Pain and Suffering in Farmed Animals

R. Eddie Clutton • Gabrielle C. Musk
Editors

Pain and Suffering in Farmed Animals

First Steps in Better Understanding and Management

Springer

Editors
R. Eddie Clutton
The Wellcome Trust Critical Care
Laboratory for Large Animals
LARIF, The Roslin Institute,
University of Edinburgh
Easter Bush, Midlothian, UK

Gabrielle C. Musk
School of Veterinary Medicine
Murdoch University
Murdoch, WA, Australia

Editorial Contact: Annette Klaus-Brusola

ISBN 978-3-032-11406-8 ISBN 978-3-032-11407-5 (eBook)
https://doi.org/10.1007/978-3-032-11407-5

This Springer imprint is published by the registered company Springer Nature Switzerland AG
The registered company address is: Gewerbestrasse 11, 6330 Cham, Switzerland

Foreword

Introduction

Animals were domesticated long before their capacity to endure pain and suffering was acknowledged. Currently, production, or farmed animals, includes mammals, fish, crustaceans, and insects, and while these animals are raised for food, many are also used in other contexts such as biomedical research, entertainment, e.g., zoos, and as beasts of burden. This book discusses a range of production animal species in the context of a range of environments in which they are kept. The chapters cover concepts associated with pain and its causes, its recognition and treatment, and species-specific detail about the current management of animals, with particular emphasis on where there is room for improvement, and what, in some instances, are the barriers to managing pain and its related suffering.

The management of pain in animals has undergone significant changes over the past few decades. Historically, veterinarians were advised to avoid treating pain, it being assumed that pain served natural protective purposes to prevent further self-injury. This perspective was exemplified by Yoxall (1978) who, in 1978, concluded a review of the recognition and control of pain in small animals by stating that it was appropriate to "minimise treatment of pain in circumstances in which, by its presence, pain is reducing the likelihood of further damage occurring", reflecting a consensus within veterinary medicine in the 1970s. However, the publication of *Pain Management in Animals* (Flecknell and Waterman-Pearson 2000) and the *Veterinary Clinics of North America Small Animal Practice*'s special issue on analgesia edited by Mathews (2000) in 2000 represented a shift in attitudes towards the importance of the recognition of animal pain (initially in laboratory animals and companion animals) and the prioritisation of its management.

Today, most veterinarians, and anyone involved with looking after or co-existing with animals, accept that pain causes suffering, compromises animal welfare, and impedes healing. In the laboratory setting, it is recognised that pain corrupts scientific data, and in production settings, unmanaged pain may decrease productivity. For veterinarians there is a professional obligation, articulated in veterinary oaths and codes of conduct, that ensures the avoidance and relief of pain in animals is a priority. For others involved with animals, such responsibility may not be embedded in the same way, but animal welfare legislation will, at least when it exists, provide societally acceptable guidance on how animals should not be treated. Yet, modern practices with respect to analgesia continue to evolve—at least when farmed species are concerned, with persistent barriers to effective pain management being outdated attitudes, economic constraints, and logistical challenges.

This book explores the physiology of pain, the connection between pain and animal welfare, advances in pain assessment in farmed animals, common myths and misconceptions, and the barriers that continue to impede progress in this field. By addressing these topics, the current state of veterinary analgesia can be better understood and opportunities for further improvement to enhance the welfare of production animals in various environments can be identified.

Historical Perspective on Analgesia in Animals

Early Views on Pain Management

Although current approaches to pain identification, assessment, and treatment in animals are in some instances comparable to those employed in the care of humans the same cannot be said for all species of animals in all parts of the world and in all environments. For example, cattle in developing countries are not managed the same way as cattle in developed countries, and even within a single region, the treatment of pain in cattle can vary. Extensively farmed cattle in the north of Western Australia will not be afforded the same access to pain relief as those in a biomedical or agricultural research project in Perth. Despite significant advances in local anaesthesia (cocaine replaced by procaine from 1905 and by lidocaine from 1943), multimodal analgesia (e.g., Crile's anoci-association described in 1913) (Crile 1913), extradural anaesthesia described in 1928 (Brook 1935; McLeod and Frank 1928), and general anaesthesia (e.g., the introduction of the barbiturate sodium pentobarbital into veterinary practice in the 1930s) (Wright and Oyler 1934), adoption and use of anaesthesia and analgesia in production animals has been a slow process compared to other species, as exemplified by the following reports.
Bemis and colleagues (1924) presented a review entitled "Anesthesia, general and local" at the 60th annual meeting of the American Veterinary Medical Association held in Montreal, Canada, August 27–31, 1923. While noting the importance of anaesthesia and improvements in use, they highlighted the need for considerably more progress:
… we wish to call your attention to the living existence of a very familiar scene to most every veterinary practitioner who attends clinics, we dare say, at practically any state, sectional, or even national association meeting. We talk, preach, demonstrate and unanimously agree upon the finer points so important to every-day practice and progress, and then call upon and permit someone of our number to step forth boldly in the arena as of old, and apply his Stone Age method of acrobatic butchery without fear and trembling, for the helpless patient is so securely confined and otherwise tortured, that the hero brandishing his modern array of instruments is at no time in any degree of danger. By permitting such crude and inhumane demonstrations at our meetings, we are encouraging the general practice of such methods when we should be using every means to suppress them.

Haskins (1987) provided advice on the use of post-operative analgesics in a small animal intensive care setting at the AVMA Colloquium on Recognition and Alleviation of Animal Pain and Distress (May 15–17, 1987). It was stated that "the administration of an analgesic is clearly indicated in some critically ill patients and not in others. The proper course of action when the clinical signs are not clear remains enigmatic. Analgesics should not be administered as a matter of course when a patient is not exhibiting clinical signs referable to pain just because a person experiencing the same disorder might routinely receive analgesics." Such references to pain management in production animals are not common, as the bias towards companion animals was, and is still, strong. However, there is no reason to suspect that pain management in livestock species was any better implemented.

In response to the apparent absence of studies describing pain management in veterinary clinical practice, Hansen and Hardie (1993) reviewed post-operative analgesic use in 243 dogs and 15 cats in a veterinary teaching hospital in the USA during the period 1983–1989. It was noted that even when moderate to severe pain was observed, analgesics were withheld from nearly half of the dogs. Only one cat received analgesics following surgery. It was suggested that an "indoctrinated fear of adverse effects" contributed to this low use of analgesics. Furthermore, the authors stated that for trauma patients with fractures "clinicians may have relied on pain to limit the dog's mobility before surgery and anticipated that surgical repair would bring relief of pain." The authors concluded that "the findings of this study suggest disturbingly inadequate care", commenting that this situation is not dissimilar to that described in humans following surgery where standard post-operative pain management was denounced as grossly inadequate (Harmer 1991). Again, such studies in production animals were not conducted as pain in these species was not considered as high a priority as that of companion animals. In response to the report by Hansen and Hardie (1993), Boge (1993) strongly advocated on behalf of all animals that the "… standard of care for the 1990s must be to provide a pain-free environment for all our patients."

For much of veterinary history, the treatment of pain was not a priority, largely due to misconceptions about its role in recovery from surgery. A.T. Yoxall (1978) articulated the prevailing view of the 1970s, suggesting that veterinarians should minimise the treatment of pain in scenarios where pain might deter animals from further injury. This recommendation, while well-intentioned, was not evidence-based and perpetuated the notion that pain was a necessary or even beneficial component of surgical recovery.

This perspective reflected a broader lack of understanding of pain physiology and its effect on animal welfare. Pain was seen as a natural part of the healing process, rather than as a condition that required active intervention. The absence of validated tools to measure pain in animals and the limited availability of veterinary-specific analgesics further hindered progress.

Nevertheless, the Brambell report (1965), published in 1965, and the result of a British parliamentary enquiry into the welfare of intensively farmed livestock marked the introduction of the Five Freedoms concept: freedom from hunger and thirst, freedom from discomfort, freedom from pain, injury and disease, freedom to express normal behaviour, and freedom from fear and distress. Although these "freedoms" provided a simple, yet holistic, way of considering the welfare state of an animal, they have evolved into the more complex but applicable Five Domains of Animal Welfare model (Mellor et al. 2020). This model is a contemporary approach to animal welfare evaluation and incorporates pain as a major contributor to animal health, the mental state of animals and their overall welfare.

Emergence of Modern Veterinary Analgesia

Despite the 1965 acknowledgement of the role of pain in the welfare of all species, it was not until the turn of the millennium that a significant shift in attitudes towards pain management in animals came about (a decade later than advocated by Boge) (Boge 1993). The release of *Pain Management in Animals* by Flecknell and Waterman-Pearson in 2000 (Flecknell and Waterman-Pearson 2000) and a special issue of *Veterinary Clinics of North America Small Animal Practice* (Mathews 2000) dedicated to analgesia are examples of publications that reflected the veterinary profession's increased interest and commitment to identifying and managing animal pain.

These publications strongly argued that:

- Pain occurs in animals and causes suffering.
- Pain relief should be prioritised as part of a veterinarian's ethical responsibility to promote animal welfare.

At that time, however, only about 50% of dogs and cats undergoing ovariohysterectomies (OHE) received analgesics, underscoring the gap between knowledge and practice. These advances also coincided with growing public expectations for veterinarians to recognise and address animal pain, aligning with societal demands for greater compassion for animals. Nonetheless, implementation of pain relief in livestock species has always been well behind pain management of companion animal species, i.e., dogs and cats.

Highlighting the paucity of approved medicines for livestock species, it is salutary to review the analgesic, local anaesthetic and sedative agents approved for use in livestock species in the USA and Great Britain, summarised in Table 1. Notably, no products are approved for use in poultry.

Table 1 Drugs approved for analgesia/local anaesthesia/general anaesthesia and sedation approved for use in livestock species in (A) the UK and (B) the USA

(A) UK. Full prescribing details available as Summary of Product Characteristics (SPC) (https://www.vmd.defra.gov.uk/ProductInformationDatabase/search)	
Cattle (10 actives)	**Route of administration and indication for use**
Carprofen	Injection To reduce clinical signs in acute infectious respiratory disease and acute mastitis
Detomidine	Injection Sedation and analgesia during various examinations and treatments. For premedication before administration anaesthetic agents

(continued)

Table 1 (continued)

Flunixin	Injection Reduction of acute inflammation associated with respiratory disease. Adjunctive therapy in the treatment of acute mastitis
Ketamine	Injection Used in combination with a sedative for immobilisation, sedation, and general anaesthesia
Ketoprofen	Injection Reducing the pyrexia and distress associated with bacterial respiratory disease. Reducing pain associated with lameness
Meloxicam	Injection For use in acute respiratory infection to reduce clinical signs. For adjunctive therapy in the treatment of acute mastitis. For the relief of post-operative pain following dehorning in calves
Metamizole +/– Hyoscine Butylbromide	Injection Treatment of smooth muscle spasms and pain associated with underlying disorders of the gastrointestinal tract, urogenital system, and bile excretory organs
Procaine + Epinephrine	Injection Local anaesthesia with an anaesthetic effect of 1–2 hours
Sodium salicylate	Water, milk (calves) Supportive treatment of pyrexia in acute respiratory disease
Xylazine	Injection Sedative with some analgesic and muscle relaxant properties
Pigs (13 actives)	
Acetylsalicylic acid	Top dress Supportive treatment for reduction of pyrexia
Azaperone	Injection A neuroleptic sedative to be used for the treatment of (i) Aggression; (ii) Stress, including transport-related stress; (iii) Obstetric conditions, e.g., cessation of parturition due to excitation, as an obstetric aid in manual delivery, inversion of the vagina, prolapse of the uterus, pathological straining; (iv) Premedication in local and general anaesthesia
Flunixin	Injection Adjunctive therapy in the treatment of swine respiratory diseases
Ketamine	Injection Used in combination with a sedative for immobilisation, sedation, and general anaesthesia
Ketoprofen	Injection, water Reducing the pyrexia and respiratory rate associated with bacterial or viral respiratory disease
Lidocaine/ Bupivacaine/ Adrenaline/ Cetrimide	Topical For use in piglets up to 7 days of age. Lidocaine and bupivacaine can form a metabolite (2,6-xylidine) in humans, which can induce carcinogenic effects at high doses in long-term toxicology studies in rats
Meloxicam	Injection, oral To reduce the symptoms of lameness and inflammation. For adjunctive therapy in the treatment of puerperal septicaemia and toxaemia
Metamizole +/– Hyoscine Butylbromide	Injection Treatment of smooth muscle spasms and pain associated with underlying disorders of the gastrointestinal tract, urogenital system, and bile excretory organs

(continued)

Table 1 (continued)

Paracetamol	Water, feed Symptomatic treatment of fever in the context of respiratory diseases
Procaine + Epinephrine	Injection Local anaesthesia with an anaesthetic effect of 1–2 hours
Sodium salicylate	Water For the treatment of inflammation
Poultry (0 actives)	
(B) USA. Of 48 analgesic, anaesthetic, and sedative drugs listed in the FDA Green Book (https://www.fda.gov/animal-veterinary/products/approved-animal-drug-products-green-book) (Acepromazine; Acetylsalicylic Acid-Methylprednisolone; Aminopentamide-Ketamine-Promazine; Atipamezole; **Azaperone**; Bcdinvetmab; Bupivacaine; Buprenorphine; Butorphanol; Carprofen; Deracoxib; Detomidine; Dexmedetomidine; Diclofenac; Diprenorphine-Etorphine; Dipyrone; Droperidol-Fentanyl; Ethylisobutrazine; Firocoxib; **Flunixin**; Frunevetmab; Grapiprant; Guaifenesin; Halothane; Isoflurane; Ketamine; **Ketoprofen**; Lidocaine-Oxytetracycline; Medetomidine; Medetomidine-Vatinoxan; Meloxicam; Mepivacaine; Naproxen; Pentobarbital; Phenylbutazone; Pregabalin; Proparacaine; Propofol; Robenacoxib; Romifidine; Sevoflurane; **Thialbarbitone**; **Thiamylal**; Thiopental; Tiletamine Hydrochloride-Zolazepam; Tolazoline; Xylazine; Yohimbine) only 5 (in bold type) are approved for use in cattle or pigs, and none in poultry.	
Full prescribing information and Freedom of Information summaries can be found at https://animaldrugsatfda.fda.gov/adafda/views/#/search	
Cattle (4 actives)	**Route of administration and indication for use**
Flunixin	Injection and pour on For the control of pyrexia associated with bovine respiratory disease and acute bovine mastitis. For the control of *pain* associated with foot rot
Ketoprofen	Injection For the control of pyrexia associated with bovine respiratory disease (BRD)
Thialbarbitone	Injection General anaesthetic in surgical procedures
Thiamylal	Injection Ultra-short-acting anaesthetic
Swine (4 actives)	
Azaperone	Injection Control of aggressiveness when mixing or regrouping weanling or feeder pigs
Flunixin	Injection For the control of pyrexia associated with swine respiratory disease
Thialbarbitone	Injection General anaesthetic in surgical procedures
Thiamylal	Injection Ultra-short-acting anaesthetic
Poultry (0 actives)	

Approved drugs are not necessarily used or available

Lessons from Human Medicine

Human medicine faced similar challenges when it came to the recognition of pain and the administration of analgesia. For example, in the 1980s, neonates undergoing surgery were often denied analgesics based on the belief that they did not experience pain or would not remember it later in life (Rodkey and Pillai Riddell 2013). The case of Jeffrey Lawson (Harmer 1991), a premature infant who underwent a thoracotomy to repair his patent ductus arteriosus, without general anaesthetic, but immobilised following administration of the neuromuscular blocking agent pancuronium, sparked community outrage and conflict within the medical community (Chamberlain 1999). Speaking to the *Los Angeles Times* (Los Angeles Times 1991), Dr Myron Yaster, a leading expert on paediatric pain at Johns Hopkins Medical Institute in Baltimore, said that "…most adults would be shocked if they saw what was done to children in hospitals without anesthetics. It's like roping and holding down a steer to brand it." Advocates for neonatal pain relief often pointed to the fact that animals received better analgesic care, highlighting the parallels between human and veterinary medicine in addressing historical misconceptions about pain. In a letter to the editor of Birth (Harmer 1991), Lawson's mother Jill wrote "…For our pets there are protective organisations and dedicated proponents to guard against surgery without anaesthesia. There appear to be none for premature babies…". In 1987, two years after Jeffrey Lawson's death, the anatomical and physiological foundation of pain perception in neonates was comprehensively described (Anand and Hickey 1987) and the American Academy of Pediatrics stated that it was no longer ethical to perform surgery on preterm babies without anaesthesia (Committee on Fetus and Newborn, Committee on Drugs, Section on Anesthesiology and Section on Surgery 1987). (See chapter "Paediatric Pain").

Pain, Welfare, and Veterinary Responsibility

The Five Freedoms Versus Five Domains Model

The Five Freedoms, long regarded as the foundation of farm animal welfare, emphasised the absence of negative experiences such as hunger, discomfort, pain, and fear. However, this model has increasingly been replaced by the Five Domains model, which takes a more nuanced view of animal welfare. Instead of striving for the elimination of all negative states—which is often unrealistic—the Five Domains focus on promoting positive experiences while managing unavoidable negatives (Mellor et al. 2020).

Under this model, pain is recognised not only as a health issue but also as a factor that affects an animal's mental state and overall welfare. Effective pain management is, it proposes, essential for improving an animal's quality of life.

The model consists of five domains that contribute to an animal's physical and/or functional state after physiological processing. Domain 1 is Nutrition, domain 2 is the physical environment, domain 3 is the animal's health (which focuses on the animal's physical state, its overall fitness, and the presence or absence of pain, injury, or disease) while domain 4 is focused on behavioural interactions. Domain 5 evaluates the animal's overall psychological well-being which is influenced by the other four domains. Domain 5 assesses the balance of positive and negative experiences, such as fear, pleasure, or frustration. These domains are used in a model to assess animal welfare by considering how factors like diet, living space, and physical health influence an animal's mental and psychological well-being.

The Bidirectional Effect of Pain on Welfare

Pain has far-reaching effects on an animal's ability to care for itself, express normal behaviour, and interact with its environment. For example:

- Animals in pain may struggle to eat, drink, or groom themselves, further compromising their health.
- Pain can prevent normal behaviours such as exploration or play, reducing an animal's overall mental stimulation.
- Chronic pain can lead to long-term health complications, including delayed healing and heightened sensitivity to pain.

Conversely, poor welfare—due to, for example, inadequate or inappropriate housing, stress, or fear—can exacerbate pain. Fear-related stress, for instance, may suppress pain behaviours in the short term but worsen chronic pain over time, as seen in human studies of anxiety and depression (Salas-González et al. 2025). Veterinarians, therefore, play a dual role in both causing and alleviating pain and distress (Carroll et al. 2022), and their ethical obligation extends to minimising pain associated with veterinary interventions, not just spontaneous disease or injury.

Recognising and Assessing Pain in Animals

Advances in Pain Recognition

Recognising pain in animals has historically been a challenge because of their inability to verbalise their experiences (in humans the gold standard of pain assessment is self-reporting). However, recent advances, e.g., facial grimace scales, have provided veterinarians with more nuanced tools to assess animal pain. Facial grimace scales are species-specific and have been validated for use in animals such as cattle, pigs, sheep, rodents, cats, and horses (Mota-Rojas et al. 2025) (see chapter "Basic Principles of Animal Pain Assessment").

Animal welfare science has contributed considerably to the ability to detect and quantify pain through behavioural changes. These advances have provided an opportunity for more accurate pain assessment by veterinarians, farmers, and those caring for farmed animals.

Challenges in Pain Recognition

Numerous challenges remain despite an increased range of pain assessment devices:

- Breed variation in facial conformation may present challenges when developing scales that rely on facial feature evaluation (i.e., a grimace scale in dogs has not yet been developed).
- Fear and stress can suppress pain behaviours, making it difficult to distinguish between a calm animal and one masking its pain. For example, a frightened animal may freeze or avoid movement, which could be misinterpreted as an absence of pain.
- Species differences in pain expression complicate efforts to generalise pain assessment tools. For instance, prey species like sheep or rabbits may hide signs of pain to avoid appearing vulnerable to predators.

Public Expectations and Veterinary Standards

The public increasingly expects veterinarians and animal carers to recognise and treat pain in animals, reflecting a growing awareness of animal welfare issues, especially in production animals. Consumers of meat are increasingly educated and informed about how the life and death of animals represents their welfare and may influence their supermarket purchases. For example, recent studies show that veterinarians are now more likely to acknowledge pain in cattle undergoing procedures such as dehorning or castration than they were a decade ago (see chapter "The Effects of Human Attitudes on Farmed Animal Pain and Its Management"). However, disparities remain—for instance, calves are less likely to receive analgesics than adult cattle, despite evidence that they experience equivalent levels of pain (Edwards-Callaway et al. 2023).

Myths and Misconceptions About Pain

Several long-held myths about animal pain have hindered its effective management in veterinary practice, either by veterinarians or their clients (Monteiro et al. 2023). These include:

1. *Pain prevents use of damaged tissues*: While this may be true in some cases, pain often leads to self-trauma, e.g., licking or biting at wounds, and delays healing.
2. *Analgesia masks physiologic indicators of deterioration*: In reality, pain is more likely to confound these indicators. Alleviating pain allows for more accurate assessment of an animal's condition.
3. *Analgesia risks unacceptable adverse effects*: While side effects are a valid concern, multimodal analgesia—using lower doses of multiple drugs—can minimise these risks.
4. *Doubts over the non-human pain experience*: The use of animals as models in human pain research justifies the use of analgesics in painful animals.

By addressing these misconceptions, veterinarians and animal carers can improve their approach to pain management and increase the likelihood that animals receive appropriate care.

Barriers to Effective Analgesia

Despite advances in veterinary analgesia, several barriers persist:

- *Time and Training*: All pain scoring systems require some degree of observer training and time to implement, which may be challenging and costly.
- *Cultural Attitudes*: Some clients and veterinarians believe that a degree of suffering is acceptable or even desirable in farmed animals.
- *Cost*: Financial constraints often limit access to analgesics, particularly for large animals or those suffering chronic pain conditions.
- *Logistical Challenges*: Adequate pain management requires round-the-clock care, including monitoring and continuous or intermittent administration of analgesia, which may be difficult to provide in certain settings.
- *Focus on Treating the Cause*: Veterinarians may prioritise resolving underlying conditions, e.g., lancing an abscess without also addressing the associated pain.
- *Access*: As is clear from Table 1, the range of licensed products that can manage pain and stress in livestock species is subject to great variation from country to country and from species to species.

Overcoming these barriers requires a multifaceted approach, including education, advocacy for affordable analgesics, and improved training in pain recognition and management.

Advances and Opportunities in Veterinary Analgesia

Lessons from Laboratory Animal Science

Laboratory animal science provides valuable insights into pain mitigation. For example, researchers are required to refine their methods to reduce pain in experimental animals, setting a standard that could be applied more broadly in veterinary and agricultural practices.

Best Practices for Pain Management

Veterinarians and farmers can improve analgesia by:

- Anticipating and monitoring the signs of pain
- Consulting veterinary anaesthetists to develop comprehensive pain management plans (Boge 1993)
- Measuring pain and evaluating responses to treatment to continuously refine protocols

Challenges in Drug Development and Registration

Developing and registering new analgesics for veterinary use is a complex and expensive process. Determining appropriate doses and the number needed to treat (NNT) requires extensive research. Regulatory barriers also vary by country; for example, xylazine is widely used in the UK but access is restricted in other regions (see Table above). Harmonising these regulations could improve access to effective analgesics.

Future Innovations to Refine the Management of Pain and Suffering

New Analgesic Drugs

While the search for new drugs with analgesic activity continues, it is unlikely that a new drug will satisfy all the needs of all animal subjects. Therefore, there is a need for a range of new molecular entities with appropriate safety and effectiveness. One potential avenue may be the carbamorphines (Akiyama et al. 2025), morphine derivatives modified to produce a pharmacological profile with retained analgesic activity but reduced respiratory depression and addictive potential.

N-of-1 Studies

While new analgesic drugs may be welcome and widely used, use of existing analgesics has much scope for major improvement. It is known that there is substantial intra- and inter-individual variation in effectiveness and safety of analgesic drugs (due to physiological, pathological, pharmacokinetic (PK), pharmacodynamic (PD), pharmacogenomic (PG), and other differences), suggesting that approaches to customisation are needed.

Nikles (2015) defined N-of-1 trials as "within-patient, randomized, double-blind, cross-over comparisons of a treatment with either a placebo or another treatment. Because the trials involve only one patient, with that patient acting as their own control, the comparison is made between multiple (usually three periods for each alternative) cross-over time intervals. N-of-1 trials can identify individual patients who respond (or do not) to medications for chronic, stable conditions." He and co-workers (2021) undertook a comprehensive review of N-of-1 trials in chronic pain management and found substantial evidence supporting their value in identifying the best available treatment in cases of refractory osteoarthritis, chronic musculoskeletal pain, and neuropathic pain. They concluded that emerging simple and facilitated strategies to carry out N-of-1 trials will improve the overall chronic pain treatment at the precision medication level. Formal N-of-1 trials do not have a long history of use in veterinary medicine, despite being a valuable and informative approach to individualised treatment. The founders of the Evidence Based Medicine revolution at McMaster University in Canada set up an N-of-1 consulting service in 1986 (Guyatt et al. 1986). Cummings et al. (2023) provided an excellent review of the value of N-of-1 data in zoo animal medicine. Such studies could readily be applied to livestock species, especially valuable individual animals requiring pain management.

Automatic Pain Assessment (APA)

In reviewing factors affecting analgesic selection in large animals, Clutton (2009) noted that "variation in an individual's response to noxious stimulation and analgesics coupled with poor agreement between pain assessors means that standardized post-operative pain management will result in some animals remaining in pain whilst others are over-dosed." It was further concluded that "a given animal's pain management requires continuous assessment and dosage re-adjustment on the basis of response; only this will balance the effects of pain and analgesics." Advances in technology, including continuous and automatic pain assessment, may increasingly enable such refinements in pain management.

Cascella (2024) provides a thorough overview of automatic pain assessment (APA) with pragmatic suggestions for clinical translation. Cascella describes how the facial action coding systems (FACS or grimace scales) developed by Ekman and Friesen in 1978 (building on the work of Charles Bell (1806) in 1806 and Darwin (1872) in 1872) allows analysis of complex patterns to identify subtle indicators of pain. Continuous monitoring permits real-time pain assessment while avoiding non-observance of critical pain events. The application of APA to the assessment of pain in cats using deep neural networks and machine learning models to predict facial landmarks inherent in the feline grimace scale has recently been described (Steagall et al. 2023). This represents a significant milestone in pain recognition and demonstrates the feasibility of extension of this approach to any species in which facial expressions or other detectable landmarks provide a valid basis of pain assessment.

Madden et al. (2021) point out that experimental pain research in humans has largely ignored the importance of intra-individual variability, which is generally "considered a nuisance, rather than a feature worthy of attention". The authors conclude that "acknowledging intraindividual variability will be key to optimising the translation of experimental research to the clinical context." The availability of validated automatic and continuous pain assessment tools may well allow refined management of such intraindividual variation in pain response as currently captured intermittently by repeated grimace scale assessment.

Real-World Data/Real-World Evidence (RWD/RWE)

Summarising the situation in human drug development, Abbasi et al. (2025) concluded that there is potential to accelerate clinical research as sources of real-world data become increasingly available. Early experience suggests that the quality and utility of the data can be enhanced by improving data capture at source (ensuring data is complete, consistent, accurate, and verifiable) and enabling longitudinal data to be collected to describe outcomes of treatment, effectiveness, and adverse events. Such a situation can equally apply to the range of livestock species encountered in clinical veterinary practice.

Golden et al. (2022) and Girolamo et al. (2022) analysed the reference sources supporting dose rates of medicines included in the *Exotic Animal Formulary* (5th Edition, 2018) and reported a high proportion of secondary sources which included narrative reviews, commentaries, viewpoints, and other non-peer-reviewed observations and reports. In view of the vast number of species encountered in veterinary practice and the large number of medicines that may be used, it is not surprising that the data available may not be comprehensive and of high quality.

In the preface to the 6th edition of Carpenter's *Exotic Animal Formulary* (Carpenter and Harms 2023) the editors emphasise that "… there are relatively few pharmacokinetic/pharmacodynamic studies on exotic companion pets for the medications we use in practice. Until more pharmacokinetic efficacy and safety studies of the drugs that we use are conducted, most dosages used in these species are based on empirical data, observations, and experience", emphasising the motivation behind the publication of the *Zoo and Wild Mammal Formulary* by Alicia Hahn (2019). Just as there are differences in effective dose regimens between species, there are also differences within a species. In the case of livestock for example, breed differences in the response to xylazine have been observed. Raptopoulos and Weaver (1984) found that Hereford steers became recumbent after injection of xylazine more readily than the Friesian steers and took longer to recover but showed fewer reactions to surgical stimulation than the Friesians. Brahman cattle have been reported to be the most sensitive to xylazine (Greene and Thurmon 1988), often responding to one-tenth the usual bovine dose. Another example of a prominent intra-species difference is that reported by Fletcher (1974) who described the Isle of Rhum red deer off the west coast of Scotland being extremely sensitive to the effect of xylazine in comparison with the red deer on the mainland.

Other causes of variation in response include temperament (Hopkins 1972). The initial demeanour of the animal patient greatly influences the sedation produced by a given dose of an alpha-2-adrenergic agonist (Abrahamsen 2008). Calm, quiet patients require smaller doses, whereas anxious or unruly patients require larger doses with unpredictable behaviour. Capturing these observations can lead to a more individualised dose regimen.

How can this unmet need for reliable dose regimens be satisfied?

The growing recognition of the value of Real-World Data (RWD) and Real-World Evidence (RWE) provides a significant opportunity for improving the evidence underpinning recommendations for dose rates.

Davies et al. (2025) demonstrate in a retrospective analysis that unstructured real-world data recorded as free-text clinical narratives can be used to identify outcomes in veterinary pharmaco-epidemiological studies. For prospective studies of pharmacotherapy in uncommon or rare animal breeds and species, a structured data template could be prepared for use by all veterinarians, providing consistent, complete, reliable and high quality RWD that could be converted to valuable RWE. This could be updated regularly as more RWD became available.

In 2021 the USA FDA Center for Veterinary Medicine (CVM) published Guidance for Industry GFI #266 (Center for Veterinary Medicine (CVM) 2021) entitled Use of Real-World Data and Real-World Evidence to Support Effectiveness of New Animal Drugs. The CVM defined Real-World Evidence (RWE) as the clinical evidence of the effectiveness of a new animal drug derived from analysis of Real-World Data (RWD).

Although it is not envisaged that regulatory approval will be sought for the thousands of potential drug-species-dose rate triads that exist, nonetheless, the rigour of the regulatory process provides an excellent guide to high quality and credible evidence.

CVM defines Real-World Data (RWD) as "data routinely collected from a variety of sources relating to the health and productivity of animals, the delivery of veterinary care, or the management of livestock/animals for food." Examples of RWD applicable to new animal drugs include: (i) data derived from health records of veterinary practices, farms, or livestock management companies (including handwritten paper records and electronic veterinary medical records (EVMRs); (ii) data from product and disease registries; and (iii) data from other sources that can inform on animal health status such as mobile and, or remote health sensing devices for animals.

Data quality, provenance, and transparency are fundamentally important in the acquisition of RWD. As the CVM emphasise, "the reliability of RWD is determined by the quality and integrity of the data and the source from which it is derived. In general, high quality data, regardless of whether it is collected manually or with electronic systems, are attributable, legible, contemporaneous, original, and accurate."

The purposes for which RWE may be potentially used includes, according to CVM, the characterisation of a dosage regimen and definition of the conditions of use.

Already substantial RWD is being collected for zoo animals. Teare and Conde (2023) described how standardised data collected from multiple sources had enormous potential to refine drug use in zoological medicine, especially as it pertains to the relief of pain and suffering. They focused on the massive database Species360 (https://species360.org/), which arose from the International Species Information System (ISIS), and provides a central storage database derived from the clinical records of zoological species worldwide, with the capacity for record analysis. It may be accessed via the Zoological Information Management System (ZIMS) and is supported by around 1300 organisations. In 2020, it received an average of 380,000 records per month. This database is an excellent example of RWD with potential (using guidance provided by CVM GFI #266) for conversion to Real-World Evidence. There is no reason why RWD could not be collected by those working with livestock to support improved pain management.

There are some important considerations in the generation of RWE from RWD. Garry et al. (2025) warn that as large datasets are formed it is necessary to critically appraise the potential effect of standardised formats, which may reduce outliers and variability and thus mask rare events that are nonetheless of clinical importance. Variation and heterogeneity are inherent in biology, only their absence should be of concern.

Physiologically Based Pharmacokinetic-Pharmacodynamic (PBPKPD) Modelling and Simulation

The availability of PBPKPD models of humans and dogs indicates an increasing likelihood that in silico models of other animal species will be developed that will allow:

1. Simulation of exposure between different formulations and routes of administration
2. Evaluation of differences in oral drug absorption due to species-specific physiological differences in gut wall structure and function
3. Evaluation of inter-subject pharmacokinetic/pharmacodynamic (PK/PD) variability associated with physiological variability
4. Prediction of PK and PD using virtual species-appropriate models
5. Translation between species, accounting for different physiological and anatomical characteristics, as well as differences in metabolising enzymes, transporters, and plasma protein levels between preclinical species and humans

The data required to build a population file for a new species is in most cases readily available and includes: (i) body mass; (ii) tissue volume/mass; (iii) cardiac output; (iv) tissue blood flow; (v) plasma protein concentration; (vi) haematocrit; (vii) glomerular filtration rate; (viii) tissue composition (intra- and extracellular water content, protein, neutral lipid, neutral and acidic phospholipid content); (ix) microsomal protein content/number of hepatocytes per gram of liver; (x) intestinal characteristics (length, fluid volume, pH, etc.).

For animal species of interest, such a model would allow improved dose determination, individualised to patient age, gender, weight, and disease status.

Conclusion

Pain management in animals has come a long way from its early days of neglect, but significant challenges remain, especially for production animal species. By recognising pain as a critical aspect of animal welfare, debunking outdated myths, and addressing barriers to effective pain management, caregivers (veterinarians, farmers, and anyone with animal care responsibilities) can continue to improve the care they provide their animals. The future of veterinary analgesia lies in advancing pain recognition, refining treatment methods and advocating for affordable access to analgesics. Ultimately, the profession must continue to prioritise the alleviation of pain to honour its commitment to animal welfare.

Stephen W. Page
Advanced Veterinary Therapeutics
Newtown, NSW, Australia

References

Abbasi AB, Curtis LH, Califf RM (2025) The promise of real-world data for research - what are we missing? N Engl J Med 393(4):318–321
Abrahamsen EJ (2008) Ruminant field anesthesia. Veterin Clin Food Anim Pract 24(3):429–441
Akiyama S, Ople RS, Kremsmair A, Ramos-Gonzalez N, Nedungadan T, Kennedy BJ, Appourchaux K, Eans SO, Tsai BA, Kraml C, Huang X-P, McLaughlin JP, Majumdar S, Sarpong R (2025) Total synthesis and biological activity of "carbamorphine": O-to-CH2 replacement in the E-ring of the morphine core structure. Proc Natl Acad Sci 122(27):e2425438122
Anand KJ, Hickey PR (1987) Pain and its effects in the human neonate and fetus. N Engl J Med 317(21):1321–1329
Bell C (1806) Essays on the anatomy of expression in painting. Paternoster Row, London, Longman, Hurst, Rees, and Orme
Bemis HE, Guard WF, Covault CH (1924) Anesthesia, general and local. J Am Vet Med Assoc 64(4):413–439
Boge P (1993) Pain control. J Am Vet Med Assoc 203(3):359
Brambell (Chairman) FWR (1965) Report of the Technical Committee to Enquire into the Welfare of Animals kept under Intensive Livestock Husbandry Systems. London Her Majesty's Stationery Office
Brook GB (1935) Spinal (epidural), in the domestic animals: a review of our knowledge at the present time. Vet Rec 25(19–23):549–553; 576–581; 597–608; 631–635; 659–667
Carpenter JW, Harms CA (2023) Carpenter's Exotic Animal Formulary, 6th edn. Elsevier, St. Louis
Carroll SL, Sykes BW, Mills PC (2022) Moving toward fear-free husbandry and veterinary care for horses. Animals (Basel) 12(21)
Cascella M (2024) Suggestions for APA research and translational perspective. In: Exploring AI in pain research and management. Springer Nature Switzerland, Cham, pp 85–103
Center for Veterinary Medicine (CVM) (2021) CVM GFI #266 use of real-world data and real-world evidence to support effectiveness of new animal drugs. Guidance for industry. Food and Drug Administration, Rockville
Chamberlain DBP (1999) Babies don't feel pain: a century of denial in medicine*. J Prenatal Perinatal Psychol Health 14(1/2):145–168
Clutton RE (2009) Clinical studies, pain and ethics. J Small Anim Pract 50(2):59–60
Clutton RE (2018) A review of factors affecting analgesic selection in large animals undergoing translational research. Vet J 236:12–22
Committee on Fetus and Newborn, Committee on Drugs, Section on Anesthesiology and Section on Surgery (1987) Neonatal Anesthesia. Pediatrics 80(3):446
Crile GW (1913) The kinetic theory of shock and its prevention through anoci-association (shockless operation). Lancet 182(4688):7–16
Cummings CO, Eisenbarth J, Krucik DDR (2023) The value of N-of-1 data in zoological medicine: a methodological review. J Zoo Wildl Med 54(3):417–427
Darwin C (1872) Chapter V. Special expressions of animals. In: Expression of the emotions in man and animals. John Murray, London, pp 116–146

Davies H, Noble PJ, Fins IS, Pinchbeck G, Singleton D, Pirmohamed M, Killick D (2025) Developing electronic health records as a source of real-world data for veterinary pharmacoepidemiology. Front Vet Sci 12:1550468

Edwards-Callaway LN, Keller KP, Oselinsky K, Johnstone E, Cramer C, Román-Muñiz N, Stallones L, Coetzee JF (2023) A nationwide survey on producer and veterinarian perceptions of the painfulness of procedures and disease states in dairy and beef cattle. Front Pain Res (Lausanne) 4:1059224

Flecknell PA, Waterman-Pearson A (2000) Pain Management in Animals. W. B. Saunders, London

Fletcher J (1974) Hypersensitivity of an isolated population of red deer (Cervus elaphus) to xylazine. Vet Rec 94(5):85–86

Garry EM, Baglivo A, Govil P, Duryea JL, Liu W, Lasky T, Chakravarty A, Rivera DR, Bradley MC (2025) Evaluating the impact of data standardization on real-world data. Pharmacoepidemiol Drug Saf 34(8):e70191

Girolamo ND, Caron M, Brandão J, Reynders RM (2022) Most veterinarians treating exotic animals use formularies to select drug dosages without consistently checking their sources. J Am Vet Med Assoc 260(8):1–11

Golden O, Kline S, Tarpalechee L, Brandão J, Di Girolamo N (2022) Evaluation of sources cited by an exotic animal formulary for supporting drug dosages and reference intervals in mammals. J Small Anim Pract 63(12):863–872

Greene SA, Thurmon JC (1988) Xylazine—a review of its pharmacology and use in veterinary medicine. J Vet Pharmacol Ther 11(4):295–313

Guyatt G, Sackett D, Taylor DW, Chong J, Roberts R, Pugsley S (1986) Determining optimal therapy—randomized trials in individual patients. N Engl J Med 314(14):889–892

Hahn A (ed) (2019) Zoo and wild mammal formulary. Wiley, Hoboken

Hansen B, Hardie E (1993) Prescription and use of analgesics in dogs and cats in a veterinary teaching hospital: 258 cases (1983-1989). J Am Vet Med Assoc 202(9):1485–1494

Harmer M (1991) Postoperative pain relief—time to take our heads out of the sand? Anaesthesia 46(3):167–168

Haskins SC (1987) Use of analgesics postoperatively and in a small animal intensive care setting. J Am Vet Med Assoc 191(10):1266–1268

He W, Cui Z, Chen Y, Wang F, Li F (2021) Status of N-of-1 trials in chronic pain management: a narrative review. Pain Ther 10(2):1013–1028

Hopkins TJ (1972) The clinical pharmacology of xylazine in cattle. Aust Vet J 48(3):109–112

Lawson JR (1986) Letter to the editor. Birth 13(2 June):124–125

Los Angeles Times (1991) "Surgeons finding even the very young feel pain. More premature babies now getting painkillers." The Baltimore Evening Sun Retrieved 8 July 2019, from https://www.baltimoresun.com/news/bs-xpm-1991-10-28-1991301120-story.html

Madden VJ, Kamerman PR, Catley MJ, Bellan V, Russek LN, Camfferman D, Lorimer Moseley G (2021) Variability in experimental pain studies: nuisance or opportunity? Br J Anaesth 126(2):e61–e64

Mathews KA (2000) Management of Pain. W. B. Saunders, Philadelphia

McLeod WM, Frank ER (1928) A preliminary report regarding epidural anesthesia in equines and bovines. J Am Vet Med Assoc 72(3):327–335

Mellor DJ, Beausoleil NJ, Littlewood KE, McLean AN, McGreevy PD, Jones B, Wilkins C (2020) The 2020 Five Domains model: including human-animal interactions in assessments of animal welfare. Animals (Basel) 10(10)

Monteiro BP, Lascelles BDX, Murrell J, Robertson S, Steagall PVM, Wright B (2023) 2022 WSAVA guidelines for the recognition, assessment and treatment of pain. J Small Anim Pract 64(4):177–254

Mota-Rojas D, Whittaker AL, Coria-Avila GA, Martínez-Burnes J, Mora-Medina P, Domínguez-Oliva A, Hernández-Avalos I, Olmos-Hernández A, Verduzco-Mendoza A, Casas-Alvarado A, Grandin T (2025) How facial expressions reveal acute pain in domestic animals with facial pain scales as a diagnostic tool. Front Vet Sci 12:1546719

Nikles J (2015) What are N-of-1 trials? In: Nikles J, Mitchell G (eds) The essential guide to N-of-1 trials in health. Springer Netherlands, Dordrecht, pp 9–17

Raptopoulos D, Weaver BM (1984) Observations following intravenous xylazine administration in steers. Vet Rec 114(23):567–569

Rodkey EN, Pillai Riddell R (2013) The infancy of infant pain research: the experimental origins of infant pain denial. J Pain 14(4):338–350

Salas-González J, Heredia-Rizo AM, Fricke-Comellas H, Chimenti RL, Casuso-Holgado MJ (2025) Patterns of pain perception in individuals with anxiety or depressive disorders: a systematic review and meta-analysis of experimental pain research. J Pain 35:105530

Steagall PV, Monteiro BP, Marangoni S, Moussa M, Sautié M (2023) Fully automated deep learning models with smartphone applicability for prediction of pain using the Feline Grimace Scale. Sci Rep 13(1):21584

Teare JA, Conde DA (2023) Chapter 7 - Harnessing standardized data and analytics to advance zoological medicine. In: Miller E, Lamberski N, Calle P (eds) Fowler' s zoo and wild animal medicine current therapy, vol 10. W.B. Saunders, New Delhi, pp 35–40

Wright JG, Oyler M (1934) Nembutal anaesthesia in the dog. Vet Rec 46(49):1463–1471

Yoxall AT (1978) Pain in small animals—its recognition and control. J Small Anim Pract 19(8):423–438

Prologue

This work resulted from a meeting with Professor Sabine Kästner (University of Veterinary Medicine, Hannover) in 2019 during which the potential of a book covering the extensive subjective of "*Pain and Suffering in Farmed Animals*" was mooted. Since then, the work has slowly become a reality as a collection of works by a committed group of passionate colleagues from around the world. Some of these contributors were already well-known to the co-editors from the field of veterinary anaesthesia, but many were new connections, coming from arguably the more appropriate fields of animal welfare science and "One-Health" medicine. Consequently, we as co-editors feel immensely privileged to have been able to combine the wisdom and enthusiasm of such a varied group of scientists under one title, with contents ranging from physiology to ethics, and from poultry to decapod crustacea. This subject variation was not entirely planned. Once key areas had been identified, i.e. food and/or fibre providers from the main farmed mammalian, avian and piscean species, a degree of "mission-creep" followed which considered the inclusion of deer, horses and rabbits. Whilst the former have been described, horses have not because their global role is predominantly that of recreation, entertainment and, or work, and like rabbits—which have also been excluded—are commonly treated as companion animals. Consequently, many veterinary resources covering the subjects of equine and lapine pain management already exist. In contrast, species selection revealed the suitability of lower taxa, i.e., fish, cephalopod molluscs, decapod crustacea and insects for inclusion: there are currently proposals to commercialize octopus farming for food, while there is an increasing appreciation that these lower taxa may be capable of experiencing pain and suffering, albeit in different qualitative forms. Their inclusion, we feel, is justified given the number of animals previously ignored under this subject title and the degree of "future-proofing" such an approach affords. For the same reason, it was felt that extending the discussion beyond those animals kept under "modern" farming systems would increase the book's appeal. Consequently, there are references to "farmed animals" in research, in zoological collections, in developing countries, and as companions throughout the book. While this may prompt some to feel that we have been over-ambitious in defining the book's scope, we would argue that in addition to identifying the current state of knowledge in the most pertinent areas, we have defined subjects for future *consideration* at least, if not for research and action.

We thank Professor Kästner for the opportunity to curate this book in the first instance and to all our contributors for their patience in seeing this project through to fruition. We hope they remain sufficiently enthused to perhaps contribute to future editions of this, or similar books, as further information becomes available.

Finally, we thank our families for their patience over the last few years. We hope that they will also consider our efforts worthwhile.

R. Eddie Clutton
Gabrielle C. Musk

Contents

Slaughter, Euthanasia and Killing of Production Animals

Production Animal Species

Contributors

Gustavo Ferlini Agne Davies Livestock Research Centre, University of Adelaide, Adelaide, SA, Australia

School of Veterinary Sciences, University of Adelaide, Adelaide, SA, Australia

Amaya Albalat Institute of Aquaculture, University of Stirling, Stirling, UK

Rachel A. Ankeny Philosophy Group, Wageningen University, Wageningen, The Netherlands

Daniel T. Barratt School of Biomedicine, University of Adelaide, Adelaide, SA, Australia

Davies Livestock Research Centre, University of Adelaide, Adelaide, SA, Australia

Australian Research Council Centre of Excellence for Nanoscale BioPhotonics, University of Adelaide, Adelaide, SA, Australia

Meghan Barrett Department of Biology, Indiana University Indianapolis, Indianapolis, IN, USA

Andrew Bell School of Biodiversity, One Health and Veterinary Medicine, University of Glasgow, Glasgow, UK

Laura Bordoni Department of Anaesthesia and Pain Medicine, Sir Charles Gairdner Hospital, QEII Medical Centre, Perth, WA, Australia

Emily A. Buddle School of Agriculture, Food, & Wine, University of Adelaide, Adelaide, Australia

Paul Chambers Associate Professor (Retired) School Of Veterinary Science, Massey University, Rowan University, Palmerston North, New Zealand

Lip Yong Choo Department of Anaesthesia and Pain Medicine, Sir Charles Gairdner Hospital, QEII Medical Centre, Perth, WA, Australia

Louise Clark Davies Veterinary Specialists, Manor Farm Business Park Higham Gobion, Hertfordshire, UK

R. Eddie Clutton The Wellcome Trust Critical Care Laboratory for Large Animals, The Roslin Institute, University of Edinburgh, Easter Bush, Midlothian, UK

Christopher J. Coates Zoology, Ryan Institute, School of Natural Sciences, University of Galway, Galway, Ireland

Robyn J. Crook Department of Biology, San Francisco State University, San Francisco, CA, USA

Andrew Crump Animal Welfare Science & Ethics, Pathobiology & Population Science, Royal Veterinary College, London, UK

Centre for Philosophy of Natural & Social Science, London School of Economics & Political Science, London, UK

Juliet Duncan The University of Edinburgh, Edinburgh, UK

Thomas Dutton The Avian Vet, Fairford, UK

Samuel Evans School of Biomedicine, University of Adelaide, Adelaide, SA, Australia

Davies Livestock Research Centre, University of Adelaide, Adelaide, SA, Australia

Australian Research Council Centre of Excellence for Nanoscale BioPhotonics, University of Adelaide, Adelaide, SA, Australia

Rob A. Gregory Deer Industry New Zealand, Wellington, New Zealand

Leisha Hewitt Livestock Welfare, Franklin, TAS, Australia

Mark R. Hutchinson School of Biomedicine, University of Adelaide, Adelaide, SA, Australia

Davies Livestock Research Centre, University of Adelaide, Adelaide, SA, Australia

Australian Research Council Centre of Excellence for Nanoscale BioPhotonics, University of Adelaide, Adelaide, SA, Australia

Craig Johnson Animal Welfare Science and Bioethics Centre, School of Veterinary Sciences Tāwharau Ora, Massey University, Palmerston North, New Zealand

Charlotte H. Johnston School of Biomedicine, University of Adelaide, Adelaide, SA, Australia

Davies Livestock Research Centre, University of Adelaide, Adelaide, SA, Australia

Kavitha Kongara Anatomy and Physiology, Shreiber School of Veterinary Medicine, Rowan University, Glassboro, NJ, USA

Benjamin Lecorps Bristol Veterinary School, University of Bristol, Bristol, UK

Danila Marini Davies Livestock Research Centre, University of Adelaide, Adelaide, SA, Australia

School of Veterinary Sciences, University of Adelaide, Adelaide, SA, Australia

School of Animal and Veterinary Science, Davies Livestock Research Centre, Adelaide University, Adelaide, SA, Australia

Fernando Martinez-Taboada 3A Animal Anaesthesia and Analgesia, Upwey, VIC, Australia

Jo Murrell Bristol Veterinary Specialists, Bristol, UK

Gabrielle C. Musk School of Veterinary Medicine, Murdoch University, Murdoch, WA, Australia

Sanam Mustafa School of Biomedicine, University of Adelaide, Adelaide, SA, Australia

Davies Livestock Research Centre, University of Adelaide, Adelaide, SA, Australia

Australian Research Council Centre of Excellence for Nanoscale BioPhotonics, University of Adelaide, Adelaide, SA, Australia

Izabella Maree Norris Animal Welfare Science and Bioethics Centre, Massey University, Palmerston North, New Zealand

Adam Powell Institute of Biological, Environmental & Rural Sciences (IBERS), Aberystwyth University, Gogerddan, Aberystwyth, Ceredigion, UK

Kenneth M. D. Rutherford Animal Behaviour and Welfare, SRUC, Edinburgh, UK

Geoff Simm Division of Global Agriculture and Food Systems, Royal (Dick) School of Veterinary Studies, University of Edinburgh, Easter Bush Campus, Midlothian, UK

Preet Singh School of Veterinary Science, Massey University, Palmerston North, New Zealand

Alison Small Armidale, Agriculture and Food, CSIRO, FD McMaster Research Laboratory, Chiswick, New England Highway, Armidale, NSW, Australia

Dinakaran Venkatachalam Independant researcher, Rowan University, Palmerston North, New Zealand

Daniel M. Weary Animal Welfare Programme, Faculty of Land and Food Systems, University of British Columbia, Vancouver, Canada

Alexandra L. Whittaker School of Veterinary Science, University of Adelaide, Adelaide, Australia

The Role and Impact of Production Animals in Society

Pain and Suffering in Farmed Animals

Benjamin Lecorps and Daniel M. Weary

Abstract

The chapter focuses on the scientific assessment of the emotional experience of pain in farmed animals. A variety of methods allow for insights regarding the issue to felt pain and how much this matters to the animal. These include simple behavioural responses (such as avoidance, body movements and vocalizations), as well as more sophisticated behavioural tests that allow for stronger inferences regarding the emotional component of pain. Some painful experiences can have more persistent effects on animals, resulting in negative mood states. These long-lasting negative affective experiences are of obvious interest when seeking to understand the impact of pain on animal welfare as negative mood means that future experiences are more likely to be experienced negatively. The expectations that animals form regarding pain experiences may make painful experiences even more unpleasant for animals, and a combination of painful experiences and other factors related to the context of the experience (including expectations) may result in suffering, especially when animals are also experiencing fear and have little control over what they experience. Negative experiences (such as pain and fear) may reset what animals expect, setting up a potential cascade of negative experiences; alternatively, the use of positive reinforcement training can mitigate negative expectations in a way that reduces the risk of suffering in farmed animals.

Keywords

Felt emotion · Analgesia · Conditioned-place avoidance · Pain expectations · Animal wellbeing

1 An Introduction to Pain in Farmed Animals

Pain is a common experience for animals kept on farms. The most obvious examples involve the direct pain caused to animals that are part of agricultural practices, such as dehorning cattle, castrating pigs and tail docking lambs. Pain is not the intended outcome of these procedures; these are carried on to reduce perceived risks for farm-

B. Lecorps
Bristol Veterinary School, University of Bristol, Bristol, UK

D. M. Weary (✉)
Animal Welfare Programme, Faculty of Land and Food Systems, University of British Columbia, Vancouver, Canada
e-mail: dan.weary@ubc.ca

R. E. Clutton, G. C. Musk (eds.), *Pain and Suffering in Farmed Animals*,
https://doi.org/10.1007/978-3-032-11407-5_1

ers and animals (as with dehorning cattle) to improve perceived product quality (as in castration of pigs) or to reduce the risk of disease (as with tail docking sheep). But, regardless of why these procedures are carried out, the animal experiences pain. In addition to these direct harms, animals on farms can experience pain resulting from indirect actions or inactions of those responsible for the farm, such as the decisions around housing and management that can increase the risk of health and behavioural problems that can cause pain, e.g. lameness, mastitis, tail biting, feather pecking, etc.

Pain can be defined as "*an unpleasant sensory and emotional experience associated with, or resembling that associated with, actual or potential tissue damage*" (Raja et al. 2020). In this chapter the focus is on the emotional component, as this is what matters to the animal, and hence what is relevant to their welfare, and is also what should matter in terms of the ethics of causing and preventing pain in these animals.

The first aim of this chapter is to describe some of the ways in which the emotional experience of pain can be scientifically assessed in farmed animals. To begin, descriptions of behavioural responses are discussed and then more sophisticated approaches to pain assessment that allow for stronger inferences regarding the emotional component are described.

As will be seen, some painful experiences may not only matter in the moment, i.e. while the pain is felt, but can have lasting consequences on the animal's mood state. These long-lasting negative affective experiences are of obvious interest when seeking to understand the impact of a procedure on animal welfare. Beyond the affective consequences, negative mood means that future experiences (even those that might otherwise be perceived as neutral) are more likely to be experienced as negative, a phenomenon known as cognitive bias (Paul et al. 2005). Thus, in addition to the immediate harms associated with acute pain, animals may experience additional harms related to the emergence of negative expectations. Following from this insight, the chapter aims to explore the role of expectations on pain experiences (and vice versa) and to understand how the combination of painful experiences and other factors related to the context of the experience (including expectations) results in suffering.

2 Scientific Approaches to Assessing the Emotional Component of Pain in Farmed Animals

Above, the argument was made that what matters to animals, and what matters ethically, is the felt, emotional component of pain. But these feelings are private, subjective states that are typically assessed using verbal reports in humans. Can science be used to meaningfully assess such states in non-verbal animals? In recent years, much progress has been made in this area (de Waal and Andrews 2022), such that researchers today draw from a wide range of techniques that allow for inferences of varying strength regarding the feelings of pain (and other emotions). In this section, some of the approaches that have been used are briefly reviewed with discussion of their strengths and weaknesses. To begin, the use of acute behavioural responses during painful procedures is described, and then techniques that more directly assess animal motivation to avoid these experiences (using approach avoidance methods) and how they remember the experience (using conditioned avoidance testing) are presented. Then the assessment of the long-lasting consequences of painful experiences on mood states (using judgement bias and anhedonia tests) is examined. Here, the focus is especially on behavioural tests, but readers should note that the emotional component of pain in farm animals has also been assessed using physiological measures, with mixed results (Box 1).

Box 1: Limitations of Physiological Methods to Assess Pain

Physiological measures have been used to study pain in farm animals, and although they are relatively easy to put in place (most are based on blood samples), they suffer from important limitations when considering their ability to reflect the emotional component of pain. Many physiological approaches are based upon changes in activation of the stress response systems. Measures of heart rate, blood pressure, hormonal responses or even changes in blood flow detected using infrared cameras have all been used to assess emotions in animals (e.g. Ede et al. 2019c). However, these approaches are more likely to capture the level of arousal or the intensity of an emotional experience rather than its valence, i.e. if it is perceived as positive or negative. An example of this limitation is that serum cortisol (a widely used physiological marker of stress) increases in response to positive events such as sexual intercourse (Colborn et al. 1991) and exercise (Stranahan et al. 2008), meaning increases are not specific to negative states. Thus, an important limitation of physiological measures is that they reveal very little about the animal's subjective experience. In contexts where affective valence is clear, measures of arousal can be of value. Some markers, such as the rate of telomere attrition, may provide insights into the cumulative affective experiences of animals (Bateson and Poirier 2019).

Acute Behavioural Responses The immediate responses to presumably painful events, including escape attempts, vocalizations, and grimaces, are often used to draw inferences regarding the affective experience of pain (Ede et al. 2019c). For instance, piglets express more kicks and high-frequency vocalizations when castrated compared to sham castration (Lou et al. 2022), and disbudded calves express more ear flicks and head shakes compared to sham-disbudded animals (Faulkner and Weary 2000). However, these behaviours are not specific to pain. For example, piglets may vocalize when being handled, and calves may flick their ears in response to flies or other disturbances.

Some behaviours are thought to be more closely associated with pain per se, including specific facial expressions. Darwin (1872) discussed the value of using facial expression as an indicator of emotional states in animals. Building upon work on non-verbal human infants, researchers have developed ways of categorizing facial expressions to assess pain in mice (Langford et al. 2010) and in a range of other species including horses (Gleerup et al. 2015b), pigs (Viscardi et al. 2017), cattle (Gleerup et al. 2015a) and sheep (McLennan et al. 2016). This work shows promise, but concerns regarding the validity of these measures remain (Fischer-Tenhagen et al. 2022).

The principal benefit of using acute behavioural measures for pain assessment is that responses are intuitively compelling, and that little special training or experimental set-up is required, so responses can be measured using simple observations. This approach means that the results are of clear practical significance, and that no animals need to be harmed simply for the purpose of studying pain. These reasons likely help explain why this approach to pain assessment is so commonly used by scientists.

Unfortunately, there are important limitations with these measures. Most notably, the strength of inference that can be made relating specifically to the felt, emotional aspect of the pain experience is weak. For example, some withdrawal or avoidance responses can be automatic, happening much more rapidly than would be possible for a process requiring conscious appraisal of an emotional response. Some responses typically assessed in the literature may not fall in this category and are more likely to be consistent with felt, emotional experiences, but in some cases, responses may occur before, or even independent to felt emotion. In addition, these types of responses are often inconsistent across animals.

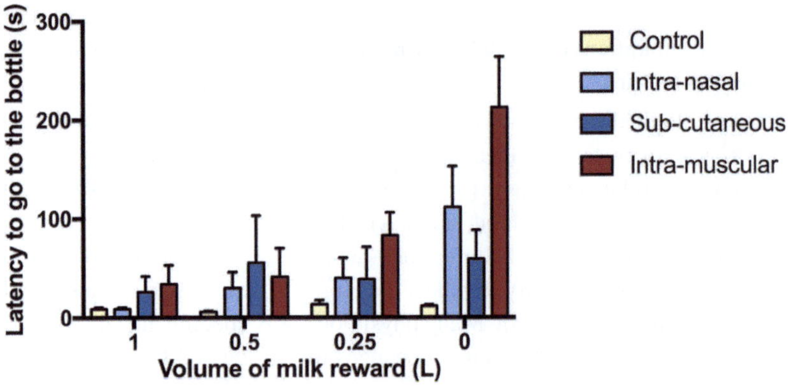

Fig. 1 Latencies of calves to approach a milk bottle (containing decreasing amounts of milk) when doing so was paired with one of four treatments: intra-muscular injection, sub-cutaneous injection, intra-nasal injection or a null control. (Figure adapted from Ede et al. 2018)

For example, only 10–20% of calves vocalize during castration (Stilwell et al. 2008), but it seems unlikely that the remaining calves do not experience pain during the procedure. It would thus seem risky to conclude that animals that express fewer acute responses are experiencing less pain. These measures may thus lack both sensitivity and specificity.

Approach/Avoidance How much pain matters to animals can be assessed using economic contingencies. These experimental approaches aim at quantifying how much animals want something, e.g. how much they are willing to work for pain relief, or how motivated they are to avoid an experience, e.g. how much they are willing to give up to avoid pain. Typically, these methods require that animals are placed in a situation in which they must trade-off rewards and punishments (Kirkden and Pajor 2006). For instance, the assessment of how motivated calves were to avoid different types of painful procedure (in this case, different injection routes) was investigated by assessing whether the calves would give up access to a milk reward to avoid the experience (Ede et al. 2018). Calves were initially trained to go to a bottle to receive a 1 L milk reward. Once calves were rapidly approaching the bottle, injection treatments began (with calves randomly assigned to one of four conditions: intra-muscular, sub-cutaneous, intra-nasal or no injection), and the volume of milk was progressively decreased. Calves continued to approach the bottle when 1 L of milk was provided, but those receiving the intramuscular injection were slower to do so, and treatment differences became more pronounced as the milk reward was reduced. These results show that some routes of injection (like sub-cutaneous) were relatively innocuous to the calves, such that they were willing to accept the procedure for only a small reward, but other methods were more unpleasant for the animals (Fig. 1).

Approach/avoidance experiments have a number of benefits for assessing pain. One is that the responses cannot be said to be reflexive; in the example cited, the calves were making an economic trade-off, and doing so based on past experiences with the painful stimulus, i.e. the animals were responding to the memory of the experience rather than to the actual experience in the moment, as well as their expectations regarding the reward. Such evidence provides a strong basis for concluding that avoiding the experience matters to the animal and can even allow for a quantitative estimate of how much this matters (in this case in terms of lost access to milk).

Approach/avoidance paradigms are especially well suited for evaluating experiences that are repeated and thus allow animals to learn the association. The approach is less suitable for studying procedures that animals would typically experience just once, e.g. castration. In the example presented, the animals required enough time to learn that the reward quantity changed over time, giving them a chance to adjust their behaviour.

Conditioned Place Avoidance Animals tend to avoid places where they experienced negative emotions (and choose to spend time in places where they experienced positive or less negative states), a phenomenon known as 'conditioned place avoidance' (or conditioned place preference). After animals have learnt an association, for example, between a brightly coloured enclosure and a painful procedure, they can be tested by providing them the option of re-entering this location versus a control location that was not associated with pain. In one study, Ede et al. (2019b) used this approach to assess how dairy calves remembered pain during the hours after hot-iron disbudding. Calves experienced a 6-hour recovery period following disbudding in one distinctively coloured pen, and experienced recovery from a sham procedure (with sedation but no disbudding) in another. In the days that followed, calves were tested by allowing them to freely move between the two test pens. The results showed that calves chose to spend more time in the pen associated with the sham procedure (or lie down there) than the one associated with disbudding (Fig. 2, left panel). On the basis of this evidence, the authors concluded that calves learnt to associate the pen with post-operative pain, and this memory of pain was sufficient to motivate the animals to avoid this location when tested on the days following the procedure.

As with the approach/avoidance paradigms, conditioned place avoidance testing has a number of important benefits for assessing the emotional component of pain. One strength of this approach is that the conditioned response reflects the entire experience in the conditioning pen, including the cumulative post-operative pain, which is likely to peak during this period. A disadvantage of this method is that the calf's overall assessment could be driven by anything else experienced in the two pens, and not just the post-operative pain. To help counter this criticism, the authors performed a second study where one pen was associated with recovery from disbudding (as in the first study) and the second pen with recovery from disbudding with an analgesic known to be effective in mitigating the post-operative pain (Ede et al. 2019a). This study showed that calves preferred the pen associated with the analgesic, indicating that preference was indeed driven by their memory of pain.

A second weakness of conditioned place preference testing is that it only shows the rank preference; we can conclude which treatment is better, but not by how much, and cannot use these results to judge how seriously the animal is affected by the pain. To better understand this impact, studies can assess how painful experi-

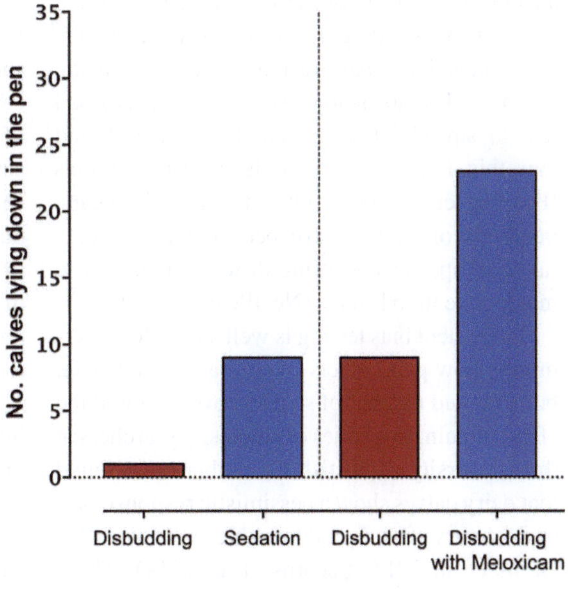

Fig. 2 Number of calves lying down in the different pen associated with either disbudding or sedation (left panel) or disbudding without or with analgesia (right panel). (Figure adapted from Ede et al. 2019a, b)

ences affect mood states as discussed in the following section.

Mood States Pain can result in changes in mood that reflect the emotional toll the experience has on an individual. Mood changes are typically assessed using judgment bias or anhedonia in animals – the former uses the close link between negative emotions and more negative judgment of ambiguous situations, and the latter explores the reduced willingness to seek pleasurable activities when experiencing negative emotions. Both phenomena are often observed in people experiencing low mood and, when persistent, are signs of mood-related disorders.

To test for judgement bias, researchers first train animals to associate a cue, e.g. a red screen, with a positive outcome (a food reward) and another cue, e.g. a white screen, with a negative outcome (a puff of air). Once animals have learnt to *approach* the positive cue and *avoid* the negative one, they are tested with one or more intermediate cues, e.g. pink screens, for which animals have not been trained to associate any outcome. If animals treat the ambiguous cue more like the positive one, e.g. by approaching the pink screen, it can be assumed that they expected the chances of being rewarded were higher than the chances of being punished; in other words, they displayed an optimistic response. However, if the animal responds by avoiding the ambiguous cue, as if it were negative, they displayed a pessimistic response. The first study testing this idea exposed rats to stressful housing conditions and found that this induced a pessimistic response bias (Harding et al. 2004). Over the past 20 years, hundreds of studies have been conducted in a range of species and using different methods to manipulate mood states (Neville et al. 2020).

Judgement bias testing is well suited for determining how pain affects an animal's mood state, but a limited number of studies have explored the effect of pain. In a series of studies, researchers at the University of British Columbia have found that dairy calves show a pessimistic response bias in the hours after disbudding (Neave et al. 2013; Daros et al. 2014; Lecorps et al. 2019). These results indicate that this pain causes a negative mood state in these animals.

Anhedonia Testing for anhedonia is typically simpler than testing for judgment bias, as it does not require as much animal training. Anhedonia can be assessed simply by measuring how much animals prefer a sweet solution versus plain water; the assumptions being that animals normally prefer the sweet solution (because of the rewarding properties of sucrose), and that this preference will wane if animals experience intense negative affective experiences (Rygula et al. 2005). This approach has been used extensively in studies using laboratory animals as animal models for depression, but only one study (Lecorps et al. 2020a) used this approach to study the implications of pain in farmed animals. The results of that study showed that dairy calves reduced their consumption of a sweet solution for several days after disbudding; these results correspond well with those from judgement bias testing, indicating pain-induced low mood in these animals.

Judgment bias and anhedonia provide a different perspective on the emotional consequences of pain in animals, by assessing the longer lasting affective consequences. Determining if, and for how long, animals experience low mood can allow a better assessment of the effect and duration of these negative experiences. These measures reflect the toll of the pain (and perhaps other impacts of these procedures) on the welfare of the animals in ways that the behavioural measures discussed at the beginning of this chapter cannot capture. More minor procedures of shorter duration or impact (like injections) are less likely to result in judgment bias or anhedonia, but further work is required to determine which procedures (under which conditions) are associated with these longer lasting effects on mood.

Although judgment bias and anhedonia testing can further our understanding of pain, both these methodologies also suffer from limitations. For example, judgment bias testing can be difficult to apply due to the need for training animals, and because this method involves rewards and punishments, the training itself may also contribute to changes in emotional states. Also, judgement bias tests may not be suitable when longer term testing is required, as animals can learn to recognize the ambiguous cues with continued exposure (Roelofs et al. 2016). There are also controversies over the validity of using sweet solutions to assess anhedonia, even in rodents, and this type of reward may be even less appropriate in animals for which sucrose may be less rewarding (such as adult dairy cattle). Methodologies assessing both motivational (wanting) and consummatory (liking) anhedonia may prove more useful (Treadway and Zald 2011), and further research is warranted to assess other types of pleasurable activities, such as play and grooming (e.g. Mandel et al. 2013; Lecorps et al. 2020b, 2021). For instance, recent results show that specific play behaviours like play-fighting are suppressed for at least 7 days post-disbudding, suggesting calves avoid this behaviour because of evoked pain or because they display long-lasting fear of pain (St John Wallis et al. 2025).

Assessing Long-Lasting Pain in Animals Most methodologies mentioned to this point have focused on assessing pain lasting from minutes to hours. There are only a few studies exploring pain lasting days to weeks, and a limited number of these used methodologies that would allow inferences about the emotional component of pain. For instance, measures of wound sensitivity or wound-directed behaviours have been used for both castration (Nogues et al. 2021) and disbudding pain (Casoni et al. 2019), but there is a difference between a wound still being sensitive to touch (provoked pain) versus suffering from pain even when the tissue is unprovoked. A potentially promising approach to this problem is determining whether animals seek pain relief or whether the provision of pain relief induces positive affect. Monitoring how long these effects are detectable may allow for inferences regarding the duration of the pain, but these methodologies are still in their infancy in farm animals, with recent evidence suggesting calves benefit from pain relief for multiple days post-disbudding (Colston et al. 2024). Considering the relevance of long-lasting pain when assessing animal welfare, and the lack of research to date, more work considering long-term consequences of painful procedures and of pathologies inducing pain is needed.

3 Understanding Pain in Context

Response to pain has long been thought of as a stimulus-driven experience. However, pain is now known to be affected by a range of contextual factors including expectations about pain. Excellent examples of the importance of expectations are placebo and nocebo analgesia; a drug's analgesic properties can either be eliminated or enhanced depending on whether subjects hold negative or positive expectations about the drug (Bingel et al. 2011). Human patients' expectations about pain before a surgery, such as fear of pain or capacity to cope with pain, can affect their pain experience after surgery (Wang et al. 2018) and how likely they are to experience chronic pain (Jackson et al. 2016). Thus, pain may be better understood when considering the context in which it is experienced (Wiech and Shriver 2018).

Most of the work to date on pain in farm animals focuses on the immediate experience and response, with little consideration given to the context in which animals experience the pain (Weary 2019; Lecorps and Weary 2024). Two important contextual factors influencing if and how much animals are affected by painful experiences are control and fear; both should be taken into account by practitioners and researchers interested in understanding and mitigating painful experiences in animals.

Control Probably the most compelling evidence that the context in which pain is experienced matters comes from studies exploring the effect of predictability and control on animals' responses. In a classic study, Weiss (1972) showed that rats given the ability to predict the occurrence of electrical shocks was less likely to suffer from stress-related diseases, compared to animals who could not predict when shocks would occur. In addition, Weiss found that rats given some control, i.e. providing them with a way to escape the shocks, were at much lower risk of disease than were rats who received the same number of shocks and could not control their occurrence. These results support the idea that it is the animal's perception of the experience that drives the negative consequences of the situation.

There is limited work exploring the effect of predictability and control on affective responses of farmed animals to pain, but work in sheep shows that giving animals the ability to predict and control negative experiences can alleviate the negative effects of stressors (Doyle et al. 2011). Similarly, a study showed that positive reinforcement and counterconditioning can make dairy cows voluntarily accept restraint and reduce their response to a subcutaneous injection (Lomb et al. 2021). These authors worked with three groups of animals: those that were *naïve* (no experiences with restraint and injections), *habituated* (repeated experiences with restraint), and *trained* (animals slowly trained to voluntarily be restrained in return for food treats). On the day of testing, all animals were restrained and received a subcutaneous injection. *Habituated* heifers reacted the most to the injection, perhaps because *trained* heifers were distracted by the food treat, and because *naïve* heifers experienced stress-induced analgesia. In the days following the injection, all the *trained* heifers voluntarily entered the restraint area, but with one exception, none of the *habituation* or *naïve* heifers returned to this area. In this example, the behavioural responses during and after a procedure are associated with both the acute stimulus, i.e. the injection that all heifers received, and also by the animals' expectations. Moreover, this example also shows how expectations can be engineered, in this case creating positive expectations via training.

Fear Pain is often associated with fear and this may affect the experience in many ways. It is fear that allows animals to avoid experiences that are associated with pain (see the conditioned place aversion example above). In some cases, acute fear may actually mitigate pain, i.e. stress-induced analgesia, as referred to in the example above. More generally, fear is likely to make the experience of pain worse or to extend the negative consequences of pain. Research on how fear affects the experience of pain is limited in farmed animals so examples from the human and laboratory animal literature that are relevant to farmed animals are considered.

Fear plays an important role in pain responses and pain recovery. The main function of fear in the context of pain is to trigger avoidance behaviours, such as to protect an individual from re-experiencing pain. It is the fear of pain that prevents us from doing movements associated with pain like putting weight on a sore ankle. However, after some time has passed, it is expected that individuals restart moving and progressively come back to previous activity levels. Although experiencing fear has a protective function, the cost of excessive and persistent fear responses is worth consideration given that "*engaging in avoidance may (paradoxically) increase rather than decrease pain-related fear*" (van Vliet et al. 2018) and thus delay recovery. In this way, individuals may be stuck in a cycle where their fear of pain triggers pain vigilance which increases negative affect potentially leading to anxiety or depression as well as creating long-term pain-related disabilities (Rogers and Farris 2022). In this way, instead of protecting individuals from pain, fear may prolong negative affect and physical disabilities associated with pain.

Similar processes have been found in laboratory animals where conditioned fear responses can persist for long periods (Meulders 2020). Thus, some animals appear to be prone to maintaining beliefs about pain-eliciting situations, leading to prolonged fear, even though they were repeatedly exposed to the unconditioned stimulus in absence of pain (Winters et al. 2021). It has also been shown that the severity of lameness in some horses does not correlate with the severity of the physical injury (Ijichi et al. 2014) perhaps because some horses developed persistent fear and continued to limp even if walking is no longer associated with pain.

Fear of pain is often directed to specific movements or external stimuli. However, the experience of pain can lead to a more dramatic change in an animal's perception of its world (Box 2). For example, work in laboratory animals shows that electrical shocks not only lead to lasting fear conditioning for the environment where they received the shocks but can also lead to conditioned aversion of any new environment. This effect, known as fear overgeneralization and "*characterized by a failure to discriminate safe from aversive environmental contexts*" (Winters et al. 2021), has implications for livestock who after a painful procedure involving humans, may consider all interactions with humans as likely to be painful. Although evidence is lacking, preliminary work suggest that castrated sheep became more fearful (assessed using a range of tests involving different types of novel situations) than intact animals (Vandenheede and Bouissou 1996). Tail-docked piglets were also found more fearful of humans as highlighted by longer latencies to contact handlers (Tallet et al. 2019), and some calves displayed persistent fear conditioning after hot-iron disbudding despite three extinction sessions (where the pen was no longer associated with pain, Ede et al. 2019b; Fig. 3). Overall, fear of pain may either persist or generalize in animals, which may mean that some procedures with relatively short-lasting acute pain may have negative consequences for animals well beyond the time in which the animal is in pain.

Box 2: Pain Is a Matter of Expectations

Pain is accompanied by changes in expectations, with animals typically appraising their world in a more pessimistic way while in pain and potentially even after the pain is gone. How the brain integrates information also appears to matter for how pain is processed and responded to. The brain does not passively integrate information coming from senses but is actively processing this information in the context of prior knowledge (Clark 2013). The brain is therefore thought to work as a sophisticated predictive (or Bayesian) machine. This means that animals generate expectations in situations where they expect to feel pain, and considering the role of expectations can therefore help understanding affective experiences of animals. There are several reasons why this cognitive process is important to consider. First, animals may not be able to predict the occurrence of pain (if this is the first time they experience a procedure). In this case, the situation would strongly violate the animals' expectations, leading to more severe affective experiences (Mendl and Paul 2020). Second, the experience of pain itself will reset the animal's expectations, such that the animal will be more likely to believe it will experience pain in future encounters. This type of effect explains why animals show shifts in perception (judgment bias) after painful procedures. If animals are expecting that their world is a dangerous place, they may experience fear more easily, e.g. every time they interacts with humans. Finally, traumatic or repeated negative experiences may disrupt the functioning of this system and lead to an incapacity to update expectations (Lecorps and Weary 2024). This form of harm has not yet been explored in the animal literature but is now gaining interest in the human literature to explain how mood-related disorders develop and persist (van de Cruys and Van Dessel 2021) and could explain why some animals show persistent and generalized fear.

4 Where Does This Leave Us?

This Chapter has examined how scientific methods can tell us about the nature, severity and duration of the felt experience of pain in farm animals. It has also highlighted that it is important to consider pain in the context in which it is experienced and that the welfare consequences of pain range far beyond pain itself. The section that follows considers what this knowledge means for practitioners and their efforts to reduce pain in the animals they work with.

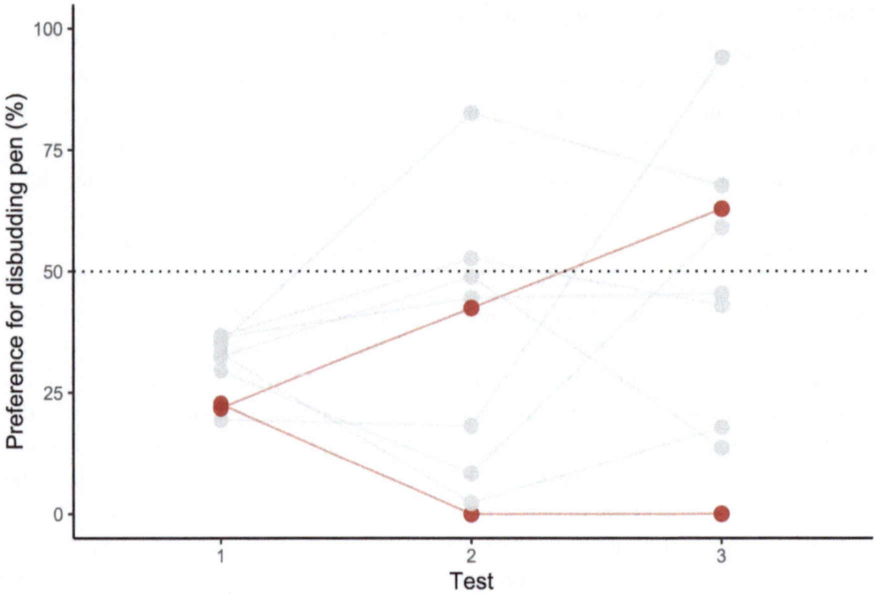

Fig. 3 Conditioned place aversion after disbudding. Calves were tested over three sessions, and whether they spent less time in the pen where they were disbudded (versus a pen where they were sham disbudded) was assessed. All calves displayed an aversion to the pen where they were disbudded in session 1 but this effect waned in sessions 2 and 3. Although some calves recovered (did not show aversion anymore) some displayed persistent aversion. All dots and lines represent individual calves. The two red lines show two contrasting 'trajectories' in terms of persistent fear conditioning. Points below 50% indicate that animals spent less time in the pen where they were previously disbudded. (Figure adapted from Ede et al. 2019b)

Phasing Out Painful Procedures Those responsible for the care of farmed animals need to find ways to avoid painful procedures such as dehorning, castration and tail docking. In some cases, alternatives may be relatively easy to adopt. Over the past decade, Canada's dairy industry eliminated the procedure of tail docking once it became clear that there was little advantage to the procedure. Similarly, dehorning is now rarer in the beef industry, given the availability of excellent genetics for animals that are genetically hornless, i.e. polled; polled genetics are now increasingly available for dairy breeds. In some cases, eliminating painful procedures will take more work, for example, if this requires changes in processing, e.g. slaughtering intact male pigs at younger ages to avoid boar taint, or management, e.g. providing more space to laying hens to reduce the risk of feather pecking. The important message is that for each of these painful procedures, potential alternatives already exist. Some jurisdictions and production standards have already phased out some of these procedures, e.g. in EU's organic standards (Duval et al. 2020).

Provide Effective Pain Control Animals feel pain, and this pain matters to them; all this can be assessed using the methods outlined in this Chapter. But this is more than just a matter of scientific interest. This knowledge brings with it the moral requirement that those caring for animals do what they can to avoid this pain. Most obviously, this includes providing adequate pain control; for many of the procedures discussed this will require a multimodal approach including the use of a sedative, a local anesthetic, and analgesics to treat the post-operative pain. In some instances, caregivers can select versions of a procedure that are less painful (as shown above for pain associated with different types of injection). Recent work has also shown that, when calves need to be castrated, surgery is associated with less pain in the weeks following the procedure, compared to the common alternative of band castration (Nogues et al. 2021). Procedures likely to trigger long-lasting pain (like band castration) should be considered especially problematic because there are no realistic pain control methods available when this technique is applied.

Consider the Big Picture The effect of painful procedures is not limited to the physical injury that is caused. Consideration must also be given to the risks of triggering long-lasting consequences including increased fear. Minimizing fear can be achieved in a variety of ways such as providing more predictability and control, habituating or counter-conditioning animals to places where they experience mild pains, and so on. Such techniques are now commonly used in zoos to treat animals that would otherwise require sedation for veterinary care. Animal training goals are achievable now and should be more widely used in the future.

Reflecting on the context in which animals experience pain will provide clues to how best to manage this from the animals' perspective. For instance, when disbudding calves, a simple way to avoid fear could be to use sedation. Another option could be to avoid imposing multiple stressors at once. For instance, perhaps disbudding should not happen at the same time as weaning, as these negative experiences may have synergistic effects, i.e. weaning may be more difficult for animals in pain.

Tailoring Practices to Individual Circumstances Animals, like humans, are not all equal with respect to pain. Some animals may be especially vulnerable to these experiences. There is limited research on individual variation in response to pain in farmed animals, but early data suggest that personality traits may play a role. For example, in one study, it was found that more pessimistic calves were more likely to show anhedonia after disbudding (Fig. 4, Lecorps et al. 2020a). Many factors, including the animal's pre-

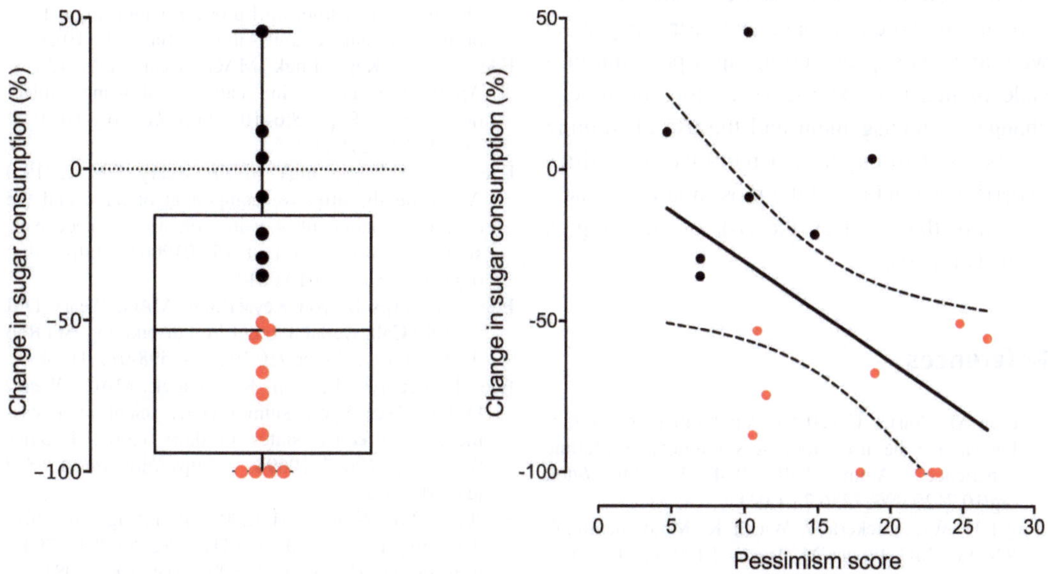

Fig. 4 Pessimistic calves are more vulnerable to pain induced by disbudding. Calves' consumption of a sweet solution was monitored before and after disbudding as a means to assess anhedonia (the more animals decreased their consumption the worst their response to pain). Percentage change in sweet solution consumption on the day after disbudding was significantly lower (left panel), and more pessimistic calves (orange dots) had stronger decreases, suggesting they were more affected by the postoperative pain (right panel). (Figure adapted from Lecorps et al. 2020a)

vious experiences, are likely to play a role in determining which animals are at heightened risk. Regardless of what makes an individual more prone to pain (or to fear pain), this vulnerability needs to be addressed. Some individuals will display much more intense responses and should be treated accordingly.

5 Conclusion

How much pain hurts, i.e. the affective component of pain, and how much it matters to animals, can be studied using a variety of scientific methods. Painful procedures can have a multitude of negative effects, beyond the acute pain. For example, painful experiences can result in the development of negative mood states that then also affect how the animals experience other aspects of their life, e.g. judging neutral experiences as negative, and no longer enjoying previous pleasures. These harms are most likely when animals are also experiencing fear and have little or no control over what they experience, factors that are common for farmed animals. Negative experiences (such as pain and fear) may reset what animals expect, setting up a potential cascade of negative experiences. More positively, changes in management and the use of training can be used to engineer a narrative of positive experiences (and avoid the most traumatic ones) in a way that reduces the risk of suffering in farmed animals.

References

Bateson M, Poirier C (2019) Can biomarkers of biological age be used to assess cumulative lifetime experience? Anim Welf 28:41–56. https://doi.org/10.7120/09627286.28.1.041

Bingel U, Wanigasekera V, Wiech K, Mhuircheartaigh RN, Lee MC, Ploner M, Tracey I (2011) The effect of treatment expectation on drug efficacy: imaging the analgesic benefit of the opioid remifentanil. Sci Transl Med 3:1–10. https://doi.org/10.1126/scitranslmed.3001244

Casoni D, Mirra A, Suter MR, Gutzwiller A, Spadavecchia C (2019) Can disbudding of calves (one versus four weeks of age) induce chronic pain? Physiol Behav 199:47–55. https://doi.org/10.1016/j.physbeh.2018.11.010

Clark A (2013) Whatever next? Predictive brains, situated agents, and the future of cognitive science. Behav Brain Sci 36:181–204. https://doi.org/10.1017/S0140525X12000477

Colston KP, Ede T, Mendl MT, Lecorps B (2024) Cold therapy and pain relief after hot-iron disbudding in dairy calves. PLOS One 19(7). https://doi.org/10.1371/journal.pone.0306889

Colborn DR, Thompson DL, Roth TL, Capehart JS, White KL (1991) Responses of cortisol and prolactin to sexual excitement and stress in stallions and geldings. J Anim Sci 69:2556–2562

Daros RR, Costa JHC, von Keyserlingk MAG, Hötzel MJ, Weary DM (2014) Separation from the dam causes negative judgement bias in dairy calves. PLoS One 9:e98429. https://doi.org/10.1371/journal.pone.0098429

Darwin C (1872) The expression of the emotions in man and animals. University of Chicago

de Waal FBM, Andrews K (2022) The question of animal emotions. Science 375:1351–1352. https://doi.org/10.1126/science.abo2378

Doyle RE, Lee C, Deiss V, Fisher AD, Hinch GN, Boissy A (2011) Measuring judgement bias and emotional reactivity in sheep following long-term exposure to unpredictable and aversive events. Physiol Behav 102:503–510. https://doi.org/10.1016/j.physbeh.2011.01.001

Duval E, von Keyserlingk MAG, Lecorps B (2020) Organic dairy cattle: do European Union regulations promote animal welfare? Animals (Basel) 10(10):1786

Ede T, Von Keyserlingk MAG, Weary DM (2018) Approach-aversion in calves following injections. Sci Rep 8:6–10. https://doi.org/10.1038/s41598-018-27669-7

Ede T, Von Keyserlingk MAG, Weary DM (2019a) Assessing the affective component of pain, and the efficacy of pain control, using conditioned place aversion in calves. Biol Lett 15:20190642. https://doi.org/10.1098/rsbl.2019.0642

Ede T, Lecorps B, von Keyserlingk MAG, Weary DM (2019b) Calf aversion to hot-iron disbudding. Sci Rep 9:5344. https://doi.org/10.1038/s41598-019-41798-7

Ede T, Lecorps B, von Keyserlingk MAG, Weary DM (2019c) Symposium review: scientific assessment of affective states in dairy cattle. J Dairy Sci 102:10677–10694. https://doi.org/10.3168/jds.2019-16325

Faulkner PM, Weary DM (2000) Reducing pain after dehorning in dairy calves. J Dairy Sci 83:2037–2041. https://doi.org/10.3168/jds.S0022-0302(00)75084-3

Fischer-Tenhagen C, Meier J, Pohl A (2022) "Do not look at me like that": is the facial expression score reliable and accurate to evaluate pain in large domestic animals? A systematic review. Front Vet Sci 9:1002681. https://doi.org/10.3389/fvets.2022.1002681

Gleerup KB, Andersen PH, Munksgaard L, Forkman B (2015a) Pain evaluation in dairy cattle. Appl Anim

Behav Sci 171:25–32. https://doi.org/10.1016/j.applanim.2015.08.023

Gleerup KB, Forkman B, Lindegaard C, Andersen PH (2015b) An equine pain face. Vet Anaesth Analg 42:103–114. https://doi.org/10.1111/vaa.12212

Harding EJ, Paul ES, Mendl M (2004) Cognitive bias and affective state. Nature 427:312. https://doi.org/10.1038/427312a

Ijichi C, Collins LM, Elwood RW (2014) Pain expression is linked to personality in horses. Appl Anim Behav Sci 152:38–43. https://doi.org/10.1016/j.applanim.2013.12.007

Jackson T, Tian P, Wang Y, Iezzi T, Xie W (2016) Toward identifying moderators of associations between presurgery emotional distress and postoperative pain outcomes: a meta-analysis of longitudinal studies. J Pain 17:874–888. https://doi.org/10.1016/j.jpain.2016.04.003

Kirkden RD, Pajor EA (2006) Using preference, motivation and aversion tests to ask scientific questions about animals' feelings. Appl Anim Behav Sci 100:29–47. https://doi.org/10.1016/j.applanim.2006.04.009

Langford DJ, Bailey AL, Chanda ML, Clarke SE, Drummond TE, Echols S, Glick S, Ingrao J, Klassen-Ross T, Lacroix-Fralish ML, Matsumiya L, Sorge RE, Sotocinal SG, Tabaka JM, Wong D, van den Maagdenberg AM, Ferrari MD, Craig KD, Mogil JS (2010) Coding of facial expressions of pain in the laboratory mouse. Nat Methods 7:447–449. https://doi.org/10.1038/nmeth.1455

Lecorps B, Weary DM (2024) Animal affect, welfare and the Bayesian brain. Anim Welf 33:e39

Lecorps B, Ludwig BR, von Keyserlingk MAG, Weary DM (2019) Pain-induced pessimism and anhedonia: evidence from a novel probability-based judgment bias test. Front Behav Neurosci 13:54. https://doi.org/10.3389/fnbeh.2019.00054

Lecorps B, Nogues E, von Keyserlingk MAG, Weary DM (2020a) Pessimistic dairy calves are more vulnerable to pain-induced anhedonia. PLoS One 15:1–19. https://doi.org/10.1371/journal.pone.0242100

Lecorps B, Weary DM, von Keyserlingk MAG (2020b) Regrouping induces anhedonia-like responses in dairy heifers. JDS Commun 1(2):45–49. https://doi.org/10.3168/jdsc.2020-0023

Lecorps B, Welk A, Weary DM, von Keyserlingk MAG (2021) Postpartum stressors cause a reduction in mechanical brush use in dairy cows. Animals 11:3031. https://doi.org/10.3390/ani11113031

Lomb J, Mauger A, von Keyserlingk MAG, Weary DM (2021) Effects of positive reinforcement training for heifers on responses to a subcutaneous injection. J Dairy Sci 104:6146–6158. https://doi.org/10.3168/jds.2020-19463

Lou M, Ventura B, Deen J, Li Y (2022) Surgical castration changes struggle behavior and vocalizations in male piglets. J Appl Anim Welf Sci 25:410–417. https://doi.org/10.1080/10888705.2021.1916938

Mandel R, Whay HR, Nicol CJ, Klement E (2013) The effect of food location, heat load, and intrusive medical procedures on brushing activity in dairy cows. J Dairy Sci 96:6506–6513. https://doi.org/10.3168/jds.2013-6941

McLennan KM, Rebelo CJB, Corke MJ, Holmes MA, Leach MC, Constantino-Casas F (2016) Development of a facial expression scale using footrot and mastitis as models of pain in sheep. Appl Anim Behav Sci 176:19–26. https://doi.org/10.1016/j.applanim.2016.01.007

Mendl M, Paul ES (2020) Animal affect and decision-making. Neurosci Biobehav Rev 112:144–163. https://doi.org/10.1016/j.neubiorev.2020.01.025

Meulders A (2020) Fear in the context of pain: lessons learned from 100 years of fear conditioning research. Behav Res Ther 131:103635. https://doi.org/10.1016/j.brat.2020.103635

Neave HW, Daros RR, Costa JHC, von Keyserlingk MAG, Weary DM (2013) Pain and pessimism: dairy calves exhibit negative judgement bias following hot-iron disbudding. PLoS One 8:e80556. https://doi.org/10.1371/journal.pone.0080556

Neville V, Nakagawa S, Zidar J, Paul ES, Lagisz M, Bateson M, Lovlie H, Mendl M (2020) Pharmacological manipulations of judgement bias: a systematic review and meta-analysis. Neurosci Biobehav Rev 108:269–286. https://doi.org/10.1101/612382

Nogues E, von Keyserlingk MAG, Weary DM (2021) Pain in the weeks following surgical and rubber ring castration in dairy calves. J Dairy Sci 104:12881–12886. https://doi.org/10.3168/jds.2021-20127

Paul ES, Harding EJ, Mendl M (2005) Measuring emotional processes in animals: the utility of a cognitive approach. Neurosci Biobehav Rev 29:469–491. https://doi.org/10.1016/j.neubiorev.2005.01.002

Raja SN, Carr DB, Cohen M, Finnerup NB, Flor H, Gibson S, Keefe FJ, Mogil JS, Ringkamp M, Sluka KA, Song XJ, Stevens B, Sullivan MD, Tutelman PR, Ushida T, Vader K (2020) The revised International Association for the Study of Pain definition of pain: concepts, challenges, and compromises. Pain 161:1976–1982. https://doi.org/10.1097/j.pain.0000000000001939

Roelofs S, Boleij H, Nordquist R, Van Der Staay FJ (2016) Making decisions under ambiguity: judgment bias tasks for assessing emotional state in animals. Front Behav Neurosci 10:1–16. https://doi.org/10.3389/fnbeh.2016.00119

Rogers AH, Farris SG (2022) A meta-analysis of the associations of elements of the fear-avoidance model of chronic pain with negative affect, depression, anxiety, pain-related disability and pain intensity. Eur J Pain 26:1611–1635. https://doi.org/10.1002/ejp.1994

Rygula R, Abumaria N, Flügge G, Fuchs E, Rüther E, Havemann-Reinecke U (2005) Anhedonia and motivational deficits in rats: impact of chronic social stress. Behav Brain Res 162:127–134. https://doi.org/10.1016/j.bbr.2005.03.009

Stilwell G, Lima MS, Broom DM (2008) Comparing plasma cortisol and behaviour of calves dehorned with caustic paste after non-steroidal-anti-inflammatory analgesia. Livest Sci 119:63–69. https://doi.org/10.1016/j.livsci.2008.02.013

St John Wallis A, Held SD, Mendl MT, von Keyserlingk MA, Weary DM, Lecorps B (2025) Pain and pessimism affect calves' play behaviours post-disbudding Discover Animals 2(1). https://doi.org/10.1007/s44338-025-00105-7

Stranahan AM, Lee K, Mattson MP (2008) Central mechanisms of HPA axis regulation by voluntary exercise. NeuroMolecular Med 10:118–127. https://doi.org/10.1007/s12017-008-8027-0

Tallet C, Rakotomahandry M, Herlemont S, Prunier A (2019) Evidence of Pain, Stress, and Fear of Humans During Tail Docking and the Next Four Weeks in Piglets (Sus scrofa domesticus). Front Vet Sci 6:1–11. https://doi.org/10.3389/fvets.2019.00462

Treadway MT, Zald DH (2011) Reconsidering anhedonia in depression: lessons from translational neuroscience. Neurosci Biobehav Rev 35:537–555. https://doi.org/10.1016/j.neubiorev.2010.06.006

van de Cruys S, Van Dessel P (2021) Mental distress through the prism of predictive processing theory. Curr Opin Psychol 41:107–112. https://doi.org/10.1016/j.copsyc.2021.07.006

van Vliet CM, Meulders A, Vancleef LMG, Vlaeyen JWS (2018) The opportunity to avoid pain may paradoxically increase fear. J Pain 19:1222–1230. https://doi.org/10.1016/j.jpain.2018.05.003

Vandenheede M, Bouissou MF (1996) Effects of castration on fear reactions of male sheep. Appl Anim Behav Sci 47:211–224. https://doi.org/10.1016/0168-1591(95)00673-7

Viscardi AV, Hunniford M, Lawlis P, Leach M, Turner PV (2017) Development of a piglet grimace scale to evaluate piglet pain using facial expressions following castration and tail docking: a pilot study. Front Vet Sci 4:1–9. https://doi.org/10.3389/fvets.2017.00051

Wang L, Chang Y, Kennedy SA, Hong PJ, Chow N, Couban RJ, McCabe RE, Bieling PJ, Busse JW (2018) Perioperative psychotherapy for persistent post-surgical pain and physical impairment: a meta-analysis of randomised trials. Br J Anaesth 120:1304–1314. https://doi.org/10.1016/j.bja.2017.10.026

Weary DM (2019) Suffering, agency and the Bayesian mind. In: FD MM (ed) Mental health and well-being in animals. CABI, Wallingford, pp 1–35

Weiss JM (1972) Psychological factors in stress and disease. Sci Am 226:104–113. https://doi.org/10.1038/scientificamerican0672-104

Wiech K, Shriver A (2018) Cognition doesn't only modulate pain perception; it's a central component of it. AJOB Neurosci 9:196–198. https://doi.org/10.1080/21507740.2018.1496180

Winters JJ, Hardy LW, Sullivan JM, Powell NA, Qutaish M, Nair S, Heimann J, Ghayoor A, Polyak I, Chaby L, Rodriguez E, Chaar D, Oscherwitz J, Liberzon I (2021) Functional deficit in hippocampal activity during fear extinction recall in the single prolonged-stress model of PTSD in male rats. Behav Brain Res 396:112902. https://doi.org/10.1016/j.bbr.2020.112902

Sociocultural and Religious Influences on Practices

Rachel A. Ankeny, Emily A. Buddle, and Alexandra L. Whittaker

Abstract

Acceptable farming practices, and associated conceptualizations of pain and suffering, differ not only between different religions of the world but also between diverse cultural backgrounds and social settings. This chapter explores the various influences on production animal management that derive from sources external to animal welfare science and that result in different attitudes to farm animal welfare. The focus is primarily on cattle, sheep, pigs and poultry, as these are the species for which there is the most evidence and information; other farmed animals (i.e. those that are used to provide food for human consumption), are also considered because these animals are amongst the most numerous in terms of scale and impacts (e.g., fish and insects). The chapter focuses on specific tensions to illustrate those issues that are likely to be most useful to those actively involved in animal care and management (e.g., veterinary professionals, farmers and scientists). It is hoped that the chapter reveals the importance of an evidence-based understanding of current worldviews, the role of animal welfare science and the evolution of cultural differences to aid such professionals to empathize and engage in dialogue with their clients, other researchers or members of the general public with regard to their approaches to pain, suffering and animal welfare.

Keywords

Animal welfare · Animal ethics · Livestock · Farm animals · Cross-cultural attitudes · Religious views

1 Introduction

With the rapid growth of understanding of the biological and psychological bases of pain and suffering in farm animals come inconsistencies in best practices. Many have become accentuated by cultural and particularly religious influences, which have been cited as key drivers associated with what are held by some to be poor and even inhumane processes. Differences in practices and attitudes about treatment of farm animals are often sources of debate and contention within and between cultures and have become especially

R. A. Ankeny (✉)
Philosophy Group, Wageningen University, Wageningen, The Netherlands
e-mail: rachel.ankeny@wur.nl

E. A. Buddle
School of Agriculture, Food, & Wine, University of Adelaide, Adelaide, Australia

A. L. Whittaker
School of Veterinary Science, University of Adelaide, Adelaide, Australia

R. E. Clutton, G. C. Musk (eds.), *Pain and Suffering in Farmed Animals*,
https://doi.org/10.1007/978-3-032-11407-5_2

evident in multicultural settings in our increasingly globalized world. Although there are occasional calls for global harmonisation of regulations associated with animal welfare, there are many barriers to this ideal.

In the recent past, the strongest voices of concern about farm animal treatment came from the abolitionists, animal rights or liberation activists whose aims were to eliminate the use of non-human animals for production-related purposes and ban the consumption of non-human animals altogether. Those voices remain currently active but have been joined by others wishing to improve processes associated with animal production whilst not necessarily eliminating these types of production. Examples of these concerns include live export, (commercial transport of livestock across national borders including long distances on sea or land), halal and kosher slaughter, and the killing of bobby calves (young, usually male, calves that are superfluous to needs in the dairy industry).

Acceptable farming practices and associated conceptualizations of pain and welfare differ not only between different religions but also between diverse cultural and social groups. The factors that influence production animal management derive from values outside of animal welfare science, as well as contentions within it, for instance the prioritization of affective versus objective measures to drive improvements in farm animal welfare. Western approaches to farmed animal pain and welfare rely heavily on an evidence base. For this reason, cattle, sheep, pigs and poultry figure predominantly in debates. However, other farmed animals, that is, those that are held under some form of control to provide food for human consumption, are also important on the basis of numbers alone (e.g., fish and insects). The scope of this chapter is not to provide an inclusive overview of all perspectives but selectively highlight those of most use to those actively involved in animal welfare and management, such as veterinary professionals, farmers and scientists working in animal welfare or with animals.

2 Attitudes About Farmed Animal Welfare

Understanding how social, cultural and religious factors can influence the way humans view pain and suffering in farm animals is facilitated by considering how humans construct their worldviews. The ways in which humans perceive and organize the world around them, including their understandings of other humans and non-human animals, are often shaped by specific cultural and historical circumstances within which they are socialised. Culturally constructed meanings come to be attached to things, and these in turn become assumptions that are not only taken for granted but form the basis of an individual's worldviews. People's ongoing experiences within the world, and the ways in which they respond to these experiences, are also fundamental to shaping perceptions, and hence their worldviews do not always align strictly to the culture from which they are raised or in which they live.

Everyday experiences can result in individuals going against their sociocultural backgrounds and lead them to reflect on their own values and actions, which in turn can lead to tensions with dominant sociocultural norms (Ingold 2000; Milton 2002). A simple example of daily experiences shifting cultural conditioning and associated values relates to the traditional dominance of meat consumption in Australia. Australia has one of the largest per capita rates of meat consumption globally, with meat eating being deeply ingrained in Australian culture (Santich 1990; Ankeny 2008). However, various sociocultural shifts such as people living in urban areas away from livestock production, increases in industrialized agriculture and the growth of climate consciousness have resulted in different daily experiences and in turn in greater reflection about the dominant role that meat has in Australian diets, resulting in reductions of, or in some cases complete abstinence from, meat consumption.

The crux of this chapter is an exploration of how the roles of culture and associated phenomena, including religion and social structures, influence practices used to farm animals and shape attitudes toward and perceptions of other cultures and their practices in this domain. In an increasingly multicultural and globalized world, veterinary professionals and animal scientists will encounter a variety of cultures that may have different understandings of animal welfare that run counter to their own sociocultural contexts and grounding, and which challenge existing worldviews and beliefs. Thus, it is critical to have a deep understanding of how personal worldviews and understandings of animal welfare are shaped, as well as how those of other cultures arose, to help such professionals empathize and engage in dialogue with their clients, other researchers or members of the general public with regard to their approaches to pain, suffering and animal welfare.

Contemporary approaches to animal welfare generally have been grounded on the relatively recent idea that non-human animals are sentient and thus consideration should be given to their well-being or suffering, especially when they are under the care of humans who control how they are used and what is done to them, as is the case with farm animals. Historically, there were diverse views on this issue: for instance, some interpret Christianity (and the Dominion Theory in the Old Testament in particular) as holding that humans are distinct from other types of animals over whom they have stewardship and are permitted to use as they see fit. Others, however, view the idea of stewardship to involve caring for non-human others, treating them as individuals rather than as a uniform group or as property, and argue that we have moral obligations to avoid causing unnecessary suffering and to treat them responsibly (Rollin 1981; Regan 2004).

Some who argue for animal rights or animal liberation contend that focus on improving animal welfare fails to bring fundamental assumptions about humans' relationships with non-human species into full view and fosters continued exploitation of these animals, with some explicitly associating Christian thought with ongoing contributions to animal cruelty and suffering due to speciesism (see Singer 1975, but compare his later capitulation on this point in Singer 2023). Other scholars have defended a Christian ethics of farmed animal welfare which holds that the lives of all creatures have value precisely because they have been created by God as ends in themselves, with no creature existing merely to foster the well-being of another and humans being obligated to enable the flourishing of fellow creatures wherever possible (Adam et al. 2019). Some interpret the notion of dominion in the Judeo-Christian tradition as resulting in strong obligations to care for animals (Scully 2002).

Similarly, other religious traditions have been variously interpreted as implying that humans have responsibilities to care for non-human animals, including Shariah law in Islam. Most belief systems associated with major religions emphasize the need for humans to be extremely respectful of non-human animals, including during rearing, production and slaughtering processes. Some commentators note how religious dietary norms related to animals (such as fasting from meat, and halal and kosher practices) can be viewed as part of these norms, and in turn underscoring the widely recognized social and cultural value of meat in particular (Tieman and Hassan 2015).

Even in those religious traditions that are associated with not consuming certain animal products, attitudes to animal welfare involve complex combinations of religion, history and politics, among other factors. For instance, in India, practices involving cruelty to animals including animal sacrifices were scrutinized by various religious or spiritual movements at different times for diverse reasons, including ideas associated with non-violence, purity and reincarnation (Berti 2019). A new way of viewing these concerns was introduced along with colonization by the British: the inherent tension at the centre of animal welfare debates in the West resulting from non-human animals both as resources for human consumption and as objects of compassion was further amplified in India due to the complex political and socio-religious dynamics of the colonial situation (Chakrabarti 2010). The British enacted legislation to prevent animal cruelty in India (the earliest in Bengal in 1869), particularly against what was viewed as barbaric local practices, while simultaneously exploiting animals for colonial economic enterprises, e.g. for military uses and in hunting, and introducing the practice of vivisection for animal experimentation to India (Chakrabarti 2010).

Legislation and activism associated with farm animal welfare reveals that many points of concern today are not in fact new: for instance the UK Society for the Prevention of Cruelty to Animals (SPCA) founded in 1824 (which became the RSPCA in 1840) campaigned against several matters of current concern, such as calf bleeding for veal, livestock transport conditions, inhumane slaughter methods, dehorning of cattle and nose-branding of sheep (Fairholme and Pain 1934), as part of a broader movement motivated by a Christian ethic of mercy, kindness and compassion that sought to civilize the working class and raise its moral standards (Ritvo 1987). The UK was amongst the earliest to enact national legislation, with the Cruelty to Animals Act passing in 1835. In contrast in the United States, there was piecemeal state legislation against animal cruelty dating back to the 1800s, but no national laws until the Animal Welfare Act of 1966. Most commentators attribute significant legislative progress around farm animal welfare to the Brambell investigation in the United Kingdom, which occurred in the mid-1960s in part as a response to Harrison's 1964 book *Animal Machines* (Woods 2011). The UK government established the Farm Animal Welfare Advisory Committee in 1967 (which later became the Farm Animal Welfare Council in 1979), and it promulgated the basic guidelines that later became the Five Freedoms. Recently, there have been efforts to improve farm animal welfare in Europe to reflect broader social values (Bock and Buller 2013; Miele et al. 2013), although it is acknowledged that there are also members of the community who do not support animal production, and others for whom it is of little concern.

3 Shifting Relationships, Shifting Views

In order to situate an exploration of sociocultural factors that influence views on farmed animal welfare, it is important to keep the major shifts in the ways in which farm animals have been raised in view. Hundreds of years ago, in many Western, and what now may be considered to be more developed locales, there was a gradual shift away from nomadic herding to enclosed farming. The Industrial Revolution in the eighteenth to nineteenth centuries had significant effects on livestock farming resulting from the mass production of farming equipment and the rise of more scientific breeding techniques.

From the second half of the twentieth century onwards in many locales but particularly Western Europe and the United States, the intensification of animal agriculture developed rapidly. Most intensively, farmed animals today have been selectively bred to maximize efficient production, with limited attention paid to the effects of such breeding on their welfare. In these sorts of industrialized settings, production animals tended to have had limited environmental stimulation and did not typically engage in species-specific behaviours. The resulting increases in animal productivity are further supported by consumer expectations about meat being a relatively low-cost protein source to be consumed in larger quantities on a daily basis. Few consumers have first-hand knowledge about animal production, and instead tend to rely on over idealized or romanticized notions of production animal lives–or alternatively accept the criticisms of an anti-farming lobby which are not always based on fact.

It is important to note that not all animal farming follows these patterns. In many other locales, particularly in Asia, Africa and South America, small or subsistence farming is the more common production system. Consequently, the reasoning associated with concerns about animal welfare may also differ across countries or cultures. In developed countries, animal welfare is often considered important due to the intrinsic value of the animals, based on their abilities to experience suffering. However, in other cultures where subsistence or small rural farming enterprises are more widespread, animal welfare may be linked in people's minds with what is formally termed the concept of 'one welfare', which promotes the critical links between non-human animal welfare, human well-being, and the physical and social environment (Pinillos et al. 2016). Thus, farm animal welfare is linked not only with greater productivity and improved animal health with reduced expenses, but also with higher qualities of human life and of the environment with less mortality and reduced veterinary medical expenses.

Attitudes toward animals are shaped by other factors that affect human interests in caring about certain species or groups. Whilst it is often assumed that the level of care that humans provide for a particular species is relatively universal and largely driven by our knowledge of their biology, notably sentient abilities and pain perception, proximity to species and cultural norms are likely to be important moderating factors. Such norms are often described as related to the nature of the human–animal bond between certain species, particularly in the context of companion animals, a group which research has shown are more valued by humans than other animals and hence popularly thought to require higher levels of animal welfare protection (Rioja-Lang et al. 2020; Sandgren et al. 2020; Morton et al. 2022).

For example, consider the koala. In a recent study (Sinclair et al. 2022), Australians ranked koala welfare more important than human welfare and the welfare of other animals with which they were presented in the study, including another native animal, the kangaroo. Dogs were the exception and were perceived to have similar importance to koalas. However, this finding was not replicated in any of the data from other countries. Indeed, koala welfare even being ranked as unimportant by those in some countries such as Pakistan and Sudan. This apparent anomaly can likely be attributed to the symbolic value of koalas in Australian culture, along with tendencies to anthropomorphize them and take a pro-conservation stance (Markwell 2021). Although the kangaroo also has symbolic value for Australians, they are perceived as pests in many locations where they are overpopulated (as in fact is the koala in some places), consumed for meat including in pet food and subject to culling programs, which probably reinforces a difference in attitudes towards the two species (Boulet et al. 2021).

This phenomenon does not only apply to charismatic megafauna. In the same research study (Sinclair et al. 2022), those in a number of countries (Bangladesh, Pakistan, Nigeria and Sudan) considered the welfare of farmed animals to be more important than that of companion animals, which went against the researchers' hypothesis. These countries have more subsistence farming and close rural living conditions, which create opportunities to develop more appreciation and respect for farm animals. In addition, proximity responses may be enhanced where there is contact with fewer numbers of animals and a greater number of interactions with individuals occur. These conditions are also less likely to generate cognitive dissonance, which is a psychological phenomenon that occurs when a person holds two contradictory beliefs at the same time, or the discomfort that a person feels when their behaviours do not align with their values or beliefs. In the case of farmed animals, cognitive dissonance is documented to arise amongst many meat eaters when they consider that animals must be killed to create the food that they are consuming, so how can they claim to care for animals, generating the so-called 'meat paradox' (Loughnan et al. 2014; Dowsett et al. 2018; Rothgerber and Rosenfeld 2021).

It has been clearly established that the nature of human–animal interactions between farmers and their animals influences productivity and welfare (Hemsworth et al. 2000). Good relationships between farmers and their animals mediate the effects of stressful or painful husbandry procedures (Hemsworth 2003). For example, tail-docked lambs showed lower stress responses when they had previously been handled positively (Tosi and Hemsworth 2002). This relationship has bidirectional effects, making the work less stressful and dangerous for farmers and improving their satisfaction with their chosen livelihood. However, there is concern that increasing modernization of farming and particularly greater levels of intensification dehumanizes the work of livestock farmers and prevents them from interacting with farm animals in the ways that they would prefer (Porcher 2006).

There is evidence that attachment and affection towards animals depend on the function of the animal and the type of production system (Wilkie 2005). As an example, hobby farmers often form a strong attachment to their small number of animals and provide them with preferential treatment, whereas farmers who rear large numbers of animals professionally are more likely to form impersonal relationships. Farmers may develop greater attachment to breeding animals as they tend to be on the farm longer and require more physical involvement of the farmer when compared to non-breeding animals. This evidence suggests that attachment is mediated by the frequency, intensity and intimacy of farmers' contact with their animals (Wilkie 2005).

Recent research has extended the psychological construct of solidarity to the field of human–animal interactions. Solidarity is defined as the sense of belonging, attachment and closeness felt toward other animals based on intergroup relations (Amiot and Bastian 2017). Solidarity allows humans to extend their thinking beyond fellow humans and develop values that can be transferred to non-human species and the environment (Stern et al. 1993). This type of solidarity with animals is postulated to be linked with a person's broader connection to nature and is more likely when a pro-ecological standpoint pre-exists. A stronger sense of social identification, or solidarity with farm animals, is more likely to lead to better care of them and a desire to reduce practices that cause them pain (Boaitey et al. 2023).

A proximity to animals, individual human–animal relationships and connectedness are important influencers of an animal's treatment at both an individual and a societal level and drive consumer choices associated with animals. In recognition of the effects that humans can have on welfare, the recently updated five domains model for welfare assessment now includes assessment of human–animal interactions (Mellor et al. 2020). However, our modern husbandry systems may moderate pro-animal attitudes due to the effects of scale and impersonalisation, as discussed above.

4 Diverse Approaches to Animal Welfare

To emphasize some of the diverse ways in which practices associated with managing pain and suffering in farm animals and their underlying concepts differ between cultures, be they associated with ethnic groups, geographic locations, types of training, diverse forms of production practices or other types of diversity, three case studies are provided to illustrate the existence of different norms and their implications.

4.1 "It's Always Been Done This Way": Mulesing, Tail Docking, Castration, and Other Routine Management

Livestock farming requires the appropriate management of animals to ensure optimal production and health. The management of sheep includes several husbandry procedures which vary in terms of the level of pain experienced by the animal, from painless "backlining" with an ectoparasiticide to control lice to shearing and hoof trimming, through to more painful procedures including castration, tail docking and mulesing[1] (Stafford 2017). Attitudes toward these types of routine husbandry procedures, particularly those that inflict pain, often raise tensions between farmers and community members. Both Australian consumers and farmers believe that good animal welfare is associated with a life free from (or with little) pain (Buddle et al. 2021,

[1]The objective of mulesing is to remove wool-bearing, wrinkled skin and increase the bare perineal area in Merino sheep which reduces the risk of flystrike. 'Mulesing' commonly involves the removal of the skin from the breech and one strip of skin along each side of the tail (Fisher 2011).

2023). However, contention arises between the community and industry when discussing which husbandry practices are considered necessary.

Farmer attitudes toward animal welfare are known to affect what husbandry practices are used on farm (Coleman et al. 2022). How pain is perceived by farmers influences how it is managed (Thompson 2016). For example, recognition of the pain that a sheep experiences during mulesing has increased since the practice began in the late 1920s (Fisher 2011). However, while some farmers say that they do not want to carry out painful procedures such as mulesing (Wells et al. 2011), many farmers consider these procedures as 'necessary evils' performed to avoid other significant welfare issues such as flystrike and injury caused by aggression in uncastrated males (Buddle et al. 2021) and are unwilling to stop such procedures until there are viable alternatives.

While an alternative for these husbandry practices has yet to be found, many Australian farmers praise the recent greater accessibility of pain relief to help manage the level of pain that an animal experiences during these procedures (Buddle et al. 2021). The increasing use of pain relief during routine husbandry procedures in Australia and elsewhere is due predominantly to community concerns about animal welfare and the resulting increase in market pressure to maintain animal welfare. However, regulation of the use of pain relief remains a contested issue. While producer advocacy organizations in most Australian states support the mandated use of pain relief when mulesing lambs, the use of pain relief for other procedures is only recommended; such organisations lobby against further mandates, arguing that cost and efficacy are barriers to greater acceptance of pain relief. Furthermore, even with multimodal pain relief (e.g., combining a local anaesthetic and a non-steroidal anti-inflammatory drug), pain is not eliminated, and thus the challenge of pain mitigation and of phasing out of painful husbandry procedures remains (Small et al. 2022).

This example shows that some procedures and practices that may appear to be inhumane or induce suffering may currently be the best available approaches to avoiding conditions that would be more inhumane or painful than the procedures themselves. However, a lack of transparency for the public about these sorts of trade-offs is inhibiting open dialogue about the real costs of farm animal production systems and where there are current gaps in best practices.

4.2 "Of Course Free Range Is Better!" Tensions in Views on the Best Production Systems

Research into attitudes about animal welfare demonstrates that perceptions of pain and suffering go beyond social and cultural contexts, with attitudes differing according to whether intensive and extensive production systems are envisioned to be the norm. For example, Australian meat consumers are known to have clear ideas about what constitutes a 'good life', that is, one free from pain and suffering, which is typically envisioned to involve animals freely grazing in paddocks as occurs in extensive production methods (Buddle et al. 2023). Animals raised in intensive systems, such as those used in caged egg production or sow stalls, are generally thought to enjoy lower welfare, as these conditions prompt the perception of animal suffering. The definitions of what are considered to be extensive as compared to intensive modes of production are in constant evolution, with some production systems arguably representing both at various points in time, for example cattle raised in barns during winter in Europe but grazing outside during the summer (Buddle and Ankeny in press).

Furthermore, scientific findings do not always align with community attitudes about preferred production methods. In recent years, free-range eggs have grown in popularity, in part due to the perception that the welfare of free-ranging chickens is better. Animal activist groups around the world have campaigned for the abolition of caged-egg production, resulting in retailers and large food companies sourcing less of these products, and governments in various locales placing restrictions on caged-egg production. Popular understandings often involve idealized and idyllic visions of 'happy chickens' romping freely in green fields, expressing their 'natural' behaviours including eating what they wish and freely roaming. In qualitative research, caged-egg production was described by many participants as 'cruel', with the desire to purchase free-range eggs often described in connection with efforts to avoid 'industrialized' food. Participants had strongly held perceptions that caged-egg production was wrong, unnatural and even disgusting, with diet and confinement being key negative aspects of conventional intensive production (Bray and Ankeny 2017).

However, the welfare of chickens in free-range systems compared with those raised in barns, cages or otherwise is not clearly better. For instance, free-range chickens are exposed to greater amounts of diseases and predation than those kept in barns or cages (Bestman and Bikker-Ouwejan 2020). In some situations, increased levels of cannibalism have been documented (Cronin et al. 2018). More generally, ethical and animal welfare issues are not absent in free-range systems (Parker and de Costa 2016). The food studies scholar Mara Miele (2011) suggests that the idea of the 'happy chicken' has been invented and is a notion that is simultaneously "ambivalent and precarious" because happiness is presented as a factor that contributes to better-tasting food. However, many aspects of animals' lives are not covered by welfare claims on food, and it may well be the case that those conditions that are easiest to convey via marketing and retailing claims are likely to come to be associated with higher animal welfare in the popular imagination even in the absence of scientific evidence.

4.3 Debates over Religious Slaughter

Religious or ritual slaughter practices (principally for halal meat) have recently attracted public concern, particularly as a result of unfavourable media coverage depicting animals in abattoirs outside of the export country subjected to sub-standard slaughter practices, involving beating, cutting of tendons and protracted dying processes (Hastreiter 2013). Vocal opponents assert that halal slaughter processes are cruel and that halal meat (and other products) and those who produce and sell them should be boycotted.

The controversy around religious slaughter methods has arisen primarily because of its association with the lack of use of stunning techniques to reduce suffering. When performed appropriately, stunning reduces suffering by producing a loss of consciousness and insensibility to pain before the throat is cut and the animal bleeds to death. Lack of stunning is associated with welfare concerns because of the stress resulting from the restraint procedures, pain from the neck cut and the respiratory distress associated with bleeding out (Gibson et al. 2009; Gregory et al. 2010). However, it is important to remember that typical Western slaughter practices are also accompanied by animal welfare concerns since lairage, restraint, poor handling and stunning failure also occur (Loyer et al. 2020). As a result, the differences between these methods should be viewed as one of degree rather than as dramatically different. More importantly, Islamic authorities in some Western countries have accepted the use of stunning methods for halal slaughter (as interpretation in Islam is a local practice), provided they leave the animal's body intact, as occurs for instance with electrical or percussive stunning (Bergeaud-Blackler et al. 2016).

In contrast, while Jewish law requires kindness to animals (Rosen 2004), the Torah does not specifically address slaughter (Silver 2011). The rules for kosher slaughter (shechita) and inspection are documented in the Talmud as well as other religious texts with an emphasis on ensuring that the meat is religiously 'clean' (Grandin and Regenstein 1994). Rules and practices associated with shechita do not vary, and pre-slaughter stunning is strictly prohibited (Rosen 2004; Silver 2011; Zivotofsky 2011; Anil 2012).

Whilst there are undoubtedly animal welfare concerns associated with religious slaughter practices in relation to pain (Mellor et al. 2009; Leffert 2021; Mota-Rojas et al. 2021), this concern highlights not only the influence of religious factors on attitudes towards animal pain and associated practices, but underscores the importance of sociocultural and related factors (Bergeaud-Blackler et al. 2016). Many discussions about halal slaughter begin from an assumption that Western practices are superior or take explicitly racist viewpoints. They also privilege a particular view on the science: for example, Jewish authorities have rejected stunning not because of antiquated religious beliefs or similar, but because they believe that the traditional method for shechita slaughter is painless (Anil 2012). Islamic holy writings contained in the Hadith emphasize that animals should not be mistreated and that pain during the killing process should be minimized (Masri 2007; Wood 2013; Caruana 2020).

Although there is scientific evidence to dispute claims that religious slaughter is pain-free, it is not clear how robust this evidence is when compared to alternative secular slaughter practices, which themselves are not without welfare concerns. For regulators looking to ban religious slaughter, as has occurred in New Zealand and the Netherlands, a key question is whether the scientific evidence justifies the threat and disrespect to culture resulting from a ban and whether such bans represent troubling forms of neocolonialism with Western views being forced on other cultures (Garcia and McGlone 2022; see also Rollin 2007; cf. Ahuja 2009). The controversies around halal slaughter in several countries also highlight a significant disparity in public scrutiny about halal versus kosher slaughter practices. Whilst these differences may be partially attributed to the relative differences in demand for these products, the higher visibility of halal products in major retailers and media interest about halal practices resulting from live export concerns, there is also an element of Islamophobia inherent in this disparity, with halal products sometimes being linked deliberately to terrorism and crime.

As this example reveals, it is often difficult to disentangle the effects of religiosity, cultural effects and country-specific factors on attitudes towards farmed animals. It is often assumed that citizens in developed and particularly Western countries care more about the welfare of animals than do those in less-developed countries (Garcia and McGlone 2022; Sinclair et al. 2022). However, several studies show that residents in low- and middle-income countries (LMICs) often value farmed animal species and their welfare as highly, if not more highly, than do those in developed countries (see e.g. Sinclair and Phillips 2017; Anderson and Tyler 2018; Sinclair et al. 2022). LMICs each face distinct challenges based on their histories, cultures and socioeconomic factors: hence, any attempts at animal welfare reforms must be carefully considered and tailored to be culturally and practically relevant, if uptake and change is to ensue.

5 Conclusions

This chapter has explored just a few of the many differences in views and attitudes about farm animal welfare by widening the usual definition of culture, which focuses on ethnicity or religion, to consider differences in production systems, relationships to animals and beliefs about what makes production 'good', among other factors. It must be noted that several of the domains that have been investigated are in flux and are evolving, such as visions of extensive versus intensive farming and the practices within them, as well as approaches to pain control. What has been constant is that producers and others seeking to foster best practices must make decisions in real time often without adequate evidence from animal welfare science, either because the questions of interest have not been investigated or because the context in which a question has been studied is not directly relevant for the context of interest, for instance because it has been done with another breed, in a location that is different in important ways such as due to weather patterns or within a different production system. There remain considerable opportunities for the focused study of pain in both farmed and other animals, and more generally, for attention to what counts as evidence and why in animal welfare science.

Rigorous processes are required for using the best scientific evidence when examining pain in animal production practices, and for this evidence to more actively inform best practice standards and regulation. However, a richer understanding of sociocultural factors that are associated with understanding farm animal welfare is also required. It is critical for those working with farmed animals, particularly those positioned between humans and non-human animals such as veterinarians and producers, to become knowledgeable about diverse cultural conceptualisations of animal pain and suffering, and to be aware that the available scientific evidence may not always provide a definitive answer about what is best for the animals concerned.

References

Adam MB, Clough DL, Grumett D (2019) A Christian case for farmed animal welfare. Animals 9(12):1116. https://doi.org/10.3390/ani9121116

Ahuja N (2009) Postcolonial critique in a multispecies world. PMLA 124:556–563. https://www.jstor.org/stable/25614298

Amiot CE, Bastian B (2017) Solidarity with animals: assessing a relevant dimension of social identification with animals. PLoS One 12:e0168184. https://doi.org/10.1371/journal.pone.0168184

Anderson J, Tyler L for Faunalytics (2018) Attitudes toward farmed animals in the BRIC countries. https://faunalytics.org/attitudes-towards-farmed-animals-bric-countries/

Anil MH (2012) Religious slaughter: a current controversial animal welfare issue. Anim Front 2:64–67. https://doi.org/10.2527/af.2012-0051

Ankeny RA (2008) The moral economy of red meat in Australia. In: Friedland S (ed) Food and morality: proceedings of the Oxford symposium on food and cookery 2007. Prospect Books, Blackawton, Totnes, pp 20–28

Bergeaud-Blackler F, Fischer J, Lever J (eds) (2016) Halal matters: Islam, politics and markets in global perspective. Routledge, London

Berti D (2019) Animals in the public debate: welfare, rights, and conservationism in India. Religion 10:475. https://doi.org/10.3390/rel10080475

Bestman M, Bikker-Ouwejan J (2020) Predation in organic and free-range egg production. Animals 10:177. https://doi.org/10.3390/ani10020177

Boaitey A, Eden M, Jette-Nantel S (2023) Too close to eat? Solidarity with animals, animal welfare and antibiotic use. J Appl Anim Welf Sci 26(4):479–492. https://doi.org/10.1080/10888705.2021.1986713

Bock B, Buller H (2013) Healthy, happy and humane: evidence in farm animal welfare policy. Sociol Rural 53:390–411. https://doi.org/10.1111/soru.12011

Boulet M, Borg K, Faulkner N, Smith L (2021) Evenly split: exploring the highly polarized public response to the use of lethal methods to manage overabundant native wildlife in Australia. J Nat Conserv 61:125995. https://doi.org/10.1016/j.jnc.2021.125995

Bray HJ, Ankeny RA (2017) Happy chickens lay tastier eggs: motivations for buying free-range eggs in Australia. Anthrozoös 30:213–226. https://doi.org/10.1080/08927936.2017.1310986

Buddle EA, Ankeny RA (in press) Raising the steaks: refining our understandings of red meat production and consumption through recognition of conflicting and shared values. In: Efstathiou S (ed) Transforming MEAT: reducing, replacing and refining meat use. Routledge, New York

Buddle EA, Bray HJ, Ankeny RA (2021) "Of course we care!": a qualitative exploration of Australian producers' understandings of farm animal welfare issues. J Rural Stud 83:50–59. https://doi.org/10.1016/j.jrurstud.2021.02.024

Buddle EA, Bray HJ, Ankeny RA (2023) Values of Australian meat consumers related to sheep and beef cattle welfare: what makes a good life and a good death? Food Ethics 8(1):5. https://doi.org/10.1007/s41055-022-00114-2

Caruana SJL (2020) Different religions, different animal ethics? Anim Front 10:8–14. https://doi.org/10.1093/af/vfz047

Chakrabarti P (2010) Beasts of burden: animals and laboratory research in colonial India. Sci Hist xlviii:125–152. https://doi.org/10.1177/007327531004800201

Coleman GJ, Hemsworth PH, Hemsworth LM, Munoz CA, Rice M (2022) Differences in public and producer attitudes toward animal welfare in the red meat industries. Front Psychol 13:875881. https://doi.org/10.3389/fpsyg.2022.875221

Cronin GM, Hocroft RL, Groves PG, Hall EJS, Phalen DN, Hemsworth PH (2018) Why did severe feather pecking and cannibalism outbreaks occur? An unintended case study while investigating the effects of forage and stress on pullets during rearing. Poult Sci 97:1484–1502. https://doi.org/10.3382/ps/pey022

Dowsett E, Semmler C, Bray H, Ankeny RA, Chur-Hansen A (2018) Neutralising the meat paradox: cognitive dissonance, gender and eating animals. Appetite 123:280–288. https://doi.org/10.1016/j.appet.2018.01.005

Fairholme EG, Pain W (1934) A century of work for animals: the history of the RSPCA, 1824–1934, 2nd ed. J Murray, London

Fisher A (2011) Addressing pain caused by mulesing in sheep. J Appl Anim Behav Sci 135(3):232–240. https://doi.org/10.1016/j.applanim.2011.10.019

Garcia A, McGlone JJ (2022) Animal welfare and the acknowledgment of cultural differences. Animals 12:474. https://doi.org/10.3390/ani12040474

Gibson T, Johnson C, Murrell J, Hulls C, Mitchinson S, Stafford K, Johnstone A, Mellor D (2009) Electroencephalographic responses of halothane-anaesthetised calves to slaughter by ventral-neck incision without prior stunning. N Z Vet J 57:77–83. https://doi.org/10.1080/00480169.2009.36882

Grandin T, Regenstein JM (1994) Religious slaughter and animal welfare: a discussion for meat scientists. Meat Focus Int 3(1):115–123

Gregory NG, Fielding H, von Wenzlawowicz M, Von Holleben K (2010) Time to collapse following slaughter without stunning in cattle. Meat Sci 85:66–69. https://doi.org/10.1016/j.meatsci.2009.12.005

Harrison R (1964) Animal machines: the new factory farming industry. Stuart, London

Hastreiter MT (2013) Animal welfare standards and Australia's live exports industry to Indonesia: creating an opportunity out of a crisis. Wash U Glob Stud Law Rev 12:181. https://openscholarship.wustl.edu/law_globalstudies/vol12/iss1/5

Hemsworth PH (2003) Human–animal interactions in livestock production. Appl Anim Behav Sci 81:185–198. https://doi.org/10.1016/S0168-1591(02)00280-0

Hemsworth P, Coleman G, Barnett J, Borg S (2000) Relationships between human–animal interactions and productivity of commercial dairy cows. J Anim Sci 78:2821–2831. https://doi.org/10.2527/2000.78112821x

Ingold T (2000) The perception of the environment: essays on livelihood, dwelling and skill. Routledge, London

Leffert S (2021) Is ongoing ritual slaughter of livestock justifiable in modern America? J Appl Anim Res 49(1):492–522. https://doi.org/10.1080/09712119.2021.2011296

Loughnan S, Bastian B, Haslam N (2014) The psychology of eating animals. Curr Dir Psychol Sci 23(2):104–108. https://doi.org/10.1177/0963721414525781

Loyer J, Whittaker AL, Buddle EA, Ankeny RA (2020) A review of legal regulation of religious slaughter in Australia: failure to regulate or a regulatory fail? Animals 10:1530. https://doi.org/10.3390/ani10091530

Markwell K (2021) Getting close to a national icon: an examination of the involvement of the koala (Phascolarctos cinereus) in Australian tourism. Tour Recreat Res 46:473–486. https://doi.org/10.1080/02508281.2020.1815411

Masri ABA (2007) Animal welfare in Islam. The Islamic Foundation, Leicestershire

Mellor D, Gibson T, Johnson C (2009) A re-evaluation of the need to stun calves prior to slaughter by ventral-neck incision: an introductory review. N Z Vet J 57:74–76. https://doi.org/10.1080/00480169.2009.36881

Mellor DJ, Beausoleil NJ, Littlewood KE, McLean AN, McGreevy PD, Jones B, Wilkins C (2020) The 2020 five domains model: including human–animal interactions in assessments of animal welfare. Animals 10:1870. https://doi.org/10.3390/ani10101870

Miele M (2011) The taste of happiness: free-range chicken. Environ Plan A 43:2076–2090. https://doi.org/10.1068/a43257

Miele M, Blokhuis H, Bennett R, Bock B (2013) Changes in farming and in stakeholder concern for animal welfare. In: Blokhuis H, Miele M, Veissier I, Jones B (eds) Improving farm animal welfare: science and society working together: the welfare quality approach. Wageningen Academic Publisher, Wageningen, pp 19–47

Milton K (2002) Loving nature: towards an ecology of emotion. Routledge, London

Morton R, Hebart ML, Ankeny RA, Whittaker AL (2022) An investigation into 'community expectations' surrounding animal welfare law enforcement in Australia. Front Anim Sci 3:991042. https://doi.org/10.3389/fanim.2022.991042

Mota-Rojas D, Napolitano F, Strappini A, Orihuela A, Ghezzi MD, Hernández-Ávalos I, Mora-Medina P, Whittaker AL (2021) Pain at the slaughterhouse in ruminants with a focus on the neurobiology of sensitisation. Animals 11:1085. https://doi.org/10.3390/ani11041085

Parker C, de Costa J (2016) Misleading the ethical consumer: the regulation of free-range egg labelling. Melb Univ Law Rev 39:895–949. http://hdl.handle.net/11343/233834

Pinillos RG, Appleby M, Manteca X, Scott-Park F, Smith C, Velarde A (2016) One welfare: a platform for improving human and animal welfare. Vet Rec 179(16):412–413. https://doi.org/10.1136/vr.i5470

Porcher J (2006) Well-being and suffering in livestock farming: living conditions at work for people and animals. Sociologie du Travail 48:e56–e70. https://doi.org/10.1016/j.soctra.2006.02.001

Regan T (2004) Christians are what Christians eat. In: Sapontzis S (ed) Food for thought: the debate over eating meat. Prometheus Books, Amherst, pp 177–185

Rioja-Lang FC, Connor M, Bacon HJ, Lawrence AB, Dwyer CM (2020) Prioritization of farm animal welfare issues using expert consensus. Front Vet Sci 6:495. https://doi.org/10.3389/fvets.2019.00495

Ritvo H (1987) The animal estate: the English and other creatures in the Victorian age. Harvard University Press, Cambridge, MA

Rollin BE (1981) Animal rights and human morality. Prometheus Books, Buffalo

Rollin BE (2007) Cultural variation, animal welfare and telos. Anim Welf 16:129

Rosen SD (2004) Physiological insights into shechita. Vet Rec 15(24):759–765. https://doi.org/10.1136/vr.154.24.759

Rothgerber H, Rosenfeld DL (2021) Meat-related cognitive dissonance: the social psychology of eating animals. Soc Personal Psychol Compass 15:e12592. https://doi.org/10.1111/spc3.12592

Sandgren EP, Streiffer R, Dykema J, Assad N, Moberg J (2020) Attitudes toward animals, and how species and purpose affect animal research justifiability, among undergraduate students and faculty. PLoS One 15:e0233204. https://doi.org/10.1371/journal.pone.0233204

Santich B (1990) Feed the man meat. In: Walker H (ed) Oxford symposium on food and cookery, 1989: staple foods: proceedings. Prospect Books, London, pp 224–227

Scully M (2002) Dominion: the power of man, the suffering of animals, and the call to mercy. St Martin's, New York

Silver J (2011) Understanding freedom of religion in a religious industry: Kosher slaughter (shechita) and animal welfare. Vic Univ Wellingt Law Rev 42(1):671–704. https://doi.org/10.26686/vuwlr.v42i4.5113

Sinclair M, Phillips CJ (2017) The cross-cultural importance of animal protection and other world social issues. J Agric Environ Ethics 30:439–455. https://doi.org/10.1007/s10806-017-9676-5

Sinclair M, Lee NYP, Hötzel MJ, de Luna MCT, Sharma A, Idris M, Derkley T, Li C, Islam MA, Iyasere OS et al (2022) International perceptions of animals and the importance of their welfare. Front Anim Sci 3:960379. https://doi.org/10.3389/fanim.2022.960379

Singer P (1975) Animal liberation: a new ethics for our treatment of animals. Random House, New York

Singer P (2023) Animal liberation now. HarperCollins, New York

Small A, Fisher AD, Lee C, Colditz I (2022) Analgesia for sheep in commercial production: where to next? Animals 11:112. https://doi.org/10.3390/ani11041127

Stafford K (2017) Husbandry procedures. In: Ferguson DM, Lee C, Fisher A (eds) Advances in sheep welfare. Woodhead Publishing, Sawston, pp 211–226. https://doi.org/10.1016/C2015-0-00983-2

Stern PC, Dietz T, Kalof L (1993) Value orientations, gender, and environmental concern. Environ Behav 25:322–348. https://doi.org/10.1177/0013916593255002

Thompson CS (2016) Assessing attitudes towards welfare and pain in farm animals. PhD thesis, University of Edinburgh

Tieman M, Hassan FH (2015) Convergence of food systems: Kosher, Christian and Halal. Br Food J 117(9):2313–2327. https://doi.org/10.1108/BFJ-02-2015-0058

Tosi M, Hemsworth P (2002) Stockperson–husbandry interactions and animal welfare in the extensive livestock industries. In: Proceedings of the Conference at the 36th Congress of the International Society for Applied Ethology, pp 6–10

Wells AED, Sneddon J, Lee JA, Blache D (2011) Farmers' response to societal concerns about farm animal welfare: the case of mulesing. J Agric Environ Ethics 24:645–658. https://doi.org/10.1007/s10806-010-9284-0

Wilkie R (2005) Sentient commodities and productive paradoxes: the ambiguous nature of human–livestock relations in Northeast Scotland. J Rural Stud 21:213–230. https://doi.org/10.1016/j.jrurstud.2004.10.002

Wood A (2013) Animal welfare under the Shari'a. Macq Law J 155(12):155–172. http://www.austlii.edu.au/au/journals/MqLawJl/2013/18.html

Woods A (2011) From cruelty to welfare: the emergence of farm animal welfare in Britain, 1964–1971. Endeavour 36:14–22. https://doi.org/10.1016/j.endeavour.2011.10.003

Zivotofsky AZ (2011) Government regulations of shechita (Jewish religious slaughter) in the twenty-first century: are they ethical? J Agric Environ Ethics 25(5):747–763. https://doi.org/10.1007/s10806-011-9324-4

A History of Pain and Suffering in Farmed Animals Used for Research

R. Eddie Clutton

Abstract

Farmed animals have been used in anatomical and physiological studies since classical times, presumably for reasons of widespread availability, the benefits of comparative studies and size-relevance. Some forms, e.g. fish and cephalopoda, were probably used for their curiosity value alone. From earliest times, observers have complained that the pain and suffering experienced by animals involved in noxious studies were immorally cruel and undermined the value of scientific observations, although these were not sufficient reasons to stop the practice. That farmed animals were reared primarily for consumption and so were destined to experience profound unpleasantness during the slaughter and butchery process probably assuaged the concerns of those conducting farmed animal vivisection but not the pain nor suffering endured by the animals involved. The introduction of anaesthetics in the mid-nineteenth century went some way to improve matters, although an increased awareness of sentience as a characteristic of an ever-increasing range of non-human animals—farmed or otherwise—and their inclusion into a moral community was probably more influential. The animal rights movement of the late twentieth century focused societal interest into the fate of all animals involved in biomedical research and through numerous strategies prompted the creation of legislation protecting laboratory animals. This chapter encourages the reader to *imagine* the experiences of farmed animals undergoing biomedical research throughout history while identifying the major developments that have arisen to minimize their pain and suffering.

Keywords

Pain · Animals · History

1 Introduction

Farmed animals have been used in "experiments" since the neolithic period and presumably (and mainly) for the same reasons as now: apart from being widely available, pigs, sheep, goats and calves are similar in size to adult human beings; pigs and humans share physiological similarities. Then as now, birds, fish and crustacea have been studied by natural philosophers for their unique features and for the insights a comparative approach offers. Humanitarian or ethical qualms with experimenting on farmed animals are readily quashed by the knowledge that most will be

R. E. Clutton (✉)
The Wellcome Trust Critical Care Laboratory for Large Animals, LARIF, The Roslin Institute, University of Edinburgh, Easter Bush, Midlothian, UK
e-mail: e.clutton@ed.ac.uk

R. E. Clutton, G. C. Musk (eds.), *Pain and Suffering in Farmed Animals*,
https://doi.org/10.1007/978-3-032-11407-5_3

butchered for the plate *anyway*. Large, farmed mammals are still used in studies for their size relevance, e.g. in the testing of implantable devices or organ explants destined for adult human patients. They are also used for agricultural and environmental research, as are farmed fish, poultry and insects. The use of genetically modified variants has increased rapidly over the last few decades. In terms of numbers used, few farmed animals are currently involved in very few experiments compared with mice, rats and zebrafish (although the latter are not "farmed" for any purpose other than domestic aquaria and to provide scientific models).

A history of farmed animal use in research is readily produced: the only requirement is evidence of what was done, to what, and in what year. In contrast, producing a history of the *pain and suffering* of laboratory animals is considerably more difficult. Putting aside the controversial subject of anthropomorphism, it is difficult to *truly* imagine the experiences of a dog in eighteenth century France, which, having heard the cries of its conspecifics for several days is brought to a room smelling of dissection and death, to be nailed by the paws to a wooden table and have one's backbone chiselled away in the study of spinal reflexes. Except for those who have undergone similar procedures without anaesthesia, then *much* must be left to the imagination—along with an acceptance that an animal's sensory and emotional—and perhaps even its evaluative capabilities—are not too different from that of a human being. However, imagination and empathy are not always necessary. Many vivisectionists and natural philosophers, i.e. early scientists, declared openly, and with varying levels of guilt, that their actions were cruel and inhumane because they *obviously* caused pain and suffering—but were *necessary.* (Although some nineteenth century vivisectionists declared no guilt whatsoever). Additionally, many early experimental reports, particularly those involving vivisection, describe animal reactions in sufficient detail (and often illustrated) to render the reader's imagination redundant. Unfortunately, then as now, many experimental reports overlooked signs of animal pain and suffering, and restricted descriptions to observations of interest to the experimenter himself.[1] Fortunately, experimental reports not infrequently—possibly inadvertently—contain information that provides convincing evidence for an animal's pain experience, e.g. alarmingly high heart rates during cardiac studies on conscious animals. Importantly, the history of animal experimentation is punctuated by reports from enraged observers containing descriptions of the animals' experience that leave few (but the hardest-hearted) in any doubt that animals suffered horribly before the introduction of anaesthetics (although many continued to suffer thereafter). Consequently, this chapter includes descriptions of pain and suffering in farmed experimental animals. The goal is not to turn stomachs but provide details that allow an appreciation of the major improvements that have occurred over time.

Throughout the history of animal experiments, the pain and suffering of the creatures involved and the corruption of any subsequently generated data did not go unrecognized by both lay observers and early scientists. As a result of this, antivivisection organizations arose in most industrialized countries, followed by legislation concerned with protecting laboratory animal welfare, almost always with a mandate to use anaesthetics in noxious procedures. This has limited the pain and suffering of all laboratory animal species since the mid-nineteenth century. However, there remains room for improvement worldwide and in particular, the reporting of anaesthetic and pain management techniques in the scientific literature remains suboptimal because amongst other things, it supports an ongoing reproducibility crisis. In the UK, all vertebrates, i.e. mammals, birds, fish and amphibians are protected under the Animals (Scientific Procedures) Act 1986. In 1993, the Act extended protection to the octopus and later, in 2013, to other edible cephalopod molluscs, i.e. squid, cuttlefish and nautilus. This is germane because the industrialized farming of octopus is currently being debated within the European Union (EU)

[1] This sexual bias is deliberate: until the mid-twentieth century, animal vivisectioners were exclusively male.

(Guerreiro 2022). It is to be hoped that within the next few decades, the decapod crustaceans, i.e. lobsters and crabs, will also receive greater protection both in the laboratory, on the "farm", and in the kitchen.

This chapter invites the reader to consider the pain and suffering endured by farmed animal species involved in "research" from earliest times. This will be done by examining primarily the involvement of pigs, sheep, goats, cattle and poultry, but also fish, cephalopods and insects, in descriptions and illustrations of the experiments in which these species have, and are currently involved in. The record is not complete: this would make the chapter too long and possibly inure the reader to the suffering described. Therefore, it relies on examples. Additionally, in reviewing key (primarily European) events in the history of animal experimentation, major social, ethical and legal events are identified that have mitigated farmed animal pain and suffering and reveal the considerable progress that has been made, as well as the improvements that are still required. The chapter is Anglocentric in approach because the UK was a leader in establishing the current situation, and again, space precludes an international analysis.

Attempts to fully appreciate the past and present experiences of farmed animals in the laboratory require the understanding that unlike traditional laboratory animals, i.e. mice, rats, fish, etc., which are purpose-bred, the larger farmed animals are usually created for agricultural, i.e. non-scientific, purposes, and are reared on and obtained from commercial farms. (Though the generation of genetically modified large animals under laboratory conditions for scientific and commercial purposes is increasing). A "farm-based" origin has four potential consequences for any animal involved. First, many will have undergone noxious husbandry procedures, e.g. castration, as neonates and may be variably sensitized to pain as adults (see Chap. 7). Second, many will be accustomed to an agricultural environment, which differs in numerous ways (both good and bad) to laboratory living conditions. Third, the necessary transport of farmed animals to the research facility by road is stressful. Fourth, and perhaps the most important, is a level of inescapable emotional stress that may exist in all modern farmed animals. According to Yuval Harari in his foreword to "Animal Liberation Now" (Singer 2023), this has arisen because modern farmed animals have, and are being denied those basic instincts which have evolved over the millennia preceding the agricultural revolution (mid-seventeenth to late nineteenth centuries) and which have ensured their survival under wild conditions. Being unable to express fundamental behaviours, e.g. nesting, roaming, socializing or caring for their young, means they are suffering not only physical harm but intense psychological distress. Considering this and the previous three concerns makes it a sobering possibility that a farmed laboratory animal's suffering is not only profound and unseen, but exceeds that of the laboratory rodent and starts well before the first experiment begins.

2 The Neolithic Age

A bovine skull found at a French Neolithic (3400–3000 BC) site, *ca* 1985, has a hole in the right frontal bone and reveals the animal underwent cranial surgery using the same techniques as those being used contemporaneously on humans. Authorities have questioned whether the surgery was done to benefit the animal (as an early act of veterinary surgery), as an experimental exercise (the first vivisection) or as the first example of surgical training on animals attempting to refine techniques of trepanation in humans (Ramirez Rozzi and Froment 2018). It is possible that the procedure may have been done simply to see what might happen, i.e. how the animal would have reacted, because animals have been vivisected ever since for all four of these reasons. Whatever the rationale, it is worth noting that trephinations were being conducted successfully on human subjects as early as 6000 BC (Verano and Finger 2010).

3 The Classical Age

Early descriptions of the effects of dissection on conscious animals, i.e. pain and suffering, are inconsistently reported in early treatises, though Galen reported being made uncomfortable by the reactions of apes to brain surgery (Andreas-Holger and Tröhler 1987). Others, e.g. the Empiricists, revealed concern for animal suffering by complaining the animals' violent reactions rendered study findings questionable (Franco 2013). In most reports, animal pain and suffering can only be surmised in descriptions of what was done in the absence of anaesthetics, analgesics and precision surgical instruments.

One of the earliest *probable* accounts of farmed animal vivisection dates from 500 BC, when the Greek Alcmaeon of Croton (Italy) is reported to have severed the optic nerves of goats and commented on their ensuing blindness. It is difficult to confirm this, and some have doubted whether he even dissected goats at all, because he believed they breathed through their ears (Codellas 1932).

Two centuries later, Erasistratus (304–250 BC) of Alexandria performed experiments involving live animals. He observed birds and demonstrated the relationship between food intake and weight gain. Using pigs, he also established that the trachea was an air tube and that the lungs were pneumatic organs. However, he and fellow physician Herophilus were infamous not for animal, but human vivisection: the so-called "butchers of Alexandria" allegedly dissected nearly 600 live prisoners (Franco 2013).

According to Rupke, the author of the Hippocratic text, "On the Heart" written in 350 BC, described cutting a pig's throat as it drank coloured water to study the act of swallowing". The same author—possibly Hippocrates himself—also described opening the chest of another (unidentified) living animal and observed how the auricles and ventricles of the heart beat alternatively (Rupke 1987).

Aristotle (384–322 BC) reported his dissections of numerous farmed animal species in "*De partibus animalium*" (On the parts of animals), including poultry and fish. In discussing "*the ox, the goat, the deer*", i.e. ruminants, he records that their "*multiplicity of stomachs*" is because "*the mouth, owing to its lack of teeth, fails to perform its due office as regards the food*" (Ogle 1882). In his other book "*Historia de animalibus*" (A History of Animals), he appeared to accept cruelty in making such observations: to better identify veins, he recommended, "*turning only to those animals that were strangled after they had been starved beforehand*". He was also aware that the post-mortem appearance of lungs depended on whether the animal was killed by strangulation or "cutting their throats" or "*removed from the animals being dissected, whose mass of blood escaped immediately*" [upon death] (Leunissen 2021). Surprisingly, the Empiricist school, which was founded on Aristotelian thinking later discredited the study of functional anatomy by vivisection on the grounds of—amongst other things—cruelty. Empiricists believed pain and death would distort normal anatomy (Franco 2013). Aristotle was also interested in lower taxa: he commented on the octopus's colour-changing abilities for camouflage and signalling.

In his treatise "*De anatomicis administrationibus*", the Graeco-Roman physician Galen (129–216) described techniques for dissecting live animals. While his interest was the human body, the Roman ban on human dissection in 150 BC forced him to vivisect pigs, goats, cattle and Barbary macaques. Focusing on the anatomy of vocalization, he discovered that the vocal protestations of vivisected animals ceased immediately upon cutting their recurrent laryngeal nerves. Carefully dissecting a live pig, he reports tracing the nerves from the spinal cord and tying off those controlling the vocal cords. This stopped the animal squealing but allowed breathing and movement to continue. On releasing the ligature, the pig squealed loudly again. Galen appeared to be sensitive to animal suffering, preferring pigs or goats over apes because the unpleasant expression of the ape discomfited him during brain dissections. Galen may have vivisected cattle, as he was aware of intracranial "*retia mirabilia*", a network of blood vessels at the base of brain that amongst mammals exist only in oxen and sheep (Rupke 1987).

The Arab physician and surgeon Avenzoar (Ibn Zuhr or Ibn Sina) (1094–1162) influenced twelfth century (and later) medicine by applying the experimental method and introducing animal testing. It is widely believed he performed procedures on animals before doing them on humans, his most notable being a tracheostomy, which he conducted on a goat (and which subsequently survived) (Abdel-Halim 2005).

4 The Renaissance

The advent of printing in the mid-fifteenth century allowed reports and illustrations of vivisectors at work. This clarified to a greater audience what their subjects were enduring (see Fig. 1). The influential anatomy book "De *humani corporis fabrica libri septem*" (On the Fabric of the Human Body in Seven Books) reveals much in text and images about the suffering of the pigs, dogs and other animals dissected by the anatomist and physician Andreas Vesalius (1514–1564) at the University of Padua. Amongst its many features, the "*Fabrica*" (published in 1543) is remarkable for its historiated capitals.[2] One historiated "Q" shows a conscious pig undergoing tracheostomy, possibly just before a tube was passed and used to intermittently inflate the animal's lungs (Slutsky 2015). However, given Galen's knowledge of vocalization, it is possible the illustration depicts Vesalius performing a recurrent laryngeal neurectomy, which would have silenced the screaming animal. Other historiated capitals show *putti*[3] hoisting the body of a small mammal (perhaps a dog) by its rear legs; another shows three medical *putti* strangling a stray dog before dissection. A third shows *putti* dissecting a pregnant dog, with one removing a puppy. Such images reveal much with respect to animal pain and suffering in Vesalius's studies and are confirmed by Vesalius himself, who describes the pregnant bitch he dissected as "cruciata", i.e. as crucified or tortured.

William Harvey's (1578–1657) thesis ("*Excercitatio anatomica de Motu cordis et sanguinis animalibus*" (An anatomical exercise on the motion of the heart and blood in living beings) published in 1628, provided the most accurate description of blood circulation and heart function of the time. His studies involved the vivisection of pigs, sheep, goats, calves, deer and other mammals, fish,[4] amphibians, reptiles and insects.[5] During early studies observing "*De Motu Cordi*" in con-

Fig. 1 Pig prepared for vivisection. As indicated by Vesalius, "In this figure thou seest a pig bound to a plank as we are accustomed to prepare one for performing a vivisection" (Translation in Lambert, 1936, p. 373). Woodcut illustration from the *Fabrica* (1543), Liber II, p. 237. (Courtesy of the U.S. National Library of Medicine)

[2] An historiated capital is an enlarged and decorated capital letter, often the first letter of a section of text, that incorporates a narrative scene or figure relevant to the text it introduces.

[3] Putti" (plural of "putto") refers to depictions of chubby, often winged, male child figures in art.

[4] Most fish used in modern research are zebrafish (*Danio rerio*), although *salmonidae* (trout & salmon), goldfish, medaka and carp are also used.

[5] *Drosophila melanogaster* (the fruit fly) is the most com-

scious animals whose ribs had been surgically removed, Harvey complained that the heart's action was often too fast, a complication which *later* prompted him to use "colder animals, such as toads, frogs and serpents, or 'dying dogs and hogs'". That excessive heart rates were probably the result of the unimaginable pain caused by thoracic surgery in conscious animals did not go unnoticed. Riolan (1648) and later O'Meara (1665) argued that, "… the miserable torture of vivisection places the body in an unnatural state and that amid the terrible pains of vivisection all the juices are brought to flow together, thus denying the validity of animal experimentation" (Ryder 2000).

French philosopher and vivisector René Descartes (1596–1650) did not share these concerns, believing that animals, "When burnt with a hot iron or cut with a knife, their writhing and screaming are like the creaking of a hinge, no more" (Regan and Singer 1989). Descartes felt no need to comment on the torments of the lower taxa, which were at this time attracting the attention of sixteenth and seventeenth century experimenters. Building on the earlier work of Zacharias Janssen (1585–pre1632) van Leeuwenhoek (1632–1723) developed single-lens microscopes which allowed the closer study of non-mammalian tissues and organs. For example, Malpighi (1628–1694) studied capillaries in fish gills and examined chicken embryos, while making discoveries about organ formation, blood vessels and neural development. Thomas Bartholin (1616–1680) and Olaus Rudbeck (1630–1702) also used chickens to establish the existence of the lymphatic circulation. William Harvey studied chicken embryos and published "*Exercitationes de Generatione Animalium*" (1651), one of the first descriptions of embryonic development (Whitteridge 1978). Fish eggs were also viewed microscopically in early (sixteenth century) embryology studies. The microscope as a means for experimental refinement did not escape Robert Hooke who in 1665 wrote,

> *The microscope enables one to look at nature 'acting according to her usual course and way, undisturbed, whereas when we endeavour to pry into her secrets by breaking open the doors upon her, and dissecting and mangling creatures whilst there is life yet within them, we find her indeed at work, but put into such disorder by the violence offered that we cannot tell if the results are of any significance.*

Some ground-breaking studies were purely observational. Francesco Redi (1626–1697) used insects to demonstrate that maggots do not arise spontaneously but come from fly eggs; he also used fish to show that only fish beget fish (Parke 2014). Observations of octopus and squid led to early theories about limb regeneration and tissue repair. Furthermore, the large eyes of cephalopod molluscs were used to study lens structure, light absorption and retina formation, while the large axons of squid and octopuses and their reaction to stimuli led to early theories on reflexes.

Not all Renaissance animal experiments necessitated vivisection or involved cruelty. For example, Francis Glisson (1597–1677) author of the *Anatomia Hepatis* (1654) used human cadavers as his primary source, and probably obtained pig, sheep and goat livers from the slaughterhouse. Nor were all renaissance experiments gory: indeed, the first published reports of animal *drug* testing have been assigned to Paracelsus, whose "*Opera Medico-chemica Sive Paradoxa*" (1605) records the action of ether on domestic fowl in experiments conducted in either 1540 or 1541. The Opera states:

> *Of all the extracts of vitriol (ether), this is the most important, being stable. It has an agreeable taste, so that even chickens take it gladly, and thereafter fall asleep for a long time, awakening unharmed.* (Davison 1958)

---

monly used insect in research. Other insects involved are crickets, black soldier flies, pollinating bees spp. and relevant infectious disease vectors.

5 The Enlightenment

5.1 The Royal Society (1660–Present)

Vivisection prospered during the eighteenth and nineteenth centuries as physiology became established as a distinct scientific discipline. Pioneers in anatomy, chemistry and physics contributed to a growing understanding of biological processes, and experimental methodology became increasingly important. In parallel came a wider appreciation of animal pain and suffering of animals. For example, Johann Brunner, Professor of Medicine at Heidelberg, described the animals he dissected as "the martyrs of the anatomists" (Andreas-Holger and Tröhler 1987) whilst early members of the Royal Society, i.e. Christopher Wren, Robert Boyle and Robert Hooke, who, recognizing the cruelty of vivisection and its corrupting effects on data, began preliminary experiments with intravenous anaesthesia. After one operation on a dog's thorax in 1664, Robert Hooke wrote to Robert Boyle:

> *I shall hardly be induc'd to make any further trials of this kind because of the torture of the creature but certainly the inquiry would be very noble if we could any way find a way soe to stupify the creature as that it might not be sensible which I fear there is hardly any opiate will performe.* (Dorrington and Poole 2013)

Contemporary artwork began to reveal both technical aspects of animal vivisection and occasionally the reactions of onlookers. In Fig. 2, (*An Experiment on a Bird in the Air Pump* 1768), a bird is being asphyxiated before a group of onlookers, which exhibit reactions ranging from grief, disbelief and dismay, through curiosity to total indifference. The central figure's scientific curiosity seems to overcome his concern for the bird's distress. To the latter, the English philosopher and social reformer Jeremy Bentham would ask,

> *The question is not, Can they reason? nor, Can they talk? but, Can they suffer?* (Bentham 1823)

By 1660—the time of the Royal Society's foundation—dogs appear to have become the preferred experimental species throughout Europe, although birds[6] were also studied. Indeed, any available species appeared to be "fair game".

5.2 The Growth of Physiology

Upon his death, some acknowledged the French vivisectionist Francoise Magendie (1783–1855) as the "father of experimental physiology". Elsewhere (mainly England), he was despised and had succeeded in fomenting more anti-vivisectionist (and Francophobic) sentiment than any other European scientist (Ryder 2000). Fortunately for the farmed animal species, his barbaric studies of nerve function involved frogs, dogs, cats and rabbits, although he also used retired army horses who had survived Waterloo. Descriptions of his vivisections, often performed publicly, abound and reveal his absolute indifference to animal suffering and to those who judged his actions cruel. John Elliotson, afterwards, Professor of Medicine at the University of London, attended his demonstrations in Paris and reported Magendie's clumsy savagery,

> *… who cut living animals here and there with no definite object but just to see what would happen.*

In 1822, the world's first legislation against animal cruelty arising by means of a Parliamentary procedure, "The Cruel Treatment of Cattle Act 1822" was established in the UK following the efforts of "Humanity Dick" (Richard) Martin MP. Its intent was to prevent the cruel and improper treatment of horses, mares, geldings, mules, asses, cows, heifers, steers, oxen, sheep and other cattle but excluded (baitable) bulls (and dogs) until 1835. It is unclear whether pigs were protected under "other cattle" (or chattels), but what is known is that the term "wanton cruelty" did not yet apply to vivisection, as this was deemed to be deliberate, ensuring experimental

[6]The bird in "An Experiment on a Bird in the Air Pump" was a cockatiel; most birds used in modern research are domestic fowl (*Gallus domesticus*), quail (*Coturnix coturnix*), ducks (*Anatidae* spp) and turkeys (*Meleagris gallopavo*).

Fig. 2 Joseph Wright 'of Derby', 1734–1797. *An Experiment on a Bird in the Air Pump*, 1768. oil on canvas, 183 × 244 cm. (Presented by Edward Tyrrell, 1863©. The National Gallery, London. With permission)

animals went unprotected by this Act (Bates 2017).

Farmed animals also escaped the attention of Claude Bernard (1813–1878), a student of Magendie whose reputation as a physiologist and vivisector came to supersede and challenge, respectively, that of his mentor. Bernard similarly used dogs to study cardiovascular, digestive, hepatic and metabolic function, whilst rabbits and frogs were used in nerve and muscle function studies (Franco 2013).

Louis Pasteur (1822–1895), a French chemist and microbiologist, used chickens in fowl cholera research, while sheep and goats were the main animal models for his successful anthrax experiments. His use of pigs and cattle in other experiments is difficult to confirm, presumably because the results were less consequential. In contrast to his French predecessors and contemporaries, Pasteur was sensitive to animal suffering: he admitted unease with some experiments and used anaesthetics to prevent unnecessary suffering. He also employed what are now recognized as humane end-points (HEPs; see below).

5.3 Farmed Animals and Early Physiological Studies

Farmed animals did not escape all eighteenth- to nineteenth-century vivisectionists. Pigs continued to be used in studies of cardiovascular, digestive and pancreatic function, in skin grafting experiments, and in the development of surgical techniques, although dogs had assumed the primacy (Gutierrez et al. 2015). The growing preference for dogs is not difficult to understand. Whilst both were freely available, pigs had additional value as comestible items. Furthermore, pigs are physically powerful, i.e. difficult to restrain, and their screaming, which can reach 110 dB (Weeks et al. 2009), is injurious to human hearing. Preferences would not have been affected by the introduction of anaesthetics: in general, difficulties with physical restraint and securing venous access (for drug injections) make pigs more troublesome to anaesthetize than dogs.

In contrast, dogs did not replace sheep nor goats in eighteenth century science. This is possibly because small ruminants often remain silent and cease to struggle when physically restrained and operated upon. This would have had welfare implications because the late nineteenth century saw sheep being used in noxious cardiovascular studies. For example, Stephen Hales (1677–1761) bled sheep to death while filling the still-beating heart with molten wax (to measure stroke volume) (Clark-Kennedy 1929). Sheep were also being increasingly used as models for skeletal research, including osteoporosis, bone grafting, fracture repair technique development and fracture healing. Studies on sheep skin as a treatment for burns and skin grafting technique development would have been only slightly less noxious. The nineteenth century also saw sheep being used (along with cats, dogs, rats, rabbits, birds and frogs) in live xenografting (inter-species pedicle flaps) experiments. In these, the donor animal was typically immobilized and affixed to the human patient for several days in an effort to promote vascularization of xenografted tissue (Gibson 1955; Ozhathil et al. 2021).

Distinct features of avian reproduction assured a role for poultry, particularly chickens, ducks and quails in embryology and early developmental studies: eighteenth century naturalists such as Karl Ernst von Baer (1792–1876) studied chicken embryos to understand organ development and postulate the germ layer theory. These studies were unlikely to have been noxious: even today, chick embryos only receive protection under the Animals (Scientific Procedures) Act 1986 (see below) 2/3rds the way through incubation, i.e. after 14 days.

Fish and their embryos featured in a surprising range of 18th–nineteenth century experiments: the transparent embryo allowed observational microscope-based studies on blood circulation and heart development. The German physician Caspar Wolff (1734–1794) used them to establish the germ layer and epigenetic theories of development. These too would not be regarded as noxious under the current Animals (Scientific Procedures) Act 1896, which protects larval forms of fish and amphibians only after they can feed independently (for zebrafish, this is 5 days after fertilization).

Early natural philosophers were unafraid of self-experimentation, but examples of humans sharing experimental animal suffering are rare. One exception is the American physician Hugh Williamson (1735–1819) who, in reporting (1775) the lethal effects of electric eels upon hapless catfish, placed his own hand in the water as the former paralyzed and killed the latter.

The French naturalist Pierre Broussonet (1761–1807) first reported (1786) the regenerative capacity of teleost fish after studying events following pectoral fin amputation in goldfish. This is pertinent because fin clipping is currently regarded as a noxious (albeit brief) experience for laboratory fish (Roques et al. 2010).

From the late nineteenth century, ophthalmic anatomy and physiology and the study of vision shifted from fish to include cephalopods. Preliminary studies involved the anatomical structure of cephalopodic eyes and optic nerves but led rapidly to the electrophysiological and biochemical examination of vision itself (Dröscher 2016).

Studies involving insects in the eighteenth and nineteenth centuries were largely innocuous, being limited to observing their association with human and animal diseases, for example, malaria and mosquitoes. That said, Pasteur was interested in the microbial causes of silkworm disease, while William Bateson (1861–1926) and early geneticists observed hereditary traits in butterflies and beetles. Jean Pierre Mégnin (1828–1905) demonstrated how fly larvae and insect activity could help determine the time of death in forensic cases.

5.4 Animal Anaesthesia

After experimenting on worms, insects and his dog, William T.G. Morton administered ether vapour to one Gilbert Abbott for the removal of a tumour and before a large audience at the Massachusetts General Hospital. The date was the 16th October 1846, and thereafter, anaesthetics would transform surgery. News of the breakthrough travelled east and saw the first general anaesthetics being delivered to humans in Paris on December 15th (1846), in London on December 19th (1846), and in Vienna on 27 January 1847. Uptake by veterinary practitioners was prompt: on 29 January 1847, *The Times* of London reported the first successful use of an anaesthetic in a horse with ether at the Royal Veterinary College, probably by William Sewell, the then principal (Jones 2023). Some 10 days later, in Vienna, Josef Seifert began anaesthetizing and reporting on ether anaesthesia in horses, goats, cattle, oxen and dogs (Moens 2025). Seifert also described the apparatus he had developed for administering ether to large animals. In 1848, *The Lancet* reported a study by Thomas Wakley comparing the effects of ether and chloroform on 100 animals—predominantly dogs, but also cats, rabbits, rats, mice, guinea pigs, a hedgehog, a pig, two sheep, a donkey, two mares and some pigeons (Wakley 1848). To all intents and purposes, 1848 was the year when animal experiments could have been conducted without the unimaginable suffering that they had hitherto caused. Unfortunately, the humane benefits that anaesthesia offered were not immediately or universally adopted, particularly in France. The history of anaesthesia and analgesia in refining animal experiments has been reviewed (Clutton 2020) but neglects cephalopod molluscs whose sentience and ability to mount physical and emotional responses to noxious stimuli have only recently been recognized (Winlow and Di Cosmo 2019).

5.5 Anti-Vivisection Movements

Throughout the nineteenth century, growing disgust at the use of vivisection in physiological studies (particularly after the introduction of effective anaesthetics) promoted the proliferation of numerous anti-vivisectionist movements across Europe. In the United Kingdom, the activities of the "Victoria Street Society for the Protection of Animals from Vivisection" led by the Anglo-Irish reformer Frances Power Cobbe, were the principal driving force for the enactment of the Cruelty to Animals Act 1876, which was the world's first legislation to regulate the use and treatment of live animals in scientific research. Amongst numerous desirable features, the Act set special store on the protection of horses, dogs and cats. This anomaly, which persists in the Animal (Scientific Procedures) Act 1986, reflected the speciesist prejudices of Cobbe herself. Legislation in other countries tends not to differentiate farmed animal species from companion animal or traditional laboratory animal species, i.e. rodents and lagomorphs, although most make special allowances for non-human primates (NHPs).

6 The Twentieth Century

Two World Wars, the Great Depression and the threat of nuclear annihilation distracted attention from animal welfare matters for much of the twentieth century, although major changes affecting farmed laboratory animal welfare occurred in the latter quarter. In industrialized countries, changes were driven by the burgeoning of medical biosciences and pressure on governments to ensure the safety of consumer items, e.g. medical devices, pharmaceuticals, tobacco, food additives and cosmetics. Increasing concerns with laboratory animal welfare led to the formulation of the 3Rs principle (see below). Public concerns with "factory" farming led to the "5 Freedoms" concept; farm animal welfare becoming a subject for scientific focus. An increasing acceptance of the moral status of animals based on—amongst other things—a growing awareness of their sentience, led to protective legislation being established. That animals underwent painful, sometimes horrific procedures in laboratories became an inescapable fact, with photographic (and later video) evidence leaving the public (rightly or wrongly) in little doubt that some scientists were cruel, dispassionate, insensitive and/or dismissive. The growth of online social media has propagated the volume, scope and diversity of those holding opinions on these matters, as well as the opinions held.

6.1 The Growth of Biomedical Research

Medical research was poorly developed at the turn of the twentieth century and relied on rudimentary techniques and limited scientific understanding. The century subsequently saw transformative advances that significantly improved human (and animal) health. Major contributors to these advances—which often involved animals—were: technological innovations, e.g. X-rays and MRI, and a greater understanding of biology and disease, e.g. infection and the discovery of penicillin. Establishing DNA's structure in 1953 led to the development of genetic engineering, gene therapy and molecular medicine. The development of technically advanced surgical procedures, e.g. organ transplantation, had been based on large animal studies. The recognition of the value of collaborative effort, and the establishment of institutions and funding sources, many involving national governments and/or pharmaceutical companies, were also important. The latter half of the twentieth century saw increasing global research efforts, with organizations like the World Health Organization initiating large-scale health projects, e.g. smallpox eradication. Farmed animals would play a role in many of these developments.

6.2 The Growth of Animal Testing

The widespread use of animals in safety testing in the early twentieth century was initiated by public health and safety requirements, which saw the development of new medical devices, pharmaceuticals, chemicals and household products whose risk to consumers (and later the environment) was unknown. The thalidomide tragedy of the 1950's and '60's in which an antiemetic drug caused severe birth defects when taken by pregnant women, highlighted the need for reliable safety testing. At this time, animals were used to assess toxicity, dosing effects and potential side effects of substances before being made publicly available. The absence of robust safety standards led to the introduction of legislation, e.g. the Medicines Act 1968, which required more systematic testing of new drugs, including the use of animal models. Currently, detailed guidelines drawn up by the European Medicines Agency's (EMA) scientific committees, the International Conference of Harmonisation (ICH) and the World Medical Association's Ethical Principles for Medical Research Involving Human Subjects (the Helsinki Declaration) require that new medicines be tested on animals before they can be given to human volunteers or patients. Current regulatory guidelines in the UK and EU recommend that tests are carried out in at least two species: one rodent and one non-rodent species. Dogs (beagles) are the default non-rodent model used in toxicology studies, although rabbits are often used for evaluating reproductive and developmental toxicity, as well as local toxicity. Non-human primates (NHPs) such as cynomolgus macaques or marmosets are used in toxicology studies, particularly for investigating biotherapeutics. Minipigs are increasingly replacing dogs and rabbits in some areas of regulatory testing (Ganderup et al. 2012). In 2023, 21% of all procedures carried out on animals in the UK were for regulatory purposes, rather than for basic or applied research, and the majority (94%) of these were undertaken to satisfy UK and/or EU legislation. In 2023, 305,385 procedures were carried out in the UK for "testing" purposes, the most commonly used species were mice (36.7%) followed by rats (30.3%). Large numbers of sheep (16.3%) were used in the production of blood-based products (a regulatory procedure under the A(SP)A 1986), while large numbers (5.8%) of non-Danio piscean species were used in ecotoxicity testing. The role of poultry (2.4%), cattle (0.4%) and pigs (0.4%) in regulatory testing was considerably less.

The degree of pain and suffering experienced by farmed animal species in drug and toxicity testing depends very much on the test substances and the test conditions. Such testing in the UK is only permissible under A(SP)A 1986, so safeguards, e.g. humane end-points, are established beforehand to prevent prolonged suffering in recovery experiments.

Many pharmaceutical studies involving minipigs (primates and dogs) frequently require that animals are instrumented for telemetry, which is the in vivo sensing of physiological data and its transmission to remote receiving, recording and storing equipment. This can involve major surgery with subsequent recovery, demanding that thought be given to post-operative analgesia. Fortuitously, telemetric data itself may assist considerably in the quantification of post-operational pain.

6.3 Growing Twentieth Century Concerns with Laboratory Animal Welfare

6.3.1 Russell and Burch (1959)

The University of London Animal Welfare Society (ULAWS), was formed in 1926 by Major Charles Hume with the aim of recruiting academics to promote animal welfare. It became the Universities Federation for Animal Welfare (UFAW) in 1938. In 1959, a UFAW-sponsored report authored by William Russell and Rex Burch, entitled "The Principles of Humane Experimental Technique" (Russell and Burch 1959) described the 3Rs principle and established an enduring ethical defence of animal use in research. The "Rs" refer to replacement, reduction and refinement, and their combined objective can be summarized thus: "When scientific goals cannot be achieved using non-animal alternatives (replacement), the fewest number (reduction) of the least sentient animals which experience a life of maximum welfare and minimum suffering (refinement) should be used. The 3Rs principle is globally recognized and adherence to it mandated throughout the process that begins with the application for scientific research funding through ethical study approval and licensing to study prosecution and report publication. Whilst focusing on laboratory rodents, The Principles of Humane Experimental Technique did not differentiate "traditional" from farmed animal species in its philosophy. Since its inception, the 3Rs principle has underpinned most, if not all of the national legislation protecting laboratory animals, such as the Animals (Scientific Procedures) Act 1986 (Anonymous 2014) and the EU Directive 2010/63/EU (Anonymous 2010). International governance, e.g. organizations such as the OECD (Organisation for Economic Co-operation and Development) and WHO (the World Health Organization), also recognize and endorse the 3Rs principle.

6.3.2 Animal Machines (1964)

Farmed animal welfare gained major societal attention in 1964 when Ruth Harrison's book, "Animal Machines", was published (Harrison 1964). Harrison visited farms intensively rearing veal calves, broilers, laying hens and pigs and recorded her findings. "Animal Machines" revealed to the public and the UK government that the post-war years had seen farm animals become production items with scant concern for their welfare. The book also informed about procedures such as "crating" veal calves, battery-caging laying hens, "stalling" pregnant sows, as well as castration, tail-docking, beak-trimming and dehorning young stock.

Reaction to the book was so great that the Government ordered an investigation. Chaired by Roger Brambell, the Brambell Committee presented its 85-page "Report of the Technical Committee to Inquire into the Welfare of Animals Kept under Intensive Livestock Husbandry Systems" in 1965—the so-called "Brambell Report". Amongst other things, the report identified five fundamental farmed animal freedoms (see below) and established the independent Farm Animal Welfare Advisory Committee to monitor the UK's livestock production sector. In 1979, when its name was changed to the Farm Animal Welfare Council, the initial 5Fs were codified.

(i) Freedom from hunger and thirst, by ready access to fresh water and a diet to maintain full health and vigour;
(ii) Freedom from discomfort, by providing an appropriate environment including shelter and a comfortable resting area;
(iii) Freedom from pain, injury or disease: by prevention or rapid diagnosis and treatment;

(iv) Freedom to express normal behaviour, by providing sufficient space, proper facilities and company of the animal's own kind; and.
(v) Freedom from fear and distress, by ensuring conditions and treatment which avoid mental suffering.

In 2011, The Farm Animal Welfare Committee (FAWC) replaced the Farm Animal Welfare Council, although its terms of reference remained similar, i.e. to maintain surveillance of farm animal welfare on agricultural land, at market, in transit and at the place of slaughter and to advise Government of any necessary changes. The "5Fs" protect large laboratory farmed animals obtained from farms or markets in the UK through the Animal Welfare Act 2006. Protection continues until animals come "under license", i.e. protected under the Animals (Scientific Procedures) Act 1986 upon admission to a licensed research establishment.

6.3.3 Animal Liberation

In 1975, "Animal Liberation", written by Australian moral philosopher Peter Singer, reenergized the anti-vivisection movement (Singer 1975). The Animal Liberation Front (ALF), founded in 1976, considered the work to be the founding philosophical statement of its *raison d'etre* and subsequently ensured that laboratory animal welfare remained a public concern. The book examines laboratory animal abuse in the chapter "Tools for Research" and the plight of intensively farmed animals in "Down on the factory farm", but the subjects are not overlapped. In both, Singer provides evidence for the inhumanity of many laboratory and intensive farm practices but says little on the efforts that were being made to improve animal welfare in either sphere. For example, he makes no reference to "The Principle of Humane Experimental Technique" and only cursorily mentions the Brambell report; the 5Fs are not acknowledged.

However, in a later work, Singer opines that there is a greater justification for the exploitation of farmed animals in laboratories than on farms. He states,

> *The knowledge gained from some experiments on animals does save lives and reduces suffering. Hence, the benefits of animal experimentation exceed the benefits of eating animals, and the former stands a better chance of being justifiable than the latter.*

He emphasizes that this only applies when animal experiments fulfil strict conditions relating to the significance of the knowledge to be gained, the unavailability of alternative animal-free techniques and the care taken to avoid pain. Under these conditions, the death of an animal in an experiment can be defended" (Singer 1980).

In 2023, "a new book, rather than a revision" (Singer's words) was published (Singer 2023). Entitled "Animal Liberation Now: The Definitive Classic Renewed", the book records Singer's latest assessment of the state of animal rights. As before, the two largest chapters describe animal abuse in scientific research and farming. In the renewed version, Singer decries the failure of veganism, the further intensification of factory farming in recently developed countries and the burgeoning numbers of animals killed in science and agriculture. However, he cautiously applauds the introduction of more stringent regulations in both laboratories and farms and quietly justifies some experiments, including the use of NHPs in Parkinson's disease research. He correctly points out that the results of many animal experiments are not replicable (see below) and account for unacceptable animal wastage. He also explains (correctly) that research continues to reveal that animals, including farmed species, possess many mental capacities similar to those of humans. The new book supports the increasing belief that consciousness may exist in lower animal orders, and that some insects and crustacean decapods may be sentient. That octopuses show signs of an extraordinary "alien-like" intelligence is acknowledged. It should be noted that such beliefs are not at odds with modern scientific thinking, and changes are already extant in regulations protecting laboratory animals of lower taxonomic orders.

6.3.4 The Animals (Scientific Procedures) Act 1986

In the UK, all research animals[7] are protected by the Animals (Scientific Procedures) Act 1986 (A(SP)A 1986). The Act allows scientific procedures (defined later) to be carried out on protected animals only by Home Office (HO) licensed individuals (with proven competency in the procedures involved), working on HO-approved and licensed projects in a HO-licensed laboratory. Project licenses are only granted if the study's benefits to people, animals or the environment are likely to exceed the harms experienced by the animals involved and where alternatives don't exist. This harm: benefit ratio (H: BR) is established by a mandatory ethical and scientific review process undertaken by an Institutional Animal Welfare and Ethical Review Board (AWERB). For each study, AWERBs approve established humane end-points (HEPs), which are clear criteria that trigger the termination of a procedure before an animal experiences harm that is not authorized or scientifically justified. The severity of licensed studies must be pro and retrospectively assessed as being terminal, mild, moderate and severe or sub-threshold (defined later). The welfare of laboratory animals is the prescribed and legal responsibility of all license holders, but within the laboratory, is assured on a day-to-day basis by designated Named Animal Care and Welfare Officers (NACWOs). The overall health and welfare of all animals within a given establishment is the responsibility of the Named Veterinary Surgeon (NVS). Whilst experimental *equidae*, dogs and cats receive special treatment in the UK under the current law (and predominantly in the study approval stage), the Act provides clear and strong protection for all other species, from mice, rats, fish and toads to pigs, sheep and cuttlefish.

Other regulations, e.g. EU Directive 2010/63/EU, have much in common with A(SP)A 1986, including a shared recognition of the special needs of NHPs. However, beyond that, the former does not afford differential protection to different species.

6.3.5 Pain, Suffering and Openness

Magendie's animal vivisections conducted publicly to large audiences left few lay and medical observers in any doubt about the pain and suffering being gleefully inflicted. In stimulating widespread anti-vivisection sentiment, he drove medical vivisectors in the UK to adopt more secretive practices. This was fuelled by a widespread belief that medical experimenters who conducted such acts could not be trusted with human patients. Consequently, one provision of the Cruelty to Animals Act 1876 was the prohibition of *public* acts of vivisection, which further drove the practice into laboratories and away from public scrutiny. This was not always successful; the observation and reporting of a mismanaged anaesthetic supervised by Earnest Starling (Professor of Physiology at University College London) on a brown terrier-like dog in 1903 by activists Louise Lind-af-Hageby and Leisa Schartau initiated the Brown Dog riots, which smouldered for 7 years. Some 60 years later, Singer's "Animal Liberation" and the rise of the ALF would ensure that animal experiments were conducted in even greater secrecy. Propagandist images of smoking beagles, the Draize test applied to rabbits and the insertion of electrodes into the brains of primates confined in stereotaxic devices ensured that a public majority would believe that scientists were deliberately cruel. More recently, efforts to increase openness and present a more balanced view have intensified. These include the mandatory publication of statistics; the promotion of humane animal research and concordats on "openness".

6.4 Home Office (UK) Statistics and ALURES

Some idea of the pain incurred by experimental farmed animals may be gleaned by analysing published data. In the UK, annual reports, i.e. "Annual Statistics of Scientific Procedures on

[7]"Protected" animals includes all vertebrates, i.e., mammals, birds, amphibians, reptiles and fish and some invertebrates, i.e., the cephalopod molluscs: octopus species; squid; cuttlefish and nautilus.

Living Animals, Great Britain", are published and provide details on the number of regulated scientific procedures involving living animals carried out in the previous calendar year along with the purpose and severity of those procedures (Anon 2024). Using 2023's reports to illustrate this, Table 1 reveals that with the exception of poultry and fish, farmed animals underwent relatively few procedures compared with mice and zebrafish.

Under the A(SP)A 1986, a regulated procedure refers to one carried out on a protected animal for a scientific or educational purpose and may cause that animal a level of pain, suffering, distress or lasting harm equivalent to, or greater than, that caused by inserting a hypodermic needle according to good veterinary practice. Based on this "lower threshold", procedures are classified as "sub-threshold", "mild", "moderate", "severe" or "non-recovery" depending upon the degree of pain, suffering, distress or lasting harm that the animal is expected to experience (see Table 2).

Home Office data reveals the number of procedures carried out in each of the five severity bands. Table 3 reveals that in 2023, farmed animals were predominantly involved in "mild" procedures. While farmed animals collectively underwent some 19% of all procedures, the majority of these were sheep, poultry and fish other than Danio species.

Beyond severity, the HO designates procedures according to their purposes (basic research, translational research or regulatory). These are further subdivided into sub-purposes. Basic research involves the 11 main organ systems, including cancers, although behaviour, development and dentistry are also covered. Translational research is also sub-purposed to 11 human organ systems, but also incorporates animal disease and welfare. Regulatory procedures are described as routine production, quality control and toxicity and safety testing. Few of the 31 sub-purposes of regulatory procedures involved farmed animals in 2023, although notably, some 49,000 sheep were used in the routine production of blood-based products, while fish were heavily involved in eco-toxicity studies, species preservation and environmental protection. Whilst the sheep numbers are large, all underwent mild procedures, probably involving little more than venous cannulation and blood collection. In contrast, 1071 and 188 non-Danio fish underwent severe procedures in studies of chronic and acute eco-toxicity, respectively. The capacity of fish to suffer in adverse environmental conditions cannot be under-estimated (Chap. 21), even if it may be more difficult to recognize.

Table 4 indicates the purposes and sub-purposes for which farmed animals experienced *severe* regulated procedures in the UK in 2023. The data reveal that cattle underwent a disproportionately greater number and range of severe procedures than other farmed animals, although large numbers of fish were involved in severe studies of ecotoxicity.

Unfortunately, not knowing *precisely* what regulated procedures are involved in each sub-purpose study makes it impossible to predict the type and extent of pain and suffering involved, beyond the definitions provided in Table 2. Thus, the 17 procedures conducted on sheep in basic, cardiovascular, blood and lymphatic system research may have ranged from heart transplantation surgeries to studies involving exsanguination. Both meet the criteria of "severe", but the nature of the pain and subsequent suffering experienced are unlikely to be the same. That said, the pre-establishment of HEPs would have ensured limits to the extent and duration of suffering irrespective of the procedures conducted. Furthermore, whilst the eventual elimination of animals from all aspects of medical research is a public expectation and the goal of the 3Rs philosophy, the ending of all *severe* experiments in all species is an initial strategic priority of many bodies in the UK, including the Royal Society for the Prevention of Cruelty to Animals.

Information on animal use in the UK is freely available (see: Statistics of scientific procedures on living animals, Great Britain: 2022—GOV.UK) and identifies how many animals, of what species, suffered (approximately) how much and for what purpose—but information is relatively onerous to glean, particularly when compared with ALURES, the EU Statistical Database (EU

Table 1 Ranked percentage involvement (by species) in the 1,468,570 regulated procedures performed in the UK in 2023 under the Animals (Scientific) Procedures Act 1986

Animal Group	Species	Total procedures	Percentage of total procedures
Mammal	Mouse (*Mus musculus*)	879,725	59.9%
Fish	Zebrafish (*Danio rerio*)	150,481	10.2%
Mammal	Rat (*Rattus norvegicus*)	143,353	9.8%
Bird	Domestic fowl (*Gallus domesticus*)	112,000	7.6%
Mammal	Sheep (*Ovis aries*)	62,201	4.2%
Fish	Other fish (other Pisces)	59,445	4.0%
Amphibian	Xenopus (*laevis and tropicalis*)	12,662	0.9%
Mammal	Horse and other equid (Equidae)	11,409	0.8%
Mammal	Rabbit (*Oryctolagus cuniculus*)	8,936	0.6%
Mammal	Pig (*Sus scrofa domesticus*)	4,811	0.3%
Mammal	Guinea-pig (*Cavia porcellus*)	3,769	0.3%
Mammal	Beagle (*Canis lupus familiaris*)	3,544	0.2%
Mammal	Cattle (*Bos taurus*)	3,482	0.2%
Bird	Other bird (other Aves)	2,907	0.2%
Mammal	Hamster (Syrian) (*Mesocricetus auratus*)	2,077	0.1%
Primate	Cynomolgus monkey (*Macaca fascicularis*)	2,032	0.1%
Amphibian	Other amphibian (other Amphibia)	1,200	0.1%
Amphibian	Rana (*temporaria and pipiens*)	1,182	0.1%
Mammal	Other carnivore (other Carnivora)	945	0.1%
Mammal	Other rodent (other Rodentia)	826	0.1%
Mammal	Other mammal (other Mammalia)	423	0.0%
Mammal	Mongolian Gerbil (*Meriones unguiculatus*)	377	0.0%
Mammal	Ferret (*Mustela putorius furo*)	231	0.0%
Mammal	Other dog (other Canis)	205	0.0%
Bird	Quail (*Coturnix coturnix*)	100	0.0%
Primate	Rhesus monkey (*Macaca mulatta*)	73	0.0%
Primate	Marmoset and tamarin	64	0.0%
Mammal	Cat (*Felis catus*)	63	0.0%
Mammal	Goat (*Capra aegagrus hircus*)	36	0.0%
Reptile	Reptile (Reptilia)	11	0.0%
Mammal	Hamster (Chinese) (*Cricetulus griseus*)	0	0.0%
Cephalopods	Cephalopod (Cephalopoda)	0	0.0%
		1,468,570	100

Data rows for the principle farmed animal species are highlighted

Commission 2025). As with project license holders in the UK, Directive 2010/63/EU requires that all member states of the EU submit information on animal use in the interests of improved transparency. In 2021, the European Commission indicated their intention to optimize transparency in animal use via the launch of a public, open-access database, containing all information from the member states. The ALURES database provides detailed information on the use of animals for scientific purposes throughout the EU. It includes complete datasets on each use of an animal and what type of procedures were carried out. The ALURES database is freely accessible to scientists, regulators, policy makers, authorities, ethical committees and the public, and

Table 2 Severity "banding" of regulated procedures under the Animals (Scientific Procedures) Act 1986; definitions and examples

Procedural severity & definitions[a]	Examples	Notes
Sub-threshold procedures refer to supra-threshold procedures that later transpire to be less noxious than anticipated	Breeding genetically altered animals under project licence authority but without a harmful phenotype Testing compounds in-feed, which the animals eat normally and suffer no consequences of dosing	>40% of all procedures
Non-recovery procedures involve the animal receiving a general anaesthetic before the procedure begins, and is humanely killed without regaining consciousness	Numerous Toxicity testing with innocuous compounds	Approximately 3% of all procedures
Mild procedures cause short-term mild pain, suffering or distress and/or minor changes in well-being or condition	An anaesthetic Intramuscular, -venous, -peritoneal injection Non-invasive imaging, e.g. CT scanning Short-term social isolation Taking a blood sample Repeated handling Superficial surgical procedure, e.g. skin biopsy	Recovery from mild procedures is immediate/or rapid with no lasting effect evident upon physical examination. Mild procedures account for >33% of all procedures
Moderate procedures cause: (i) short-term moderate pain, suffering or distress; or (ii) long-lasting mild pain, suffering or distress; and/or (iii) moderate impairment in well-being or condition	Surgery under general anaesthetic Inducing cancer Diet modification Exposing animals to stimuli they would normally avoid but preventing them from doing so	Moderate procedures cause significant and easily detectable disturbances to an animal's normal state. The disturbance is enough for animals to show discomfort, abnormal behaviours, significant weight loss or other indicators of poor welfare, but does not prevent normal feeding and drinking or other normal activities, other than for short periods or to a limited extent for longer periods. Accounts for approximately 15%
Severe procedures are likely to cause animals: (i) severe pain, suffering or distress, or (ii) long-lasting moderate pain, suffering or distress; and/or (iii) severe impairment of the animal's well-being or general condition	Any test where death is the end-point or fatalities are expected Spinal cord injury Testing a device that would cause pain/death if it were to fail Pancreatitis studies Inescapable electric shocks Breeding genetically disordered animals that are expected to manifest severe & persistent impairment of general condition, e.g. Huntington's disease, muscular dystrophy	Severe procedures cause major departure from the animal's usual state of health and well-being. It would usually include long-term disease processes where assistance with normal activities, such as feeding and drinking, are required or where significant deficits in behaviours/activities persist. This would include any state that a person would find difficult to tolerate or disease where clinical signs have progressed to such an extent that it threatens the life of an animal. Severe procedures currently account for approximately 3% of all procedures

[a]Paraphrased from: Advisory notes on recording and reporting the actual severity of regulated procedures and Annex VIII Animals Directive 2010/63 EU

Table 3 The number of regulated procedures undergone by farmed animals in the United Kingdom (2023) according to severity banding

Species	Non-recovery	Sub-threshold	Mild	Moderate	Severe	Total procedures	% of total procedures
Cattle	0	7	3051	408	16	3482	0.20
Sheep	6	101	61,778	298	18	62,201	4.20
Goat	0	0	17	19	0	36	0.00
Pig	330	149	3124	1184	24	4111	0.30
Other fish	3	10,232	34,468	13,366	1376	59,445	4.00
Birds (Fowl)	85	951	109,975	750	239	112,000	7.60
Birds (Quail)	0	4	91	5	0	100	0.00
Cephalopods	0	0	0	0	0	0	0.00
Total farmed animals	424	11,444	212,504	16,030	1673	241,375	16.3
All other species	63,576	284,556	511,496	393,970	53,327	1227,195	83.7
[a]	0.67	4.02	41.55	4.07	3.14	18.85	
All species	64,000	296,000	724,000	410,000	55,000	1,468,570	100

Based on published Home Office Figures

[a]Proportion (percentage) of farmed animals undergoing procedures in each band relative to total numbers

Table 4 The purposes and sub-purposes for which farmed animals experienced *severe* regulated procedures in the United Kingdom in 2023. Home Office data

Species				Cattle	Sheep	Goats	Pigs	Poultry	Fish
Basic research			Cardiovascular Blood and Lymphatic System	0	17	0	0	0	15
			Gastrointestinal System including Liver	11	0	0	0	0	0
			Musculoskeletal System	83	0	0	0	0	0
			Immune System	438	0	0	0	0	3
			Multisystemic	154	0	0	0	0	0
			Developmental	0	0	0	0	0	1
			Other	12	0	0	0	0	0
Translational or applied research			Human Sensory Organ Disorders (skin, eyes and ears)	0	0	0	1	0	0
			Animal Diseases and Disorders	819	0	0	16	217	82
			Nutrition	192	0	0	0	0	0
			Animal Welfare	120	0	0	0	4	0
			Diagnosis of diseases	132	0	0	0	0	0
			Non-regulatory toxicology and ecotoxicology	0	0	0	0	0	15
Regulatory	Routine production		Other	70	0	0	0	0	0
	Quality control		Batch safety testing	10	0	0	0	0	0
			Batch potency testing	618	0	0	0	0	0
			Other efficacy and tolerance testing	527	1	0	1	18	0
	toxicity and other safety testing including pharmacology		Acute and sub-toxicity testing (Non lethal methods)	0	0	0	1	0	0
		Other types of regulatory procedure	Repeated dose toxicity < 28 days	0	0	0	3	0	0
			Repeated dose toxicity > 90 days	0	0	0	2	0	0
			Target animal safety	145	0	0	0	0	0
			Kinetics	2	0	0	0	0	0
		Ecotoxicity	Ecotoxicity; acute toxicity	0	0	0	0	0	188
			Ecotoxicity; chronic toxicity	0	0	0	0	0	1,071
			Ecotoxicity; bioaccumulation	0	0	0	0	0	1
			Protection of the natural environment	149	0	0	0	0	0
			total procedures	3,482	18	0	24	239	1,376
			% of total procedures	0.23%	0.00%	0.00%	0.00%	0.01%	0.09%

allows details of the severity of farmed animal experiments conducted in the EU to be identified rapidly.

6.4.1 Promoting "Open" Humane Animal Research

Numerous organizations currently exist to promote the open, humane and ethical use of animals in research, e.g. Understanding Animal Research, Animal Research Tomorrow, the National Centre for the Replacement, Refinement and Reduction of Animals in Research (UK), Norway's National Platform for the advancement of the 3Rs (NORECOPA), the Danish 3R-Center, the Association of Medical Research Charities, the European Animal Research Association and the Laboratory Animal Science Association. Whilst these organizations differ in several ways, their broad common remits are: the ethical treatment of animals in research; unconditional implementation of the 3Rs; proactive and fact-based interactive dialogue with society; increased transparency, accountability and trust through open science.

6.4.2 The Non-technical Summary

The non-technical summary (NTS) is a recent initiative designed to better inform the public about the nature of animal experimentation (and the suffering involved). In the EU and the UK, the provision of an acceptable NTS is a condition for a project license award. The NTS must provide a clear, concise and accessible summary of the research project's aims, benefits and implications, without technical scientific jargon. It must explain firstly why animals are necessary and then justify why a particular species is being used. Importantly, details of what exactly happens to the animals during and after the experiments, and details of the procedures that the animals will experience, must be provided. Finally, details of the expected level of harm or pain that the animals may undergo and how any discomfort will be managed and minimized must be provided. Such information is made freely available to the public and in theory overcomes any remaining requirement for imagination in understanding the animal's experience.

6.4.3 Limitations of Current Methods for Evaluating Pain and Suffering

Pain experiences amongst humans vary considerably (Coghill 2010). Such variability can only increase when observers attempt to evaluate animal pain. Variability in assessment is undesirable as decision making over humane end-points—a legal and scientific requirement—requires a high degree of consensus. Evaluating farmed animal pain is complicated by a general lack of relevant information arising from inadequate reporting in the scientific literature, and a limited number of relevant scoring systems.

6.4.4 Lack of Relevant Information

Pain recognition is less straightforward in farmed compared to traditional laboratory animals for several reasons. The relatively few number of the former undergoing regulated procedures compared to mice and rats (Table 1) limits the opportunities to collect (and publish) data on suspected behavioural, physiological and endocrinological indicators of pain and suffering. (A corollary of the fewer farmed animal numbers involved in this is the limited funding available to study farmed laboratory animal pain). This is manifest in the fewer number of publications dedicated to pain assessment in the farmed laboratory animal species (Fig. 3).

Furthermore, information on farmed animal pain assessment focuses mainly on pain experienced by animals under farm rather than laboratory conditions; scientific reports describing pain assessment in young animals undergoing routine procedures such as castration, teeth clipping and tail docking or adults suffering environmental pain, e.g. lameness, mastitis, predominate. A recent review article (Ison et al. 2016) entitled "A Review of pain assessment in pigs" cited 206 articles whose differing objectives confirm this point. Figure 4 shows the prevalence of scoring systems devised for farm, not laboratory use. This is important because the pain experiences (and signs thereof) of young animals undergoing painful husbandry procedures are not the same as adults with chronic pain conditions. Neither is comparable with the acute pain and suffering endured by sheep recovering from major orthopaedic surgery, or pigs recovering from thoracotomy operations, without effective analgesia.

6.4.5 Pain Scoring Systems

For traditional laboratory animal species, adequate consensus in pain evaluation may be achieved using pain scales (Chap. 10). These are based on collective experiences, experimental studies and, more recently, the analysis of video-recorded behavioural pain-related aberrations using artificial intelligence. In all cases, behavioural aberrations are gauged against normal behaviours and pain is quantified in approximate proportion to the degree of aberration from the norm.

Numerous factors affect the validity of pain scoring systems: beyond non-animal factors (observer bias and training, scoring system design, validation and reliability testing), the most important variable is species. The behavioural response of pigs to severe procedures is not those of sheep. Related factors include: strain (or breed); age; sex; function; source (and level of domestication); health status; previous experiences. The nature of the procedure conducted is critically important: the pain experienced after major orthopaedic surgery will not be the same as that occurring after liver transplantation, although the degree of suffering may be similar. Contextual influences may also be important: environmental factors (noise, temperature, handling) may affect behaviour and physiological responses, confounding evaluation. Consequently, a pain scoring system devised for piglet castration will be useless when applied to adult pigs recovering from thoracic surgery.

6.4.6 Inadequate Reporting of Animal Studies

The subject of what constitutes "effective analgesia" in laboratory animals is another major impediment to gaining a general appreciation of laboratory animal pain and suffering because for some time, the description of laboratory animal pain management, i.e. recognition, quantification and importantly treatment, in scientific reports has been inadequate—if it was described at all.

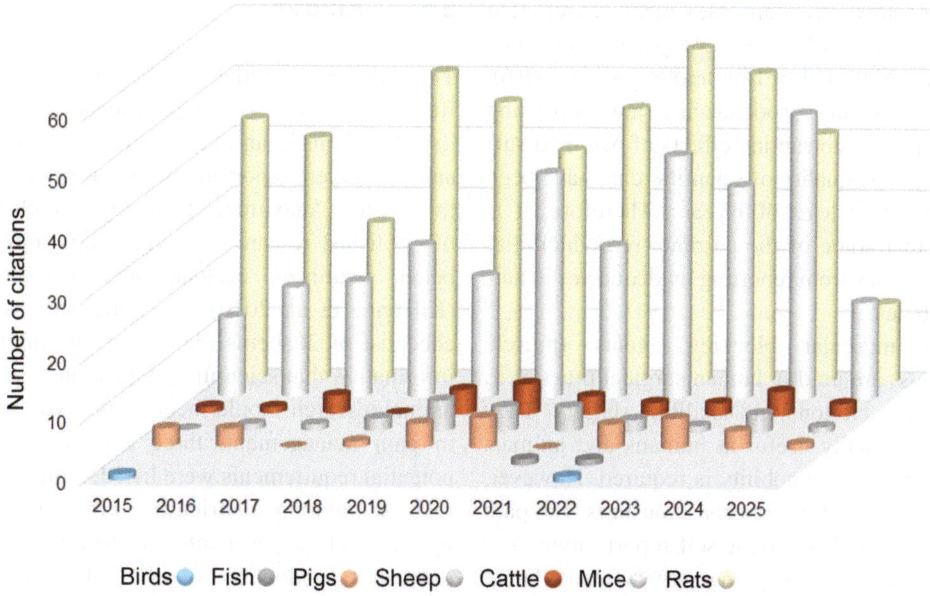

Fig. 3 The frequency of publications devoted to pain scoring in farmed and traditional laboratory animal species from 2015–2025. (PubMed search terms: Pain; Scoring; (species))

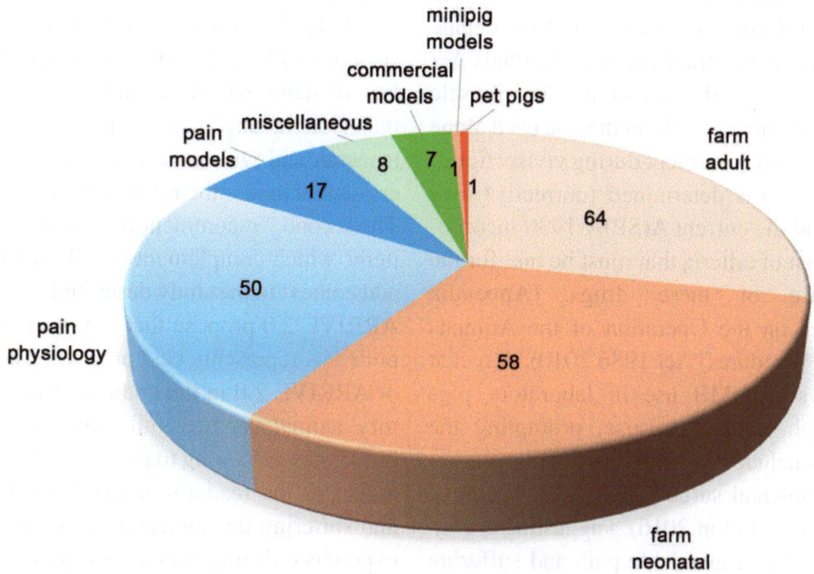

Fig. 4 Pie-chart showing subject categories of 206 references cited in "Pain assessment in Pigs" Ison et al. 2016. The figure reveals the preponderance of data arising from farm rather than laboratory sources

One structured literature review (Coulter et al. 2009) compared the extent of analgesic administration in larger laboratory animals, including pigs and sheep, in 74 studies between 2000 and 2001 and 75 studies between 2005 and 2006. Despite a modest increase (from 50% to 63%) in the reported administration of analgesics, it was concluded that the reporting of analgesic administration to "large" laboratory species was suboptimal. Another work focusing entirely on pigs found postoperative pain assessment was reported in only 23 of 233 (10%) of papers published between 2012 and 2014 (Bradbury et al. 2016). These reporting inadequacies are surprising given that the corrupting effects of pain and suffering on the quality of scientific data have been known for at least 400 years. Therefore, it is likely that some of the current "reproducibility crisis[8]" arises from reporting inadequacies in animal studies.

Neuromuscular blocking agents (muscle relaxants) are used in human surgical practice to reduce muscle tone and facilitate surgery. They are occasionally useful in humans and animals when absolute immobility is required. However, they have no effect on consciousness nor pain mitigation, and horrifying self-reports abound of human patients undergoing surgery whilst being aware but unable to protest. Ironically, Claude Bernard conducted much animal research on neuromuscular blocking agents in dogs—without anaesthetics—and in the full knowledge of what he was doing. Consequently, much of the debate in the formulation of the Cruelty to Animals Act 1876 was given to the question, "Are muscle relaxants anaesthetics, and can they be used alone in animals to stop movement during vivisection?" Fortunately, it was determined (correctly) they were not, and the current A(SP)A 1986 incorporates a full list of criteria that must be met for the humane use of these drugs (Appendix H. Guidance on the Operation of the Animals (Scientific Procedures) Act 1986 2010). A recent study analysing NMB use in laboratory pigs found a high level of misuse, prompting the authors to conclude that awareness in pigs undergoing experimental surgery may be widespread (Bradbury and Clutton 2016), suggesting a considerable degree of avoidable pain and suffering may be experienced by many pigs undergoing experimental surgery.

[8]The "reproducibility crisis" refers to a growing concern that many published research findings are difficult or impossible to replicate, potentially undermining the credibility and advancement of scientific knowledge.

6.5 ARRIVE

The ARRIVE guidelines (Animal Research: Reporting of In Vivo Experiments) were published in 2010 in an attempt to improve the full and transparent reporting of animal research and thereby to (i) maximize the quality and reliability of published research and (ii) enable others to better scrutinize, evaluate and reproduce it (Kilkenny et al. 2010). The guidelines were a checklist of 20 items identifying the minimum information that scientific publications reporting animal research should incorporate. With respect to pain management, three explicit and four potential requirements were listed, indicating the relative importance attributed to these in reporting animal experiments. Unfortunately, the enthusiastic adoption of the ARRIVE 2010 guidelines by research institutions, funding bodies and publishers did not translate into improved practice, necessitating a review of requirements 10 years later.

ARRIVE 2.0 (Percie du Sert et al. 2020) consists of two lists: The first, "essential" list identifies 10 items which are the basic minimum for inclusion in any manuscript describing animal research and without which readers and reviewers cannot assess the reliability of study findings. The second, "recommended" list also contains 10 items which complement the "Essential 10" and add context to the study described. The authors of ARRIVE 2.0 propose that reporting the items in both sets represents best practice. The relevance of ARRIVE 2.0 in the context of farmed laboratory animals is that full compliance with the guidelines pertaining to pain management greatly enhances the reader's appreciation for the pain and suffering the subject animals did–or did not experience during a given study. It also reflects (and arguably encourages) the scientists' commitment to animal welfare. Tables 5a, b lists the *relevant* provisions of ARRIVE 2.0.

6.5.1 The Use of Farmed Animals in Twentieth Century Research

Table 6 indicates some of the research areas involving farmed animal species during the twentieth century. While an absence of transpar-

Table 5 (a) Items required under ARRIVE 2.0 that: (i) allow estimation of pain and suffering experienced by animals described in scientific reports; (ii) encourage scientists to comply with animal welfare requirements; and (iii) improve experimental reproducibility. (b) Recommended items adding context to the study report

(a) *ARRIVE 2.0; Essential 10*
(9) *Experimental procedures*
(9a) In surgical procedures (including sham surgery) what was done, how it was done, and what was used;
Describe the surgical procedure Detail the anaesthetic used (including dose and other information listed in pharmacological procedures) Detail the pre and post analgesia regimen Describe pre-surgery procedures such as fasting Describe aseptic techniques involved Describing associated monitoring, e.g. assessment anaesthetic plane during surgery Indicate whether the procedure is terminal or not Describe post-surgery procedures Indicate the duration of the procedure and duration of anaesthesia Detail the physiological variables measured
(9b) When and how often?
Describe the timing and frequency of welfare assessments, taking into consideration the animal's normal activity patterns
(9c) Where?
Indicate where studies were performed, e.g. dedicated laboratory space or animal facility, home cage, open field arena, water maze, and whether periods of physiological or procedural acclimatization were included in the study protocol, including type and duration
(b) *ARRIVE 2.0; Recommended 10*
(16) *Animal Care and monitoring*
(16a) Describe any interventions or steps taken in the experimental protocols to reduce pain, suffering, and distress. Clearly describe pain management strategies, including:
Specific analgesics used Administration method, e.g. formulation, route, dose, concentration, volume, frequency, timing, and equipment used. Rationale for the choice, e.g. animal model, disease/pathology, procedure, mechanism of action, pharmacokinetics, personnel safety Protocol modifications to reduce pain, suffering and distress, e.g. changes to the anaesthetic protocol, increased frequency of monitoring, procedural modifications, habituation, etc. If analgesics or other welfare measures, reasonably expected for the procedure performed, are not performed for experimental reasons, report the scientific justification.
(16b) Report any expected or unexpected adverse events
Report any events that adversely affected animal welfare in the study, e.g. cardiovascular and respiratory depression, CNS disturbance, hypothermia, reduction of food intake. Indicate whether they were expected or unexpected. If adverse events were not observed, or not recorded during the study, explicitly state this
(16c) Describe the humane endpoints (HEPs) established for the study, the signs that were monitored, and the frequency of monitoring. If the study did not have HEPs, state this
Detail the HEPs established for the specific study, species and strain. Include clear criteria of the clinical signs monitored and clinical signs that led to euthanasia or other defined actions. Include details such as general welfare indicators. e.g. weight loss, reduced food intake, abnormal posture, and procedure-specific welfare indicators, e.g. tumour size in cancer studies. Report the timing and frequency of monitoring, taking into consideration the normal circadian rhythm of the animal and timing of scientific procedures, as well as any increase in the frequency of monitoring, e.g. post-surgery recovery, critical times during disease studies, or following the observation of an adverse event. Publishing score sheets of the clinical signs that were monitored can help guide other researchers to develop clinically relevant welfare assessments, particularly for studies reporting novel procedures

With permission from the NC3Rs

ency—the working norm—and poor scientific reporting make it difficult to gauge the degree of pain and suffering experienced by the animals involved in the first 80 years of that century, the changes mentioned above ensured major improvements occurred throughout the industrialized world in the last twenty. Tables 6 and 7 (below) have been synthesized from the following articles: pigs (Gutierrez et al. 2015), sheep (Banstola and Reynolds 2022), goats (Alvites et al. 2021), birds (Kumar et al. 2024), fish (Sarasamma et al. 2017), cephalopods (Fiorito et al. 2014), insects (Baqri 2023), general (Polejaeva et al. 2016).

6.5.2 The Twenty-First Century

Table 7 shows the key areas in which farmed animal species are currently involved. Whilst it

Table 6 Farmed animal species involvement indicated by "+" in twentieth century research

	Pigs	Sheep	Goats	Cattle	Birds	Fish	Ceph.	Insects
Discovery of Rous sarcoma virus and cancer research (1910)					+			+
Discovery of chromosomes & gene Inheritance (1910–20)								
Insulin production (1920)	+							
Influenza virus discovery & vaccine development (1930)					+			
Embryology & developmental biology (1930–present)					+			+
Identification of genetic mutations (1930–50)								+
Production of yellow fever vaccine (1930)					+			
Discovery of the action potential in nerve cells (1930–1950)							+	
Development of nerve conduction studies (1940–1960)							+	
Development of polio vaccine (1950)				+				
Discovery of the immune system's Bursa of Fabricius (1956)					+			
Development of artificial heart valves (1950–60)		+	+	+				
Antibody production for infectious diseases (1950–present)			+					
Heart valve replacement (1960–90)	+	+	+	+				
Organ transplantation (1960–present)	+	+		+				
Xenotransplantation (1960–80)	+			+				
Pulmonary & cardiovascular research (1960–present)		+	+					
Fetal surgery & neonatal medicine (1960–present)		+						
Blood coagulation research; haemophilia therapy (1960–present)				+				
Fundamental studies on vertebrate development (1960–1980s)						+		
Advances in neurotransmitter research (1960–1980)							+	
Early research on memory and learning (1980–1990)							+	
Advances in immunology & vaccines (1970)	+							
Artificial lung development (1970)		+						
Surfactant therapy for premature babies (1970–80)		+						
Osteoporosis & bone research (1970–present)			+					
Vaccine development for infectious diseases (1970–present)				+				
Marek's disease vaccine & breakthroughs in viral oncology (1970)					+			
Toxicology & environmental health research (1970–present)						+		
Development of insect-borne disease control strategies (1970)								+
Burn treatment & skin grafts (1980)	+							
Bone grafting & orthopaedic implants (1980–present)		+						
Monoclonal antibody research (1980)				+				

Avian leukosis virus & Human Leukaemia Research (1980)					+			
First use of zebrafish in genetic research (1980)						+		
Discovery of circadian rhythms (1980)								+
Understanding neurological disorders (1980–1990)								+
Transgenic animals for protein production (1990)			+					
Cloning (1996)		+						
Corneal transplants & eye surgery innovations (1990)				+				
Bovine collagen for wound healing & cosmetic surgery (1990)				+				
Use of chick embryos in stem cell research (1990)					+			
Discovery of cancer genes using zebrafish (1990)						+		
Insights into camouflage and skin regeneration (1990)							+	
Breakthroughs in cancer research (1990)								+

Not all research areas are listed. The absence of "+" does not indicate absolute species exclusion from the corresponding research area. Years refer to the decade research began or was accomplished

Ceph cephalopod

Table 7 Farmed animal species involvement indicated by "+" in twenty first century research

	Pigs	Sheep	Goats	Cattle	Birds	Fish	Ceph.	Insects
Advances in stem cell therapy (2000–present)	+	+	+					
Xenotransplantation & tissue engineering (2000–present)	+	+	+	+				
Gene-edited sheep for human disease research (2000–present)		+						
Transgenic animals producing human antithrombin (2000)			+					
Diabetes treatment innovations (2000–present)	+							
Genetic modification for environmental adaptation				+				
Genetic modification for disease resistance	+			+	+			
Cardiac research & pacemaker development (2000–present)	+							
Production of human therapeutic proteins in calf milk (2000)				+				
Tissue engineering & regenerative medicine (2000–present)				+			+	
Neuroplasticity and brain regeneration research (2000–present)							+	
Obesity & metabolic syndrome studies (2000–present)	+				+			
Cancer research (2000–present)	+				+			
Avian Flu Vaccine & Pandemic Preparedness (2000)					+			
Human disease modelling (zebrafish) (2000–present)						+		
Bionic & artificial limbs testing (2010–present)		+						
Cartilage & joint regeneration (2010–present)		+						
Heart failure & artificial heart research (2010–present)		+						
3D tissue bioprinting (2010–present)		+		+				
Use of chicken embryos in 3D bioprinting (2010–present)					+			
Bone grafting & orthopaedic Implants (2010–present)								
Human collagen development for wound healing (2010–present)			+					
Neurological disease research (2010–present)			+					
Cancer research & immunotherapy (2010–present)			+					
Liver disease & cirrhosis studies (2010–present)			+					
Milk-based production of human pharmaceuticals (2010–present)			+					
Gene therapy for haemophilia (2010–present)			+					
First use of CRISPR gene editing in insects (2010)								+
CRISPR gene editing for organ transplants (2010–present)	+							
CRISPR gene editing for disease research (2010–present)			+	+				
CRISPR gene editing in chickens (2010)					+			
Development of artificial organs (2010–present)	+							
Monoclonal antibody production in chickens (2010–present)					+			
Chick models of neural & brain disease research (2010–present)					+			
Liver disease & hepatitis research (2010–present)	+							
Neuroscience & Alzheimer's disease (2010–present)	+			+		+		
Personalized medicine & drug testing (2010–present)	+					+		
Gene therapy advances (2010–present)	+							
Microbiome & gut health research (2010–present)	+							+
Gene therapy & disease modelling (2010–present)								+
Mosquito-based malaria & zika virus research (2010–present)								+
Insect-derived antimicrobial peptides (2010–present)								+
Regenerative medicine & wound healing (2010–present)						+		+
Drug discovery for cancer & aging (2010–present)								+

(continued)

Table 7 (continued)

	Pigs	Sheep	Goats	Cattle	Birds	Fish	Ceph.	Insects
Insect-inspired robotics & prosthetics (2010–present)								+
Studying the effects of space travel on biology (2010–present)								+
Biocompatible material for medicine development (2010–present)							+	
Cephalopod-inspired robotics for prosthetics (2010–present)							+	
Bio-inspired drug delivery systems (2010–present)							+	
Breakthroughs in skin regeneration & burns (2010–present)							+	
Squid ink for cancer diagnosis & treatment (2010–present)							+	
Zika virus research using chick models (2016)					+			
CRISPR gene editing in fish (2013–present)						+		
Understanding autism & neurological disorders (2010–present)						+		
Modelling heart & cardiovascular disorders (2010–present)						+		
Bird flu research & public health surveillance (2020)					+			
Advanced vaccine research (2020)			+					
Antibody production for emerging infectious diseases (2020)			+					
Development of artificial milk for infant nutrition (2020)			+					
Muscular dystrophy & genetic disease research (2020)			+					
Chickens as bioreactors for drug production (2020)					+			
COVID-19 & respiratory disease research (2020)	+	+		+	+	+		
Spinal cord injury research (2020)		+					+	
Artificial blood vessel development (2020)	+	+						
Gene therapy for muscular dystrophy (2020)		+						
Artificial uterus & preterm birth research (2020)		+						
Breakthroughs in vision & eye disease research (2020)								+
Using insects for biodegradable drug delivery (2020)								+
Dementia & memory loss research (2020)								+
Future of synthetic biology & genetic engineering (2020)								+
Artificial blood (and oxygen carrier) development (2020)				+				+
Cancer immunotherapy & antibody research (2020)				+				
Development of bioengineered skin & burn treatments (2020)				+				
Advanced wound healing using cephalopod biomaterials (2020)							+	
Studying sleep & brain activity for human sleep disorders (2020)							+	
Biodegradable implants & surgical adhesives (2020)							+	
Cephalopod-inspired vision for blindness treatments (2020)							+	
Camouflage-inspired medical textiles & bandages (2020)							+	
Brain-machine interfaces (2020)							+	
Cancer immunotherapy research (2020)						+		
Studying the effects of aging (2020)						+		
Bone & skeletal disease research (2020)						+		
Regeneration of human retinal cells (2020)						+		
Studying rare genetic disorders (2020)						+		
Anti-inflammatory drug development (2020)						+		
Cardiovascular regeneration & stem cell therapy (2020)						+		

Not all research areas are listed. The absence of a "+" does not indicate absolute species exclusion from the corresponding research area. Years refer to the decade research began or was accomplished
Ceph cephalopod

would be naïve to believe that these no longer involve pain and suffering, major improvements have clearly been made and continue to occur. The most serious "driver" for improvement is replacement; several national 3Rs organizations are now considering the year animal experimentation will finally end in their jurisdictions. However, it is difficult to see how large animal models can be replaced by organoids, "bodies-on-a-chip", AI-based models and 3D bioprinting or indeed humans for safety and efficacy testing heart-assist and similar devices. The replacement drive comes from numerous bodies, including the NC3Rs and Replacing Animal Research (formerly the Fund for the Replacement of Animals in Medical Experiments). Until they achieve success, efforts to improve the welfare of farmed animals in the laboratory continue. The Federation of European Laboratory Animal Science Associations (FELASA) has commissioned working groups on subjects relevant to farmed laboratory animal welfare. For example, the FELASA Working Group on Farm Animals Involved in Research: "Anaesthesia and Analgesia" and "The Health and Welfare of Ruminants and Pigs". Recently (2025), the Research Animal Anaesthesia Network received formal support from the Association of Veterinary Anaesthetists; the former exists to promote the involvement of veterinary anaesthetists in animal experiments by fostering collaboration between experienced laboratory animal anaesthetists and scientists requiring assistance or advice in animal anaesthesia and analgesia.

6.5.3 Public Understanding

The efforts of numerous bodies to eliminate pain and suffering in laboratory animals of all species are not universally recognized by the public. In 2018, Ipsos Mori[9] published the findings of a United Kingdom survey into the public acceptance, awareness and attitudes towards the use of animals in scientific research (Ipsos 2018), which showed that most (71%) of the public accept the use of animals in scientific research, providing: it is for medical research; there is no unnecessary suffering; and there is no animal-free alternative. The survey also revealed the public's views on the acceptability of using different species in different types of research (see Fig. 5). The use of farmed animal species was favoured over dogs and cats, while the use of rodents and fish caused the least concern. A significant proportion of respondents thought the involvement of great apes in both medical and animal research was justified, even though this had been banned in the United Kingdom since 1998 (except for ethically justifiable studies, e.g. species-preservation). Furthermore, 41% of the 38% of respondents who believed that animal cosmetics testing was still permissible in the United Kingdom, considered themselves to be "very well-informed" on the subject of animal experiments. Clearly, feeling informed about animal use in research is not necessarily grounded in objective knowledge about the subject: testing cosmetics on animals has been banned in the United Kingdom since 1998 and across the EU since 2013.

7 Conclusion

In terms of numbers, farmed animals (with the exception of fish and poultry) have never figured prominently in scientific research, but when they have, many have suffered grievously. However, the last 30 years has seen major improvements in the welfare of all laboratory animal species in the industrialized world and never has there been more openness, allowing the interested public to determine for themselves how great and/or unnecessary that pain and suffering continues to be. Those who continue to doubt are encouraged to visit the NORECOPA (Smith 2025) and/or ALURES websites.

[9] Ipsos MORI, now Ipsos, is the UK-based arm of the global Ipsos group, which specializes in research across areas of social and public policy. It abides by the Market Research Society Code of Conduct.

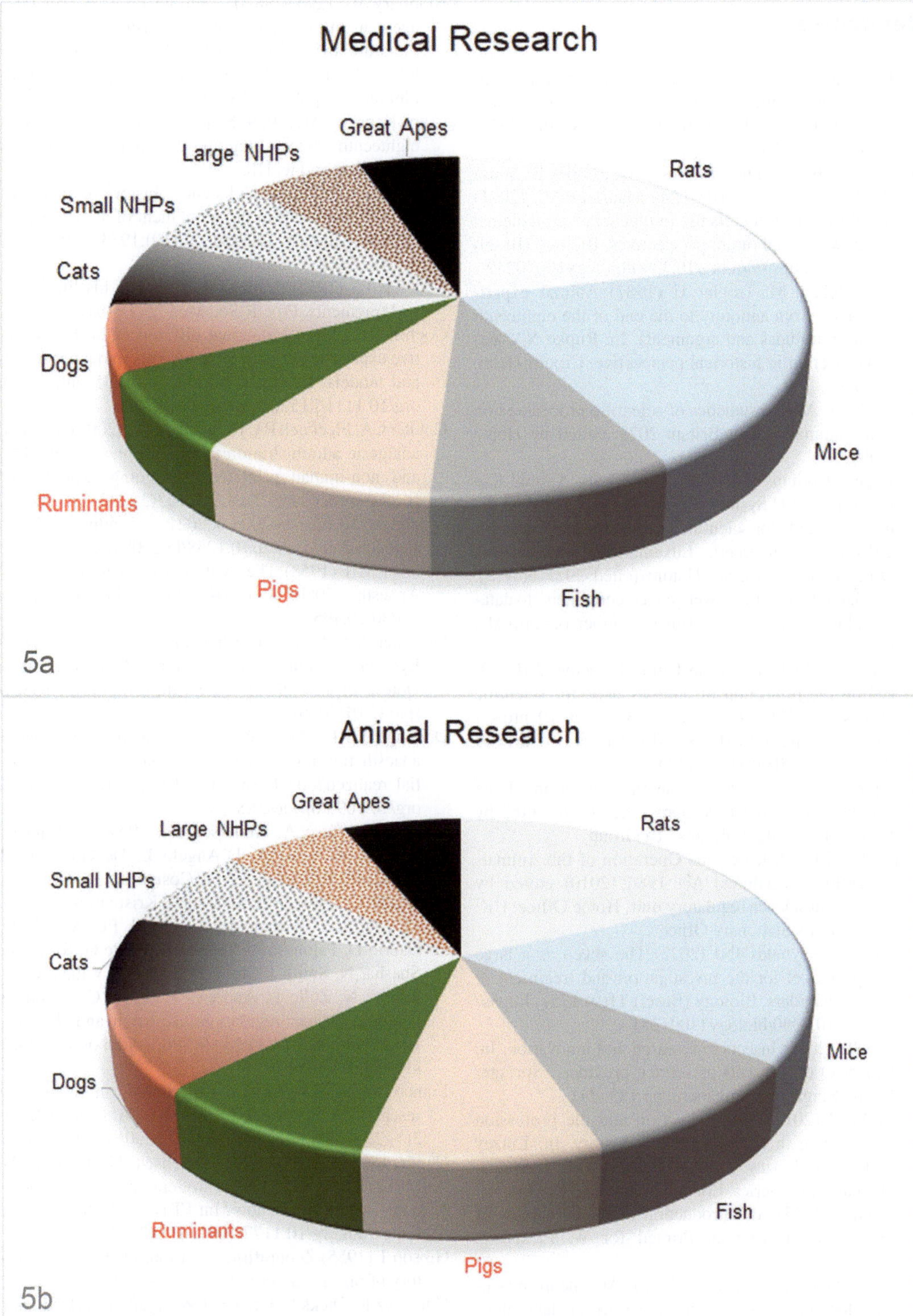

Figs. 5 Pie-charts showing the relative proportion of survey respondents accepting the use of different species in medical (**a**) compared with animal (**b**) research. Birds, amphibians and small mammals, e.g. rabbits and ferrets, have been excluded. NHPs: non-human primates. (Data taken (with permission) from the 'Public attitudes to animal research in 2018' Ipsos; Department for Business, Energy & Industrial Strategy)

References

Abdel-Halim RE (2005) Contributions of Ibn Zuhr (Avenzoar) to the progress of surgery: a study and translations from his book Al-Taisir. Saudi Med J 26(9):1333–1339

Alvites RD, Branquinho MV, Sousa AC, Lopes B, Sousa P, Mendonça C, Atayde LM, Maurício AC (2021) Small ruminants and its use in regenerative medicine: recent works and future perspectives. Biology (Basel) 10(3):249. https://doi.org/10.3390/biology10030249

Andreas-Holger M, Tröhler U (1987) Animal experimentation from antiquity to the end of the eighteenth century: attitudes and arguments. In: Rupke NA (ed) In vivisection in historical perspective. Croom Helm, London

Anon (2024) Annual statistics of scientific procedures on living animals, Great Britain 2023. edited by Home Office

EEuropean Commission (2025) ALURES – Animal Use Reporting – EU System: EU statistics database on the use of animals for scientific purposes under Directive 2010/63/EU [Internet]. Luxembourg: Publications Office of the European Union; [cited 2025 Nov 5]. Available from: https://webgate.ec.europa.eu/envdataportal/content/alures/section1_number-of-animals.html

Anonymous (2010) European Union Directive 2010/63/EU on the protection of animals used for scientific purposes. Off J Eur Union: L 276/33-L 276/79. https://eur-lex.europa.eu/LexUriServ/LexUriServ.do?uri=OJ:L:2010:276:0033:0079:en:PDF

Anonymous (2014) Guidance on the Operation of the Animals (Scientific Procedures) Act 1986 edited by Home Office. UK: Williams Lea Group

Appendix H Guidance on the Operation of the Animals (Scientific Procedures) Act 1986 (2010) edited by Animals in science regulatory unit; Home Office. UK: Her Majest's Stationary Office

Banstola A, Reynolds JNJ (2022) The sheep as a large animal model for the investigation and treatment of human disorders. Biology (Basel) 11(9):1251. https://doi.org/10.3390/biology11091251

Baqri SSR (2023) Insects in research and innovation. In: Omkar O (ed) Insects as service providers. Springer Nature Singapore, Singapore, pp 183–241

Bates AWH (2017) Anti-vivisection and the profession of medicine in Britain; a social history. In: Linzey A, Cohn P, Linzey C (eds) The Palgrave Macmillan Animal Ethics Series. London, Plagrave MacMaillan

Bentham J (1823) An introduction to the principles of morals and legislation. Printed for W. Pickering, London

Bradbury AG, Clutton RE (2016) Are neuromuscular blocking agents being misused in laboratory pigs? Br J Anaesth 116(4):476–485. https://doi.org/10.1093/bja/aew019. https://www.ncbi.nlm.nih.gov/pubmed/26934943

Bradbury AG, Eddleston M, Clutton RE (2016) Pain management in pigs undergoing experimental surgery; a literature review (2012-4). Br J Anaesth 116(1):37–45. https://doi.org/10.1093/bja/aev301. https://www.ncbi.nlm.nih.gov/pubmed/26433866

Clark-Kennedy AE (1929) Stephen Hales, D.D., F.R.S: an eighteenth century biography. Cambridge University Press, Cambridge, UK

Clutton RE (2020) An anglocentric history of anaesthetics and analgesics in the refinement of animal experiments. Animals (Basel) 10(10):1933. https://doi.org/10.3390/ani10101933

Codellas PS (1932) Alcmaeon of Croton: his life, work, and fragments. Proc R Soc Med 25(7):1041–1046

Coghill RC (2010) Individual differences in the subjective experience of pain: new insights into mechanisms and models. Headache 50(9):1531–1535. https://doi.org/10.1111/j.1526-4610.2010.01763.x

Coulter CA, Flecknell PA, Richardson CA (2009) Reported analgesic administration to rabbits, pigs, sheep, dogs and non-human primates undergoing experimental surgical procedures. Lab Anim 43(3):232–238. https://doi.org/10.1258/la.2008.008021. https://journals.sagepub.com/doi/pdf/10.1258/la.2008.008021

Davison MH (1958) The evolution of anaesthesia. Br J Anaesth 30(10):495–504. https://doi.org/10.1093/bja/30.10.495

Dröscher A (2016) Pioneering Studies on Cephalopod's Eye and Vision at the Stazione Zoologica Anton Dohrn. Front Physiol 7:618. https://doi.org/10.3389/fphys.2016.00618

Dorrington KL, Poole W (2013) The first intravenous anaesthetic: how well was it managed and its potential realized? Br J Anaesth 110(1):7–12. https://doi.org/10.1093/bja/aes388

Fiorito G, Affuso A, Anderson DB, Basil J, Bonnaud L, Botta G, Cole A, D'Angelo L, De Girolamo P, Dennison N, Dickel L, Di Cosmo A, Di Cristo C, Gestal C, Fonseca R, Grasso F, Kristiansen T, Kuba M, Maffucci F, Manciocco A, Mark FC, Melillo D, Osorio D, Palumbo A, Perkins K, Ponte G, Raspa M, Shashar N, Smith J, Smith D, Sykes A, Villanueva R, Tublitz N, Zullo L, Andrews P (2014) Cephalopods in neuroscience: regulations, research and the 3Rs. Invert Neurosci 14(1):13–36. https://doi.org/10.1007/s10158-013-0165-x

Franco NH (2013) Animal experiments in biomedical research: a historical perspective. Animals (Basel) 3(1):238–273. https://doi.org/10.3390/ani3010238

Ganderup NC, Harvey W, Mortensen JT, Harrouk W (2012) The minipig as nonrodent species in toxicology—where are we now? Int J Toxicol 31(6):507–528. https://doi.org/10.1177/1091581812462039

Gibson T (1955) Zoografting: a curious chapter in the history of plastic surgery. Br J Plast Surg 8:234–242

Gutierrez K, Dicks N, Glanzner W, Agellon L, Bordignon V (2015) Efficacy of the porcine species in biomedical research. Front Genet 6:293. https://doi.org/10.3389/fgene.2015.00293

Guerreiro (2022) is EU parliamentary question E-001498/2022 Parliamentary question I Octopus farming in the EU I E-001498/2022 I European Parliament

Harrison R (1964) Animal Machines. Vincent Stuart Publishers, London

Ipsos (2018) Public attitudes to animal research in 2018. In: Energy & industrial strategy department of business. Ipsos MORI, London

Ison SH, Clutton RE, Di Giminiani P, Rutherford KM (2016) A review of pain assessment in pigs. Front Vet Sci 3:108. https://doi.org/10.3389/fvets.2016.00108

Jones BV (2023) The long story but short history of veterinary anaesthesia. In Improve veterinary practice. https://www.veterinary-practice.com/article/the-long-story-but-short-history-of-veterinary-anaesthesia

Kilkenny C, Browne WJ, Cuthill IC, Emerson M, Altman DG (2010) Improving bioscience research reporting: the ARRIVE guidelines for reporting animal research. PLoS Biol 8(6):e1000412. https://doi.org/10.1371/journal.pbio.1000412. https://spiral.imperial.ac.uk/bitstream/10044/1/53981/2/Improving%20bioscience%20research%20reporting%3a%20the%20ARRIVE%20guidelines%20for%20reporting%20animal%20research.pdf

Kumar R, Kumar S, Gupta RK, Yadav I (2024) Avian models for biomedical research. In: Sreelatha HV, Patel S, Nagarajan P (eds) Animal models in research: principles and practice. Springer Nature Singapore, Singapore, pp 169–190

Leunissen M (2021) Empiricism and Hearsay in Aristotle's zoological collection of facts. In: Connell SM, In Cambridge companions to philosophy (eds) The Cambridge companion to Aristotle's biology. Cambridge University Press, Cambridge, pp 64–82

Moens Y (2025) Joseph Seifert, veterinarian at the Imperial Court in Austria, a forgotten pioneer in veterinary anaesthesia in Vienna. Vet Anaesth Analg 52(6):969–975. https://doi.org/10.1016/j.vaa.2025.01.013

Ogle W (1882) Aristotle on the parts of animals. Kegan, Paul, Trench & Co., London. Translated by Ogle. W.

Ozhathil DK, Tay MW, Wolf SE, Branski LK (2021) A narrative review of the history of skin grafting in burn care. Medicina (Kaunas) 57(4):380. https://doi.org/10.3390/medicina57040380

Parke EC (2014) Flies from meat and wasps from trees: Reevaluating Francesco Redi's spontaneous generation experiments. Stud Hist Philos Biol Biomed Sci 45:34–42. https://doi.org/10.1016/j.shpsc.2013.12.005

Percie du Sert N, Hurst V, Ahluwalia A, Alam S, Avey MT, Baker M, Browne WJ, Clark A, Cuthill IC, Dirnagl U, Emerson M, Garner P, Holgate ST, Howells DW, Karp NA, Lazic SE, Lidster K, MacCallum CJ, Macleod M, Pearl EJ, Petersen OH, Rawle F, Reynolds P, Rooney K, Sena ES, Silberberg SD, Steckler T, Würbel H (2020) The ARRIVE guidelines 2.0: updated guidelines for reporting animal research. PLoS Biol 18(7):e3000410. https://doi.org/10.1371/journal.pbio.3000410

Polejaeva IA, Rutigliano HM, Wells KD (2016) Livestock in biomedical research: history, current status and future prospective. Reprod Fertil Dev 28(1-2):112–124. https://doi.org/10.1071/rd15343

Ramirez Rozzi F, Froment A (2018) Earliest animal cranial surgery: from cow to man in the neolithic. Sci Rep 8(1):5536. https://doi.org/10.1038/s41598-018-23914-1

Regan T, Singer P (1989) Animals are machines. In: Animal rights and human obligations. Prentice Hall, Upper Saddle River, pp 13–19

Roques JA, Abbink W, Geurds F, van de Vis H, Flik G (2010) Tailfin clipping, a painful procedure: Studies on Nile tilapia and common carp. Physiol Behav 101(4):533–540. https://doi.org/10.1016/j.physbeh.2010.08.001

Rupke NA (1987) Vivisection in historical perspective. In: Bynum WF, Porter R (eds) The Wellcome Institue Series in the history of medicine, edited by N.A. Rupke. Croom Helm, London

Russell WMS, Burch RL (1959) The principles of humane experimental technique. Methuen, London

Ryder RD (2000) Animal revolution: changing attitudes towards speciesism. Berg Publishers, Oxford, p 101–104, 106

Sarasamma S, Varikkodan MM, Liang ST, Lin YC, Wang WP, Hsiao CD (2017) Zebrafish: a premier vertebrate model for biomedical research in Indian scenario. Zebrafish 14(6):589–605. https://doi.org/10.1089/zeb.2017.1447

Singer P (1975) Animal liberation. Pimlico, London

Singer P (1980) Animals and the value of life. In: Beauchamp TL, Regan T (eds) Matters of life and death. Temple University Press, Philadelphia

Singer P (2023) Animal liberation now: the definitive classic renewed. Harper, an imprint of HarperCollins Publishers, New York

Slutsky AS (2015) History of Mechanical Ventilation. From Vesalius to Ventilator-induced Lung Injury. Am J Respir Crit Care Med 191(10):1106–1115. https://doi.org/10.1164/rccm.201503-0421PP

Smith A (2025) The path to better science: practical advice on implementing the 3Rs and related principles in animal research [oral presentation]. Presented at: 13th World Congress on Alternatives and Animal Use in the Life Sciences; Rio de Janeiro

Verano JW, Finger S (2010) Chapter 1: ancient trepanation. Handb Clin Neurol 95:3–14. https://doi.org/10.1016/s0072-9752(08)02101-5

Wakley TH (1848) A record of one hundred experiments on animals, with ether and chloroform. Lancet 51:19–25

Weeks CA, Brown SN, Lane S, Heasman L, Benson T, Warriss PD (2009) Noise levels in lairages for cattle, sheep and pigs in abattoirs in England and Wales. Vet Rec 165(11):308–314. https://doi.org/10.1136/vr.165.11.308

Whitteridge G (1978) William Harvey on the circulation of the blood and on generation. Am J Med 65(6):888–890. https://doi.org/10.1016/0002-9343(78)90737-4

Winlow W, Di Cosmo A (2019) Editorial: sentience, pain, and anesthesia in advanced invertebrates. Front Physiol 10:1141. https://doi.org/10.3389/fphys.2019.01141

Environmental and Ecological Impacts of Livestock Farming

Geoff Simm

Abstract

There has been a dramatic expansion of global livestock production over the past 60 years or so. This increase in production has been driven partly by growth in the human population but also by changing demand associated with growing affluence in some countries, lower relative cost of many livestock products, urbanisation and increasingly global food systems. While this expansion of production has contributed to global food and nutritional security and livelihoods, it has also had significant negative environmental and ecological effects. Biodiversity loss and greenhouse gas emissions are especially important, with global food systems ranking as the first and second most important drivers of these, respectively. This chapter briefly reviews the evidence for the adverse and beneficial effects of livestock production and explores opportunities to improve the sustainability of livestock systems.

Keywords

Livestock production · Biodiversity · Climate change · Greenhouse gases · Food security · Sustainable livestock

1 Introduction

The fortunes of the human race have been closely intertwined with livestock farming since it began between 5000 and 10,000 years ago in various parts of the world. For much of this time, in many cultures, different types of livestock farming have been regarded as making positive net contributions to livelihoods, food and nutritional security, as well as being of cultural importance. (Exceptions to these uses of animals on religious or ethical grounds are discussed in other chapters of this book.) However, the significant expansion of livestock production globally over the last 60 years or so, and the adoption of some new farming practices has prompted growing attention to environmental, ecological and other impacts of livestock production. The introduction of more intensive systems in some countries in the mid-1900s—initially, and in particular in poultry and pig systems—led to concerns over the adverse effects on animal welfare (see Harrison 1964) and the local environment. More recently, the Food and Agriculture Organisation (FAO) of the United Nations' (UN) report

G. Simm (✉)
Division of Global Agriculture and Food Systems, Royal (Dick) School of Veterinary Studies, University of Edinburgh, Easter Bush Campus, Midlothian, UK
e-mail: geoff.simm@ed.ac.uk

R. E. Clutton, G. C. Musk (eds.), *Pain and Suffering in Farmed Animals*,
https://doi.org/10.1007/978-3-032-11407-5_4

Livestock's Long Shadow (FAO 2006) intensified the debate, reflecting on the global expansion of livestock production, and documenting other, broader environmental impacts of livestock production, e.g. on biodiversity, and arising from greenhouse gas emissions, water use and pollution. Many other influential studies have followed, reporting more detailed evidence on the impacts of food systems in general, and livestock production and consumption in particular, on land and other resource use, biodiversity, greenhouse gas emissions and human health (see later sections).

In this chapter, the drivers of expansion of livestock farming globally, and its environmental and ecological impacts, are reviewed. With the debate on the future of livestock farming becoming increasingly polarised, the positive contributions of livestock farming, as well its negative effects, are explored. Lastly, some of the opportunities to develop more sustainable approaches and systems that aim to harness some of the benefits of livestock farming while reducing adverse consequences are examined.

2 Drivers of Expansion of Livestock Farming

The growth in the global human population is one of the major drivers of increasing demand for, and production of, animal-sourced foods. Figures 1 and 2 show how the world population has changed over the last 10,000 years, and how it is projected to change up to 2100. The fourfold increase in population in the last century or so is especially notable, following relatively slow growth in the preceding millennia. This relatively recent and dramatic increase in the human population puts enormous pressure on food systems and the natural systems on which they depend.

However, it is not only population growth that affects demand for food. Figures 3 and 4 show the increase in meat production by region and type. These charts show a 4.7-fold increase in meat production over a period of 60 years, during which the population grew by around 2.6-fold. One of the key reasons for this disproportionate increase is the tendency for average meat and other animal-sourced foods consumption to rise—at least at a national level—as wealth increases (see Fig. 5). Income growth in poorer countries in particular leads to larger increases in meat consumption (Milford et al. 2019). Also, the relative cost of animal-sourced foods fell compared to other foods over the period shown in the graphs, making them more affordable. These factors largely explain the dramatic growth of production and consumption of meat in Asia. There were 6-fold and 2.6-fold increases in global egg and milk production, respectively, over the same period. Social factors such as urbanisation, and increasingly global food commodity markets, also have an impact on demand (Alexander et al. 2016; Milford et al. 2019).

3 Environmental and Ecological Impacts of Livestock Farming

3.1 Ecosystems, Land Use and Livestock Farming

Human activity has unintentionally (and intentionally) affected ecosystems for millennia, leading to *"species extinctions, invasions, introductions and domestications, increased soil erosion, altered fire frequency and hydrology, and [inciting] profound changes in primary productivity and other key biogeochemical and ecosystems processes"* (Ellis et al. 2010).

The relatively recent significant expansion of agriculture to meet the needs of the growing human population—including the expansion of global livestock production—has had a particularly dramatic effect. For example, Ellis et al. (2010) estimated that in 1700, almost half of the terrestrial biosphere was wild, with no human settlement or significant land use, with most of the rest being in a semi-natural state. By 2000, the reverse was true, with most of the terrestrial biosphere under agriculture or settlement, less than 20% semi-natural and only 25%

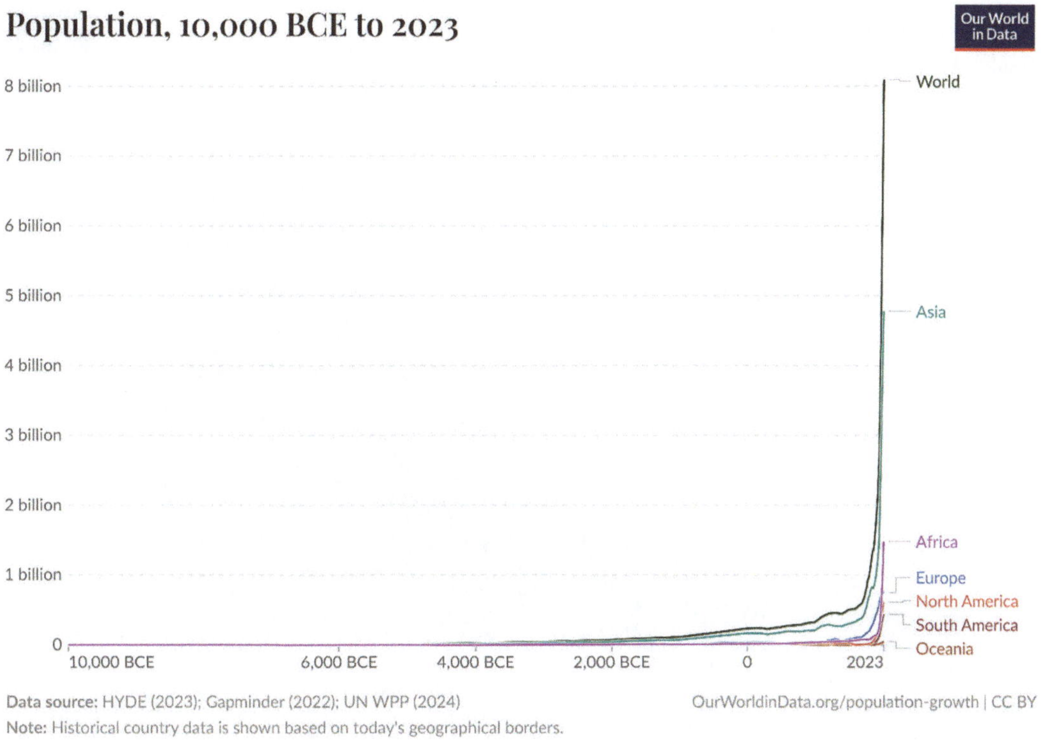

Fig. 1 World population from 10,000 BCE to 2021. (Roser and Ritchie 2023; Our World in Data 2025, CC-BY, based on data from HYDE, Gapminder and the UN World Population Prospects)

wild. Bar-On et al. (2018) estimate that the biomass of farmed animals on earth now far exceeds that of wild animals (see visualisations in Ritchie 2019 and National Food Strategy 2021).

Over the last few decades, there has been increasing awareness and concern over the global decline in biodiversity and the role of human food systems in this. The Intergovernmental Science-Policy Platform on Biodiversity and Ecosystem Services (Brondizio et al. 2019) estimates that a million species of plants and animals now face extinction. Ritchie and Roser (2019) reported that agriculture was a threat to 24,000 of the 28,000 species then on the IUCN Red List threatened with extinction. The global food system—in particular the conversion of natural ecosystems to pasture or cropping—is believed to be the primary driver of this trend (see, for example, Benton et al. 2021). Overuse or inappropriate use of fertilisers and pesticides in some places, repeated cropping of one or a few species, and farming practices that deplete soil health exacerbate the problem.

Recent FAO data show that over three-quarters of global agricultural land is used for livestock grazing or for feed production for livestock, with less than a quarter used for crop production for direct human consumption (see Fig. 6). As livestock are often fed co-products from other agrifood systems, or crops of a quality unsuitable for processing, this distinction between crop production for feeding humans versus livestock is less absolute than it may seem—see Mottet et al. (2017). Nonetheless, the growth in global livestock production over recent decades is a major contributor to the loss of biodiversity and other ecosystem effects.

There are parallel concerns in aquatic environments. The weight of wild-caught fish and seafood more than doubled between 1960 and 1990, threatening the status of many wild fish stocks (FAO 2022). Since the early 1990s, there has been a dramatic expansion of aquaculture—the

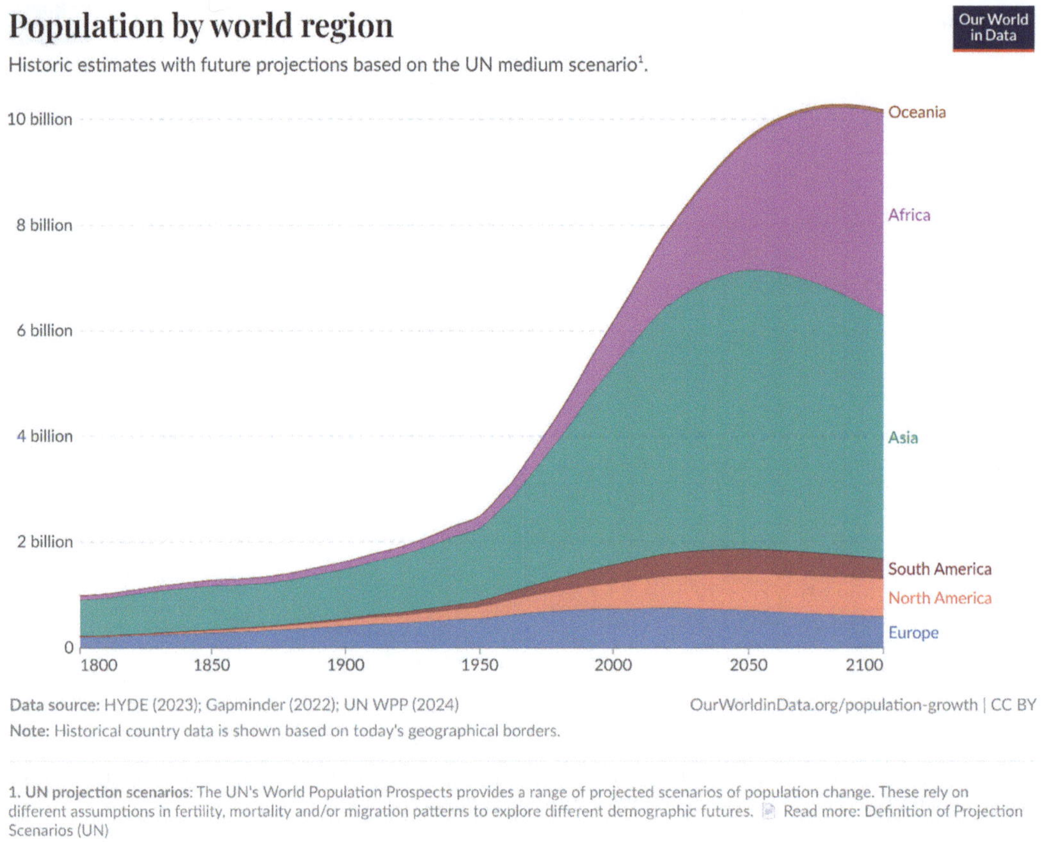

Fig. 2 World population from 1800 to 2100. (Roser and Ritchie 2023, Our World in Data 2025, CC-BY, based on data from HYDE, Gapminder and the UN World Population Division)

farming of fish and other seafood—such that global aquaculture production now exceeds wild-caught fish production: the latter "plateaued" in the 1990s. This expansion of aquaculture has helped reduce pressure on wild fish stocks, but it creates environmental impacts of its own. These include at least localised environmental effects of excess feed and excreta near fish farms, risk of disease and parasite exchange with wild stocks, water contamination from drug treatment, and the risk of "genetic contamination" of wild fish from escapees in some systems (e.g. Bergqvist and Gunnarsson 2013; SAMS 2018).

There are other problematic interactions between farming and the environment. Environmental animal pathogens can pose a direct risk to human health (in addition to the risks from direct contact with animals or from handling or consuming livestock products). The depletion of natural habitats, especially tropical forests, and the closer juxtaposition of wild and farmed animals, increases the risk of the emergence and spread of zoonotic diseases (those affecting humans and other animals; UNEP and ILRI 2020). The high or inappropriate use of antimicrobial treatments in livestock farming in some systems or regions increases the prevalence of antimicrobial resistance in livestock. This is important in its own right but can also contribute to the spread of resistance in the environment, humans and wildlife, though the relative importance of different pathways remains unclear (Graham et al. 2019; Larsson and Flach 2022).

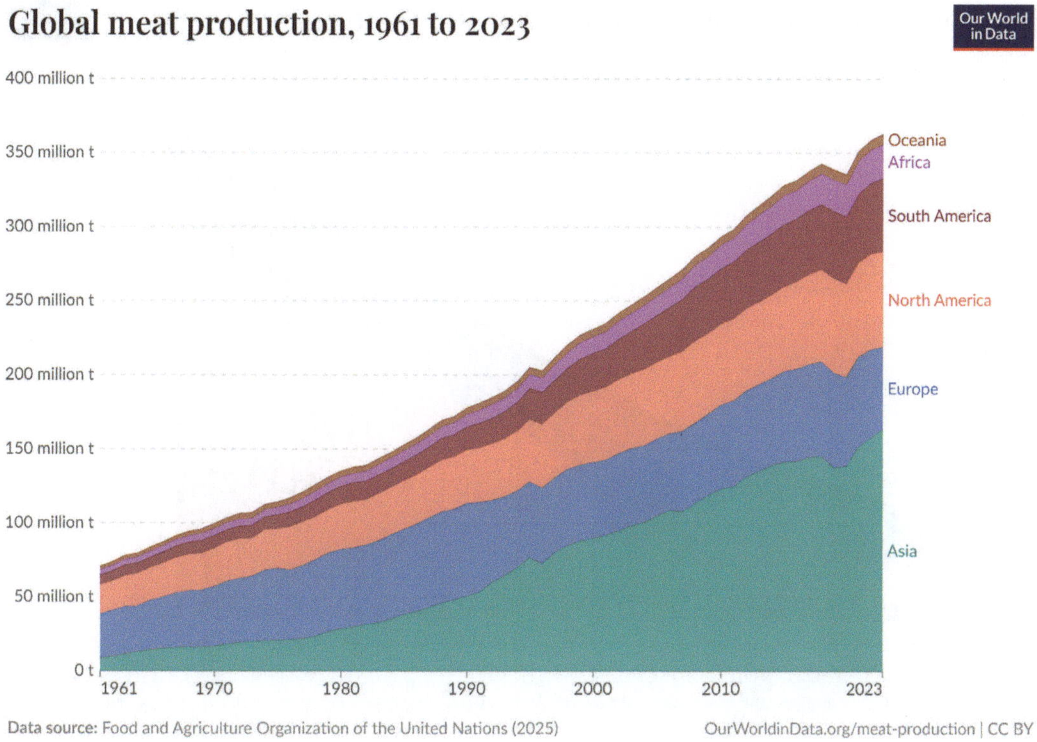

Fig. 3 Global meat production by region. (Ritchie et al. 2023, Our World in Data 2025, CC-BY, based on data from FAO)

3.2 Biogeochemical Cycles

Human activity in general, and farming in particular, has a major impact on biogeochemical cycles—the pathways through which key elements or compounds essential for life circulate through different compartments of the earth, including various solid layers of the earth, the oceans and other waters and the atmosphere.

Fresh water accounts for a small fraction of water on the planet but is vital for most forms of life. FAO (2017b) estimates that agriculture uses 70% of this freshwater globally. A growing human population and a warming climate make access to freshwater increasingly contentious in many parts of the world. Farming has an important effect on water quality too. Effluent or run-off from manure, slurry or silage stores, and run-off from fertilised land, leads to nitrogen and phosphorus entering water courses or groundwater. This accumulation of nutrients, or eutrophication, can lead to excessive and harmful algal growth or blooms. Pesticides, veterinary medicines and other treatments can also contaminate water.

Farming has affected the nitrogen cycle, which is critical for the survival of living organisms as well as the productivity of ecosystems and farming itself. While abundant in the atmosphere, it is the conversion of dinitrogen gas to other forms of nitrogen that makes it biologically available to most organisms. This process happens naturally through fixation of nitrogen by some microorganisms, including those associated with the roots of legumes, and when lightening occurs. It also happens through industrial activity, such as the so-called “Haber–Bosch” process of manufacturing nitrogen fertiliser invented in the early twentieth century, and through the combustion of fossil fuels (see e.g. Bernhard 2010). Synthetic fertilisers produced by the Haber–Bosch process have become extensively used in agriculture and helped support the “Green Revolution”. Excess nitrogen, i.e. that fraction that cannot be used by

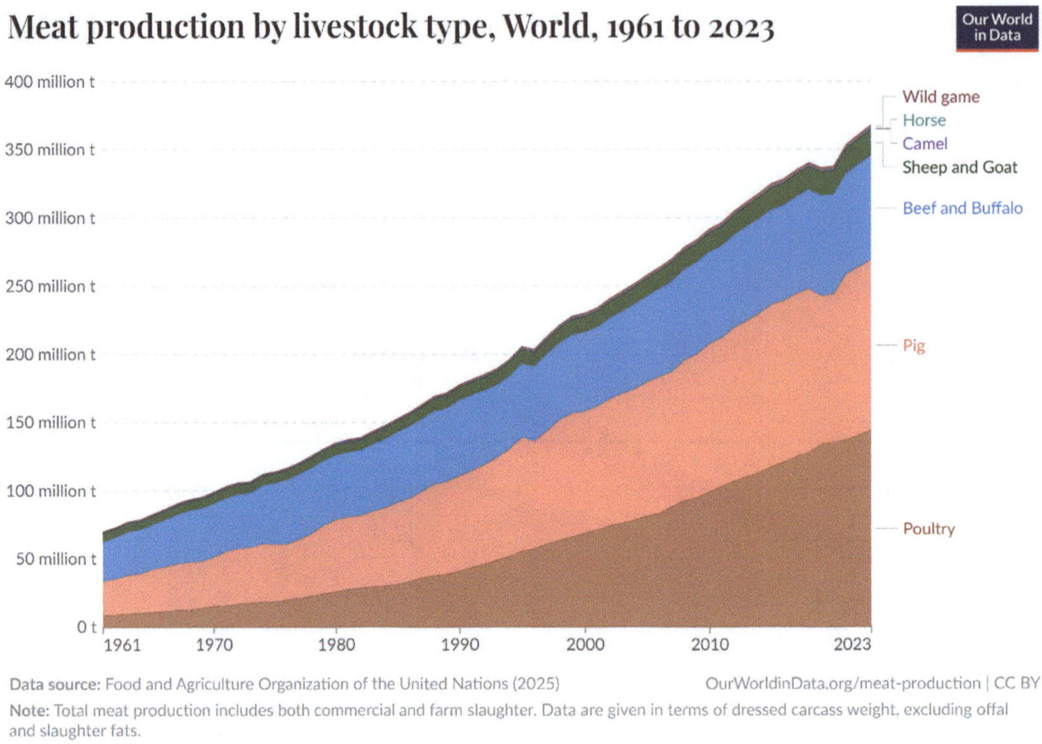

Fig. 4 Global meat production by livestock. (Ritchie et al. 2023, Our World in Data 2025, CC-BY, based on data from FAO)

plants, or immediately converted back to its unreactive atmospheric form by microbes, pollutes water courses or enters the atmosphere either as nitrous oxide (one of the most important greenhouse gases) or as other active nitrogen compounds such as ammonia and nitrogen oxides (Fowler et al. 2013).

Phosphorus is essential for plant and animal growth, so it has been intensively mined for fertiliser production. Human activity has led to both a concerning depletion of natural phosphorus sources as well as its accumulation in areas of intense crop or livestock production. The natural processes that move phosphates through land and oceans are very slow, prompting efforts to improve their recovery and reuse from agricultural and human waste.

The carbon cycle—specifically the fast component which affects its flow between land, atmosphere and oceans (see e.g. NASA 2023)—has been greatly affected by human activity. This has involved the burning of fossil fuels, which releases greenhouse gases into the atmosphere, and the clearance of land, which leads to loss of plant biomass and carbon therein, and oxidation of soil carbon through soil disturbance. Farming is an important energy user, and hence carbon dioxide emitter, but more importantly, it also contributes other potent greenhouse gases. These include methane from enteric fermentation in the digestive tracts of ruminant livestock and from manure and plant residue management) and nitrous oxide (from livestock manure, soil tillage and fertiliser application).

The entire global food system is estimated to produce around a third of all greenhouse gases (IPCC 2019; Crippa et al. 2021). Agriculture is responsible for around 12% of net anthropogenic emissions (in CO_2 equivalents), with forestry and other land use including deforestation,

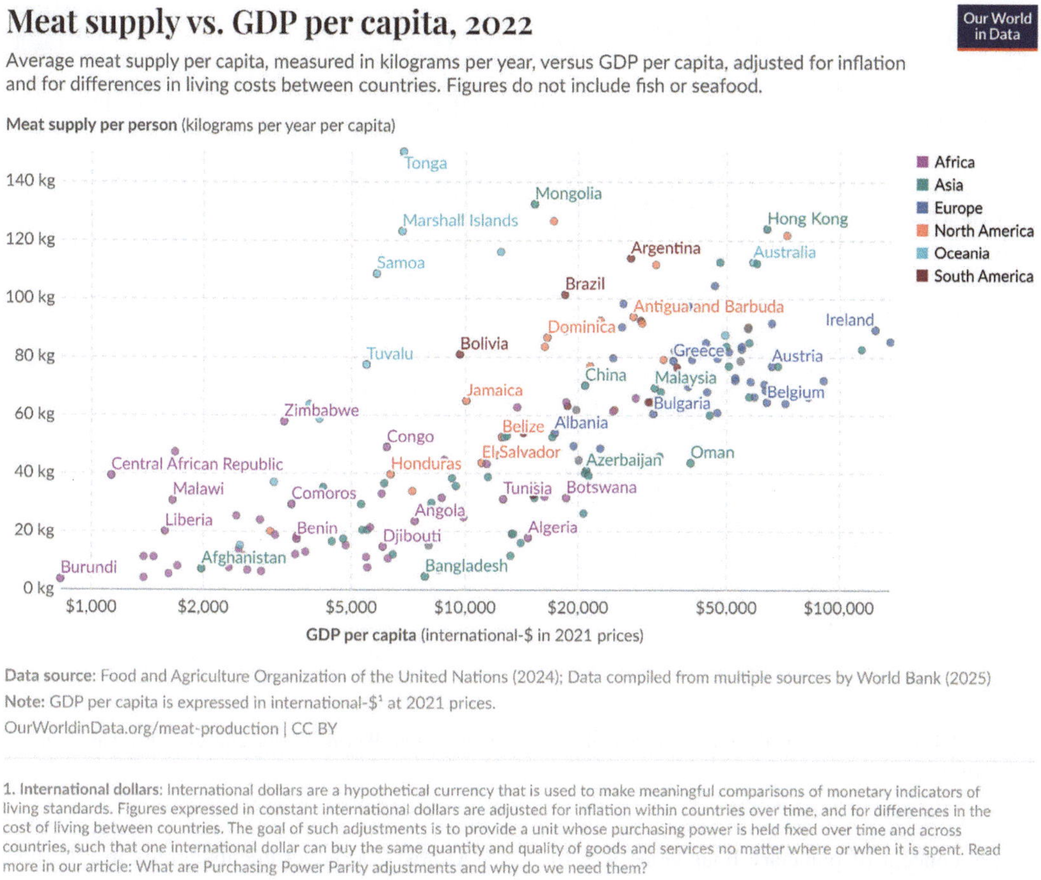

Fig. 5 Annual meat consumption versus GDP per capita in 2022. (Ritchie et al. 2023, Our World in Data 2025, CC-BY, based on data from World Bank)

accounting for a further 11% (IPCC 2019). The production phase of agriculture, rather than processing, transport, retail phases etc., is responsible for most emissions (see, for example, Poore and Nemecek 2018). Protein-rich foods, especially animal-sourced foods, tend to be associated with higher emissions. Around 14.5% of global emissions are from livestock; cattle are responsible for about two-thirds of the livestock emissions, largely methane from rumen fermentation (FAO 2017a). For monogastric livestock, the majority of emissions are from feed production, including soya associated with deforestation. It is important to note that expressing methane emissions as CO_2 equivalents overstates the effect of constant or reducing emissions on global warming and underestimates the impact of new methane emissions (Lynch et al. 2020). Consequently, it is best to consider methane separately from longer lived greenhouse gases when seeking to control emissions from farming through technical or policy interventions—see Sect. 5.5 for more on these.

3.3 Planetary Boundaries

As well as being of major intrinsic importance, the effect of human activities on natural systems poses an existential threat to humans themselves. Humans depend on healthy ecosystems for provisioning (providing food, wood, fibre, water, fuel, etc.), regulating (climate regulation, flood management, water filtration, pollination, etc.), cultural and supporting "services" (nutrient cycles, soil formation, habitats for biodiversity; see e.g. NatureScot 2023).

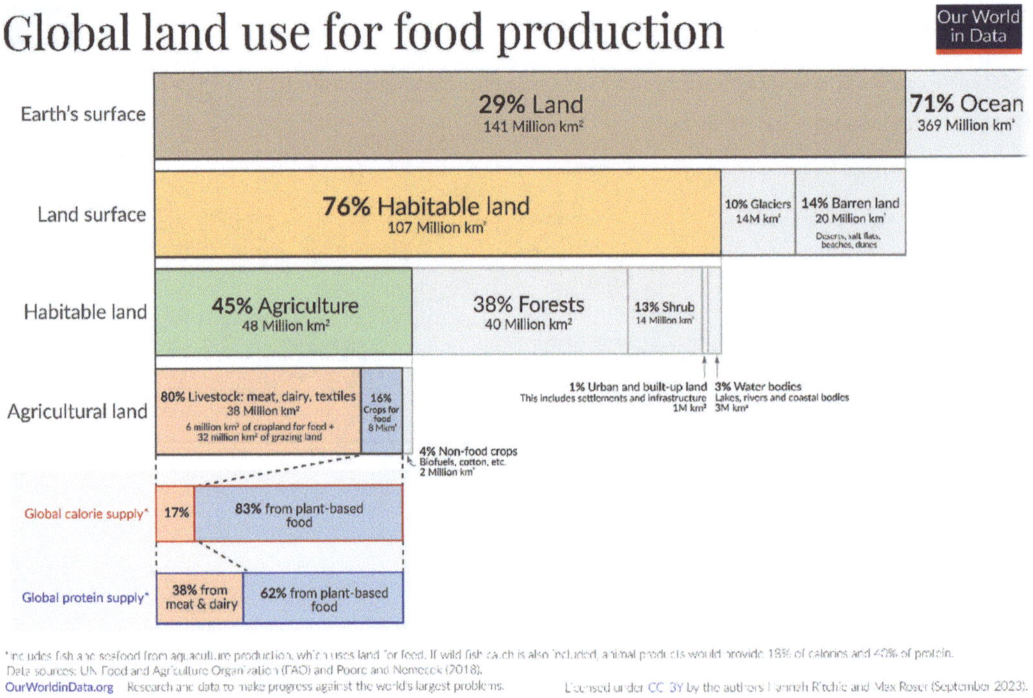

Fig. 6 Global land use for food production. (Ritchie and Roser 2024, Our World in Data, 2025, CC-BY, based on data from FAO)

The concept of planetary boundaries was formulated to recognise the interrelationships between Earth's biophysical processes and to help identify a "safe operating space" for human development without destabilising these processes (Rockström et al. 2009; Steffen et al. 2015). The latter study identified the particular risk of human activity transgressing boundaries related to climate change and biodiversity and "[driving] the Earth system into a new state". A recent update from Richardson et al. (2023) reported that six of the nine planetary boundaries have now been transgressed, and that Earth was now "…well outside of the safe operating space for humanity". Raworth's "Doughnut Economics" (Raworth 2017) extended this approach to include a social foundation as well as an ecological ceiling for the safe operating space. Similarly, Dasgupta (2021) highlighted that our economies are embedded in nature, and that we urgently need to reform our measures of economic success to reflect this and to reform our institutions and systems to support such changes.

Almost 200 governments agreed in 2022 that at least 30% of terrestrial and inland water, and coastal and marine areas—especially areas of particular importance for biodiversity and ecosystem functions and services—should be effectively conserved and managed by 2030 (UN 2022). This commitment highlights the need for more sustainable and responsible food production and consumption to ensure space for nature (see Sect. 5.5).

4 The Costs and Benefits of Livestock Production and Consumption

While this chapter has focused thus far on the environmental costs of livestock production, it is important to recognise its benefits. The costs and benefits in economic, social and environmental domains are summarised in Table 1, drawing on Simm et al. (2021).

Table 1 Summary of the costs and benefits of livestock production and consumption

Domain	Benefits	Costs
Economic	Major global economic importance	Cost of liabilities/ externalities
Social	Central to livelihoods, including 1.7 billion poorest people, 70% women. Co-products such as fertiliser, transport, draught, capital	
	Valued dietary component for many important source of dietary protein, energy, highly bioavailable micronutrients	Link between (high) consumption of some products & disease risk Risk of zoonotic disease. Animal welfare concerns in some systems
Environment	Some systems support biodiversity/C sequestration Farmed animals themselves are part of (agricultural) biodiversity	Loss of natural habitat for food production Source of environmental pollutants, locally and globally, including greenhouse gases
	Ability to use feed resources not of direct human use (grassland, co-products)	Competition for land, water and other resources for grazing or livestock feed production

The benefits include the economic value of livestock production—the livestock sector accounts for around 40% of the value of global agricultural output (FAO 2019a). However, there are hidden, or not-so-hidden, costs on human health, the environment and elsewhere in the economy. Some argue that the costs of global food and land use systems exceed their market value (Food and Land Use Coalition 2019).

Livestock also make broader contributions to livelihoods. They are crucial in around 60% of rural households in developing countries—including those of smallholder farmers, agro-pastoralists and pastoralists, contributing to the livelihoods of about 1.7 billion poor people, 70% of whom are women (FAO 2019b). In many cultures, livestock are important as a source of economic and social capital, fibre, hides, fertiliser, draught power or transport, as well as food. They also have important economic and cultural roles in other communities, including many rural communities, those that are socially or economically fragile, or those that are in protracted conflict (Defra 2006; FAO 2015, 2016a).

Livestock products are a valued part of many peoples' diet. They provide a particularly valuable source of protein, energy and highly bio-available micronutrients, such as calcium, iron and vitamin B12. Farmed seafood also provides a major source of long chain fatty acids considered essential for human development and health. Small amounts of livestock-derived foods can make a significant contribution to the well-being of pregnant and lactating mothers, the physical and cognitive development of children in early life, and to other nutritionally vulnerable groups (HLPE 2016; Grace et al. 2018; Zerfu et al. 2023). Hence, it is particularly important that the increasingly polarised debate in the Global North on the future of livestock production does not overshadow the important contribution of livestock to nutrition, livelihoods and development in the Global South. Conversely, livestock keeping in low and middle-income countries is associated with a higher risk of disease transmission (Zerfu et al. 2023) and high consumption of some animal-sourced foods is associated with a higher incidence of some diseases (Dietary Guidelines Advisory Committee 2020). Some livestock production systems give rise to animal welfare or other ethical concerns, which are explored in more detail in other chapters of this book.

Well-managed livestock farming systems, especially grazing or mixed farming systems with intermediate stocking rates, contribute to biodiversity, e.g. by maintaining habitats suitable for particular plant species, and providing feed sources for particular insects, and in turn, those bird species that feed off them (FAO 2016b; HNV 2023). For example, many of Europe's most endangered habitats and species depend on "High Nature Value" farming practices that have evolved in specific regions, and are based on semi-natural pastures and meadows (HNV 2023). Conversely, overgrazing is a major cause of biodiversity loss, land and soil degradation.

In some regions and systems, grazing contributes to carbon sequestration, though often not enough to fully offset the methane emissions of grazing animals (Garnett et al. 2017). However, restoring degraded grasslands can be effective in sequestering carbon and increasing agricultural productivity, thereby reducing pressure for further natural habitat loss (Smith 2014; Conant et al. 2017; de Oliveira et al. 2017). Grazing systems also often include hedgerows and trees, which are beneficial for biodiversity and carbon sequestration. Integration of grazing and forestry can deliver multiple benefits (Perks et al. 2018). Conversely, as discussed previously, livestock farming globally is a driver of habitat loss, pollution and global greenhouse gas emissions.

The ability of livestock to use feed resources unsuitable for direct human consumption is a major asset and is discussed further below. However, the growth in livestock production in recent decades has driven competition for land for grazing and livestock feed production and caused adverse ecosystem effects, including the loss of natural habitat.

Livestock production is directly linked—positively, negatively or both—to at least 10 of the 17 United Nations Sustainable Development Goals (Goals 1, 2, 3, 5, 8, 10, 12, 13, 14, 15; UN 2019), and indirectly to most or all of the rest.

5 Towards More Sustainable Livestock Systems

5.1 Guiding Frameworks

The environmental, ecological and food system challenges outlined thus far are complex and interrelated. They require urgent and systemic responses. Livestock production is a major contributor to many of the challenges. However, as well as being part of the problem, there are many opportunities for innovation in livestock systems to contribute to solutions.

There are many concepts and frameworks available to guide a systemic response, from global to farm level. These include the concept of sustainability, defined by the UN Brundtland Commission in 1987 as "meeting the needs of the present without compromising the ability of future generations to meet their own needs" (UN 1987). More recently, the United Nations formulated the 17 Sustainable Development Goals intended to be achieve by 2030—a task that is proving very challenging for most signatories. One Health and Planetary Health are similar, holistic concepts, which increasingly recognise the interconnectedness of human, animal, plant and ecosystem health (OHHLEP et al. 2022; Planetary Health Alliance 2023). As alluded to earlier, Planetary Boundaries and the related Doughnut Economics attempt to define safe operating spaces for human development. Similarly, the Circular Economy concept is based on eliminating waste and pollution, circulating products and materials at their highest value, and regenerating nature (Ellen MacArthur Foundation 2023). While most of these frameworks are high level, a growing range of tools is available or under development to help evaluate and improve sustainability at the food system or farm level (see e.g. FAO 2014; Aquaculture Stewardship Council 2023; Farm Animal Investment Risk and Return 2023; Sustainable Food Trust 2003; Trinity AgTech 2023; World Benchmarking Alliance 2023).

5.2 Sustainable Livestock Systems

An overarching aim for future livestock systems might be defined as to breed and manage for optimal performance in economically, socially and environmentally sustainable systems (Simm 2023). While the specific challenges and priorities will vary from country to country, there are some common themes. Economic priorities will usually include profitability, productivity and efficiency but, depending on the country and context, livestock may be valued mainly as financial and cultural assets. There is a particular need to improve livestock productivity in regions or systems where this is low, and the environmental impact per unit of food produced is relatively high (e.g. low-producing ruminants may produce little human food, but they still emit methane). This approach is often termed sustainable intensification. While increasing productivity could increase total emissions if the livestock population remains constant, it would reduce emissions per unit product and potentially allow for land sparing or other land uses. Social priorities include the contribution of livestock farming to livelihoods; the affordability of the products; high standards of animal welfare; and the wider "social licence" of systems, including the technologies used, e.g. gene editing, and other new breeding technologies. Key environmental issues include greenhouse gas emissions; carbon sequestration potential; the use of land, water and other natural resources; the circularity of systems; and effects on nature and biodiversity, whether by sparing or sharing land for nature. Balmford (2021) concludes that the impact of farming on nature is minimised by maximising yields in appropriate places, so leaving as much land as possible for nature. The National Food Strategy (2021) suggests that a modified "three compartment" model is more appropriate in countries like the United Kingdom with long-established farming systems, with some land farmed intensively, some spared for nature, and some with lower intensity farming more conducive to nature.

5.3 Improving Resource Use Efficiency

Around a third of all food produced globally is lost during production, harvest, storage or processing, or wasted in households (FAO 2023). Tackling food loss and waste is clearly beneficial for system sustainability, food security, and reducing greenhouse gas emissions, so it is a key target for technical innovation and societal change.

It is often assumed that competition for land and other resources for growing human food versus livestock feed is minimised when there is no livestock production. However, Van Zanten et al. (2018) showed that livestock's ability to use co-products and other "low opportunity cost feeds" means that arable land use is minimised when between 16% and 40% of the human diet comes from animal-sourced protein. Prioritising the use of low opportunity cost feed by livestock is an important way of reducing competition and increasing circularity (WWF 2022). Lee et al. (2021) show that when considering overall nutrient value supplied, the demand for land for human-edible crops is smaller with ruminant than monogastric systems. Wilkinson and Lee (2018) discuss the optimal use of human-edible feeds by ruminant livestock. It is important that future breeding programmes and production systems continue to focus on attributes that enhance the complementary role of livestock.

Competition for resources in growing human food rather than livestock feed could also be reduced by developing novel livestock feed sources—such as insects, algae or novel microbial proteins (see e.g. Saadaoui et al. 2021; WWF 2021a, b; Deep Branch 2023). Similarly, modification of the fatty acid profile of crops to better meet the needs of carnivorous farmed fish, such as salmon, has potential to improve efficiency and reduce the dependence on wild-caught fish or their co-products in feed ingredients (Betancor et al. 2015).

Improvements in livestock breeding, nutrition, health and management can also make major contributions to resource use efficiency. For example, Capper et al. (2009) estimated that, compared to U.S. dairy systems in 1944, those in 2007 required 21% of the animals, 23% of the feedstuffs, 35% of the water and only 10% of the land per billion kg of milk produced. Similarly, 2007 dairy systems produced 24% of the manure, 43% of methane and 56% of the nitrous oxide per billion kg of milk, compared with 1944 systems.

Genetic improvement is a particularly effective way of improving resource use efficiency. Typically, well-designed breeding programmes achieve cumulative annual rates of change of

1–3% in traits of economic importance. While these annual changes are small, they are cumulative with ongoing selection, so the changes can be dramatic over a few decades. For example, much of the over four-fold increase in the average yield of dairy cows in the United Kingdom since the 1930s is down to selection between and within dairy breeds (Simm et al. 2021). As illustrated above, such changes in productivity typically drive improvements in resource use efficiency. However, it is important that breeding programmes target improved performance in relevant, sustainable systems (including appropriate feeds) and that breeding goals do not overemphasise productivity traits but put appropriate emphasis on functionality, health and welfare-related traits.

Since around 2010, genomic selection has begun to revolutionise livestock breeding. Genomic selection involves augmenting the selection of breeding animals based on the use of genome-wide genetic markers to identify individuals and families carrying markers known from previous studies to be associated with traits of interest (Simm et al. 2021). Genomic selection can allow earlier identification of elite breeding stock, and also allow selection for traits that are difficult or expensive to measure in commercial herds or flocks—like feed intake, methane emissions or disease resistance. Hence, it is especially valuable in addressing resource use efficiency. Advances in statistical, genetic and reproductive technologies will continue to offer new opportunities.

A wide range of innovations has been reported to further improve overall resource use efficiency of livestock systems, and reduce environmental and ecological impact, including precision farming approaches, using data and new technologies to better manage system inputs and outputs (see e.g. CIEL 2022; EA-PLF 2023).

5.4 Climate Change Mitigation and Adaptation

Climate change poses twin challenges for food systems. On the one hand the contribution food systems, especially animal-sourced foods, make to greenhouse gas emissions must be reduced. On the other hand, food systems must adapt to unavoidable climate change.

With respect to mitigation, avoiding further exploitation of natural habitats for agriculture and continuing innovation in resource use efficiency are key priorities. In ruminant systems, reducing methane emissions is especially important. A range of studies has investigated the ability of dietary changes, feed additives, vaccines and breeding approaches to do so. The use of feed additives for housed cattle and breeding approaches is probably most likely to have an effect in the short term (see Simm et al. 2021; Defra et al. 2022). Breeding for improved productivity already delivers indirect benefits in terms of reduced greenhouse gas emissions per unit of product (emission intensity). Including direct or proxy measures of methane emissions in breeding programmes could reduce emissions intensity further (e.g. by 29% in dairy cattle by 2050; de Haas et al. 2021). There appears to be genetic variation among cattle in the population of rumen microbes they host, and so the levels of methane produced. This offers a promising approach for future breeding schemes, which could further reduce emissions (see Martínez-Álvaro et al. 2022). Better tools are required that allow both breeders and their customers to identify, and be credited in the supply chain, for producing breeding stock with lower environmental impact.

Adaptation to unavoidable climate change is also critical, with temperature rises, changes in rainfall, and more frequent extreme weather events challenging current food production systems in many regions. For livestock, the biggest impacts are on feed availability and quality, the growing risk of heat stress in many parts of the world, and new disease risks (CCAFS 2022). Thornton et al. (2021) illustrate that the increase in extreme heat stress risk this century in many parts of the tropics and some temperate zones will prove very challenging to livestock production and welfare.

5.5 Demand-Side Changes

Previous sections have focused primarily on supply-side opportunities to reduce the effects of livestock production. While research and innovation are helping to reduce the environmental footprint of livestock production, achieving sustainable food production globally requires demand-side changes too. Targets include reducing waste, and prioritising sustainable diets, including a reduction of animal-sourced food consumption where it is high. For example, the National Food Strategy (2021) proposed a basket of measures including reducing meat consumption by 30%, halving food waste, and increasing crop yields by 15%, in order to feed the United Kingdom population with a third less land.

6 Conclusion

Our food and farming systems have responded remarkably well to the challenge of feeding a dramatically growing global population over recent decades. However, the global expansion of farming in general, and livestock production in particular, has had major unintended environmental and ecological consequences. Innovation in livestock systems is needed to help enhance their benefits, while reducing the economic, environmental and social costs. Science and technology can make major contributions to improving the environmental and ecological footprint of livestock farming. However, it seems unlikely that they alone can meet the challenge of equitably feeding an expected 10.5 billion people well without further damaging the natural systems on which humans and animals depend. This problem will also need social, economic and political innovation and interventions.

Acknowledgements I am grateful to Professor Marie Haskell, Professor Jon Hillier and Carys Redman-White for advice on sections of the text and sources.

References

Alexander P, Brown C, Arneth A, Finnigan J, Rounsevell MD (2016) Human appropriation of land for food: the role of diet. Glob Environ Chang 41:88–98. http://www.sciencedirect.com/science/article/pii/S0959378016302370

Aquaculture Stewardship Council (2023) Farm standards. https://www.asc-aqua.org/what-we-do/our-standards/. Accessed 10 Feb 2023

Balmford A (2021) Concentrating vs. spreading our footprint: how to meet humanity's needs at least cost to nature. J Zool 315(2021):79–109. https://doi.org/10.1111/jzo.12920

Bar-On YM, Phillips R, Milo R (2018) The biomass distribution on Earth. Proc Natl Acad Sci U S A 115(25):6506–6511. http://www.pnas.org/content/115/25/6506

Benton TG, Bieg C, Harwatt H, Pudasaini R, Wellesley L (2021) Food system impacts on biodiversity loss. Three levers for food system transformation in support of nature. Research Paper Chatham House. https://www.chathamhouse.org/sites/default/files/2021-02/2021-02-03-food-system-biodiversity-loss-benton--et-al_0.pdf. Accessed 10 Feb 2023

Bergqvist J, Gunnarsson S (2013) Finfish aquaculture: animal welfare, the environment, and ethical implications. J Agric Environ Ethics 26:75–99. https://doi.org/10.1007/s10806-011-9346-y

Bernhard A (2010) The nitrogen cycle: processes, players, and human impact. Nat Educ Knowl 3(10):25

Betancor M, Sprague M, Usher S et al (2015) A nutritionally-enhanced oil from transgenic Camelina sativa effectively replaces fish oil as a source of eicosapentaenoic acid for fish. Sci Rep 5:8104. https://doi.org/10.1038/srep08104

Brondizio ES, Settele J, Díaz S, Ngo HT (eds) (2019) Global assessment report on biodiversity and ecosystem services of the Intergovernmental Science-Policy Platform on Biodiversity and Ecosystem Services. https://doi.org/10.5281/zenodo.3831673. Accessed 10 Feb 2023

Capper JL, Cady RA, Bauman DE (2009) The environmental impact of dairy production: 1944 compared with 2007. J Anim Sci 87:2160–2167. https://doi.org/10.2527/jas.2009-1781

CCAFS (2022) Climate impact on production: Livestock. https://ccafs.cgiar.org/bigfacts/#theme=climate-impacts-production&subtheme=livestock. Accessed 20 Feb 2023

CIEL (2022) Net zero and livestock. How farmers can reduce emissions. https://cielivestockcouk/expertise/net-zero-carbon-uk-livestock/report-april-2022/. Accessed 10 Feb 2023

Conant RT, Cerri CEP, Osborne BB, Paustian K (2017) Grassland management impacts on soil carbon stocks:

a new synthesis. Ecol Appl 27:662–668. https://esajournals.onlinelibrary.wiley.com/doi/pdf/10.1002/eap.1473
Crippa M, Solazzo E, Guizzardi D et al (2021) Food systems are responsible for a third of global anthropogenic GHG emissions. Nat Food 2:198–209
Dasgupta P (2021) The economics of biodiversity: the Dasgupta review. HM Treasury, London. https://assets.publishing.service.gov.uk/government/uploads/system/uploads/attachment_data/file/962785/The_Economics_of_Biodiversity_The_Dasgupta_Review_Full_Report.pdf. Accessed 10 Feb 2023
de Haas Y, Veerkamp RF, de Jong G, Aldridge MN (2021) Selective breeding as a mitigation tool for methane emissions from dairy cattle. Animal 15(Suppl):1. https://doi.org/10.1016/j.animal.2021.100294
de Oliveira SR, Barioni LG, Hall JAJ, Moretti AC, Fonseca Veloso R, Alexander P, Crespolini M, Moran D (2017) Sustainable intensification of Brazilian livestock production through optimized pasture restoration. Agric Syst 153:201–211
Deep Branch (2023) Clean ingredients for climate-friendly food. https://deepbranch.com/. Accessed 10 Feb 2023
Defra (2006) UK National action plan on farm animal genetic resources. https://assets.publishing.service.gov.uk/government/uploads/system/uploads/attachment_data/file/69397/pb12190-fangr-actionplan.pdf. Accessed 10 Feb 2023
Defra et al. (2022) Call for evidence on methane suppressing feed products. https://consult.defra.gov.uk/agriclimate/methane-supressing-feed-products/supporting_documents/Methane%20Suppressing%20Feed%20Products%20Call%20for%20Evidence.pdf. Accessed 10 Feb 2023
Dietary Guidelines Advisory Committee (2020) Scientific report of the 2020 dietary guidelines advisory committee: advisory report to the secretary of agriculture and the secretary of health and human services. US Department of Agriculture, Agricultural Research Service, Washington, DC. https://www.dietaryguidelines.gov/sites/default/files/2020-07/ScientificReport_of_the_2020DietaryGuidelinesAdvisoryCommittee_first-print.pdf. Accessed 10 Feb 2023
EA-PLF (2023) European association for precision livestock farming. Conference proceedings. https://www.eaplf.eu/. Accessed 10 Feb 2023
Ellen MacArthur Foundation (2023) What is a circular economy? https://ellenmacarthurfoundation.org/topics/circular-economy-introduction/overview#principles. Accessed 10 Feb 2023
Ellis EC, Klein Goldewijk K, Siebert S, Lightman D, Ramankutty N (2010) Anthropogenic transformation of the biomes, 1700 to 2000. Glob Ecol Biogeogr 19:589–606. https://doi.org/10.1111/j.1466-8238.2010.00540.x
FAO (2006) Livestock's long shadow. Environmental issues and options. http://www.fao.org/3/a-a0701e.pdf. Accessed 10 Feb 2023
FAO (2014) Sustainability assessment of food and agricultural systems guidelines. Version 3.0. Food and Agriculture Organization of the United Nations, Rome. https://www.fao.org/3/i3957e/i3957e.pdf. Accessed 10 February 2023
FAO (2015) The second report on the state of the world's animal genetic resources for food and agriculture. Commission on Genetic Resources for Food and Agriculture, Food and Agriculture Organization of the United Nations, Rome. http://www.fao.org/3/a--i4787e.pdf. Accessed 10 February 2023
FAO (2016a) Livestock in protracted crises. Guidance note. http://www.fao.org/3/a-i6637e.pdf. Accessed 10 Feb 2023
FAO (2016b) Principles for the assessment of livestock impacts on biodiversity. http://www.fao.org/3/a--i6492e.pdf. Accessed 10 Feb 2023
FAO (2017a) Livestock solutions for climate change. https://www.fao.org/3/i8098e/i8098e.pdf. Accessed 10 Feb 2023
FAO (2017b) Water for sustainable food and agriculture. FAO. http://www.fao.org/3/i7959e/i7959e.pdf. Accessed 10 Feb 2023
FAO (2019a) Animal production. http://www.fao.org/animal-production/en/. Accessed 10 Feb 2023
FAO (2019b) Decent rural employment. http://www.fao.org/rural-employment/agricultural-sub-sectors/livestock/en/. Accessed 10 Feb 2023
FAO (2022) The state of world fisheries and aquaculture. Towards blue transformation. https://www.fao.org/3/cc0461en/cc0461en.pdf. Accessed 10 Feb 2023
FAO (2023) Tackling food loss and waste: a triple win opportunity. https://www.fao.org/newsroom/detail/FAO-UNEP-agriculture-environment-food-loss-waste-day-2022/en. Accessed 10 Feb 2023
Farm Animal Investment Risk and Return (2023) Coller FAIRR protein producer index. https://www.fairr.org/index/. Accessed 10 Feb 2023
Food and Land Use Coalition (2019) Growing better: ten critical transitions to transform food systems and land use. https://www.foodandlandusecoalition.org/wp-content/uploads/2019/09/FOLU-GrowingBetter-GlobalReport.pdf. Accessed 10 Feb 2023
Fowler D, Coyle M, Skiba U, Sutton MA, Cape JN, Reis S, Sheppard LJ, Jenkins A, Grizzetti B, Galloway JN, Vitousek P, Leach A, Bouwman AF, Butterbach-Bahl K, Dentener F, Stevenson D, Amann M, Voss M (2013) The global nitrogen cycle in the twenty-first century. Phil Trans R Soc Lond B Biol Sci 368(1621):20130164. https://doi.org/10.1098/rstb.2013.0164
Garnett T, Godde C, Muller A, Röös E, Smith P, de Boer IJM, Ermgassen E, Herrero M, van Middelaar C, Schader C and van Zanten H (2017) Grazed and Confused? Ruminating on cattle, grazing systems, methane, nitrous oxide, the soil carbon sequestration question—and what it all means for greenhouse gas emissions. FCRN, University of Oxford. Available at: https://www.oxfordmartin.ox.ac.uk/downloads/reports/fcrn_gnc_report.pdf. Accessed 20 Feb 2023

Grace D, Dominguez-Salas P, Alonso S, Lannerstad M, Muunda E, Ngwili N, Omar A, Khan M, Otobo E (2018) The influence of livestock-derived foods on nutrition during the first 1,000 days of life ILRI Research Report 44. International Livestock Research Institute, Nairobi. https://cgspace.cgiar.org/handle/10568/92907. Accessed 10 Feb 2023

Graham DW, Bergeron G, Bourassa MW, Dickson J, Gomes F, Howe A, Kahn LH, Morley PS, Scott HM, Simjee S, Singer RS, Smith TC, Storrs C, Wittum TE (2019) Complexities in understanding antimicrobial resistance across domesticated animal, human, and environmental systems. Ann N Y Acad Sci 1441(1):17–30. https://doi.org/10.1111/nyas.14036. PMID: 30924539; PMCID: PMC6850694

Harrison R (1964) Animal machines. Vincent Stuart Ltd, London

HLPE (2016) Sustainable agricultural development for food security and nutrition: a report by the high level panel of experts on food security and nutrition. http://www.fao.org/fileadmin/user_upload/hlpe/hlpe_documents/HLPE_Reports/HLPE-Report-10_EN.pdf. Accessed 10 Feb 2023

HNV (2023) High nature value farming. http://www.high--nature-value-farming.eu/. Accessed 10 Feb 2023

IPCC (2019) Summary for policymakers. In: Shukla PR, Skea J, Calvo Buendia E, Masson-Delmotte V, Pörtner H-O, Roberts DC, Zhai P, Slade R, Connors S, van Diemen R, Ferrat M, Haughey E, Luz S, Neogi S, Pathak M, Petzold J, Portugal Pereira J, Vyas P, Huntley E, Kissick K, Belkacemi M, Malley J (eds) Climate change and land: an IPCC special report on climate change, desertification, land degradation, sustainable land management, food security, and greenhouse gas fluxes in terrestrial ecosystems. doi:https://doi.org/10.1017/9781009157988.001

Larsson DGJ, Flach CF (2022) Antibiotic resistance in the environment. Nat Rev Microbiol 20:257–269. https://doi.org/10.1038/s41579-021-00649-x

Lee MRF, Domingues JP, McAuliffe GA, Tichit M, Accatino F, Takahashi T (2021) Nutrient provision capacity of alternative livestock farming systems per area of arable farmland required. Sci Rep 11:14975. https://doi.org/10.1038/s41598-021-93782-9

Lynch J, Cain M, Pierrehumbert R, Allen M (2020) Demonstrating GWP*: a means of reporting warming equivalent emissions that captures the contrasting impacts of short- and long-lived climate pollutants. Environ Res Lett 15(4):044023. https://doi.org/10.1088/1748-9326/ab6d7e

Martínez-Álvaro M, Auffret MD, Duthie C-A, Dewhurst RJ, Cleveland MA, Watson M, Roehe R (2022) Bovine host genome acts on rumen microbiome function linked to methane emissions. Commun Biol 5:350. https://doi.org/10.1038/s42003-022-03293-0

Milford AB, Le Mouël C, Bodirsky BL, Rolinski S (2019) Drivers of meat consumption. Appetite 141:104313. https://doi.org/10.1016/j.appet.2019.06.005

Mottet A, de Haan C, Falcucci A, Tempio G, Opio C, Gerber P (2017) Livestock: on our plates or eating at our table? A new analysis of the feed/food debate. Glob Food Sec 14:1–8. https://doi.org/10.1016/j.gfs.2017.01.001

NASA (2023) NASA earth observatory. The carbon cycle. https://earthobservatory.nasa.gov/features/CarbonCycle. Accessed 10 Feb 2023

National Food Strategy (2021) National food strategy. Independent review. The plan. https://wwwnationalfoodstrategyorg/the-report/. Accessed 10 Feb 2023

NatureScot (2023) What are ecosystem services? https://www.nature.scot/scotlands-biodiversity/scottish-biodiversity-strategy-and-cop15/ecosystem-approach/ecosystem-services-natures-benefits#:~:text=Ecosystem%20Services%20are%20the%20direct,as%20reducing%20stress%20and%20anxiety. Accessed 10 Feb 2023

OHHLEP, Adisasmito WB, Almuhairi S, Behravesh CB, Bilivogui P, Bukachi SA et al (2022) One health: a new definition for a sustainable and healthy future. PLoS Pathog 18(6):e1010537. https://doi.org/10.1371/journal.ppat.1010537

Perks M, Khomik M, Bathgate S, Chapman S, Slee B, Yeluripati J, Roberts D, Morison J (2018) Agroforestry in Scotland—potential benefits in a changing climate. https://www.climatexchange.org.uk/media/3312/agroforestry-in-Scotland-potential-benefits-in-a-changing-climate.pdf. Accessed 23 Feb 2023

Planetary Health Alliance (2023) Planetary Health. https://www.planetaryhealthalliance.org/planetary-health. Accessed 10 Feb 2023

Poore J, Nemecek T (2018) Reducing food's environmental impacts through producers and consumers. Science 360:987–992

Raworth K (2017) Doughnut economics. Random House Business, London

Richardson K, Steffen W, Lucht W, Bendtsen J, Cornell SE, Donges JF, Drüke M, Fetzer I, Bala G, von Bloh W, Feulner G, Fiedler S, Gerten D, Gleeson T, Hofmann M, Huiskamp W, Kummu M, Mohan C, Nogués-Bravo D, Petri S, Porkka M, Rahmstorf S, Schaphoff S, Thonicke K, Tobian A, Virkki V, Wang-Erlandsson L, Weber L, Rockström J (2023) Earth beyond six of nine planetary boundaries. Sci Adv 9:eadh2458. https://www.science.org/doi/epdf/10.1126/sciadv.adh2458. Accessed 8 April 2024

Ritchie H (2019) Humans make up just 0.01% of Earth's life - what's the rest? Our world in data. https://ourworldindata.org/life-on-earth#:~:text=Livestock%20accounts%20for%20more%20biomass,by%20a%20factor%20of%20ten. Accessed 10 Feb 2023

Ritchie H, Roser M (2019) Land use. Our world in data. https://ourworldindataorg/land-use#breakdown-of-global-land-use-today. Accessed 10 Feb 2023

Ritchie H, Roser M (2024) Half of the world's habitable land is used for agriculture. Published online at OurWorldinData.org. Retrieved from: 'https://ourworldindata.org/global-land-for-agriculture' [Online Resource]. Accessed 23 April 2025

Ritchie H, Rosado P, Roser M (2023) Meat and dairy production. Published online at OurWorldinData.org. Retrieved from: https://ourworldindata.org/meat-production [Online Resource]. Accessed 23 April 2025

Rockström J, Steffen W, Noone K, Persson Å, Chapin FSIII, Lambin E, Lenton TM, Scheffer M, Folke C, Schellnhuber HJ, Nykvist B, de Wit CA, Hughes T, van der Leeuw S, Rodhe H, Sörlin S, Snyder PK, Costanza R, Svedin U, Falkenmark M, Karlberg L, Corell RW, Fabry VJ, Hansen J, Walker B, Liverman D, Richardson K, Crutzen P, Foley J (2009) Planetary boundaries: exploring the safe operating space for humanity. Ecol Soc 14:32. http://www.ecologyandsociety.org/vol14/iss2/art32/

Roser M, Ritchie H (2023) How has world population growth changed over time? Published online at OurWorldinData.org. Retrieved from: https://ourworldindata.org/population-growth-over-time [Online Resource]. Accessed 23 April 2025

Saadaoui I, Rasheed R, Aguilar A et al (2021) Microalgal-based feed: promising alternative feedstocks for livestock and poultry production. J Anim Sci Biotechnol 12(1):76. https://doi.org/10.1186/s40104-021-00593-z

SAMS (2018) Review of the environmental impacts of salmon farming in Scotland. https://archive2021.parliament.scot/S5_Environment/General%20Documents/20180125_SAMS_Review_of_Environmental_Impact_of_Salmon_Farming_-_Report.pdf. Accessed 10 Feb 2023

Simm G (2023) Food sustainability—where does beef fit into feeding the world? British Cattle Breeders Club Conference. https://www.cattlebreeders.org.uk/digests/78/papers/3647/

Simm G, Pollott G, Mrode R, Houston R, Marshall K (2021) Genetic improvement of farmed animals. CABI, Wallingford

Smith P (2014) Do grasslands act as a perpetual sink for carbon? Glob Chang Biol 20:2708–2711. https://doi.org/10.1111/gcb.12561

Steffen W, Richardson K, Rockström J, Cornell SE, Fetzer I, Bennett EM, Biggs R, Carpenter SR, de Vries W, de Wit CA, Folke C, Gerten D, Heinke J, Mace GM, Persson LM, Ramanathan V, Reyers B, Sörlin S (2015) Planetary boundaries: guiding human development on a changing planet. Science 347:1259855. https://doi.org/10.1126/science.1259855

Sustainable Food Trust (2003) Measuring sustainability. https://sustainablefoodtrust.org/our-work/measuring-sustainability/. Accessed 10 Feb 2023

Thornton P, Nelson G, Mayberry D, Herrero M (2021) Increases in extreme heat stress in domesticated livestock species during the twenty-first century. Glob Chang Biol 27:5762–5772. https://doi.org/10.1111/gcb.15825

Trinity AgTech (2023) Case studies. https://www.trinity-agtech.com/case-studies. Accessed 10 Feb 2023

UN (1987) Report of the World Commission on Environment and Development. https://digitallibrary.un.org/record/139811?ln=en. Accessed 20 Feb 2023

UN (2019) United Nations Sustainable Development Goals. https://sustainabledevelopment.un.org/sdgs. Accessed 10 Feb 2023

UN (2022) United Nations Environment Programme. Convention on biological diversity. Decision adopted by the conference of the parties to the convention on biological diversity. https://www.cbd.int/doc/decisions/cop-15/cop-15-dec-04-en.pdf. Accessed 10 Feb 2023

UNEP and ILRI (2020) Preventing the next pandemic: zoonotic diseases and how to break the chain of transmission. https://www.unep.org/resources/report/preventing-future-zoonotic-disease-outbreaks-protecting-environment-animals-and. Accessed 10 Feb 2023

Van Zanten HHE, Herrero M, Van Hal O, Röös E, Muller A, Garnett T, Gerber PJ, Schader C, De Boer IJM (2018) Defining a land boundary for sustainable livestock consumption. Glob Chang Biol 24:4185–4194

Wilkinson JM, Lee MRF (2018) Review: use of human-edible animal feeds by ruminant livestock. Animal 12:1735–1743. https://doi.org/10.1017/S175173111700218X

World Benchmarking Alliance (2023) Food and agriculture benchmark. https://www.worldbenchmarkingalliance.org/food-and-agriculture-benchmark/. Accessed 10 Feb 2023

WWF (2021a) Freshwater systems. World Wildlife Fund. https://wwwworldwildlifeorg/industries/freshwater-systems. Accessed 10 Feb 2023

WWF (2021b) The future of feed: a WWF roadmap to accelerating insect protein in UK feeds. https://www.wwf.org.uk/sites/default/files/2021-06/The_future_of_feed_July_2021.pdf. Accessed 10 Feb 2023

WWF (2022) The future of feed: how low opportunity cost livestock feed could support a more regenerative UK foods system. https://www.wwf.org.uk/sites/default/files/2022-06/future_of_feed_full_report.pdf. Accessed 10 Feb 2023

Zerfu T, Nguyen G, Duncan A, Baltenweck I, Brown F, Iannotti L, McNeill G (2023) Associations between livestock keeping, morbidity and nutritional status of children and women in low- and middle-income countries: a systematic review. Nutr Res Rev 36(2):526–543. https://doi.org/10.1017/S0954422422000233

The Effects of Human Attitudes on Farmed Animal Pain and Its Management

Juliet Duncan

Abstract

Pain reduces an animal's quality of life (QoL), is a welfare concern (Belshaw Z, Yeates J. Vet J 239:59–64, 2018; Muir WW. Equine Vet J 45:653–655, 2013) and in many circumstances, alleviating pain in farmed animals is an important element in improving their welfare. Effective pain mitigation in any farmed species begins with an appreciation of their capacity to feel pain and to suffer and then, an ability to recognise species-specific pain signs. Being unable to verbally communicate means animals predominantly demonstrate their pain experiences to observers through altered behaviour, and only under biologically appropriate conditions. Consequently, many pain behaviours may be subtle rather than ostentatious. Conscientious caregivers, i.e. owners, paid stockpersons and veterinarians, must be able to recognise these behavioural indicators of pain. The communication of pain and suffering to human observers may elicit emotional and caring responses (Hadjistavropoulos T, Craig KD, Duck S, Cano A, Goubert L, Jackson PL, Mogil JS, Rainville P, Sullivan MJL, Williams ACDC, Vervoort, Fitzgerald. Psychol Bull 137:910–939, 2011), but how humans contextualise animal pain will be influenced by numerous factors, including their broader attitudes to animals, personal experiences and attitudes to pain in general.

Internal biases and attitudes towards pain may influence how a caregiver responds to observing a farm animal in pain. Caregivers with more negative or less useful attitudes towards pain in farm animals may not be motivated to alleviate pain and suffering. In these caregivers, it is important to understand how caregiver attitudes can be influenced to support positive changes in behaviour and subsequent improvements in farm animal welfare.

To elucidate the processes whereby observer bias may arise and the influence of caregiver attitude on treatment of pain in farm animals, this chapter will consider: (1) What are attitudes? (2) Animal characteristics which influence attitude to pain in farm animals; (3) Caregiver characteristics that influence attitude to pain in farm animals.

Keywords

Attitude · Pain · Farm animals

1 Introduction

Pain reduces an animal's quality of life (QoL), is a welfare concern (Belshaw and Yeates 2018; Muir 2013) and in many circumstances, allevi-

J. Duncan (✉)
The University of Edinburgh, Edinburgh, UK
e-mail: juliet.duncan@ed.ac.uk

R. E. Clutton, G. C. Musk (eds.), *Pain and Suffering in Farmed Animals*,
https://doi.org/10.1007/978-3-032-11407-5_5

ating pain in farmed animals is an important element in improving their welfare. Effective pain mitigation in any farmed species begins with an appreciation of their capacity to feel pain and to suffer and then, an ability to recognise species-specific pain signs. Being unable to verbally communicate means animals predominantly demonstrate their pain experiences to observers through altered behaviour, and only under biologically appropriate conditions. Consequently, many pain behaviours may be subtle rather than ostentatious. Conscientious caregivers, i.e. owners, paid stockpersons and veterinarians, must be able to recognise these behavioural indicators of pain. The communication of pain and suffering to human observers may elicit emotional and caring responses (Hadjistavropoulos et al. 2011), but how humans contextualise animal pain will be influenced by numerous factors, including their broader attitudes to animals, personal experiences and attitudes to pain in general.

Internal biases and attitudes towards pain may influence how a caregiver responds to observing a farm animal in pain. Caregivers with more negative or less useful attitudes towards pain in farm animals may not be motivated to alleviate pain and suffering. In these caregivers, it is important to understand how caregiver attitudes can be influenced to support positive changes in behaviour and subsequent improvements in farm animal welfare.

To elucidate the processes whereby observer bias may arise and the influence of caregiver attitude on treatment of pain in farm animals, this chapter will consider: (1) What are attitudes? (2) Animal characteristics which influence attitude to pain in farm animals; (3) Caregiver characteristics that influence attitude to pain in farm animals.

2 What Are Attitudes?

The nature of 'attitude' is debated: some regard it as a singular construct, while others hold it as multidimensional model with at least three components, i.e. feelings, beliefs and behaviours (Franzoi 2000; Eagley and Chaiken 1993). Attitudes are commonly examined using these affective, behavioural and cognitive components in the 'tripartite' model (or the 'ABC of attitudes'). It was previously believed that everyone's attitudes contained all three components, but this theory is now questioned: some do not contain all three, while others are mutually inconsistent (Miller and Tesser 1986). Consequently, defining 'attitude' is not straightforward and the differences between the terms 'attitude', 'belief' and 'perception' are unclear (Ung et al. 2016; Jones 2005).

While the term 'attitude' is widely used (implying an agreed meaning), there is no formal definition. At its simplest, an attitude is an assessment of oneself, other people, ideas and objects in the world (Petty et al. 1997). Essentially, an attitude reflects one's feelings, opinions and beliefs about numerous aspects of the world.

Importantly, our attitudes can influence behaviour, decision-making and interactions with others. Whether positive or negative, attitudes play a significant role in shaping an individual's responses and actions. Value attitudes are of considerable societal interest because they strongly influence social thought and can predict an individual's actions. Most people believe that knowing how others think and feel about something will provide insights into how they process acquired information and act upon it. The way people think influences behaviour, while attitudes colour how incoming information is perceived.

The 'strength' of an attitude is important because strong attitudes possessing well-recognised characteristics are more likely to affect the thoughts and behaviours of others, are resistant to change and more stable over time. The predictability of someone's behaviour is also directly proportional to the strength of their attitudes.

In this chapter, the term 'attitude' will be used to describe a learned tendency to evaluate things in a certain way, differing in strength and direction or valency (Eagley and Chaiken 1993).

2.1 Attitude Change

The potential for pain to adversely affect welfare applies to all animals but is of particular societal interest when farmed animals are involved. To alleviate pain in animals much time has been devoted to studying pain, and the vast literature includes many facets—for example management, education, neurobiology, psychology, physiology, pharmacology and assessment. However, a farm animal caregiver's attitude towards pain will affect their motivation to manage it (Strong et al. 2002; Jones 2005). Considerable work has focused on the attitudes of veterinarians (practitioners and students), farmers and owners towards animal pain (Ellingsen et al. 2010; Wikman et al. 2013; Becker et al. 2013; Bell et al. 2014; Capner et al. 1999; Coleman and Slingsby 2007; Lorena et al. 2013; Perret-Gentil et al. 2014; Raekallio et al. 2003). Promoting pro-welfare attitudes in farm animal caregivers is desirable for moral, economic and legal reasons. Attitudes are likely to influence the energy and scale with which farmed animal pain (and its welfare consequences) are approached in the future (Rodkey and Pillai Riddell 2013; Darlow et al. 2012; Schechter and Allen 1986; Hellyer et al. 1999; Becker et al. 2013, 2014; Bell et al. 2014; Capner et al. 1999; Coleman and Slingsby 2007; Dohoo and Dohoo 1996; Hewson et al. 2007; Hugonnard et al. 2004; Huxley and Whay 2006; Ison and Rutherford 2014; Keown et al. 2011; Lascelles et al. 1999; Lim et al. 2014; Lorena et al. 2013, 2014).

While the literature devoted to the psychology of *attitude change* (or attitudinal malleability) is extensive, its importance in the ongoing process of improving farmed animal welfare means a brief overview is necessary. The process of communicating persuasive messages can change attitudes (Eagley and Chaiken 1993; Franzoi 2000) with persuasion defined as, '*the process of consciously attempting to change attitudes through the transmission of some message*' (Franzoi 2000). Numerous theories about how attitudes change have led to the development of theoretical models, with the most predominant being Petty and Cacioppo's Elaboration Likelihood Model (ELM) and Chaiken et al's Heuristic-Systematic Model (HSM) (Eagley and Chaiken 1993) which focus on the way people process information.

The heuristic-systematic model of attitude change suggests that there are two main ways in which attitudes can be changed. The first way is by people assimilating available information and consideration of known 'facts', which leads to conclusions and potential changes of attitude. This is similar to the process a student may experience where naive attitudes may develop into more sophisticated attitudes through learning about and research of a particular topic. However, this type of attitude change requires effort and time. It is more common for attitudes to be changed by simpler methods which require less cognitive effort; for example, a person may adopt the attitudes of the majority or consensus. This requires little cognitive effort and appeals to the human desire to 'fit in'.

When considering attitudes to pain in farm animals, numerous factors influence how caregivers receive and accept information, and whether they assimilate this information in a way which would lead to a change in their pre-existing attitudes. These factors include characteristics of the person receiving the information (recipient factors), characteristics of the person delivering the information (source factors) and the way in which information is framed (message factors) (Shaver 1987).

2.2 Recipient Factors Which Influence Attitude Change

2.2.1 Intelligence

Recipient factors include intelligence and self-esteem, because these are associated with a susceptibility to persuasive messages. Those of greater intelligence and self-esteem are less inclined to attitude change (Petty et al. 1997). Persuasive messages like those already held exert a reinforcing effect. That said, conflicting persuasive messages engender resistance to attitude change (Petty et al. 1997). This implies that those with no established attitudes to animal pain are more readily influenced than those with embedded attitudes. This concept has consequences when attempting to improve welfare attitudes to animals.

2.2.2 Age

There is conflicting evidence that a recipient's age may influence attitudinal malleability. Franzoi (2000) claims people are most susceptible to attitude change during adolescence and early adulthood. Tyler and Schuller (1991) found little difference in attitudinal malleability between age groups, while Jones (2005) has suggested personal experiences, as much as age, are important. However, there may be more opportunities to engage the younger generation of animal caregivers through dissemination of information through formal education and even through social media. Engaging with the younger generation of animal caregivers and promoting early attitude change to encourage useful attitudes towards farm animal pain may result in a generational attitude shift and improvements in farm animal pain in the future.

2.2.3 Conflict or Concurrence

The extent to which persuasive messages concur or conflict with pre-existing attitudes is central to cognitive dissonance theory (Shaver 1987). This is based on the principle that people are motivated to be consistent in their attitudes, thoughts (cognitive elements) and behaviour (Jones 2005). People experiencing conflict between two cognitive elements, or between a cognitive element and a behaviour become emotionally uncomfortable, and are so motivated to alter their thinking (Shaver 1987; Wood 2000). A commonplace example of cognitive dissonance pertinent to farmed animal welfare involves caring farmers who avoid inflicting pain on their animals but who, being aware of the advantages of painful (though legal) husbandry procedures, e.g. castration, tail docking, are prepared to commit them. In these cases, dissonance reduction might involve the belief that the animal is exaggerating its pain and/or that the procedure is not unacceptably painful.

Cognitive dissonance is a complex phenomenon that could explain why some aspects of farmed animal pain are dismissed or not fully acknowledged by caregivers. Cognitive dissonance in situations where farm animals are experiencing pain causes mental discomfort and stress for the caregiver. Education and providing farm animal caregivers with financially viable and practical alternative solutions to farm animal pain will improve both farm animal welfare and caregiver well-being.

3 Animal Characteristics Which Influence Attitude to Pain in Farm Animals

3.1 Animal Species and Purpose

Speciesism influences attitudes towards the welfare of animals used in research. When asked to justify the defensibility of animal experimentation, students and faculty at the University of Wisconsin strongly supported the use of fish, followed by rats or mice, pigs or sheep, monkeys and dogs or cats (Sandgren et al. 2020). Singer (2009) argued that the value of a life should be considered on a graded spectrum based on an organism's cognitive capacities, rather than on the simple framework of speciesism, which values an organism (and therefore its welfare status) on the species alone.

Research into attitudes towards animal pain has focused on both domesticated and non-domesticated species, including dogs, cats, horses, farm animals, reptiles, fish and amphibians (Vaisanen et al. 2008; Bradbury et al. 2016; Machin 2001; Becker et al. 2013, 2014; Bell et al. 2014; Capner et al. 1999; Dohoo and Dohoo 1998; Dujardin and van Loon 2011; Flaherty 2012; Hewson et al. 2006, 2007; Hugonnard et al. 2004; Ison and Rutherford 2014; Lascelles et al. 1999; Lorena et al. 2013; Perret-Gentil et al. 2014, Price et al. 2002; Raekallio et al. 2003). Most have investigated attitudes towards pain management in the clinical contexts of acute post-surgical or chronic (orthopaedic or oncological) pain. Very few studies have investigated the influence of species, in particular farm animal species on caregiver attitude to pain.

The perceived purpose of a farm animal can influence a caregiver's attitude toward welfare. An Australian survey found that people were less willing to contribute financially to improvements in animal welfare for farm animals compared to pet animals (Taylor and Signal 2009). This may reflect how farm animals are valued within the wider society and the value of a farm animal compared to companion animals, which are sometimes afforded a status more akin to family members.

In another study, pig farmers and non-farming individuals were asked to ascribe 'capacity to experience pain' to various species, including dogs, cats, pigs and cows (Peden et al. 2020). The researchers anticipated that pig farmers would ascribe pigs with a lower capacity to feel pain due to 'psychological defence' or cognitive dissonance of breeding pigs for meat production. However, this was not the case and pig farmers attitudes to pain in pigs were comparable to other groups of people (Peden et al. 2020).

3.2 Sentience

Animal sentience (at its simplest, consciousness and self-awareness) has been described more fully as an animal's capacity to subjectively feel or experience an array of sensory inputs, which are processed by a nervous system into outputs of sensations, emotions and decisions about appropriate responses, communications with others, and the generation of goal-directed behaviours' (Turner 2021). Its possession by different animal species appears to strongly influence the public's attitudes to animal welfare. In 2013, Kupsala et al., investigating attitudes to welfare in farmed animal species, found that people rated salmon lower on their mental capabilities and ability to experience pain, compared to other farmed animals, such as cattle and sheep. This finding supports the notion that attitudinal differences to animal welfare are based—at least in part—on perceptions of sentience and concurs with previous research that indicates an animal's perceived sentience affects human attitudes to their welfare (Kupsala et al. 2013; Mather 2011; Singer 2009; Proctor 2012).

3.3 The 'Cuteness' Factor

Ethologist Konrad Lorenz (1943) suggested certain physical characteristics in infants, such as large head, high and protruding forehead, large eyes, small nose and mouth, trigger an evolutionary caregiving response in parents. These caregiving responses include actions which would be considered beneficial or positive in terms of animal welfare, for example providing comfort.

There is evidence to suggest that animals which possess 'baby- like' features are more likely to evoke a positive caregiver response from humans (Kruger 2015). In the context of farm animal pain, this raises the possibility of animals which are deemed to be 'cute' or possess baby-like features being more likely to have pain recognised and treated by their caregivers.

4 Caregiver Characteristics Which Influence Attitude to Pain in Farm Animals

How people perceive a farm animal's capacity to experience pain appears to depend on their own characteristics. For example, human gender, age, occupation, cultural & demographic differences and pet ownership (Vaisanen et al. 2008; Cornish et al. 2018).

4.1 Influence of Caregiver Gender

The gender of human patients affects their *willingness* to report their own pain as well as *how* they report it (Robinson et al. 2001; Wise et al. 2002; Racine et al. 2012). An individual's gender also influences how they perceive pain in other humans (Robinson and Wise 2003). For example, human health worker characteristics—including gender—may be a contributing factor to unequal pain management within hospitals (Bartley et al. 2015; Boissoneault et al. 2016).

Attitudes to animal pain have been studied extensively (Becker et al. 2013, 2014; Bell et al. 2014; Capner et al. 1999; Coleman and Slingsby 2007; Dohoo and Dohoo 1996, 1998; Hewson et al. 2006, 2007; Hugonnard et al. 2004). However, few have focused on the effect of observer gender on assessment of pain in farm animals.

An exception was a questionnaire-based study by Kielland et al. (2009), which invited respondents to rank painful conditions in cattle ranging in severity from fractures to skin calluses. Intended for use by cattle farmers, the questionnaire was 'piloted' on 171 Norwegian veterinary students. The students assigned pain scores to the various conditions using both Numerical Rating Scores (NRS) and Visual Analogue Scale (VAS). The resulting data identified gender of the students as a strong predictor of how they rated the painful conditions in the cattle. Similarly to studies examining observer perception of human pain, female students tended to award greater VAS and NRS scores for the same conditions in cattle compared to the scores awarded by male students. More recent studies have also found that gender influences attitudes towards pain in cattle in both qualified veterinarians and also famers (Tschoner et al. 2021; Van Dyke et al. 2021).

4.2 Influence of Caregiver Family

A study involving veterinary students found an association between the number of siblings a respondent possessed and the award of greater pain scores to a range of painful conditions found in cattle (Kielland et al. 2009). Students with multiple siblings tended to award higher scores for VAS and NRS, with the authors speculating whether having siblings promoted the development of empathy in these individuals and therefore enhanced their ability to empathise across species.

Observing pain in others has a profound and lasting effect on peoples' attitudes to pain (Goubert et al. 2011). There are two potential pathways of observational learning: (i) classical conditioning, in which initially neutral events are associated with (positive or negative) stimuli (Askew and Field 2007) or; (ii) operant conditioning, in which a behaviour is followed by positive or negative effects (Lousberg et al. 1996). People who observe the pain expressions in others elicited by specific painful stimuli later mimic the same pain expression when exposed to the same stimulus (Bandura 1965). This demonstrates a human tendency to adopt cues on how to respond to painful stimuli, i.e. information is acquired about situations through observing another's behaviour in that situation.

In the context of farm animal pain, caregivers who have experience of observing pain in other humans and farm animals may be more attuned to recognising pain cues and behaviour in farm animals under their care in the future.

4.3 Influence of Caregiver Background and Education

The influence of background on attitudes to farm animal pain is poorly understood. As veterinary students' progress through their training, their perceptions of animal sentience appear to develop (Paul and Podberscek 2000). This has important welfare implications as animals which are viewed as sentient are more likely to treated appropriately for pain. It is difficult to determine whether attitudinal changes regarding animal sentience are associated with veterinary education per se, or represent a maturation of developing attitudes as students age from late teens to adulthood. However, other studies have failed to demonstrate a relationship between age and attitude towards pain in farm animals (Kielland et al. 2010; Remnant et al. 2017).

Post-graduation experience with farm animal species seems to influence veterinarians' attitude towards pain. Huxley and Whay (2006) found that British veterinarians with postgraduate training or qualifications assigned significantly higher pain scores for surgery, and higher pain scores for claw amputation, caesarean section, digital dermatitis and fracture of a distal limb in cattle compared to those without postgraduate qualifications. There also seems to be a difference in attitudes to farm animal pain between veterinarians and farmers (who are the primary caregivers), with farmers rating diseases such as mastitis and interdigital dermatitis as more painful than veterinarians (Thomsen et al. 2012). This may reflect that as primary caregivers, the farmers have greater experience in observing the behaviour of these animals in pain along with the associated distress and welfare implications of these diseases.

4.4 Influence of Caregiver Empathy

Most people experience empathy and undergo sensory processing when they witness others in pain (Giummarra et al. 2014). The extent of the empathetic response depends (not unsurprisingly) on their previous pain experience, as well as gender and any perceived similarity with the observed person (Preis and Kroener-Herwig 2012). Empathic concern refers to the motivation to care for someone in distress or pain (Gleichgerrcht and Decety 2014).

Norring et al. (2014) examined the sensitivity, empathy and attitudes of Finnish veterinary students to cattle pain using a web-based questionnaire. Participants were asked to score thirteen conditions, ranging in severity from a neck callus to a distal limb or tuber coxae fracture, for their painfulness. Participant empathy was determined by the analysis of Likert scale responses and an empathy measure called the 'Davies Interpersonal Reactivity Index' (IRI). The authors modified the empathic concern scale of the IRI by changing the terms used, thus allowing empathy towards animals, rather than humans, to be gauged. Seventy-nine veterinary students (the majority of whom were female) completed the questionnaire. Analyses revealed a positive association with how respondents scored pain in cattle and their empathy towards humans.

The study also examined the effects of gender, education, experience and empathy for humans and found a direct association between family size and empathy for animals. This result conflicted with previous work which found: i) having a child in the house was not associated with increased empathy for animals; and ii) an inverse relationship between larger households and empathy towards animals Paul (2000), Ellingsen et al. (2010). However, Kielland et al. (2009) had also found those with more siblings gave higher pain scores as described earlier in this chapter.

Green et al. (2009) studied the relationship between observer empathy and their ability to perceive pain in others. Psychology undergraduate students were asked to rate the facial pain expressed by humans undergoing a cold pressor test using the Davies empathic concern scale (reflecting an individual's tendency to show compassion and concern for another's distress). Students who scored highly for empathic concern were more likely to award higher pain scores than those who were less empathetic.

A review by Goubert et al. (2005) of the effects of empathy on the perception of pain in others discussed how empathy may foster distress and prompt an inclination to help when observing others in pain. The review also described the ability to accurately perceive another's pain or distress as empathic accuracy (Goubert et al. 2005). The authors proposed that this approach is fundamental to the process of delivering effective care to people in pain. Importantly, it is suggested that models of pain need to consider the personal and interpersonal processes of empathy, which occur when observing another in pain. These ideas were reiterated in a later review by Craig et al. (2010) who also suggested that in order to understand pain in another person, it is important that non-verbal communication is taken in consideration.

It is difficult to ascertain fully what the impact of caregiver empathy is on pain in farm animals. A recent study explored caregiver empathy and farm animals, but in the context of euthanasia (Alves et al. 2024). Further research is needed to investigate how empathy interacts with other characteristics to influence caregiver attitudes to pain in farm animals.

5 Summary

Attitudes to farm animal pain are complex, influenced by many factors and developed over time. It is likely that caregivers of farm animals have a range of attitudes towards pain in the animals under their care, and these may differ depending on the species involved. To improve the identification and treatment of pain in farm animals, exploring new ways to influence attitude change in caregivers who do not acknowledge sentience in these species is vital. However, it is important to acknowledge that changing attitudes of caregivers is challenging and requires an understanding of how those attitudes have been developed in order to devise methods to successfully change attitudes towards pain for the benefit of future generations of farm animals.

References

Alves LKS, Pairis-Garcia MD, Arruda AG, De Melo CAF, Gomes NAC, Hoshino RY, Garbossa CAP (2024) Perceptions of swine euthanasia among Brazilian caretakers from non-integrated swine farms. Front Vet Sci 11:1513141

Askew C, Field AP (2007) Vicarious learning and the development of fears in childhood. Behav Res Ther 45:2616–2627

Bandura A (1965) Influence of model's reinforcement contingencies on the acquisition of imitative responses. J Pers Soc Psychol 1:589–595

Bartley EJ, Boissoneault J, Vargovich AM, Wandner LD, Hirsh AT, Lok BC, Heft MW, Robinson ME (2015) The influence of health care professional characteristics on pain management decisions. Pain Med 16:99–111

Becker J, Reist M, Friedli K, Strabel D, Wuthrich M, Steiner A (2013) Current attitudes of bovine practitioners, claw-trimmers and farmers in Switzerland to pain and painful interventions in the feet in dairy cattle. Vet J 196:467–476

Becker J, Reist M, Steiner A (2014) Factors influencing the attitudes of cattle veterinarians, farmers, and claw trimmers towards the pain associated with the treatment of sole ulcers and the sensitivity to pain of dairy cows. Vet J 200:38–43

Bell A, Helm J, Reid J (2014) Veterinarians' attitudes to chronic pain in dogs. Vet Rec 175:428

Belshaw Z, Yeates J (2018) Assessment of quality of life and chronic pain in dogs. Vet J 239:59–64

Boissoneault J, Mundt JM, Bartley EJ, Wandner LD, Hirsh AT, Robinson ME (2016) Assessment of the influence of demographic and professional characteristics on health care providers' pain management decisions using virtual humans. J Dent Educ 80:578–587

Bradbury AG, Eddleston M, Clutton RE (2016) Pain management in pigs undergoing experimental surgery; a literature review (2012-4). Br J Anaesth 116:37–45

Capner CA, Lascelles BDX, Waterman-Pearson AE (1999) Current British veterinary attitudes to perioperative analgesia for dogs. Vet Rec 145:95–99

Coleman DL, Slingsby LS (2007) Attitudes of veterinary nurses to the assessment of pain and the use of pain scales. Vet Rec 160:541–544

Cornish A, Wilson B, Raubenheimer D, Mcgreevy P (2018) Demographics regarding belief in non-human animal sentience and emotional empathy with animals: a pilot study among attendees of an animal welfare symposium. Animals (Basel) 8

Craig KD, Versloot J, Goubert L, Vervoort T, Crombez G (2010) Perceiving pain in others: automatic and controlled mechanisms. J Pain 11:101–108

Darlow B, Fullen BM, Dean S, Hurley DA, Baxter GD, Dowell A (2012) The association between health care professional attitudes and beliefs and the attitudes and beliefs, clinical management, and outcomes of patients with low back pain: a systematic review. Eur J Pain 16:3–17

Dohoo SE, Dohoo IR (1996) Postoperative use of analgesics in dogs and cats by Canadian veterinarians. Can Vet J 37:546–551

Dohoo SE, Dohoo IR (1998) Attitudes and concerns of Canadian animal health technologists toward postoperative pain management in dogs and cats. Can Vet J 39:491–496

Dujardin CL, Van Loon JP (2011) Pain recognition and treatment in the horse: a survey of equine veterinarians in The Netherlands and Belgium. Tijdschr Diergeneeskd 136:715–724

Eagley AH, Chaiken S (1993) The psychology of attitudes. Harcourt, Brace, & Janovich, Fort Worth

Ellingsen K, Zanella AJ, Bjerkas E, Indrebo A (2010) The relationship between empathy, perception of pain and attitudes toward pets among Norwegian dog owners. Anthrozoös 23:231–243

Flaherty D (2012) Acute pain in small animals: how far have we come and where are we going? J Small Anim Pract 53:369–370

Franzoi SL (2000) Social psychology. McGraw Hill, Boston

Giummarra MJ, Fitzgibbon BM, Georgiou-Karistianis N, Beukelman M, Verdejo-Garcia A, Blumberg Z, Chou M, Gibson SJ (2014) Affective, sensory and empathic sharing of another's pain: the empathy for pain scale. Eur J Pain 19:807

Gleichgerrcht E, Decety J (2014) The relationship between different facets of empathy, pain perception and compassion fatigue among physicians. Front Behav Neurosci 8:243

Goubert L, Craig KD, Vervoort T, Morley S, Sullivan MJ, De CWAC, Cano A, Crombez G (2005) Facing others in pain: the effects of empathy. Pain 118:285–288

Goubert L, Vlaeyen JW, Crombez G, Craig KD (2011) Learning about pain from others: an observational learning account. J Pain 12:167–174

Green AD, Tripp DA, Sullivan MJ, Davidson M (2009) The relationship between empathy and estimates of observed pain. Pain Med 10:381–392

Hadjistavropoulos T, Craig KD, Duck S, Cano A, Goubert L, Jackson PL, Mogil JS, Rainville P, Sullivan MJL, Williams ACDC, Vervoort T, Fitzgerald TD (2011) A biopsychosocial formulation of pain communication. Psychol Bull 137:910–939

Hellyer PW, Frederick C, Lacy M, Salman MD, Wagner AE (1999) Attitudes of veterinary medical students, house officers, clinical faculty, and staff toward pain management in animals. J Am Vet Med Assoc 214:238–244

Hewson CJ, Dohoo IR, Lemke KA (2006) Factors affecting the use of postincisional analgesics in dogs and cats by Canadian veterinarians in 2001. Can Vet J 47:453–459, Ottawa, Canadian Veterinary Medical Association

Hewson CJ, Dohoo IR, Lemke KA, Barkema HW (2007) Factors affecting Canadian veterinarians' use of analgesics when dehorning beef and dairy calves. Can Vet J 48:1129–1136

Hugonnard M, Leblond A, Keroack S, Cadore JL, Troncy E (2004) Attitudes and concerns of French veterinarians towards pain and analgesia in dogs and cats. (special issue: feline analgesia.). Vet Anaesth Analg 31:154–163

Huxley JN, Whay HR (2006) Current attitudes of cattle practitioners to pain and the use of analgesics in cattle. Vet Rec 159:662–668

Ison SH, Rutherford KM (2014) Attitudes of farmers and veterinarians towards pain and the use of pain relief in pigs. Vet J 202:622–627

Jones D (2005) Attitudes and beliefs about chronic noncancer pain and the role of education in preparing community based health and social care professional to undertake pain management. PhD, Queen Margaret University

Keown AJ, Farnworth MJ, Adams NJ (2011) Attitudes towards perception and management of pain in rabbits and Guinea pigs by a sample of veterinarians in New Zealand. N Z Vet J 59:305–310

Kielland C, Skjerve E, Zanella AJ (2009) Attitudes of veterinary students to pain in cattle. Vet Rec 165:254–258

Kielland C, Skjerve E, Osteras O, Zanella AJ (2010) Dairy farmer attitudes and empathy towards animals are associated with animal welfare indicators. J Dairy Sci 93:2998–3006

Kruger DJ (2015) Non-mammalian infants requiring parental care elicit greater human caregiving reactions than superprecocial infants do. Ethology 121:769–774

Kupsala S, Jokinen P, Vinnari M (2013) Who cares about farmed fish? Citizen perceptions of the welfare and the mental abilities of fish. (special section: fish welfare: challenge for science and ethics.). J Agri Environ Ethics 26:119–135

Lascelles BDX, Capner CA, Waterman-Pearson AE (1999) Current British veterinary attitudes to perioperative analgesia for cats and small mammals. Vet Rec 145:601–604

Lim MY, Chen HC, Omar MA (2014) Assessment of post-operative pain in cats: a case study on veterinary students of Universiti Putra Malaysia. J Vet Med Educ 41:197–203

Lorena SE, Luna SP, Lascelles BD, Corrente JE (2013) Attitude of Brazilian veterinarians in the recognition and treatment of pain in horses and cattle. Vet Anaesth Analg 40:410–418

Lorena SE, Luna SP, Lascelles BD, Corrente JE (2014) Current attitudes regarding the use of perioperative analgesics in dogs and cats by Brazilian veterinarians. Vet Anaesth Analg 41:82–89

Lorenz K (1943) Die angeborenen Formen moglicher Arfahrung [The inate forms of potential experience.]. Zietschrift für Tierpsychologie 5:234–409

Lousberg R, Groenman NH, Schmidt AJ, Gielen AA (1996) Operant conditioning of the pain experience. Percept Mot Skills 83:883–900

Machin KL (2001) Fish, amphibian, and reptile analgesia. Vet Clin Exotic Anim Pract 4:19–33

Mather JA (2011) Philosophical background of attitudes toward and treatment of invertebrates. ILAR J 52:205–212

Miller MG, Tesser A (1986) Effects of Affective and Cognitive Focus on the Attitude–Behavior Relation. J Pers Soc Psychol 51(2):270–276

Muir WW (2013) Stress and pain: their relationship to health related quality of life (HRQL) for horses. Equine Vet J 45:653–655

Norring M, Wikman I, Hokkanen AH, Kujala MV, Hanninen L (2014) Empathic veterinarians score cattle pain higher. Vet J 200:186–190

Paul ES (2000) Empathy with animals and with humans: are they linked? Anthrozoös 13:194–202

Paul ES, Podberscek AL (2000) Veterinary education and students' attitudes towards animal welfare. Vet Rec 146:269–272

Peden RSE, Camerlink I, Boyle LA, Loughnan S, Akaichi F, Turner SP (2020) Belief in pigs' capacity to suffer: an assessment of pig farmers, veterinarians, students, and citizens. Anthrozoös 33(1):21–36. https://doi.org/10.1080/08927936.2020.1694304

Perret-Gentil F, Doherr MG, Spadavecchia C, Levionnois OL (2014) Attitudes of Swiss veterinarians towards pain and analgesia in dogs and cats. Schweiz Arch Tierheilkd 156:111–117

Petty RE, Wegener DT, Fabrigar LR (1997) Attitudes and attitude change. Annu Rev Psychol 48:609–647

Preis MA, Kroener-Herwig B (2012) Empathy for pain: the effects of prior experience and sex. Eur J Pain 16:1311–1319

Price J, Marques JM, Welsh EM, Waran NK (2002) Pilot epidemiological study of attitudes towards pain in horses. Vet Rec 151:570–575

Proctor H (2012) Animal sentience: where are we and where are we heading? Animals (Basel) 2:628–639

Racine M, Tousignant-Laflamme Y, Kloda LA, Dion D, Dupuis G, Choiniere M (2012) A systematic literature review of 10 years of research on sex/gender and experimental pain perception – part 1: are there really differences between women and men? Pain 153:602–618

Raekallio M, Heinonen KM, Kuussaari J, Vainio O (2003) Pain alleviation in animals: attitudes and practices of Finnish veterinarians. Vet J 165:131–135

Remnant JG, Tremlett A, Huxley JN, Hudson CD (2017) Clinician attitudes to pain and use of analgesia in cattle: where are we 10 years on? Vet Rec 181:400

Robinson ME, Wise EA (2003) Gender bias in the observation of experimental pain. Pain 104:259–264

Robinson ME, Riley JL 3rd, Myers CD, Papas RK, Wise EA, Waxenberg LB, Fillingim RB (2001) Gender role expectations of pain: relationship to sex differences in pain. J Pain 2:251–257

Rodkey EN, Pillai Riddell R (2013) The infancy of infant pain research: the experimental origins of infant pain denial. J Pain 14:338–350

Sandgren EP, Streiffer R, Dykema J, Assad N, Moberg J (2020) Attitudes toward animals, and how species and purpose affect animal research justifiability, among undergraduate students and faculty. PLoS One 15:e0233204

Schechter NL, Allen D (1986) Physicians' attitudes toward pain in children. J Dev Behav Pediatr 7:350–354

Shaver KG (1987) Principles of social psychology. Lawrence Erlbaum Association, Hillsdale

Singer P (2009) Speciesism and moral status. Metaphilosophy 40:567–581

Strong J, S J, Unruh AM, Vicenzino B (2002) Pain assessment and measurement. In: Strong J, U A, Wright A, Baxter Gd (ed). Churchill Livingstone, Edinburgh

Taylor N, Signal TD (2009) Willingness to pay: Australian consumers and "on the farm" welfare. J Appl Anim Welf Sci 12:345–359

Thomsen PT, Anneberg I, Herskin MS (2012) Differences in attitudes of farmers and veterinarians towards pain in dairy cows. Vet J 194:94–97

Tschoner T, Sauter-Louis C, Peinhofer V, Feist M (2021) Exploring the attitudes of Bavarian farmers towards pain in cattle and how they differ from the attitudes of bovine veterinarians. Vet Rec 189:e515

Turner PV (2021) Moving beyond the absence of pain and distress: focusing on positive animal welfare. ILAR J 60:366–372

Tyler TR, Schuller RA (1991) Aging and attitude change. J Pers Soc Psychol 61:689–697

Ung A, Salamonson Y, Hu W, Gallego G (2016) Assessing knowledge, perceptions and attitudes to pain management among medical and nursing students: a review of the literature. Br J Pain 10:8–21

Vaisanen MA, Tuomikoski-Alin SK, Brodbelt DC, Vainio OM (2008) Opinions of Finnish small animal owners about surgery and pain management in small animals. J Small Anim Pract 49:626–632

Van Dyke R, Connor M, Miele A (2021) An investigation into the perceptions of veterinarians towards perioperative pain management in Calves. Animals (Basel) 11

Wikman I, Hokkanen AH, Pastell M, Kauppinen T, Valros A, Hanninen L (2013) Dairy producer attitudes to pain in cattle in relation to disbudding calves. J Dairy Sci 96:6894–6903

Wise EA, Price DD, Myers CD, Heft MW, Robinson ME (2002) Gender role expectations of pain: relationship to experimental pain perception. Pain 96:335–342

Wood W (2000) Attitude change: persuasion and social influence. Annu Rev Psychol 51:539–570

Production Animal Pain and Its Management

Acute Pain

Andrew Bell

Abstract

Pain is an unpleasant sensation and may arise in production animals as a result of disease, injury and routine husbandry practices. Across species, pain is a complex multidimensional experience generated by activity in neural circuits. The molecules, cells and neural circuits that drive pain sensation are present at multiple levels of the nervous system, including within peripheral nerves, the spinal cord and brain. Understanding the mechanisms underlying pain signalling at each of these levels is an essential prerequisite for rational prevention and treatment of pain in animals. This chapter addresses each of the components of "pain pathway" and how they contribute to the different dimensions of the pain experience.

Keywords

Acute pain · Pain dimensions · Nociception · Sensory neurons · Spinal cord · Neuronal circuits

A. Bell (✉)
School of Biodiversity, One Health and Veterinary Medicine, University of Glasgow, Glasgow, UK
e-mail: Andrew.Bell@glasgow.ac.uk

1 Introduction

Acute Pain is an unpleasant sensation that usually arises because of damaging or intense stimuli. Pain can have many causes and in production animal systems these can include disease, traumatic injury, routine husbandry practices, e.g. castration, and natural processes such as parturition (McLennan 2018).

Acute pain usually persists only until the stimulus or injury has resolved, unlike chronic pain which persists for longer. Despite its unpleasant nature, acute pain is a sensation of immense evolutionary importance. It serves to protect from injury and preserve the integrity of the body. Humans and animals born with rare genetic insensitivities to pain tend to have dramatically shortened life expectancies as they are not sufficiently motivated to avoid noxious stimuli and protect damaged tissues after injury (Cox et al. 2006). However, despite the essential protective features pain offers, it should be recognised that it is a profoundly unpleasant and aversive sensation. In the context of civilised society, contemporary medical practice and indeed within production animal systems, pain should therefore not be viewed as protective, but as something to prevent and treat.

Pain arises because of coordinated activity across multiple levels of the nervous system (Fig. 1). Painful stimuli are encoded into electrical activity by specific pain-sensing nerves

R. E. Clutton, G. C. Musk (eds.), *Pain and Suffering in Farmed Animals*,
https://doi.org/10.1007/978-3-032-11407-5_6

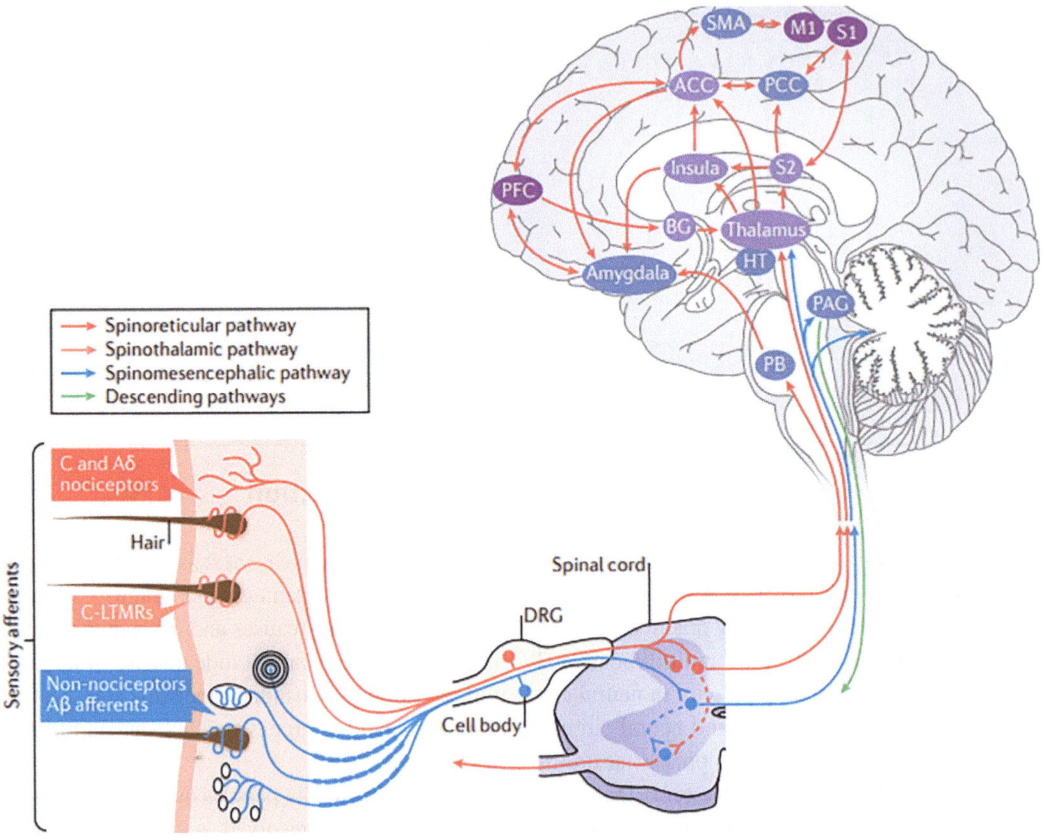

Fig. 1 Pain results from coordinated activity in neurons across the nervous system. Ultimately, pain is felt because of activity across multiple brain centres. However, the nociceptive signals that drive this usually arise in the tissues due to activity in pain-sensing nociceptors (red). This pain information is relayed to the central nervous system alongside touch information (blue) via the dorsal horn of the spinal cord. From the spinal cord, various ascending pathways project to different brain areas. Of note, the brain can also regulate spinal nociceptive transmission in a top-down manner via descending pathways arising from brainstem nuclei. (C-LTMR C-low threshold mechanoceptor, DRG dorsal root ganglia, PB parabrachial nucleus, PAG periaqueductal grey, HT hypothalamus, BG basal ganglia, PFC prefrontal cortex, ACC/PCC anterior/posterior cingulate cortex, S1 primary somatosensory cortex). (Source https://doi.org/10.1038/nrn.2016.162)

termed nociceptors. These nerves innervate the skin, joints and viscera across the body and activity in these nerves is the primary driver of most kinds of pain. While one end of the nociceptor innervates the periphery and senses pain, the other end connects to second-order neurons in the dorsal horn of the spinal cord. Hence, electrical activity caused by noxious stimulation reaches the central nervous system rapidly. At the spinal cord level, second-order neurons within the dorsal horn are responsible for modulating the "pain" signals and then relaying them to higher centres in the brain via specific nerve tracts.

Ultimately, it is neural activity across multiple areas of the brain that leads to the perception of pain in an animal. It is important to recognise that activity driven by painful stimuli in the sensory system is only pain when it is perceived as such at the level of the brain. Electrical activity in the pain-sensing neurons of the periphery and spinal cord is strictly termed nociception. Pain and nociception are different phenomena, and while nociception will often lead to pain, this relationship is not fixed. Multiple cerebral influences, for example, expectation or distraction, may modify the qualitative nature of the pain percept. As an extreme example of the disconnect between nociception and pain, an animal cannot feel pain while anaesthetised despite high levels of activity in the nociceptive pathway.

2 Classifying Pain

2.1 Pain Can Be Classified According to the Process by Which it Arises

Diverse types of stimuli can elicit pain and hence pain is commonly classified into 3 categories by cause: nociceptive, neuropathic and nociplastic. Nociceptive pain is caused by the activation of peripheral nociceptors by a noxious stimulus. This stimulus may be thermal (heat/cold), mechanical or chemical. Arguably, most pain in production animals is nociceptive. As examples, the application of heat as part of routine husbandry procedures, like disbudding in cattle, causes acute thermal nociceptive pain. Conducting surgical procedures drives mechanical nociception via damage to the skin, integument and internal organs. Furthermore, the inflammation that results from surgery or chronic disease processes generates inflammatory mediators that activate a multitude of peripheral receptors on nociceptors driving chemical pain.

Neuropathic pain, the second category, is caused by an injury or lesion of the nervous system itself. In humans, primary neuropathic pain is a common and debilitating condition with multiple aetiologies such as diabetic peripheral neuropathy, traumatic nerve damage or spinal cord injury. It may be particularly distressing as it is associated with a unique set of symptoms which may include sharp electric shock-like pain, numbness, pins and needles (paraesthesia) and burning or cold sensations. Primary neuropathic pain with causes akin to those in humans is arguably quite rare in production animals. However, a neuropathic component contributing to the overall pain experienced by an animal is possible in many circumstances where nerves are damaged by trauma or surgery. For example, in procedures such as tail or digit amputation, where nerves are traumatised or cut as part of the surgical procedure, long-lasting neuropathic symptoms may be a feature (Di Giminiani et al. 2017).

Nociplastic pain is a relatively newly established third category of pain. Pain of this nature is generated neither by nociceptive nor neuropathic mechanisms, but rather by a dysfunction of the central nervous system that serves to distort the processing of normal sensory stimuli. In humans, the classic example of this type of pain would be fibromyalgia, where widespread pain is felt but without any obvious cause. Mechanistically, nociplastic pain is poorly understood, and primary examples of this type of pain in veterinary species are limited. However, secondary nociplastic mechanisms also underlie the increased sensitivity of the central nervous system and consequent amplification of pain often seen in chronically painful diseases such as osteoarthritis or chronic inflammatory lesions.

2.2 Somatic and Visceral Pain

Pain may also be classified by its site of origin into somatic or visceral categories. Somatic pain is initiated by the stimulation of nociceptors in the skin and is well-defined and clearly located. Somatic pain may also be elicited by activation of nociceptors in deeper structures such as ligaments, tendons, bones and muscles, e.g. by fractures, and in these cases, pain tends to be duller and more poorly localised. In contrast, pain arising from internal organs is termed visceral pain. Visceral pain is of a very different nature to somatic pain, not least because the stimuli that initiate it are different to those that elicit somatic pain. Visceral nociceptors respond poorly to burning and cutting but are very sensitive to stretch, distention and inflammation. Secondly, viscera are innervated in such a way that stimuli are difficult to localise, and this results in dull pains that may even be referred to other structures. For example, people having a heart attack often feel the pain from their myocardium in their left arm and jaw. Lastly, visceral nociception drives a qualitatively different pain percept to that of somatic pain, with predominant symptoms in humans described as dull, sickening or cramping. Despite differences in aetiology, visceral pain still leads to unpleasant affective states that can affect welfare. Many examples of visceral pain in production animal systems exist, for example, it commonly arises as a result of gastrointestinal disease ("colic") or uterine disease, e.g. metritis.

2.3 The "Dimensions" of the Pain Experience

Irrespective of the cause, pain is a multidimensional experience. The International Association for the Study of Pain defines pain as, "An unpleasant sensory and emotional experience associated with, or resembling that associated with, actual or potential tissue damage". Of relevance to veterinary medicine, this definition has a number of notes, and one states that, "Verbal description is only one of several behaviours to express pain; inability to communicate does not negate the possibility that a human or a nonhuman animal experiences pain".

The IASP definition refers to pain being a multidimensional experience with both sensory and emotional aspects. However, classically, the pain experience is divided into three dimensions (or aspects): Sensory/discriminatory; affective/motivational; and cognitive/evaluative (Melzack and Casey 1968).

2.4 Sensory/Discriminatory Aspect

The sensory/discriminatory aspects of pain allow animals to recognise the intensity, nature and location of a painful stimulus. That might be hot or cold, sharp or dull, stinging or aching with a location in a specific bodily part. While this pain dimension allows appreciation of the pain, the resultant feelings are not necessarily associated with an emotional valence, i.e. the sensory/discriminatory aspect of pain alone is not necessarily unpleasant. However, in the majority of instances, sensory discriminatory aspects of pain are associated with negative affective and motivational aspects.

2.5 Affective/Motivational Aspects

The affective/motivational dimension encompasses the unpleasant nature of the pain and the urge to escape that unpleasantness. Crucially, this element of pain has an emotional valence, that is to say that it causes negative feelings. This is the element of the pain experience that results in suffering and has an impact on animal welfare. Other affective states, such as stress, fear or those caused by longstanding disease, may also contribute to the overall unpleasantness and the consequences in terms of the animal's welfare. Perceived pain is more severe in animals that are subjected to negative affective states and these states can arise as a result of prolonged disease, such as lameness in cattle (Steagall et al. 2021).

2.6 Cognitive/Evaluative Aspects

The cognitive dimension allows for higher cognitive appraisal activities to modify the sensory percept or its associated unpleasantness. This is a dimension that is easier to understand in humans, where, for example, worries about the effects of pain on ability to work or socialise might worsen the overall experience. Both the sensory/discriminatory aspects and the affective/motivational aspects can be altered by cognitive processes, for example, during games, competition or fighting, powerful cognitive processes can serve to inhibit both the sensation of pain and its unpleasantness for a period of time. It is undeniable that the capacity of animals to engage in high levels of cognition (such as concern about pain affecting the future) is more limited than it is in humans. That said, a number of recent behavioural studies have shed new light on the intricacies of social and cognitive interactions in laboratory rodents and the effect of these on pain (Mogil 2019a, b). As such, it is worthwhile including an appraisal of potential cognitive aspects when considering pain, at least in mammalian production animals.

As described briefly above and in Fig. 1, pain is an experience that arises due to coordinated activity within the somatosensory nervous system, spanning nerves from the periphery to the brain. There accumulates an ever-increasing knowledge of the neural architecture responsible for pain sensation and its different dimensions. It is worth noting, however, that much of our knowledge about how pain is generated is derived from studies involving either laboratory rodents or human beings. Bespoke studies examining whether peculiarities exist in the pain-generating systems of most production animal species are lacking, and many of the pathophysiological mechanisms described in this chapter are derived from work in rodents. That said, given the evolutionary importance of pain, it is highly likely that many important pain molecules, cells and neuronal circuits are conserved across species. In the following chapter, the individual components of the pain pathway, their functions and how they interact to generate the multiple dimensions of pain perception will be considered.

3 Nociceptors

Most pain begins as a peripheral stimulus that is sensed by a nociceptor, the first link in the pain pathway. Anatomically, the cell bodies of nociceptors and other sensory neurons are located in the dorsal root ganglia (DRG) adjacent to the spinal cord, or the trigeminal ganglia in the case of sensory information arising from the face. The cells have an axon (the portion of the cell that serves as the conduit for sensory information) that has both a peripheral branch, which innervates the tissues and a central branch, which arborises in the dorsal horn of the spinal cord. Despite the fact that all sensory neurons share this anatomical arrangement, there is a tremendous diversity in the molecular makeup and function of nociceptors and this underlies how qualitatively different pains are generated. For many years, pain neuroscientists have sought to understand this heterogeneity and its functional implications. This has traditionally been done using anatomical and electrophysiological means but more recently, the advent of molecular and advanced genetic sequencing techniques has revolutionised the way nociceptors and other components of the nervous system are classified.

3.1 Classical Sensory Neuron Classification Using Size and Myelination

Sensory neurons have been traditionally classified by their cell body size and degree of axonal myelination, both of which determine their conduction velocity. In contrast to the large myelinated (Aβ) fibres which convey touch, nociceptors are medium to small-diameter fibres which may be myelinated (Aδ) or unmyelinated (C-fibre). The Aδ afferents transmit well localised "fast" or "first" pain, whereas the C-fibres conduct more slowly and convey poorly localised "slow" pain (Fig. 2). A further small class of C-fibre exists which does not transmit nociceptive information; rather these fibres encode pleasant or gentle touch sensations, and these are termed C-low threshold mechanoreceptors (C-LTMRs) (Abraira and Ginty 2013). Each class of primary afferent described above may have distinct and characteristic endings in the skin that serve to sense specific stimuli. Nociceptors and C-LTMRs terminate in the skin simply as free nerve endings are associated with hair follicles. However, Aβ-touch fibres may be associated with a number of specialised end-organs such as Merkel cell–neurite complexes, Meissner corpuscles, Pacinian corpuscles and Ruffini corpuscles. These structures do not transduce painful stimuli but rather sense different qualities of touch (Handler and Ginty 2021).

3.2 Classifying Neurons Using Receptors, Ion Channels & Neurochemicals

Nociceptors may be further defined in a molecular sense, i.e. based on the receptors, ion channels

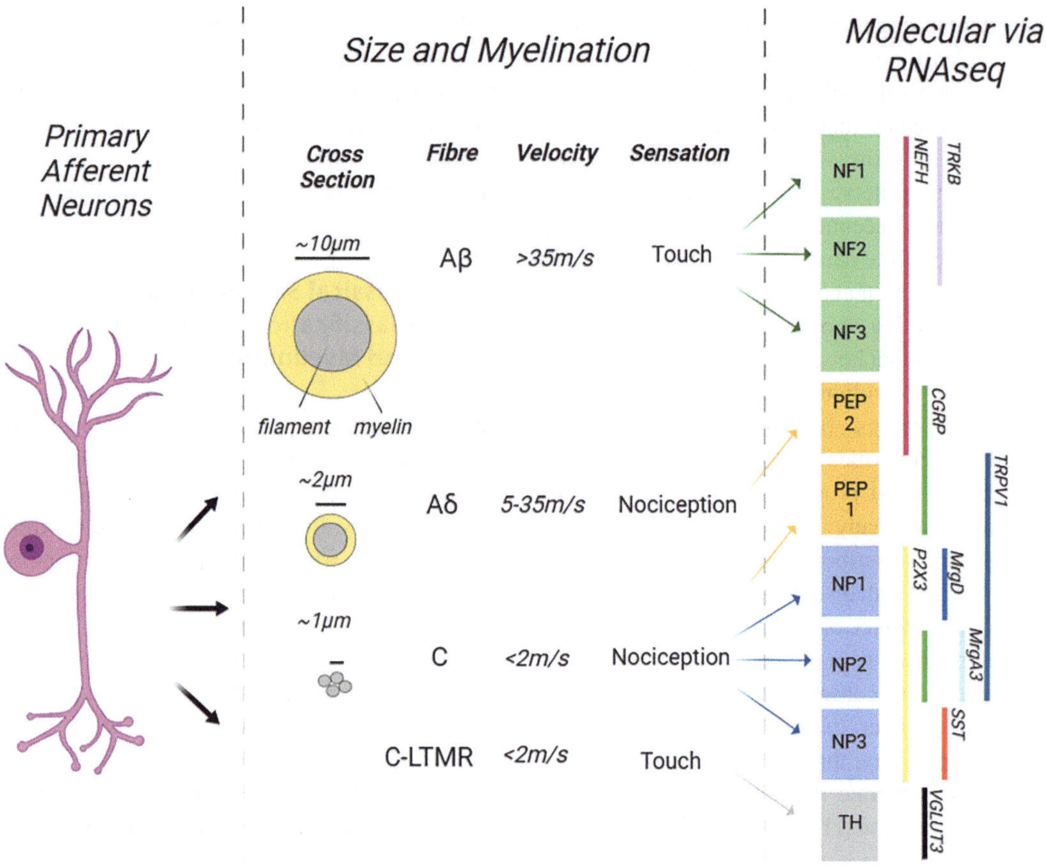

Fig. 2 Primary afferent neurons may be divided into classes based on their size and myelination, which both determine their conduction velocity. Furthermore, these cells can be classified based on their molecular make up and a scheme from Usoskin et al. (2015) is shown. Nine classes of afferent neuron are shown with some limited examples of how different molecular markers are distributed across the classes and how they correspond to existing classifications. Neurofilament Heavy (NEFH) expression indicates myelination. Abbreviations not in main text—TrkB Tropomyosin receptor kinase B, MrgD/A3 Mas-related G-protein coupled receptor, SST Somatostatin

and neurochemicals that they express, rather than their functional or conduction properties. The use of gene and protein expression to define functional populations amongst nociceptors is becoming increasingly popular. For example, C-fibre nociceptors are commonly classified as either peptidergic or non-peptidergic based on substances they express. Peptidergic nociceptors express substance P, calcitonin gene-related peptide (CGRP) and high levels of the TRPV1 ion channel, while non-peptidergic cells bind the plant lectin IB4 and express the ATP-gated ion channel P2X3. It is thought that most nociceptors irrespective of class, are essentially polymodal, i.e. transduce and transmit more than one stimulus modality, e.g. heat, mechanical pain, itch, etc. There is some evidence that these different peptidergic and non-peptidergic classes have unique functions and preferentially transduce certain stimuli. While peptidergic fibres are generally considered crucial for pain sensation, several studies have shown that a non-peptidergic class marked by the murine G-protein receptor MrgprD may actually transmit non-painful stimuli that do not cause aversive behaviours. However, this changes in the context of injury and resultant chronic pain, where these non-peptidergic fibres begin to transmit information that is perceived as noxious (Huang et al. 2019).

However, many molecular schemes to define sensory neurons and nociceptors have been discovered serendipitously and may not truly represent the functional diversity of these cells. Recent advances in science, including the ability to rapidly sequence genetic material, have revolutionised the ability to measure cell types at the scale and complexity required. One such technology is single-cell RNA sequencing, and this has been employed extensively in the previous few years to characterise DRG neuron subtypes. This approach involves measuring the complete gene expression profile of each individual cell in the dorsal root ganglion. Because functional cell-type identity is reflected in the gene expression profile of individual cells, this sequencing, followed by unsupervised grouping of similar expression profiles reveals meaningful cell populations. Several research groups have generated single-cell transcriptomic atlases of rodent, primate and human sensory neurons with broadly similar results. These studies highlight the phenomenal complexity of sensory neurons and allow nociceptors to be subdivided into multiple functional types. One of these schemes defines three types of non-peptidergic neurons, two peptidergic subclasses, four low-threshold categories and two classes of proprioceptors (that convey the locations of limbs to the brain) (Usoskin et al. 2015). Defining the molecular complexity of the somatosensory system in this unbiased manner has numerous advantages, principally, the sheer volume of data that has become available can help better understand the function of these cells in both health and disease, and even predict the action of analgesic drugs. For example, powerful methods have been developed that take data from RNA sequencing experiments and use the newly discovered receptor and cellular pathways to highlight existing drugs that could be repurposed as therapies. While this has not yet been fruitful in pain research, the concept is an exciting one.

3.3 The Validity of Rodent Models

One note of caution in terms of this approach relates to the fact that much of the work defining primary afferents (and other components of the nociceptive pathway) has been conducted in rodent models. While there appears to be a high degree of similarity between species in this regard, some important differences do exist; for example, not all dogs express the kappa opioid receptor in primary afferents while humans and rodents do (Iadarola et al. 2018). However, although rodents may seem far detached from humans and other larger mammals, they share an overwhelming majority of genes. It is likely that the fundamental molecular basis of sensation at the level of the DRG is largely similar across species making rodent findings valuable. Currently, there are no single-cell atlases of sensory neurons derived from common production animal species and an understanding of any species-specific adaptations, if indeed these exist, is limited. However, atlases generated by single-cell sequencing have been acquired from other production animal tissues; porcine immune cells, for example. As RNA-sequencing becomes more accessible and affordable, it is likely that the molecular understanding of sensory neurons in production animals will grow.

4 The Complexity of Sensation

Sensory neurons—including nociceptors—which innervate tissues throughout the body, are able to sense and discriminate different stimuli: humans can sense a myriad of painful and non-painful stimuli from the burning of a hot iron to the feeling of a light breeze. The mechanisms whereby the mechanical and thermal properties of these stimuli are ultimately converted into electrical impulses in specific classes of sensory nerves lie in the multitude of specific receptors and ion channels in the nerve terminals, which are tuned to specific stimuli. When these Ion channels are activated by the appropriate stimulus, they generate an electrical current by opening and allowing an influx of sodium or calcium ions.

4.1 TRP Ion Channels

The 2021 Nobel prize for physiology and medicine was awarded to two researchers who have been instrumental in defining the molecules underlying sensory transduction. David Julius received the award for his discovery of members of the transient receptor potential (TRP) family of ion channels. Of these, the TRP Vanilloid 1 (TRPV1) channel has a fundamental role in the detection of noxious heat (Vandewauw et al. 2018). TRPV1 is the receptor for capsaicin, the active ingredient in chilli peppers, and acts as a receptor for noxious heat. It is only found it certain populations of nociceptors and not in low-threshold touch fibres or in proprioceptors. Other TRP channels also have roles in sensing specific thermal and chemical stimuli. For example, TRPM8, the receptor for menthol, is proposed to have major roles in noxious cold sensation.

4.2 Piezo Channels

The Nobel prize was also awarded to Ardem Patapoutian for the parallel discovery of a novel channel called Piezo2. He demonstrated that Piezo2 is the major mechanical transducer in sensory nerves and is responsible for our sense of touch (Ledford and Callaway 2021). However, it is an oversimplification to state that TRPV1, TRPM8 and Piezo are wholly responsible for the transduction of all stimuli because a plethora of other receptors and ion channels are also integral to sensing the varying noxious and non-noxious aspects of the world animals and humans inhabit. Nevertheless, the principle that specific receptors and channels underlie the transduction of specific stimuli and that these are expressed selectively in different types of sensory nerve cells represents the fundamental basis of somatosensation.

4.3 A Comparative Anatomy of Transduction

Some species have receptors that respond differently to painful stimuli that have arisen in response to evolutionary pressures. For example, some birds that eat chilli peppers have evolved TRPV1 receptors that are less sensitive to capsaicin. The naked mole rat, a small burrowing mammal, has evolved to tolerate acidic environments and capsaicin. These animals also respond much less to inflammation than one would expect. This startling phenotype arises from mutations to the TrKA receptor in nociceptors, which acts alongside TRPV1 to generate electrical activity in the nerves (Omerbašić et al. 2016). These kinds of adaptations in the way pain signals are transduced are very much the exception rather than the rule. Unsurprisingly, animals that feel less pain but also survive as a species are uncommon due to the evolutionary importance of pain itself. For the most part, and including production animals, it is believed that the molecular basis of signal transduction in nociceptors is generally conserved across species.

Following transduction of the stimulus, the electrical activity is conducted along the afferent nerve as an action potential. An influx of sodium through voltage-gated sodium channels is primarily responsible for this and different classes of sodium channels, such as Nav 1.1, Nav 1.6, Nav 1.7, Nav 1.8 and Nav 1.9, are present in sensory neurons, with the latter three being predominantly expressed in nociceptors. The different channels have different response properties, and Nav 1.7 is widely regarded as the crucially important "pain" channel because it sets the "gain" in cells on which it is present (Emery et al. 2016). Mutations within this channel have been shown to underlie dramatic insensitivities to pain in human subjects, while gain of function mutations causing hyperactivity in this channel have been shown to cause painful disorders. As a result, there has been significant interest in developing analgesics that act to block this channel, although none have yet succeeded.

5 Distinguishing Touch from Pain in the Central Nervous System

The sensations of touch and pain share many features in terms of how they are generated in the periphery. Both sensations are transduced by sensory neurons, and this information is carried to the spinal cord as electrical impulses. Different sensory neuron types, namely nociceptors and low-threshold touch fibres, allow touch and pain to be sensed and segregated in the periphery, but how these distinctions persist within the spinal cord is of considerable interest. The possibility of crosstalk between these two modalities is currently mooted, as this kind of phenomenon could lead to the undesirable situation in which touch becomes pain.

Sensory afferents from the periphery terminate at synapses with second-order neurons in the dorsal horn of the spinal cord (see Fig. 1). By convention, the dorsal horn is divided into ten layers or laminae; these divisions being traditionally based on the appearance of the cells in each layer. However, the termination pattern of sensory neurons in the dorsal horn is determined by fibre function. While low-threshold afferents generally terminate and synapse in lamina III and deeper laminae, nociceptive afferents synapse in the superficial layers (laminae I and II, Fig. 1 in the Chronic pain chapter) (Todd 2010). However, fibres carrying touch information and synapsing in the deeper laminae also branch and project directly to brainstem nuclei via the dorsal columns. In this way, touch information can reach "higher" centres directly, rapidly and without synapsing via the spinal cord; some touch information is not modulated at the spinal level, unlike pain (see below). The specific projection patterns of sensory neurons mean that touch and pain processing areas are spatially segregated in the spinal dorsal horn under normal conditions. However, in many chronically painful conditions, this segregation may be absent, which is important because it means non-painful stimuli may become painful. Both touch and pain processing areas of the spinal cord contain second-order neurons that act as the next link in the pathways that convey touch and pain information to the brain. Nociceptive-specific projections are located in the most superficial lamina of the dorsal horn, with some present in deeper laminae (III, IV & V). These cells have axons that travel in various white matter spinal tracts to higher centres in the brain. In human beings, these tracts are termed the anterolateral system, although the term "ventrolateral system" is more appropriate in quadrupedal animals, as this better describes their location in the white matter. Some species differences are thought to exist in the targets of nociceptive tracts, considered the most important for pain sensation. The spinothalamic tract seems to be the most important in human beings and primates, while the spinocervicothalamic and the spinoparabrachial tracts may predominate in carnivora and rodents, respectively (Dostrovsky and Craig 2013).

Projection neurons constitute a very small proportion of cells in the dorsal horn of the spinal cord; the majority of cells in this area are local interneurons. These cells may be excitatory or inhibitory and within each class exhibit considerable molecular and functional diversity. This has been revealed in single-cell sequencing studies involving both rodents and human beings (Häring et al. 2018). While these interneurons form complex circuits and have numerous functions in spinal pain modulation (that exceed this chapter's scope), it is their presence that partly explains some spinally mediated interactions between touch and pain. Of clinical significance is the centrally mediated phenomenon known as allodynia, in which "touch" information accesses pain-specific projection neurons in chronic pain disease states causing normal touching to feel painful. The mechanisms underlying this are discussed in the chronic pain chapter of this book. This and related phenomena illustrate the long-held belief that the spinal cord is the site of considerable processing of "pain" signals, much of which is influenced by incoming touch signals.

5.1 The Gate Theory of Pain

The seminal "Gate Theory" of pain, which was proposed in 1965 by Melzack and Wall holds that while the extent to which a stimulus produces pain depends on the magnitude of the signal in nociceptive-specific primary afferents, this activity can be modulated at spinal cord level by non-nociceptive afferent signals. Thus, rubbing a painful area after an injury reduces the pain sensation at the level of the dorsal horn by non-nociceptive signals rather than cognitive mechanisms in the brain. Melzack and Wall proposed a spinal circuit to explain this, which began with separate pain and touch afferents and ended with a projection neuron (Fig. 3), At the centre of the gate control circuit lies an inhibitory interneuron, which can be activated by activity in large touch fibres resulting in feed-forward inhibition of the projected output. Current knowledge of the complexity of spinal interneurons and the circuits they engage (Peirs and Seal 2016) reveals the original gate theory to be an oversimplification (Fig. 3). However, what is clear is that interneurons in the spinal cord have critical roles in pain behaviour by both preventing cross-talk between modalities and facilitating modulation of pain signals.

6 Pain Is Not "Hard Wired"

An early illustration of nociceptive transmission from the periphery to the brain was drawn by the seventeenth-century philosopher René Descartes. He envisaged painful information as "fast-moving particles of fire" that moved up a wire from the foot to the brain (Fig. 4). Predictably, viewing the nociceptive axis as a simple wire has become untenable: multiple synapses and different neuronal classes exist between the periphery and the brain. What has also become evident is that the system itself is capable of plasticity, i.e. capable of anatomical and functional change under certain conditions, including prolonged pain. Nociceptive pathways are particularly plastic and have the capacity to amplify signals, resulting in the phenomena of: i) hyperalgesia, were painful stimuli are perceived as even more painful; and ii) allodynia, where normally non-noxious stimuli generate the perception of pain. Pain hypersensitivity, which can be distressing to animals, may be generated because of either peripheral or central sensitisation. The mechanisms underlying central sensitisation, i.e. spinal cord and brain-mediated amplification of the signal, are most commonly seen in chronically painful conditions, and are described further in the chronic pain chapter of this book.

7 Peripheral Sensitisation

Peripheral sensitisation occurs commonly in acute pain. The term "peripheral" sensitisation describes changes to nociceptors that occur at the site of disease or injury. These changes sensitise the nociceptor resulting in hyperalgesia and allodynia. When tissues are injured or damaged, inflammatory mediators are generated by the injured tissue or by the inevitable influx of immune cells that sensitise nociceptors. This is exemplified by the sensation of tenderness that arises with sunburn, when simply touching the skin is painful. This pain is the inevitable accompaniment of redness and inflammation in the skin causing peripheral sensitisation. Significant peripheral sensitisation almost inevitably accompanies husbandry procedures such as castration, or tail docking in production animals, or when inflammatory disease is present.

7.1 Peripheral Sensitisation and Mechanisms

Peripheral sensitisation is caused by the release of inflammatory mediators from numerous non-neuronal cells, including fibroblasts, keratinocytes, platelets and immune cells. Released

Fig. 3 (continued) cells (T). Advances in our knowledge of dorsal horn circuitry have resulted in much more complex circuit diagrams being proposed, as shown here. (Modified from Melzack and Wall (1965) and Peirs and Seal (2016). Adapted from Peirs & Seal https://doi.org/10.1126/science.aaf8933)

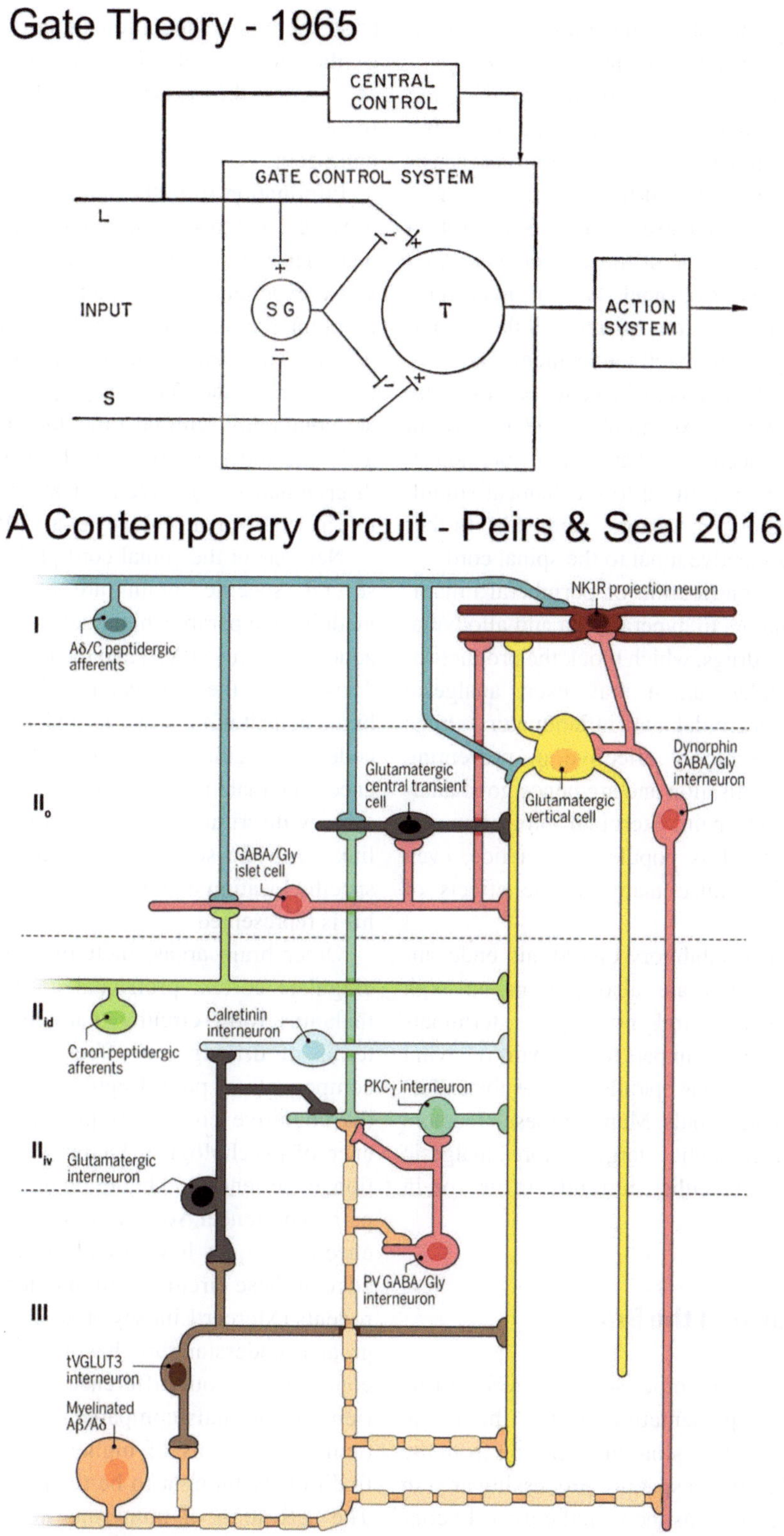

Fig. 3 Increasing knowledge of spinal cord circuitry over the past 50 years. The original gate theory proposed that primary afferent input to the dorsal horn, comprising large diameter fibres (L) and small diameter fibres (S), is modulated by the actions of a single inhibitory substantia gelatinosa interneuron (SG) before passing to central transmission

inflammatory mediators constitute the "inflammatory soup", which contains a range of signalling molecules (Fig. 5). In addition to attracting immune cells, fighting infection and promoting healing, these inflammatory mediators also sensitise nociceptors such that the nociceptor can be activated by less intense stimuli. Sensitisation may then occur by one of three mechanisms: (i) direct activation of channels causing nociceptor activation; (ii) activation of intracellular regulatory pathways to indirectly alter membrane proteins; or (iii) alterations to the genes expressed by the cell such that excitability is increased. In addition, nociceptors that were previously "silent" can be sensitised to mechanical stimuli by inflammatory mediators, and this further increases nociceptive input to the spinal cord.

Given the contribution of peripheral inflammatory mediators to hyperalgesia and allodynia, it follows that drugs, which block the products of peripheral inflammation may exert analgesic effects. Non-steroidal anti-inflammatory drugs (NSAIDs) are highly effective in preventing peripheral sensitisation and are hence good analgesics for acute pain. Steroidal anti-inflammatory agents are less popular in practice, even though they produce many of the effects of NSAIDs.

A number of inhibitory opioid and endocannabinoid receptors are also present on both peripheral and central nociceptive terminals (Fig. 5) and account in part for the pain-relieving effects of endogenous opioids, such as the endorphins and cannabinoids. Many of these receptors are also therapeutic targets for analgesic drugs, which are either currently in use or in development.

8 Pain and the Brain

Pain is unlike many other senses in terms of its somatotopic representation within the brain. For example, with vision, information from the retina is relayed to a specific processing area in the brain identified as the visual cortex. In contrast, there is no specific pain cortex, and none of the areas involved in pain perception are specific to the sensation; in fact, many are involved in other aspects of perception and behaviour.

Information from the spinal cord is projected to various centres in the brain, and the resultant neuronal activity gives rise to the multidimensional experience that is pain. The combined set of brain regions which are activated during a painful experience, and which can be investigated using functional imaging (such as MRI), are commonly referred to as the "pain matrix". This encompasses areas involved with sensory-discriminative, affective-motivational and cognitive aspects of pain (Tracey et al. 2019).

Neurons in the spinal cord project directly to several specific brain areas, including the medulla, the parabrachial nucleus (PB), the periaqueductal grey (PAG) and the thalamus (Fig. 1). These areas then project to other areas of the brain, constituting networks which are thought to code for specific dimensions of the pain experience. The thalamus has been associated with the sensory-discriminatory aspects of pain due to its links with the somatosensory cortex where the specific location and nature of the painful stimulus is represented.

Other brain areas, including the amygdala, cingulate cortex, prefrontal cortex and hypothalamus form circuits that are important in terms of driving the affective and cognitive components of pain. Regulation of pain within the cognitive circuits helps explain the influence of psychological factors such as expectation, experience and belief on the shaping of the pain experience. Advances in modern neuroscience techniques have greatly increased knowledge of these circuits both in human beings and rodents (Mercer Lindsay et al. 2021). However, greater understanding has failed to establish conclusions about differences in the pain experience of animals compared with humans, particularly in terms of higher cognitive aspects traditionaly thought to be the preserve of man. The activation of numerous neuronal circuits

Fig. 4 René Descartes envisaged painful information as "fast-moving particles of fire" that moved up a wire from the foot to the brain. (From Traite de l'homme (1664) by René Descartes)

and brain areas are similar in the two mammalian extremes of rodents and humans, and while the complexity of the latter is greater, this neuroanatomical homology can be used to argue that higher level cognitive capabilities may exist in some animals. Several well-designed behavioural rodent studies have demonstrated that pain may be affected by cognitive factors such as memory and social empathy (Mogil 2019a, b). And whilst the interpretation of these kinds of study may be tinged by anthropomorphism, a renewed attention to the cognitive, emotional and social abilities of animals and how this affects their pain will undoubtedly continue to drive further research. That rodents can show diverse and complex responses to pain beyond simple aversion makes it likely that pain-activated brain networks can produce negative affective states. This in turn, means that pain in production animal species, when present, almost certainly contributes to their burden of suffering and a lower quality of life.

8.1 The Brain Exerts "Top-Down Control" Over "Ascending" Nociceptive System Information

In addition to formulating the pain experience, the brain also exerts a powerful "top-down" control on spinal pain processing via descending nerve control mechanisms. This mechanism allows cognitive, emotional or autonomic nervous factors to regulate spinal nociceptive processing in both facilitatory or inhibitory ways, i.e. to heighten or suppress the ultimate pain experience. Such control mechanisms probably explain situations in which the protective role of acute pain is over-ridden; for example, when racehorses sustain limb fractures but complete the race, only pulling up lame after crossing the finish line. In these situations, motivating factors from "higher" levels effectively abolish the pain experience by inhibiting spinal nociceptive transmission.

The descending neurons responsible for pain suppression project from two brain areas: the ros-

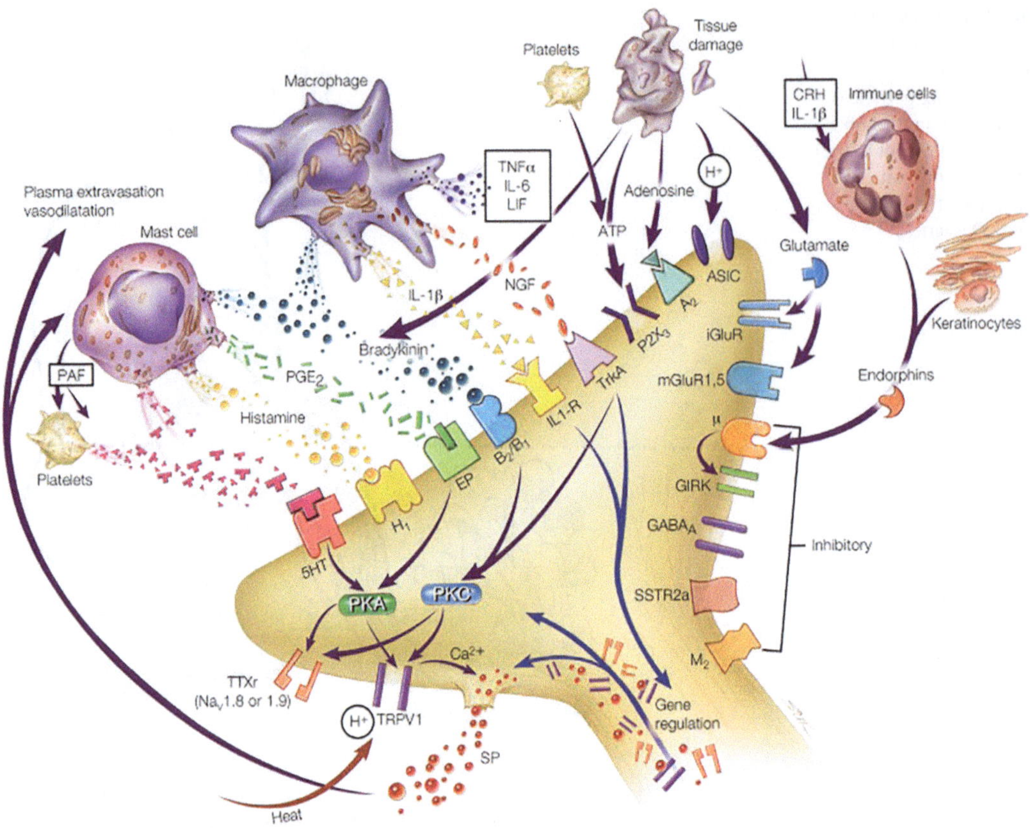

Fig. 5 Mechanisms of peripheral sensitisation. Following tissue injury, a multitude of immune cells release inflammatory mediators such as nerve growth factor (NGF), bradykinin and histamine. Each of these molecules acts via specific receptors to increase the sensitivity of the nociceptor terminal. Of note, while peripheral sensitisation involves the action of substances that sensitise the nociceptor, other inhibitory receptors are also present on peripheral nociceptor terminals, and these can cause analgesia by reducing the excitability of the nerve. (From 5th Edition Melzack & Wall's Textbook of Pain)

troventral medulla (RVM) and the locus coeruleus (LC). These project to the dorsal horn where they release 5-hydroxytryptamine (5-HT) and noradrenaline (NA). Noradrenaline acts via α_2-ceptors, resulting in antinociceptive effects. 5-HT release can produce anti- or pro-nociceptive effects through its action at the 5-HT$_7$ or 5-HT$_3$ receptors, respectively.

In summary, pain is a multidimensional experience generated by activity in specialised pain networks at multiple levels of the nervous system. The introduction of new high-resolution scientific techniques, such as gene sequencing has led to an ever-increasing knowledge of the identities of the component parts of the system. In some cases, it is possible to assign specific dimensions of the pain experience to specific molecules, cells and circuits within the nociceptive and pain pathways. Continued research in the workings of this complex system is required for continued progress in the diagnosis and treatment of acute pain in human beings and animals.

References[1]

*Abraira VE, Ginty DD (2013) The sensory neurons of touch. Neuron 79:618–639. https://doi.org/10.1016/j.neuron.2013.07.051

Cox JJ, Reimann F, Nicholas AK et al (2006) An SCN9A channelopathy causes congenital inability to experience pain. Nature 444:894–898. https://doi.org/10.1038/nature05413

Di Giminiani P, Edwards SA, Malcolm EM et al (2017) Characterization of short- and long-term mechanical sensitisation following surgical tail amputation in pigs. Sci Rep 7:4827. https://doi.org/10.1038/s41598-017-05404-y

Dostrovsky J, Craig B (2013) Ascending projection systems. In: Wall & Melzack's textbook of pain, 6th edn. Saunders, Philadelphia, pp 182–197

Emery EC, Luiz AP, Wood JN (2016) Nav1.7 and other voltage-gated sodium channels as drug targets for pain relief. Expert Opin Ther Targets 20:975–983. https://doi.org/10.1517/14728222.2016.1162295

Handler A, Ginty DD (2021) The mechanosensory neurons of touch and their mechanisms of activation. Nat Rev Neurosci 22:521–537. https://doi.org/10.1038/s41583-021-00489-x

*Häring M, Zeisel A, Hochgerner H, et al (2018) Neuronal atlas of the dorsal horn defines its architecture and links sensory input to transcriptional cell types. Nature Neuroscience 21:869-880. https://doi.org/10.1038/s41593-018-0141-1

Huang T, Lin S-H, Malewicz NM et al (2019) Identifying the pathways required for coping behaviours associated with sustained pain. Nature 565:86–90. https://doi.org/10.1038/s41586-018-0793-8

Iadarola M, Sapio M, Raithel S et al (2018) Long-term pain relief in canine osteoarthritis by a single intra-articular injection of resiniferatoxin, a potent TRPV1 agonist. Pain 159:2105–2114. https://doi.org/10.1097/j.pain.0000000000001314

Ledford H, Callaway E (2021) Medicine Nobel goes to scientists who discovered biology of senses. Nature 598:246–246. https://doi.org/10.1038/d41586-021-01283-6

McLennan KM (2018) Why pain is still a welfare issue for farm animals, and how facial expression could be the answer. Agriculture 8:127. https://doi.org/10.3390/agriculture8080127

Melzack R, Casey K (1968) Sensory, motivational, and central control determinants of pain. In: The skin senses. Charles C Thomas, pp 423–439

Melzack R, Wall PD (1965) Pain mechanisms: a new theory. Science 150:971–979

*Mercer Lindsay N, Chen C, Gilam G, et al (2021) Brain circuits for pain and its treatment. Sci Transl Med 13:eabj7360. https://doi.org/10.1126/scitranslmed.abj7360

Mogil JS (2019a) The translatability of pain across species. Philos Trans R Soc Lond B Biol Sci 374:20190286. https://doi.org/10.1098/rstb.2019.0286

Mogil JS (2019b) Mice are people too: Increasing evidence for cognitive, emotional and social capabilities in laboratory rodents. Can Psychol 60:14–20. https://doi.org/10.1037/cap0000166

Omerbašić D, Smith ESJ, Moroni M et al (2016) Hypofunctional TrkA accounts for the absence of pain sensitization in the African naked mole-rat. Cell Rep 17:748–758. https://doi.org/10.1016/j.celrep.2016.09.035

*Peirs C, Seal RP (2016) Neural circuits for pain: Recent advances and current views. Science 354:578–584. https://doi.org/10.1126/science.aaf8933

Steagall PV, Bustamante H, Johnson CB, Turner PV (2021) Pain management in farm animals: focus on cattle, sheep and pigs. Animals 11:1483. https://doi.org/10.3390/ani11061483

Todd AJ (2010) Neuronal circuitry for pain processing in the dorsal horn. Nat Rev Neurosci 11:823–836. https://doi.org/10.1038/nrn2947

*Tracey I, Woolf CJ, Andrews NA (2019) Composite pain biomarker signatures for objective assessment and effective treatment. Neuron 101:783–800. https://doi.org/10.1016/j.neuron.2019.02.019

*Usoskin D, Furlan A, Islam S, et al (2015) Unbiased classification of sensory neuron types by large-scale single-cell RNA sequencing. Nat Neurosci 18:145–153. https://doi.org/10.1038/nn.3881

Vandewauw I, De Clercq K, Mulier M et al (2018) A TRP channel trio mediates acute noxious heat sensing. Nature 555:662–666. https://doi.org/10.1038/nature26137

[1] References marked with "*" are of outstanding interest.

Chronic Pain

Louise Clark

Abstract

There is increasing awareness of, and concerns with the effect chronic pain may have on farmed animals. Chronic pain is probably the product of evolutionary processes that make the individual less vulnerable to attack and so enhance survival. However, it may simultaneously compromise both welfare and production (Crook RJ, Dickson K, Hanlon RT, Walters ET. Curr Biol 24:1121–1125. https://doi.org/10.1016/j.cub.2014.03.043, 2014). Chronic pain is not simply a temporal continuum of acute pain but the perceived result of a series of peripheral, spinal and central mechanisms. Physiological changes at multiple levels within the nociceptive pathway may facilitate or mitigate the development and perception of chronic pain. Understanding how chronic pain develops is the first step in creating strategies to mitigate it in farmed animals.

Keywords

Chronic pain · Peripheral sensitisation · Central sensitisation · Brain

1 Chronic Pain as an Entity in Farm Animals

Suffering, rather than pain, brings people into doctors` surgeries seeking relief. Chronic pain is more than a sensory process, while suffering is a property of the human brain and depends upon consciousness (Loeser 2000). In animals as in humans, pain and suffering are inextricably linked (*see acute pain, chapter* "Pain in Parturition"). Animals are sentient and therefore can suffer (UK Government 2021); an important cause of suffering in farmed animals is chronic or persistent pain.

Many production animals undergo surgical procedures as part of routine husbandry and management. Whilst the literature in this field generally concentrates on the alleviation and mitigation of acute pain, there are some data, which infer that these procedures have a long-term effect on the individuals involved (Clark 2020). Alvarez et al. (2019) followed goat kids that were disbudded with a hot-iron technique and noted that the kids were painful throughout the 7 weeks it took for the wounds to heal, raising welfare concerns about this procedure. This study does not demonstrate the development of chronic pain, but the persistence of the post-procedural pain may be relevant to the genesis of chronic pain. Similar concerns have been raised about the long-term welfare implications of velvet removal in deer (Wilson and Stafford 2002) and tail docking in

L. Clark (✉)
Davies Veterinary Specialists, Manor Farm Business Park Higham Gobion, Hertfordshire, UK
e-mail: Louise.Clark@vetspecialists.co.uk

R. E. Clutton, G. C. Musk (eds.), *Pain and Suffering in Farmed Animals*,
https://doi.org/10.1007/978-3-032-11407-5_7

pigs (Tallet et al. 2019). Tail docking modifies behaviours indicative of pain during the procedure that persist until weaning (4 weeks later), again raising questions about the post-procedural pain trajectory. When younger calves were disbudded (at 3 days compared to 35 days), sensitivity at a second site, a marker for generalised pain sensitivity, was increased (Adcock and Tucker 2018). Performing painful procedures near birth can cause lasting changes in pain perception, extending into adulthood (Walker et al. 2009; Walker 2017). Whilst these husbandry procedures are potential sources of chronic pain in all farm animal species, mastitis and foot rot are highly prevalent, well-documented sources of chronic pain in dairy cows and sheep, respectively.

2 Overview of Chronic Pain

Much of the following chapter is 'generic' insofar that the research relates to both human patients and animal models (Kuner and Flor 2017). Chronic pain is that which persists for prolonged periods and often extends beyond the period of healing of the original injury. The International Association for the Study of Pain (IASP), the world's largest pain-focused professional organisation, defines chronic pain as pain that persists or recurs for longer than 3 months (IASP 2023a). Chronic pain may persist despite successful management of the condition that initiated it, or because the underlying condition cannot be effectively treated.

Chronic pain is generally considered 'maladaptive' because it provides no apparent protective or recuperative benefits. It is also felt that it represents a dysregulation of pain mechanisms, although current hypotheses relating to aetiology are more nuanced. A commonly accepted theory proposes that chronic pain is driven by mechanisms that evolved to reduce enhanced vulnerability to attack, where aggressors target prey showing physical impairment after injury. Pain and anxiety may persist after severe injury enhances vigilance and behavioural caution, which in turn decreases heightened vulnerability to attack caused by motor impairment and disfigurement. These culminate in increased survival chances and the opportunity to reproduce (Walters 2019). In a seminal study involving squid, Crook et al. (2014) described how post-injury sensitisation changed their behaviour to an extent that resulted in less predation. A second, related theory proposes that 'excessive' pain is normal because the cost of *more* pain is often less than the cost of too *little* (the smoke-detector principle). The severe injuries occurring in people with diminished capacity to feel pain reveal the usefulness of pain in motivating escape from and avoidance of situations causing tissue damage (Nesse and Schulkin 2019).

The sensitisation of damage-sensing processes is an adaptation that facilitates tissue healing. It may also provide extra protection in dangerous environments (Crook et al. 2014). Unfortunately, such protection carries the risk of there being no system to mitigate the effect, in which case the experience of pain lowers the pain threshold and causes chronic pain (Nesse and Schulkin 2019). In summarising recent developments, Price and Dussor (2014) questioned whether this plasticity resolves when the injury has healed. There is evidence that pain plasticity may persist long after an injury has healed. This does not mean that every injury causes chronic pain, because it is surmised that endogenous analgesic mechanisms have developed to mask persistent nociceptive plasticity. This process involves adaptations in the spinal cord's opioid systems and brain stem centres, e.g. the periaqueductal grey (PAG) and the rostroventromedial medulla (RVM), which are constituents of the descending pain modulatory system. It is, therefore, possible that a propensity to develop chronic pain is an evolutionarily encoded feature of complex neural systems. Growing evidence supports the belief that it is the (in)ability to mask this persistent nociceptive plasticity that may determine whether an injury leads to chronic pain or not.

3 Classification of Chronic Pain

Chronic pain is not a temporal continuum of acute pain, although the mechanisms driving pain probably change on a continuum from acute nociceptive pain to when chronic pain begins. Pain is commonly classified by cause: nociceptive, neuropathic and nociplastic. Nociceptive pain arises when peripheral nociceptors are activated by noxious thermal (heat/cold), mechanical or chemical stimuli. Some chronic pain seen in production animals is nociceptive in aetiology, e.g. foot-rot in sheep. This, the most common cause of lameness in the United Kingdom sheep flock, is considered extremely painful. The pain arising from foot rot can become chronic or recurrent. In the latter case, inflammatory mediators activate numerous peripheral receptors, driving chemical pain. Neuropathic pain results from injuries or lesions of the nervous system and contributes to the injured animal's overall pain experienced. Neuropathic pain is likely in some chronic disease states, e.g. in tail or digit amputation or debeaking in poultry, where nerves that are traumatised or cut during the procedure propagate persistent pain impulses with neuropathic features. Nociplastic pain, which has only recently been recognised, is generated by CNS dysfunction that distorts the processing of normal sensory stimuli, rather than by nociceptive nor neuropathic mechanisms. Primary examples of nociplastic pain in animals are limited. However, secondary nociplastic mechanisms underlie increased CNS sensitivity and consequent pain amplification often seen in chronically painful diseases such as chronic mastitis. This classification system reveals that clinical pain is often a combination of all three 'types'.

4 The Transition from Acute to Chronic Pain

The transition from acute to chronic pain is incompletely understood despite the considerable research conducted in this area. Recent evidence suggests that the immune and nervous systems communicate to protect the host from injury, and that chronic pain conditions often result when such 'conversations' go 'wrong' (McMahon et al. 2015). On one level, chronic pain may be considered as a neuroinflammatory disorder mediated by neuronal and non-neuronal cells. Immune cells, e.g. macrophages, T cells and mast cells, become activated at the site of inflammation and release pro-inflammatory cytokines and chemokines (Davis et al. 2020). When these cells enter the peripheral and central nervous systems, they express numerous inflammatory mediators that influence pain transmission and the transition from acute to chronic pain. At each level of the nervous system, immune cells can render nociceptive circuitry dysfunctional. The physiological changes that may lead to chronic pain are probably best considered as events occurring along the nociceptive pathway (Fig. 1 Acute Pain chapter). The mechanisms leading to persistent pain are many and varied, and nociceptive pathways may become dysfunctional at one or several sites simultaneously.

4.1 Peripheral Sensitisation

The term 'peripheral sensitisation' (see chapter "Pain in Parturition") describes functional changes to nociceptors that occur at disease or injury sites. It occurs because inflammatory mediators are released from numerous non-neuronal cells, including fibroblasts, keratinocytes, platelets and immune cells. These inflammatory mediators, along with numerous signalling molecules constitute the 'inflammatory soup' (Fig. 5 Acute Pain chapter), which attracts immune cells, fights infection, promotes healing but also sensitises nociceptors, which become activated by less intense stimuli. Sensitisation may occur in three ways: (i) direct channel activation—causing nociceptor activation; (ii) activation of intracellular regulatory pathways that indirectly alter membrane proteins; or (iii) alterations to cellular gene expression increasing excitability. Additionally, inflammatory mediators can sensitise nociceptors that were previously 'silent', which further increases nociceptive input to the spinal cord.

Altered membrane proteins and gene expression tend to cause long-lasting peripheral sensitisation because of changes to the cellular machinery. The overall effects of peripheral sensitisation are that the same input produces a much greater ascending output, and the animal will display a greater behavioural response.

4.2 Spinal Cord Plasticity

Spinal cord anatomy is detailed in chapter "Pain in Parturition", acute pain. Sensory afferent nerve fibres from the periphery terminate at synapses with second-order neurons in the dorsal horn of the spinal cord. Low-threshold afferents generally terminate and synapse in lamina III and lamina anatomically deep to this layer, and nociceptive afferents synapse in the superficial dorsal horn (laminae I and II, Fig. 3, Acute Pain chapter and Fig. 1 below). Fibres carrying touch information not only synapse in the deeper laminae, but also branch and project directly to nuclei in the brainstem via the dorsal columns. In this way, touch information reaches higher centres directly, i.e. without synapsing in the spinal cord. Under normal conditions, touch and pain processing areas are spatially segregated in the spinal dorsal horn. Each of these areas contains second-order neurons that act as the next link in the pathways that convey touch and pain information to the brain.

4.3 Central Sensitisation and Allodynia

Woolf described central sensitisation as a 'process through which a state of hyperexcitability is established in the CNS leading to enhanced processing of nociceptive inputs' (Woolf 1983). Sensory afferent inputs can trigger prolonged, but reversible increases in the excitability and synaptic efficacy of central nociceptive neurons. Central sensitisation manifests as pain hypersensitivity, particularly dynamic mechanical allodynia (see below), secondary punctate or pressure hyperalgesia, after-sensations and enhanced temporal summation (Woolf 2011).

Allodynia arises when pain or unpleasant sensations are felt in response to normally innocuous stimuli, such as light touch (mechanical allodynia) or modest temperature changes (mostly cold allodynia). Allodynia can be defined as 'pain due to a stimulus that does not normally provoke pain' (IASP 2023b). An example would be a light feather touch or a few drops of cold water (that should only produce sensation), causing pain. Allodynia is a common feature of chronic pain syndromes in people; it is unpleasant and can be difficult to manage. Sufferers usually describe the pain as sharp, stinging or burning. Some describe it as if they have a bad sunburn. The most common causes of allodynia in people include diabetes, shingles, fibromyalgia and migraine headaches (Cleveland Clinic 2024). We have little data in respect to animals.

Whilst allodynia can be rapidly evoked by the acute absence of endogenous inhibitory control without structural changes in the spinal cord, structural reorganisation, called plasticity, clearly contributes to its role in chronic pain (see Fig. 1). This involves a reorganisation of synapses, cells and circuits (Kuner and Flor 2017). A desegregation of touch and pain may arise at the level of spinal processing through multiple distinct, but not mutually exclusive mechanisms. These include structural remodelling of circuits to enable physical links between nociceptive and non-nociceptive neurons or disinhibition of existing physical links, for which there is currently more evidence (Kuner and Flor 2017). Put another way, the pathways for pain and touch that should be separate are able to interact with each other.

A good example of this lack of desegregation is seen when the 'gate control' mechanism malfunctions. Normally, touch sensations, e.g. rubbing an injured part, *inhibit* the resulting pain through a feed-forward inhibitory circuit within the superficial laminae of the dorsal horn. In mechanical allodynia, injury impairs the feed-forward inhibitory circuit, allowing light touch to engage the nociceptive network. Rodent studies have shown that low-threshold Aβ-fibre (touch)

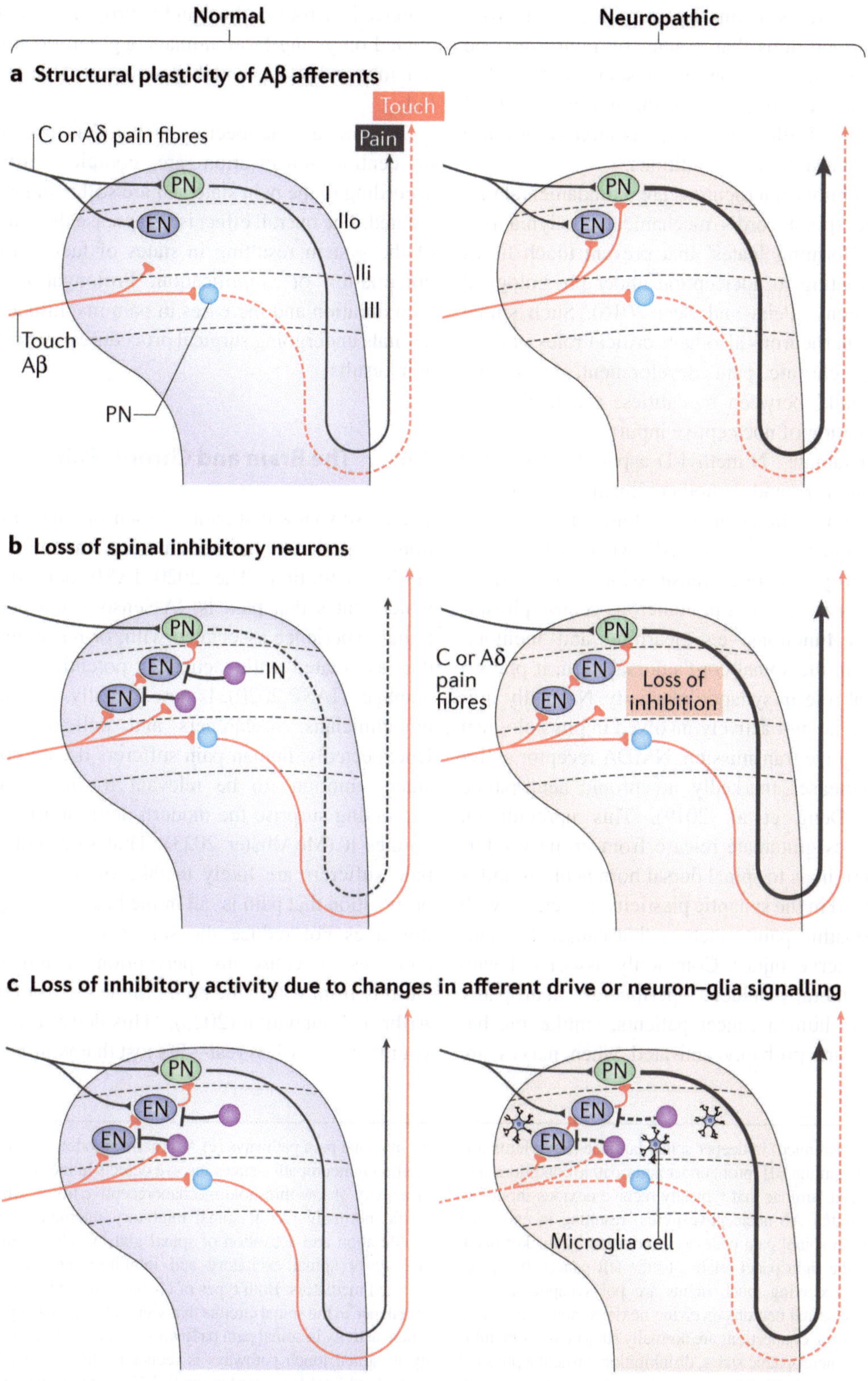

Fig. 1 Remodelling of spinal circuits. Models of potential changes in spinal circuitry segregating pain from innocuous touch in neuropathic mechanical allodynia. (Reproduced from Kuner and Flor (2017) with permission (including figure legend)) (**a**) Touch-sensitive low threshold mechanoreceptive fibres (Aβ fibres) that normally terminate

input activates lamina I pain projection neurons when conditions that mimic the injury-induced decrease in inhibition are present. Such studies have also established that the normally 'silent' mechanical allodynia circuit is already in place under physiological conditions.

Inhibitory interneurons are a fundamental part of the spinal cord's mechanical allodynia network, forming 'gates' that prevent touch fibres contributing to nociception under physiological conditions (Peirs and Seal 2016). Such spinal cord interneurons also have critical roles in mitigating chronic pain development, preventing 'crosstalk' between modalities, and facilitating modulation of nociceptive input.

Glutamate N-methyl-D-aspartate (NMDA) receptors provide another illustration of this complexity. Glutamate is the dominant excitatory neurotransmitter in the CNS, where it facilitates excitatory synaptic transmission. The NMDA receptor is involved in numerous central physiological functions, e.g. learning and memory, while in the spinal cord's dorsal horn, it plays a pivotal role in synaptic plasticity. Normally quiescent, i.e. not actively involved in physiological nociceptive transmission, NMDA receptor activity increases markedly in chronic neuropathic pain (Deng et al. 2019). This upregulation increases glutamate release from primary afferent terminals to spinal dorsal horn neurons and is involved in the synaptic plasticity associated with neuropathic pain—such as that caused by traumatic nerve injury. Commonly associated with chemotherapy-induced peripheral neuropathy seen in human cancer patients, similar mechanisms are probably activated when nerves are damaged in routine husbandry procedures conducted on young farm animals, e.g. tail docking in lambs and piglets and disbudding in calves and kids.

In summary, the specific mechanisms involved in central sensitisation are complex, differ according to the pain state and are still being elucidated. The overall effect is to increase the 'gain' of the system resulting in states of facilitation, potentiation or amplification. Prolongation of pain duration and increases in pain magnitude in animals undergoing surgical procedures are probable results.

4.4 The Brain and Chronic Pain

The IASP's view that pain is 'a sensory and emotional experience' has not changed since its first (1979) definition. The 2020 IASP definition, which states that pain is, 'A sensory and emotional experience associated with, or resembling that associated with, actual or potential tissue damage' (IASP 2020). Is now globally accepted by clinicians, researchers and policymakers. Unexpectedly, human pain sufferers do not consider 'emotion' to be relevant to their pain, expressing surprise the modern definition incorporates it (McAllister 2023). That said, human pain sufferers are likely to take offence to the proposition that pain is 'all in the head', although this does not reduce the statement's veracity: pain, as a conscious perception, originates entirely from within the brain. In the words of the Arthritis Foundation (2023), 'This doesn't mean your pain is any less real—it's just that your brain

Fig. 1 (continued) in deeper spinal laminae (inner lamina II (IIi) and lamina III) sprout under neuropathic conditions into superficial laminae that typically receive noxious inputs (C fibres and/or Aδ nociceptive fibres) resulting in enhanced activity in spinal pain pathways (indicated by thicker black line in the right panel relative to the left panel) (**b**) Spinal neurons receiving touch inputs are polysynaptically connected to spinal neurons receiving noxious inputs, and these polysynaptic connections are normally subject to strong inhibition. In neuropathic states, disinhibition through a physical loss of spinal inhibitory neurons can activate crosstalk between touch and pain circuits and result in increased activity in spinal pain pathways (**c**) Alternative models for disinhibition in neuropathic states involve defects in the structure or activity of low-threshold mechanoreceptive fibres, which would normally recruit spinal inhibitory interneurons, or proliferation and activation of spinal glia, which modulate activity of spinal excitatory and inhibitory neurons via secreted mediators. Both types of changes would lead to an imbalance in the spinal circuits that segregate touch and pain, where activity in spinal pain pathways is increased and activity in spinal touch pathways is reduced (indicated by the dashed red line) IIo outer lamina II, EN excitatory interneuron, IN inhibitory interneuron, PN projection neuron

literally creates what your body feels, and in cases of chronic pain, your brain helps perpetuate it'. Current neurobiological understanding provides no evidence that a farmed animal's pain experience is any different.

Pain is unlike many other senses in terms of its somatotopic representation within the brain. Vision, for example, involves impulses being relayed from the retina directly to the visual cortex. There is no 'corresponding' pain cortex, and none of the areas involved in pain perception are devoted to that role alone, many being involved in other aspects of perception and behaviour. For example, the amygdala, part of the limbic system of the brain, plays a key role in processing emotions and emotional reactions in general, and is also involved in the regulation of pain behaviour in numerous species.

Nociception is initiated in the periphery and ascends through the hind and midbrains to the thalamus. The latter two structures contribute to the pain 'neuromatrix', a dynamic, ever-changing set of neural networks that determine conscious pain perception. The pain neuromatrix was conceptualised by Ronald Melzack (1999) to explain the nature of pain. The pain matrix encompasses areas involved with the sensory-discriminative, affective-motivational and cognitive–evaluative dimensions of pain. It is anatomically and functionally diverse (Tracey et al. 2019; Kano et al. 2020). Spinal cord neurons project directly to several specific brain areas, including the medulla, the parabrachial nucleus (PB) the periaqueductal grey (PAG) and the thalamus. These then project to 'higher' areas, constituting networks coding for specific dimensions of the pain experience. The thalamus has been associated with sensory-discriminatory aspects of pain because of links with the posterior insula, a part of the somatosensory cortex. The insula (insular cortex) is involved in formulating the location, intensity, salience (prominence) and quality of noxious stimuli (Baliki et al. 2009; Wager et al. 2013) and believed to process convergent information to produce an emotionally relevant context for sensory experience (Luan Phan et al. 2002).

The limbic system, cingulate cortex and hypothalamic pituitary axis (HPA) are important in driving the affective (emotional) and cognitive components of pain. Complementing findings from human brain-imaging studies, rodent studies have focused on plasticity in particular regions of the pain matrix (Peirs and Seal 2016) and have tended to support the notion that individual brain areas have multi-functional roles, and that emotional aspects of pain, such as anxiety and fear are linked to the anterior cingulate cortex (ACC). Rodent studies have also indicated an important role for ACC synapses in the mutually enhancing interactions of anxiety and chronic pain. The limbic system, which includes the amygdala, hippocampus and prefrontal cortex, is involved in the regulation of pain behaviour in numerous species. For some time, it has been clear that an individual's emotional and motivational state before noxious stimulation can affect the pain experience, including its evaluation. This has recently been supported by brain imaging studies in humans. The anticipation of pain can increase the intensity of perceived pain and activate regions involved in sensory and emotional pain processing, i.e. the prefrontal and insular cortex (Bingel et al. 2011). The complexity of regulation within and between brain circuits helps explain the influence of psychological factors such as expectation, experience and belief on the pain experience (Mercer Lindsay et al. 2021).

For reasons both ethical and methodological, the changes that occur within the brain when acute pain transitions to the chronic form are poorly understood. Most brain processes are considered complex because multiple brain networks (and thus multiple brain areas) discharge and work synchronously to execute specific individual tasks. Similarly, numerous brain areas are simultaneously activated when pain is experienced. Dysfunction at any level can produce unregulated, persistent pain. One of the challenges in understanding pain neurobiology is that there is no single cortical structure or pathway responsible for the overall pain experience.

Understanding neural networks has been facilitated by technologies such as functional Magnetic Resonance Imaging (fMRI) and advanced software because both have allowed the generation of connectivity models.

In humans, seven main brain 'networks' exist under the control of two dominant control networks known as the central executive network (CEN) and the default mode network (DMN). The CEN focuses on external goals and is responsible for the active creation of memory associated with external activities. The DMN exhibits the highest overlap in its structural and functional connectivity (Horn et al. 2013) and controls the brain's passive, internal processes for thinking and evaluating ideas (Androulakis et al. 2018). The DMN is a large-scale network based mainly on the medial prefrontal cortex, posterior cingulate cortex/precuneus and the angular gyrus (Buckner et al. 2008). In healthy brains, the activity of a third, the salience network (SN), ensures that the CEN and DMN are not active simultaneously. The SN determines the controlling network at any given moment by acting like a meeting moderator, i.e. letting only one person speak at a time. It operates to ensure information is not missed or messages become mixed. The SN moderates the activity of the DMN and the CEN by constantly monitoring the external world, and in assessing how other brain networks react to new information and stimulation (Sughrue 2022). The main functional areas, or nodes of the SN are in the anterior midcingulate cortex, the anterior insula, the pre-supplementary motor areas, and the amygdala (Crawford 2022).

Importantly, and in connection with the limbic system, the SN registers the volume of signals involving emotional response and so plays critical roles in pain processing, emotion, reward and motivation (Menon 2015). When the SN is damaged through injury, dementia or other degenerative diseases, the dominant control networks malfunction. Furthermore, physical damage to the SN affects pain perception. When the SN becomes over or under-active, network switching and emotional regulation become mishandled leading to general brain malfunctioning which features in depression, anxiety, post-traumatic stress disorder and schizophrenia. Current research is focusing on the SN's role in irregularly moderating the brain's response to pain.

The suffering of chronic pain projects beyond pain; depression and anxiety, sleep disturbances and decision-making abnormalities significantly diminish the subject's quality of life. Unexpectedly, chronic pain harms cortical areas unrelated to pain (Apkarian et al. 2004). The fMRIs of chronic back pain sufferers taken during simple visual attention tasks showed reduced deactivation in several key DMN regions compared to those from healthy pain-free individuals. A comparison of three different chronic pain syndromes showed common DMN changes, i.e. decreased connectivity of the medial prefrontal cortex with posterior DMN elements, and increased connectivity with the insular cortex, and in direct proportion to the pain intensity. Thus, DMN function seems to be reorganised in chronic pain, which may reflect the maladaptive physiology of different types of chronic pain.

Nociceptive information is further modulated by the descending pain modulatory system (DPMS). This complex, top-down system allows cognitive, emotional or autonomic nervous factors to inhibit or facilitate nociceptive processing; thereby heightening or reducing the eventual pain experience. Descending neurons project from the rostroventral medulla (RVM) and locus coeruleus (LC) to the spinal dorsal horn. This mechanism is discussed in more detail in the acute pain chapter. Being influenced by both sensory and affective brain circuits, the DPMS can either facilitate or inhibit pain perception according to the individual's emotional state, as well as the extent of injury. Plasticity in the DPMS pathways is a key factor in the transition from acute to chronic pain, and understanding DMPS dysfunction and plasticity in chronic pain is a current research goal (Kucharczyk et al. 2023).

5 A Synopsis of the Mechanisms Involved in 'The Brain in Chronic Pain'

The mechanisms associated with chronic pain perception in the brain are incompletely understood, multifold and complex. They include: (i) expansion and shifts of cortical representations; (ii) impaired descending inhibitory control; (iii) altered structural integrity and connectivity of white matter; (iv) alterations in grey-matter volume; (v) altered resting-state and pain-evoked functional connectivity (network function); and (vi) altered glial activity.

6 Summary

It is widely felt that chronic pain has an inestimable—potentially major—and adverse effect on animal welfare. Therefore, until the processes whereby acute pain becomes chronic are better understood, and form the basis of rational therapy, it behoves the humanitarian to ensure acute pain is managed as appropriately and aggressively as possible (Rosenberger and Pogatzki-Zahn 2022).

References

Adcock S, Tucker CB (2018) The effect of disbudding age on healing and pain sensitivity in dairy calves. J Dairy Sci 101:10361–10373

Alvarez L, Adcock SJ, Tucker CB (2019) Sensitivity and wound healing after hot-iron disbudding in goat kids. J Dairy Sci 102:10152–10162

Androulakis XM, Krebs KA, Jenkins C et al (2018) Central executive and default mode network intranet work functional connectivity patterns in chronic migraine. J Neurol Disord 6(5):393

Apkarian AV, Sosa Y, Krauss BR, Thomas PS et al (2004) Chronic pain patients are impaired on an emotional decision-making task. Pain 108:129–136

Arthritis Foundation (2023) https://www.arthritis.org/health-wellness/healthy-living/managing-pain/understanding-pain/pain-brain-connection. Accessed 12 Jan 23

Baliki MN, Geha PY, Apkarian AV (2009) Parsing pain perception between nociceptive representation and magnitude estimation. J Neurophysiol 101(2):875–887. https://doi.org/10.1152/jn.91100.2008. Epub 2008 Dec 10. PMID: 19073802; PMCID: PMC3815214

Bingel U, Wanigasekera V, Wiech K, Ni Mhuircheartaigh R, Lee MC, Ploner M, Tracey I (2011) The effect of treatment expectation on drug efficacy: imaging the analgesic benefit of the opioid remifentanil. Sci Transl Med 3(70):Article 70ra14. https://doi.org/10.1126/scitranslmed.3001244

Buckner RL, Andrews-Hanna JR, Schacter DL (2008) The brain's default network: anatomy, function, and relevance to disease. Ann N Y Acad Sci 1124(1):1–38

Clark L (2020) Chronic (or persistent) postsurgical pain: a veterinary problem? Vet Anaesth Analg 48(1):4–6. https://doi.org/10.1016/j.vaa.2020.08.007

Cleveland Clinic (2024) https://my.clevelandclinic.org/health/symptoms/21570-allodynia. Accessed 3 Jan 2024

Crawford L (2022) https://www.o8t.com/blog/salience-network. Accessed 3 Jan 2024

Crook RJ, Dickson K, Hanlon RT, Walters ET (2014) Nociceptive sensitization reduces predation risk. Curr Biol 24:1121–1125. https://doi.org/10.1016/j.cub.2014.03.043

Davis KD, Aghaeepour N, Ahn AH, Angst MS, Borsook D, Brenton A, Burczynski ME, Crean C, Edwards R, Gaudilliere B, Hergenroeder GW, Iadarola MJ, Iyengar S, Jiang Y, Kong JT, Mackey S, Saab CY, Sang CN, Scholz J, Segerdahl M, Tracey I, Veasley C, Wang J, Wager TD, Wasan AD, Pelleymounter MA (2020) Discovery and validation of biomarkers to aid the development of safe and effective pain therapeutics: challenges and opportunities. Nat Rev Neurol 16(7):381–400. https://doi.org/10.1038/s41582-020-0362-2

Deng M, Shao-Rui C, Hui-Lin P (2019) Presynaptic NMDA receptors control nociceptive transmission at the spinal cord level in neuropathic pain. Rev Cell Mol Life Sci 76(10):1889–1899

Horn A, Ostwald D, Reisert M, Blankenburg F (2013) The structural-functional connectome and the default mode network of the human brain. NeuroImage 102:142–151

IASP (2020) https://www.iasp-pain.org/publications/pain-research-forum/papers-of-the-week/paper/146236-revised-international-association-study-pain-definition-pain-concepts-challenges-and/. Accessed 24 Jan 23

IASP (2023a) https://www.iasp-pain.org/advocacy/definitions-of-chronic-pain-syndromes/. Accessed 24 Jan 23

IASP (2023b) https://www.iasp-pain.org/resources/terminology/. Accessed 24 Jan 23

Kano M, Grinsvall C, Ran Q et al (2020) Resting state functional connectivity of the pain matrix and default mode network in irritable bowel syndrome: a graph theoretical analysis. Sci Rep 10:11015

Kucharczyk MW, Di Domenico F, Bannister K (2023) A critical brainstem relay for mediation of diffuse noxious inhibitory controls. Brain 146:2259–2267

Kuner R, Flor H (2017) Structural plasticity and reorganisation in chronic pain. Nat Rev Neurosci 18(1):20–30. Erratum in: Nat Rev Neurosci. 2017 Feb;18(2):158. Erratum in: Nat Rev Neurosci. 2017 18, 2, 113

Loeser JD (2000). Review) Clin J Pain 16(2 Suppl):S2–S6

Luan Phan K, Wager T, Taylor SF, Liberzon I (2002) Functional neuroanatomy of emotion: a meta-analysis of emotion activation studies in PET and fMRI. Neuroimage 16(2):331–348. https://doi.org/10.1006/nimg.2002.1087

McAllister MJ (2023) https://www.instituteforchronicpain.org/understanding-chronic-pain/what-is-chronic-pain/what-is-pain. Accessed 3 Jan 2024

McMahon SB, La Russa F, Bennett DL (2015) Crosstalk between the nociceptive and immune systems in host defence and disease. Nat Rev Neurosci 16:389–402

Melzack R (1999) From the gate to the neuromatrix. Pain S6:S121–S126

Menon V (2015) Large-scale functional brain organization. Brain Mapping 2:449–459

Mercer Lindsay N, Chen C, Gilam G et al (2021) Brain circuits for pain and its treatment. Sci Transl Med 13:eabj7360. https://doi.org/10.1126/scitranslmed.abj7360

Nesse RM, Schulkin J (2019) An evolutionary medicine perspective on pain and its disorders. Philos Trans R Soc B 374:20190288

Peirs C, Seal RP (2016) Neural circuits for pain: Recent advances and current views. Science 354(6312):578–584

Price TJ, Dussor G (2014) Evolution: the advantage of 'maladaptive' pain plasticity. Curr Biol 24(10):R384–R386

Rosenberger DC, Pogatzki-Zahn EM (2022) Chronic postsurgical pain – update on incidence, risk factors and preventive treatment options. BJA Educ 22(5):190–196. https://doi.org/10.1016/j.bjae.2021.11.008. Epub 2022 Feb 24. PMID: 35496645; PMCID: PMC9039436

Sughrue M (2022) https://www.o8t.com/blog/important-brain-networks. Accessed 21 Jan 23

Tallet C, Rakotomahandry M, Herlemont S, Prunier A (2019) Evidence of pain, stress, and fear of humans during tail docking and the next four weeks in piglets (Sus scrofa domesticus). Front Vet Sci 6:462

Tracey I, Woolf CJ, Andrews NA (2019) Composite pain biomarker signatures for objective assessment and effective treatment. Neuron 101:783–800. https://doi.org/10.1016/j.neuron.2019.02.019

UK government (2021) https://www.gov.uk/government/news/animals-to-be-formally-recognised-as-sentient-beings-in-domestic-law. Accessed 5 Dec 2022

Wager TD, Atlas LY, Lindquist MA, Roy M, Woo CW, Kross E (2013) An fMRI-based neurologic signature of physical pain. N Engl J Med 368(15):1388–1397. https://doi.org/10.1056/NEJMoa1204471. PMID: 23574118; PMCID: PMC3691100

Walker SM (2017) Translational studies identify long-term impact of prior neonatal pain experience. Pain 158(Suppl 1):S29–S42. https://doi.org/10.1097/j.pain.0000000000000784

Walker SM, Tochiki KK, Fitzgerald M (2009) Hindpaw incision in early life increases the hyperalgesic response to repeat surgical injury: critical period and dependence on initial afferent activity. Pain 147:99–106

Walters ET (2019) Adaptive mechanisms driving maladaptive pain: how chronic ongoing activity in primary nociceptors can enhance evolutionary fitness after severe injury. Philos Trans R Soc Lond Ser B Biol Sci 374:1785

Wilson PR, Stafford KJ (2002) Welfare of farmed deer in New Zealand. 2. Velvet antler removal. NZ Vet J 50:221–227

Woolf CJ (1983) Evidence for a central component of post-injury pain hypersensitivity. Nature 306:686–688

Woolf CJ (2011) Central sensitization: implications for the diagnosis and treatment of pain. Pain 152(3 Suppl):S2–S15. https://doi.org/10.1016/j.pain.2010.09.030. Epub 2010 Oct 18

Pain in Parturition

Lip Yong Choo and Laura Bordoni

Abstract

The experience of labour pain is multifaceted and differs greatly between and within species. Labour pain in the human parturient is especially complex, being influenced by not only the degree of nociceptive input but by an individual's psychological, emotional, cognitive, social, cultural and environmental circumstances, as well. In view of the limited evidence available exploring pain in the labouring animal, it is hoped that through examination of evidence and common practices in humans, some inferences may be drawn.

Keywords

Pain · Labour · Parturition · Delivery · Epidural

1 Historical Attitudes to Pain in Labour

The management of pain in labour has changed markedly throughout the years, influenced not only by medical advancements but also importantly by society's changing attitudes towards how labour pain is subjectively and objectively perceived.

For centuries, the belief that pain was necessary for the progress of labour, compounded by religious ideals and misinterpretations of biblical scripture, resulted in the limited exploration of analgesic options for the labouring woman. In fact, history has documented that pre-fifteenth-century midwives were burned at the stake for offering pain relief during labour (Lurie 2004). These ideals slowly began to change in the nineteenth century, which saw the adoption of novel methods for managing labour pain, including the inhalational agents ether, chloroform and nitrous oxide. With this shift in attitude came the early movement promoting the right of the parturient to pain relief, spurring the development of alternative options such as the 'twilight sleep', a morphine and scopolamine sedation technique used for decades before falling out of favour due to the significant level of morbidity and mortality that was associated with its use (Barnett 2005). The early 1900s also saw the development of regional analgesic techniques, with the founding father of modern pain medicine, John Bonica, famously using epidural analgesia for his labouring wife following a complication she had suffered with the use of ether (Skowronski 2015).

Reduced uptake of many of these analgesic methods due to safety concerns coincided with another stark change in societal attitudes towards

L. Y. Choo (✉) · L. Bordoni
Department of Anaesthesia and Pain Medicine, Sir Charles Gairdner Hospital, QEII Medical Centre, Perth, WA, Australia
e-mail: lip.choo@health.wa.gov.au

R. E. Clutton, G. C. Musk (eds.), *Pain and Suffering in Farmed Animals*,
https://doi.org/10.1007/978-3-032-11407-5_8

how labour pain was viewed and therefore managed. In the 1940s, the now controversial English obstetrician Dick-Read (1943) advocated that labour was a physiological process and as such, should not be inherently painful. He posited that 'primitive' women experienced little pain during childbirth and claimed that 'civilised' western women were more pain-intolerant due to ignorance, fear and the long-held cultural belief that related suffering in childbirth to punishment for sin. He advocated for women to receive education to target these causative factors and was a proponent for relaxation and reassurance over pharmacological techniques. The French obstetrician Lemaze built on this approach by popularising psychoprophylaxis techniques such as breathing and relaxation techniques (Hrešanová 2016). The natural childbirth movement followed, with many supporters believing that the birthing process had become too paternalistic and over-medicalised (Rothman 1989), with many of the non-pharmacological methods described by Lemaze and his advocators still being popular today (Howard 2018).

1.1 A Contemporary View of Pain in Labour

1.1.1 Pain in Labour Can be Severe and Intolerable

In contrast to the views of Dick-Read, it is now commonly acknowledged that the pain experienced by the human parturient during labour can be severe and intolerable. In Melzack's 'The myth of painless childbirth' it was documented that labouring nulliparous women with no prepared childbirth training compared their pain in labour to that of amputating a digit (Melzack 1984). Furthermore, Dick-Read's assumption that 'primitive' women do not feel significant pain during childbirth has also been rebuked. The American College of Obstetricians and Gynecologists (ACOG) recently stated that there was no other circumstances in which it is considered acceptable for an individual to experience severe pain, amenable to safe pain relief, while under a physician's care (ACOG 2019). The right to pain relief in labour has been similarly echoed by other bodies and the contemporary move away from paternalistic to a greater focus on patient autonomy has seen the creation of modern birthing plans for labour pain management that focus on shared decision-making and more patient-centric care.

2 Adverse Effects of Pain in Labour

Providing options for pain relief in labour is not only humane but may also confer specific benefits to the vulnerable mother or foetus. Evidence shows that the pain of labour may lead to detrimental physiological effects for both the parturient and her offspring and may be associated with the development of long-term negative psychological effects.

2.1 Physiological Effects

2.1.1 Hyperventilation

Periods of intermittent hyperventilation during painful contractions cause a reduction in maternal arterial carbon dioxide tension. If the resulting respiratory alkalosis is severe, as may be associated with extreme levels of pain and distress, uteroplacental vasoconstriction with a reduced blood flow and oxygen delivery to the foetus may occur (Tomimatsu et al. 2013). Maternal alkalosis also produces a leftward shift in the oxyhaemoglobin-dissociation curve, increasing the affinity of oxygen for maternal haemoglobin, reducing oxygen off-loading to the foetus and further reducing oxygen delivery. The hyperventilation-induced reduction in arterial carbon dioxide also reduces ventilatory drive, causing a period of compensatory hypoventilation, which in the absence of supplemental oxygen, can lead to maternal and foetal hypoxaemia (Rooth 1982; Minnich et al. 1990). Epidural analgesia has been shown to limit hyperventilation and its associated increase in oxygen consumption (Hagerdal et al. 1983), improving both maternal and foetal oxygen tension (Bergmans et al. 1996; Reynolds et al. 2002).

2.1.2 Neuroendocrine Effects

Pain and stress during labour cause an increase in plasma catecholamine levels, in turn increasing cardiac output and peripheral vascular resistance. While these effects are usually well tolerated in healthy women, those with specific pathologies such as significant cardiac disease or pre-eclampsia will be at increased risk of morbidity. The increase in peripheral vascular resistance can produce foetal hypoxaemia and acidosis via a reduction in uteroplacental blood flow, as shown in both animal and human studies (Morishima et al. 1979; Hood et al. 1986; Vogl et al. 2006). Morishima et al. showed that when a stressful stimulus was applied to maternal baboons, a reduction in uterine blood flow resulted, which was accompanied by a reduction in foetal heart rate and arterial oxygenation status. Once the stressful stimulus was abated or removed, these changes were reversed (Morishima et al. 1979). In turn, many of these effects can be reversed by the delivery of effective analgesia, with neuraxial local anaesthetic and neuraxial opioid having been shown to reduce maternal plasma catecholamine levels by up to half (Shnider et al. 1983; Cascio et al. 1997).

The stress response to a painful labour can also result in an increased production and release of gastrin and reduce gastrointestinal motility, increasing gastric acidity and volume and therefore the risk of pulmonary aspiration, especially if general anaesthesia is required for emergent operative delivery of the foetus.

2.2 Psychological Effects

Studies examining the psychological sequelae of unrelieved pain during labour have found an association with postpartum depression (Hiltunen et al. 2004; Suhitharan et al. 2016) and in some instances, posttraumatic stress disorder (Soet et al. 2003). One study found that the severity of acute postpartum pain was a better correlate for postpartum depression than whether the baby was born via vaginal or caesarean delivery (Eisenach et al. 2008), suggesting that the pain response plays a greater role in the development of psychological sequelae than the degree of tissue trauma itself.

While low-quality evidence suggests epidural analgesia may be protective against developing postpartum depression (Ding et al. 2014), systematic reviews highlight the difficulty that many studies have in controlling for bias and confounding factors, reflected by the heterogeneity of the data (Orbach-Zinger et al. 2021; Wang et al. 2022). Conversely, other studies have found that a small portion of women who received labour epidural analgesia described feelings of 'deprivation' at the loss of a more 'natural' childbirth experience (Billewicz-Driemel and Milne 1976). While none of these women required psychiatric counselling as a result of this experience, older case reports suggest that this response may not be the same for all (Stewart 1982). A more recent study similarly found an increased incidence of postpartum depression in women who had not anticipated receiving a labour epidural but did have that procedure performed (Orbach-Zinger et al. 2018).

While this conflicting data indicates that more robust research is required to better identify and understand the possible psychological effects that labour pain and its management can have, what is clearly demonstrated is the complexity and variability in the responses to pain in labour, emphasising the need for individualised care.

2.3 Development of Chronic Pain

Chronic postdelivery pain has an estimated incidence of 10%, 1.8% and 0.3% at 2, 6 and 12 months, respectively (Eisenach et al. 2013). While the severity of acute postoperative pain has been linked to the development of persistent postsurgical pain (Kehlet et al. 2006), targeted research determining whether a similar link exists in the obstetric population is required. Such research should aim to control for the multiple confounding factors that make this situation unique from the non-obstetric population, such as care of the new-born, so that risk factors that influence the development of chronic postdelivery pain can be more accurately identified and stratified.

3 Factors Contributing to Pain in Labour

The tremendous variability of the labour pain experience observed between women, despite seemingly similar biological processes, can only be explained by the presence of other contributing factors. Melzack's neuromatrix theory of pain (Melzack 2005) introduced in the early twenty-first century is a useful contemporary model of pain that is highly applicable to labour. The theory describes the affective-emotional, cognitive-evaluative and sensory-discriminative processing areas in the brain that interact with one another to shape an individual's unique pain response. It is important to not only consider the anatomical and physiological basis of labour pain, but to have a deep appreciation of these other determinants so that a more holistic, individualised patient-centred approach can be taken.

3.1 The Pain Pathways

Over 50 years ago, Melzack and Wall first used the gate-control theory of pain to explain how the pain signal is transmitted and regulated from the peripheral nerve to the spinal cord (Melzack and Wall 1965). Modern pharmacological and interventional management of labour pain has largely focused on interrupting the nociceptive signal along this pathway.

3.1.1 Peripheral Pain Pathway

First Stage of Labour

The first stage of labour is from the onset of regular uterine contractions to full cervical dilatation. This characteristically visceral-type pain, often described as cramping or aching, originates from the uterus, chiefly the lower uterine segment, and the cervix. Each uterine contraction aims to stretch the cervix so it can accommodate the foetus for delivery. The distention of uterine and cervical mechanoreceptors and local tissue ischaemia caused by these contractions results in depolarisation of the A-delta and C primary afferent nerve terminals, resulting in propagation of an action potential along them. These afferent nerves pass through the paracervical region, the hypogastric nerve plexus and the lumbar sympathetic chain to enter the spinal cord through the dorsal roots of T10 to L1. As such, analgesia in the first stage of labour could theoretically include peripheral blockade anywhere along this pathway, such as with a paracervical, paravertebral, lumbar sympathetic or epidural (T10 to L1 dermatome) block, discussed in more detail below.

Second Stage of Labour

The second stage of labour is from full cervical dilatation to delivery of the foetus and is said to be more painful than the first stage of labour (Brownridge 1995). In addition to the visceral-type pain described during the first stage of labour, pain in the second stage also includes somatic-type pain originating from vaginal and perineal distention, ischaemia or trauma. Activation of A-delta and C nociceptive fibres in these tissues results in the transmission of the signal via the pudendal nerve dorsal root ganglions from S2 to S4. As such, effective pain relief during this stage might include a pudendal nerve block or extension of the epidural block from T10 down to S4.

3.1.2 Central Pain Pathway

The organisation and transmission of the nociceptive signal is like that of other acute pain situations. Upon entering the spinal cord, the primary afferent neurons terminate in the ipsilateral laminae of the dorsal horn, where they synapse with secondary afferent neurons. These secondary afferents ascend the spinal cord largely via the ventral spinothalamic tract, going on to stimulate neurons located in the thalamus. Neurons also project to the somatosensory cortex, where pain is consciously interpreted, as well as to the spinoreticular and spinomesencephalic tracts. From these latter two tracts, projections to the areas of vigilance (the locus coeruleus and reticular formation) and areas of cardiorespiratory control (caudal medulla and nucleus tractus solitarus) result in the increased sympathetic outflow observed during labour. The descending inhibitory pathway is activated from the periaqueductal grey locus coeruleus, nucleus raphe magnus, rostral medial medulla and the cerebellum, acting to dampen the ascending transmission of the nociceptive signal. The neuromatrix is further comprised of the M1 motor cortex, the prefrontal cortex (area for memory), and the insular cortex and the anterior cingulate cortex (areas for emotional response), all of which are connected via a complex arrangement of interconnecting axons. While the management of labour pain has largely focused on anatomical targets of nociception, many of these central pain pathways, focusing on the neurophysiological basis of labour pain, might provide an alternative avenue for research and hence increase the number of analgesic options for the future labouring parturient.

3.2 Physiological Factors

3.2.1 Progress of Labour

The progression of labour is characterised by the increasing frequency and tonicity of uterine contractions and their associated cervical dilatation, and the engagement of the foetal head onto the pelvic ring. This progress, and the subsequent vaginal and perineal stretching during the second stage of labour, can contribute towards the type, duration and intensity of pain experienced as labour progresses. Early studies have demonstrated an increasing severity of pain with cervical dilatation (Hardy and Javert 1949; Javert and Hardy 1951), however, other studies have demonstrated that the degree of cervical dilatation only accounted for a small portion of the variability in pain scores observed between parturient (Conell-Price et al. 2008; Debiec et al. 2009), again demonstrating the multidimensional nature of labour pain.

As labour progresses, increasing levels of oxytocin are released from the posterior pituitary gland, peaking around the time of birth. Oxytocin is not only responsible for causing the uterine contractions of labour but also acts as a potent neuromodulator, altering the activity of other neurons within the central nervous system of mammals (Walter et al. 2021). Its action on the hippocampus, amygdala and nucleus accumbens contributes to the feeling of euphoria, likely positively shaping the way pain is perceived, especially during the later stages of labour when it is at high concentrations (Shnol et al. 2014; Walter et al. 2021). Oxytocin also exerts an inhibitory action on the glutamate-mediated sensory synaptic transmission between the primary afferent fibres and dorsal horn neurones, as well as mediating the descending inhibitory pathway, both regulating spinal nociceptive transmission (Robinson et al. 2002). The positive behavioural and analgesic effects are likely to be sustained due to the delayed degradation of oxytocin found within the central nervous system.

In women receiving epidural analgesia, the serum oxytocin concentration has been found to be reduced, presumably due to the administration of local anaesthetics and opioids impeding the transmission of afferent neural signals (Goodfellow et al. 1983). This phenomenon is also known as the inhibition of the Ferguson reflex, referring to the physiological neuroendocrine response whereby cervical distension due to foetal descent results in oxytocin production. As such, the duration of epidural analgesia administration has a theoretical potential association with the duration and progress of labour, which is also due to the extent of the associated motor blockade. The clinical significance of this theory has been questioned, especially in several more recent (post 2004) randomised trials that showed that the use of epidural analgesia was not associated with an increased duration of the first stage of labour (Wong et al. 2005; Wang et al. 2009), instrumental delivery rate (Wang et al.

2017; Anim-Somuah et al. 2018), and caesarean section rate (Anim-Somuah et al. 2018). Importantly, these trials used lower concentrations of local anaesthetic agents (as is common contemporary practice) versus older studies which used higher concentrations, hence resulting in differing degrees of neuronal blockade, including motor blockade which can interfere with the parturient's ability to push. This paralysis has been shown to prolong the second stage of labour (Cheng et al. 2014); however, the extent of this effect is markedly reduced with lower concentrations of local anaesthetic (Anim-Somuah et al. 2018).

Labour pain can also be influenced by several other maternal physiological factors, such as the presence of pre-existing chronic pain, opioid tolerance or addictive behavioural traits. These factors have the potential to increase pain perception in the labouring parturient due to modulation and dysregulation of nociceptors and chemical mediators (Krans et al. 2015; Soens et al. 2019).

3.3 Pathological Factors

Pathological processes such as chorioamnionitis, endometritis, cervicitis, vaginitis, infections of the urinary tract, and sepsis can occur concurrently during labour. These processes can evoke a complex cellular response and immunological cascade resulting in inflammation and the formation of neurotoxins and chemical mediators which activate nociceptors, producing a hyperalgesic state (Guan et al. 2016). Labour pain can also be exacerbated by the presence of uterine rupture, placenta abruption or penetrating or blunt trauma to the gravid uterus. Bulky tumours, benign or malignant, may exert excessive pressure within the abdomen to aggravate the painful sensations already experienced during labour. Differentiating between labour pain with or without a coinciding pathological process is difficult, and therefore, other signs indicating a pathology that should not be ignored during the perinatal period should be elicited.

3.4 Psychosocial Factors

Psychological predisposition, behavioural priming and any pre-existing mental health issues can shape the overall labour pain experience of the parturient (Spiby et al. 2003; Mestrovic et al. 2015). As human parturients develop certain expectations of their pain levels prior to the actual labour, many might also experience feelings of fear and anxiety. Some parturients are able to adopt various coping strategies and confidence in managing the pain of labour through adequate acquisition of childbirth knowledge and preparation (Lang et al. 2006; Siyoum and Mekonnen 2019). This approach is highly beneficial, since it is known that poor coping mechanisms, compounded by maternal stress and anxiety, can adversely affect labour experience through exaggerated pain perception (Spiby et al. 2003).

The sociocultural environment of the parturient also plays a pivotal role in shaping the labour pain perception. In particular, the level of education, socioeconomic status, personal beliefs and values, family traditions, cultural heritage, local and regional customs, and societal expectations determine, to varying extents, the way pain is expressed, perceived, assessed and managed (Navarro-Prado et al. 2022). The presence of a support person, such as the birthing partner or midwife, during pregnancy may also modify the labouring experience of the parturient by providing ongoing positive reassurances, promoting confidence and reinforcing positive coping strategies (Schug et al. 2016).

3.5 Foetal Factors

Abnormal uterine distension due to increased foetal weight, multiple gestations and abnormal presentation such as breech or horizontal lie, may result in excessive uterine stretching, visceral pressure and bony impaction, causing unpleasant and painful sensations in the parturient. Foetal death in utero can also trigger an inflammatory cascade from the release of toxins causing uterine stimulation, systemic sepsis and coagulopathy in the parturient, not to mention the additional psychological impact contributing to the pain response.

4 Assessment of the Human Parturient Experiencing Labour Pain

4.1 Assessment of Pain in Labour

Pain assessment during labour is vital to ensure effective pain management strategies can be individually tailored to the needs of each parturient. Periodic pain evaluation and trending of pain scores using validated assessment tools provide meaningful and valuable information to allow clinicians to gauge the effectiveness of the analgesia initiated, progress of labour and the presence of pathological processes. Holistic pain assessment should ideally encompass all of the biopsychosocial elements of labour pain, but may not be practical in the acute setting. This section aims to provide an overview of labour pain assessment, comprising qualitative and quantitative assessment tools.

4.1.1 Qualitative Assessment

Qualitative assessment relies on the use of language to describe pain, which is by self-reporting in humans. In animals, this approach requires observation and interpretation of behaviours and responses of animals to interactions and interventions. Labour pain in the human parturient is emotive or emotion-triggering, involving the use of feelings to describe pain, for example, agonising, irritating, crippling, and traumatising. A useful pain assessment framework used by clinicians for evaluating pain in the humans is the mnemonic *SOCRATES*, which is an acronym for site, onset, character, radiation, associated features/symptoms, timing/duration, exacerbating/relieving factors and severity/intensity (Table 1).

4.1.2 Quantitative Assessment

Pain scoring tools are unidimensional and designed to evaluate pain by determining the level or degree of pain experienced by the parturient (Jones et al. 2015). The numerical rating scale (NRS) is an applied score that requires the parturient to provide a score from 0 to 10 or 100 to rate their pain level, with 0 being no pain and 10 or 100 being the worst-ever pain experienced.

Table 1 The SOCRATES pain assessment framework

Site	The site indicates the locality and distribution of pain. The pain could be localised or diffuse, unilateral or bilateral, symmetrical or asymmetrical, and superficial or deep. In the labouring parturient, the locality of pain depends on the stage of labour. In the first stage of labour, pain occurs in the abdomen due to uterine contractions. As labour progresses towards the second stage, pelvic and lower back pain gradually develop as the foetus engages with the pelvic inlet and commences its descent.
Onset	The onset describes the commencement, including the events preceding the noxious episode, and trajectory of pain. For instance, the pain could be an acute new onset pain, or a chronic pre-existing pain. During labour, pain is experienced when uterine contractions develop and improves once uterine contractions cease.
Character	The character describes the nature of pain. Somatic pain, due to active tissue damage involving A-delta pain fibres, is often described as sharp and localised. Visceral pain involving C fibres is dull, achy and diffuse. Neuropathic pain is an electric-shock type or shooting sensation due to direct neural tissue injury. Mixed pain is a combination of different pain types. Visceral pain is experienced during the first stage of labour and is often diffuse. In the second stage of labour, somatic pain occurs due to vaginal and perineal distension.
Radiation	Radiation describes the referral of pain to a site adjacent to or distant from the primary site of pain. As labour progresses, pain experienced in the groin or perineum can radiate to the thighs due to femoral nerve entrapment during extreme hip flexion in the lithotomy position.

(continued)

Table 1 (continued)

Associated features/ symptoms	Associated features/symptoms are clinical manifestations that are experienced during the painful episode and could include nausea, vomiting or dyspnoea. The labouring parturient might develop nausea and diaphoresis when the sensation of pain intensifies.
Timing/ duration	The timing/duration describes the temporal change and pattern of pain, such as episodic, constant, colicky, or crescendo. The pain experienced during labour comes in 'waves', corresponding to each uterine contraction.
Exacerbating/ relieving factors	Exacerbating/relieving factors are those that aggravate or reduce the pain such as positional changes and/ or the use of pharmacological agents. For instance, the parturient might be able to relieve her labour pain by immersing in water, or pain may be aggravated when the support person is absent.
Severity/ intensity	The degree of pain could be described using certain words such as mild or severe, or a value or derived score using a quantitative tool. The intensity of pain escalates with increasing frequency and tonicity of uterine contraction.

Much like the NRS, the Verbal Rating Scale (VRS) provides an objective measure of pain by asking the human parturient to rate their pain level using an adjective ranging from absent to severe to the worst imaginable pain on a similar graduated scale. These self-assessment tools for humans can be used in animals but rely upon careful and adequate observation and interpretation of animal behaviour, mental state and response to interaction and intervention.

The Visual Analogue Scale (VAS) and adaptations of it, offers more utility in the non-human parturient. The original form of this assessment tool uses a straight line with one end meaning no pain and the other end meaning the worst pain imaginable. The parturient indicates along that scale where their pain level sits at the time of asking. Used more commonly in paediatric medicine and more applicable to the non-human parturient is the Wong-Baker FACES® tool (Wong and Baker 1988): an elaborated version of the VAS, incorporating a series of graphic imagery depicting facial expressions ranging from teary and frowning to smiling, corresponding to pain levels and a pre-determined value (Fig. 1). It may be possible for the various facial grimace scales developed for animals to be integrated into a species-specific VAS.

Novel pain assessment tools have been developed and trialled in the human parturient to improve existing tools and address the growing concerns that the human parturient's capacity for self-evaluation can vary as birth progresses, thereby inaccurately reflecting the actual experience of the birthing woman. One such tool, ESVADOPA, is a pain assessment tool covering six items across two domains (facial muscle, body response, verbal response, restlessness, ability to relax and vegetative symptoms). This instrument, yet to be validated for widespread use, was designed with its key feature being the ability to assess the expression of childbirth pain based on the carer's observation without interfering in the childbirth process.

4.1.3 Assessment of Epidural Analgesia Block Efficacy

Pain assessment in the labouring parturient must be ongoing and should include an informed assessment of the efficacy of any instigated therapies. Should neuraxial analgesia be performed (most commonly epidural analgesia) ongoing assessment of the efficacy and development of any immediate or delayed unwanted effects should be regularly performed. In addition to the routine qualitative and quantitative pain assessments described, it is important to ascertain the dermatomal height, density and quality of the block. The sensory, motor and sympathetic components, as well as maternal satisfaction, are also assessed periodically and documented.

As previously described, neuraxial blockade in the form of epidural analgesia established for labour should cover the T10 to L1 nerve root levels during uterine contractions and cervical dilatation of the first stage of labour, and later include the S2 to S4 nerve routes during the second stage

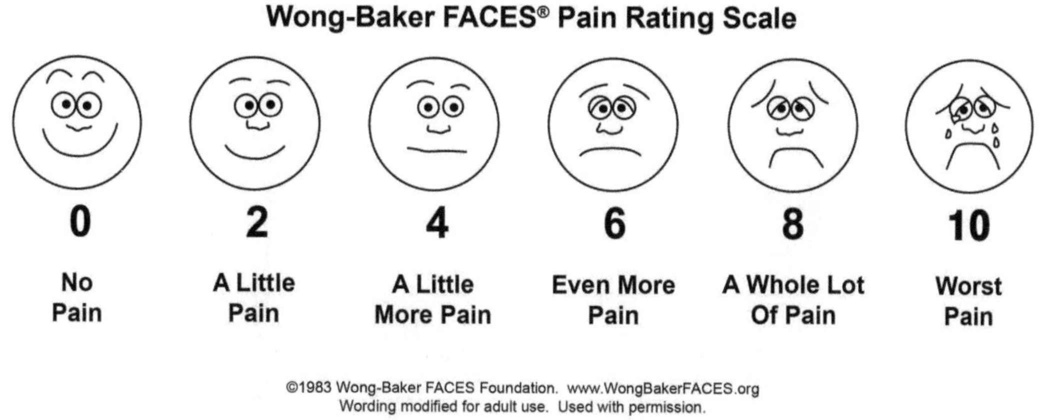

Fig. 1 Diagram illustrating a series of schematic facial expressions corresponding to a pain score (from 0 to 10) and a brief description. (Picture reproduced with permission from the publisher and authors)

of labour resulting from the stretch and distension of the vagina and perineum. The ideal epidural block will cover the sensory components of these levels, with minimal motor block. This strategy not only aims to aid maternal comfort and satisfaction but will also assist the parturient to push during the second stage of labour. A 'light' block using a high-volume low-concentration local anaesthetic aims to achieve this. Practical assessment of the block height and density includes assessment of the dermatomal block height to cold, since the neurons carrying the nociceptive signal to the spinal cord also carry the temperature signal (Fig. 2).

A 'dense' block, as may be observed with a higher concentration of local anaesthetic, is often characterised by reduced motor function and should be avoided due to safety reasons and its resulting effects on the progress of labour and delivery, including prolonged second stage of labour (Cheng et al. 2014; Anim-Somuah et al. 2018) and increased rate of instrumental delivery (Sultan et al. 2013). The Modified Bromage Scale (Breen et al. 1993), revised from the original Bromage Scale (Bromage 1965), is a tool used to evaluate the degree of motor blockade resulting from neuraxial analgesia or anaesthesia. A receding block, characterised by an escalation of pain associated with a reduction in the dermatomal level covered by the block, requires a dose 'top-up' or evaluation of the catheter for malposition or dislodgement. Since it would be difficult to evaluate the dermatomal level in the non-human parturient, increasing pain in a previously comfortable parturient following epidural analgesia will be the main indication of a receding block.

Other unwanted effects resulting from epidural analgesia can include an inadvertent dural puncture, characterised by a constellation of clinical manifestations such as debilitating and postural headache, visual disturbances, tinnitus and neck stiffness, commonly occurring over 24 h *after* the dural puncture. The exact pathophysiology of the symptoms is not fully understood, but they are thought to occur due to persistent leakage of cerebrospinal fluid, causing intracranial hypotension and meningeal irritation. Dural puncture can be recognised (free flow of cerebrospinal fluid back through the epidural needle), or unrecognised. The latter will often result in a high local anaesthetic block once the local anaesthetic is administered, characterised by reduced upper limb strength, dyspnoea, and, in higher doses of local anaesthetic, respiratory and cardiovascular collapse requiring ventilatory and cardiovascular support. In addition, epidural analgesia can be associated with urinary retention, nausea and vomiting, cardiovascular compromise including bradycardia and hypotension,

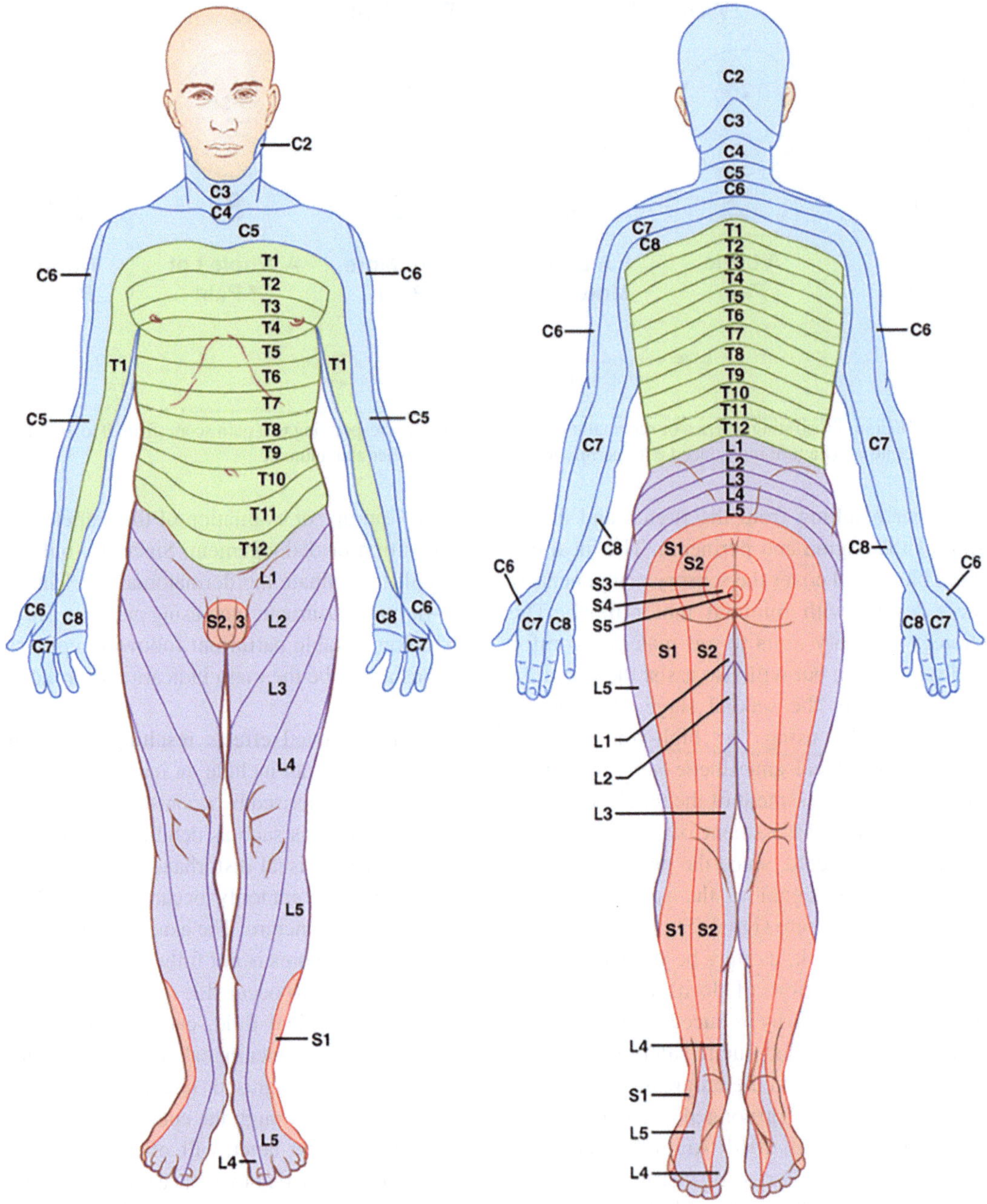

Fig. 2 Diagram illustrating spinal dermatomes in a human both anterior (left) and posterior (right). The dermatomal levels are colour-coded according to the respective spinal segments—cervical (blue), thoracic (green), lumbar (purple) and sciatic (red). (Turnbull and Aleshi 2015). (Picture reproduced with permission from Springer Nature (License number 5601250794585))

bleeding such as epidural haematoma, infection such as epidural abscess and temporary or permanent nerve injury.

Additional details regarding epidural analgesia on its benefits, patient selection and techniques are described in the next section.

5 Labour Pain Management in the Human Parturient

The contemporary view of pain in labour acknowledges that the experience can be severe and intolerable and when amenable to pain ameliorating measures, timely and effective relief should be offered, providing a much-improved experience for the labouring parturient as well as conferring physiological and psychological benefits to mother and foetus. This is supported by numerous obstetric and anaesthesia professional bodies (ACOG 2019; Bisson et al. 2019) who also propose the empowerment of the parturient with robust knowledge of the labour journey and options of holistic pain relief, further supporting the contemporary shift away from the traditional paternalistic approach towards managing pain in labour. This allows women to actively participate in their own care by gaining some control over their own labour experience. Studies have shown that a lack of relevant information has been associated with increased anxiety, which can be detrimental to the overall labour experience (Brewin and Bradley 1982; Waldenstrom et al. 1996; Gibbins and Thomson 2001).

For most healthy women, the selected method for pain relief during labour is dependent on personal choice and the modalities available. Since the introduction of chloroform for labour analgesia in 1847, different methods and medications have been studied and developed to relieve the pain of labour, with pharmacological interventions primarily aiming to relieve this by reducing nociceptive transmission, and non-pharmacological interventions encouraging improved empowerment and self-efficacy. This section will explore the evidence surrounding the use of these strategies to manage labour pain in the human parturient.

5.1 Pharmacological Approaches

There is a broad range of pharmacological agents available for labour analgesia and these can be subdivided into their respective routes of administration. The choice of agent is guided by its efficacy and safety profiles, availability and patient preference.

Drug safety profiles are outlined by individual national governing bodies such as the Australian Drug Evaluation Committee (ADEC) and the United States Food and Drug Administration (FDA). Many of these use the following pregnancy categories: A, B, C, D and X. Category A refers to drugs without reported adverse effects in humans based on conducted studies, B refers to no effect in humans with adverse effects seen in animals or no effects in animals without human data, C refers to adverse effects in animals without human data or no data available for both animals and humans, D is adverse effects demonstrated in humans or adverse effects in animals with strong mechanistic expectation of effects in humans, and X referring to drugs that exert adverse effects in humans or animals without indication for use during pregnancy. Importantly, the safety profile of a drug should always be balanced by its potential benefits.

5.1.1 Nitrous Oxide

Historically, nitrous oxide has been a popular medication for labour analgesia since the late 1800s due to its ease of use and widespread availability. Nitrous oxide is an inorganic gas administered with oxygen, the mixture otherwise known as Entonox, via the inhalational route and can be self-administered safely using a breathing system with a demand valve fitted to prevent the parturient from 'overdosing' on the agent. Its mechanism of action is postulated to involve N-methyl D-aspartate (NMDA) antagonism, pain perception modulation via the alpha-2 receptors in the dorsal horn of the spinal cord and the release of endogenous opioids in the central nervous system, hence reducing pain sensitivity and providing analgesia. The onset of action occurs within 60 s after administration and is rapidly washed out leading to a quick offset. The efficacy of nitrous oxide for labour describes improved (but not complete) pain relief for many (Klomp et al. 2012), but this effect is at the expense of common side effects including nausea, vomiting, dizziness and dysphoria which can limit its use.

5.1.2 Opioids

Opioids characteristically act on central G-protein mu opioid receptors to exert their antinociceptive effects and can be administered via the oral, intravenous (IV) or intramuscular (IM) routes, as well as administered directly around the neural tissues using central neuraxial or locoregional techniques. Each opioid exhibits a unique pharmacokinetic and pharmacodynamic profile, giving rise to different speeds of onset, durations of action and adverse effects. These characteristics, along with availability, often dictate institutional or physician preference. A recent systematic review found that parenteral opioids were associated with moderate satisfaction scores, with pain relief and maternal satisfaction decreasing after only 1–2 h (Smith et al. 2018b). This effect should be balanced by the potential dose-dependent adverse effects of nausea, vomiting, drowsiness and respiratory depression, as well as the risk:benefit profile of the alternative options for labour pain management.

The classes of opioids range from naturally occurring compounds such as morphine and diamorphine to synthetic agents such as oxycodone, codeine, fentanyl, pethidine, buprenorphine and methadone, with availability being subject to geopolitical variations across the globe. The use of parenteral pethidine (1 mg kg^{-1} IM), also known as meperidine, and morphine (2.5–5 mg IV or 5–10 mg IM) in labour are perhaps the most extensively used and studied opioids worldwide; however, the use of remifentanil patient-controlled analgesia (PCA) has been gaining popularity in some regions. PCA involves a self-administered dose via a push-button with a 'lock-out' period designed to only allow additional doses to be administered after the drug has reached its peak effect. Due to its metabolism by non-specific plasma and tissue esterases, remifentanil is an ultra-short-acting opioid, making it a safer alternative than other opioids for PCA use, such as fentanyl, which is commonly used in the postoperative setting. It has a fast onset of action, approximately between 30 and 60 s and reaches peak effect between 2 and 3 min (Babenco et al. 2000). A commonly administered dose is between 20 and 30 μg with a lockout period of 2–3 min. While remifentanil crosses the placenta, ultra-fast metabolism also takes place in the foetus, reducing the risk of neonatal adverse effects such as postnatal respiratory depression. However, continuous monitoring of maternal oxygen saturation and respiratory rate is recommended as case reports describe incidents of maternal respiratory and cardiac arrest related to its use in labour (Muchatuta and Kinsella 2013; Logtenberg et al. 2019). A trial comparing remifentanil PCA to 100 mg of IM pethidine in labour found that supplemental oxygenation was required in over 40% of remifentanil PCA cases, versus only 1% of those receiving IM pethidine (Wilson et al. 2018). Of note, however, was that the remifentanil PCA group were significantly less likely to subsequently request epidural analgesia, highlighting the importance of balancing potential side effects with the efficacy profile of each option.

Pethidine is approximately thirty times more lipid soluble than morphine and is metabolised to the active metabolite, norpethidine. It crosses the placenta readily by passive diffusion and accumulates in the acidaemic foetal blood due to ion-trapping. As the concentration in the foetus peaks 2–3 h after maternal dosing, pethidine is ideally administered four or more hours before delivery to avoid peak accumulation of the drug in the foetus at birth, as this may adversely affect neonatal behaviour and breastfeeding (Caldwell et al. 1978; Kuhnert et al. 1985; Nissen et al. 1997). Norpethidine, which also has convulsant properties and is contraindicated in parturients with severe pre-eclampsia and epilepsy, is not reversed by naloxone and could potentially increase the risk of a serotonergic crisis, neurotoxicity, and significant adverse drug interactions.

Morphine was a popular labour analgesic option in the 1920–30s when it was used with hyoscine to induce 'twilight sleep' (Harries and Collis 2020). Compared to most other opioids, it is longer acting with a duration of action lasting between 3 and 6 h when administered intravenously. Morphine is metabolised to morphine-6-glucuronide, an opioid agonist and active metabolite, and morphine-3-glucuronide, an opioid antagonist, in varying quantities. These metabolites do not have the same pro-convulsant activity as that of pethidine. Although it crosses the placenta rapidly, a relatively smaller amount is detected in the foetal circulation as it diffuses back into the maternal circulation passively due to rapid maternal elimination, creating a diffusion gradient. Despite this movement out of the placenta, systemically administered morphine has been associated with both maternal and neonatal side effects at the higher, more efficacious doses for labour analgesia, and as such, has fallen out of favour in many centres where superior analgesic options, such as epidural analgesia, are available (Olofsson et al. 1996; Ullman et al. 2010).

5.1.3 Simple Analgesics

The literature regarding the efficacy of non-opioid analgesics such as paracetamol and non-steroid anti-inflammatory drugs (NSAIDs) is limited. Nevertheless, the evidence presented by smaller studies has demonstrated the usefulness of paracetamol (or acetaminophen) as a simple, safe, inexpensive and widely available agent that could be used as an adjunct to labour analgesia (Abd-El-Maeboud et al. 2014; Anter et al. 2022), especially in early labour. NSAIDs are not used due to the theoretical risk of premature closure of the ductus arteriosus in the foetus (Koren et al. 2006). The use of NSAIDs has been extensively studied in a variety of parturient farmed animal species, principally because of their availability, legality and affordability, rather than efficacy.

5.1.4 Adjuncts

The use of other adjunctive medications for labour analgesia has also been described in medical literature, but overall these approaches lack strong evidence to support their routine use, in part due to the foetal effects resulting from placental drug transfer (Othman et al. 2012).

5.2 Regional Analgesic Techniques

A separate mode of labour analgesia, which involves the direct deposition of pharmacological agents around neural tissues, can be achieved via lumbar epidural, spinal, combined spinal and epidural, or peripheral nerve blockades. Local anaesthetic agents, with or without adjuncts, are the primary agent administered for these techniques, acting to directly inhibit the action potential signal transmission along the nerve axon, theoretically offering complete attenuation of pain in the distribution covered.

5.2.1 Epidural Analgesia

Epidural analgesia is a common technique employed for managing labour pain in the human parturient and has been shown to provide unparalleled and consistently effective labour analgesia compared to other options (Jones et al. 2012; Schug et al. 2016). Epidural analgesia provides physiological benefits through minimisation of the sensory component of labour thus promoting better maternal rest, limiting excessive release of catecholamines and prostaglandins during painful uterine contractions and cervical dilatation, and facilitating more coordinated uterine contractions (Harries and Collis 2020). It is also known to minimise the excessive oxygen consumption associated with labour pain and maternal effort and can reduce maternal and foetal acidosis through the enhancement of placental and intervillous blood flow by arteriolar vasodilatation (Harries and Collis 2020).

In assessing a parturient for the suitability of epidural analgesia, the indications and contraindications should be thoroughly considered. Some of the specific indications for labour epidural include maternal request, anticipated exaggerated painful response due to augmentation of labour or abnormal foetal lie, hypertensive disease or pre-eclampsia, cardiorespiratory disease and cerebrovascular diseases. Since an epidural used for labour analgesia can also later be employed for anaesthesia, for example, for unplanned caesarean section, other indications can include those, which obviate the avoidance of general anaesthesia, such as morbid obesity, predicted difficult airway or history of anaesthetic complications such as malignant hyperthermia. Absolute contraindications to performing an epidural are patient refusal, allergy to drugs or material used in the siting of an epidural, maternal sepsis, raised intracranial pressure, severe clotting disorders, a lack of appropriately trained staff or equipment and the presence of an infective process in proximity to the proposed injection site. Some medical conditions, such as hypovolaemia with ongoing haemorrhage, mild thrombocytopaenia, raised inflammatory markers, specific anatomical anomalies, cardiovascular diseases and active neurological conditions are considered relative contraindications of epidural analgesia.

Access to the epidural space is gained with the aid of an epidural needle (Fig. 3), most commonly using a loss of resistance to saline or air technique (Fig. 4.1). The epidural catheter is inserted and local anaesthetic drugs, with or without adjuncts, are deposited via this conduit into the epidural space (Fig. 4.2). The volume of drug administered spreads within the potential space of the epidural cavity in directions that can be difficult to determine precisely and would often depend on the parturient body position, allowing gravity to partially carry the medication towards dependent areas. The local anaesthetic works directly on nerve roots at the levels above and below the site of insertion with minimal direct effect on the spinal cord, aiming to provide an effective sensory blockade without significant motor involvement. Epidural analgesia is maintained for the duration of the labour via a programmed pump delivery of a dilute local anaesthetic agent. A common regime used at the author's institution includes the use of a 0.0625% (or 0.625 mg mL^{-1}) bupivacaine with 2.5 μg mL^{-1} fentanyl premix solution administered as a 10 mL bolus every hour and the ability for the parturient to self-administer an additional 5 mL boluses every 15 min to cover any breakthrough pain experienced during labour.

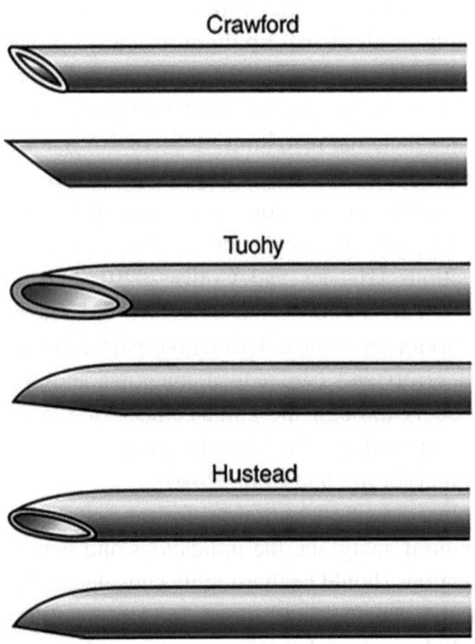

Fig. 3 A pictorial display of different epidural needles highlighting differences in the design. (Picture reproduced with permission Springer Nature (License number 5601250794585))

As mentioned, a labour epidural can also be used as a modality for surgical anaesthesia in the event of a caesarean delivery, or for postpartum analgesia, for example, using pethidine patient-controlled epidural analgesia.

The assessment of epidural analgesia block efficacy is described in greater detail in the previous section.

5.2.2 Spinal (Subarachnoid) Analgesia

In comparison to epidural analgesia, a spinal block achieves analgesia more rapidly (commonly 1–2 min versus 10–20 min) using a significantly smaller volume of local anaesthetic, with or without adjuncts such as fentanyl or clonidine. Using a much smaller gauge needle, these drugs are injected directly into the intrathecal or subarachnoid space after breaching the dura mater, the meningeal layer encasing the spinal cord and cerebrospinal fluid (Fig. 5). An epidural injection, in contrast, does not breach

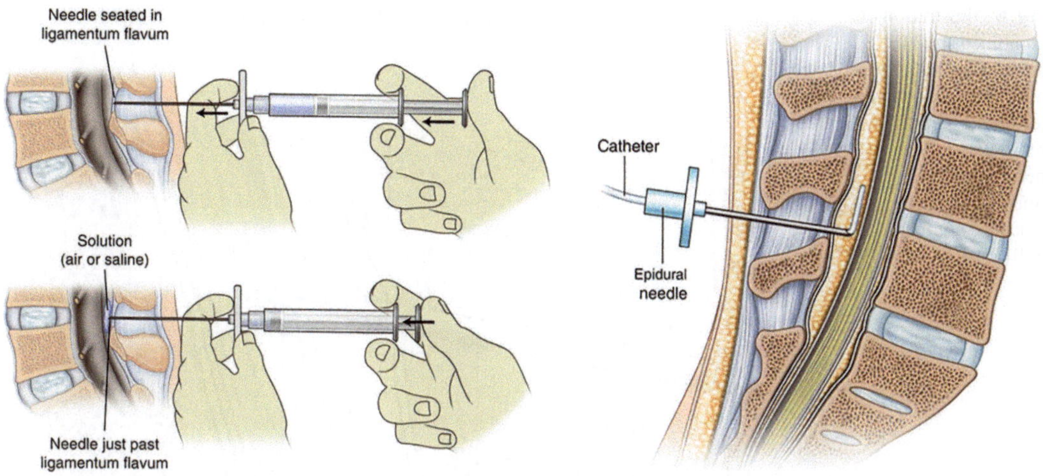

Fig. 4 Illustration demonstrating access to the epidural space is achieved using the epidural needle attached to a loss-of-resistance syringe (Fig. 4.1) and the insertion of an epidural catheter via the epidural needle once the epidural space is found (Fig. 4.2). The epidural space is a potential space located between the ligamentum flavum and dura mater. (Picture reproduced with permission from Springer Nature (License number 5601250794585))

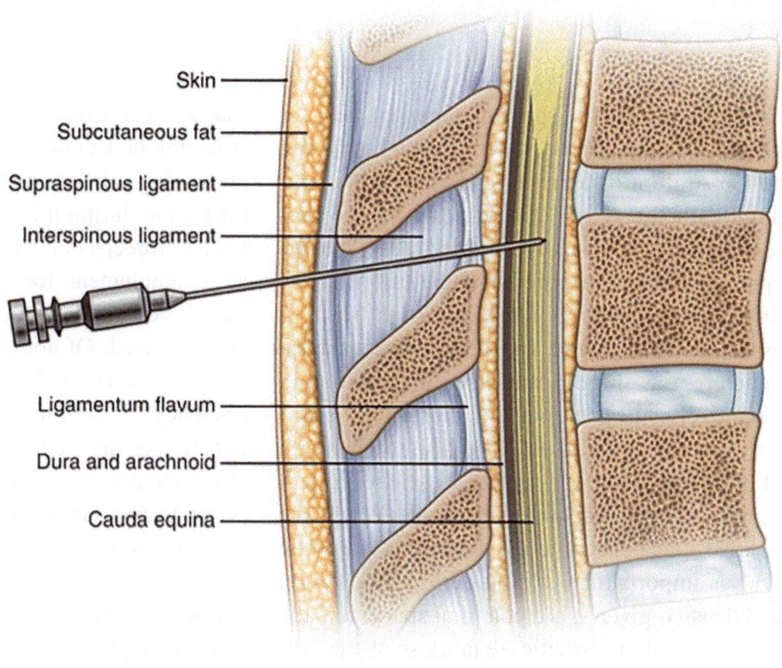

Fig. 5 Illustration of a spinal needle crossing all the ligamentous structures with its tip lodged within the intrathecal or subarachnoid space near the spinal cord. (Picture reproduced with permission from Springer Nature (License number 5601250794585))

the dura (Fig. 4). The chosen drugs spread within the cerebrospinal fluid and act directly on the ascending and descending tracts of the spinal cord via diffusion. When higher doses of local anaesthetic are used, such as that required for surgical anaesthesia, dense sensory (touch, temperature, pain and proprioception) and motor block results. Where spinal blockade is employed in the setting of labour pain, it is commonly performed using a combined spinal-epidural (CSE) needle-through-needle technique (Fig. 6), allowing for pain relief to be achieved

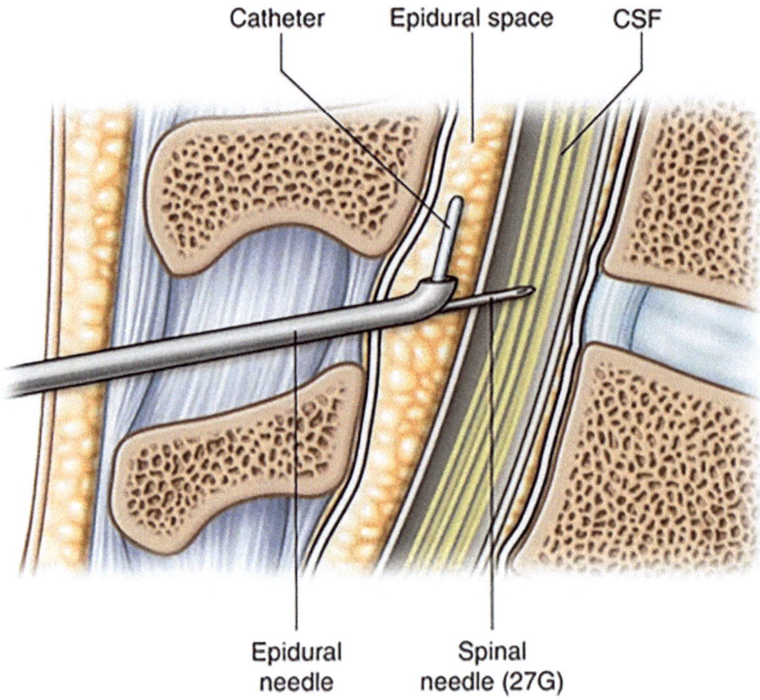

Fig. 6 Illustration of a combined spinal epidural using a needle-through-needle technique. (Picture reproduced with permission from Springer Nature (License number 5601250794585))

more quickly via the spinal component, and for prolonged pain relief achieved through ongoing administration of local anaesthetic via the epidural catheter. A spinal block can also be performed alone, usually immediately before epidural catheter insertion in situations whereby the latter insertion is likely to be difficult or associated with a high risk of inadvertent dural puncture, e.g. if the parturient is unable to hold a relatively still position for its insertion. A spinal dose of approximately one fifth to one tenth of that required for surgical anaesthesia will usually achieve adequate analgesia. While the volume and concentration of local anaesthetic drug is important for the epidural block height and density, given it is a potential space, it is the dose of local anaesthetic, regardless of the volume or concentration that is more important in achieving a spinal block. As such, the same local anaesthetic solution that is administered for epidural analgesia can also be used for the spinal analgesic component.

5.2.3 Pudendal Nerve Block

Bilateral pudendal nerve blocks are useful for alleviating pain arising from vaginal and perineal distension during the second stage of labour and for low forceps delivery. This block may be used as a supplement for epidural analgesia if the sacral nerves (S2 to S4) are not sufficiently anaesthetised. Of note, anterior perineum sensation is not abolished as this region is supplied by branches of the ilioinguinal and genitofemoral nerves. As the pudendal nerve block does not affect the cervix or upper vagina, it is not appropriate to provide analgesic coverage for uterine manipulation and in mid-forceps delivery.

5.2.4 Paracervical Block

This block provides some relief from the pain of uterine contractions and cervical dilatation or manipulation during the active phase of labour. It is however not effective during the second stage of labour because the sensory nerves from the perineum are not blocked. Prior to the 1970s, the paracervical block was commonly performed for labour analgesia but owing to the high incidence of foetal bradycardia and the superior safety profile of other techniques, it is no longer popular.

5.3 Non-pharmacological Approaches

Unlike the pharmacological interventions, non-pharmacological strategies are aimed at helping the labouring parturient cope with labour pain and provide a sense of control over the birthing process. Over the course of the twentieth century, numerous interventions have been trialled and experimented on the labouring parturient, most of which are supported by low to moderate level evidence, suggesting that more ongoing robust clinical trials would assist in further determining their efficacy (Jones et al. 2012; Schug et al. 2016). Despite limited evidence advocating their regular use, many interventions are still performed as they are deemed low-risk, inexpensive, simple to administer and readily available.

Parturients may benefit from having a support person such as a birthing partner, midwife, or doula to create a calm birthing environment, offering physical touch and ongoing psychological encouragement to allay anxiety (Tarkka and Paunonen 1996; Taheri et al. 2018). Simple distraction techniques such as breathing and relaxation exercises (Smith et al. 2018a), water immersion (Cluett et al. 2018), music or audiotherapy (Smith et al. 2018a) and aromatherapy (Smith et al. 2011) are also thought to be helpful in improving the labouring experience. A few studies have explored the engagement of a therapist to equip the parturient with effective coping strategies through goal setting, hypnosis (Madden et al. 2016), mindfulness exercises, cognitive therapy, and biofeedback therapy (Barragan Loayza et al. 2011) such as eye movement desensitisation and reprocessing (EMDR). Overall, studies employing the above measures as described have yielded inconclusive findings.

The muscular pain experienced during labour can partially be alleviated by the direct application of a heat or cool pad onto specific sites of the body (Smith et al. 2018a). This treatment works via stimulation of the skin thermal receptors, which activate wide dynamic range (WDR) neurons located within the spinal cord, thus modulating nociceptive transmission. The same mechanism is also applied in the use of the transcutaneous electrical nerve stimulation (TENS), whereby small pulsatile electrical currents are administered to the skin via a low-impedance gel pad to stimulate nociceptors and modulate the spinal cord signal transmissions. Several meta-analyses, however, have not found an appreciable effect on labour pain (Dowswell et al. 2009). In addition, tissue massage (Smith et al. 2018a), body movements and yoga exercises promote improved joint flexibility and relieve muscle tensions and spasms likely to develop during labour. Adopting certain postures with the use of adjuncts such as the 'birth ball' (Makvandi et al. 2015; Delgado et al. 2019), sacroiliac belts or uterine support pillows are other low-risk alternatives.

Semi-invasive interventions have also been explored. Acupuncture, which has a more than a thousand-year history of use in Asia, is a form of alternative medicine involving the insertion of fine needles at different locations and depths along the human body to regulate the body's "energy" (or meridian), improve specific organ functions and relieve pain. A Cochrane review found that even with the limited low-certainty evidence, targeted acupuncture may increase satisfaction with pain management and reduce the use of pharmacological analgesia (Smith et al. 2020). Sterile water injection is a more novel technique, which involves the administration of intracutaneous injections of 0.5 mL sterile water in the lumbosacral region. Although evidence supporting its widespread use is currently lacking (Derry et al. 2012), further research is currently underway (Lee et al. 2022).

5.4 Pain in Animals During Parturition

In the acute setting of parturition, pain can be especially difficult to identify, assess and prevent or treat. Nevertheless, in many cases, pending parturition may be expected and extra surveillance, and preparation to treat pain in these animals is prudent. Pain associated with parturition is poorly defined in animals in general, but there is adequate evidence, not just by extrapolation across species from humans to animals, but in animals directly, that parturition may be painful (Scott 2005; Mainau and Manteca 2011; Proudfoot 2019). There are circumstances where pain may be more likely, and these include prolonged labour; large foetuses; pre-existing osteoarthritis; and prior surgery such as caesarean or vulval surgery.

Most research on the impact of parturition in animals focuses on the endocrine changes associated with gestation, parturition and lactation (Aurich et al. 1990). Endogenous opioids are known to increase during parturition in cattle (Aurich et al. 1990), suggesting that physiologically, parturition is painful. In sows, nociceptive thresholds increase during late pregnancy, again, representing an endogenous defence against labour pain (Jarvis et al. 1997). However, cortisol, ACTH and beta-endorphin concentrations did not change during piglet expulsion, suggesting that expulsion of the piglet foetus is not painful (Jarvis et al. 1999).

More subjectively, cattle vets rank dystocia as one of the most painful conditions experienced by cattle (Huxley and Whay 2006). For sheep, it is similarly acknowledged that parturition is likely to be associated with pain that should be treated (Scott 2005). However, there is little research into the safety and efficacy of analgesic strategies for the periparturient period in production animals. Except caesarean procedures in cattle and sheep, analgesia is rarely administered (Scott 2005).

References

Abd-El-Maeboud KH, Elbohoty AE, Mohammed WE, Elgamel HM, Ali WA (2014) Intravenous infusion of paracetamol for intrapartum analgesia. J Obstet Gynaecol Res 40(11):2152–2157

ACOG (2019) American College of Obstetricians and Gynecologists' committee on practice bulletins—obstetrics. ACOG practice bulletin no. 209: obstetric analgesia and anesthesia. Obstet Gynecol 133(3)

Anim-Somuah M, Smyth RM, Cyna AM, Cuthbert A (2018) Epidural versus non-epidural or no analgesia for pain management in labour. Cochrane Database Syst Rev 5(5):CD000331

Anter ME, Abdel Attey Saleh S, Shawkey Allam S, Mohamed Nofal A (2022) Efficacy and safety of intravenous paracetamol in management of labour pains in a low resource setting: a randomized clinical trial. J Matern Fetal Neonatal Med 35(25):6320–6328

Aurich JE, Dobrinski I, Hoppen HO, Grunert E (1990) Beta-endorphin and met-enkephalin in plasma of cattle during pregnancy, parturition and the neonatal period. J Reprod Fertil 89(2):605–612

Babenco HD, Conard PF, Gross JB (2000) The pharmacodynamic effect of a remifentanil bolus on ventilatory control. Anesthesiology 92(2):393–398

Barnett R (2005) A horse named 'Twilight Sleep': the language of obstetric anaesthesia in 20th century Britain. Int J Obstet Anesth 14(4):310–315

Barragan Loayza IM, Sola I, Juando Prats C (2011) Biofeedback for pain management during labour. Cochrane Database Syst Rev (6):CD006168

Bergmans MG, van Geijn HP, Hasaart TH, Weber T, Nickelsen C (1996) Fetal and maternal transcutaneous PCO2 levels during labour and the influence of epidural analgesia. Eur J Obstet Gynecol Reprod Biol 67(2):127–132

Billewicz-Driemel AM, Milne MD (1976) Long-term assessment of extradural analgesia for the relief of pain in labour. II: sense of "deprivation" after extradural analgesia in labour: relevant or not? Br J Anaesth 48(2):139–144

Bisson DL, Newell SD, Laxton C, O. Royal College of and Gynaecologists (2019) Antenatal and postnatal analgesia: scientific impact paper no. 59. BJOG 126(4):e114–e124

Breen TW, Shapiro T, Glass B, Foster-Payne D, Oriol NE (1993) Epidural anesthesia for labor in an ambulatory patient. Anesth Analg 77(5):919–924

Brewin C, Bradley C (1982) Perceived control and the experience of childbirth. Br J Clin Psychol 21(Pt 4):263–269

Bromage PR (1965) A comparison of the hydrochloride salts of lignocaine and prilocaine for epidural analgesia. Br J Anaesth 37(10):753–761

Brownridge P (1995) The nature and consequences of childbirth pain. Eur J Obstet Gynecol Reprod Biol 59(Suppl):S9–S15

Caldwell J, Wakile LA, Notarianni LJ, Smith RL, Correy GJ, Lieberman BA, Beard RW, Finnie MD, Snedden W (1978) Maternal and neonatal disposition of pethidine in childbirth–a study using quantitative gas chromatography-mass spectrometry. Life Sci 22(7):589–596

Cascio M, Pygon B, Bernett C, Ramanathan S (1997) Labour analgesia with intrathecal fentanyl decreases maternal stress. Can J Anaesth 44(6):605–609

Cheng YW, Shaffer BL, Nicholson JM, Caughey AB (2014) Second stage of labor and epidural use: a larger effect than previously suggested. Obstet Gynecol 123(3):527–535

Cluett ER, Burns E, Cuthbert A (2018) Immersion in water during labour and birth. Cochrane Database Syst Rev 5(5):CD000111

Conell-Price J, Evans JB, Hong D, Shafer S, Flood P (2008) The development and validation of a dynamic model to account for the progress of labor in the assessment of pain. Anesth Analg 106(5):1509–1515. table of contents

Debiec J, Conell-Price J, Evansmith J, Shafer SL, Flood P (2009) Mathematical modeling of the pain and progress of the first stage of nulliparous labor. Anesthesiology 111(5):1093–1110

Delgado A, Maia T, Melo RS, Lemos A (2019) Birth ball use for women in labor: a systematic review and meta-analysis. Complement Ther Clin Pract 35:92–101

Derry S, Straube S, Moore RA, Hancock H, Collins SL (2012) Intracutaneous or subcutaneous sterile water injection compared with blinded controls for pain management in labour. Cochrane Database Syst Rev 1:CD009107

Dick-Read G (1943) Childbirth without fear. Originally published as: Revelation of Childbirth. Heinemann, London

Ding T, Wang DX, Qu Y, Chen Q, Zhu SN (2014) Epidural labor analgesia is associated with a decreased risk of postpartum depression: a prospective cohort study. Anesth Analg 119(2):383–392

Dowswell T, Bedwell C, Lavender T, Neilson JP (2009) Transcutaneous electrical nerve stimulation (TENS) for pain relief in labour. Cochrane Database Syst Rev (2):CD007214

Eisenach JC, Pan PH, Smiley R, Lavand'homme P, Landau R, Houle TT (2008) Severity of acute pain after childbirth, but not type of delivery, predicts persistent pain and postpartum depression. Pain 140(1):87–94

Eisenach JC, Pan P, Smiley RM, Lavand'homme P, Landau R, Houle TT (2013) Resolution of pain after childbirth. Anesthesiology 118(1):143–151

Gibbins J, Thomson AM (2001) Women's expectations and experiences of childbirth. Midwifery 17(4):302–313

Goodfellow CF, Hull MG, Swaab DF, Dogterom J, Buijs RM (1983) Oxytocin deficiency at delivery with epidural analgesia. Br J Obstet Gynaecol 90(3):214–219

Guan Z, Hellman J, Schumacher M (2016) Contemporary views on inflammatory pain mechanisms: TRPing over innate and microglial pathways. F1000Res 5:2425

Hagerdal M, Morgan CW, Sumner AE, Gutsche BB (1983) Minute ventilation and oxygen consumption during labor with epidural analgesia. Anesthesiology 59(5):425–427

Hardy JD, Javert CT (1949) Studies on pain: measurements of pain intensity in childbirth. J Clin Invest 28(1):153–162

Harries S, Collis R (2020) Chapter 9: Regional analgesia for labour. In: Collis R (ed) Obstetric anaesthesia, 2nd edn. Oxford University Press, pp P237–P272

Hiltunen P, Raudaskoski T, Ebeling H, Moilanen I (2004) Does pain relief during delivery decrease the risk of postnatal depression? Acta Obstet Gynecol Scand 83(3):257–261

Hood DD, Dewan DM, James FM 3rd (1986) Maternal and fetal effects of epinephrine in gravid ewes. Anesthesiology 64(5):610–613

Howard ED (2018) Promoting maternal confidence, coping, and comfort in latent labor: evidence-based strategies and resources. J Perinat Neonatal Nurs 32(4):291–294

Hrešanová E (2016) The psychoprophylactic method of painless childbirth in socialist Czechoslovakia: from state propaganda to activism of enthusiasts. Med Hist 60(4):534–556

Huxley JN, Whay HR (2006) Current attitudes of cattle practitioners to pain and the use of analgesics in cattle. Vet Rec 159(20):662–668

Jarvis S, McLean KA, Chirnside J, Deans LA, Calvert SK, Molony V, Lawrence AB (1997) Opioid-mediated changes in nociceptive threshold during pregnancy and parturition in the sow. Pain 72(1–2):153–159

Jarvis S, Lawrence AB, McLean KA, Deans LA, Chirnside J, Calvert SK (1999) The effect of piglet expulsion in the sow on plasma cortisol, adrenocorticotropic hormone and β-endorphin. Reprod Domest Anim 34(2):89–94

Javert CT, Hardy JD (1951) Influence of analgesic on pain intensity during labor; with a note on natural childbirth. Anesthesiology 12(2):189–215

Jones L, Othman M, Dowswell T, Alfirevic Z, Gates S, Newburn M, Jordan S, Lavender T, Neilson JP (2012) Pain management for women in labour: an overview of systematic reviews. Cochrane Database Syst Rev 2012(3):CD009234

Jones LE, Whitburn LY, Davey MA, Small R (2015) Assessment of pain associated with childbirth: Women's perspectives, preferences and solutions. Midwifery 31(7):708–712. https://doi.org/10.1016/j.midw.2015.03.012. PMID:25886966

Kehlet H, Jensen TS, Woolf CJ (2006) Persistent post-surgical pain: risk factors and prevention. Lancet 367(9522):1618–1625

Klomp T, van Poppel M, Jones L, Lazet J, Di Nisio M, Lagro-Janssen AL (2012) Inhaled analgesia for pain management in labour. Cochrane Database Syst Rev (9):CD009351

Koren G, Florescu A, Costei AM, Boskovic R, Moretti ME (2006) Nonsteroidal antiinflammatory drugs during third trimester and the risk of premature clo-

sure of the ductus arteriosus: a meta-analysis. Ann Pharmacother 40(5):824–829
Krans EE, Cochran G, Bogen DL (2015) Caring for opioid-dependent pregnant women: prenatal and postpartum care considerations. Clin Obstet Gynecol 58(2):370–379
Kuhnert BR, Linn PL, Kennard MJ, Kuhnert PM (1985) Effects of low doses of meperidine on neonatal behavior. Anesth Analg 64(3):335–342
Lang AJ, Sorrell JT, Rodgers CS, Lebeck MM (2006) Anxiety sensitivity as a predictor of labor pain. Eur J Pain 10(3):263–270
Lee N, Gao Y, Martensson LB, Callaway L, Barnett B, Kildea S (2022) Sterile water injections for relief of labour pain (the SATURN trial): study protocol for a randomised controlled trial. Trials 23(1):155
Logtenberg SLM, Vink ML, Godfried MB, Beenakkers ICM, Schellevis FG, Mol BW, Verhoeven CJ (2019) Serious adverse events attributed to remifentanil patient-controlled analgesia during labour in The Netherlands. Int J Obstet Anesth 39:22–28
Lurie S (2004) Euphemia Maclean, Agnes Sampson and pain relief during labour in 16th century Edinburgh. Anaesthesia 59(8):834–835
Madden K, Middleton P, Cyna AM, Matthewson M, Jones L (2016) Hypnosis for pain management during labour and childbirth. Cochrane Database Syst Rev 2016(5):CD009356
Mainau E, Manteca X (2011) Pain and discomfort caused by parturition in cows and sows. Appl Anim Behav Sci 135(3):241–251
Makvandi S, Latifnejad Roudsari R, Sadeghi R, Karimi L (2015) Effect of birth ball on labor pain relief: a systematic review and meta-analysis. J Obstet Gynaecol Res 41(11):1679–1686
Melzack R (1984) The myth of painless childbirth (the John J. Bonica lecture). Pain 19(4):321–337
Melzack R (2005) Evolution of the neuromatrix theory of pain. The Prithvi Raj Lecture: presented at the third world congress of World Institute of Pain, Barcelona 2004. Pain Pract 5(2):85–94
Melzack R, Wall PD (1965) Pain mechanisms: a new theory. Science 150(3699):971–979
Mestrovic AH, Bilic M, Loncar LB, Mickovic V, Loncar Z (2015) Psychological factors in experience of pain during childbirth. Coll Antropol 39(3):557–565
Minnich ME, Brown M, Clark RB, Miller FC, Thompson DS (1990) Oxygen desaturation in women in labor. J Reprod Med 35(7):693–696
Morishima HO, Yeh MN, James LS (1979) Reduced uterine blood flow and fetal hypoxemia with acute maternal stress: experimental observation in the pregnant baboon. Am J Obstet Gynecol 134(3):270–275
Muchatuta NA, Kinsella SM (2013) Remifentanil for labour analgesia: time to draw breath? Anaesthesia 68(3):231–235
Navarro-Prado S, Sanchez-Ojeda MA, Marmolejo-Martin J, Kapravelou G, Fernandez-Gomez E, Martin-Salvador A (2022) Cultural influence on the expression of labour-associated pain. BMC Pregnancy Childbirth 22(1):836
Nissen E, Widstrom AM, Lilja G, Matthiesen AS, Uvnas-Moberg K, Jacobsson G, Boreus LO (1997) Effects of routinely given pethidine during labour on infants' developing breastfeeding behaviour. Effects of dose-delivery time interval and various concentrations of pethidine/norpethidine in cord plasma. Acta Paediatr 86(2):201–208
Olofsson C, Ekblom A, Ekman-Ordeberg G, Hjelm A, Irestedt L (1996) Lack of analgesic effect of systemically administered morphine or pethidine on labour pain. Br J Obstet Gynaecol 103(10):968–972
Orbach-Zinger S, Landau R, Harousch AB, Ovad O, Caspi L, Kornilov E, Ioscovich A, Bracco D, Davis A, Fireman S, Hoshen M, Eidelman LA (2018) The relationship between women's intention to request a labor epidural analgesia, actually delivering with labor epidural analgesia, and postpartum depression at 6 weeks: a prospective observational study. Anesth Analg 126(5):1590–1597
Orbach-Zinger S, Heesen M, Grigoriadis S, Heesen P, Halpern S (2021) A systematic review of the association between postpartum depression and neuraxial labor analgesia. Int J Obstet Anesth 45:142–149
Othman M, Jones L, Neilson JP (2012) Non-opioid drugs for pain management in labour. Cochrane Database Syst Rev (7):CD009223
Proudfoot KL (2019) Maternal behavior and design of the maternity pen. Vet Clin N Am Food Anim Pract 35(1):111–124
Reynolds F, Sharma SK, Seed PT (2002) Analgesia in labour and fetal acid-base balance: a meta-analysis comparing epidural with systemic opioid analgesia. BJOG 109(12):1344–1353
Robinson DA, Wei F, Wang GD, Li P, Kim SJ, Vogt SK, Muglia LJ, Zhuo M (2002) Oxytocin mediates stress-induced analgesia in adult mice. J Physiol 540(Pt 2):593–606
Rooth G (1982) Maternal hypoxaemia during labor. Padiatr Padol 17(2):231–236
Rothman BK (1989) Motherhood: beyond patriarchy. Nova Law Rev 13(2):481–486
Schug SA, Palmer GM, Scott DA, Halliwell R, Trinca J (2016) Acute pain management: scientific evidence, fourth edition, 2015. Med J Aust 204(8):315–317
Scott PR (2005) The management and welfare of some common ovine obstetrical problems in the United Kingdom. Vet J 170(1):33–40
Shnider SM, Abboud TK, Artal R, Henriksen EH, Stefani SJ, Levinson G (1983) Maternal catecholamines decrease during labor after lumbar epidural anesthesia. Am J Obstet Gynecol 147(1):13–15
Shnol H, Paul N, Belfer I (2014) Labor pain mechanisms. Int Anesthesiol Clin 52(3):1–17
Siyoum M, Mekonnen S (2019) Labor pain control and associated factors among women who gave birth at Leku primary hospital, Southern Ethiopia. BMC Res Notes 12(1):619

Skowronski GA (2015) Pain relief in childbirth: changing historical and feminist perspectives. Anaesth Intensive Care 43(Suppl):25–28

Smith CA, Collins CT, Crowther CA (2011) Aromatherapy for pain management in labour. Cochrane Database Syst Rev (7):CD009215

Smith CA, Levett KM, Collins CT, Armour M, Dahlen HG, Suganuma M (2018a) Relaxation techniques for pain management in labour. Cochrane Database Syst Rev 3(3):CD009514

Smith LA, Burns E, Cuthbert A (2018b) Parenteral opioids for maternal pain management in labour. Cochrane Database Syst Rev 6(6):CD007396

Smith CA, Collins CT, Levett KM, Armour M, Dahlen HG, Tan AL, Mesgarpour B (2020) Acupuncture or acupressure for pain management during labour. Cochrane Database Syst Rev 2(2):CD009232

Soens MA, He J, Bateman BT (2019) Anesthesia considerations and post-operative pain management in pregnant women with chronic opioid use. Semin Perinatol 43(3):149–161

Soet JE, Brack GA, DiIorio C (2003) Prevalence and predictors of women's experience of psychological trauma during childbirth. Birth 30(1):36–46

Spiby H, Slade P, Escott D, Henderson B, Fraser RB (2003) Selected coping strategies in labor: an investigation of women's experiences. Birth 30(3):189–194

Stewart DE (1982) Psychiatric symptoms following attempted natural childbirth. Can Med Assoc J 127(8):713–716

Suhitharan T, Pham TP, Chen H, Assam PN, Sultana R, Han NL, Tan EC, Sng BL (2016) Investigating analgesic and psychological factors associated with risk of postpartum depression development: a case-control study. Neuropsychiatr Dis Treat 12:1333–1339

Sultan P, Murphy C, Halpern S, Carvalho B (2013) The effect of low concentrations versus high concentrations of local anesthetics for labour analgesia on obstetric and anesthetic outcomes: a meta-analysis. Can J Anaesth 60(9):840–854

Taheri M, Takian A, Taghizadeh Z, Jafari N, Sarafraz N (2018) Creating a positive perception of childbirth experience: systematic review and meta-analysis of prenatal and intrapartum interventions. Reprod Health 15(1):73

Tarkka MT, Paunonen M (1996) Social support and its impact on mothers' experiences of childbirth. J Adv Nurs 23(1):70–75

Tomimatsu T, Kakigano A, Mimura K, Kanayama T, Koyama S, Fujita S, Taniguchi Y, Kanagawa T, Kimura T (2013) Maternal carbon dioxide level during labor and its possible effect on fetal cerebral oxygenation: mini review. J Obstet Gynaecol Res 39(1):1–6

Turnbull JH, Aleshi P (2015) Spinal and epidural anesthesia. In: Sikka P, Beaman S, Street J (eds) Basic clinical anesthesia. Springer, New York. https://doi.org/10.1007/978-1-4939-1737-2_21

Ullman R, Smith LA, Burns E, Mori R, Dowswell T (2010) Parenteral opioids for maternal pain relief in labour. Cochrane Database Syst Rev (9):CD007396

Vogl SE, Worda C, Egarter C, Bieglmayer C, Szekeres T, Huber J, Husslein P (2006) Mode of delivery is associated with maternal and fetal endocrine stress response. BJOG 113(4):441–445

Waldenstrom U, Borg IM, Olsson B, Skold M, Wall S (1996) The childbirth experience: a study of 295 new mothers. Birth 23(3):144–153

Walter MH, Abele H, Plappert CF (2021) The role of oxytocin and the effect of stress during childbirth: neurobiological basics and implications for mother and child. Front Endocrinol (Lausanne) 12:742236

Wang F, Shen X, Guo X, Peng Y, Gu X, G. Labor Analgesia Examining (2009) Epidural analgesia in the latent phase of labor and the risk of cesarean delivery: a five-year randomized controlled trial. Anesthesiology 111(4):871–880

Wang TT, Sun S, Huang SQ (2017) Effects of epidural labor analgesia with low concentrations of local anesthetics on obstetric outcomes: a systematic review and meta-analysis of randomized controlled trials. Anesth Analg 124(5):1571–1580

Wang J, Zhao G, Song G, Liu J (2022) Association between neuraxial labor analgesia and postpartum depression: a meta-analysis. J Affect Disord 311:95–102

Wilson MJA, MacArthur C, Hewitt CA, Handley K, Gao F, Beeson L, Daniels J, R. T. C. Group (2018) Intravenous remifentanil patient-controlled analgesia versus intramuscular pethidine for pain relief in labour (RESPITE): an open-label, multicentre, randomised controlled trial. Lancet 392(10148):662–672

Wong DL, Baker CM (1988) Pain in children: comparison of assessment scales. Pediatr Nurs 14(1):9–17

Wong CA, Scavone BM, Peaceman AM, McCarthy RJ, Sullivan JT, Diaz NT, Yaghmour E, Marcus RJ, Sherwani SS, Sproviero MT, Yilmaz M, Patel R, Robles C, Grouper S (2005) The risk of cesarean delivery with neuraxial analgesia given early versus late in labor. N Engl J Med 352(7):655–665

Pain in 4D: How Do Early Life Events Create Susceptibilities to Pain

Charlotte H. Johnston, Samuel Evans, Danila Marini, Daniel T. Barratt, Sanam Mustafa, Gustavo Ferlini Agne, and Mark R. Hutchinson

Abstract

Pain in humans and animals is a complex experience that is influenced by numerous biological, psychological and social factors. It is increasingly recognised that contributors to pain exist across multiple dimensions, including time and space. This chapter examines the concept of four-dimensional contributors to pain and emphasises the multiscale nature of biopsychosocial factors that contribute to chronic pain development in particular.

C. H. Johnston
School of Biomedicine, University of Adelaide, Adelaide, SA, Australia

Davies Livestock Research Centre, University of Adelaide, Adelaide, SA, Australia

S. Evans · D. T. Barratt · S. Mustafa · M. R. Hutchinson (✉)
School of Biomedicine, University of Adelaide, Adelaide, SA, Australia

Davies Livestock Research Centre, University of Adelaide, Adelaide, SA, Australia

Australian Research Council Centre of Excellence for Nanoscale BioPhotonics, University of Adelaide, Adelaide, SA, Australia
e-mail: mark.hutchinson@adelaide.edu.au

D. Marini · G. F. Agne
Davies Livestock Research Centre, University of Adelaide, Adelaide, SA, Australia

School of Veterinary Sciences, University of Adelaide, Adelaide, SA, Australia

At the nanoscale, there are molecular and cellular processes that contribute to chronic pain development across all species. Importantly contributors to pain also exist across multiple timescales, from nanoseconds to years. For example, early life molecular events, stress and trauma can exert long-lasting effects on the developing brain and increase the risk of developing chronic pain later in life. Similarly, social support, coping strategies and expectations can also play important roles in shaping the pain experience.

Understanding the multidimensional nature of pain is critical for developing effective interventions that address the diverse factors that contribute to chronic pain. By taking a multiscale and multidimensional approach to pain, targeted interventions can be developed that address the unique needs of individual patients. Mainly focusing on pain studies conducted in rodents and humans, this chapter will focus on the role of early life events in shaping the experience of pain later in life. Although similarities in the mechanisms underlying pain are expected across species, unique contributors to pain (e.g. prior experience, environmental and genetic influences) are likely to result in nuanced differences. Nevertheless, a convergent understanding of pain across species will rapidly advance knowledge and technology transfer between

R. E. Clutton, G. C. Musk (eds.), *Pain and Suffering in Farmed Animals*,
https://doi.org/10.1007/978-3-032-11407-5_9

human medicine and animal veterinary industries to address the unmet challenges of underdiagnosed and untreated pain.

Keywords

Acute Pain · Chronic pain · Inflammation · Sensitisation · Glia · Epigenetics · Immune system · Plasticity · Descending modulation · Early life events

1 Introduction

Pain is a complex and multifaceted phenomenon that transcends species to impact both human and animal kind. As a subjective, unpleasant sensory and emotional experience, pain involves a range of biological, psychological and social factors, and can have a profound effect on an individual's wellbeing in the presence of actual or perceived tissue damage. Hence, the biopsychosocial construct is the accepted state-of-the-art model that exists today (Raja et al. 2020). Pain can be acute or chronic, and can be influenced by a range of factors, including genetics, environment and prior experiences. Acute pain can serve a critical purpose for the organism, warning of danger, impending tissue injury and illness. However, current understanding suggests chronic pain serves no adaptive purpose and, as such, can be considered a dysfunctional state. The experience of pain is unique to each individual and can be difficult to quantify or objectively measure. In animals, as in humans, pain is a similarly complex phenomenon and can have significant effects on an individual's health, well-being, and behaviour. This pain can have a secondary profound effect on the community in which the human or animal lives. In humans and animals, the objective measurement of pain is challenging owing to the difficulties in communicating it verbally. Measuring and then understanding the molecular and cellular mechanisms that underlie pain is crucial for developing effective management strategies and improving clinical outcomes for both humans and animals.

Pain results from numerous causes in humans and animals, including surgical procedures, injuries, burns, infection, disease and poor living conditions. Understanding the shared and distinct mechanisms that underlie these different aetiologies is crucial for developing effective treatments and management strategies. However, the complex aetiology of acute and chronic pain means that the molecular origins and treatments of pain are far from straightforward and are inherently multi-dimensional. Moreover, the distinction between the experience of pain, and the ascending communication of body events to the brain—termed nociception—is critical. Here the brain can receive nociceptive data, and then contextualise this with other inputs to integrate a painful experience. For example, pain experienced early in life has different molecular and cellular nociceptive contributors compared to pain experienced later in life. Early life events, including prenatal development, birth and the neonatal period, are critical periods in which the developing nervous system is highly sensitive to environmental influences (Fitzgerald 2005). These early life events can have profound and long-lasting effects on the journey of a nociceptive signal, the subsequent susceptibility to pain experiences, and the possible development of chronic pain later in life. While current understanding of early-life events on pain perception in farmed animals is comparatively limited, there is a unique opportunity to learn from decades of research in humans and laboratory rodents. This will allow knowledge and technology transfer between species for the benefit of both human medicine and animal veterinary industries. Such an approach will create an awakening of precision interventions and precision management for pain to address the immense societal burden of underdiagnosed and untreated pain.

Early life experiences can 'prime' the nervous system for future pain experiences (Beggs et al. 2012). During early development across all species, the nervous system is highly plastic and capable of adapting to environmental changes. Experimental rodent and animal data have demonstrated that early life events, such as maternal separation, social isolation, or exposure to noxious stimuli, can lead to long-lasting changes in the nervous system, resulting in an increased risk of chronic pain (Beggs et al. 2012; Grace et al. 2014). Early life stressors, including maternal separation and social isolation, have been shown to result in changes to the stress response system, including the hypothalamic-pituitary-adrenal (HPA) axis (Bilbo and Schwarz 2012; Cirulli et al. 2003). The HPA axis is a key mediator of the stress response, and changes to this system can result in alterations to pain sensitivity and chronic pain development (Burke et al. 2017). Additionally, early life events can lead to changes in the structure and function of the nervous system, including the development of peripheral sensitisation, central sensitisation, and altered pain modulation, to be discussed in detail later in this chapter (Fitzgerald 2005). There is excellent correlational evidence that these early life cellular and molecular events are mirrored across different species and are corroborated by longitudinal human data.

Understanding the specific mechanisms by which early life events prime the nervous system for future pain experiences is crucial for developing effective management strategies for pain (Grace et al. 2014, 2021). Examining the latest rodent-to-human research on early-life pain provides insights into the specific factors that contribute to chronic pain development and the factors that can be targeted for prevention and treatment. In the following sections of this chapter, the specific biological, psychological, and social factors that contribute to the development of pain susceptibilities will be explored, with a particular focus on the role of early life events arising as the brain is developing. The latest research on the molecular events associated with chronic pain development will be explored, along with the sensory and emotional elements of pain, the challenges of behavioural quantification of chronic pain, and the opportunities that new technologies and approaches offer for pain quantification and management. By examining these different factors, we can gain a deeper understanding of the complexities of pain and develop more effective strategies for its prevention and management.

2 Pain as a Biopsychosocial Construct

The biopsychosocial model of pain acknowledges that biological, psychological and social factors can all contribute to the development and maintenance of pain (Raja et al. 2020). This model emphasises the importance of understanding the interaction between these different factors to effectively manage pain. It is crucial that what is meant by this interaction is understood. For the biopsychosocial construct of pain to be fully embodied, this requires information contained in nanoscale molecular interactions, occurring at the atomic scale, to converge with cellular signalling pathways and cellular systems at the micron level. These signals then orchestrate coordinated cellular system responses that occur over millimetre distances, to cause organ system events at centimetre resolution. All of this is occurring within an organism (animal to human) that is living and behaving in a dynamic environment at metre resolution beyond the proprioceptive and anatomical boundaries of the body. Moreover, these 3D events occur and change with time, over scales of nanoseconds to years, some precipitating immediate consequences, and others laying foundations for future experiences later in life (Fig. 1). All of this must be acknowledged when pain is viewed as biopsychosocial: this is a four-dimensional challenge that human brains are unsuited to conceptualise or to handle computationally. This human computational failing has hindered the study and understanding of the complexity of pain.

Despite the limitations in human understanding, pain research has identified several molecular and cellular systems that are hypothesised to be involved in the development and maintenance of pain. Importantly, when reading the relevant literature, each target often appears to be explained as a sole trigger or cause of pain; the 4D nature of pain dismisses this. Consequently, each target must be approached with a systems-level understanding of where such knowledge must fit within the specific contextual view. For example, it is clear that key molecules such as nerve growth factor (NGF) play a crucial role in

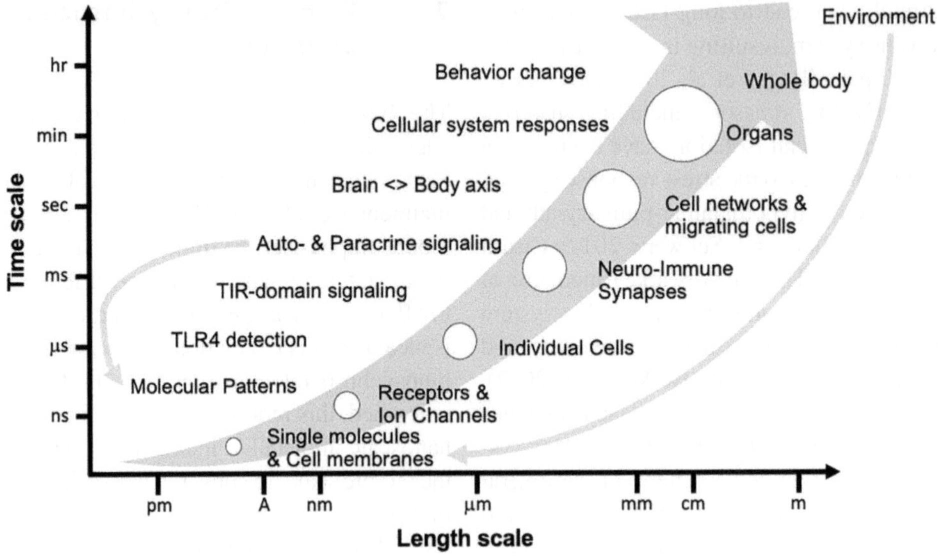

Fig. 1 The 4D concept of pain as a biopsychosocial construct that spans vast length and time scales

the development and maintenance of pain (Costigan et al. 2009). Here, NGF acts at the nanoscale by binding to another molecular target, the tropomyosin receptor kinase A (TrkA) receptor. This molecular interaction occurs uniquely on the sensory neurons that express the TrkA receptor, promoting neuronal survival and sensitising these neurons to noxious stimuli (Costigan et al. 2009). Crucially, these events are sufficient to trigger pain, but are not always necessary. Moreover, molecular 'information processing' occurs in a paralleled fashion across multiple ligand and receptor systems working in concert.

Other diverse molecules that have been implicated in the development and maintenance of pain include the pro-inflammatory cytokines interleukin-1 beta (IL-1β), tumour necrosis factor-alpha (TNF-α), and interleukin-6 (IL-6) (Grace et al. 2014). These classical immune molecules are produced in response to tissue damage and inflammation and can promote the sensitisation of nociceptive neurons, contributing to chronic pain development. Such signalling of immune factors to neuronal sites was previously not acknowledged and, in some cases, actively dismissed (Mustafa et al. 2023). It is now clear that molecular danger signals of diverse origins can act at these nanoscale systems to alert discrete cellular pathways to current or impending threats. This may warrant the organism to acknowledge this presence with conscious attention, and hence the expression of pain. This expansion of biological factors that contribute to pain has profoundly deepened and accelerated the appreciation of the complexity of biopsychosocial pain mechanisms. However, an understanding and continued research exploration of this complexity has yet to develop at the same rate as this complexity itself. This is of specific importance as these signalling events at the nanoscale can then be scaled to the understanding of multicellular functional units and, eventually behaviours.

In order for this molecular data to be transmitted, cellular systems need to be recruited across millimetres to centimetres. At the cellular level, research has focused on understanding the changes that occur in sensory neurons, glial cells, and other cells in the nervous system in response to tissue damage and inflammation. One of the key cellular changes that occurs in response to tissue damage is peripheral sensitisation. Peripheral sensitisation refers to the process by which nociceptive neurons become sensitised to noxious stimuli, resulting in an increased responsiveness to stimuli that would not normally be perceived as painful (von Hehn et al. 2012; Woolf 2011). This process is mediated by the previously introduced nanoscale molecular events, such as the release of inflammatory molecules and NGF from damaged tissues, which then act to sensitise nociceptive neurons by causing changes in receptor expression and protein functions. In addition to peripheral sensitisation, central sensitisation is another cellular process that is hypothesised to play a key role in chronic pain development. Central sensitisation refers to the process by which the neurons in the spinal cord and brain become sensitised to noxious stimuli (Woolf 2011). This process is thought to be mediated by the release of excitatory neurotransmitters, such as glutamate, and the activation of N-methyl-D-aspartate (NMDA) receptors on neurons in the central nervous system. These central nervous system level changes occur at the single neuron level and across multi-neuron systems, whereby 're-wiring' of the spinal cord and brain occurs that causes facilitation of nociception and hence pain experience (von Hehn et al. 2012). As such, these events require hundreds of molecular switches to be activated and cause structural changes that can persist long after the initial acute precipitating events (Grace et al. 2012; Wangzhou et al. 2021). These changes are occurring at the level of mRNA transcription right through to proteomic and microanatomical cellular structural changes. As such, biopsychosocial elements of pain can become 'hard-wired' and profoundly difficult to 'reset', further contributing to the complexity of pain.

The biopsychosocial model of pain also acknowledges the role of psychological factors in the development and maintenance of pain. Psychological factors, such as anxiety, depression and catastrophising, can all contribute to the experience of pain in humans (Crombez et al. 2013). These factors can influence the perception of pain, the ability to cope with pain, and the likelihood of developing chronic pain. Social factors, such as social support and cultural beliefs, can also influence the experience of pain in humans. For example, research has shown that social support can buffer against the negative effects of pain (Meints and Edwards 2018). This type of social level event has been demonstrated across species with social buffering reported in lambs undergoing tail docking in the presence of an undocked conspecific (Larrondo et al. 2021; Guesgen et al. 2014). In addition, cultural beliefs about pain can influence how pain is perceived and managed (Peacock and Patel 2008). For example, some cultures may view pain as a normal part of life, while others may view pain as a sign of weakness. Collectively, drawing on the 4D nature of pain, these types of psychological and social factors enable mechanistically driven broadening of the range of stimuli and states that contribute to pain and, in and of themselves, are sufficient to trigger pain. Therefore, at the root of this is the acknowledgement that *how* humans 'think' and *what* humans think and experience contribute mechanistically to the pain experience. Once again, these are not always individually necessary, yet are persistent physiological systems that must be considered.

The biopsychosocial model of pain recognises that pain is a complex phenomenon that involves the interaction of multiple biological, psychological and social factors. Understanding these factors is crucial for the future creation and delivery of effective pain management in all species. The development of new pain treatments requires a better understanding and harnessing of the molecular and cellular systems that are involved in pain development and maintenance. This will require a multidisciplinary convergent science approach involving researchers from several disciplines, including neurobiology, psychology and sociology.

3 Acute to Chronic Pain Transition

The transition from acute to chronic pain involves a complex interplay of molecular and cellular events that ultimately lead to maladaptive pain. While acute pain serves as an essential protective mechanism, chronic pain can be debilitating and have a significant effect on an individual's quality of life. Understanding the molecular and cellular mechanisms that underlie the transition from acute to chronic pain is crucial for developing effective therapies for chronic pain (Hutchinson and Terry 2019).

One of the key factors in the transition from acute to chronic pain is the persistence of immune signalling at and below the classical inflammation threshold. Classical inflammation is initiated in response to tissue damage or injury, and its resolution is necessary for the restoration of tissue homeostasis. However, in some cases, the inflammatory response becomes chronic with sub-inflammation immune signalling persisting, leading to persistent pain. This persistent immune signalling is driven by a range of immune cells, including macrophages, T cells, and mast cells, which release pro-inflammatory cytokines and chemokines (Davies et al. 2020). These mediators can directly activate nociceptors, leading to sensitisation and chronic pain development. At the heart of these processes is the maladaptive function that occurs at a newly appreciated nano- to micro-anatomical structure, termed the neuroimmune synapse. At the neuroimmune synapse, normal homeostatic balance is lost, with increased allostatic load first burdening and then exhausting the cellular control systems. Allostatic load is the accumulated burden of maintaining homeostasis by continually escalating of more on and off feedback loops attempting to stabilise the system. Such an accumulated burden presents with identical phenotypic presentations of pain, with profoundly different scales of mechanistic cellular and molecular systems engaged. This eventually results in exhaustion of the systems and a sudden watershed moment that results in uncontrolled and unremitting pain.

The transition from acute to chronic pain is also driven by changes in the neuronal systems side of the equation. This includes changes in the properties of nociceptors (defined in detail in the next section), alterations in synaptic transmission within the spinal cord, and changes in the activity of immune-like cells in the brain and spinal cord, called glia. One of the key changes in the nervous system in chronic pain is the sensitisation of nociceptors, which results in the lowering of the threshold for activation and the propagation of nociceptive signals. This sensitisation can be driven by the release of inflammatory mediators, including prostaglandins and bradykinin, which can directly activate nociceptors (von Hehn et al. 2012). Sensitisation also occurs as a result of changes in the expression of ion channels, including TRPV1 and Nav1.7, which play critical roles in nociceptor function.

Changes in synaptic transmission within the spinal cord are also crucial in chronic pain development. In particular, changes in the activity of glial cells, including microglia and astrocytes, have been implicated in chronic pain development. Glial cells can release a range of inflammatory mediators, including cytokines and chemokines, which can modulate synaptic transmission within the spinal cord. For example, microglia have been shown to release the cytokines TNF-α and IL-1β, which can enhance synaptic transmission in the dorsal horn of the spinal cord and contribute to chronic pain development (Watkins et al. 2007). All of these collectively lead to a maladaptive neuroimmune synapse function. The drivers of these changes in neuroimmune synapse function often require changes in gene expression. These transcriptional changes can occur in both neurons and non-neuronal cells, including immune cells and glial cells (Wangzhou et al. 2021). For example, changes in gene expression in macrophages and T cells have been shown to play a critical role in chronic pain development. In particular, the expression of pro-inflammatory cytokines, including IL-1β and TNF-α, are upregulated in immune cells in the peripheral nervous system, contributing to the maintenance of chronic pain.

Therefore, the transition from acute to chronic pain involves a complex interplay of molecular and cellular events that can occur in both the peripheral and central nervous system, as well as in the immune system.

4 Sensory and Emotional Elements That Can Culminate in the Experience of Pain

Pain is a complex and multifaceted phenomenon that involves both sensory and emotional processing within the central nervous system. Sensory processing involves the detection of noxious stimuli, the transmission of nerve impulses (action potentials) to the brain, and the subsequent interpretation of these signals by the brain. Emotional processing, in contrast, involves an individual's emotional state and its interaction with their response to multiple psychological inputs that can include pain, including the affective and motivational aspects of pain. The interaction of these different pain components is thought to be important for the overall pain experience involving complex molecular and cellular systems within the nervous system (Raja et al. 2020).

At the molecular and cellular level, the primary system involved in the sensory processing of pain is the nociceptive system. This system includes specialised sensory neurons called nociceptors, which detect noxious stimuli and transmit nerve impulses to the spinal cord and brain (Ji et al. 2016; von Hehn et al. 2012). Nociceptors are activated by a range of stimuli, including heat, cold, pressure, and chemical irritants. Once activated, they transmit nerve impulses to the spinal cord and brain, where the signals are interpreted as pain. Prior to this perception of danger as pain (or pain as danger), the neural impulse transmission along nociceptive pathways is nociception. This is a critical distinction, as the pain of an unconscious being may be altered by their loss of higher cortical controls, but the nociceptive consequences may persist unabated.

The transmission of nociceptive signals is modulated by a range of molecular and cellular systems. These include neurotransmitters, such as glutamate and substance P, and ion channels, such as NMDA receptors and TRPV1 channels (Grace et al. 2014, 2021; Wangzhou et al. 2021). These systems are thought to play a role in the development and maintenance of chronic pain. One of the key brain regions involved in the sensory processing of pain is the somatosensory cortex. This is responsible for processing the location, intensity, and quality of noxious stimuli (Apkarian et al. 2005). In addition to the somatosensory cortex, brain-imaging studies have identified the insular cortex as a key brain region involved in sensory pain processing. The insular cortex is thought to be involved in the sensory-discriminative aspects of pain, including the location, intensity and quality of noxious stimuli (Wager et al. 2013).

The emotional processing of pain, on the other hand, involves the limbic system and the hypothalamic-pituitary-adrenal (HPA) axis, which is involved in the stress response. While not wishing to anthropomorphise the animal state of emotion, comparative biology allows the inference of affective state to defined brain regions. For example, the limbic system across species includes several brain regions, including the amygdala, hippocampus and prefrontal cortex (Price 2000). These regions are involved in the affective and motivational aspects of pain, including the emotional response to pain and the regulation of pain behaviour across species. The inverse is also true, i.e. where the emotional and motivational state of the individual in the lead up to noxious stimulation is sufficient to change the experience and evaluation of the painful insult. For example, the amygdala is involved in the emotional response to pain, while the prefrontal cortex is involved in the regulation of pain behaviour. Brain imaging studies have shown that emotional factors can influence the experience of pain. For example, in human trials the anticipation of pain can lead to an increase in pain intensity (Moseley 2005). Here the use of functional magnetic resonance imaging (fMRI) to measure brain activity in response to painful stimuli showed that the anticipation of pain activated the prefrontal cortex and insular cortex, which are key brain regions involved in sensory and emotional pain processing.

These complex CNS responses and adaptations can trigger physiological responses through the activation of circulating hormone systems like the hypothalamus-pituitary-adrenal axis

(HPA axis). The HPA axis is a key mediator of the stress response, and is thought to play a role in the development and maintenance of chronic pain across species. The HPA axis includes the hypothalamus, pituitary gland, and adrenal gland. In response to stress, the hypothalamus releases corticotropin-releasing hormone (CRH), which stimulates the release of adrenocorticotropic hormone (ACTH) from the pituitary gland. ACTH then stimulates the release of cortisol from the adrenal gland. Cortisol is a stress hormone that is thought to play a role in the development and maintenance of chronic pain (Hannibal and Bishop 2014). Of note, these events of the HPA axis are anatomically distributed across the whole body from the adrenal gland (just proximal to the kidneys) right up to the hypothalamus in the brain. The signalling factors that comprise the HPA axis are highly sensitive to—but are not specific to stress. As such, the recruitment of the HPA axis by painful experiences has wide-ranging consequences for numerous physiological systems. In contrast, and reciprocally, non-stress-triggered HPA axis controls can affect the biopsychosocial experience of pain.

The molecular and cellular systems involved in the sensory and emotional processing of pain are complex and multifaceted. They involve a range of neurotransmitters, ion channels and brain regions, and they are thought to play a role in the development and maintenance of chronic pain. Contextually understanding these systems and their interactions in 4D is crucial for the development of new pain treatments.

5 The Crucial Distributed Anatomical Cellular Events that Can Culminate in Pain

The processes that culminate in pain perception can begin peripherally and progress in series to brain-level activity, or can occur in parallel from multiple anatomical compartments, leading to a multidimensional 4D experience. From the extremities, nociception can be triggered in peripheral tissues, where damage or injury leads to the release of inflammatory mediators that activate nociceptors (Ji et al. 2016). Nociceptors are sensory neurons that are specialised to detect noxious stimuli and transmit this information to the spinal cord. In the periphery, non-neuronal cells, such as immune cells, also play a critical role in pain modulation and future pain processing. These cells release a range of inflammatory mediators that can directly sensitise nociceptors and increase their responsiveness to noxious stimuli. For example, cytokines and chemokines released by immune cells can bind to nociceptor receptors, increasing their activity and sensitivity to stimuli (Grace et al. 2014). Wangzhou et al. (2021) exemplified this through the curation of a ligand-receptor interactome platform, where nociceptor receptors and immune factor ligands were paired, based on known nanoscale molecular affinities, to allow conceptual and computational modelling of the vast array of parallel molecular events that drive nociception. These data demonstrate that hundreds of options exist for the potential mass parallel processing of nociceptive information at the molecular to cellular level, any of which may be sufficient to cause a painful experience for the individual.

The importance of non-neuronal cells such as mast cells, keratinocytes, dendritic cells, and macrophages in the peripheral mechanisms that underlie pain development is being increasingly recognised. This realisation is also extending to non-host cells such as bacteria that are now understood to directly communicate with the nociceptive system. Mast cells, for example, have been shown to play a crucial role in pain development by releasing a range of mediators such as histamine, cytokines and proteases, which can sensitise nociceptors and contribute to neurogenic inflammation (Mai et al. 2021). In addition, bacteria can also play a role in pain processing, as has been demonstrated in studies investigating the role of microbes in chronic pain (Diogenes et al. 2011). These animal and human studies have demonstrated that a gram-negative cell wall factor, lipopolysaccharide (LPS), enhances the function of the Transient Receptor Potential Vanilloid 1 (TRPV1) ion channel, which is involved in pain sensation, by activating Toll-like receptor 4 (TLR4) in trigeminal sensory neurons.

This process leads to an increased responsiveness of TRPV1 to painful stimuli, thereby contributing to the development of inflammatory pain. The consequences of this cross-sensitisation is detailed later in this chapter.

Keratinocytes—cells that make up the outermost layer of the skin—are also implicated in the peripheral events leading to pain. These cells have been shown to express a range of nociceptive ion channels and receptors, and are capable of producing a range of inflammatory mediators that can sensitise nociceptors and contribute to the development of pain (Koivisto et al. 2022). Dendritic cells and macrophages are immune cells that are present in peripheral tissues, and have also been shown to participate in pain processing by releasing pro-inflammatory cytokines and chemokines that can sensitise nociceptors and contribute to neurogenic inflammation (Ji et al. 2016). These findings highlight the importance of non-neuronal cells in the peripheral mechanisms that contribute to the development of pain. Moreover, this concept of neurogenic inflammation creates a brain and spinal to skin level of descending nociceptive sensitisation and facilitation. Such mechanisms create a completion of a circular feed-forward loop of signalling, such that the originating molecular event that culminates in pain cannot be solely attributed to the location of the symptomatic peripheral inflammation. Specifically, it is plausible for an event such as this to arise from descending brain and spinal cord activity, causing neurogenic inflammatory factor release and the establishment of a centrally facilitated but peripherally mediated inflammatory state.

The ascending transmission of nociceptive signals from the periphery to the spinal cord occurs through the dorsal root ganglia (DRG), which contain the cell bodies of sensory neurons. The DRG is a critical site for the integration of nociceptive information, as it contains a range of signalling molecules and receptors that can modulate nociceptor activity. Changes in DRG function have been implicated in chronic pain development. For example, Grace et al. (2021) highlighted the role of neuroimmune signalling in the DRG and suggested that targeting these signalling pathways may lead to new chronic pain treatments. Macrophages and natural Killer (NK) cells are also found within the DRG and have been shown to participate in the development of neuropathic pain. Studies have shown that depletion of NK cells leads to a reduction in neuropathic pain (Davies et al. 2020). These data suggest NK cell cytotoxic function is dysregulated in pain states and is sufficient to exaggerate pain. The presence of satellite glial cells in the DRG is another important cellular presence that contributes to pain processing. These cells play a role in regulating nociceptive signalling within the DRG and can contribute to chronic pain development (Lee and Kim 2020). T cells and B cells have also been identified in the DRG, and are thought to play a role in the immune response to injury and inflammation within the DRG itself. Dysregulation of these cells has been linked to the development of neuropathic pain (Totsch and Sorge 2017).

Once nociceptive signals reach the spinal cord, they are transmitted to higher brain regions for processing. The dorsal horn of the spinal cord is a critical site for the integration of nociceptive information. Here, synaptic connections between sensory neurons and second-order neurons are subject to modulation by a range of signalling molecules and receptors, including glutamate, GABA, and neuropeptides. These signalling molecules can increase or decrease the activity of nociceptive circuits, leading to changes in pain sensitivity. A wide range of non-neuronal cells in the spinal cord also play a crucial role in the development and maintenance of chronic pain. Microglia, for example, the resident immune-like cells of the central nervous system, play a key role in the pathogenesis of chronic pain. Microglial reactivity and proliferation occur following peripheral nerve injury, with these cells releasing a range of pro-inflammatory cytokines, chemokines and proteases that contribute to the development of pain hypersensitivity (Grace et al. 2014; Nicotra et al. 2012).

Oligodendrocytes, which are responsible for myelinating axons within the central nervous system, also play a role in chronic pain development. Oligodendrocyte precursor cells (OPCs)

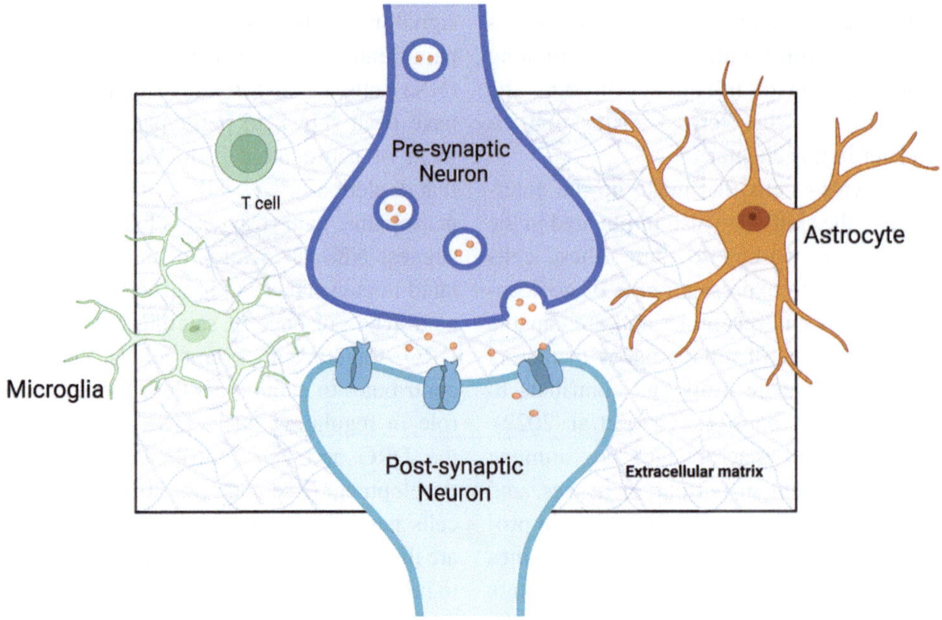

Fig. 2 Schematic diagram of the microglial penta-partite synapse consisting of the pre- and post-synaptic neurons, astrocytic processes (orange), microglial processes (light green), and the extracellular matrix (background). T cells may also influence synaptic function in the presence of disease. (Created in Biorender.com. Adapted from Aramideh et al. (2021))

proliferate and differentiate into oligodendrocytes in response to nerve injury, with these cells playing a role in remyelination and the restoration of axonal function. However, excessive OPC differentiation can lead to the formation of non-myelinating Schwann cells, which contribute to the development of neuropathic pain (Gritsch et al. 2014). Astrocytes are another type of glial cell that play an important role in chronic pain development. Astrocyte reactivity has been shown to contribute to pain hypersensitivity, with these cells releasing a range of pro-inflammatory cytokines, chemokines and other mediators that sensitise nociceptors and contribute to reactive immune signalling (Xu et al. 2021). Moreover, astrocytes can lose their excitatory neurotransmitter homeostatic functions, owing to reductions in their expression of critical reuptake transporter enzymes. This culminates in the abundance of glutamate within synapses and hyperexcitability. Importantly, all of these adaptations do not occur distal to critical neuronal synapses. Rather, it is now clear that these non-neuronal cells are intimately co-located anatomically with the critical nociceptive neuronal systems in cellular structures, originally termed the tripartite synapse. Here, an astrocyte end-foot encapsulates the pre and postsynaptic terminals. This relationship has now been extended to encompass multiple cell types, thus creating the pentapartite synapse (Aramideh et al. 2021) (Fig. 2).

The conceptualisation of the pentapartite synapse has triggered a revolution in neuroscience, as it breaks the previous dogma that the brain is immune-privileged. It has emphasised the importance of the ability of peripheral immune cells to be able to migrate into the spinal cord and contribute to nociception. T cells, which are classically viewed only as an important component of the adaptive immune system, infiltrate the spinal cord following nerve injury and contribute to the development of neuropathic pain through the release of pro-inflammatory cytokines and chemokines (Totsch and Sorge 2017). Likewise, peripheral macrophages also play a role in chronic pain development because these cells infiltrate the spinal cord following peripheral nerve injury and promote pain hypersensitivity by releasing pro-inflammatory cytokines and chemokines (Grace et al. 2011).

These findings highlight the importance of examining multiple anatomical compartments and a range of cells and molecules in the assessment of pain perception and inflammatory signalling, which could potentially improve diagnosis, treatment, and management of chronic pain. The role of non-neuronal cells in the development and maintenance of pain is becoming increasingly recognised, and future research could focus on the interplay between these cells and neurons in the development of pain. Additionally, targeting non-neuronal cells may lead to new treatments for chronic pain, providing novel avenues for drug development.

6 Primed for Pain: There Is No Biological Reset Button in Life

Chronic pain is a complex condition that arises from numerous underlying mechanisms, including persistent immune signalling, peripheral sensitisation and changes in the central nervous system. The last few decades have seen researchers come to recognise the important role played by non-neuronal cells, particularly glial cells, in the development and maintenance of chronic pain. Glial cells are the most abundant cells in the central nervous system, and they play a critical role in maintaining the homeostasis of the neuronal environment. This section will examine the critical role of glial priming and changes in cellular reactivity set points in the transition from acute to chronic pain, and how early life events can lead to lifelong pain consequences.

Glial cells, including microglia and astrocytes, play important roles in the development and maintenance of chronic pain. These cells are involved in the regulation of neuronal excitability, synaptic plasticity and the immune response. They are critical to the nano- and micro-anatomical structures such as the pentapartite synapse (Aramideh et al. 2021). As such, these cells are intimately connected to neuronal processing and transition of nociceptive information. Microglia and astrocytes can become 'primed', or sensitised, following first exposure to inflammatory signals, leading to a long-lived susceptibility to increased cytokine and chemokine expression, excessive production of reactive oxygen species, and other signalling molecules upon subsequent insults (O'Neil et al. 2022). This priming is thought to contribute to the transition from acute to chronic pain by increasing the sensitivity of neurons to noxious stimuli and altering the cellular reactivity set points of glial cells (Mika et al. 2013).

Glial cells, particularly microglia, play a critical role in chronic pain development. Exposure to an inflammatory stimulus during early life results in the priming of microglia, leading to an enhanced response to subsequent inflammatory challenges in adulthood (Bilbo and Schwarz 2009) (Fig. 3). This priming was associated with changes in the expression of a number of proteins, including fractalkine and its receptor, which plays a critical role in the communication between neurons and microglia (Watkins et al. 2007). Other studies have shown that early life exposure to stress can lead to a persistent increase in the expression of proinflammatory cytokines in microglia, leading to increased sensitivity to subsequent inflammatory stimuli (Schwarz and Bilbo 2012).

Astrocytes also play a critical role in the development and maintenance of chronic pain. For example, exposure to an inflammatory stimulus during early life results in a persistent increase in the expression of the cytokine IL-1β in astrocytes in rats (Hsieh et al. 2020). This increase is associated with an enhanced response to subsequent inflammatory challenges and increased sensitivity to pain. Other studies have shown that astrocytes also play a critical role in the regulation of neuronal excitability and synaptic plasticity, with changes in astrocyte function contributing to chronic pain development.

The transition from acute to chronic pain is also associated with changes in the cellular reactivity set points of glial cells. In particular, chronic pain is associated with an increase in the expression of proinflammatory cytokines

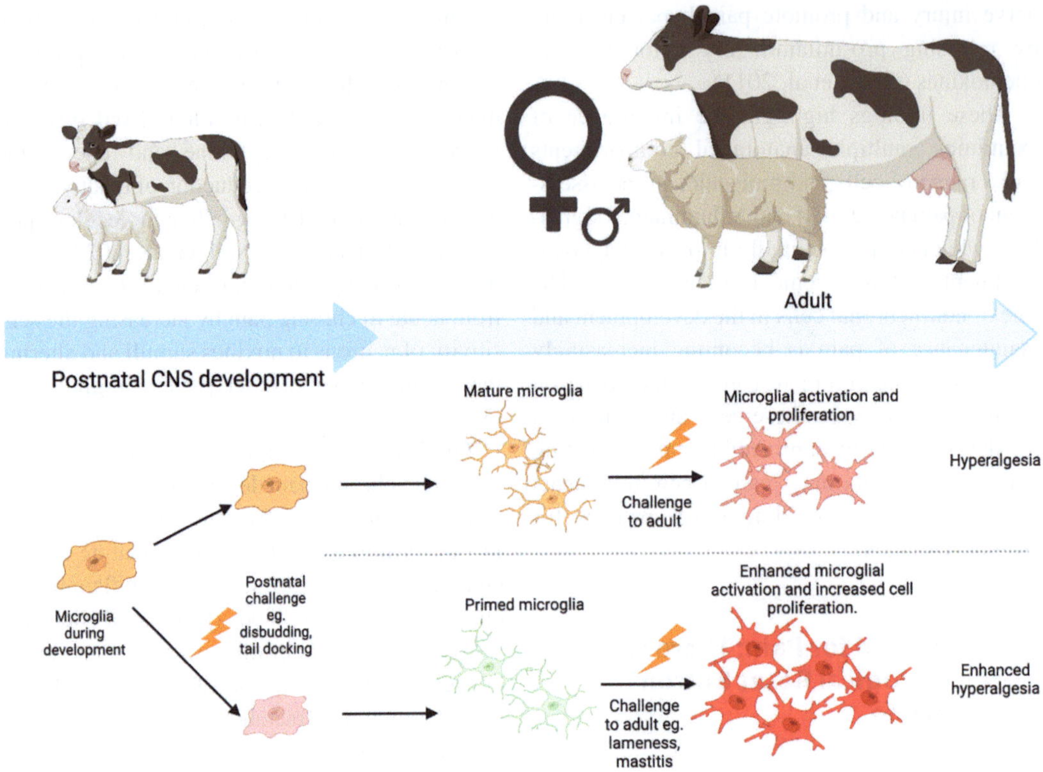

Fig. 3 Disruption to normal CNS development during early life can lead to priming of microglia and result in an enhanced glial response to challenges in later life. One example may be disbudding or tail docking in young livestock. The greater prevalence of exaggerated pain in human females may have considerable implications within the livestock sector, owing to the predominance of female animals in the adult population. (Created in Biorender.com. Adapted from Burke et al. (2016))

and chemokines, as well as an increase in the production of reactive oxygen species. Changes in cellular reactivity set points contribute to the maintenance of chronic pain by altering the response of glial cells to subsequent inflammatory stimuli. Changes in neurodevelopment and the somatosensory system also contribute to the development and maintenance of chronic pain. Exposure to early life stressors can result in long-lasting changes in the structure and function of the somatosensory system, increasing risk of chronic pain development. In particular, changes in the structure and function of the dorsal root ganglia, spinal cord and brainstem have been associated with chronic pain development.

7 Cross Sensitisation—the Innocuous Challenge That Can Create Pain

Cross-sensitisation is a phenomenon that occurs when two receptor systems interact and result in an exaggerated pain response, e.g. the interaction between TLR4 and TRPV1 channels results in increased pain sensitivity. This interaction plays a role in chronic pain development, rendering it an important study area for research. TRPV1 is a non-selective neuronal-ion channel and is activated by heat (> 43 °C) amongst other numerous endogenous and exogenous chemicals, including capsaicin (the pungent component of chilli peppers; (Caterina et al. 1997). Situated throughout the nervous system, from peripheral nociceptors, spinal cord and higher brain regions, TRPV1 is ideally positioned to integrate signals from multiple systems and affect neuronal outcomes (Mezey et al. 2000).

Extensively studied for its role in nociception, TRPV1 activation on peripheral nociceptors results in action-potential generation and an immediate 'pain' response. This is followed by a brief period of hypersensitivity in the tissue surrounding the site of activation (Caterina and Julius 2001). Interestingly, the influence of other parallel receptor signalling systems alters TRPV1 signalling and subsequent painful responses. The innate immune receptor TLR4, and activation of its signalling by bacterial lipopolysaccharide (LPS) enhances the release of the neurotransmitter, calcitonin gene-related peptide (CGRP) following TRPV1 activation (Hua et al. 1996). This effect of immune signalling on TRPV1 is also observed behaviourally, with pain behaviours in a headache model enhanced 5 hour following LPS administration (Kemper et al. 1998). In a preclinical colitis model (2,4,6-trinitrobenzine sulfate (TNBS)-induced colitis), capsaicin-induced currents in peripheral neurons were significantly potentiated, an effect attenuated in TLR4 knockout mice (Wu et al. 2019). It is now known that several pro-inflammatory mediators sensitise or upregulate transcription and membrane trafficking of TRPV1, enhancing intracellular cation ingress and subsequent neurotransmitter release. These include cytokines (IL-6, IL-1β, TNF-α), NGF, prostaglandins (PGE_2, PI_2) and bradykinin. Receptor sensitisation involves phosphorylation and dephosphorylation of the receptor by various protein kinases, including protein kinase A and C (PKA/C), phospholipase C (PLC), Src tyrosine kinase (Src) and phosphoinositide 3-kinase (PI3K) (Ji et al. 2016).

Potentiated capsaicin responses following LPS application are observed over much shorter time periods in vitro. For example, a 15-minute LPS application increases capsaicin-induced inward current amplitude and calcium accumulation in TLR4/TRPV1 expressing HEK cells (Min et al. 2014). The same outcome is observed in rat trigeminal neurons, where the response is attenuated by the TLR4 antagonist LPS-RS (Diogenes et al. 2011). LPS-RS blocks potentiated capsaicin-induced intercellular calcium accumulation in cells excised from a rat chemotherapy-induced neuropathic pain (CINP) model (Li et al. 2015b). The same effect is observed in ex vivo spinal cord slices; an increased rate of miniature excitatory postsynaptic currents (mEPSCs) to a second application of capsaicin following paclitaxel treatment, an effect blocked by LPS-RS (Li et al. 2015a). The exact mechanism driving the relatively rapid changes to receptor function requires further investigation. TLR4-mediated kinase activity is a likely candidate. PKA and PKC, *Src* and calcium-dependent increases in phospholipase C (PLC) and PI3K are reported 5–60 minutes after TLR4 stimulation (Cabral et al. 2015). The timing of these increases makes this a plausible explanation of TLR4-dependent sensitisation of TRPV1.

TLR4 and TRPV1 have been identified within the same neurons in human and rat trigeminal ganglia (TG), and DRG. This in vivo co-expression and supporting evidence from immunoprecipitation studies have led to investigations of direct protein–protein interaction between these receptors. Such an interaction is not unique, as a TRPA1/TLR7 interaction has been reported, where TLR7 activity potentiates TRPA1 induced inward current. In HEK cells co-expressing TRPV1 and TLR4, response amplitude and calcium accumulation are increased following capsaicin compared to cells expressing TRPV1 only (Min et al. 2014). The authors of this observation propose a link between TRPV1 and the TIR domain of TLR4, whereby a physical nanoscale interaction prevents desensitisation and internalisation of the TRPV1 (Min et al. 2018). Li and colleagues found that TLR4 may facilitate receptor desensitisation from repeated capsaicin doses. However, more investigation is needed to clarify this proposed link (Li et al. 2015b). Another theory involves indirect relationships with scaffolding proteins (Min et al. 2018). Cytoplasmic scaffolding protein A-kinase anchoring protein 79 (AKAP79) is important to the sensitising effect of PKA and PKC on TRPV1 and is proposed as a potential candidate.

Importantly, a human trial by Hutchinson et al. (2013) shows that the TLR4/TRPV1 relationship has clinical relevance. Potentiation of mechanical allodynia and hyperalgesia was observed at 3 hours, but not 2 hours or directly following administration of intravenous low dose (0.4 ng kg^{-1}) LPS (Hutchinson et al. 2013). The timing of potentiation correlated with peak levels of serum IL-6, suggesting inflammatory-mediated receptor sensitisation (Hutchinson et al. 2013). There is evidence to suggest that the interaction between TLR4 and TRPV1 has the potential to affect clinical pain outcomes. It is important to clarify important regulators of this interaction in inflammatory models, and if potential protein-protein interactions are relevant to clinical pain outcomes.

The co-expression and interaction of TLR4 and TRPV1 in single neuronal cells provides one example of crosstalk between receptor systems leading to cross-sensitisation and exaggerated pain responses. The cross-sensitisation of receptor systems, like the relationship between TLR4 and TRPV1 has profound clinical relevance, and there is evidence to suggest that this interaction has the potential to affect clinical pain outcomes. When this type of cross sensitisation and receptor signalling modulation is viewed in the context of the mass parallel processing possible within the pain interactome created by Wangzhou et al. (2021), a simple view of approximately 80 signalling pairs operating in parallel, any of which being sufficient to trigger a painful response, means that there are $(2^{80})^{-1} = 1.2 \times 10^{24}$ potential molecular pathways to pain (almost all of which could not be blocked by targeting a single signalling pair); this is greater than the total number of stars in the universe. And this is only taking into account the 'simple' primary afferent events; it does not account for the immense complexity at the psychological and social levels of the pain experience and modulation. This complexity of the systems at play underscores the need to appreciate this challenge, not necessarily understanding all of it, in order to address the pain experience.

8 Early Life Events that Can Drive Chronic Pain

The concept of early life infection and stress creating changes that promote pain later in life is a complex and important area of study in pain research. Early life stressors, such as inflammation, surgical incision, and maternal separation, changes central nociceptive signalling and pain outcomes in adult rodents (Beggs et al. 2012; Walker et al. 2015, 2016). These stressors reveal altered baseline mechanical and thermal sensitivity, as well as altered response to reinjury, in adults. Increased excitatory and reduced inhibitory spinal input suggests that developmental changes alter the nociceptive environment. Critically, the types of physiological insults used in these experimental models are commonplace in animal husbandry practices and human life experience. As such, an appreciation of how these life events contribute to pain later in life is vital if industrial, organisational and societal systems that are unintentionally priming for later life pain consequences are to be addressed.

Common models of neonatal stress are surgical incision injury and paw inflammation. In adult animals who have been exposed to neonatal incision, increased thermal and mechanical nociceptive thresholds are observed. Likewise, in a carrageenan paw inflammatory model, rats who received carrageenan between post-natal days 0 and 5 showed increased thermal and mechanical thresholds in adulthood (Laprairie et al. 2008). Ipsi- and contra-lateral threshold increases were observed in one instance. In carrageenan-treated neonates, increased descending inhibition from the rostral ventromedial medulla (RVM) is reported as a potential contributor. Maternal separation, which is a common model of neonatal stress results in higher mechanical thresholds in mice compared to controls (Salberg et al. 2020). Note, that these experimental models of separating rodents from dams during critical phases of neurodevelopment are the same as the production practices used during animal weaning.

In agreement with these behavioural observations, dorsal horn recordings from an early life mid-thigh incision model show decreased miniature excitatory postsynaptic currents (mEPSCs) at day 19 (Li et al. 2009). While RVM activity was altered in a rat model, both ipsi- and contralateral biceps femoris show altered reflex excitability, revealing altered RVM control favouring inhibition (Walker et al. 2015). In contrast, altered inhibitory signalling is reported in a mouse incision model. Decreased inhibition within the spinal dorsal horn is observed by a reduction in glycine receptor-mediated inputs (Li et al. 2013). Neonatal inflammatory models using carrageenan complete Freund's adjuvant (CFA), and LPS also report reduced thermal and mechanical thresholds in adult mice and rats when administered at day 1 and 14 (Blom et al. 2006; Boissé et al. 2005; Roizenblatt et al. 2010). Dorsal horn neuronal recordings also show increased responsiveness to mechanical and thermal stimuli at 8–10 weeks in CFA-treated rat neonates, with an increased receptive field on the rat hindpaw, suggesting increased afferent input/activity (Peng et al. 2003). In contrast, one study using a neonatal-CFA model reports no thermal threshold differences in adulthood. However, this study does report animals displayed increased spontaneous activity, increased responsiveness to innocuous brushing, and increased density of sciatic nerve primary afferents (Ruda et al. 2000). Multiple studies also report no threshold changes following neonatal CFA or carrageenan administration, despite the expansion of saphenous and sciatic nerve input to superficial dorsal horn laminae (Chu et al. 2007; Walker et al. 2003). While baseline thresholds in adult animals exposed to neonatal stress vary, the same cannot be said of results upon re-exposure to nociceptive stress.

In incision models, upon re-incision, mechanical and thermal thresholds are significantly lower than in animals not exposed to neonatal incision (Beggs et al. 2012; Moriarty et al. 2018, 2019). Interestingly, the extent of mechanical and thermal sensitivity is altered depending on the site of neonatal incision, suggesting changes may be unique to specific spinal regions (Moriarty et al. 2019). Secondary carrageen or CFA challenge in adult rats reduces thermal and mechanical thresholds in neonatally treated animals (Blom et al. 2006; Lim et al. 2009; Ruda et al. 2000). Similarly, CFA-treated neonates show decreased capsaicin-induced thermal and mechanical thresholds as adults when CFA is administered at day 1, and not when it is administered at day 14. This was correlated with increased dorsal horn capsaicin-induced c-Fos activation (Hohmann et al. 2005), an experimental marker of neuronal activity. In a non-invasive maternal separation model, increased anxiety behaviours and mechanical threshold were observed, but not hot/cold sensitivity, following a paw incision surgery in adolescent rats. This correlated with changes in gene expression within the brain and altered serum inflammatory markers (TNF-α, IL-6, CRP) (Salberg et al. 2020). Despite these examples, the pre-clinical literature strongly implies that early-life nociceptive challenge affects later life responses to re-injury and/or inflammation.

Cellular and neuronal elements of the CNS have been identified as important factors in re-incision hypersensitivity. For example, the microglial inhibitor minocycline administered during neonatal incision blocks re-incision hypersensitivity in male, but not female rats (Moriarty et al. 2019). This adds to the evidence suggesting that microglia play an important role in exaggerated re-incision hypersensitivity, as microglial responses are primed to secondary incision (Beggs et al. 2012). In contrast, targeting purely neural signals with morphine in mice during neonatal incision had no effect on enhanced re-incision responses (Moriarty et al. 2018). Multiple changes in neuronal excitability are reported in animals exposed to neonatal incision and subsequent re-injury, including increased primary afferent input and action potential discharge in projection neurons, and altered production of long-term potentiation, both of which promote pronociceptive environments in re-injured adults (Li and Baccei 2016; Li et al. 2015a, b).

This greater understanding of the long-term consequences of early life pain in rodents and humans has spurred growing interest from the livestock sector. There are substantial welfare and productivity implications for the industry given the ubiquity of surgical procedures being performed in young production animals and maternal separation resulting from weaning (Adcock 2021). The incision models discussed above were adapted to a real-world farm situation by McCracken et al. (2010) in their study on tail docking and castration in male lambs. The lambs were castrated with a rubber ring at either one or ten days of age. At one month of age, these lambs were tail-docked using a rubber ring. Lambs castrated at one day of age showed significantly higher pain behaviours following tail docking compared to those castrated at 10 days old. Another study adapting pre-clinical research to production-specific practices found that female lambs tail docked with a rubber ring or administered LPS at 3–4 days of age showed greater pain-related behaviours prior to parturition at 24 months of age compared to handled controls (Clark et al. 2014). In the same study, the lambs from the ewes exposed to LPS had reduced pain sensitivity after tail docking with or without castration at 3 days of age, demonstrating transgenerational effects of early-life inflammation. The findings of these studies are consistent with previous studies in rodents, indicating altered pain experience in later life following neonatal tissue injury and inflammation (Adcock 2021; Beggs et al. 2012; Fitzgerald and McKelvey 2016). There may also be other changes in stress responsiveness and resilience that are recognised in humans and rodents that are yet to be revealed in livestock (Adcock 2021).

Overall, the evidence suggests that early life stressors and infections can have lasting effects on pain processing and sensitivity in adulthood and that the fourth dimension of pain, time, can be critically important. These findings have important implications for the prevention and treatment of chronic pain, particularly in vulnerable populations such as animals, infants and young children. By understanding the mechanisms underlying the effects of early life stressors and infections on pain processing, it may be possible to develop new treatments and preventative measures for chronic pain conditions. Furthermore, identifying the specific stressors and infections, and timings of those insults that are most likely to contribute to chronic pain in adulthood, can help to inform targeted interventions and preventative strategies.

9 The Complexity of Genomic Contributions to Pain

Genomic and epigenetic factors can play a role in the development of pain conditions. One of the most well-known examples is the very rare gain-of-function mutation in the *SCN9A* gene, which encodes the voltage-gated sodium channel Nav1.7, that is responsible for inherited erythromelalgia, a condition characterised by burning pain and redness in the extremities (Fertleman et al. 2006). This mutation leads to hyperexcitability of nociceptors and increased pain sensitivity. Conversely, nonsense mutations in *SCN9A* result in loss of Nav1.7 function and insensitivity to pain (Emery et al. 2015). The contribution of more common genetic factors to pain susceptibility has been widely studied, and several genetic polymorphisms have been identified as potential modifiers of nociceptive signalling, analgesic response, or as risk factors for the development of pain disorders (Pain genes database Lacroix-Fralish et al. (2007)). One widely studied example is the *COMT* gene, which encodes the catechol-O-methyltransferase enzyme that metabolises catecholamines, including dopamine and norepinephrine. Polymorphisms in this gene have been associated with increased pain sensitivity and altered responses to pain medications (Diatchenko et al. 2005).

However, it is important to note that the contribution of individual genetic polymorphisms to pain susceptibility is often modest, and it is the interaction of multiple genetic and environmental factors that ultimately determines an individual's pain susceptibility. For example, multiple immune signalling gene polymorphisms (e.g. in *TLR2*, *TLR4*, *IL1B*, *IL1R2*, *IL2*, *IL6*, *IL6R*, *IL10*, *CXCL8*, *TNF* and *BDNF*) have been identified to work in concert to influence pain sensitivity, analgesic response and chronic pain risk (Barratt et al. 2021; Somogyi et al. 2016; Stephens et al. 2014, 2017). This conceptually makes sense given the pain interactome discussed previously and the known complex parallel signalling events that occur in immunology. However, most data analysis and research methods employed previously have not accounted for these types of appreciated physiological states. Therefore, innovations in the collective approach to research methods are needed to tackle this complex pain challenge.

Innovations across the omics fields are further contributing to pain research, and epigenetic mechanisms may also affect pain susceptibility. Epigenetic modifications such as DNA methylation, histone modification, and non-coding RNA expression can influence gene expression and alter pain sensitivity. For example, a recent study found that DNA methylation of the *FKBP5* gene, which regulates the stress response, was associated with increased pain sensitivity in patients with fibromyalgia (Gerra et al. 2021). The evidence for differential blood DNA methylation of pain-relevant genes in human chronic pain (including chronic postsurgical pain) patients has grown substantially in the last few years (Achenbach et al. 2022; Aroke et al. 2022; Chidambaran et al. 2019; Ciampi de Andrade et al. 2017; Menzies et al. 2013; Montesino-Goicolea et al. 2020; Paoloni-Giacobino et al. 2020; Rubino et al. 2022; Stenz et al. 2022; Takenaka et al. 2020), including correlations with symptom severity (Rubino et al. 2022; Takenaka et al. 2020), differentiation of chronic nociceptive from chronic neuropathic pain (Stenz et al. 2022), and the potential influence of childhood trauma on pain sensitivity-related gene methylation (Achenbach et al. 2019). Other studies have found that histone modifications, particularly histone acetylation, play a role in the development of neuropathic pain (Luo et al. 2021).

The heritability of pain is also an important consideration. Family and twin studies have consistently demonstrated that a significant proportion of the variance in pain susceptibility is attributable to inherited factors. For example, a study of twins found that the heritability of chronic pain was estimated to be approximately 50% (Nielsen et al. 2012).

In conclusion, the contribution of genomic and epigenetic factors to pain susceptibility across all species is complex and multifactorial. While individual genetic polymorphisms can contribute to pain susceptibility, it is the interaction of multiple genetic and environmental factors that ultimately determines an individual's pain sensitivity. Furthermore, the heritability of pain underscores the critical importance of genetic and potentially inherited epigenetic factors in the development of pain conditions. Understanding the interplay between genetic, epigenetic, and environmental factors is crucial to fully conceptualising the multidimensional biopsychosocial nature of pain and the potential intergenerational impacts.

10 The Complexity of Species Comparisons of Pain

The complex and multifaceted characteristics of pain complicates its study in different species. Despite the many shared neural mechanisms that underlie pain perception in different animals, there are significant differences in the way that pain is experienced, expressed and managed. These differences are driven by variations in genetics, anatomy and behaviour, as well as by the biopsychosocial contexts in which pain occurs.

One of the primary challenges in studying pain across species is the significant differences in the cellular and anatomical features of the nervous system. For example, the expression of ion channels and receptors involved in pain signalling can differ significantly between species. One well-known example of this is the Nav1.7 ion channel, which plays a critical role in the transmission of pain signals. In humans, Nav1.7 is critical for nociceptive signalling. However, while Nav1.7 is present in most mammalian species, its expression and function vary widely. For example, while mice express Nav1.7 in nociceptive neurons, the channel is not essential for acute pain perception in these animals (Nassar et al. 2004). Similarly, while Nav1.7 plays a significant role in human pain perception (and in other non-human primates), it is not required for pain sensation in some non-primate species, e.g. chickens and fish (Bourinet et al. 2014).

TRPV1 channels provide another example of relevant interspecific differences. TRPV1 is present in a variety of vertebrate species, including mammals, birds (chickens), amphibians (tropical clawed frogs), and teleost fishes (zebrafish). These channels are generally activated by heat, making this a conserved trait across these species. However, recent research has shown that the heat-evoked activity of TRPV1 is reduced in certain animals, such as ground squirrels and camels (Saito and Tominaga 2017). While heat activation is a common trait, sensitivity to capsaicin varies among species. For example, TRPV1 channels in humans, rodents and dogs are highly sensitive to capsaicin, whereas those in rabbits, chickens, tropical clawed frogs and zebrafish have lower capsaicin sensitivity. This variation explains the scientific basis for coating bird feeders with capsaicin or chili powder: this deters squirrels from consuming the birdseed while leaving birds undeterred (Saito and Tominaga 2017).

Other cellular differences can also contribute to variations in pain perception between species. For example, the expression of different subtypes of opioid receptors and their associated endogenous opioid peptides differ significantly between species. In some cases, these differences can explain species-specific responses to analgesic drugs. For example, mice and rats have lower expression of the mu-opioid receptor in the brain than humans and primates, which may contribute to the reduced effectiveness of opioid analgesics in these animals (Hutchinson et al. 2011).

Anatomical differences can also play a significant role in shaping the experience and expression of pain in different species. The presence or absence of specialised nociceptors, as well as variations in their density and distribution, can contribute to differences in pain perception. In some species, such as reptiles and fish, nociceptors are less specialised and less densely distributed than in mammals; this can affect the quality and intensity of pain perception (Sneddon 2015). Similarly, the size and structure of the spinal cord and brain can also affect pain perception. Some species, such as fish and amphibians, have relatively simple spinal cord structures and limited brain function, which can affect their ability to perceive and respond to nociception and pain (Sneddon 2015).

In addition to cellular and anatomical differences, variations in neuroendocrine systems can also play a role in forming pain perception and its management in different species. For example, the HPA axis, which is responsible for the release of stress hormones such as cortisol, can affect pain perception and modulation. In some species, such as fish and reptiles, the HPA axis is less well developed than in mammals, which can affect their responses to painful stimuli.

Finally, the social and behavioural contexts in which pain occurs can also contribute to significant variations in pain perception and expression between species. While many species display overt signs of pain, such as vocalisation, body language and altered behaviours, some animals may exhibit more subtle or atypical responses to painful stimuli. In some cases, these responses may be influenced by the social dynamics within a given species, such as the need to maintain social hierarchy or avoid attracting predators. Additionally, the complex social and environmental contexts in which animals live can affect how they experience pain and their willingness to express pain-related behaviours. For example, animals that are living in unpredictable or dangerous environments may be more likely to hide signs of pain in order to avoid attracting predators or to continue to forage for food. On the other hand, animals that are living in more stable or controlled environments may be more likely to display pain-related behaviours, as they do not have the same pressures to conceal their pain. These factors can further complicate the measurement and interpretation of pain in different species and highlight the need for more nuanced and species-specific approaches to studying pain.

Understanding the biopsychosocial nature of pain is a complex and multifaceted task, particularly when attempting to compare pain experiences between different species or even between predator and prey animals. Differences in neuroanatomy, neurophysiology and social context can all contribute to differences in pain perception and expression, making it challenging to develop a unified understanding of pain across the animal kingdom. However, with advances in technology and a growing awareness of the importance of studying animal pain, there is hope that we can continue to refine our understanding of the mechanisms underlying pain and develop more effective strategies for treating it in both humans and animals.

11 The New Technologies Frontier—Biomarkers of Pain

Pain is a complex phenomenon that remains difficult to identify and quantify. Pain biomarkers, i.e., biological indicators of pain and measurements of the physiological state of the body, are of great current interest in the field of pain diagnosis and measurement. By using appropriate biomarkers, pain identification, quantification, and treatment can be improved, especially when multiple layers of species-specific neuroimmune modulatory pathways are taken into consideration. The specificity of the biomarker in different species, however, poses a challenge and makes it necessary to understand more about the molecular events that underpin the biomarker mechanism and how this is associatively linked, or causally implicated, in the pain processes. This section will explore the complexities of using pain biomarkers in different species, particularly between predator and prey animals, and how cellular and anatomical differences, neuroendocrine systems, and social and behavioural contexts can affect pain perception and expression.

Whilst the development of pain biomarkers has emerged as a promising strategy to improve the diagnosis and treatment of chronic pain, there is a blend of challenges and opportunities involved in discovering and validating these for pain therapeutics (Davis et al. 2020). First, there is the need for rigorous validation, to ensure biomarker specificity and sensitivity for pain. Another challenge is the need for standardised and reproducible methods and technologies for sample collection, processing and analysis, as well as appropriate statistical methods for data analysis. Importantly, these approaches need to be designed from the outset to be economical and time-efficient, given the prevalence of pain and the desire to diagnose and measure it. A third challenge is the complexity of pain and the need for a multimodal approach that incorporates both subjective and objective measures (Davis et al. 2020). Despite these challenges, there are several opportunities for the development of pain biomarkers, including the use of new technologies such as multiomics, imaging and wearables, as well as the integration of reported outcomes and machine learning algorithms. All of this raises medical, legal and ethical issues surrounding the use of 'objective tests' for pain diagnosis and treatment (Davis et al. 2017). Moving forward, continued research, collaboration and innovation will be needed to overcome these challenges and fully realise the potential of pain biomarkers for chronic pain.

This chapter has intended to make clear that the immune and nervous systems are tightly interconnected, with bidirectional communication and cellular movement between immune cells and neurons. This communication takes place through the neuroimmune synapse, where neurotransmitters, neuropeptides and cytokines modulate immune cell function and vice versa. As a result, there is a growing interest in developing blood-based biomarkers of pain that reflect the dynamic interplay between the immune and nervous systems (Evans et al. 2021; Kwok et al. 2012, 2013). The development of these biomarkers would enable researchers to quantify the molecular activity at the neuroimmune synapse and facilitate the discovery of new pain therapeutics. Several studies have demonstrated that peripheral blood mononuclear cells (PBMCs) from individuals with chronic pain exhibit increased responsiveness to TLR ligands, which are known to induce immune cell activation and inflammation (Evans et al. 2021; Kwok et al. 2012, 2013). Additionally, studies have identified changes in androgen levels, along with TLR4 responsiveness of PBMCs, in young women with dysmenorrhoea, providing insight into the link between hormonal factors and pain (Evans et al. 2021). Given the mechanistic links of these cellular blood fractions to pain and the importance of the receptor systems that have been targeted, these approaches appear to have good face validity as possible sensitive and specific pain biomarkers. Moreover, they highlight the value of quantifying the accessible functional units of the neuroimmune synapse in the rapid development of future safe and effective pain therapeutics.

Devising strategies to rapidly develop pain solutions is of paramount concern. Given the complexity of the biopsychosocial pain state introduced up to now, it may be asked 'how can a single measurement identify *and* quantify pain?'. Clearly, the complexity of pain identification is further enhanced when multiple layers of species-specific neuroimmune modulatory pathways are taken into consideration. In the example provided earlier, the specific biomarker Kwok et al. (2012) identified was the expression of the proinflammatory cytokine interleukin-1β (IL-1β) following stimulation of toll-like receptors (TLR) 2 and 4. Exposing PBMCs to LPS and Pam(3)CysK (a tri-acylated lipopeptide) caused an increase in IL-1β. PBMCs isolated from humans suffering from chronic pain secreted significantly higher levels of IL-1β compared to those isolated from pain-free controls. This response was recreated in rats in a follow-up study using a graded model of chronic pain to further confirm that the IL-1β increase was the result of pain and not aberrant inflammation (Kwok et al. 2013). The assay used in the latter studies demonstrated similar neuro-immune modulatory behaviour between humans and rodents when both were experiencing chronic pain, but the relative sensitivities to the LPS innate immune challenge is profoundly different with different concentrations of the LPS needed to achieve a similar output. Nevertheless, the assay still distinguished PBMCs isolated from pain versus no-pain backgrounds. This highlights the need for a species-specific understanding of how these types of assays function and possibly point to different susceptibilities to pain.

12 Innovation of Novel Pain Treatments

The search for new pain treatments for both humans and animals is ongoing and requires a comprehensive understanding of the biopsychosocial nature of pain, as well as the neuronal and non-neuronal mechanisms that underlie pain perception and expression. The development of new drugs and non-drug therapies, as well as the optimisation of early life periods for optimal brain development, are all critical areas of research for pain treatment.

One promising area of research is the targeting of the neuroimmune interface in pain. This represents a potentially untapped field for the development of new pain treatments, particularly in chronic pain conditions. Targeting this interface involves identifying and targeting the cellular and molecular pathways that underpin the interactions between the immune and nervous systems in the context of pain perception and modulation, i.e. the neuroimmune synapse.

A key target in the neuroimmune interface is the glial pain mechanism. Glial cells, such as microglia and astrocytes, play critical roles in modulating pain perception and transmission. The reactive activation of glial cells can lead to the release of proinflammatory mediators and signalling primers that contribute to pain hypersensitivity. The targeting of glial cells and their downstream consequences, such as the release of proinflammatory cytokines and chemokines, represents a promising area of research for developing new pain therapies. Moreover, sex-based differences in pain mechanisms may mean that sex-specific pain interventions may be mechanistically viable and have improved efficacy.

In addition to drug therapies, drug-free treatments such as physical therapy, cognitive-behavioural therapy and mindfulness-based interventions have shown promise in the treatment of pain in humans. These approaches focus on the mind-body connection and can help patients manage their pain through a combination of physical and mental techniques. How these might be applied to animals is not clear at this stage, but perhaps can be linked to creating controllability of biopsychosocial threats. For example, simple access to a wooden dowel during a stressful experience is sufficient to give medial prefrontal cortex 'control' of a stressful experience. Moreover, safety cues are trainable and, through conditioning can be used to provide beneficial physiological resolution of stressful events (Hutchinson and Terry 2019). Furthermore, early life periods represent a critical period for brain development and the establishment of pain pathways. Optimisation of early life periods through early-life pain management could have long-lasting effects on pain perception and modulation throughout the lifespan (Hutchinson and Terry 2019).

The search for new pain treatment approaches for both humans and animals involves a multidisciplinary approach that incorporates both drug and non-drug therapies, as well as optimisation of early life periods for optimal brain development. Targeting the neuroimmune interface and glial pain mechanisms represents a promising area of research for the development of new pain therapies. The integration of these approaches could ultimately lead to the development of more effective and personalised pain management strategies.

13 Conclusion

Pain is a complex phenomenon that is difficult to identify and quantify. This chapter has explored the intricacies of pain from a variety of perspectives, including the underlying mechanisms, the interaction between pain and mental health, and the complexities of comparing pain across different species. The most recent advances in pain research, including the development of new technologies for identifying and quantifying pain, as well as the ongoing search for effective pain treatments, have also been examined. At the heart of the discussion is the fundamental understanding that pain is not simply a sensory experience, but a complex and multifaceted phenomenon that is shaped by a variety of biological, psychological, and social factors. From the interaction of immune and nervous systems, through the role of genetics and epigenetics in pain susceptibility, to the effect of social and environmental factors on pain perception and expression, the exploration of pain has highlighted the vast network of factors that contribute to pain in different individuals and species.

It is critical that a deeper understanding of the biopsychosocial nature of pain remains an important future research objective. This concept underscores the complexity of pain and highlights the need for a multidimensional approach to pain diagnosis and treatment that takes into account the interplay between biological, psychological, and social factors. Truly exploring pain in 4D. In this context, the development of pain biomarkers has emerged as a promising strategy for improving the diagnosis and treatment of chronic pain. By using these biomarkers, pain identification, quantification, and treatment can be improved, especially when multiple layers of species-specific neuroimmune modulatory pathways are considered. However, the development of pain biomarkers is not without its challenges. One of the primary issues is the need for rigorous validation to ensure their specificity and sensitivity for pain. Additionally, the need for standardised, reproducible and scalable methods and technologies for sample collection, processing and analysis, are needed. The development of new appropriate statistical methods for data analysis is also required because current approaches fail to account for the complexities within biology, and the appropriate constraints to known physiological systems. These approaches must be designed from the outset to be economical and time-efficient given the prevalence of pain and the desire to diagnose and measure it.

Progress in understanding the nature of pain and the subsequent development of more effective strategies for its treatment will depend on developing new technologies and approaches to pain research, including the use of pain biomarkers. The study of pain is a rapidly evolving and multidisciplinary field that requires a comprehensive and nuanced approach. Continued research, collaboration, and innovation will be essential to overcome the many challenges of pain research and to fully realise a future where pain is still experienced but the burden of unwanted and unproductive pain is eliminated.

Acknowledgements MRH is the recipient of an ARC Future Fellowship (FT180100565), and CHJ is the recipient of an associated PhD scholarship. The Davies Livestock Research Centre has also supported this work.

References

Achenbach J, Rhein M, Gombert S, Meyer-Bockenkamp F, Buhck M, Eberhardt M, Leffler A, Frieling H, Karst M (2019) Childhood traumatization is associated with differences in TRPA1 promoter methylation in female patients with multisomatoform disorder with pain as the leading bodily symptom. Clin Epigenetics 11(1):126

Achenbach J, Rhein M, Glahn A, Frieling H, Karst M (2022) Leptin promoter methylation in female patients with painful multisomatoform disorder and chronic widespread pain. Clin Epigenetics 14(1):13

Adcock SJJ (2021) Early life painful procedures: long-term consequences and implications for farm animal welfare. Front Anim Sci 2. https://doi.org/10.3389/fanim.2021.759522

Apkarian AV, Bushnell MC, Treede R-D, Zubieta J-K (2005) Human brain mechanisms of pain perception and regulation in health and disease. Eur J Pain 9(4):463–484

Aramideh JA, Vidal-Itriago A, Morsch M, Graeber MB (2021) Cytokine signalling at the microglial Penta-partite synapse. Int J Mol Sci 22(24). https://doi.org/10.3390/ijms222413186

Aroke EN, Jackson P, Meng L, Huo Z, Overstreet DS, Penn TM, Quinn TL, Cruz-Almeida Y, Goodin BR (2022) Differential DNA methylation in black and white individuals with chronic low back pain enrich different genomic pathways. Neurobiol Pain 11:100086

Barratt DT, Sia AT, Tan E-C, Somogyi AA (2021) Innate immune and neuronal genetic markers are highly predictive of postoperative pain and morphine patient-controlled analgesia requirements in Indian but not Chinese or Malay hysterectomy patients. Pain Med. https://doi.org/10.1093/pm/pnab172

Beggs S, Currie G, Salter MW, Fitzgerald M, Walker SM (2012) Priming of adult pain responses by neonatal pain experience: maintenance by central neuroimmune activity. Brain 135(Pt 2):404–417

Bilbo SD, Schwarz JM (2009) Early-life programming of later-life brain and behavior: a critical role for the immune system. Front Behav Neurosci 3:14

Bilbo SD, Schwarz JM (2012) The immune system and developmental programming of brain and behavior. Front Neuroendocrinol 33(3):267–286

Blom JMC, Benatti C, Alboni S, Capone G, Ferraguti C, Brunello N, Tascedda F (2006) Early postnatal chronic inflammation produces long-term changes in pain behavior and N-methyl-D-aspartate receptor subtype gene expression in the central nervous system of adult mice. J Neurosci Res 84(8):1789–1798

Boissé L, Spencer SJ, Mouihate A, Vergnolle N, Pittman QJ (2005) Neonatal immune challenge alters nociception in the adult rat. Pain 119(1–3):133–141

Bourinet E, Altier C, Hildebrand ME, Trang T, Salter MW, Zamponi GW (2014) Calcium-permeable ion channels in pain signaling. Physiol Rev 94(1):81–140

Burke NN, Fan CY, Trang T (2016) Microglia in health and pain: impact of noxious early life events. Exp Physiol 101(8):1003–1021. https://doi.org/10.1113/eph.2016.101.issue-810.1113/EP085714

Burke NN, Finn DP, McGuire BE, Roche M (2017) Psychological stress in early life as a predisposing factor for chronic pain development: clinical and pre-clinical evidence and neurobiological mechanisms. J Neurosci Res 95(6):1257–1270

Cabral JM, Grácio D, Soares-da-Silva P, Magro F (2015) Short- and long-term regulation of intestinal Na+/H+ exchange by toll-like receptors TLR4 and TLR5. Am J Physiol Gastrointest Liver Physiol 309(8):G703–G715

Caterina MJ, Julius D (2001) The vanilloid receptor: a molecular gateway to the pain pathway. Annu Rev Neurosci 24:487–517

Caterina MJ, Schumacher MA, Tominaga M, Rosen TA, Levine JD, Julius D (1997) The capsaicin receptor: a heat-activated ion channel in the pain pathway. Nature 389(6653):816–824

Chidambaran V, Zhang X, Geisler K, Stubbeman BL, Chen X, Weirauch MT, Meller J, Ji H (2019) Enrichment of genomic pathways based on differential DNA methylation associated with chronic post-surgical pain and anxiety in children: a prospective, pilot study. J Pain 20(7):771–785

Chu Y-C, Chan K-H, Tsou M-Y, Lin S-M, Hsieh Y-C, Tao Y-X (2007) Mechanical pain hypersensitivity after incisional surgery is enhanced in rats subjected to neonatal peripheral inflammation: effects of N-methyl-D-aspartate receptor antagonists. Anesthesiology 106(6):1204–1212

Ciampi de Andrade D, Maschietto M, Galhardoni R, Gouveia G, Chile T, Victorino Krepischi AC, Dale CS, Brunoni AR, Parravano DC, Cueva Moscoso AS, Raicher I, Kaziyama HHS, Teixeira MJ, Brentani HP (2017) Epigenetics insights into chronic pain: DNA hypomethylation in fibromyalgia-a controlled pilot-study. Pain 158(8):1473–1480

Cirulli F, Berry A, Alleva E (2003) Early disruption of the mother-infant relationship: effects on brain plasticity and implications for psychopathology. Neurosci Biobehav Rev 27(1–2):73–82

Clark C, Murrell J, Fernyhough M, O'Rourke T, Mendl M (2014) Long-term and trans-generational effects of neonatal experience on sheep behaviour. Biol Lett 10(7). https://doi.org/10.1098/rsbl.2014.0273

Costigan M, Scholz J, Woolf CJ (2009) Neuropathic pain: a maladaptive response of the nervous system to damage. Annu Rev Neurosci 32:1–32

Crombez G, Van Ryckeghem DML, Eccleston C, Van Damme S (2013) Attentional bias to pain-related information: a meta-analysis. Pain 154(4):497–510

Davies AJ, Rinaldi S, Costigan M, Oh SB (2020) Cytotoxic immunity in peripheral nerve injury and pain. Front Neurosci 14:142

Davis KD, Flor H, Greely HT, Iannetti GD, Mackey S, Ploner M, Pustilnik A, Tracey I, Treede R-D, Wager TD (2017) Brain imaging tests for chronic pain: medical, legal and ethical issues and recommendations. Nat Rev Neurol 13(10):624–638

Davis KD, Aghaeepour N, Ahn AH, Angst MS, Borsook D, Brenton A, Burczynski ME, Crean C, Edwards R, Gaudilliere B, Hergenroeder GW, Iadarola MJ, Iyengar S, Jiang Y, Kong J-T, Mackey S, Saab CY, Sang CN, Scholz J et al (2020) Discovery and validation of biomarkers to aid the development of safe and effective pain therapeutics: challenges and opportunities. Nat Rev Neurol 16(7):381–400

Diatchenko L, Slade GD, Nackley AG, Bhalang K, Sigurdsson A, Belfer I, Goldman D, Xu K, Shabalina SA, Shagin D, Max MB, Makarov SS, Maixner W (2005) Genetic basis for individual variations in pain perception and the development of a chronic pain condition. Hum Mol Genet 14(1):135–143

Diogenes A, Ferraz CCR, Akopian AN, Henry MA, Hargreaves KM (2011) LPS sensitizes TRPV1 via activation of TLR4 in trigeminal sensory neurons. J Dent Res 90(6):759–764

Emery EC, Habib AM, Cox JJ, Nicholas AK, Gribble FM, Woods CG, Reimann F (2015) Novel SCN9A mutations underlying extreme pain phenotypes: unexpected electrophysiological and clinical phenotype correlations. J Neurosci Off J Soc Neurosci 35(20):7674–7681

Evans SF, Kwok Y, Solterbeck A, Pyragius C, Hull ML, Hutchinson MR, Rolan P (2021) The relationship between androgens and days per month of period pain, pelvic pain, headache, and TLR4 responsiveness of peripheral blood mononuclear cells in young women with dysmenorrhoea. J Pain Res 14:585–599

Fertleman CR, Baker MD, Parker KA, Moffatt S, Elmslie FV, Abrahamsen B, Ostman J, Klugbauer N, Wood JN, Gardiner RM, Rees M (2006) SCN9A mutations in paroxysmal extreme pain disorder: allelic variants underlie distinct channel defects and phenotypes. Neuron 52(5):767–774

Fitzgerald M (2005) The development of nociceptive circuits. Nat Rev Neurosci 6(7):507–520

Fitzgerald M, McKelvey R (2016) Nerve injury and neuropathic pain—a question of age. Exp Neurol 275:296–302. https://doi.org/10.1016/j.expneurol.2015.07.013

Gerra MC, Carnevali D, Ossola P, González-Villar A, Pedersen IS, Triñanes Y, Donnini C, Manfredini M, Arendt-Nielsen L, Carrillo-de-la-Peña MT (2021) DNA methylation changes in fibromyalgia suggest the role of the immune-inflammatory response and central sensitization. J Clin Med Res 10(21). https://doi.org/10.3390/jcm10214992

Grace PM, Hutchinson MR, Bishop A, Somogyi AA, Mayrhofer G, Rolan PE (2011) Adoptive transfer of peripheral immune cells potentiates allodynia in a graded chronic constriction injury model of neuropathic pain. Brain Behav Immun 25(3):503–513

Grace PM, Hurley D, Barratt DT, Tsykin A, Watkins LR, Rolan PE, Hutchinson MR (2012) Harnessing pain heterogeneity and RNA transcriptome to identify blood-based pain biomarkers: a novel correlational study design and bioinformatics approach in a graded chronic constriction injury model. J Neurochem 122(5):976–994

Grace PM, Hutchinson MR, Maier SF, Watkins LR (2014) Pathological pain and the neuroimmune interface. Nat Rev Immunol 14(4):217–231

Grace PM, Tawfik VL, Svensson CI, Burton MD, Loggia ML, Hutchinson MR (2021) The Neuroimmunology of chronic pain: from rodents to humans. J Neurosci Off J Soc Neurosci 41(5):855–865

Gritsch S, Lu J, Thilemann S, Wörtge S, Möbius W, Bruttger J, Karram K, Ruhwedel T, Blanfeld M, Vardeh D, Waisman A, Nave K-A, Kuner R (2014) Oligodendrocyte ablation triggers central pain independently of innate or adaptive immune responses in mice. Nat Commun 5:5472

Guesgen MJ, Beausoleil NJ, Minot EO, Stewart M, Stafford KJ (2014) Social context and other factors influence the behavioural expression of pain by lambs. Appl Anim Behav Sci 159:41–49. https://doi.org/10.1016/j.applanim.2014.07.008

Hannibal KE, Bishop MD (2014) Chronic stress, cortisol dysfunction, and pain: a psychoneuroendocrine rationale for stress management in pain rehabilitation. Phys Ther 94(12):1816–1825

von Hehn CA, Baron R, Woolf CJ (2012) Deconstructing the neuropathic pain phenotype to reveal neural mechanisms. Neuron 73(4):638–652

Hohmann AG, Neely MH, Piña J, Nackley AG (2005) Neonatal chronic hind paw inflammation alters sensitization to intradermal capsaicin in adult rats: a behavioral and immunocytochemical study. J Pain 6(12):798–808

Hsieh C-T, Lee Y-J, Lee JW, Lu S, Tucci MA, Dai X, Ojeda NB, Lee HJ, Fan L-W, Tien L-T (2020) Interleukin-1 receptor antagonist ameliorates the pain hypersensitivity, spinal inflammation and oxidative stress induced by systemic lipopolysaccharide in neonatal rats. Neurochem Int 135:104686

Hua X-Y, Chen P, Fox A, Myers RR (1996) Involvement of cytokines in lipopolysaccharide-induced facilitation of CGRP release from capsaicin-sensitive nerves in the trachea: studies with interleukin-1β and tumor necrosis factor-α. J Neurosci Off J Soc Neurosci 16(15):4742–4748

Hutchinson MR, Terry R (2019) Review: what innovations in pain measurement and control might be possible if we could quantify the neuroimmune synapse? Animal 13(12):3000–3008

Hutchinson MR, Shavit Y, Grace PM, Rice KC, Maier SF, Watkins LR (2011) Exploring the neuroimmunopharmacology of opioids: an integrative review of mechanisms of central immune signaling and their implications for opioid analgesia. Pharmacol Rev 63(3):772–810

Hutchinson MR, Buijs M, Tuke J, Kwok YH, Gentgall M, Williams D, Rolan P (2013) Low-dose endotoxin potentiates capsaicin-induced pain in man: evidence for a pain neuroimmune connection. Brain Behav Immun 30:3–11

Ji R-R, Chamessian A, Zhang Y-Q (2016) Pain regulation by non-neuronal cells and inflammation. Science 354(6312):572–577

Kemper RHA, Spoelstra MB, Meijler WJ, Ter Horst GJ (1998) Lipopolysaccharide-induced hyperalgesia of intracranial capsaicin sensitive afferents in conscious rats. Pain 78(3):181–190

Koivisto A-P, Belvisi MG, Gaudet R, Szallasi A (2022) Advances in TRP channel drug discovery: from target validation to clinical studies. Nat Rev Drug Discov 21(1):41–59

Kwok YH, Hutchinson MR, Gentgall MG, Rolan PE (2012) Increased responsiveness of peripheral blood mononuclear cells to in vitro TLR 2, 4 and 7 ligand stimulation in chronic pain patients. PLoS One 7:e44232. https://www.ncbi.nlm.nih.gov/pmc/articles/PMC3429430/

Kwok YH, Tuke J, Nicotra LL, Grace PM, Rolan PE, Hutchinson MR (2013) TLR 2 and 4 responsiveness from isolated peripheral blood mononuclear cells from rats and humans as potential chronic pain biomarkers. PLoS One 8(10):e77799

Lacroix-Fralish ML, Ledoux JB, Mogil JS (2007) The pain genes database: an interactive web browser of pain-related transgenic knockout studies. Pain 131(1–2):3.e1–3.e4

Laprairie JL, Johns ME, Murphy AZ (2008) Preemptive morphine analgesia attenuates the long-term consequences of neonatal inflammation in male and female rats. Pediatr Res 64(6):625–630

Larrondo C, Orihuela A, Strappini A, Acosta-Jamett G, Mota-Rojas D, Gallo C (2021) Provision of straw and the presence of undocked lambs reduce the behavioural and physiological expressions of pain and stress associated with tail docking in lambs: a preliminary study. Anim Prod Sci 61(4):423–431. https://doi.org/10.1071/AN20237

Lee JH, Kim W (2020) The role of satellite glial cells, astrocytes, and microglia in oxaliplatin-induced neuropathic pain. Biomedicines 8(9):324. https://doi.org/10.3390/biomedicines8090324

Li J, Baccei ML (2016) Neonatal tissue damage promotes spike timing-dependent synaptic long-term potentiation in adult spinal projection neurons. J Neurosci 36(19):5405–5416. https://doi.org/10.1523/JNEUROSCI.3547-15.2016

Li J, Xie W, Zhang J-M, Baccei ML (2009) Peripheral nerve injury sensitizes neonatal dorsal horn neurons to tumor necrosis factor-alpha. Mol Pain 5:10

Li J, Blankenship ML, Baccei ML (2013) Deficits in glycinergic inhibition within adult spinal nociceptive circuits after neonatal tissue damage. Pain 154(7):1129–1139

Li Y, Adamek P, Zhang H, Tatsui CE, Rhines LD, Mrozkova P, Li Q, Kosturakis AK, Cassidy RM, Harrison DS, Cata JP, Sapire K, Zhang H, Kennamer-Chapman RM, Jawad AB, Ghetti A, Yan J, Palecek J, Dougherty PM (2015a) The cancer chemotherapeutic paclitaxel increases human and rodent sensory neuron responses to TRPV1 by activation of TLR4. J Neurosci 35(39):13487–13500

Li Y, Zhang H, Kosturakis AK, Cassidy RM, Zhang H, Kennamer-Chapman RM, Jawad AB, Colomand CM, Harrison DS, Dougherty PM (2015b) MAPK signaling downstream to TLR4 contributes to paclitaxel-induced peripheral neuropathy. Brain Behav Immun 49:255–266

Lim EJ, Back SK, Kim MA, Li C, Lee J, Jeong KY, Na HS (2009) Long-lasting neonatal inflammation enhances pain responses to subsequent inflammation, but not peripheral nerve injury in adult rats. Int J Develop Neurosci 27(3):215–222

Luo D, Li X, Tang S, Song F, Li W, Xie G, Liang J, Zhou J (2021) Epigenetic modifications in neuropathic pain. Mol Pain 17:17448069211056767

Mai L, Liu Q, Huang F, He H, Fan W (2021) Involvement of mast cells in the pathophysiology of pain. Front Cell Neurosci 15:665066

McCracken L, Waran N, Mitchinson S, Johnson CB (2010) Effect of age at castration on behavioural response to subsequent tail docking in lambs. Vet Anaesth Analg 37(4):375–381. https://doi.org/10.1111/j.1467-2995.2010.00547.x

Meints SM, Edwards RR (2018) Evaluating psychosocial contributions to chronic pain outcomes. Prog Neuro-Psychopharmacol Biol Psychiatry 87(Pt B):168–182

Menzies V, Lyon DE, Archer KJ, Zhou Q, Brumelle J, Jones KH, Gao G, York TP, Jackson-Cook C (2013) Epigenetic alterations and an increased frequency of micronuclei in women with fibromyalgia. Nurs Res Pract 2013:795784

Mezey E, Tóth ZE, Cortright DN, Arzubi MK, Krause JE, Elde R, Guo A, Blumberg PM, Szallasi A (2000) Distribution of mRNA for vanilloid receptor subtype 1 (VR1), and VR1-like immunoreactivity, in the central nervous system of the rat and human. Proc Natl Acad Sci USA 97(7):3655–3660

Mika J, Zychowska M, Popiolek-Barczyk K, Rojewska E, Przewlocka B (2013) Importance of glial activation in neuropathic pain. Eur J Pharmacol 716(1–3):106–119

Min H, Lee H, Lim H, Jang YH, Chung SJ, Lee CJ, Lee SJ (2014) TLR4 enhances histamine-mediated pruritus by potentiating TRPV1 activity. Mol Brain 7:59

Min H, Cho W-H, Lee H, Choi B, Kim Y-J, Lee HK, Joo Y, Jung SJ, Choi S-Y, Lee S, Lee SJ (2018) Association of TRPV1 and TLR4 through the TIR domain potentiates TRPV1 activity by blocking activation-induced desensitization. Mol Pain 14:1744806918812636

Montesino-Goicolea S, Sinha P, Huo Z, Rani A, Foster TC, Cruz-Almeida Y (2020) Enrichment of genomic pathways based on differential DNA methylation profiles associated with chronic musculoskeletal pain in older adults: an exploratory study. Mol Pain 16:1744806920966902

Moriarty O, Harrington L, Beggs S, Walker SM (2018) Opioid analgesia and the somatosensory memory of neonatal surgical injury in the adult rat. Br J Anaesth 121(1):314–324

Moriarty O, Tu Y, Sengar AS, Salter MW, Beggs S, Walker SM (2019) Priming of adult incision response by early-life injury: neonatal microglial inhibition has persistent but sexually dimorphic effects in adult rats. J Neurosci 39(16):3081–3093

Moseley GL (2005) Distorted body image in complex regional pain syndrome. Neurology 65(5):773

Mustafa S, Bajic JE, Barry B, Evans S, Siemens KR, Hutchinson MR, Grace PM (2023) One immune system plays many parts: the dynamic role of the immune system in chronic pain and opioid pharmacology. Neuropharmacology 228:109459

Nassar MA, Stirling LC, Forlani G, Baker MD, Matthews EA, Dickenson AH, Wood JN (2004) Nociceptor-specific gene deletion reveals a major role for Nav1.7 (PN1) in acute and inflammatory pain. Proc Natl Acad Sci USA 101(34):12706–12711

Nicotra L, Loram LC, Watkins LR, Hutchinson MR (2012) Toll-like receptors in chronic pain. Exp Neurol 234:316. https://www.sciencedirect.com/science/article/pii/S001448861100358X.

Nielsen CS, Knudsen GP, Steingrímsdóttir ÓA (2012) Twin studies of pain. Clin Genet 82(4):331–340

O'Neil SM, Hans EE, Jiang S, Wangler LM, Godbout JP (2022) Astrocyte immunosenescence and deficits in interleukin 10 signaling in the aged brain disrupt the regulation of microglia following innate immune activation. Glia. https://doi.org/10.1002/glia.24147

Paoloni-Giacobino A, Luthi F, Stenz L, Le Carré J, Vuistiner P, Léger B (2020) Altered BDNF methylation in patients with chronic musculoskeletal pain and high biopsychosocial complexity. J Pain Res 13:1289–1296

Peacock S, Patel S (2008) Cultural influences on pain. Rev Pain 1(2):6–9

Peng YB, Ling QD, Ruda MA, Kenshalo DR (2003) Electrophysiological changes in adult rat dorsal horn neurons after neonatal peripheral inflammation. J Neurophysiol 90(1):73–80

Price DD (2000) Psychological and neural mechanisms of the affective dimension of pain. Science 288(5472):1769–1772

Raja SN, Carr DB, Cohen M, Finnerup NB, Flor H, Gibson S, Keefe FJ, Mogil JS, Ringkamp M, Sluka KA, Song X-J, Stevens B, Sullivan MD, Tutelman PR, Ushida T, Vader K (2020) The revised International Association for the Study of Pain definition of pain: concepts, challenges, and compromises. Pain 161(9):1976–1982

Roizenblatt S, Andersen ML, Bignotto M, D'Almeida V, Martins PJF, Tufik S (2010) Neonatal arthritis disturbs sleep and behaviour of adult rat offspring and their dams. Eur J Pain 14(10):985–991

Rubino E, Boschi S, Giorgio E, Pozzi E, Marcinnò A, Gallo E, Roveta F, Grassini A, Brusco A, Rainero I (2022) Analysis of the DNA methylation pattern of the promoter region of calcitonin gene-related peptide 1 gene in patients with episodic migraine: an exploratory case-control study. Neurobiol Pain 11:100089

Ruda MA, Ling QD, Hohmann AG, Peng YB, Tachibana T (2000) Altered nociceptive neuronal circuits after neonatal peripheral inflammation. Science 289(5479):628–631

Saito S, Tominaga M (2017) Evolutionary tuning of TRPA1 and TRPV1 thermal and chemical sensitivity in vertebrates. Temperature (Austin, Tex) 4(2):141–152

Salberg S, Noel M, Burke NN, Vinall J, Mychasiuk R (2020) Utilization of a rodent model to examine the neurological effects of early life adversity on adolescent pain sensitivity. Dev Psychobiol 62(3):386–399

Schwarz JM, Bilbo SD (2012) Sex, glia, and development: interactions in health and disease. Horm Behav 62(3):243–253

Sneddon LU (2015) Pain in aquatic animals. J Exp Biol 218(Pt 7):967–976

Somogyi AA, Sia AT, Tan E-C, Coller JK, Hutchinson MR, Barratt DT (2016) Ethnicity-dependent influence of innate immune genetic markers on morphine PCA requirements and adverse effects in postoperative pain. Pain 157(11):2458–2466

Stenz L, Carré JL, Luthi F, Vuistiner P, Burrus C, Paoloni-Giacobino A, Léger B (2022) Genome-wide epigenomic analyses in patients with nociceptive and neuropathic chronic pain subtypes reveals alterations in methylation of genes involved in the neuro-musculoskeletal system. J Pain 23(2):326–336

Stephens K, Cooper BA, West C, Paul SM, Baggott CR, Merriman JD, Dhruva A, Kober KM, Langford DJ, Leutwyler H, Luce JA, Schmidt BL, Abrams GM, Elboim C, Hamolsky D, Levine JD, Miaskowski C, Aouizerat BE (2014) Associations between cytokine gene variations and severe persistent breast pain in women following breast cancer surgery. J Pain 15(2):169–180

Stephens KE, Levine JD, Aouizerat BE, Paul SM, Abrams G, Conley YP, Miaskowski C (2017) Associations between genetic and epigenetic variations in cytokine genes and mild persistent breast pain in women following breast cancer surgery. Cytokine 99:203–213

Takenaka S, Sukenaga N, Ohmuraya M, Matsuki Y, Maeda L, Takao Y, Hirose M (2020) Association between neuropathic pain characteristics and DNA methylation of transient receptor potential ankyrin 1 in human peripheral blood. Medicine (Baltimore) 99(8):e19325

Totsch SK, Sorge RE (2017) Immune system involvement in specific pain conditions. Mol Pain 13:1744806917724559

Wager TD, Atlas LY, Lindquist MA, Roy M, Woo C-W, Kross E (2013) An fMRI-based neurologic signature of physical pain. N Engl J Med 368(15):1388–1397

Walker SM, Meredith-Middleton J, Cooke-Yarborough C, Fitzgerald M (2003) Neonatal inflammation and primary afferent terminal plasticity in the rat dorsal horn. Pain 105(1–2):185–195

Walker SM, Fitzgerald M, Hathway GJ (2015) Surgical injury in the neonatal rat alters the adult pattern of descending modulation from the rostroventral medulla. Anesthesiology 122(6):1391–1400

Walker SM, Beggs S, Baccei ML (2016) Persistent changes in peripheral and spinal nociceptive processing after early tissue injury. Exp Neurol 275 Pt 2(0 2):253–260

Wangzhou A, Paige C, Neerukonda SV, Naik DK, Kume M, David ET, Dussor G, Ray PR, Price TJ (2021) A ligand-receptor interactome platform for discovery of pain mechanisms and therapeutic targets. Sci Signal 14(674). https://doi.org/10.1126/scisignal.abe1648

Watkins LR, Hutchinson MR, Ledeboer A, Wieseler-Frank J, Milligan ED, Maier SF (2007) Glia as the "bad guys": implications for improving clinical pain control and the clinical utility of opioids. Brain Behav Immun 21(2):131–146

Woolf CJ (2011) Central sensitization: implications for the diagnosis and treatment of pain. Pain 152(3 Suppl):S2–S15

Wu Y, Wang Y, Wang J, Fan Q, Zhu J, Yang L, Rong W (2019) TLR4 mediates upregulation and sensitization of TRPV1 in primary afferent neurons in 2,4,6-trinitrobenzene sulfate-induced colitis. Mol Pain 15:1744806919830018

Xu Q, Ford NC, He S, Huang Q, Anderson M, Chen Z, Yang F, Crawford LK, Caterina MJ, Guan Y, Dong X (2021) Astrocytes contribute to pain gating in the spinal cord. Sci Adv 7(45):eabi6287

Basic Principles of Animal Pain Assessment

Jo Murrell

Abstract

Pain can only be managed if it can be identified and assessed. However, pain assessment in animals is challenging, especially in minimally domesticated species such as some production animals. In production animal species, various approaches are taken to assess pain, including observations of behaviour, measurement of various physiological parameters, infrared thermography, activity levels, gait analysis, nociceptive threshold testing and electroencephalography. Furthermore, scoring systems, which attribute number to the severity or intensity of pain, can be utilised and currently involve the use of uni-dimensional pain scales, composite pain scales and facial grimace scales. Pain assessment for both acute and chronic pain is more comprehensive in companion animal species but given the role of many production animal species in food production, biomedical research, exhibitions and as pets, further research is warranted.

Keywords

Acute pain · Chronic pain · Pain assessment

J. Murrell (✉)
Bristol Veterinary Specialists, Bristol, UK
e-mail: Jo.Murrell@Bristolvetspecialists.co.uk

1 Introduction

Accurate recognition and quantification of pain are prerequisites of effective pain management. Unless pain is recognized, there is a risk that it will go untreated and animals will suffer. Quantification of pain is important if under- or over-treatment of pain with analgesics are to be avoided. The under-treatment of pain is undesirable while 'over' administration of analgesic drugs may have adverse consequences. All analgesic drugs have side effects, and so tailoring of analgesic drug administration to required need is imperative. This balance is accomplished by pain quantification. The side-effects of some analgesic drugs, e.g. morphine—an opioid agonist, may be more apparent when given to animals that are not in pain and so do not require opioid medication: morphine is most likely to cause excitation when given to pain-free dogs and horses (Hamamoto-Hardman et al. 2019). Other analgesics, such as the non-steroidal anti-inflammatory drugs (NSAIDs) can have serious adverse effects including gastro-intestinal ulceration (Monteiro-Steagall et al. 2013). Consequently, limiting their use to when they are strongly indicated is prudent. Recognizing and quantifying pain is the key to implementing an effective and safe pain management strategy for individual animals, herds or flocks.

A major problem with pain assessment in animals is that it can only be accomplished by

R. E. Clutton, G. C. Musk (eds.), *Pain and Suffering in Farmed Animals*,
https://doi.org/10.1007/978-3-032-11407-5_10

'proxy'. Effective pain management in human patients is usually based on assessments made by the painful subject themselves, i.e. there is self-reporting of the pain experience—which is the 'gold standard' method of pain assessment. However, in animals (and in non-verbal children and adults), pain assessment must be conducted by a third party, which leads to a significant possibility of error in the assessment process. The profound complexity of an individual's pain experience inescapably complicates its accurate comprehension by a third party. At present, there is no way to overcome this limitation, although the use of cognitive bias as a pain assessment tool (see later) offers some promise in surmounting the challenge.

Currently, researching animal pain assessment in the published literature reveals a predominance of companion animal, i.e. dog and cat, studies, although a substantial literature also exists on the subject in laboratory rodents. This emphasis on particular species is accompanied by a focus on behavioural pain studies and the development of quantitative assessment methods. Large animal (mainly sheep and cattle) studies generally focus on the physiological effects of pain in castration and tail docking models of juvenile pain. More research is needed into the behavioural effects of pain in all large animal species, particularly in adult animals.

2 Recognizing Signs of Animal Pain

2.1 Behavioural Signs

Behavioural changes are probably the best means of identifying animal pain because behavioural analysis can detect changes revealing both the unpleasant sensory and emotional elements, which constitute the overall pain experience (Raja et al. 2020). The manifestation of emotional signs, such as dullness or depression, anxiety, fearfulness or stress depends upon the species and the individual animal's ability to cope. The sensory aspect of pain is identified by locating its source and intensity by gentle manipulation of the animal and careful palpation around the suspected lesion.

While behavioural changes associated with abdominal pain have been described in rodents, cats, dogs and horses, there is a paucity of data describing such signs in farmed animals. In rodents, abdominal pain is associated with 'cat-like back-arching' (where the animal arches its back upwards); writhing, described as a lateral contortion of the abdominal muscles and twitching of any body part (Roughan and Flecknell 2004). In cats, abdominal pain after ovariohysterectomy is manifest as restlessness, an unwillingness to curl up and sleep, resentment or even aggression associated with abdominal palpation, and occasionally increased licking of the abdominal area (Väisänen et al. 2007). In dogs, cranial abdominal pain is manifest as adoption of the 'praying position', in which the thoracic limbs are extended and lowered to the ground while the pelvic limbs are in the standing position. Horses with abdominal pain (colic) are commonly restless, may thrash about their stable, lie on their backs and writhe, and, or 'paw' [sic] the ground while standing (Sutton et al. 2013). In most of these species, characteristic facial expression changes have been associated with abdominal pain (and are discussed under pain quantification, below). In lambs, behavioural changes associated with pain are best described after tail docking and castration. Acute pain-related behaviours of restlessness, foot stamping, lying and kicking have been described after both procedures (Molony et al. 1993) and may be ameliorated by the administration of analgesics before these interventions (Small et al. 2020). In calves, behaviours associated with castration and disbudding or dehorning are also well-documented. Following disbudding, ear flicks, head shaking and head rubbing are reported, as well as tail swishing, foot stamping and vocalization (Stilwell et al. 2012; Adcock et al. 2020). These acute pain-related behaviours are probably more likely to reflect the sensory, rather than the emotional component of pain. Calf behaviours after castration include tail swishing and pelvic limb stamping, with reduced grazing and lying activities (Nogues et al. 2021). Again, some of these behaviours are mitigated by analgesic drug administration (Van der Saag et al. 2019; Martin et al. 2022). It is possible that the repertoire of behavioural changes shown by calves and lambs differs from adults due to age-dependent differences in emotional maturity, although this is an area of research that requires much further exploration. Sheep in severe pain may head press and hunch their back (Fig. 1).

Behavioural changes caused by chronic pain are likely to differ from those caused by acutely painful conditions and typically are more subtle, particularly when they result from longstanding pain. Probably the best descriptors of behavioural changes caused by chronic pain are those arising from osteoarthritis in dogs and cats. Lameness is most commonly noted but there are also changes in playfulness, toileting (in cats) (Monteiro 2020) and sociability (dogs) (Belshaw et al. 2020). Although lameness is common in dairy cattle, knowledge of other behavioural changes associated with chronic pain—including that caused by mastitis is scant in farm animals. This scarcity may reflect a greater societal importance placed on companion animal pain, compared to farmed animal pain.

Behaviours, and behavioural changes resulting from pain may be quantified using an ethogram. This method of pain assessment is a record (often graphical) of an array of behaviours a particular animal species may show over a defined period. Ethograms are usually split into frequency-dependent behaviours, where the frequency of defined events, such as foot stamping per unit time, can be counted, or 'state' behaviours, where the duration of time spent doing a particular behaviour, e.g. lying down, is calculated. This approach has been used extensively in quantifying the effects of analgesic interventions on husbandry procedures in calves, cows, sheep and lambs (Molony and Kent 1997; Calderon and Cook 2011; Silva et al. 2020; Nogues et al. 2021).

There are limitations to using behaviour for pain recognition in animals. Importantly, it is generally acknowledged that predated species, such as cattle, sheep and horses, are less likely to exhibit overt pain signs because of associated evolutionary disadvantages, i.e. that exhibiting pain-related behaviour will invite predation (Hudson et al. 2008). Consequently, detecting behavioural change in these species is challenging and may underestimate the severity and/or duration of pain they are experiencing. Cats are less likely to show pain-related behaviours than dogs: being solitary animals, there is less reason for cats to exhibit signs of pain. This personality trait contrasts with dogs, which, in evolutionary terms, are pack animals.

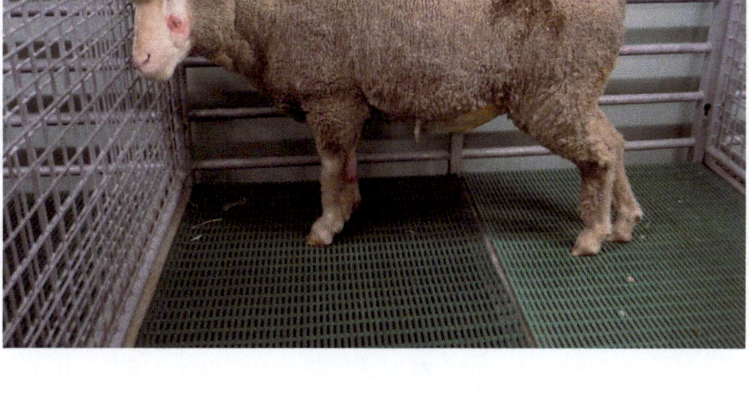

Fig. 1 Head pressing and back hunching in a sheep following a laparotomy (photo Dr G. Musk)

Direct observation restricts the manifestation of pain-related behaviours in all animals if they are aware they are being observed. However, this effect may be partially overcome by acclimatising the subjects before the direct observation period. Remote video recording of behaviour may also reduce the effects of observer presence. Systems are available to monitor home-cage behaviour in rodents and subsequently detect behavioural changes due to pain (Burand Jr et al. 2023). Such technology overcomes the problems of direct observer presence and the considerable time required for observers to watch and document video data. Using behaviour to recognize pain is rendered more accurate if the behaviours to be assessed are rigidly defined in advance, and observers (either directly or by retrospective video review) are trained to recognize them. In general, large numbers of animals are needed for behavioural studies because of significant individual variation in the behavioural manifestation of pain.

3 Physiological Measures

Changes in physiological variables such as heart and respiratory rate, blood pressure, and neuroendocrine variables such as plasma cortisol and substance P concentration have been widely used in farm animal studies investigating the painful effects of common husbandry interventions (Stafford and Mellor 2005; Coetzee 2013). In general, pain is associated with increases in heart and respiratory rate, blood pressure and plasma concentrations of neuroendocrine markers. However, while physiological changes are generally regarded as being objective measures of pain, there are several limitations in their use. First and foremost, physiological measures are not specific to pain and most likely reflect the stress that accompanies a painful experience. Therefore, interventions such as handling and isolation during husbandry procedures will also cause changes in physiology that cannot be separated from the accompanying pain. Although heart and respiratory rate and blood pressure (measured indirectly) can be registered with relative ease, measuring cortisol and substance P is more challenging. Cortisol is most commonly measured in blood, which requires invasive and stressful blood sampling (Fig. 2), although it can now be measured in saliva (Hernandez et al. 2014), faeces (Pesenhofer et al. 2006), milk (Gygax et al. 2006) and hair (Braun et al. 2017). However, blood cortisol concentrations also show profound diurnal variation, so that the timing of analgesia studies using cortisol as a biomarker is pivotal (Greco and Stabenfeldt 1997). Substance P is measured in blood or saliva and is considered by some to be more specific for pain compared to cortisol, such that when both are measured together, pain and distress can be dissociated from each other (Coetzee et al. 2008). However, there are also limitations associated with substance P measurement, e.g. there are large age and individual variations in substance P concentrations (Dockweiler et al. 2013; Tschoner et al. 2018) and measurement in the field is not straightforward. Blood tubes must contain protease inhibitors and samples must be kept on ice and promptly centrifuged to separate plasma after collection. Other pro-inflammatory biomarkers such as cytokines,e.g. IL-1β, IL-2, C-reactive protein and Tumour Necrosis Factor-α, have also been measured in studies in order to quantify inflammation and from which pain may be inferred (Mirra et al. 2018). In addition, the measurement of acute phase proteins such as fibrinogen, haptoglobin, serum amyloid A and substances such as iron has been used to assess pain in cattle (Musk et al. 2017).

Other physiological measures include feed intake (either concentrate or milk) and liveweight gain. These have been most widely used in cattle, where automatic feeders are commonly used to record feed or milk intake. However, feed intake is probably more closely linked to disease status or stress levels rather than pain per se, and so is not recommended as a sole measure (Sutherland et al. 2018). Liveweight gain can also be measured easily and automatically, but data linking decreases in liveweight gain with pain are contentious. It is important to recognize that castration itself can decrease live weight gain, so using this measure to assess medium-term pain

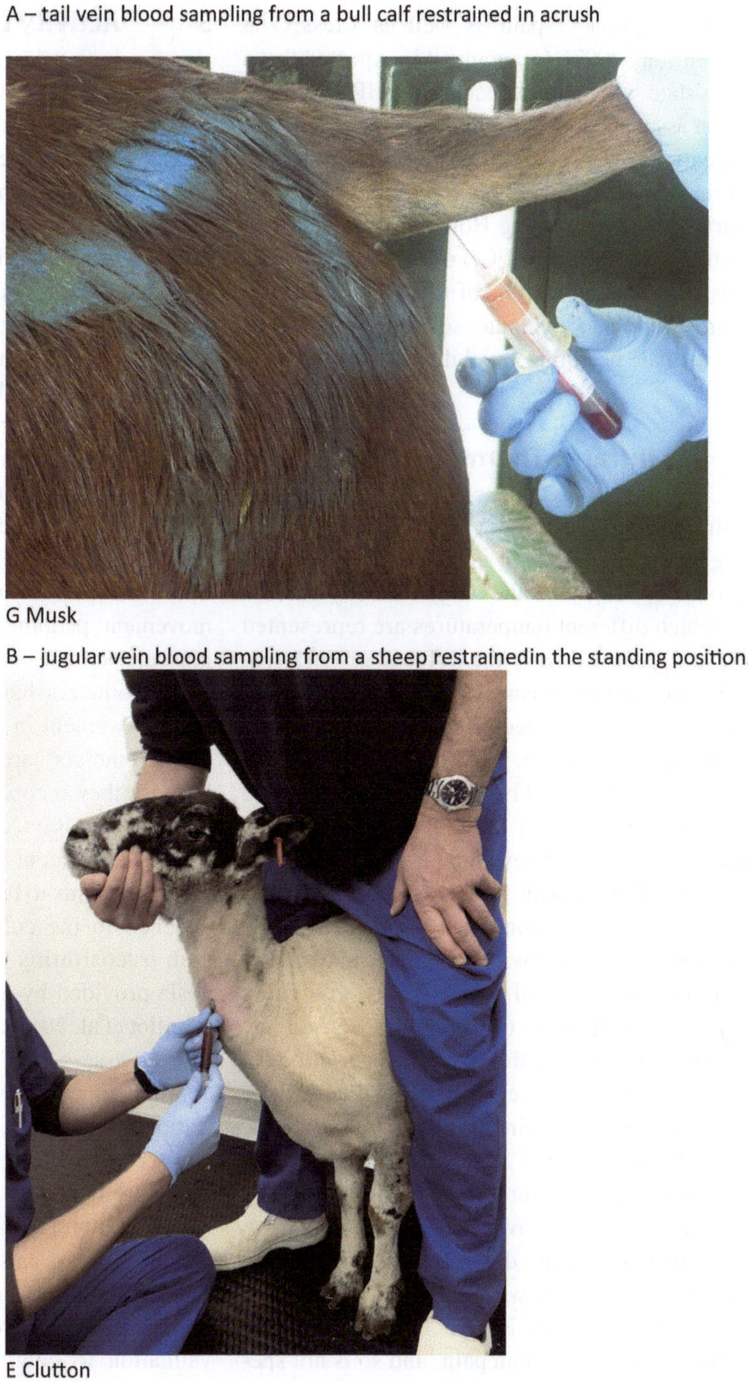

Fig. 2 Blood sampling can be stressful for the animal as it usually requires restraint and isolation. (**a**) Tail vein blood sampling from a bull calf restrained in a crush. (**b**) Jugular vein blood sampling from a sheep restrained in the standing position

after castration may be inaccurate (González et al. 2010).

Heart rate variability (HRV) has been widely investigated as a biomarker for nociception in a range of animal species. Heart rate variability describes the range of changes in the time interval between subsequent heartbeats and depends on the balance between activity in the sympathetic and parasympathetic branches of the autonomic nervous system. This balance is affected by nociception or pain as well as stress (von Borell et al. 2007). Generally, it is expected that a pain state will cause a decrease in HRV because of an associated increase in sympathetic nervous activity (Stewart et al. 2010). The electrocardiogram (ECG) can readily be measured in a wide variety of species using Holter monitors, which digitize and record the ECG over prolonged periods. Subsequent analysis of the digitized record using proprietary computer software is then used to calculate heart rate variability.

4 Infrared Thermography

Infrared thermography is in vivo digital imaging of an animal using an infrared camera. In producing a map of the animal's body surface in which different temperatures are represented by different colours, the effects of different interventions or disease states on surface temperatures can be detected easily and remotely (Fig. 3). For example, it has been used to detect inflammation caused by mastitis in dairy cattle (Berry et al. 2003; Metzner et al. 2015). Recently, changes in eye temperature have been used to identify pain in calves caused by disbudding or castration (Stewart et al. 2009; Stewart et al. 2010); with a decrease in eye temperature documented immediately following these interventions (presumably caused by sympathetically mediated vasoconstriction of the ocular blood vessels as part of the 'fight or flight response' to pain. Using the eye as a thermal biomarker also has the advantage that changes in temperature are not confounded by the presence of fur over the area. It is also an objective technique that is non-invasive and easy to carry out. However, it is important to be aware that changes in eye temperature may reflect stress rather than pain, and so is not specific to pain states (Stewart et al. 2008); using thermal imaging in combination with other pain biomarkers may give a higher degree of certainty that pain is being measured rather than stress. Infrared cameras are expensive to purchase, which may limit widespread applicability of the technique.

5 Activity Levels

The objective measurement of animal activity can be a useful tool in documenting changes in behaviour with different pain states, or after the administration of analgesic drugs (Currah et al. 2009). Counting steps is difficult to do accurately when observing animals in real-time or on video recordings. Pedometers can be used to quantify activity and can be attached directly to the legs of larger animals (Coetzee et al. 2013) (Fig. 4). They calculate the step-count from raw data using an algorithm (Theurer et al. 2013) and, in larger animals, such as calves, show good accuracy (Currah et al. 2009). Pedometers have also been attached to the tails of calves to 'count' the number of tail movements after interventions such as castration. In measuring movement, pedometers can be useful in grading musculoskeletal disease or lameness. Accelerometers have also been used to document movement in a range of species. Tri-axial accelerometers are most used in research because they record acceleration along the x, y and z axes (Ravi et al. 2005). This approach enables time spent lying, walking and standing, and step counts to be calculated. Accelerometers attached to the collar have been used in cats with osteoarthritis to objectively evaluate analgesia provided by administration of an NSAID (Guillot et al. 2013). They have also been widely used to assess pain caused by husbandry interventions in calves, such as disbudding and castration. In dogs, commercially available accelerometers, e.g. Fitbark, that attach to the collar are widely available and contain an algorithm allowing downloaded movement data analysis that reveals the time the animal spends running, playing, sleeping, etc. However, these accelerometers have not undergone objective validation to confirm that the algorithms are correct. Pedometers and accelerometers have several advantages: they are relatively inexpensive and easy to use, while the data are easy to interpret. They are also usually small and lightweight, so their attachment is unlikely to change the animal's behaviour. With appropriate storage capacity, they can also record data over a

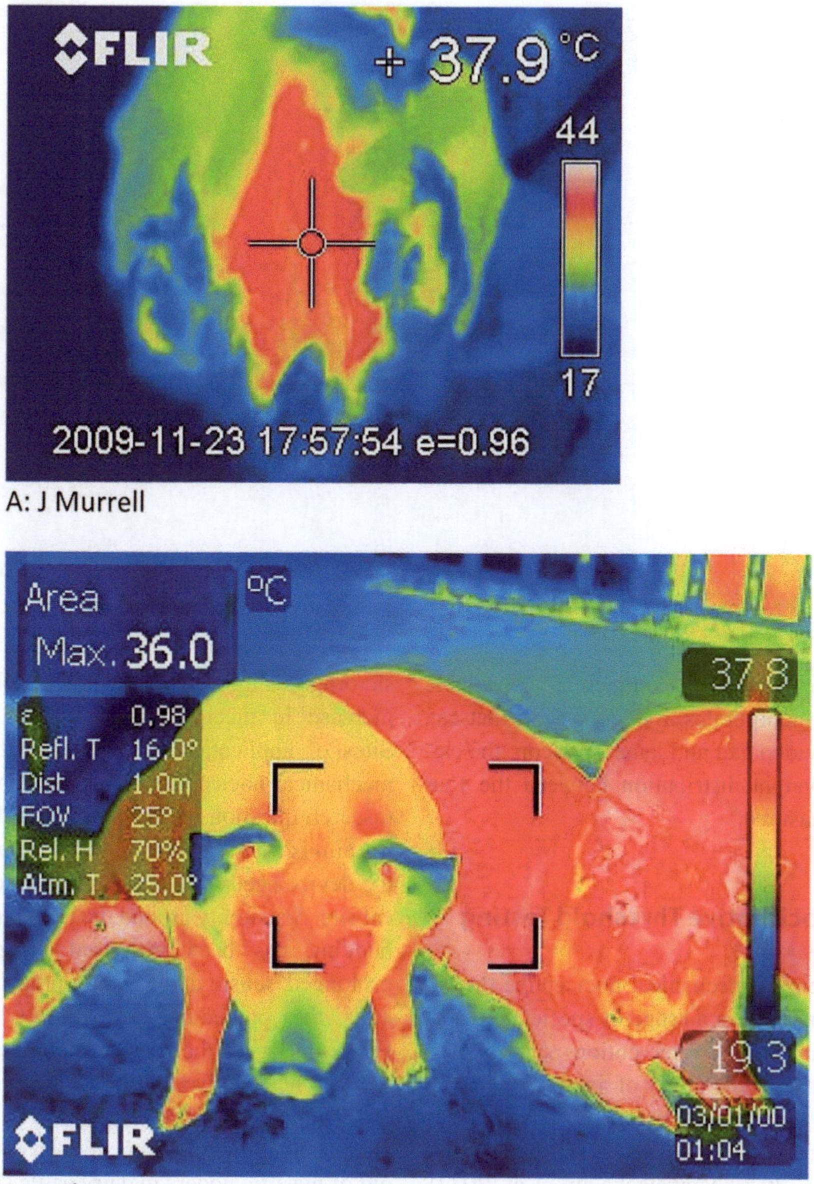

Fig. 3 (**a**) Thermal image of the ventrum of a laying hen. The keel is the red colour in the image. Thermal imaging has been used to investigate inflammation associated with keel bone fractures in laying hens (Nasr 2020). (**b**) Inappetence and reduced fluid intake associated with acute post-operative pain possibly account for the generalised vasoconstriction apparent in the pig on the left of the thermogram

relatively long period of time. Alternatively, wireless accelerometers may be used, though wireless technology restricts the area over which animals can move during a given recording period.

5.1 Pressure Mats

Pressure mats are not widely available, being specialist equipment. However, they have been used to assess lameness in dairy cattle and calves

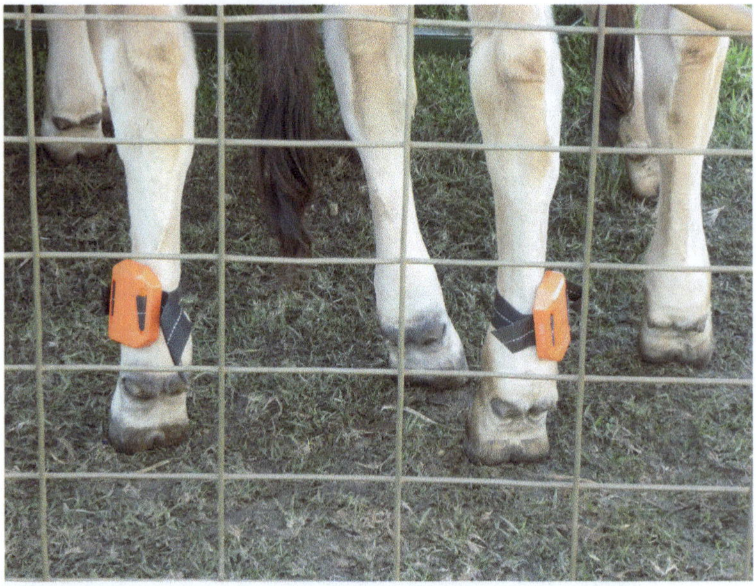

G Musk

Fig. 4 Pedometers attached to the hindlimbs of cattle to measure activity levels (photo Dr G. Musk)

(Wheeler et al. 2013). Commercially available equipment incorporates a video system that can be used to video gait and lameness score the cattle while weight distribution between the four limbs is measured.

5.2 Nociceptive Threshold Testing

Measurement of mechanical nociceptive threshold has been used in a range of species to detect pain associated with interventions such as surgery or to assess the effect of analgesic therapy, particularly the administration of opioids (Ley et al. 1995; Love et al. 2012; Musk et al. 2014; Frahm et al. 2020; Venkatachalam et al. 2022). Devices that measure mechanical nociceptive threshold effectively quantify the amount of force (in Newtons) needed to produce a behavioural response indicative of pain. Measuring mechanical nociceptive thresholds is particularly helpful in detecting the secondary mechanical hyperalgesia surrounding an area of tissue injury, where the mechanical hyperalgesia will cause a decrease in mechanical threshold. This technique effectively quantifies sensitivity to the application of manual pressure around the same area. When assessing the effect of analgesic interventions, then analgesic drugs such as opioids will generally cause an increase in mechanical nociceptive threshold. Sites of application of the device to measure mechanical nociceptive threshold depend on the research question being answered. Obviously, if the aim is to assess sensitivity after surgery, then the device needs to be applied close to the surgical site. Sensitivity around the horn bud after disbudding has been measured in goats, with prolonged sensitivity found for up to 7 weeks after the procedure (Alvarez et al. 2019). Hyperalgesia of the pastern (demonstrated by a reduced nociceptive threshold) in lame cattle has also been found (Whay et al. 1997), as well as alterations in the mechanical nociceptive threshold of mastitic udders in dairy cattle (Kemp et al. 2008). To measure the effect of analgesic drugs, a non-compressive area of the body is used to deliver the stimulus, such as the distal tibia or hock or metacarpus/metatarsus (Fig. 5) or on the rump (Fig. 6). This approach aims to increase the robustness and repeatability of the measurement. It is important to have a safe maximal value of applied force or 'cut-off' when measuring mechanical nociceptive threshold to prevent tissue damage, particularly in analgesia studies where the nociceptive threshold is likely to increase. There are many variables that can

Fig. 5 Cuff housing a pin to measure mechanical nociceptive threshold in sheep. The cuff is attached to a pneumatic device which uses air to advance the pin against the metatarsus. The end point of the test is a leg lift

J Murrell

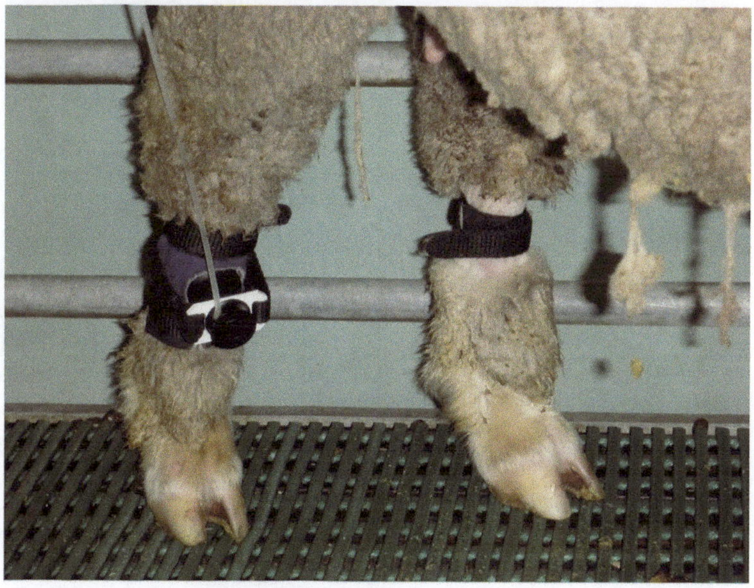

G Musk

change mechanical nociceptive thresholds, including diameter of the pin through which the force is applied, position of the device on the animal and position of the animal (lying or sitting). In dogs, the individual animal was also found to be a significant source of variability, making it important to take baseline measures before painful or analgesic interventions. The response to the mechanical stimulus may also vary from withdrawal of the area, e.g. limb, skin flick or conscious movement or turning of the head, with each of these endpoints probably signifying different things. For example, a reflex withdrawal of the limb may not always reflect conscious perception of the stimulus, whereas a head turn may be more indicative of a pain response associated with conscious stimulus perception. Using the same endpoint in each animal may be more robust in terms of repeatability of thresholds. Handheld pressure algometers designed to mea-

Fig. 6 Hand held device to measure mechanical nociceptive threshold in a dog (**a**) and a bull (with an extension tube (**b** and **c**))

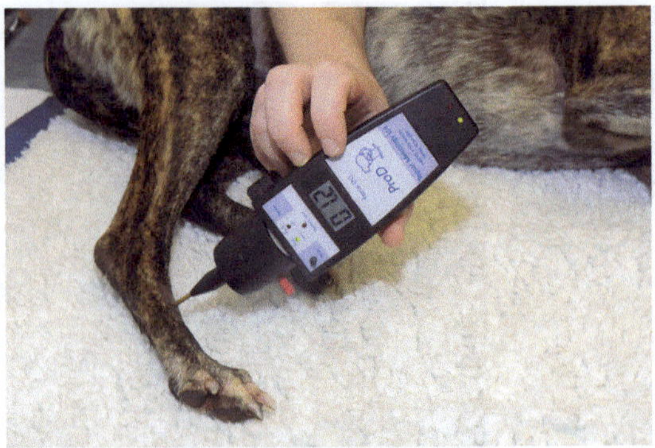

A - J Murrell

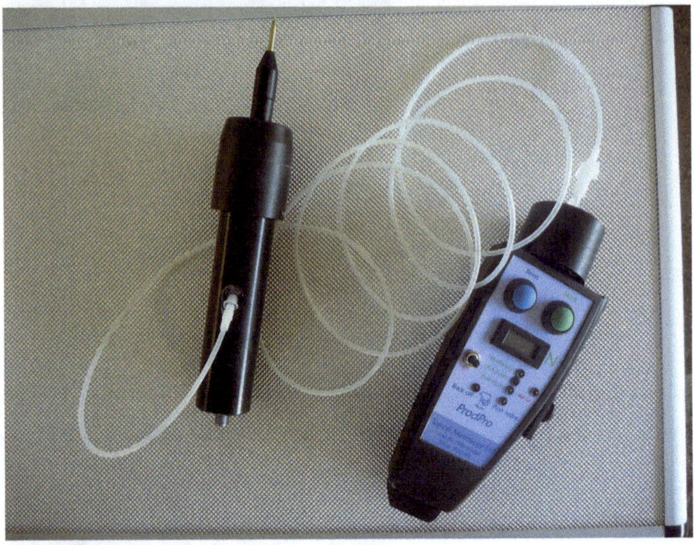

B - G Musk

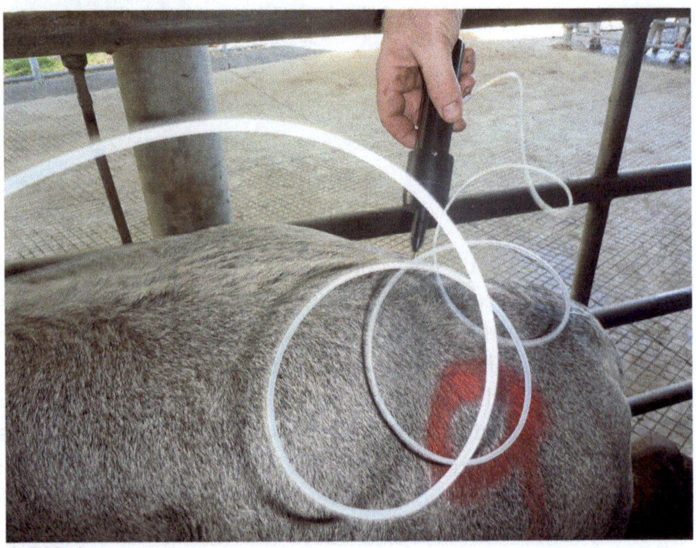

C - G Musk

sure mechanical nociceptive thresholds are relatively accessible and inexpensive, making them a useful resource to assess pain in animals.

Thermal nociceptive threshold has also been measured in studies in many species to assess the efficacy of analgesics such as opioids (Love et al. 2011; Slingsby et al. 2016; Adami and Spadavecchia 2020). Commonly, a thermode is applied to the animal's skin (after shaving to maximise contact) and heated at a constant rate once the device is activated remotely by a handheld switch (Fig. 7). Heat is applied until the animal responds, either with a skin twitch or conscious head turn or limb withdrawal, depending on where the heat stimulus is applied and the type of response being measured in the study. Thermode-based devices are designed to ensure that heating is immediately stopped once the animal has responded, and the thermode is cooled rapidly to avoid skin damage. Skin temperature at the point at which the animal responds is the outcome measure in this test. As with mechanical nociceptive thresholds, it is important that there is a cut-off applied, the temperature above which the thermode will not heat, should the animal not make a visible response to the stimulus. This is to avoid thermal injury. Thermal sensitivity is usually only increased, i.e., thermal hyperalgesia develops, in the area of surgery or tissue damage itself, as opposed to mechanical hyperalgesia, which is found in the area of surrounding tissue, so thermal hyperalgesia is less popular as an indicator for pain than mechanical hyperalgesia. Unlike pressure algometers, which are readily available and cheap to buy, the equipment to measure thermal hyperalgesia is more specialised and expensive.

5.3 Electroencephalography

Pain perception is accepted to arise in the brain, predominantly in the cerebral cortex. Therefore, there has been significant interest in electroencephalographic (EEG) measures of pain perception (Murrell and Johnson 2006). Measuring the EEG in awake, moving animals is technically challenging because of difficulties in maintaining adequate electrode contact with the skin (cutaneous electrodes) or subcutaneous tissues (needle electrodes). This results in an EEG signal which is very 'noisy', i.e. contains significant artefact, making the EEG signal difficult to interpret. Johnson and colleagues have developed a minimal anaesthesia model in which animals are anaesthetised with a low concentration of halo-

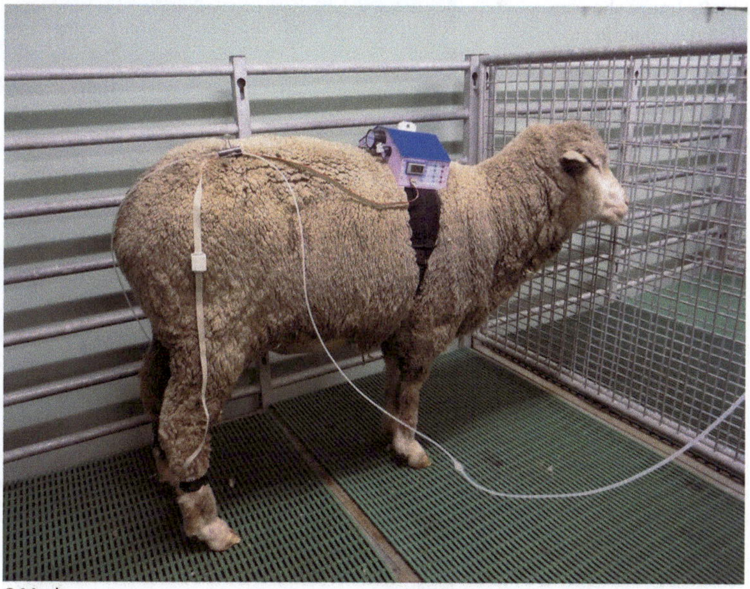

G Musk

Fig. 7 Wireless device for measuring thermal nociceptive threshold in a sheep. The probe is positioned on the lateral aspect of the hindlimb below the hock. The heat is controlled remotely and generated in the housing secured around the thorax (photo Dr G. Musk)

thane sufficient to induce unconsciousness, and the EEG recorded during the application of noxious stimuli, such as castration in horses (Murrell et al. 2003) and cattle (Lehmann et al. 2017), velvet antler removal in deer (Johnson et al. 2005) or tail docking in lambs. The EEG is recorded using three or four subcutaneous needle electrodes placed in a specific montage to record electrical cerebrocortical activity from the left and right sides of the cerebral cortex (Fig. 8). Other equipment required for EEG recording is pre-amplifiers to amplify the signal, main amplifiers, which filter the signal to specific frequencies, an analogue to digital converter and a computer. Changes in the EGG, which have been documented in the minimal anaesthesia model in response to a noxious stimulus, such as severance of the spermatic cord during castration, include an increase in median frequency of the EEG and a decrease in total power (Murrell et al. 2003). These changes represent a switch in EEG pattern from low-frequency high-amplitude activity to high-frequency low-amplitude activity in response to nociception. Application of analgesia before the noxious stimulus, e.g. by application of a lidocaine ring block around the base of the antler before velvet removal in deer, has been shown to abolish these EEG changes (Johnson et al. 2005). The use of the EEG as a biomarker for nociception in the minimal anaesthesia model has several advantages, particularly in farmed animals. It is an objective measure of nociception that does not rely on the detection of behavioural changes, which can be challenging in farmed animals. Furthermore, each animal is anaesthetised and analgesia can be provided at the end of the experiment, i.e. once all EEG data has been collected, before the animal is allowed to recover from general anaesthesia. This is advantageous from an ethical standpoint. Halothane is currently used as the inhalant agent to maintain general anaesthesia in the model because it causes less EEG depression than other inhalant agents (Murrell et al. 2008a), but it is becoming increasingly difficult to source, which may require changes to the anaesthesia protocol used for this model in the future.

5.4 Cognitive Bias

Techniques for measuring cognitive biases have been applied to animals after adaptation of human psychological methods (Lagisz et al. 2020). Cognitive bias tasks show that depressed humans judge ambiguous events negatively, i.e. regard

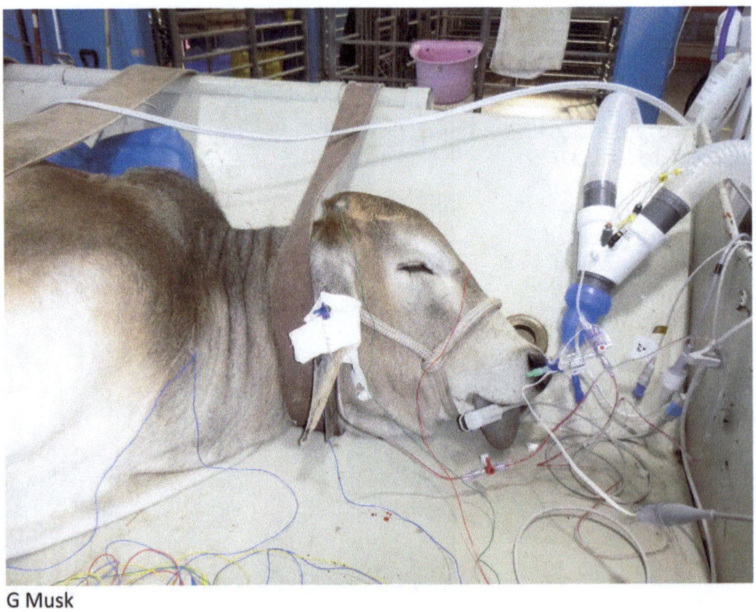

G Musk

Fig. 8 Image showing EEG recording in an anaesthetized bull calf undergoing castration in a research context (photo Dr G. Musk)

their glass as half-empty. The same principle has been applied to assess emotional states in animals. Rats housed in unpredictable housing conditions were more likely to judge ambiguous stimuli negatively compared to rats housed in predictable conditions, suggesting that the former may be in a negative emotional state (Parker et al. 2014). Recently, the same principle was applied to calves after hot iron disbudding (Neave et al. 2013) whereby calves were trained to either touch or not touch different coloured screens by associating them with either positive (e.g. milk) or negative (e.g. aversive sound and no milk cues). After hot iron disbudding the calves were presented with a variety of ambiguous coloured screens to determine whether the calf would interpret the cue positively and touch the screen with their nose, or not approach the screen and interpret it negatively. For 22 hours after disbudding, calves were more likely to interpret ambiguous cues negatively. This was the first evidence of judgement bias in animals due to pain and indicates that pain induces a negative emotional state in animals. It was also the first time that the emotional impact of pain on animals had been investigated with any certainty. Cognitive bias experiments are complex to carry out, and careful consideration must be given to potential confounders. Animals also require a significant amount of time-consuming training. However, the technique allows the emotional aspect of pain to be interrogated in animals.

5.5 Quantification of Pain

Quantifying pain is imperative to avoid the under- or over-administration of analgesic drugs. It allows tailored analgesia plans to be made that meet the demands of the individual animal, which frequently reveals the individual nature of the pain experience. Another important property of pain scales is that they can define a point for interventional (or 'rescue') analgesia, where additional rescue or 'top-up' analgesia is given. This approach is useful in both clinical practice and in research settings to standardise analgesic drug administration. A variety of systems have been developed to quantify or 'score' pain in different species. Before discussing these, it is important to recognize their need for validation: it is critical that measurement devices measure what they are designed to measure, i.e. pain, that they show good repeatability between and within observers, and that they are responsive.

6 Validity of Scoring Systems

Numerous forms of validity have been described but face, construct, content and criterion validity are commonly required for pain measurement methods (Reid et al. 2018). Face validity is whether a pain questionnaire or tool 'looks' like it measures what it is meant to measure. Face validity can be rated by asking a pool of individuals ranging from pain experts to associated parties, e.g. farmers, whether in their opinion the pain questionnaire is measuring pain. Construct validity is the extent to which the questionnaire measures the theoretical construct it is intended to measure, i.e. pain. Construct validity is tested by examining how the system scores a hypothetical construct. For example, a method's construct validity would be assured if its scores bore a linear relationship with the dose of analgesics given to animals undergoing the same procedure. A pain scoring system for post-castration pain in piglets would have high construct validity if it consistently awarded high, median, and low pain scores to animals receiving no, NSAID and NSAID with local anaesthetics, respectively. Content validity refers to the completeness of the items comprising the instrument and can be assessed formally by asking relevant experts to rate the relevance and clarity of items using a scale or more informally using subjective expert judgement. Criterion validity is the agreement of a new instrument with an existing 'gold standard' method but this comparison is difficult to execute for pain questionnaires in animals where no universally accepted gold standard exists. In verbal humans the existing gold standard is usually 'self-report', but this method is not possible in animals. Reliability is also important to measure in pain scales, where both intra-rater and inter-rater reliability must be known. Intra-rater reliability is when an unchanging subject is measured at two time points by the same individual. In practice, this process is performed using video recordings of the same animal at the same time point and assessing them at given time intervals. Inter-rater reliability is measured when two individuals rate pain in the same animal at the same time point. It is an important characteristic of any measurement system because pain scales are likely to be used to assess animals across time by different personnel; therefore, minimal inter- and intra-rater variability is highly desirable. Other important properties of pain assessment tools are responsiveness and clinical utility. Responsiveness is the ability of the tool to be sensitive enough to measure changes in pain that are relevant to the clinician and to the animal. Therefore, it is meaningful for pain scales to be able to differentiate between mild, moderate and severe pain. Clinical utility is the practical applicability of the scale. A scale needs to be easy to apply and interpret to be clinically useful. A scale that is time-consuming to complete and difficult to interpret will not be widely adopted in clinical practice.

6.1 Pain Scales for Acute Pain

Pain scales can be broadly divided into two forms: unidimensional and composite. In general, uni-dimensional scales only consider the sensory component of pain, i.e. where, and how much does it hurt? (Von Baeyer and Spagrud 2007). In contrast, composite pain scales aim to also assess the emotional component of pain, which is desirable (Murrell et al. 2008b). There has been an evolution in pain scales towards the development of composite pain scales.

6.2 Unidimensional Pain Scales

Unidimensional pain scales are typically simple descriptive scales, numerical rating scales and visual analogue scales. Simple descriptive scales (SDS) are generally divided into three or four different levels of pain, each with a descriptor and a different score assigned to each descriptor, for example from zero (no pain) to four (severe pain) (de Oliveira et al. 2014). The descriptors are generally species-specific, so that a different SDS is necessary for each species being studied. Simple descriptive scales have been used to describe lameness in farm animals. These scales are simple and quick to use and develop but are generally not validated and no gold standard SDS exists for either lameness or pain. It is also assumed (though established) that a score of 2 indicates double the pain awarded a score of 1, which is misleading. Furthermore, they are insensitive to different levels of pain and there is no validated intervention level for any SDS at which intervention analgesia administration is triggered. Numerical rating scales (NRSs) generally have a scale of 0–10 where 0 is no pain and 10 is the maximal possible pain. They have also been used for lameness assessment in several species including dogs and cattle, where 0 would be sound and 10 would be non-weight bearing on the affected limb (Plüss et al. 2021). As with SDSs, they are largely unvalidated, and no universally accepted intervention level exists for the provision of interventional analgesia. However, they are quick and simple to use and easy to understand. The NRS is also more sensitive than an SDS by virtue of having more scoring points, e.g. 10 versus 3 or 4. Visual analogue scales (VAS) comprise a 10 cm line with two anchor points, 0 at one end, representing no pain and 100 mm at the other end representing the worst pain possible with respect to the procedure being applied. The assessor marks the line corresponding to the level of pain that they believe the animal is experiencing, and the distance measured between the 0 anchor point and the mark on the line gives a quantifiable measure of pain. Typically, the VAS is completed without interacting with the animal, although a dynamic interactive visual analogue scale (DIVAS) has also been described, whereby the assessor interacts with the animal and generally applies gentle pressure around any wound or painful area before the pain assessment is made (Swaffield et al. 2020). The DIVAS is thought to be more accurate than the

VAS because of the extra information gained from the interaction. The VAS or DIVAS is very sensitive to different levels of pain—indeed, almost too sensitive because two or more assessors seldom mark the line similarly, even if they believe pain levels to be comparable. However, VASs are quick and easy to use. While there is no validated intervention level for rescue analgesia, many studies in dogs and cats have incorporated arbitrary cut-offs of 40 mm (Slingsby and Waterman-Pearson 2002; Slingsby et al. 2006). Like other unidimensional scales, an assumption that a score of 20 mm represents half as much pain as a 40 mm score is untested, i.e. the linearity of VAS is unknown. In general, composite pain scales or grimace scales have superseded unidimensional scales, although NRSs are still widely used for evaluation and quantification of lameness in different species (Welsh et al. 1993; Thompson et al. 2019; Aulakh et al. 2020).

6.3 Composite Pain Scales

Composite pain scales (CPSs) have the advantage that they attempt to capture both the sensory and emotional components of pain. The importance of the emotional component of the pain experience makes this attempt laudable: CPSs represent a significant advance in the clinical utility of pain scales. Examples of early composite pain scales are the Colorado Acute Pain Scales for dogs and cats (Palmeira et al. 2022; Lovell et al. 2022). These scales are both simple but try and incorporate psychological and behavioural aspects of the animal's behaviour as well as the response to wound palpation and body tension. The scales have only five summary levels of pain depending on the outcome of the assessment in the three categories: psychological and behavioural, response to palpation and body tension. Each category has several different descriptors that accompany it, and each category can be scored from 0 to 4. The appropriateness of the descriptors for a particular animal determines the assigned score. However, these scales are unvalidated and have since fallen into disuse with the development of validated composite pain scales. The canine Glasgow Composite Measure Pain Scale (Short Form) was developed more recently using psychometric principles (Testa et al. 2021). This technique utilises an established process, taken from human medicine, to select items for the scale, construct the questionnaire followed by subsequent testing for validity, reliability and sensitivity. The scale comprises six questions each with a number of different answers that are rated with a score from 0 (not present) to 4 or 5. The first part of the questionnaire (section A) asks the assessor to examine the animal from outside of its cage, section B involves interaction with the animal and section C asks questions pertaining to the emotional state of the animal. For section B, lameness and the response to wound palpation are assessed. Including the response to gentle palpation around the wound is critical because it allows the detection of secondary mechanical hyperalgesia, which is present when stimuli that are normally painful cause a heightened pain sensation in an area of uninjured tissue. Secondary mechanical hyperalgesia is indicative of central sensitisation and/or upregulation of the pain pathways. The maximum possible score for the scale is 20 if the animal cannot be walked and one question from section B is omitted, or 24 if the animal can be walked and lameness assessed. The defined intervention level is a score of 5 or greater out of 20 or a score of 6 or greater out of 24. This intervention level was validated in a subsequent study separate from those validating the scale itself. There are disadvantages with the canine Glasgow Composite Measure Pain Scale-Short Form: dogs that are profoundly sedated score highly on section C, which can push total scores towards the threshold for intervention analgesia. For this reason, it is recommended that scoring be delayed until at least 30 minutes after tracheal extubation to minimise this effect. Dogs that show other behavioural changes e.g. dysphoria caused by drugs, or restlessness arising from a full bladder, can also have falsely elevated scores. It is important that a 'common sense' check is applied when using any pain scale to determine whether high scores are really due to pain. Other composite pain scales have been developed and validated for rabbits (Benato et al. 2021), cats

(Calvo et al. 2014), horses (van Loon and van Dierendonck 2018), donkeys (van Dierendonck et al. 2020), cattle (de Oliveira et al. 2014), sheep (Silva et al. 2020) and pigs (Luna et al. 2020). As an example, the UNESP-Botucatu composite pain scale for pain assessment in cattle (UCAPS) (de Oliveira et al. 2014) was developed following observation of behaviours in adult cattle undergoing open orchiectomy (castration) under sedation, with local anaesthetic injected into the spermatic cord. The observer scores behaviour in five different categories: locomotion, interaction with the environment, activity, food intake and presence of different behaviours such as wagging the tail abruptly and repeatedly, licking the affected area, kicking and foot stamping or head below the line of the vertebral column. Each category can be scored from 0 (no pain) to 2 (greatest possible pain), with the presence of different behaviours scored as 0 (no behaviours are present to 2) (2 or more of the behaviours are present). Therefore, the maximum possible score for any animal is 10, with a score of 4 or greater deemed to indicate the requirement for rescue analgesia. In contrast to the Glasgow Composite Pain Scale-Short Form, the UCAPS does not require the assessor to interact directly with the animal, which would seem more appropriate for cattle pain assessment, given that cattle may be unused to human interaction.

Generally, all composite pain scales rely on behavioural evaluation of the animal, combined with varying degrees of interaction and an assignment of scores in different categories. Higher scores denote an increasing level of pain. Most composite pain scales include a threshold when the authors recommend additional analgesia be given. This is useful for clinicians treating pain or when it is necessary to standardise analgesia administration in a research study.

7 Facial Grimace Scales

There is ongoing interest in developing grimace scales in numerous species to assess acute pain. This has stemmed from human paediatric practice where robust changes in facial expression have been documented in neonates undergoing painful procedures such as heel lancing for blood sample collection (Taddio et al. 2002). The first grimace scale to be developed in animals was the Mouse Grimace Scale (Langford et al. 2010). This scale was created by taking still images of facial expression in mice that had been injected intra-peritoneally with acetic acid, thereby causing visceral pain. Changes in so-called ‘facial action units’ (FAUs) were identified: ears, eyes, cheek bulge, nose bulge, and whisker change. Subsequently, grimace scales have been developed for rats (Sotocinal et al. 2011), rabbits (Keating et al. 2012), cats (Evangelista et al. 2019), horses (Dalla Costa et al. 2014), donkeys (Orth et al. 2020), sheep (Hager et al. 2017) (Fig. 9), lambs (Guesgen et al. 2016), piglets (Viscardi et al. 2017) and ferrets (Reijgwart et al. 2017), with each of these undergoing varying degrees of validation. The Feline Grimace Scale (www.felinegrimacescale.com, accessed 15th December 2022) is notable in having undergone multiple studies to evaluate the effects of anaesthesia, variability in the scores with different observers—veterinary surgeons, owners, veterinary students and veterinary nurses and to evaluate differences in scores dependent on whether assessment of the cat is done in real time or from still images of the face. The feline grimace scale assesses 5 FAUs: orbital tension, ear position, whisker position, muzzle tension and head position, with a score assigned from 0 to 2 depending on whether the action unit is present or not, hence the maximum possible score for a cat in severe pain is 10. There is also a defined intervention score when additional analgesia is needed, which is a score of 4 or greater out of 10. Fischer-Tenhagen and colleagues conducted a recent (2022) systematic review to evaluate whether facial expression scoring is sufficiently reliable and accurate to evaluate pain in large domestic animals (Fischer-Tenhagen et al. 2022). Most of the studies examined involved horses and donkeys, although the review did consider studies of other farm animals. From the 30 studies evaluated, 18 reported considerable inter-observer variability, with five of these indicating that inter-observer variability was adequate (ICC > =0.7).

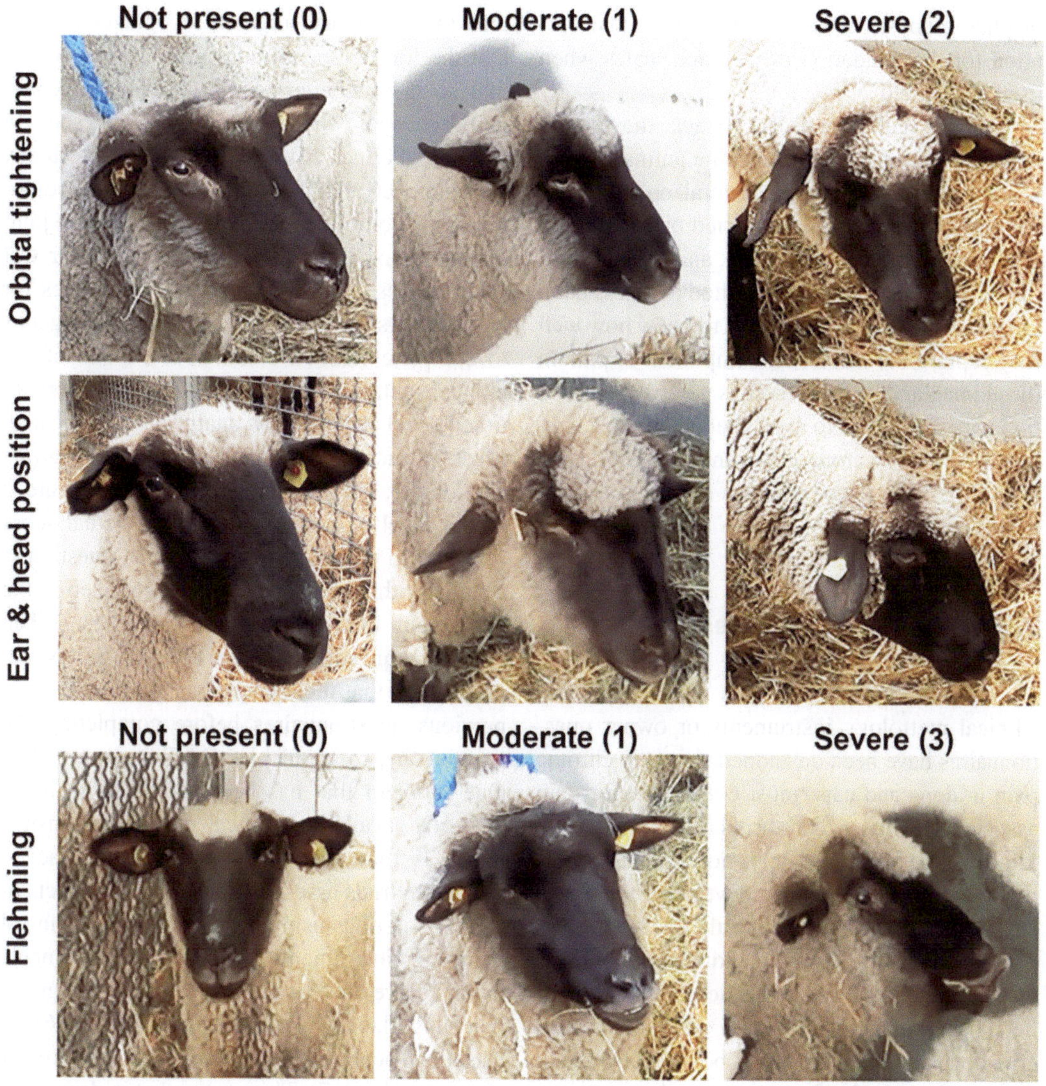

Fig. 9 The Sheep Grimace Scale as an indicator of post-operative distress and pain in laboratory sheep (Häger et al. 2017)

This suggests that most studies suffer from substantial interobserver variability. Training observers in the use of scoring systems has been suggested as best practice for grimace scales. Differences in the ability of observers to score different FAUs have also been documented. Ear position seems to be relatively easy to score in horses, whereas 'tension above the eye', 'strained mouth' and 'pronounced chin' were found to be more difficult. Agreement between observers was also low for certain FAUs in lambs undergoing castration. In contrast, agreement was good for all FAUs in sheep suffering painful clinical disease. This observation suggests that age and the nature of the pain may influence the reliability of scoring different FAUs within species. Differences in total scores for grimace scales in large animals have also been found in different species, depending on whether the scoring was done 'live' or using photographs or video recordings, with scores being generally lower when performed 'live' in real time. This disparity may be due in part to selection bias when choosing images for retrospective evaluation using video or photographs, but it is an important consideration for the validation of grimace scales when used in the field.

Grimace scales are useful because they do not require direct interaction with the animal, which is important if the animal is feral or cannot be handled. They can also be carried out quickly to give a rapid assessment of pain, and they are promoted as an 'easy to learn' method to assess pain. On the other hand, it is difficult to know how well they capture the emotional component of pain in the animal, and because there is no direct interaction with the animal to assess secondary mechanical hyperalgesia around a wound or painful area, it is difficult to know whether central sensitisation is present or not.

7.1 Pain Scales to Assess Chronic Pain

Clinical metrology instruments or owner questionnaires have been developed to assess chronic pain in dogs and cats, most commonly for pain associated with osteoarthritis or degenerative joint disease. Unfortunately, no clinical metrology instruments have been developed for use in farm animals and may reflect the fact that companion animals are treated as individuals, whereas farm animals tend to be considered on a flock or herd basis, making individual assessments less common. The human–animal bond is also structured differently for farm, compared to companion animals. Furthermore, farm animals with chronic diseases are commonly culled. In dogs, the most notable and validated clinical metrology instruments are the Helsinki Chronic Pain Index (Hielm-Björkman et al. 2009), the Liverpool Osteoarthritis in Dogs (LOAD) (Walton et al. 2013), and the Canine Brief Pain Inventory (Brown et al. 2013), all of which have been validated for osteoarthritis in dogs. The Canine Brief Pain Inventory has also been validated for bone cancer pain (Brown et al. 2009). The questionnaires generally ask for information about mobility as well as some questions targeting quality of life, e.g. does the animal engage in play? Asking the owner to complete the questionnaire (rather than the veterinary surgeon) recognizes the fact that the former spend more time with their animal, so are in a better position to evaluate behavioural changes when chronic pain is present. In cats, a Feline Musculoskeletal Pain Index (Enomoto et al. 2022) has been developed for owners to complete to evaluate degenerative joint disease pain. The Montreal Instrument for Cat Arthritis Testing has separate questionnaires for both owners/caregivers and veterinary surgeons to complete (Klinck et al. 2018). None of the clinical metrology instruments has an interventional level at whereby analgesia is indicated. Most of these instruments are available *gratis* and online; in most, higher scores indicate a higher level of pain. However, the actual score assigned to each answer for a given question is usually withheld from the owner completing the instrument to avoid bias in responses. When sequential questionnaires are completed, it is recommended that the owner does not look back at previous questionnaires before completing the current one, to avoid bias (Muller et al. 2016). One problem that has been identified with the currently available clinical metrology instruments is that they are susceptible to placebo effects. In trials assessing the analgesic effects of NSAIDs in dogs using owner-completed clinical metrology instruments, up to 50% improvements in pain score have been documented in the group receiving placebo (Rausch-Derra et al. 2016). In cats, the placebo effect appears to be even greater (Gruen et al. 2017). This means that in drug trials using clinical metrology instruments as an outcome measure, it can be difficult to show the benefit of drug administration relative to placebo. This may also lead to overestimation of treatment effects when using a clinical metrology instrument to monitor the clinical effects of analgesia in companion animals. There are currently no clinical metrology instruments to monitor chronic pain in farm animals.

7.2 Scales to Assess Quality of Life

Quality of life (QOL) is a somewhat nebulous term with no universally accepted definition when applied to animals. It has been defined as those aspects of an animal's life that make life better or worse for that specific animal—a definition that promotes the idea that QOL is very much individual-dependent. In human subjects, health-related QOL is a slightly different and narrower construct, which only considers the effect of a specific health condition on quality of life. Chronic pain and QOL are inextricably linked, with it being widely accepted that chronic pain has an adverse influence on QOL. Several different scales or instruments have been developed to try and measure QOL in dogs and cats with different health conditions, e.g. (cardiac disease (Freeman et al. 2005), osteoarthritis (Roberts et al. 2021), skin disease (Noli et al. 2011) and diabetes mellitus (Niessen et al. 2010). However, most of these instruments have undergone limited validation and are often more suited to research studies than to clinical practice. There are no instruments developed for farm animals to assess QOL, although farm assurance schemes often include questionnaires that attempt to collect information on QOL and welfare in different species.

7.3 Conclusions

Significant advances in the recognition and quantification of pain have been made across different species over the last few decades. While the behavioural assessment of pain remains the cornerstone of pain recognition, newer physiological measures such as infrared thermography and heart rate variability have come to the fore. The importance of developing validated pain scales to measure acute and chronic pain is now recognised, with many validated composite pain scales available to measure acute pain across a wide range of species. Instruments to measure chronic pain and quality of life in farm animals have lagged well behind similar developments in companion animals, despite major research interests in this area.

References

Adami C, Spadavecchia C (2020) Use of nociceptive threshold testing in cats in experimental and clinical settings: a qualitative review. Vet Anaesth Analg 47(4):419–436

Adcock SJJ, Cruz DM, Tucker CB (2020) Behavioral changes in calves 11 days after cautery disbudding: effect of local anesthesia. J Dairy Sci 103(9):8518–8525

Alvarez L, Adcock SJ, Tucker CB (2019) Sensitivity and wound healing after hot-iron disbudding in goat kids (2019). J Dairy Sci 102(11):10152–10162

Aulakh KS, Dongaonkar KR, Barnes K, Gines AJ, Bordelon JT, Hulse D, Aulakh HK, Liu CC (2020) Influence of orthopedic examination on lameness scores and interobserver and intraobserver agreement in dogs with naturally occurring elbow osteoarthritis. Vet Surg 49(3):455–462

Belshaw Z, Dean R, Asher L (2020) Could it be osteoarthritis? How dog owners and veterinary surgeons describe identifying canine osteoarthritis in a general practice setting. Prev Vet Med 185:105198

Benato L, Murrell J, Knowles TG, Rooney NJ (2021) Development of the Bristol Rabbit Pain Scale (BRPS): a multidimensional composite pain scale specific to rabbits (Oryctolagus cuniculus). PLoS One 16(6):e0252417

Berry RJ, Kennedy AD, Scott SL, Kyle BL, Schaefer AL (2003) Daily variation in the udder surface temperature of dairy cows measured by infrared thermography: potential for mastitis detection. Can J Anim Sci 83:687–693

Braun U, Michel N, Baumgartner MR, Hässig M, Binz TM (2017) Cortisol concentration of regrown hair and hair from a previously unshorn area in dairy cows. Res Vet Sci 114:412–415

Brown DC, Boston R, Coyne JC, Farrar JT (2009) A novel approach to the use of animals in studies of pain: validation of the canine brief pain inventory in canine bone cancer. Pain Med 10(1):133–142

Brown DC, Boston RC, Farrar JT (2013) Comparison of force plate gait analysis and owner assessment of pain using the Canine Brief Pain Inventory in dogs with osteoarthritis. J Vet Intern Med 27(1):22–30

Burand AJ Jr, Waltz TB, Manis AD, Hodges MR, Stucky CL (2023) HomeCageScan analysis reveals ongoing pain in Fabry rats. Neurobiol Pain 13:100113

Calderon DF, Cook NB (2011) The effect of lameness on the resting behavior and metabolic status of dairy cattle during the transition period in a freestall-housed dairy herd. J Dairy Sci 94(6):2883–2894

Calvo G, Holden E, Reid J, Scott EM, Firth A, Bell A, Robertson S, Nolan AM (2014) Development of a behaviour-based measurement tool with defined inter-

vention level for assessing acute pain in cats. J Small Anim Pract 55(12):622–629
Coetzee JF, Lubbers B, Toerber SE, Gehring R, Thomso DU, White BJ, Apley MD (2008) Plasma concentrations of substance P and cortisol in beef calves after castration or simulated castration. Am J Vet Res 69:751–762
Coetzee JF (2013) Assessment and management of pain associated with castration in cattle. Vet Clin North Am Food Anim Pract 29(1):75–101
Coetzee JF, Lechtenberg KF, Stock ML, KuKanich B (2013) Pharmacokinetics and effect of intravenous nalbuphine in weaned Holstein calves after surgical castration. J Vet Pharmacol Ther 37:169–177
Currah JM, Hendrick SH, Stookey JM (2009) The behavioral assessment and alleviation of pain associated with castration in beef calves treated with flunixin meglumine and caudal lidocaine epidural anesthesia with epinephrine. Can Vet J 50:375–382
Dalla Costa E, Minero M, Lebelt D, Stucke D, Canali E, Leach MC (2014) Development of the Horse Grimace Scale (HGS) as a pain assessment tool in horses undergoing routine castration. PLoS One 9(3):e92281
de Oliveira FA, Luna SP, do Amaral JB, Rodrigues KA, Sant'Anna AC, Daolio M, Brondani JT (2014) Validation of the UNESP-Botucatu unidimensional composite pain scale for assessing postoperative pain in cattle. BMC Vet Res 6(10):200
Dockweiler J, Coetzee J, Edwards-Callaway L, Bello N, Glynn H, Allen K, Theurer M, Jones M, Miller K, Bergamasco L (2013) Effect of castration method on neurohormonal and electroencephalographic stress indicators in Holstein calves of different ages. J Dairy Sci 96:4340–4354
Enomoto M, Lascelles BDX, Robertson JB, Gruen ME (2022) Refinement of the Feline Musculoskeletal Pain Index (FMPI) and development of the short-form FMPI. J Feline Med Surg 24(2):142–151
Evangelista MC, Watanabe R, VSY L, Monteiro BP, O'Toole E, Pang DSJ, Steagall PV (2019) Facial expressions of pain in cats: the development and validation of a FelineGrimace Scale. Sci Rep 9(1):19128
Fischer-Tenhagen C, Meier J, Pohl A (2022) "Do not look at me like that": is the facial expression score reliable and accurate to evaluate pain in large domestic animals? A systematic review. Front Vet Sci 9:1002681
Frahm S, Di Giminiani P, Stanitznig A, Schoiswohl J, Krametter-Frötscher R, Wittek T, Waiblinger S (2020) Nociceptive threshold of calves and goat kids undergoing injection of clove oil or isoeugenol for disbudding. Animals Basel 10(7):1228
Freeman LM, Rush JE, Farabaugh AE, Must AJ (2005) Development and evaluation of a questionnaire for assessing health-related quality of life in dogs with cardiac disease. Am Vet Med Assoc 226(11):1864–1868
González LA, Schwartzkopf Genswein KS, Caulkett NA, Janzén E, McAllister TA, Fierheller E, Schaefer AL, Haley DB, Stookey JM, Hendrick S (2010) Pain mitigation after band castration of beef calves and its effects on performance, behavior, Escherichia coli, and salivary cortisol. J Anim Sci 88:802–810
Greco D, Stabenfeldt GH (1997) Endocrine glands and their function. In: Cunningham JG (ed) Textbook of veterinary physiology, 2nd edn. W. B. Saunders Company, Philadelphia, PA, pp 404–439
Gruen ME, Dorman DC, Lascelles BDX (2017) Caregiver placebo effect in analgesic clinical trials for cats with naturally occurring degenerative joint disease-associated pain. Vet Rec 180(19):473
Guesgen MJ, Beausoleil NJ, Leach M, Minot EO, Stewart M, Stafford KJ (2016) Coding and quantification of a facial expression for pain in lambs. Behav Processes 132:49–56
Guillot M, Moreau M, Heit M, Martel-Pelletier J, Pelletier JP, Troncy E (2013) Characterization of osteoarthritis in cats and meloxicam efficacy using objective chronic pain evaluation tools. Vet J 196(3):360–367
Gygax L, Neuffer I, Kaufmann C, Hauser R, Wechsler B (2006) Milk cortisol concentration in automatic milking systems compared with auto-tandem milking parlors. J Dairy Sci 89:3447–3454
Häger C, Biernot S, Buettner M, Glage S, Keubler LM, Held N, Bleich EM, Otto K, Müller CW, Decker S, Talbot SR, Bleich A (2017) The Sheep Grimace Scale as an indicator of post-operative distress and pain in laboratory sheep. PLoS One 12(4):e0175839
Hamamoto-Hardman BD, Steffey EP, Weiner D, McKemie DS, Kass P, Knych HK (2019) Pharmacokinetics and selected pharmacodynamics of morphine and it's active metabolites in horses after intravenous administration of four doses. J Vet Pharmacol Ther 42(4):401–410
Hernandez CE, Thierfelder T, Svennersten-Sjaunja K, Berg C, Orihuela A, Lidfors L (2014) Time lag between peak concentrations of plasma and salivary cortisol following a stressful procedure in dairy cattle. Acta Vet Scand 56:61
Hielm-Björkman AK, Rita H, Tulamo RM (2009) Psychometric testing of the Helsinki chronic pain index by completion of a questionnaire in Finnish by owners of dogs with chronic signs of pain caused by osteoarthritis. Am J Vet Res 70(6):727–734
Hudson C, Whay H, Huxley J (2008) Recognition and management of pain in cattle. In Pract 30:126–134
Johnson CB, Wilson PR, Woodbury MR, Caulkett NA (2005) Comparison of analgesic techniques for antler removal in halothane-anaesthetized red deer (Cervus elaphus): electroencephalographic responses. Vet Anaesth Analg 32(2):61–71
Keating SC, Thomas AA, Flecknell PA, Leach MC (2012) Evaluation of EMLA cream for preventing pain during tattooing of rabbits: changes in physiological, behavioural and facial expression responses. PLoS One 7(9):e44437
Kemp MH, Nolan AM, Cripps PJ, Fitzpatrick JL (2008) Animal-based measurements of the severity of mastitis in dairy cows. Vet Record 163(6):175–179
Klinck MP, Monteiro BP, Lussier B, Guillot M, Moreau M, Otis C, Steagall PV, Frank D, Martel-Pelletier

J, Pelletier JP, Del Castillo JR, Troncy EJ (2018) Refinement of the Montreal instrument for cat arthritis testing, for use by veterinarians: detection of naturally occurring osteoarthritis in laboratory cats. Feline Med Surg 20(8):728–740

Lagisz M, Zidar J, Nakagawa S, Neville V, Sorato E, Paul ES, Bateson M, Mendl M, Løvlie H (2020) Optimism, pessimism and judgement bias in animals: a systematic review and meta-analysis. Neurosci Biobehav Rev 118:3–17

Langford DJ, Bailey AL, Chanda ML, Clarke SE, Drummond TE, Echols S, Glick S, Ingrao J, Klassen-Ross T, Lacroix-Fralish ML, Matsumiya L, Sorge RE, Sotocinal SG, Tabaka JM, Wong D, van den Maagdenberg AM, Ferrari MD, Craig KD, Mogil JS (2010) Coding of facial expressions of pain in the laboratory mouse. Nat Methods 7(6):447–449

Lehmann HS, Musk GC, Laurence M, Hyndman TH, Tuke J, Collins T, Gleerup KB, Johnson CB (2017) Mitigation of electroencephalographic and cardiovascular responses to castration in Bos indicus bulls following the administration of either lidocaine or meloxicam. Vet Anaesth Analg 44(6):1341–1352

Love EJ, Murrell J, Whay HR (2011) Thermal and mechanical nociceptive threshold testing in horses: a review. Vet Anaesth Analg 38(1):3–14

Love EJ, Taylor PM, Murrell J, Whay HR (2012) Effects of acepromazine, butorphanol and buprenorphine on thermal and mechanical nociceptive thresholds in horses. Equine Vet J 44(2):221–225

Lovell S, Simon B, Boudreau EC, Mankin J, Jeffery N (2022) Randomized clinical trial comparing outcomes after fentanyl or ketamine-dexmedetomidine analgesia in thoracolumbar spinal surgery in dogs. J Vet Intern Med 36(5):1742–1751

Ley SJ, Waterman AE, Livingston A (1995) A field study of the effect of lameness on mechanical nociceptive thresholds in sheep. Vet Rec 137(4):85–87

Luna SPL, de Araújo AL, da Nóbrega Neto PI, Brondani JT, de Oliveira FA, Azerêdo LMDS, Telles FG, Trindade PHE (2020) Validation of the UNESP Botucatu pig composite acute pain scale (UPAPS). PLoS One 15(6):e0233552

Martin MS, Kleinhenz MD, Viscardi AV, Curtis AK, Johnson BT, Montgomery SR, Lou ME, Coetzee JF (2022) Effect of bupivacaine liposome suspension administered as a local anesthetic block on indicators of pain and distress during and after surgical castration in dairy calves. J Anim Sci 100(1):skab378

Metzner M, Sauter-Louis C, Seemueller A, Petzl W, Zerbe H (2015) Infrared thermography of the udder after experimentally induced Escherichia coli mastitis in cows. Vet J 204:360–362

Molony V, Kent JE, Robertson IS (1993) Behavioural responses of lambs of three ages in the first three hours after three methods of castration and tail docking. Res Vet Sci 55(2):236–245

Molony V, Kent JE (1997) Assessment of acute pain in farm animals using behavioral and physiological measurements. J Anim Sci 75(1):266–272

Mirra A, Spadavecchia C, Bruckmaier R, Gutzwiller A, Casoni D (2018) Acute pain and peripheral sensitization following cautery disbudding in 1- and 4-week-old calves. Physiol Behav 184:248–260

Monteiro BP (2020) Feline chronic pain and osteoarthritis. Vet Clin North Am Small Anim Pract 50(4):769–788

Monteiro-Steagall BP, Steagall PV, Lascelles BD (2013) Systematic review of nonsteroidal anti-inflammatory drug-induced adverseeffects in dogs. J Vet Intern Med 27(5):1011–1019

Muller C, Gaines B, Gruen M, Case B, Arrufat K, Innes J, Lascelles BDX (2016) Evaluation of clinical metrology instrument in dogs with osteoarthritis. J Vet Intern Med 30(3):836–846

Murrell JC, Johnson CB, White KL, Taylor PM, Haberham ZL, Waterman-Pearson AE (2003) Changes in the EEG during castration in horses and ponies anaesthetized with halothane. Vet Anaesth Analg 30(3):138–146

Murrell JC, Johnson CB (2006) Neurophysiological techniques to assess pain in animals. J Vet Pharmacol Ther 29(5):325–335

Murrell JC, Waters D, Johnson CB (2008a) Comparative effects of halothane, isoflurane, sevoflurane and desflurane on the electroencephalogram of the rat. Lab Anim 42(2):161–170

Murrell JC, Psatha EP, Scott EM, Reid J, Hellebrekers LJ (2008b) Application of a modified form of the Glasgow pain scale in a veterinary teaching centre in the Netherlands. Vet Record 162(13):405–410

Musk GC, Murdoch FR, Tuke J, Kemp MW, Dixon MJ, Taylor PM (2014) Thermal and mechanical nociceptive threshold testing in pregnant sheep. Vet Anaesth Analg 41(3):305–311

Musk GC, Jacobsen S, Lehmann HS, Laurence M, Hyndman TH, Tuke SJ, Collins T, Gleerup KB, Johnson CB (2017) Objective measures for the assessment of post-operative pain in Bos indicus bull calves following castration. Animals 7(10):76. https://doi.org/10.3390/ani7100076

Nasr M (2020) The effects of keel bone fractures on the welfare and production of laying hens. PhD thesis,. University of Bristol

Neave HW, Daros RR, Costa JH, von Keyserlingk MA, Weary DM (2013) Pain and pessimism: dairy calves exhibit negative judgement bias following hot-iron disbudding. PLoS One 8(12):e80556

Niessen SJ, Powney S, Guitian J, Niessen AP, Pion PD, Shaw JA, Church DB (2010) Evaluation of a quality-of-life tool for cats with diabetes mellitus. J Vet Intern Med 24(5):1098–1105

Nogues E, von Keyserlingk MAG, Weary DM (2021) Pain in the weeks following surgical and rubber ring castration in dairy calves. J Dairy Sci 104(12):12881–12886

Noli C, Minafò G, Galzerano M (2011) Quality of life of dogs with skin diseases and their owners. Part 1: development and validation of a questionnaire. Vet Dermatol 22(4):335–343

Orth EK, Navas González FJ, Iglesias Pastrana C, Berger JM, Jeune SSL, Davis EW, McLean AK (2020) Development of a donkey grimace scale to recognize pain in donkeys (*Equus asinus*) post castration. Animals (Basel) 10(8):1411

Palmeira I, Fonseca MJ, Lafont-Lecuelle C, Pageat P, Cozzi A, Asproni P, Requicha JF, de Oliveira J (2022) Dental pain in cats: a prospective 6-month study. J Vet Dent 39(4):369–375

Parker RM, Paul ES, Burman OH, Browne WJ, Mendl M (2014) Housing conditions affect rat responses to two types of ambiguity in a reward-reward discrimination cognitive bias task. Behav Brain Res 274:73–83

Pesenhofer G, Palme R, Pesenhofer RM, Kofler J (2006) Comparison of two methods of fixation during functional claw trimming- walk-in crush versus tilt table-in dairy cows using faecal cortisol metabolite concentrations and daily milk yield as parameters. Wien Tierarztl Monatsschr 93:288–294

Plüss J, Steiner A, Alsaaod MJ (2021) Short communication: claw block application improves locomotion and weight-bearing characteristics in cattle with foot diseases. Dairy Sci 104(2):2302–2307

Raja SN, Carr DB, Cohen M, Finnerup NB, Flor H, Gibson S, Keefe FJ, Mogil JS, Ringkamp M, Sluka KA, Song XJ, Stevens B, Sullivan MD, Tutelman PR, Ushida T, Vader K (2020) The revised International Association for the Study of Pain definition of pain: concepts, challenges, and compromises. Pain 161(9):1976–1198

Roughan JV, Flecknell PA (2004) Behaviour-based assessment of the duration of laparotomy-induced abdominal pain and the analgesic effects of carprofen and buprenorphine in rats. Behav Pharmacol 15(7):461

Rausch-Derra L, Huebner M, Wofford J, Rhodes LJ (2016) A prospective, randomized, masked, placebo-controlled multisite clinical study of grapiprant, an EP4 prostaglandin receptor antagonist (PRA), in dogs with osteoarthritis. J Vet Intern Med 30(3):756–763

Ravi N, Dandekar N, Mysore P, Littman ML (2005) Activity recognition from accelerometer data. Aaai 5:1541–1546

Reid J, Nolan AM, Scott EM (2018) Measuring pain in dogs and cats using structured behavioural observation. Vet J 236:72–79

Reijgwart ML, Schoemaker NJ, Pascuzzo R, Leach MC, Stodel M, de Nies L, Hendriksen CFM, van der Meer M, Vinke CM, van Zeeland YRA (2017) The composition and initial evaluation of a grimace scale in ferrets after surgical implantation of a telemetry probe. PLoS One 12(11):e0187986

Roberts C, Armson B, Bartram D, Belshaw Z, Capon H, Cherry G, Gonzalez Villeta L, McIntyre SL, Odeyemi I, Cook AJC (2021) Construction of a conceptual framework for assessment of health-related quality of life in dogs with osteoarthritis. Front Vet Sci 8:741864

Silva NEOF, Trindade PHE, Oliveira AR, Taffarel MO, Moreira MAP, Denadai R, Rocha PB, Luna SPL (2020) Validation of the Unesp-Botucatu composite scale to assess acute postoperative abdominal pain in sheep (USAPS). PLoS One 15(10):e0239622

Slingsby LS, Waterman-Pearson AE (2002) Comparison between meloxicam and carprofen for postoperative analgesia after feline ovariohysterectomy. J Small Anim Pract 43(7):286–289

Slingsby LS, Taylor PM, Waterman-Pearson AE (2006) Effects of two doses of buprenorphine four or six hours apart on nociceptive thresholds, pain and sedation in dogs after castration. Vet Rec 159(21):705–711

Slingsby LS, Sear JW, Taylor PM, Murrell JC (2016) Effect of intramuscular methadone on pharmacokinetic data and thermal and mechanical nociceptive thresholds in the cat. J Feline Med Surg 18(11):875–881

Small AH, Jongman EC, Niemeyer D, Lee C, Colditz IG (2020) Efficacy of precisely injected single local bolus of lignocaine for alleviation of behavioural responses to pain during tail docking and castration of lambs with rubber rings. Res Vet Sci 133:210–218

Sotocinal SG, Sorge RE, Zaloum A, Tuttle AH, Martin LJ, Wieskopf JS, Mapplebeck JC, Wei P, Zhan S, Zhang S, McDougall JJ, King OD, Mogil JS (2011) The Rat Grimace Scale: a partially automated method for quantifying pain in the laboratory rat via facial expressions. Mol Pain 7:55

Stafford KJ, Mellor DJ (2005) The welfare significance of the castration of cattle: a review. N Z Vet J 53(5):271–278

Stewart M, Schaefer AL, Haley DB, Colyn J, Cook NJ, Stafford KJ, Webster J (2008) Infrared thermography as a non-invasive method for detecting fear-related responses of cattle to handling procedures. Anim Welf 17:387–393

Stewart M, Stookey J, Stafford K, Tucker C, Rogers A, Dowling S, Verkerk G, Schaefer A, Webster J (2009) Effects of local anesthetic and a nonsteroidal antiinflammatory drug on pain responses of dairy calves to hot-iron dehorning. J Dairy Sci 92:1512–1519

Stewart M, Verkerk G, Stafford K, Schaefer A, Webster J (2010) Non invasive assessment of autonomic activity for evaluation of pain in calves, using surgical castration as a model. J Dairy Sci 93:3602–3609

Stilwell G, Lima MS, Carvalho RC, Broom DM (2012) Effects of hot-iron disbudding, using regional anaesthesia with and without carprofen, on cortisol and behaviour of calves. Res Vet Sci 92(2):338–341

Sutherland M, Lowe G, Huddart F, Waas J, Stewart M (2018) Measurement of dairy calf behavior prior to onset of clinical disease and in response to disbudding using automated calf feeders and accelerometers. J Dairy Sci 101:8208–8216

Sutton GA, Dahan R, Turner D, Paltiel O (2013) A behaviour-based pain scale for horses with acute colic: scale construction. Vet J 196(3):394–401

Swaffield MJ, Molloy SL, Lipscomb VJ (2020) Prospective comparison of perioperative wound and pain score parameters in cats undergoing flank vs midline ovariectomy. J Feline Med Surg 22(2):168–177

Taddio A, Shah V, Gilbert-MacLeod C, Katz J (2002) Conditioning and hyperalgesia in newborns exposed to repeated heel lances. JAMA 288(7):857–861

Testa B, Reid J, Scott ME, Murison PJ, Bell AM (2021) The short form of the Glasgow composite measure pain scale in post-operative analgesia studies in dogs: a scoping review. Front Vet Sci 8:751949

Theurer ME, Amrine DE, White BJ (2013) Remote non-invasive assessment of pain and health status in cattle. Vet Clin North Am Food Anim Pract 29:59–74

Thompson AJ, Weary DM, Bran JA, Daros RR, Hötzel MJ, von Keyserlingk MAG (2019) Lameness and lying behavior in grazing dairy cows. J Dairy Sci 102(7):6373–6382

Tschoner T, Behrendt-Wipperman M, Rieger A, Metzner M, Knubben-Schweizer G, Reichmann F, Feist M (2018) Course of plasma substance P concentrations during umbilical surgery in calves. Berl Munch Tierarztl Wochenschr 11–12:522–528

Väisänen MA, Tuomikoski SK, Vainio OM (2007) Behavioral alterations and severity of pain in cats recovering at home following elective ovario-hysterectomy or castration. J Am Vet Med Assoc 231(2):236–242

Van der Saag D, Lomax S, Windsor PA, Hall E, White PJ (2019) Effect of lignocaine and a topical vapocoolant spray on pain response during surgical castration of beef calves. Animals (Basel) 9(4):126

Venkatachalam D, Kells N, Chambers P, Jacob A, Ward N, Singh P (2022) Pharmacokinetics and efficacy of a novel long-acting bupivacaine formulation for cornual nerve block in calves. Front Vet Sci 1(9):1060951

van Loon JPAM, Van Dierendonck MC (2018) Objective pain assessment in horses (2014-2018). Vet J 242:1–7

van Dierendonck MC, Burden FA, Rickards K, van Loon JPAM (2020) Monitoring acute pain in donkeys with the equine Utrecht University scale for donkeys composite pain assessment (EQUUS-DONKEY-COMPASS) and the equine Utrecht University scale for donkey facial assessment of pain (EQUUS-DONKEY-FAP). Animals (Basel) 10(2):354

Viscardi AV, Hunniford M, Lawlis P, Leach M, Turner PV (2017) Development of a piglet grimace scale to evaluate piglet pain using facial expressions following castration and tail docking: a pilot study. Front Vet Sci 4:51

Von Baeyer CL, Spagrud LJ (2007) Systematic review of observational (behavioral) measures of pain for children and adolescents aged 3 to 18 years. Pain 127(1–2):140–150

Von Borell E, Langbein J, Després G, Hansen S, Leterrier C, Marchant-Forde J, Marchant-Forde R, Minero M, Mohr E, Prunier A, Valance D, Veissier I (2007) Heart rate variability as a measure of autonomic regulation of cardiac activity for assessing stress and welfare in farm animals—a review. Physiol Behav 92:293–316

Walton MB, Cowderoy E, Lascelles D, Innes JF (2013) Evaluation of construct and criterion validity for the 'Liverpool Osteoarthritis in Dogs' (LOAD) clinical metrology instrument and comparison to two other instruments. PLoS One 8(3):e58125

Welsh EM, Gettinby G, Nolan AM (1993) Comparison of a visual analogue scale and a numerical rating scale for assessment of lameness, using sheep as a model. Am J Vet Res 54(6):976–983

Whay HR, Waterman AE, Webster AJ (1997) Associations between locomotion, claw lesions and nociceptive threshold in dairy heifers during the peri-partum period. Vet J 154(2):155–161

Wheeler CA, White BJ, Anderson DE, Amrine DE, Larson RL (2013) Assessment of biometric tools for quantitative gait analysis in Holstein calves. Am J Vet Res 74(11):1443–1449

Local Anaesthetics and Adjuvant Drugs

Fernando Martinez-Taboada
and Gabrielle C. Musk

Abstract

Local anaesthetics are a class of drugs that can be used to entirely block pain signals during surgical procedures. Although this group of drugs is diverse, local anaesthetics produce predictable, reliable and temporary suppression of conduction through sensory nerves, motor nerves and autonomic nerves. Local anaesthetics can be used on their own in animals that can be safely and comfortably restrained, or they can be used in conjunction with sedation or general anaesthesia to contribute to preventive and multimodal analgesia. Lignocaine is the most widely used local anaesthetic drug and is registered for use in many production animal species in many countries. It has a fast onset of action of action (1–2 min) and a duration of action of 1–2 h. In some jurisdictions, it is the only analgesic drug registered for production animals.

Keywords

Local anaesthesia · Production animals · Desensitisation · Analgesia · Lignocaine

Local anaesthetics (LAs) have an almost inconceivable potential to attenuate pain in farmed animal species; suitable formulations administered at an appropriate dose, accurately, i.e., in the right location, are capable of completely blocking all nerve conduction to and from any area of the animal's body. Blockade of sensory nerves eliminates sensation; motor nerve blockade produces paralysis, while pre- or postganglionic sympathetic nervous blockade produces vasodilatation. Thus, two components of the anaesthetic triad (Dugdale 2010), i.e., analgesia and muscle relaxation, can be achieved by the accurate injection of these versatile (and often inexpensive) compounds. When sensory and motor blockade is complete, the term 'local anaesthesia' applies. When local anaesthetic injection achieves complete or partial sensory blockade, then 'local analgesia' is more appropriate (Campoy and Schroeder 2013). With local analgesia, motor and autonomic nerve function is preserved, so there is no effect on movement or blood supply.

Local anaesthetics are a chemically heterogeneous group of drugs that produce predictable and reversible suppression of nerve conduction through the blockade of voltage-gated sodium (Na^+) channels, and consequently Na^+ influx dur-

F. Martinez-Taboada
3A Animal Anaesthesia and Analgesia,
Upwey, VIC, Australia

G. C. Musk (✉)
School of Veterinary Medicine, Murdoch University,
Murdoch, WA, Australia
e-mail: g.musk@murdoch.edu.au

R. E. Clutton, G. C. Musk (eds.), *Pain and Suffering in Farmed Animals*,
https://doi.org/10.1007/978-3-032-11407-5_11

ing the generation of nerve impulses (or action potentials).

1 History

The analgesic and hallucinatory effects derived from chewing coca leaves (*Erythroxylon coca L.*) were recognized by the Incas and other indigenous South American cultures (Gay et al. 1975). In 1855, the German chemist Friedrich Gaedcke isolated the alkaloid cocaine (Gaedcke 1855), but it remained unnamed until 1860 when Albert Niemann named the molecule and described its analgesic effects on the oral mucosae (Niemann 1860).

In 1884, Carl Koller desensitised the eye for surgery using cocaine as an ophthalmological anaesthetic (Ruetsch et al. 2001). And then in 1885, William Halsted used it to produce regional or perineural anaesthesia (Halsted 1885). James Corning delivered the first spinal anaesthetic when he accidentally injected cocaine into the subarachnoid space of a dog (Corning 1885).

Most of the LAs in current use, e.g., lidocaine, prilocaine, mepivacaine, and bupivacaine, were developed during the 1940s and 1950s. In 1989, ropivacaine was introduced, becoming clinically available in the late 1990s, with levobupivacaine becoming available in the early 2000s (Fig. 1) (Cartwright and Fyhr 1988). New local anaesthetic drugs have not been identified in the last 25 years.

2 Pharmacology

2.1 Chemical Structure and Isomerism

Chemically, all LAs consist of an aromatic lipid-soluble ring and a water-soluble amine group joined by a chemical bond whose identity forms the traditional basis for classifying these drugs, i.e., esters, amides, ketones, and ethers. While all such molecules exert local anaesthetic effects, only the esters and amides are used clinically (Lagan and McLure 2004). Interestingly, there are other drugs which have local anaesthetic properties, but do not possess the aforementioned chemical structure. Such drugs are often used for different therapeutic purposes, e.g. atropine, propranolol, clonidine, pethidine, amitriptyline and on some occasions are used as adjuvants of the LAs (McLure and Rubin 2005).

Some amide-linked molecules—bupivacaine, prilocaine, ropivacaine, mepivacaine and etidocaine, are enantiomers (or optical isomers), i.e., two stereoisomers related to each other by reflec-

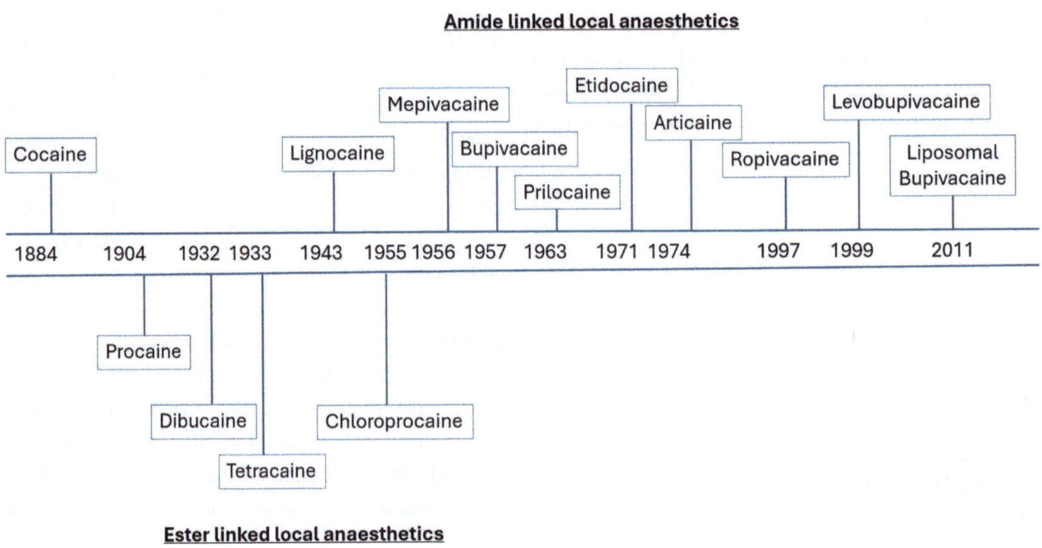

Fig. 1 Timeline of local anaesthetic discovery and development

tion. Put another way, the two molecules are mirror images of each other, so are non-superposable, like human hands. Enantiomers have similar physical properties except for the direction they rotate polarized light and the way they interact with other compounds. Because of this property, enantiomers with near-identical chemical structures have very different pharmacological properties. Clinically, LAs are usually available as a racemic mixture (of enantiomers) except for levobupivacaine and ropivacaine.

2.2 Mechanism of Action

Local anaesthetics are weak basic compounds (molecules not fully ionised or protonated in an aqueous solution) that exist in two states simultaneously, i.e. acidic (protonated or ionised)—in the form BH^+ and basic (non-ionised), taking the form B. The equilibrium between the two states is determined by the Henderson–Hasselbalch equation and relates to the drug's pK_a (the strength an acid exercises to hold a proton, or, in other words, the pKa reflects how the molecules tend to dissociate in an aqueous solution) and the pH of local tissue.

$$\mathrm{pH} = \mathrm{pKa} + \log_{10}\frac{[B]}{[BH^+]}$$

The Henderson–Hasselbalch equation relates the pH of a solution to the dissociation constant and the concentrations of the protonated and non-protonated molecules in equilibrium.

Local anaesthetics are generally formulated as hydrochlorides to increase their solubility in water. Most LA solutions, before injection, are relatively acidic, i.e. have a (low) pH which is considerably different (numerically) from the high pKa of the specific drug. Consequently, most of the drug molecules are in the form BH^+. When the solution is injected into tissue, the ambient ('local') pH is more physiological, i.e. approaches a value of 7.4, which is similar to the drug's pKa. Therefore, more molecules become non-ionised, i.e. take the form B. The non-ionised molecules are more lipid soluble, meaning they more readily cross nerve cell membranes (which are composed of a phospholipid bilayer). In the nerve cell's cytoplasm, the pH is more acidic again so the BH^+ form re-predominates. It is this protonated molecule that blocks the sodium channel and prevents the neuronal depolarization upon which nerve transmission relies.

The sodium channel blockade (Fig. 2) is reversible via weak chemical bonds (e.g. Van der Waals forces) (Hollmann et al. 2001). These interactions block the influx of sodium into the neuron and, if the blockade is extensive enough, the action potential cannot be generated. It is speculated that the molecules in the non-ionised state can also contribute to the effectiveness of the blockade by inducing a non-coordinated expansion of the cellular membrane. Local anaesthetics also act on other receptors; however, the clinical relevance of these effects remains unclear (Xiong and Strichartz 1998; Hollmann et al. 2001; Olschewski et al. 1998).

The affinity of LAs for Na^+ channels is maximal when the channel is open (active or inactive phases), and it is minimal when it is closed (deactivated or refractory phases) (Fig. 3). Therefore, when LAs are present, the faster the Na^+ channel activation, the greater the degree of blockade that ensues. Affinity for and dissociation from the Na^+ channel depend on the drug's molecular configuration and occasionally the enantiomer present. This property is an important determinant of the cardiovascular toxicity of local anaesthetics (see below) (Clarkson and Hondeghem 1985).

3 Physicochemical Properties of LAs Affecting Pharmacological Activity

3.1 Relative Molecular Mass

Other factors being equal, smaller molecules diffuse through membranes more rapidly and exert a prompter effect (Smith 2009).

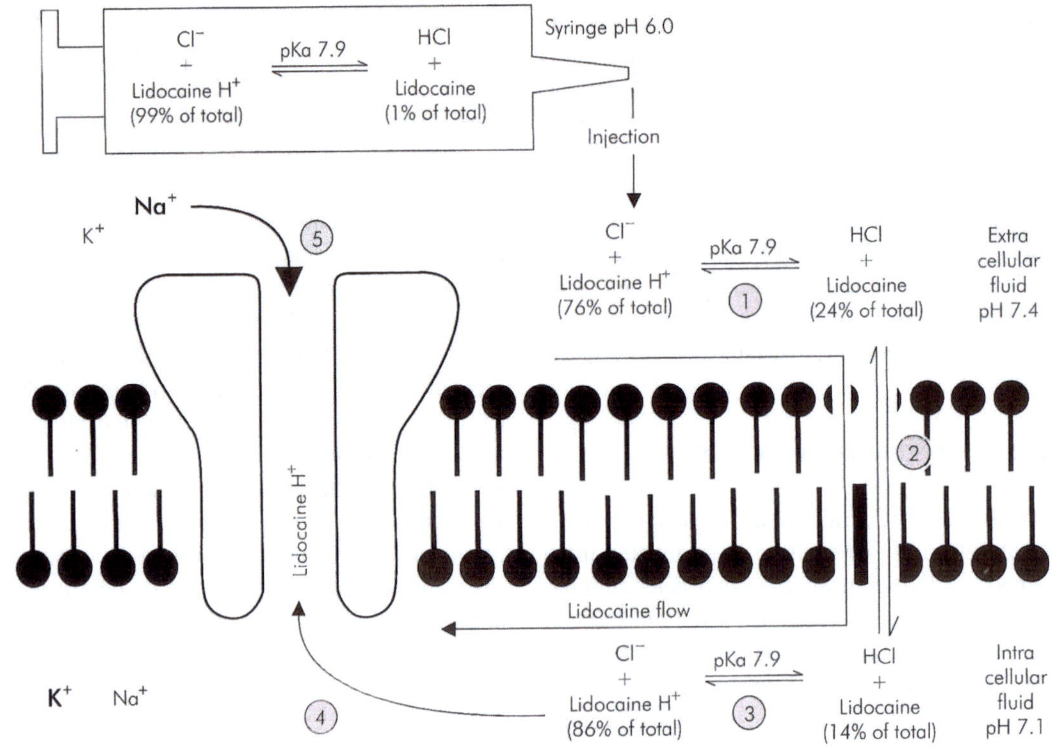

Fig. 2 Mechanism of action of local anaesthetics using lidocaine as an example

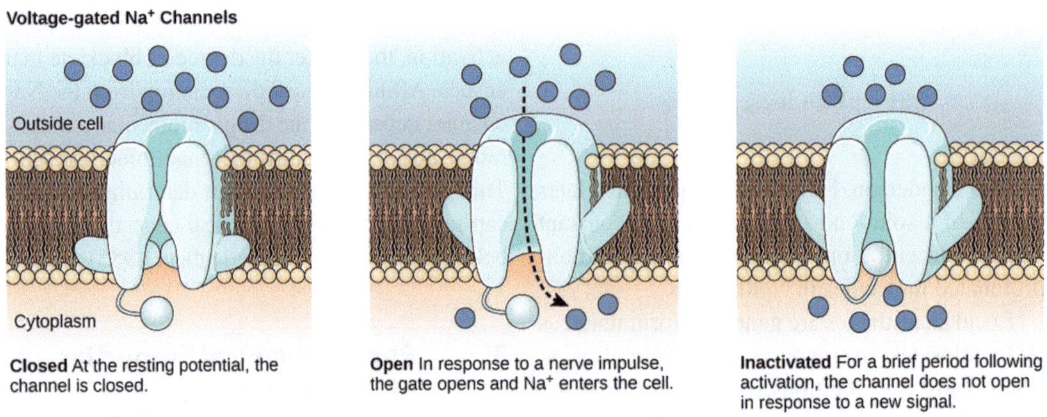

Fig. 3 The voltage-gated sodium channel opens in response to a change in the membrane voltage. After opening, it becomes inactivated for a brief period of time

Table 1 Physicochemical characteristics of local anaesthetic drugs (MW molecular weight)

	Structural classification	MW	p*K*a	Protein binding (%)	Partition coefficient	Onset	Elimination half-life (min)
Cocaine	Ester	311	8.6	95	–	Fast	100
Chloroprocaine	Ester	271	9.1	–	17	Fast	6
Prilocaine	Ester	220	7.7	55	50	Fast	100
Lidocaine	Amide	234	7.8	70	110	Fast	100
Mepivacaine	Amide	246	7.7	77	42	Fast	115
Bupivacaine	Amide	288	8.1	95	560	Moderate	210
Ropivacaine	Amide	274	8.1	94	230	Moderate	120
Levobupivacaine	Amide	288	8.1	95	–	Moderate	210

3.2 Lipid Solubility

While all LAs are lipid soluble those with greater lipid solubility diffuse more rapidly through nerve sheaths, and neural membranes of the individual axons comprising the nerves. The aromatic ring mostly determines molecular solubility although this property can also be enhanced by aliphatic chains. While lipid solubility (or lipophilicity) correlates directly with drug potency it also prolongs the onset of action (because of the aqueous nature of cytoplasm). Therefore, bupivacaine which is more lipid soluble than lidocaine, is also more potent and so its commercial presentation is often 0.5% (5 mg mL^{-1}) as opposed to 2% (20 mg mL^{-1}) for lidocaine. Hence the dose for bupivacaine is lower in milligrams per kilogram than lidocaine for a comparable effect, but the onset of action of bupivacaine is slower than lidocaine because bupivacaine is more lipid soluble than lidocaine. The ability of lipophilic drugs to cross cell membranes more rapidly than hydrophilic ones also explains the direct proportionality between lipid solubility and toxicity, both local and systemic (Heavner 2007). Drug lipid solubility is usually expressed as the heptane/buffer partition coefficient with molecules with greater coefficients being more lipid soluble (Table 1).

3.3 Ionisation (pK_a and pH)

The relative proportion of drug molecules in the BH^+ and B states depends on the drug-specific dissociation coefficient (pK_a) and the local (tissue) pH. Most LAs have a $pK_a > 7.4$, which means that at physiological pHs (7.45) or in more acidic environments, e.g. those affected by inflammation or infection, most molecules are in the protonated (BH^+) state and incapable of affecting nerves (Edgcombe and Hocking 2006).

3.4 Distribution and Absorption

Local anaesthetics bind to plasma proteins (albumin and α_1-acid-glycoprotein), and tissue proteins. Being weakly basic, LAs have high affinity for α_1-acid-glycoprotein, but albumin is also important because of its high plasma concentrations. Consequently, albumin is regarded as a high volume-low affinity system, while α_1-acid-glycoprotein binding is considered a low volume-high affinity system. Protein binding is closely related to the duration of drug action, as drugs that are highly protein-bound have a longer duration of action due to a relatively low concentration of free drug being available for metabolism.

Most LAs have a dose-dependent vasodilatory effect with vasoconstriction at low concentrations, and vasodilatation at greater (clinical) doses. The exception is cocaine, which causes intense vasoconstriction through its effect on norepinephrine reuptake (Neal 2007). Enantiomer-specific effects on vascular tone have also been observed with ropivacaine and bupivacaine (Kopacz et al. 1989; Aps and Reynolds 1978). Absorption at the injection site depends on factors discussed previously. The extent of systemic absorption also depends on the anatomical locus of the injection site. For instance, the greatest plasma drug concentrations follow intercostal and caudal injections, followed by lumbosacral epidural, brachial plexus, and sciatic and femoral nerve blocks (Tucker et al. 1972).

A considerable proportion of blood-borne LA undergoes clearance during passage through the pulmonary circulation (Jorfeldt et al. 1979). This uptake results from ion retention by lung tissue, which is acidic relative to plasma. and attenuates large changes in LA plasma concentration.

Local anaesthetics are less protein-bound during pregnancy, which increases the concentration of free unbound drug. This phenomenon, along with the lipid solubility of LAs facilitates transplacental transfer. Within the foetus, low pH levels and a relative lack of glycoproteins 'trap' the LA by ion retention (Biehl et al. 1978; Johnson et al. 1995; Ueki et al. 2009). This situation can be clinically significant when high LA doses are used, e.g. during intravenous infusion or inadvertent intravenous injection, as the risk of toxicity is increased. The transfer of LAs in milk is less well-examined, and the clinical importance of the levels detected is questionable (Ortega et al. 1999; Matsota et al. 2009).

3.5 Metabolism

The metabolism of ester and amide-linked LAs differs, which affects their allergic potential. Ester-linked LAs are hydrolysed rapidly in plasma by pseudocholinesterases. The metabolites are largely inactive, although one such metabolite, ρ-aminobenzoic acid (PABA), can be highly allergenic. This metabolism is rapid (except in cases of genetically based atypical cholinesterase). Cocaine is an exception in this group, as it is entirely metabolized by the liver.

Amide-linked LAs are metabolised in the liver in a complex, biphasic biotransformation involving cytochrome P450. This metabolism is dependent on liver perfusion and the LA in question. With fixed liver perfusion, the rate of metabolism of amide-linked LAs is: prilocaine and etidocaine (most rapid) > mepivacaine and lidocaine > ropivacaine and bupivacaine. The metabolism of prilocaine exceeds liver metabolism, indicating extrahepatic clearance processes. Although pulmonary metabolism is a possible explanation for this, evidence is inconclusive (Geng et al. 1995).

3.6 Toxicity and Anaphylaxis

Local anaesthetics have been associated with anaphylactoid, as well as local and systemic reactions. Local tissue 'toxicity' with LAs is not uncommon and is associated with the repeated use of high concentrations of long-lasting drugs. For example, bupivacaine and ropivacaine produce a regenerative necrotic myositis in pigs when they are used long term, but the clinical relevance is unclear (Zink et al. 2005). This myotoxic effect is common to all LAs, though the magnitude of damage is agent-specific; procaine has the least, and bupivacaine the greatest effect (Zink and Graf 2004). Local anaesthetics are toxic to chondrocytes (Park et al. 2011) and, on occasion, nerves themselves (Kroin et al. 1986). Bupivacaine and ropivacaine are less neurotoxic than lidocaine (Sakura et al. 2005), although this risk depends on the route of administration. The risk of bupivacaine and ropivacaine toxicity is greater with subarachnoid, rather than epidural injection (Kirihara et al. 2003).

Prilocaine, benzocaine and articaine can induce *methaemoglobinemia* in multiple species such as humans, dogs and cats, although this side effect is rare.

Overdosage with LAs may occur through accidental intravenous injection or dose miscalculation. Typically, central nervous system signs precede cardiovascular anomalies, although both are related to plasma drug concentrations (Fig. 4). Some effects may not be apparent during general anaesthesia. If the context allows, the treatment for LA toxicity is fundamentally supportive with oxygen therapy, intravenous fluid therapy, vasopressors, inotropic drugs, sedatives/anticonvulsant drugs, and antiarrhythmic drugs, depending on the severity of the case. Lipid emulsion therapy has also been found beneficial in the treatment of sys-

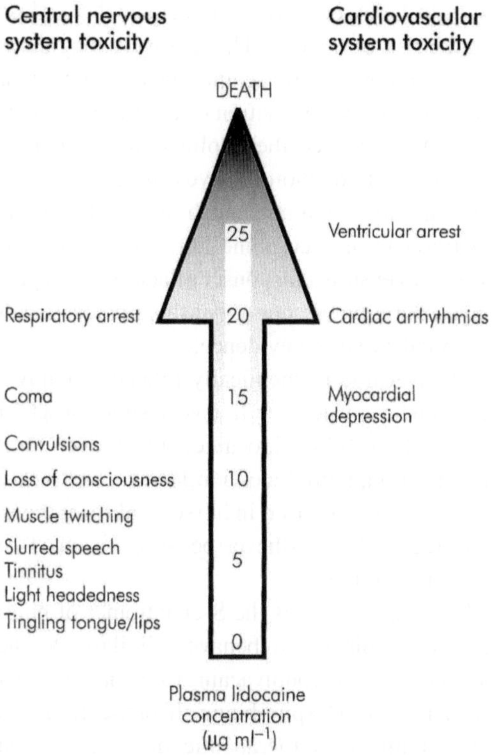

Fig. 4 Lidocaine toxicity on the central nervous and cardiovascular systems in the human

temic toxicity of LA. Bupivacaine is highly cardiotoxic due to a high affinity to the cardiac cell Na^+ and slow dissociation from the receptor. Bupivacaine cardiovascular collapse is difficult to manage and can be refractory to many treatments. Currently, the use of lipid emulsion therapy is recommended, but a combination of adrenaline-vasopressin has also been successful in experimental models (Association of A 2023).

Allergic reactions are typically associated with ester-linked LAs; ρ-aminobenzoic acid (PABA) is potentially very allergenic. Previous sensitisation is normally required, but cases of cross-sensitivity with cosmetics, food preservatives, and sulphonamides have been reported in humans. For several decades, the use of amide-linked LAs has predominated, and so ester-related allergic reactions are rare (Finucane 2003).

In general, it can be concluded that LAs are safe drugs, but they can potentially cause local and systemic damage. To minimise these risks, some recommendations are:

- Be familiar with the anatomy of the area to be injected.
- Avoid inadvertent intravenous or intra-arterial injection of LA by aspirating before injecting LA, and repeating this aspiration during the injection of the full volume of LA.
- Inject a 'test dose' of approximately a quarter of the volume/dose to be injected. Then, assess the lack of side effects before continuing the administration.
- Follow the correct time intervals (based on duration of action) between doses.
- Follow the correct dose based on species-specific therapeutic and toxic doses.
- Use the least toxic LA available.
- Use adjuvants to decrease the LA dose (see Sect. 4.6).

4 Drugs

4.1 Ester-Linked Local Anaesthetics

Cocaine, an ester of benzoic acid, is naturally present in the leaves of *Erythroxylum coca* and *E. novogranatense*. It has unique characteristics within this group such as causing vasoconstriction and undergoing hepatic metabolism. Currently, it is exclusively used for desensitizing the nasal mucosae of humans for procedures such as endoscopy. Cocaine is not used in animals.

Procaine was the first synthetic LA. This drug has a wider safety margin i.e., the range of doses within which undesirable side effects occur is larger, than cocaine. Furthermore, being ester-linked, procaine retains the propensity to cause allergies. It has a very brief duration of action and a relatively long onset time ($pK_a = 9$). Today it is used only in humans for dental procedures, and more recently as an antiaging medication (Sheikh and Dua 2023).

Tetracaine has a rapid onset of action and—uncharacteristic of ester-linked LAs—a prolonged duration of action, i.e., 4–6 h. It is considered the most toxic of all the ester-linked LAs, so currently is only used topically. It was very popular alongside cocaine and adrenaline for Ear-Nose-Throat surgery in humans, but nowadays it is mostly used in ophthalmology in humans (Smith 2009).

Benzocaine is a relatively impotent LA but has a rapid onset of action. It is an unusual ester LA in that the side chain is an ethyl group with no amine component, so it persistently exists in the non-ionised (*B*) state, which explains its rapid onset of action. It is widely used in the topical treatment for oral and throat pain, sunburn, haemorrhoids, etc., in people. Despite its impotence, it can cause allergies and methemoglobinemia. It is used in aqueous solutions as a general anaesthetic for fish and amphibians (Smith 2009).

4.2 Amide-Type Local Anaesthetics

Lidocaine has a rapid onset and a medium duration of action. The duration of action is predictable because of its plasma protein binding and hepatic metabolism. It is considered to have a wide safety margin. The toxic dose of intravenous lidocaine in dogs is 22 mg kg^{-1} (Liu et al. 1983) and 11 mg kg^{-1} in cats (Chadwick 1985). These doses are higher than therapeutic doses in both species. Toxic doses of lidocaine are not as well defined in production animals, but are anecdotally slightly lower than the toxic doses in dogs and cats. Lidocaine is also widely used as an antiarrhythmic drug (class 1B).

Bupivacaine has a prolonged onset time in part because its $pK_a = 8.1$, and a prolonged duration of action (98% plasma protein bound). The toxic (intravenous) dose in cats and dogs is 5 mg kg^{-1} (Liu et al. 1983; Chadwick 1985), which is only 2–3 times greater than the therapeutic dose (2 mg kg^{-1}). Bupivacaine also causes direct cardiovascular depression—which is difficult to manage—because of fast Na^+ channel blockade occurring in both cardiac nerves and conduction systems (causing arrhythmias) and cardiac tissue (causing myocardial depression). Previously, a 0.75% preparation was available, but currently it is more common to use the 0.25 and 0.5% solutions. These formulations are less likely to be associated with accidental overdose and the risk of toxicity. The relatively long duration of action of bupivacaine makes it a popular choice when desensitisation of the surgical site is desirable. However, the implications of the prolonged effects of motor nerves must be considered in larger animals such as cattle. Some practitioners may combine lidocaine with bupivacaine to ensure a fast onset of action and a prolonged duration of action, but this practice is not supported by strong evidence.

Mepivacaine is chemically related to bupivacaine and ropivacaine. It has a short onset of action (pK_a 7.9) like lidocaine, but a longer duration of action; it produces a mild vasoconstrictive effect. It is widely used in horses and for intravenous regional anaesthesia because its onset of action is so rapid.

Levobupivacaine is the S-enantiomer of bupivacaine. Clinically, it behaves similarly to the racemic mixture (bupivacaine) but has a wider safety margin. Despite being theoretically safer, the recommended therapeutic dose remains at 2 mg kg^{-1} (the same as bupivacaine).

Ropivacaine was synthesised in the 1950s but was not commercialised for 40 years. During its initial development, it was considered to be highly cardiotoxic, though this is now known to be an effect of the R enantiomer. Chemically related to bupivacaine, it has a similar onset and duration of action and a slightly lower potency. At low concentrations, it produces a clearly differentiated sensory blockade, so it preserves motor function. This property makes it desirable for use in production animals, especially when it is necessary to maintain them in a standing position. This selective blockade results from ropivacaine's lower lipid solubility in comparison to bupivacaine. The purification of S-ropivacaine increased the enantiomeric mixture's safety, especially when compared to bupivacaine. However, the fairest comparison would be to compare ropivacaine and levobupivacaine (both being very safe drugs).

Prilocaine is derived from lidocaine and has a similar onset and duration of action. However, it is approximately 40 times less toxic than lidocaine due to its lower lipid solubility, more rapid hepatic, renal and pulmonary metabolism, and greater tissue uptake. Therefore, it has been extensively used in intravenous regional anaesthesia in people. In combination with lidocaine, it forms a eutectic mixture: the proprietary product is known as EMLA cream (Eutectic Mixture of Local Anaesthetics), which is used for desensitisation of the skin or mucous membranes. One of prilocaine's metabolites (o-toluidine) induces methemoglobinemia when prilocaine is used in high doses. This effect is more likely in neonates and newborn babies (as foetal haemoglobin is more readily oxidised than the adult form) (Smith 2009).

4.3 New Local Anaesthetics

Current interest in LAs has two major foci: (i) prolonging the duration of action to several days (for postoperative use) and for weeks to months (for palliative use); and (ii) obtaining a more selective sensory blockade. Consequently, Exparel® (liposomal preparation of bupivacaine) was approved in 2011 for treating acute pain in humans in both the USA and Europe (Lafont et al. 1996). Sustained-release formulations of bupivacaine, such as SABER-bupivacaine (bupivacaine in a biodegradable biolayer) and HTX-011 (bupivacaine combined with a low dose meloxicam in a bio-erodible polymer), are also undergoing Phase I trials: they appear to provide up to 3 days of local anaesthesia (Prabhakar et al. 2019). Neosaxotoxin is produced by freshwater cyanobacteria and blocks the Na^+ channel at the 'site 1', producing a reversible and dose-dependent blockade. Having low affinity for cardiac Na^+ channels and being unable to cross the blood-brain barrier means it has limited cardiac or central nervous toxicity (Rathmell et al. 2015). Finally, n-butyl-p-aminobenzoate is being rediscovered as it appears to be able to produce prolonged selective sensory nerve blockade (Van Den Berg et al. 1996).

4.4 Use in Animals

For production animal species the drug most utilised is lidocaine, which is the only local anaesthetic approved for use in many countries for animals. The relatively fast onset of action, wide safety margin and short duration of action make it useful in a production context on a farm where prolonged observation of animals after a procedure is not possible. In research settings longer acting LAs such as bupivacaine and ropivacaine may be used as the ability to monitor and manage any side effects is much greater. These drugs are desirable because their duration of action is relatively long.

4.5 Other Properties

Some local anaesthetics have anti-inflammatory effects, which may be greater than non-steroidal anti-inflammatory drugs or steroids (Cassuto et al. 2006). These drugs all interact at multiple levels of the inflammatory cascade: decreasing leukocyte adhesion, migration, activation, priming, and phagocytosis. Additionally, LAs inhibit the phospholipase A_2 inhibitory effects on prostaglandin (even at low doses in the case of lidocaine), which decreases the production of thromboxanes and leukotrienes, thereby decreasing platelet aggregation and the risk of deep venous thrombosis and decreasing oedema formation during inflammation, respectively. Local anaesthetics also decrease histamine release and vascular hyperpermeability, which inhibits the oedema associated with inflammation. In addition, LAs decrease the release of interleukin (IL)1, IL8, IL6 and tumour necrosis factor α. Finally, they decrease oxygen free radical production and have a direct free radical scavenging effect (Cassuto et al. 2006; Soto et al. 2018).

Recently, the antitumoral effects of LAs have gained attention. These effects may be indirect (due to the anti-inflammatory effect) or occur directly due to the expression of Na^+ channels by the tumoral cells. Various carcinomatous cells, breast, prostate, cervix, colon, etc., express voltage-dependant Na^+ channels and their activation increases excitability and metastatic activity (this phenomenon is known as cellular excitability or CELEX). There is a current debate on the use of LAs to block these channels to limit metastasis, cellular motility, and invasiveness (Soto et al. 2018; Zhang et al. 2021).

Local anaesthetics also have antimicrobial properties. The exact mechanism is unclear, but it may be related to direct interactions with the bacterial wall or with macromolecules on the wall surface. Electrostatic interactions with membrane proteins may induce K^+ leakage from the bacterial cytoplasm and promote cell lysis. Research in this field supports the premise that antibacterial effects are related to LA concentration (even at clinically used doses) and independent of the chemical structure (ester and amide types are equally effective) except for ropivacaine. Ropivacaine has weak or no antibacterial effects at clinical concentrations, which has triggered some discussion about its intrathecal use. Finally, LAs have some antiviral and antifungal properties, but only at high concentrations (Cassuto et al. 2006).

4.6 Adjuvants

Epinephrine (adrenaline) is usually combined with LAs to decrease blood flow at the injection site to retard systemic absorption, i.e. prolong blockade (Sakura et al. 1999). It is not recommended to exceed 5 $\mu g\ mL^{-1}$ (1:200000) of adrenaline with the LA to prevent severe local vasoconstriction. Care must be taken to avoid the intravascular injection of this combination and when injecting near a terminal artery without collateral vascularisation, e.g. the dorsal artery of the penis, central retinal artery, etc., as the risk of necrosis of the extremity due to lack of blood supply is increased in these locations.

Commercially available LA solutions are slightly hypobaric (i.e. they have a lower density than cerebrospinal fluid (CSF) at 37 °C and physiological pH). As the specific gravities (SG) of bupivacaine and cerebrospinal fluid (CSF) are 1.0013 and 1.005, respectively, bupivacaine 'floats' on top of CSF during, for example, intrathecal injections. Adding glucose to the LA solution increases its SG and ensures that the mixture 'sinks'. Changing the baricity of LA solutions with glucose coupled with deliberate manipulation of the subject's body position allows precise drug distribution, e.g. limiting blockade to certain dermatomes or to only one side of the body (Phelan and Macevilly 1984).

Hyaluronidase and related enzymes temporally degrade hyaluronic acid, allowing a greater distribution of LAs throughout connective tissue at the injection site. This effect extends the area of local anaesthetic effect. However, their use is not widespread, and there is limited and conflicting evidence on the benefit of combining LAs with these enzymes. The exception is when these combinations are used for retrobulbar blocks (Kallio et al. 2000).

The alkalinization of LA solutions with sodium bicarbonate delays the onset time of the LAs and prolongs their duration of effect. This effect is due to the modification of the pH by bicarbonate, which also causes less pain at the injection site of the LA (Ririe et al. 2000). Preparations of LAs with sodium bicarbonate degrade rapidly, so it is recommended that they be mixed immediately before administration.

Opioids may also increase the duration of LA when the two drugs are mixed. This effect seems to be related to the presence of opioid receptors in the spinal cord, peripheral nerves and joint capsule (Hendrix et al. 1996; Ng et al. 2006; Mays et al. 1987).

Agonists of α_2-adrenoreceptors also increase the duration of LAs. This increase in duration seems to be related to the presence of these receptors in the central and peripheral nervous systems (Souza et al. 2010; Esmaoglu et al. 2010), but some drugs in this group (e.g. xylazine) may have a direct effect on the Na^+ channels (Aziz and Martin 1978).

5 Local Anaesthetic Use in Production Animals

Surgical pain management in all species—including production animals—is most effective when analgesics are administered: (i) pre-emptively, i.e. before painful procedures are begun; (ii) for an adequate duration; and (iii) as part of a multi-modal or polypharmacological technique. While incorporating LAs into multimodal techniques has numerous advantages, LAs are often used alone in production animals (Metcalf et al. 2024). This practice is popular in these species as the accurate injection of LA at a suitable dose provides complete sensory blockade, i.e., for brief procedures, drugs from other pharmacological classes are simply not required if physical restraint can be performed safely. Furthermore, LAs are considered safe if used at appropriate doses. Unfortunately, LAs aren't always administered preventively (or pre-emptively) in production animals under commercial conditions, where the efficient throughput of animals may be prioritised over optimal timing of administration of the LA and ultimately, over ideal pain management strategies. The time to peak effect of LAs is not instantaneous, and in an ideal environment, the LA would be administered at least 1–2 min before the painful procedure is performed. This approach requires either a slow production line of animals undergoing a procedure or double handling, where animals are injected on the first pass and the procedure is performed on the second round. The compromise in many instances is to administer the LA at the time of the procedure and accept that desensitisation of the site has not occurred, but that post-operative effects will still be beneficial and decrease pain (Small et al. 2020).

In many countries, LAs may be the only analgesic registered for use in production animal species. There are several reasons for this: most are inexpensive; they have low abuse potential, and they do not require stringent, e.g., refrigerated, storage conditions. Consequently, the use of LAs has been extensively reported in numerous production animal species (Metcalf et al. 2024). With appropriate physical restraint or sedation, local anaesthesia techniques are frequently preferable to general anaesthesia under production conditions: animals will recover sooner and return to their normal environment and activities; remain conscious throughout the procedure, experience fewer complications and the procedure will be less expensive (Edmondson 2008, 2016).

The use of analgesia for husbandry procedures in a farming context is not universal. While veterinarians, producers and consumers recognise that various procedures and conditions cause pain in production animals, the provision of pain relief, including LA, remains incomplete (van Dyke et al. 2021). To ensure that more farmed animals receive analgesia, acknowledging and understanding the barriers to the administration of pain relief is important. The main issues are the costs of analgesia, including LA, along with limited appreciation of the capacity of farmed animals to experience pain. There are also challenges with access to licensed products (van Dyke et al. 2021). Local anaesthesia is considered cheap compared to other pain relief options but the cost per animal may not be justifiable to a farmer. In some cases, LA administration has been shown to be cost effective as the potential gains in animal production offset the expense of the drug (Steagall et al. 2021).

The biggest limitation of LA for productions animals is the relatively short duration of action of these drugs (Metcalf et al. 2024; Small et al. 2021). Until longer acting drugs are developed, are accessible, and cost effective, it is likely that production animals will continue to be undertreated. In the interim, LAs offer a good option for these species. The details of various strategies for LA administration are described in the species-specific chapters in this book. In brief, in cattle, LAs have been used for local, regional, and spinal anaesthesia for a range of painful procedures such as disbudding (Herskin and Nielsen 2018), castration (Steagall et al. 2021; Musk et al. 2017; McCarthy et al. 2016), caesarean delivery (Schultz et al. 2008) and other elective and emergency procedures (Desrochers et al. 1995). In sheep, the use of LAs for tail docking (Small et al. 2020), castration (Small et al. 2020), mulesing (Lomax et al. 2013) and caesarean delivery (Scott 2005) has been extensive. The use of LAs in pigs is less extensive than in ruminants, with their use limited to castration (Yun et al. 2019) and veterinary procedures (Skarda 1996), although advanced techniques have been described. In goats, LAs are used for various procedures, including disbudding (Ajuda et al. 2020; Cuttance et al. 2023) and castration (Ajadi et al. 2012; Venkatachalam et al. 2021). For deer, articaine has been used for velveting and dehorning (Sahebjam et al. 2020; Venkatachalam et al. 2019). In chickens, LAs have been used during cloacal surgery (Khamisabadi et al. 2021), but their use in poultry production is minimal.

The use of LAs is rare in non-mammalian production animals and in invertebrates. This observation reflects the greater understanding of mammalian pain, a paucity of information on pharmacological evidence in non-mammalian species, and the challenges of pain assessment in non-mammalian species. If pain cannot be identified the motivation to provide pain relief is diminished. In-depth detail is provided in the species-specific chapters. However, in brief, in insects there has been work to characterise the target of LAs (the voltage-gated sodium channels) to understand the interactions of LAs with their receptors (Song et al. 2011). This work is ongoing as it appears that the receptors in mammals and insects may diverge (Song et al. 2011). Further detail is included in the chapter about insects in this book. Local anaesthetics have been used to investigate olfactory memory of honeybees by blocking certain areas of the brain (Devaud et al. 2007). The results of this work demonstrated that LA injections can be used to map learning tasks onto specific areas of the insect brain (Devaud et al. 2007). The pharmacological properties, safety and efficacy of LAs in fish require investigation to ensure an evidence-based approach to their use in these animals (Chatigny et al. 2017). Lidocaine appears to be an effective local anaesthetic in cephalopod molluscs (Butler-Struben et al. 2018). Finally, in decapod crustaceans, local anaesthesia has been used to desensitise eyestalks and antennae (de Souza Valente 2022).

References

Ajadi RA, Owanikin AO, Martins MM, Gazal OS (2012) Effect of epidural tramadol and lignocaine on physiological and behavioural changes in goats subjected to castration with a high tension band. N Z Vet J 60(6):344–348

Ajuda I, Battini M, Mattiello S, Arcuri C, Stilwell G (2020) Evaluation of pain mitigation strategies in goat kids after cautery disbudding. Animals (Basel) 10(2):277

Aps C, Reynolds F (1978) An intradermal study of the local anaesthetic and vascular effects of the isomers of bupivacaine. Br J Clin Pharmacol 6:63–68

Association of A (2023) Management of severe local anaesthetic toxicity. Association of Anaesthetists

Aziz MA, Martin RJ (1978) Alpha agonist and local anesthetic properties of xylazine. Zentralbl Veterinarmed 25:181–188

Biehl D, Shnider SM, Levinson G (1978) Placental transfer of lidocaine: effects of fetal acidosis. Anesthesiology 48:409–412

Butler-Struben HM, Brophy SM, Johnson NA, Crook RJ (2018) In vivo recording of neural and behavioral correlates of anesthesia induction, reversal, and euthanasia in cephalopod Molluscs. Front Physiol 9:109

Campoy L, Schroeder K (2013) General considerations. In: Campoy L, Raw RM (eds) Small animal regional Anaesthesia and analgesia. Wiley-Blackwell, pp 11–18

Cartwright PD, Fyhr P (1988) The manufacture and storage of local anesthetics. Reg Anesth 13:1–12

Cassuto J, Sinclair R, Bonderovic M (2006) Anti-inflammatory properties of local anesthetics and their present and potential clinical implications. Acta Anaesthesiol Scand 50(3):265–282

Chadwick HS (1985) Toxicity and resuscitation in lidocaine- or bupivacaine-infused cats. Anesthesiology 63:385–390

Chatigny F, Kamunde C, Creighton CM, Stevens ED (2017) Uses and doses of local anesthetics in fish, amphibians, and reptiles. J Am Assoc Lab Anim Sci 56(3):244–253

Clarkson C, Hondeghem L (1985) Mechanism for bupivacaine depression of cardiovascular conduction: fast block of sodium channels during the action potential with slow recovery from block during diastole. Anesthesiology 62:396–405

Corning JL (1885) Spinal anaesthesia and local medication of the cord. NY Med J 42:483–485

Cuttance EL, Mason WA, Laven RA, Ruddy BP, Taberner AJ, JW MK et al (2023) Investigating the use of local nerve blocks and general anaesthesia in reducing pain during and after disbudding procedure in goat kids. J Appl Anim Welf Sci:1–18

de Souza Valente C (2022) Anaesthesia of decapod crustaceans. Vet Anim Sci 16:100252

Desrochers A, St-Jean G, Anderson DE (1995) Surgical management of preputial injuries in bulls: 51 cases (1986-1994). Can Vet J 36(9):553–556

Devaud J-M, Blunk A, Podufall J, Giurfa M, Grünewald B (2007) Using local anaesthetics to block neuronal activity and map specific learning tasks to the mushroom bodies of an insect brain. Eur J Neurosci 26(11):3193–3206

Dugdale AHA (2010) Concepts of general anaesthesia. In: Veterinary Anaesthesia principles to practice. Blackwell Publishing Ltd, pp 1–4

Edgcombe H, Hocking G (2006) Local anaesthetic pharmacology 11/07/05
Edmondson MA (2008) Local and regional anesthesia in cattle. Vet Clin N Am Food Anim Pract 24(2):211–226
Edmondson MA (2016) Local, regional, and spinal anesthesia in ruminants. Vet Clin North Am Food Anim Pract 32(3):535–552
Esmaoglu A, Yegenoglu F, Akin A (2010) Dexmedetomidine added to levobupivacaine prolongs axillary brachial plexus block. Anesth Analg 111:1548–1551
Finucane BT (2003) Allergies to local anesthetics – the real truth. Can J Anaesth 50:869–874
Gaedcke F (1855) Ueber das Erythroxylin, dargestellt aus den Blättern des in Südamerika cultivirten Strauches Erythroxylon Coca. Arch Pharm 132:141–150
Gay GR, Inaba DS, Sheppard CW, Newmeyer JA (1975) Cocaine: history, epidemiology, human pharmacology, and treatment. A perspective on a new debut for an old girl. Clin Toxicol 8:149–178
Geng WP, Ebke M, Foth H (1995) Prilocaine elimination by isolated perfused rat lung and liver. Naunyn Schmiedeberg's Arch Pharmacol 351:93–98
Halsted WS (1885) Practical comments on the use and abuse of cocaine. N Y Med J 42:294–295
Heavner JE (2007) Local anesthetics. Curr Opin Anaesthesiol 20(4):336–342
Hendrix PK, Raffe MR, Robinson EP (1996) Epidural administration of bupivacaine, morphine, or their combination for postoperative analgesia in dogs. J Am Vet Med Assoc 209:598–607
Herskin MS, Nielsen BH (2018) Welfare effects of the use of a combination of local anesthesia and NSAID for disbudding analgesia in dairy calves—reviewed across different welfare concerns. Front Vet Sci 5:117
Hollmann MW, Wieczorek KS, Berger A (2001) Local anesthetic inhibition of G protein-coupled receptor signaling by interference with Galpha (q) protein function. Mol Pharmacol 59:294–301
Johnson RF, Herman N, Arney TL (1995) Bupivacaine transfer across the human term placenta. A study using the dual perfused human placental model. Anesthesiology 82:459–468
Jorfeldt L, Lewis DH, Lofstrom JB (1979) Lung uptake of lidocaine in healthy volunteers. Acta Anaesthesiol Scand 23:567–574
Kallio H, Paloheimo M, Maunuksela EL (2000) Hyaluronidase as an adjuvant in bupivacaine-lidocaine mixture for retrobulbar/peribulbar block. Anesth Analg 91:934–937
Khamisabadi A, Kazemi-Darabadi S, Akbari G (2021) Comparison of anesthetic efficacy of lidocaine and bupivacaine in spinal anesthesia in chickens. J Avian Med Surg 35(1):60–67
Kirihara Y, Saito Y, Sakura S (2003) Comparative neurotoxicity of intrathecal and epidural lidocaine in rats. Anesthesiology 99:961–968
Kopacz D, Carpenter R, Mackey D (1989) Effect of ropivacaine on capillary blood flow in pigs. Anesthesiology 71:69–74
Kroin JS, Penn RD, Levy FE (1986) Effect of repetitive lidocaine infusion on peripheral nerve. Exp Neurol 94:166–173
Lafont ND, Legros FJ, Boogaerts JG (1996) Use of liposome-associated bupivacaine in a cancer pain syndrome. Anaesthesia 51:578–579
Lagan G, McLure HA (2004) Review of local anaesthetic agents. Curr Anaesth Crit Care 15(4–5):247–254
Liu PL, Feldman HS, Giasi R (1983) Comparative CNS toxicity of lidocaine, etidocaine, bupivacaine and tetracaine in awake dogs following rapid intravenous administration. Anaesth Analg 62:375–379
Lomax S, Sheil M, Windsor PA (2013) Duration of action of a topical anaesthetic formulation for pain management of mulesing in sheep. Aust Vet J 91(4):160–167
Matsota PK, Markantonis SL, Fousteri MZ (2009) Excretion of ropivacaine in breast milk during patient-controlled epidural analgesia after cesarean delivery. Reg Anesth Pain Med 34:126–129
Mays KS, Lipman JJ, Schnapp M (1987) Local analgesia without anesthesia using peripheral perineural morphine injections. Anesth Analg 66:417–420
McCarthy D, Lomax S, Windsor PA, White PJ (2016) Effect of a topical anaesthetic formulation on the cortisol response to surgical castration of unweaned beef calves. Animal 10(1):150–156
McLure HA, Rubin AP (2005) Review of local anaesthetic agents. Minerva Anestesiol 71:59–74
Metcalf L, Lomax S, Van der Saag D, Garg S, White PJ (2024) Pain relief interventions in Australian livestock husbandry: a review of animal welfare and pain duration. Animals (Basel) 14(13):1901
Musk GC, Jacobsen S, Hyndman TH, Lehmann HS, Tuke SJ, Collins T et al (2017) Objective measures for the assessment of post-operative pain in Bos indicus bull calves following castration. Animals (Basel) 7(10):76
Neal MJ (2007) Local anaesthetics. In: Medical pharmacology. Wiley, Chichester
Ng HP, Nordström U, Axelsson K (2006) Efficacy of intra-articular bupivacaine, ropivacaine, or a combination of ropivacaine, morphine, and ketorolac on postoperative pain relief after ambulatory arthroscopic knee surgery: a randomized double-blind study. Reg Anesth Pain Med 31:26–33
Niemann A (1860) Ueber eine neue organische Base in den Cocablättern. Arch Pharm 153:129–256
Olschewski A, Hempelmann G, Vogel W (1998) Blockade of Na+ and K+ currents by local anesthetics in the dorsal horn neurons of the spinal cord. Anesthesiology 88:172–179
Ortega D, Viviand X, Lorec AM (1999) Excretion of lidocaine and bupivacaine in breast milk following epidural anesthesia for cesarean delivery. Acta Anaesthesiol Scand 43:394–397
Park J, Sutradhar BC, Hong G (2011) Comparison of the cytotoxic effects of bupivacaine, lidocaine, and mepivacaine in equine articular chondrocytes. Vet Anaesth Analg 38:127–133

Phelan DM, Macevilly M (1984) A comparison of hyper- and isobaric solutions of bupivacaine for subarachnoid block. Anaesth Intensive Care 12:101–107

Prabhakar A, Ward CT, Watson M (2019) Liposomal bupivacaine and novel local anesthetic formulations. Best Pract Res Clin Anaesthesiol 33:425–432

Rathmell JP, Strichartz G, Wanderer J (2015) Neosaxitoxin versus traditional local anesthetics: mechanism of action and sites of notable effect. Anesthesiology 123:A23

Ririe DG, Walker FO, James RL (2000) Effect of alkalinization of lidocaine on median nerve block. Br J Anaesth 84:163–168

Ruetsch YA, Böni T, Borgeat A (2001) From cocaine to ropivacaine: the history of local anesthetic drugs. Curr Top Med Chem 1(3):175–182

Sahebjam F, Kongara K, Chambers JP, Walker RE, Naffa R, Lopez-Villalobos N et al (2020) Pressure algometry validation and determination of efficacy of Articaine hydrochloride ring block in antler removal in Red Deer (Cervus elaphus). Animals (Basel) 10(11)

Sakura S, Sumi M, Marimoto N (1999) The addition of epinephrine increases intensity of sensory block during epidural anesthesia with lidocaine. Reg Anesth Pain Med 24:541–546

Sakura S, Kirihara Y, Muguruma T (2005) The comparative neurotoxicity of intrathecal lidocaine and bupivacaine in rats. Anesth Analg 101:541–547

Schultz LG, Tyler JW, Moll HD, Constantinescu GM (2008) Surgical approaches for cesarean section in cattle. Can Vet J 49(6):565–568

Scott PR (2005) The management and welfare of some common ovine obstetrical problems in the United Kingdom. Vet J 170(1):33–40

Sheikh NK, Dua A (2023) Procaine: StatPearls Publishing

Skarda RT (1996) Local and regional anesthesia in ruminants and swine. Vet Clin North Am Food Anim Pract 12(3):579–626

Small AH, Jongman EC, Niemeyer D, Lee C, Colditz IG (2020) Efficacy of precisely injected single local bolus of lignocaine for alleviation of behavioural responses to pain during tail docking and castration of lambs with rubber rings. Res Vet Sci 133:210–218

Small A, Fisher AD, Lee C, Colditz I (2021) Analgesia for sheep in commercial production: where to next? Animals 11(4):1127

Smith TC (2009) Local anaesthetic agents. Fundamentals of Anaesthesia

Song W, Silver KS, Du Y, Liu Z, Dong K (2011) Analysis of the action of lidocaine on insect sodium channels. Insect Biochem Mol Biol 41(1):36–41

Soto G, González N, Calero M (2018) Perfusión de lidocaína intravenosa. Rev Esp Aneste Reani 65:269–274

Souza SS, Intelisano TR, Biaggi D (2010) Cardiopulmonary and isoflurane-sparing effects of epidural or intravenous infusion of dexmedetomidine in cats undergoing surgery with epidural lidocaine. Vet Anaesth Analg 37:106–115

Steagall PV, Bustamante H, Johnson CB, Turner PV (2021) Pain management in farm animals: focus on cattle, sheep and pigs. Animals (Basel) 11(6):1483

Tucker GT, Moore DC, Bridenbaugh PO (1972) Systemic absorption of mepivacaine in commonly used regional block procedures. Anethesiology 37:277–287

Ueki R, Tatara T, Kariya N (2009) Comparison of placental transfer of local anesthetics in perfusates with different pH values in a human cotyledon model. J Anesth 23:526–529

Van Den Berg RJ, Wang Z, Grouls RJ (1996) The local anesthetic, n-butyl-p-aminobenzoate, reduces rat sensory neuron excitability by differential actions on fast and slow Na+ current components. Eur J Pharmacol 316:87–95

van Dyke R, Connor M, Miele A (2021) An investigation into the perceptions of veterinarians towards perioperative pain Management in Calves. Animals (Basel) 11(7):1882

Venkatachalam D, Chambers JP, Kongara K, Singh P (2019) Analgesic efficacy of articaine hydrochloride for velvet antler removal in red deer (Cervus elaphus) and analysis of drug residues in the harvested velvet antlers. N Z Vet J 67(5):228–233

Venkatachalam D, Chambers JP, Kongara K, Ward N, Jacob A, Singh PM (2021) Pharmacokinetics, efficacy and convulsive dose of articaine hydrochloride in goat kids. Vet Anaesth Analg 48(2):264–271

Xiong Z, Strichartz GR (1998) Inhibition by local anesthetics of Ca2+ channels in rat anterior pituitary cells. Eur J Pharmacol 363:81–90

Yun J, Ollila A, Valros A, Larenza-Menzies P, Heinonen M, Oliviero C et al (2019) Behavioural alterations in piglets after surgical castration: effects of analgesia and anaesthesia. Res Vet Sci 125:36–42

Zhang Y, Jing Y, Pan R, Ding K, Chen R, Meng Q (2021) Mechanisms of cancer inhibition by local anesthetics. Front Pharmacol 12:12

Zink W, Graf BM (2004) Local anesthetic myotoxicity. Reg Anesth Pain Med 29:333–340

Zink W, Bohl JR, Hacke N (2005) The long term myotoxic effects of bupivacaine and ropivacaine after continuous peripheral nerve blocks. Anesth Analg 101:548–554

Slaughter, Euthanasia and Killing of Production Animals

Slaughter, Euthanasia and Killing of Mammals

Craig Johnson

Abstract

Animals are killed in different circumstances for different reasons. Culling is used to describe the killing of animals or groups of animals to reduce or eliminate a population in a particular place. Euthanasia is derived from the Greek and means 'a good death'. The evaluation of killing methods must consider the animals to be killed and the personnel involved. In general, the extent to which an animal's death affects its welfare is estimated by the duration and severity of any adverse affective states that the animal experiences before the point at which it loses consciousness for the last time. After this point, the animal cannot experience anything. Methods for killing animals are of three types, and in each category, death is brought about by extreme and permanent alterations to either the structure and/or function of the central nervous system. Death may be caused by physical methods such as shooting, cervical dislocation and decapitation; by functional disruption of the central nervous system using methods that don't physically destroy it, such as electrocution or chemical disruption; or by deprivation of oxygen required by the central nervous system, such as exsanguination, carbon monoxide inhalation and low atmospheric pressure stunning.

Keywords

Slaughter · Euthanasia · Killing · Humane

1 Introduction

Human beings have had close relationships with non-human animals for as long as *Homo sapiens* have existed. Initially, these relationships centred around the hunting of free-living animals. Later, some species were domesticated and used as working, production or companion animals. A common factor in relationships has been the human's responsibility to choose both the timing and the manner of the animals' deaths (Allen et al. 2023). Understanding animal sentience and the need for animal welfare has prompted the evaluation of killing methods in terms of their potential effect on the animals to which they are applied (Kasperbauer and Sandøe 2015). Public opinion appears to be influenced by factors such as the age and 'cuteness' of the animal to be killed, the aesthetics of the killing method, the extent to which the need for killing is understood, and the level of publicity surrounding the killing (Sinclair et al. 2023). While these are important

C. Johnson (✉)
Animal Welfare Science and Bioethics Centre, School of Veterinary Sciences Tāwharau Ora, Massey University, Palmerston North, New Zealand
e-mail: c.b.johnson@massey.ac.nz

R. E. Clutton, G. C. Musk (eds.), *Pain and Suffering in Farmed Animals*,
https://doi.org/10.1007/978-3-032-11407-5_12

from the point of view of maintaining social license, the animal itself is the principal consideration in this process.

From the animal's perspective, the only important feature of a killing method is the effect on its welfare that will arise during its death. Some methods of killing completely protect the animal from perceiving any sensory input that may occur before the onset of death, whilst others may not or may even be aversive in their own right. Evaluations of these effects require a theoretical understanding of the processes of death and how different methods of killing work, as well as a practical understanding of the advantages and disadvantages of the killing methods themselves. This chapter will focus on these aspects and is divided into section 1, which considers the theoretical aspects involved in killing animals, and section 2, which discusses a selection of killing methods whose usefulness depends on circumstances.

2 Theoretical Aspects of Killing Animals

2.1 Circumstances in Which Animals Are Killed

Animals are killed in different circumstances for different reasons. Consequently, several terms are used to differentiate the act of killing animals in different contexts, some of which may be euphemistic and misleading. Terms such as 'sacrifice' and 'harvest' have specific meanings in other contexts and so should be avoided. The following terms are used in this chapter to describe the act of killing an animal in particular circumstances:

Kill A general term used in any circumstances where an animal is killed by any technique for any reason.

Slaughter To kill an animal so that its carcass or parts of it will enter the food chain.

Cull To kill an animal or a group of animals to reduce or eliminate a population in a particular place. Animals are culled because of overpopulation, at the end of their productive life, as a means of disease prevention or control, or because they are pests.

Euthanasia The term euthanasia is derived from the Greek and means 'a good death'. The term has two distinct uses in different sources. Taken to mean 'ending the life of an individual animal in a way that minimises or eliminates pain and distress' is used by the American Veterinary Medical Association (Leary et al. 2020). Alternatively, euthanasia can be taken to mean 'painless killing to relieve suffering' and is used in this context by the Royal College of Veterinary Surgeons (Anon 2022). This second use encompasses the first but also requires that the act of killing be in the animal's best interest in order to limit adverse welfare effects on that animal. This second use will be adopted in this chapter. The killing of an animal in circumstances that fit the first usage, but not the second, will be referred to as culling, or an alternative term if this more appropriately describes the particular circumstances.

Emergency This term is appended to others describing animal killing, to indicate urgency in the circumstances described. Thus, the killing of an injured animal where the carcass is intended for consumption is referred to as 'emergency slaughter'.

2.2 What Is Death?

Dying is usefully viewed as a process which ends in death. Some considerations of killing identify the 'point of death', but this is impossible to identify. The process of dying differs according to cause, while components of the process can occur at different times, last for different durations, and occur in different sequences. Figure 1 illustrates the processes that can occur during an animal's dying and notes how these may influence the animal's affective state, i.e., the emotions it experiences, such as fear or anxiety, during that part of the process.

2.3 Evaluating the Welfare Cost of Death

The evaluation of killing methods must consider the animals to be killed and the personnel involved. Currently, there is debate, but no agreement as to whether the loss of an animal's life per se represents an adverse welfare event beyond any unpleasant affective states the animal may experience as it dies. The debate about this aspect of dying is philosophical and focuses on an animal's 'intrinsic versus extrinsic' value. If animals are considered to have extrinsic value only, it is considered that there are no welfare implications to killing them provided that they experience no aversive affective states during the course of their death. Conversely, if animals are considered to have both extrinsic and intrinsic value, it is considered that killing them has implications for their welfare, irrespective of their perception of affective states. This is because the act of killing deprives them of the opportunity for continued existence. This debate is beyond the scope of the present chapter, so the subject of intrinsic value will not be referred to hereafter. Feldman (1991) is a good starting point for a more detailed consideration of the potential evil of death.

In general, the extent to which an animal's death affects its welfare is estimated by the duration and severity of any adverse affective states that the animal experiences before the point at which it loses consciousness for the last time. After this point, the animal cannot experience anything. Ideally, animals should not experience any adverse effects before losing consciousness, but other features of the killing method may render this impossible. Personnel who undertake or witness killing are also affected by these and other features of the killing method, e.g., its aesthetics, loudness of gun reports, etc. While the assessment of killing methods must consider the people involved, the animal's welfare is the most

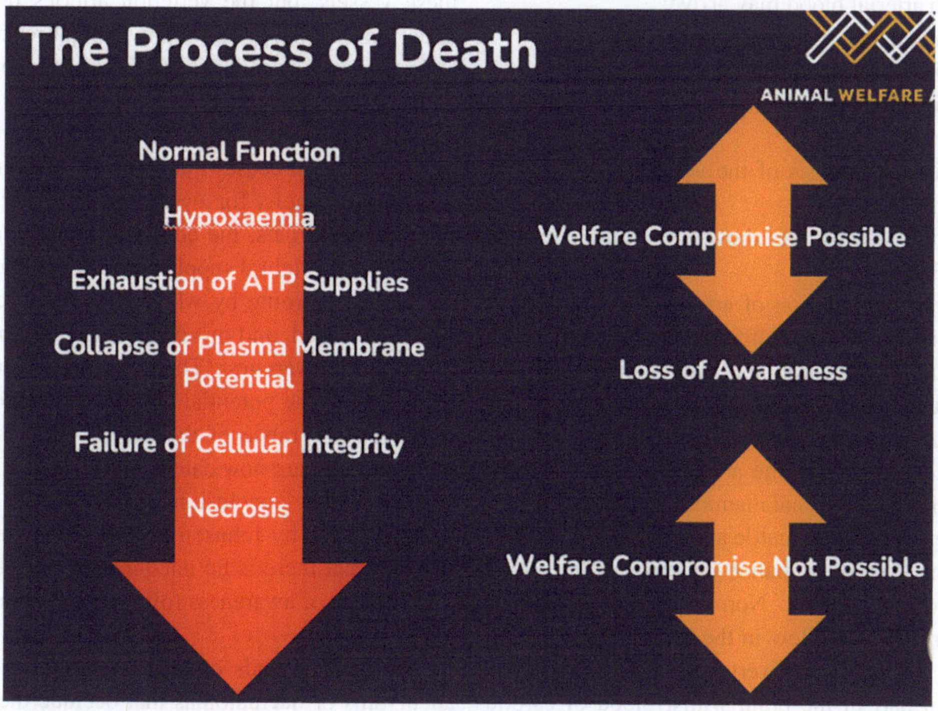

Fig. 1 Steps In the process of death

important factor in the selection of killing methods.

2.4 Physiology and Anatomy Related to Killing Methods

2.4.1 Blood Supply to the Central Nervous System

The central nervous system has a high metabolic rate and so requires an adequate supply of oxygen from arterial blood. In the context of slaughter, death is often caused by exsanguination, which is brought on by severance of major arteries in the neck (common carotid arteries) or cranial thoracic cavity (brachycephalic trunk). There are distinct species differences in the latency between severance of the common carotid arteries and loss of awareness, which are due to differences in the anatomy of the arterial supply to the arterial circle at the base of the brain and from which the majority of the brain's arterial blood is supplied.

It can be seen from a schematic diagram of the blood supply to the arterial circle in the brain (Fig. 2) that there are four possible routes by which arterial blood may arrive:

1. Via the internal carotid artery
2. Via the basilar artery
3. Via the anastomosing ramus of the maxillary artery
4. Via anastomosis of the vertebral and carotid arteries

These routes are shown in more detail in Fig. 3.

The configuration of arterial blood supply is species dependent; differences between them account for the duration of continued brain perfusion and hence the capacity to remain aware following the severance of the common carotid arteries. Details of the arterial supply are illustrated in Fig. 4 (usual mammalian pattern), Fig. 5 (sheep) and Fig. 6 (cattle and goats).

The usual mammalian pattern supplies blood via routes 1 and 2. Normal pressure gradients mean that blood flow in the basilar artery is craniad, and so blood of vertebral origin supplies the caudal part of the brain whilst blood of carotid origin supplies the rostral part. Some animals with this pattern also have an anastomosing artery connecting the carotid and the maxillary arteries, but this is small and relatively insignificant.

In sheep, the proximal section of the internal carotid artery atrophies and the lumen becomes obliterated in the weeks or months after birth, meaning that there is no possibility of anastomosis between the vertebral and carotid supplies. Blood flow in the basilar artery is caudad, and so blood of vertebral origin supplies the spinal cord and caudal brain stem, whilst the majority of the brain is supplied by carotid blood via the anastomosing ramus of the maxillary artery.

In cattle and goats, the proximal internal carotid artery also atrophies and is no longer functional after the first year of life. In this instance, there is a connection of blood from the vertebral artery with blood from the maxillary anastomosing ramus, both supplies feeding into the distal portion of the internal carotid artery. This means that the whole of the central nervous system can be supplied by blood of both carotid and vertebral arterial origins.

Severance of the common carotid arteries at slaughter interrupts the blood supply through these vessels, but the vertebral arteries remain intact. In the sheep, this supply cannot reach the arterial circle because there is no connection between the vertebral artery and the internal carotid artery, and so perfusion of the higher centres of the brain ceases immediately, resulting in loss of the capacity for awareness in 20 s or less. In cattle and goats, the connection between the carotid and vertebral supplies means that there is an alternative route by which the arterial circle can be perfused, and this can result in prolonged capacity for awareness. There are several studies investigating the potential duration of awareness following severance of the carotid arteries, but many of these are now dated and make it difficult to draw precise conclusions. These sources have been reviewed by Johnson et al. (2015), where it is noted that there is no unequivocal estimate of the duration of awareness following severance of blood vessels.

The severed ends of arteries can form false aneurysms or haematomas that occlude the vessel's end and prevent further blood loss. This is

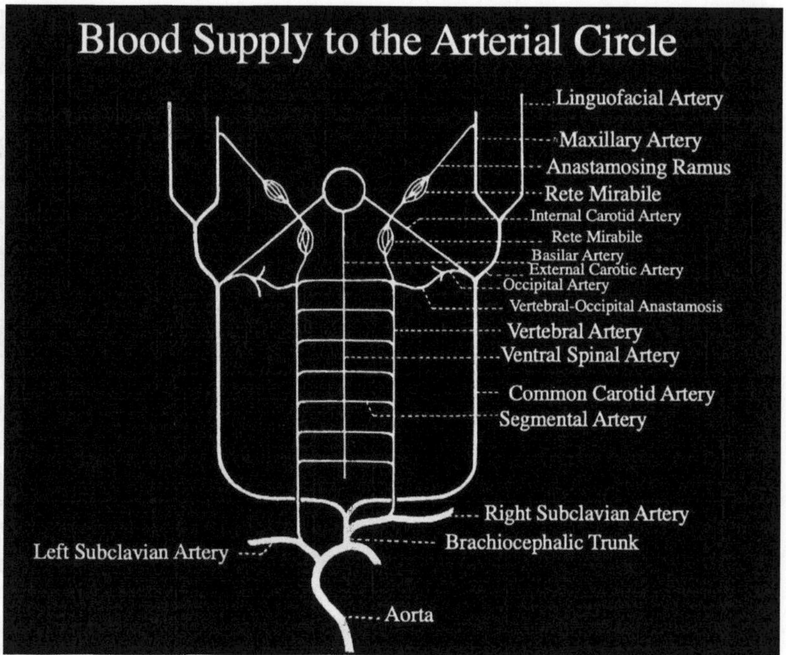

Fig. 2 Detailed blood supply to the arterial circle. (Adapted from King 1987)

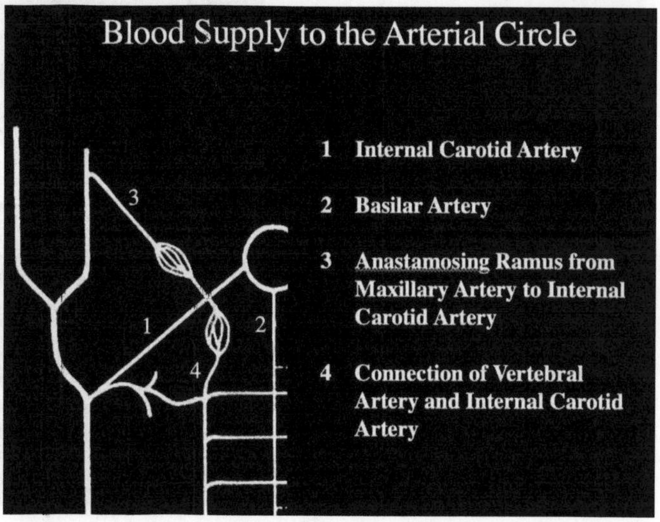

Fig. 3 Blood supply to the arterial circle. (Adapted from King 1987)

relatively common in cattle. When false aneurysms occlude the distal end of the severed carotid artery, the vertebral arterial supply may maintain perfusion to the arterial circle and prevent the animal from losing consciousness. The incidence of false aneurysm formation is similar to the incidence of prolonged times to collapse—60 s or more following halal slaughter (Gregory et al. 2010)—so may account for this phenomenon.

2.4.2 Anatomy of the Skull and Central Nervous System

The mammalian central nervous system (CNS) is housed in a supporting bath of cerebrospinal fluid within the bony structures of the skull and vertebral column. This affords it a very high degree of physical protection that must be breached when physical killing methods based on brain destruction are employed. Physical disruption of central nervous system tissues can be achieved either by

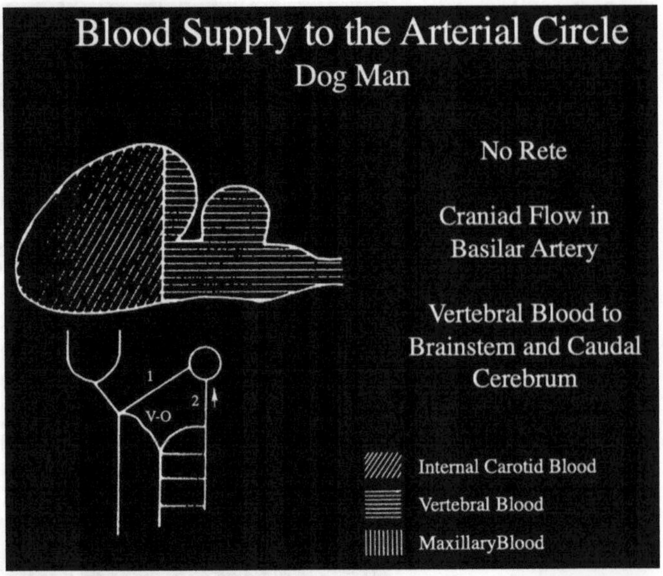

Fig. 4 Blood supply to the arterial circle in dogs and humans. (Adapted from King 1987)

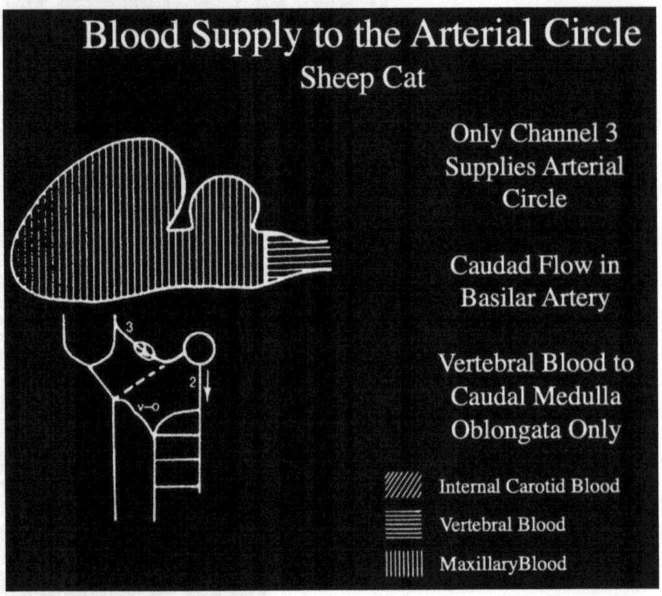

Fig. 5 Blood supply to the arterial circle in cats and sheep. (Adapted from King 1987)

physical penetration of a foreign body or by the propagation of a pressure wave of sufficient energy to disrupt function. Some physical killing methods, e.g. shooting a bullet into the brain, combine both of these elements.

Penetration of the CNS by objects with low kinetic energy causes direct damage that is restricted to the area destroyed. For this injury to kill the animal humanely, the penetrating object must destroy an area of the CNS that is critical for the maintenance of awareness and vital functions such as breathing. The penetrating object usually takes the form of a sharp implement introduced through the skull or spinal cord, e.g. a bullet or poleaxe, but similar damage can be achieved by using the tissues of the body themselves, such as in cervical dislocation, where the spinal cord is damaged due to distraction of the vertebrae of the neck.

Objects with high kinetic energy entering the body propagate a pressure wave that travels through the tissues, causing direct tissue disrup-

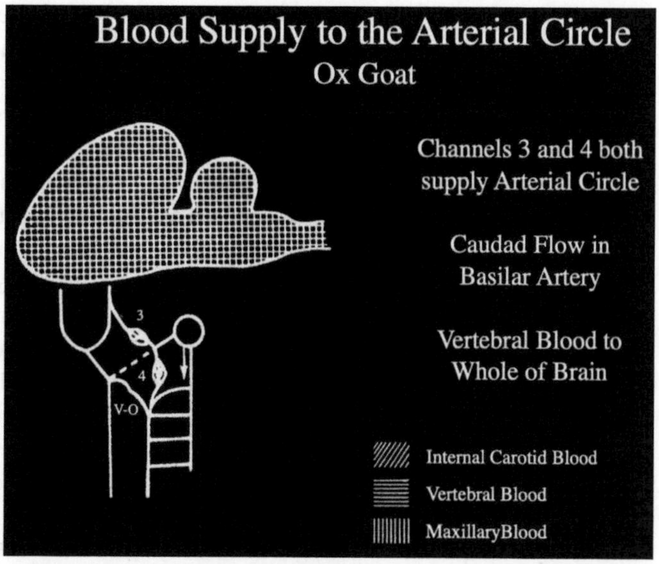

Fig. 6 Blood supply to the arterial circle in oxen and goats. (Adapted from King 1987)

tion. This pressure wave is usually linear and propagated in the general direction of travel of the originating object, but when the object contacts rigid elements, e.g., the skull, the pressure wave becomes more complex. For example, the rotational movement of the skull seen with some captive bolt placements can propagate rotational waves in the brain, causing widespread damage that is not limited to the general direction of travel of the bolt itself.

2.4.3 Hypoxia and Hypercarbia

Oxygen is essential to most forms of advanced life because it oxidises sugars, generating chemical energy by the conversion of adenosine monophosphate (AMP) into adenosine triphosphate (ATP). Intracellular chemical processes that require energy are driven by the conversion of ATP back into AMP. Failure of oxygen supply to the cells results in catastrophic breakdown of this cycle, followed by collapse of cellular processes and irreversible cellular changes, which lead to death. This process can occur very rapidly, especially in cells such as neurones, which have high metabolic demands and cannot generate energy by anaerobic mechanisms.

Hypoxic damage to cells can occur by one of four different categories of failure of oxygen delivery or utilisation (Mallat et al. 2022):

1. *Hypoxic hypoxia*—The arterial partial pressure of oxygen is insufficient to produce fully saturated haemoglobin and so the amount of oxygen in the blood is reduced. This may be due to the animal breathing an atmosphere that does not contain sufficient oxygen, such as during low atmospheric pressure stunning used in the slaughter of pigs.
2. *Hypaemic hypoxia* –There is insufficient functional haemoglobin in the blood to carry sufficient oxygen to the tissues. This may be due to an absolute reduction in haemoglobin, e.g. anaemia, or to a reduction in the function of haemoglobin, such as in carbon monoxide poisoning. Carbon monoxide poisoning is usually accidental, but this gas has occasionally been used to cull unwanted animals in a shelter context or where large numbers of animals are killed in situ. The high toxicity of carbon monoxide means that these techniques are extremely dangerous to operators and should be avoided if at all possible.
3. *Stagnant hypoxia*—There is insufficient tissue perfusion to deliver sufficient oxygen despite there being adequate blood oxyhaemoglobin. This occurs when the cardiovascular system collapses, e.g. during exsanguination as part of the slaughter process in many animals, or cardiac arrest.

4. *Histotoxic hypoxia*—Despite adequate cellular oxygen delivery, cells are unable to use it because toxic agents are interfering with cellular transport systems. This occurs with potassium cyanide poisoning, when the function of cytochrome oxidase within mitochondria is inhibited. This technique is not used for the slaughter of production animals, but may be applied to the control of free-living pest species such as possums in New Zealand.

Carbon dioxide (CO_2) is one of the main products of tissue energy production and is produced by all living cells. Carbon dioxide dissolves in water to form carbonic acid, which means that CO_2 elimination from the body is essential for maintaining normal body pH—a prerequisite for normal enzymatic processes throughout the body. The link between body pH and CO_2 removal means that CO_2 plays a major role in the regulation of respiratory, cardiovascular, and renal physiology, with these organ systems working together to maintain a normal CO_2 concentration in the animal.

Breathlessness, sometimes classified as dyspnoea, can cause an affective state, or feeling, that alerts an animal to the possibility of threatened homeostasis due to impaired respiratory function. In humans, it is perceived as increased inspiratory effort, tightness of the chest, and air hunger and is due to the mismatch between ventilatory function and the physiological drive to breathe (Coccia et al. 2016). Humans report breathlessness as one of the most distressing of the affective states. Its perception heightens anxiety and can cause feelings of panic (Williams and Carel 2018). Breathlessness has been highlighted as a significant animal welfare issue by Beausoleil and Mellor (2015).

Carbon dioxide causes aversion by two mechanisms: First, inhaling concentrations greater than 46% can cause pain because carbonic acid is formed on moist surfaces such as the membranes of the nasal passage and conjunctivae (Anton et al. 1992); Second, concentrations of CO_2 in the range of 7–10% cause breathlessness (Sechzer et al. 1960). Increases in CO_2 lead to increased ventilatory drive and a feeling of breathlessness, altered cellular function and stimulation of the sympathetic nervous system, resulting in a characteristic 'fight or flight response' that in humans is often accompanied by a sense of panic.

The physiological stimuli that cause breathlessness vary between species. In pigs (and most other mammals), physiological receptors are more sensitive to increased inspired CO_2 than decreased inspired oxygen. Under conditions of normal CO_2 concentrations, an increase in ventilatory drive is not seen until oxygen reaches approximately 50% of normal. Under conditions of normal inspired oxygen concentration ($F_IO_2 = 0.21$), an increase in ventilatory drive is seen whenever CO_2 levels increase from normal (Verbrugghe et al. 1982).

2.4.4 Electrical Nature of Excitable Tissues

The ability of the CNS to support cognitive processes depends upon the electrical activity of its neurones. Neurones use ATP to generate a membrane potential, allowing them to conduct information in the form of action potentials. The membrane potential is generated by the action of sodium potassium ATPase (Na^+-K^+ ATP-ase). This transmembrane protein uses energy from ATP to catalyse the transfer of Na^+ and K^+ across the plasma membrane, resulting in increased intracellular K^+ and increased extracellular Na^+. The membrane potential develops because of the concentration gradient of Na^+ and K^+ across the plasma membrane. The membrane is more permeable to K^+ than to Na^+, and so more K^+ leaks from the cell than Na^+ leaks in. This disparity results in more positively charged ions being outside the cell than inside, and so an electrical charge develops across the membrane: the membrane potential. Disruption of this electrical activity results in loss of coordinated neuronal activity and can result in loss of consciousness ranging from transient to permanent (Uhlhaas and Singer 2006).

Several conditions must be met for the maintenance of coordinated CNS electrical activity. These include: the presence of adequate oxygen (see hypoxia above); suitable tissue pH (see hypercarbia above); a temperature within appropriate limits; suitable concentration gradients of ions across plasma membranes; the absence of high current flux densities within tissues; the absence of toxins blocking transmembrane proteins.

Sodium—potassium ATPase activity is temperature dependent and can only maintain sufficient transfer to support coordinated neuronal activity at temperatures between about 32 and 40 °C (Dow et al. 2019; Hancock and Vasmatzidis 2003). When tissue temperatures move out of this range, coordinated electrical activity is disrupted and cognitive impairment ensues. This may cause a range of effects from impaired mentation to complete loss of awareness, depending on the severity of the deviation.

Ionic imbalances in the vicinity of plasma membranes alter the transmembrane balance between concentration and charge gradients and disrupt the membrane potential. This is most commonly seen when there is excessive extracellular potassium, although other ions, such as magnesium, have similar effects.

The delicate transmembrane ion balance that generates the membrane potential is also disrupted by strong extraneous currents. When voltages are applied across a tissue block, a current flux flows between the points of application. The path the current takes and its magnitude along the path depend on the relative resistivity at all points within the tissue and complicates the precise calculation of current flux. This flux creates gradients of potential, causing the movement of ions. Small current fluxes have no practical effect on the functioning of the tissues through which they flow, but once a sufficient magnitude is reached, they can disrupt membrane potential, making normal neuronal function impossible. Low currents produce a brief disruption while the current is present, but further increases in current flux can result in more prolonged dysfunction. Very high currents can permanently damage tissues and result in permanent loss of function. Overall, the effects of artificially induced electrical currents in tissues, such as during electrical stunning of cattle, sheep or pigs, can range from none through stunning of various durations to permanent loss of function. The effect of a particular current will depend on the magnitude and frequency of the current as well as the impedance and location of its connection to the tissues. For a more detailed treatment of current fluxes in tissues, see Vučenović et al. 2017.

Many toxins interfere with the maintenance of membrane potentials either by disrupting the energy requirements of the system (example: potassium cyanide), which disrupts cellular respiration—used in culling of free-living pest species) or by altering the ion permeability of the plasma membrane itself (example: tetrodotoxin, which selectively blocks voltage-dependent sodium channels).

2.5 Categorising Killing Methods

Methods for killing animals are of three types, as illustrated with examples in Table 1. In each category, death is brought about by extreme and permanent alterations to either the structure and/or function of the CNS.

Death caused by physical methods requires that parts of the CNS required for the continuation of life be physically destroyed. Methods in this category include shooting (free bullet or captive bolt), cervical dislocation and decapitation.

Death by functional disruption requires that CNS function be disrupted using methods that don't physically destroy it. The disruption must persist until death ensues. Methods in this category include electrocution and chemical disruption (exposure to toxins, e.g. pentobarbital (injected or orally administered) and CO_2 (administered by inhalation).

Death can also be caused by deprivation of essential substrates required by the CNS. Methods in this category include exsanguination, carbon monoxide inhalation and low atmospheric pressure stunning (LAPS), which all deprive the CNS of oxygen.

2.6 Non-welfare Considerations During Animal Slaughter

The animal's welfare must always be prioritised when they are killed, for whatever reason.

Table 1 Killing methods utilised in mammals

Category of killing method	Examples
Physical destruction	Free bullet Penetrative captive bolt Non-penetrative captive bolt Cervical dislocation Blunt force trauma
Functional disruption	Carbon dioxide inhalation Diathermic syncope Barbiturate overdose Inhaled Anaesthetic overdose Electrocution Potassium and magnesium salts
Dcprivation	Low atmospheric pressure stunning Carbon monoxide inhalation Inert gas inhalation Exsanguination
Adjuncts to killing method	Foam containment

However, other factors, which may be important in different circumstances, include:

Slaughter When animals are slaughtered for their carcasses to be consumed by humans the slaughter methods chosen must ensure the flesh is acceptable in terms of food safety (has not been contaminated with microbes), that there are no residual chemicals in the carcass, and that the meat continues to look appetising. Animals to be slaughtered are often transported to an abattoir and, after stunning, are killed by exsanguination, as this produces good carcass quality with reduced potential for microbial contamination. Whilst the transport of live animals may have greater welfare costs than home-killing, the former option is preferred to ensure the safety and quality of the meat for the consumer. Exsanguination of a conscious animal is painful because of the sensitive tissues (such as the skin and muscles in the neck) that must be incised before major blood vessels are opened. Consequently, animals are often rendered insensible by stunning (e.g. electrical, percussive stunning) before the throat is cut.

Culling of Pest Animals *Pest* control, an important part of conservation and biosecurity, frequently requires animals such as feral goats or pigs to be killed. Access to animals that are free roaming may be difficult, especially in remote areas. Traps and poisons are commonly used to kill pests, but their use requires that the pest species be targeted in ways that prevent other animals or the environment from being harmed. Traps and poisons should also be used in ways that minimise any unpleasant effects in the period between exposure and loss of consciousness. Traps aim to accomplish this by killing the animal immediately after triggering. It is important, therefore, that a particular trap be tested on the target species to avoid scenarios that do not result in rapid death as well as accidental targeting of other species. Poisons are often chosen for features other than their effects on the animal's welfare, such as rapid degradation in the environment and the availability of an antidote (contributing to the safety of a poison for non-target species), but do not affect the experience of animals that are the intended target of the poison. For example, brodifacoum is often used as a poison for rodents because it takes longer to kill them than other poisons. This results in a more severe impact on the welfare of the rodents, but means that if there is an instance of accidental ingestion (in a dog or toddler for instance) there is a greater chance of successfully treating the poisoning than there would be with more rapidly acting toxins.

Disposal of Unwanted Farm Animals Some animal production systems generate surplus animals, e.g. bull calves born to dairy cattle, male birds in laying flocks and female birds in the broiler industry. While these animals can be used for different purposes, many are disposed of as surplus and are culled in large numbers. In some systems, the animal becomes a by-product rather than the production item, so it is culled at the end of the production cycle. Examples include end-of-lay chickens and spent dairy cattle. Some may be sent for slaughter, but when economic factors preclude this option, the animals may be culled on farm. The need to cull large numbers of animals requires killing methods selected for cost and efficiency that require little or no handling of individual animals. Selected methods must consider how the remains will be disposed of to ensure harmful chemicals do not enter the food chain or the environment. An example of this is the use of pentobarbitone on animals, there are several reported instances of deaths to other animals after consumption of carcasses that they had gained access to by digging into offal pits (Kaiser et al. 2010; Payne et al. 2015).

2.7 Stunning Before Killing

In some countries and cultures, legislation or customs require that animals of some species be stunned before being slaughtered. This is usually done when the killing method is painful but may also be chosen for other reasons. During slaughter, death is usually caused by exsanguination because bleeding the carcass improves food safety and quality. Death by exsanguination is painful because in severing the necessary blood vessels the incision must be made through sensitive tissues; stunning is undertaken to render the animal unaware before the incision. This technical elaboration has implications in religious slaughter, whereby the process of slaughter is guided by scripture or other religious teachings. Some religious or cultural groups allow some types of stunning where the process meets certain criteria.

Discussions on the advantages and disadvantages of different methods of religious slaughter are complicated by the fact that definitions of death and stunning vary between different religious groups. Some Islamic authorities that allow stunning require that the stun does not kill the animal (using cessation of blood circulation or regular heartbeat as a definition of death) and that the stun effects be reversible with no long-lasting effects. This usually means the animal must be able to stand within a given period after the stun if it's not subsequently slaughtered. These definitions form an important part of halal practices in many countries, even though there is no single Islamic position on them, and variation in slaughter practice occurs between different legislatures.

2.8 Assessment of Outcomes

The effectiveness of killing methods is assessed after animals have been killed by examination of the body or by using a secondary killing method. Confirming death by examination requires that sufficient time elapse after the killing method to allow the animal to die. The lack of a heartbeat, spontaneous ventilation and corneal reflex are indicative of death. Death invariably involves a loss of awareness, but depending on the method, other physiological functions may be unchanged when this point, i.e., consciousness, is lost. The sooner the animal is examined after becoming unconscious, the more difficult it is to confirm the animal is dead, stunned or anaesthetised. When there is doubt concerning the animal's condition, the body should be held back for re-examination so that death can be confirmed before disposal or carcase processing commences. In first-world processing facilities, i.e. abattoirs, the production line moves at a pace to ensure there is adequate time to confirm death before the animal progresses to the next stage of processing.

When it is difficult to confirm death by examination, a secondary killing method should be used after the animal has lost awareness. This may be because the technique typically takes an extended period of time to kill or because the animal is being killed by someone with insufficient skills to be able to confirm death beyond doubt. Ideally, this second method should be categorically different from the first. For example, where killing involves physical destruction (e.g. free bullet), the secondary method should, for example, deprive the CNS of oxygen, e.g., exsanguination.

Comparing or measuring the effectiveness of killing methods must be conducted in context as there are multiple factors that influence the feasibility, efficiency and suitability of a method. For example, an assessment technique used in an abattoir should not affect the routine in the plant. Assessment of killing methods should provide information on the method's overall effectiveness under real world conditions. In an abattoir, the proportion of animals that require repeat stunning is used to indicate the effectiveness of the stunning process. Pain and suffering prior to death are not routinely assessed in most abattoirs although there are jurisdiction specific requirements for transport of animals to the site, holding conditions (including duration, shelter, access to feed and water), movement to the killing area and restraint. These activities present risk to animals to experience pain and suffering.

2.9 Care of Personnel Involved in Animal Killing

The killing of animals can be physically and emotionally unpleasant. Whenever people are involved in killing animals, especially when a large number of animals or an extended period of time is involved, care must be taken to ensure that they receive adequate emotional support for the work that they are undertaking. Prolonged or unprepared contact with environments where animals are killed can lead to a range of serious emotional and psychological symptoms, including anxiety, depression, dissociation and suicidal ideation (Leibler et al. 2017). These negative feelings are often compounded by feelings of stigma associated with work that involves killing animals (Baran et al. 2009). The progression of these feelings can lead to compassion fatigue, which is a secondary traumatic stress disorder with the potential for very sudden onset (Cocker and Joss 2016). Competent authorities are increasingly recognising the need to protect workers in these environments and in many jurisdictions, health and safety legislation is beginning to encompass the need for staff to be emotionally and psychologically as well as physically safe in the workplace.

2.10 Legislative Control of Killing

Meeting the welfare needs of animals whilst ensuring efficient and practicable functioning of a process may be difficult to reconcile, but expectations of the public for standards of animal care that protect the welfare of the animals are increasingly being expressed in legislation. In developing countries, local practices and expectations vary and depend on a complex interaction of cultural, religious and legislative requirements. These cultural variations in expectations are well illustrated with respect to slaughter by Sinclair et al. (2023) who compared attitudes to various aspects of slaughter including the need for stunning prior to exsanguination in 14 countries across the globe. Their findings suggest that whilst there are many differences in local practice with regard to slaughter, concern for the reduction of animal suffering is a universal human tendency rather than a cultural development to be found only in some regions. The great variations found across different legislatures can be difficult to compare directly due to differences in the way that laws are constructed and industries managed in different countries. One point of similarity is that the vast majority of countries are member nations of the World Organisation for Animal Health and so have agreed to the provisions of the Terrestrial Animal Health Code (Anon 2024). Chapter 7.5 of this code deals with the welfare of animals during slaughter and chapter 7.6 with welfare aspects of killing animals in other contexts.

3 Specific Methods of Killing Animals

The following section is intended to indicate some of the techniques that can be used to kill animals and some of the circumstances in which they may be appropriate. It is intended to be indicative only and is neither exhaustive nor contains all the information required to be able to use a particular technique safely and effectively. It is essential that operators make sure that they are fully familiar with all safety requirements and limitations before a killing method is implemented for the first time. Use of a killing method that is unfamiliar to the operator may result in poor application of the method that can contribute to needless welfare compromise for the animal despite the method itself being suitable for the circumstances in which it is employed. Further details of many of these killing methods can be found in the guidelines produced by the AVMA (Leary et al. 2020). Practical details of how to safely and effectively implement many of these methods are available from the Humane Slaughter Association (https://www.hsa.org.uk/publications/online-guides).

3.1 Physical Destruction Methods

To be confident that parts of the CNS are effectively destroyed rapidly enough to ensure a painless death requires that the animal loses consciousness before any sensations originating from the killing method can be processed. The time taken to process this information varies depending upon the sensation involved and the degree to which processing needs to progress in order that the animal's welfare becomes compromised. Under most circumstances, the time required for this processing will be in the range of tens of milliseconds to a few seconds and so practically speaking, destruction needs to be sufficiently rapid as to be regarded as instantaneous.

To optimise efficacy of these techniques, the animal must be appropriately restrained, and have its head fixated so that the target area is hit with precision in every animal on every occasion. The operator must be familiar with the anatomical landmarks that allow the technique to be correctly implemented. The process of physical restraint and head fixation may be stressful for animals, especially wild animals, which have not been previously restrained. Because of this required contact with the animal physical destruction methods may be unsuited to situations where large numbers need to be killed. Technique failure or equipment malfunction may result in painful, non-lethal injuries. For these reasons, back-up killing methods must always be available to ensure a prompt death in such circumstances.

Physical destruction methods include:

Penetrative and Non-penetrative Captive Bolt (Figs. 7 and 8) These devices impart lethal energy and cause tissue destruction in the CNS (as with free bullets), but this equipment is held directly on the restrained animal's forehead to ensure reliability, efficacy and safety for operators. Penetrative captive bolts extend a projection or bolt that is designed to penetrate the skull and enter the brain. These devices impart energy to the skull's surface which is transmitted to the brain. Some captive bolts operate pneumatically, while others use a small explosive charge, i.e. 'blanks'. Explosive captive bolts are considered to be firearms in some legislatures. When used properly, captive bolts and free bullets render animals unaware with instantaneous effect (Gibson et al. 2009a), although physical immobilisation of the head may cause the animal distress.

Cervical Dislocation This technique is restricted to small animals, e.g. rabbits and poultry. The head is held extended and pulled in such a way as to dislocate one or more cranial cervical vertebrae, rupture the spinal cord and disrupt blood flow to the brain (https://www.hsa.org.uk/neck-dislocation/manual). This can be achieved manually or using a range of commercially available devices that are designed for this technique. Cervical dislocation may not always cause an immediate loss of awareness (Hernandez et al. 2019) and so should not be used unless more reliable methods are unavailable.

Manually Applied Blunt Force Trauma This technique is functionally equivalent to the use of a non-penetrative captive bolt and, when properly applied, provides rapid loss of awareness and is 'high welfare'. The manual application of blunt force trauma is more technically demanding that other killing methods and must only be undertaken by adequately trained competent personnel. Manual application of blunt force trauma requires a more intimate connection with the ani-

Fig. 7 Two examples of penetrating captive bolt guns. (Photo credit Dr. Jordan Hampton)

mal than other killing methods and carries greater risk of emotional harm to those undertaking it (Rault et al. 2015). Many authorities now limit the circumstances in which application of blunt force trauma can be considered appropriate, for example Regulation 8 of New Zealand's Animal Welfare (Care and Procedures) Regulations (Anon 2018) prohibits the killing of any calf by blunt force trauma unless the calf is in severe pain or distress and there is no alternative available.

Free Bullet Shooting animals requires competent operators who are familiar with the safe use of firearms in the particular circumstances that the animals are in. They must be able to select and use an appropriate gun and ammunition for the animal to be killed. Free bullet shooting requires that the bullet's course runs within the animal and exits along a path avoiding people, other animals and hard surfaces—where unpredictable ricochets may occur (Leary et al. 2020). Many legislations limit the availability of fire-

Fig. 8 Non-penetrating captive bolt gun. (Photo credit Dr. Jordan Hampton)

arms and publish regulations concerning their licensing, storage and safe use.

3.2 Functional Disruption Methods

Carbon Dioxide Carbon dioxide induces narcosis by increasing concentrations of neurotransmitters such as glutamate and gamma-aminobutyric acid and acidification of the cerebrospinal fluid (Drechsler and Morris 2023). Increases in blood carbon dioxide levels result in marked increases in respiratory drive and sympathetic nervous activity. Consequently, before the loss of awareness, there are changes associated with feelings of air hunger and panic in humans. The process appears to be very aversive to animals (Kells et al. 2018b, c). This method is commonly used for the stunning of pigs prior to slaughter.

Diathermic Syncope Exposing the brain to microwave radiation to increase temperatures beyond the maximum supporting awareness (Rault et al. 2014) has been developed as a stunning method before killing sheep and cattle. This method is thought not to cause permanent CNS damage, so it may be a suitable stunning method when such damage would render the carcass unfit for consumption for cultural or religious reasons. Musk and Johnson (2024) reviewed this method and compared it to other reversible stunning methods.

Pentobarbital Pentobarbital is an injectable anaesthetic agent belonging to the barbiturates. Concentrated solutions are commonly used at increased doses as a means of killing animals. This drug is usually given by intravenous injection, but can be administered by any route. Injection by some routes, e.g., the peritoneal cavity, can be painful (Kells et al. 2018a). When animals have been killed using barbiturates, they must not enter human or animal food chains. Barbiturates are controlled substances in many legislatures and are subject to regulations concerning their storage and use. This method is used for individual animals such as those in zoos or on hobby farms whose disposal can be overseen to ensure there is no risk of consumption of the carcase.

Inhaled Anaesthetics Inhaled anaesthetic agents can be used to kill or stun animals before the use of more aversive killing methods, such as carbon dioxide inhalation or intravenous potassium or magnesium salts. Inhaled anaesthetics are medicines in many legislatures and are subject to regulations concerning storage and use. This method is used for individual animals, such as those in zoos or on hobby farms and is only appropriate for smaller animals that can be restrained to administer the anaesthetic gas.

Electrical Stunning (Electrocution) Electrical stunning is used in various forms in the slaughter of pigs, sheep and cattle. Depending on electrode placement, the size, duration and frequency of the current may cause cardiac arrest and represent a killing method in its own right or may preserve the circulation and so be reversible and permissible for some forms of religious slaughter (Sabow et al. 2017). This method of slaughter should target the brain before the heart. If the heart is affected by the electrical current first death by fibrillation is likely to be painful.

Potassium and Magnesium The intravenous injection of potassium or magnesium salts kills by terminating cardiac contraction. Magnesium crosses the blood—brain barrier but does not cause significant sedation before loss of awareness from circulatory failure ensues (Stanger et al. 2019). These methods should not be used without prior stunning or deep sedation/anaesthesia (Leary et al. 2020). They have been used in a wide variety of animals and have the advantage that they do not contaminate carcasses with dangerous chemicals that make disposal difficult.

3.3 Deprivation Methods

Low Atmospheric Pressure Stunning This technique involves depriving animals of oxygen by placing them in a chamber where the pressure is reduced to a point below that required for adequate oxygen delivery. The method is licensed for killing poultry in the USA, Canada and the EU, but its use in pigs was stressful, unaesthetic and probably associated with considerable pain arising from eardrum rupture and bowel distension (Bouwsema and Lines 2019).

Carbon Monoxide Carbon monoxide binds irreversibly to haemoglobin and prevents it from transporting oxygen. Exposure to this gas reduces oxygen availability to the tissues and results in death at concentrations as low as 0.32% (Bloom 1972). The welfare effects of this killing technique are low in many species. However, the irreversible nature of the haemoglobin binding makes the technique hazardous for personnel and should only be undertaken with extreme care in an environment that is safe for personnel (Leary et al. 2020).

Exsanguination Through severance of major arteries supplying the brain, killing by exsanguination deprives the brain of oxygen first (causing unconsciousness) and then kills the animal by failure of oxygen delivery to the heart muscles. The loss of awareness is not instantaneous and varies from species to species (King 1987; Johnson et al. 2015). Arterial severance requires an incision to be made through sensitive tissues and is considered to be painful in animals that are aware (Gibson et al. 2009b). For this reason, exsanguination is usually carried out as a secondary killing method after stunning by other techniques. Some religious slaughter practices require that animals be killed by exsanguination without prior stunning and many legislatures make specific allowance for this in their slaughter regulations.

Inert Gas Inert gases such as nitrogen or argon displace air in the inspired gas and reduce the availability of oxygen. In some species and circumstances, this appears to be less aversive than carbon dioxide inhalation, but is still associated with significant welfare impact (Sindhøj et al. 2021). Adult pigs, for example, do not appear to find hypoxia due to replacement of air with argon aversive (Raj et al. 1997), whereas rats do find this aversive (Makowska et al. 2008).

Foam Containment Whilst not a killing method per se, foam expanded with either inert gas or carbon dioxide can be used as a killing method where many animals need to be killed at once (Raj et al. 2008). This method is commonly undertaken when depopulating indoor poultry production units in the face of disease outbreaks.

References

Allen BJ, Bobier C, Dawson, Fleming PJS, Hampton J, Jachowski D, Kerley GIH, Linnell JDC, Marnewick K, Minnie L, Muthersbaugh M, O'Riain MJ, Parker D, Proulx G, Somers MJ, Titus K (2023) Why humans kill animals and why we cannot avoid it. Sci Total Environ. https://doi.org/10.1016/j.scitotenv.2023.165283

Anon (2018) New Zealand Animal Welfare (care and procedures) regulations. https://www.legislation.govt.nz/regulation/public/2018/0050/latest/whole.html#LMS22828

Anon (2022) Code of professional conduct for veterinary surgeons. Royal College of Veterinary Surgeons. https://www.rcvs.org.uk/setting-standards/advice-and-guidance/code-of-professional-conduct-for-veterinary-surgeons

Anon (2024) Terrestrial animal health code. World Organisation for Animal Health. https://www.woah.org/en/what-we-do/standards/codes-and-manuals/terrestrial-code-online-access/

Anton F, Euchner I, Handwerker HO (1992) Psychophysical examination of pain induced by defined CO_2 pulses applied to the nasal mucosa. Pain 49:53–60

Baran BE, Allen JA, Rogelberg SG et al (2009) Euthanasia-related strain and coping strategies in animal shelter employees, vol 103. Psychology Faculty Publications. https://digitalcommons.unomaha.edu/psychfacpub/103

Beausoleil NJ, Mellor DJ (2015) Introducing breathlessness as a significant animal welfare issue. N Z Vet J 63:44–51

Bloom JD (1972) Some considerations in establishing divers' breathing gas purity standards for carbon monoxide. Aerosp Med 43:633–636

Bouwsema J, Lines J (2019) Could low atmospheric pressure stunning (LAPS) be suitable for pig slaughter? A review of available information. Anim Welf 28(4):421–432. https://doi.org/10.7120/09627286.28.4.421

Coccia CBI, Palkowski GH, Schweitzer B, Motsohi T, Ntusi NAB (2016) Dyspnoea: pathophysiology and a clinical approach. S Afr Med J 106:32–36

Cocker F, Joss N (2016) Compassion fatigue among healthcare, emergency and community service workers: a systematic review. Int J Environ Res Public Health 13:618–636

Dow J, Giesbrecht GG, Danzl DF, Brugger H, Sagalyn EB, Walpoth B, Auerbach PS, McIntosh SE, Nemethy M, McDevitt M, Schoene RB, Rodway GW, Hackett PH, Zafren K, Bennett BL, Grissom CK (2019) Wilderness medical society clinical practice guidelines for the out-of-hospital evaluation and treatment of accidental hypothermia: 2019 update. Wilderness Environ Med 30:S47–S69

Drechsler M, Morris J (2023) Carbon dioxide narcosis. In: StatPearls [Internet]. https://www.ncbi.nlm.nih.gov/books/NBK551620/#

Feldman F (1991) Some puzzles about the evil of death. Philos Rev 2:205–227

Gibson TJ, Johnson CB, Murrell JC, Mitchinson SL, Stafford KJ, Mellor DJ (2009a) Electroencephalographic response to concussive non-penetrating captive bolt stunning in halothane anaesthetised calves. N Z Vet J 57:90–95

Gibson TJ, Johnson CB, Murrell JC, Hulls CM, Mitchinson SL, Stafford KJ, Johnstone AC, Mellor DJ (2009b) Electroencephalographic responses of calves to slaughter by ventral neck incision without prior stunning. N Z Vet J 57:77–83

Gregory NG, Fielding HR, von Wenzlawowicz M, von Holleben K (2010) Time to collapse following slaughter without stunning in cattle. Meat Sci 85:66–69

Hancock PA, Vasmatzidis I (2003) Effects of heat stress on cognitive performance: the current state of knowledge. Int J Hyperth 19:355–372

Hernandez E, James F, Torrey S et al (2019) Evaluation of brain death in laying hens during on-farm killing by cervical dislocation methods or pentobarbital Na+ injection. Front Vet Sci. https://doi.org/10.3389/fvets.2019.00297

Johnson CB, Mellor DJ, Hemsworth PH, Fisher AD (2015) A scientific comment on the welfare of domesticated ruminants slaughtered without stunning. N Z Vet J 63:58–65

Kaiser A, McFarland W, Siemion R, Raisbeck M (2010) Secondary pentobarbital poisoning in two dogs: a cautionary tale. J Vet Diagn Invest 22:632–634. https://doi.org/10.1177/104063871002200423

Kasperbauer TJ, Sandøe P (2015) Killing as a welfare issue. In: Višak T, Garner R (eds) The ethics of killing animals, online edn. Oxford Academic, New York. https://doi.org/10.1093/acprof:oso/9780199396078.003.0002

Kells NJ, Beausoleil NJ, Sutherland MA, Johnson CB (2018a) Electroencephalographic responses of anaesthetised pigs to intraperitoneal injection of Na+ pentobarbital. Anim Welf 27:205–214

Kells NJ, Beausoleil NJ, Johnson CB, Sutherland MA (2018b) Evaluation of different gases and gas combinations for on-farm euthanasia of pre-weaned pigs. Animals. https://doi.org/10.3390/ani8030040

Kells NJ, Beausoleil NJ, McIlhone AE, Johnson CB (2018c) Electroencephalographic responses of anaesthetised rats to carbon dioxide inhalation. Anim Welf 27:215–223

King AS (1987) Physiological and clinical anatomy of the domestic mammals. Blackwell, Oxford

Leary et al (2020) AVMA guidelines for the euthanasia of animals. American Veterinary Medical Association. https://www.avma.org/sites/default/files/2020-02/Guidelines-on-Euthanasia-2020.pdf

Leibler JH, Janulewicza PA, Perry MJ (2017) Prevalence of serious psychological distress among slaughterhouse workers at a United States beef packing plant. Work 57:105–109

Makowska IJ, Niel L, Kirkden RD, Weary DM (2008) Rats show aversion to argon-induced hypoxia. Appl Anim Behav Sci 114:572–581. https://doi.org/10.1016/j.applanim.2008.04.005

Mallat J, Rahman N, Hamed F, Hernandez G, Fischer MO (2022) Pathophysiology, mechanisms, and managements of tissue hypoxia. Anaesthesia Critical Care and Pain Medicine 41. https://doi.org/10.1016/j.accpm.2022.101087

Musk GC, Johnson CB (2024) Commentary: a comparison of the methods of the pre-slaughter stunning of cattle in Australia—mechanical, electrical, and diathermic syncope. Animals 14. https://doi.org/10.3390/ani14213141

Payne J, Farris R, Parker G, Bonhotal J, Schwarz M (2015) Quantification of sodium pentobarbital residues from equine mortality compost piles. J Anim Sci 93:1824–1829. https://doi.org/10.2527/jas.2014-8193

Raj AB, Johnson SP, Wotton SB, McInstry JL (1997) Welfare implications of gas stunning pigs: 3. The time to loss of somatosensory evoked potentials and spontaneous electrocorticogram of pigs during exposure to gases. Vet J 153(3):329–339. https://doi.org/10.1016/s1090-0233(97)80067-6. PMID: 9232122

Raj M, Smith C, Hickman GAW (2008) Novel method for killing poultry in houses with dry foam created using nitrogen. Vet Rec 162:722–723

Rault JL, Hemsworth P, Cakebread P, Mellor DJ, Johnson CB (2014) Evaluation of microwave energy as a humane stunning technique based on electroencephalography (EEG) of anesthetized cattle. Anim Welf 23:391–400

Rault J-L, Kells NJ, Johnson CB, Dennis, Sutherland M, Lay DC Jr (2015) Nitrous oxide as a humane method for piglet euthanasia: behavior and electroencephalography (EEG). Physiol Behav 151:29–37

Sabow AB, Nakyinsige K, Adeyemi KD, Johnson CB, Webster J, Farouk MM (2017) High frequency pre-slaughter electrical stunning in ruminants and poultry for halal meat production: a review. Livest Sci 202:124–134

Sechzer PH, Egbert LD, Linde HW, Cooper DY, Dripps RD, Price HL (1960) Effect of CO_2 inhalation on arterial pressure, ECG and plasma catecholamines and 17-OH corticosteroids in normal man. J Appl Physiol 15:454–458

Sinclair M, Hötzel MJ, Lee NYP, de Luna MCT, Sharma A, Idris M, Islam MA, Iyasere OS, Navarro G, Ahmed AA, Burns GL, Curry M, Marchant JN (2023) Animal welfare at slaughter: perceptions and knowledge across cultures. Front Anim Sci. https://www.frontiersin.org/journals/animal-science/articles/10.3389/fanim.2023.1141789

Sindhøj E, Lindahl C, Bark L (2021) Potential alternatives to high-concentration carbon dioxide stunning of pigs at slaughter. Animal 15. https://doi.org/10.1016/j.animal.2020.100164

Stanger KJ, Kells NJ, Fisher AD, Jubb T, Rault J-L, Johnson C (2019) Evaluation of euthanasia of sheep with intravenous saturated salt solutions to enable the collection of whole, intact brains. Anim Welf 28:397–406

Uhlhaas PJ, Singer W (2006) Neural synchrony in brain review disorders: relevance for cognitive dysfunctions and pathophysiology. Neuron 52:155–168

Verbrugghe C, Laurent, Bouverot P (1982) Chemoreflex drive of ventilation in the awake miniature pig. Respir Physiol 47:379–391

Vučenović SM, Šetrajčić JP, Vojnović M, Šetrajčić-Tomić AJ, Džambas LD (2017) Physiological processes when an electrical current passes through the tissues and organs. RAD Conf Proc 2:290–295

Williams T, Carel H (2018) Breathlessness: from bodily symptom to existential experience. In: Aho K (ed) Existential medicine. Rowman and Littlefield International, London

Euthanasia and Slaughter of Poultry

Leisha Hewitt

Abstract

Billions of poultry are produced and processed globally each year to provide meat and eggs for human consumption. During poultry production and processing, the need to kill birds arises in various contexts, including euthanasia on the farm and slaughter for human consumption in the processing plant. Regardless of circumstances, whether it involves killing individual birds or the commercial processing of thousands, animal welfare remains a critical concern. During euthanasia and slaughter processes, pain can result from prior injuries, improper handling, restraint, incorrect stunning or neck cutting (if stunning is not used or is ineffective). This chapter explores the challenges associated with euthanasia and slaughter, acknowledging the potential for birds to endure pain. Furthermore, it aims to identify ways to eliminate or reduce the risk of potentially painful experiences and improve animal welfare.

Keywords

Birds · Poultry · Slaughter · Euthanasia · Stunning

L. Hewitt (✉)
Livestock Welfare, Franklin, TAS, Australia

1 Introduction

Billions of poultry are produced and processed globally each year, making poultry one of the most significant sectors in the global food supply chain. Chickens, ducks, turkeys and other poultry species are raised to meet the growing demand for meat and eggs, which are essential protein sources for billions of people worldwide. According to the Food and Agriculture Organization (FAO), global poultry production has expanded steadily over the past decades (Fig. 1), driven by population growth, urbanisation and shifting dietary preferences.

This industry not only plays a crucial role in ensuring food security but also makes a significant contribution to the economies of many countries, particularly through exports. However, the scale of poultry production also raises important considerations, including animal welfare. According to the ABS's 'Livestock Products, Australia' report for the December 2023 quarter, approximately 182.6 million chickens were slaughtered during that period, resulting in a total chicken meat production of 361,378 tonnes (Fig. 2).

The term 'poultry' is typically used to describe domestic birds that are raised to produce meat, eggs and or feathers. Chickens, turkeys, ducks and geese are of primary commercial importance and are considered in detail here. Other poultry species that are slaughtered in smaller numbers

R. E. Clutton, G. C. Musk (eds.), *Pain and Suffering in Farmed Animals*,
https://doi.org/10.1007/978-3-032-11407-5_13

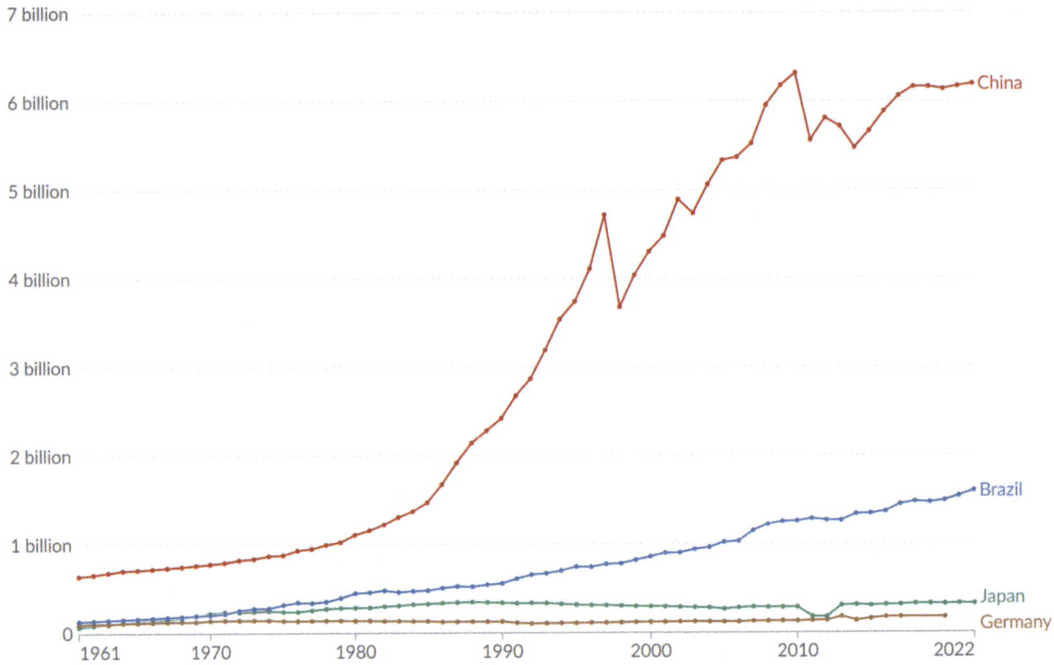

Fig. 1 Number of poultry slaughtered, 1961–2022. Food and Agriculture Organization of the United Nations (2023)—with major processing by Our World in Data

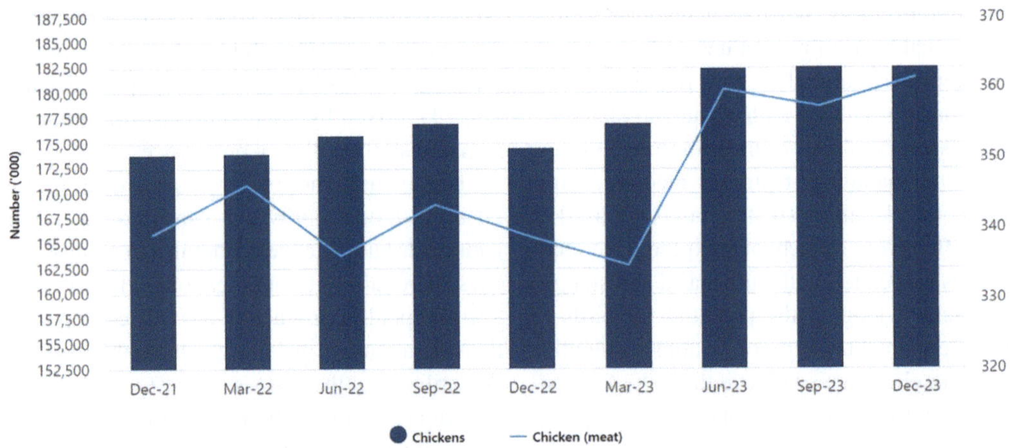

Fig. 2 Chickens slaughtered and meat produced in Australia. Australian Bureau of Statistics Livestock Products, Australia December 2023

include guinea fowl, quail, pigeons and pheasants. These species represent niche markets or are regionally significant but contribute relatively little to global poultry production compared to the primary species. While poultry may experience painful events throughout the whole production period, this chapter focuses on the welfare concerns associated with the killing process. The need to kill poultry ranges from the on-farm euthanasia of individual birds to the commercial slaughter of tens of thousands in processing plants.

Fundamental to any discussion of euthanasia and slaughter is the meaning of 'consciousness' and 'unconsciousness'. In relation to killing animals, consciousness is generally associated with the ability to feel emotions and control voluntary movement. Unconsciousness is often defined as a state of unawareness caused by the temporary or permanent disruption of brain function. As a result of this disruption, unconscious animals are unable to respond to normal stimuli, including pain (EFSA 2006). When animals are conscious, or they regain consciousness during euthanasia or slaughter, they can experience pain, fear and distress. Fear is an unpleasant emotional state induced by the perception of danger or potential danger that threatens the integrity of the animal (Von Holleben et al. 2010). It involves physiological and behavioural changes that prepare the animal to cope with danger. This can be problematic during the slaughter process when animals encounter new or unfamiliar stimuli while being held in lairage, handled or restrained. The terms 'stress' and 'distress' are often used interchangeably to describe the response to and effect of the conditions that animals are exposed to. Stress is a physiological disturbance, imposed by a stressor such as a threatening situation. It involves the activation of both the hypothalamic-pituitary-adrenal (HPA)-axis and the sympathetic nervous system (SNS), which are common to all vertebrates. The World Organisation for Animal Health (WOAH 2023) defines distress as the state of an animal that has been unable to adapt to stressors, and that manifests as abnormal physiological or behavioural responses. Distress can be acute or chronic and may result in pathological changes.

2 Sources of Pain During Euthanasia and Slaughter

The description of pain used in most animal welfare studies aligns with the definition established (and recently revised) by the International Association for the Study of Pain (IASP), i.e. 'an unpleasant sensory and emotional experience associated with, or resembling that associated with, actual or potential tissue damage' (IASP 2020). The nociceptive system, which regulates pain perception in animals, functions similarly across all mammalian species and birds. Pain perception involves an interaction between receptors, nerves, the spinal cord and the brain. Pain receptors are located in the skin, muscles, joints, periosteum, some internal organs and blood vessels. The types of receptors and the anatomical structures involved both influence the experience of pain (see Chapters Bell, Clarke and Dutton).

To prevent pain, fear and distress, euthanasia and slaughter methods should immediately induce general unconsciousness that persists until death. Under practical conditions, the Scientific Panel on Animal Health and Welfare of the European Food Safety Authority (EFSA 2004, 2006) has defined 'immediate unconsciousness' as 'unconsciousness occurring within 1 second' of the intervention being applied. Alternative methods that do not produce immediate unconsciousness must ensure the absence of pain, distress and suffering until consciousness is lost. Methods involving the use of inhalational agents such as carbon dioxide in Controlled Atmosphere Stunning (CAS) systems, fall into this category, as unconsciousness can be induced gradually and without stress before the subject dies (Gerritzen and Raj 2009).

Measurements of brain electrical function (EEG) have been used to objectively quantify the state of unconsciousness; consciousness normally ceases at some point between behavioural-unresponsiveness and brain death (AVMA 2020). However, EEG data alone cannot provide definitive information on the exact onset of unconsciousness. EEG-based brain function monitors are limited in their ability to directly indicate the presence or absence of unconsciousness, particularly around the transition point (Lambooij and Korte 2011).

Brain-stem reflex activity, including rhythmic breathing and the nictitating membrane reflex, are used as a proxy indicators of unconsciousness during and after euthanasia and slaughter. However, for some electrical stunning methods, direct observations of rhythmic breathing and signs interpreted as epilepsy (such as muscle spasms or convulsions) are not sufficiently reliable indicators of unconsciousness (Anastasov and Wotton 2012). These researchers concluded that poultry assessed after leaving the water bath appeared to be stunned but could have been in a state of electrically induced paralysis. Consequently, assessing the effectiveness of some euthanasia and slaughter methods in a commercial setting can be challenging.

During the euthanasia and slaughter of birds, pain can result from injuries caused, for example, by inappropriate handling and or restraint, incorrect stunning or neck cutting (if stunning is not used or is ineffective), and the overall effect on avian welfare is cumulative. For example, leg injuries, such as fractures sustained during catching and loading into transport containers, can worsen the pain during subsequent handling at the processing plant, particularly when the birds are restrained in metal shackles for electrical stunning (Fig. 3).

Animal-based measures related to the experience of pain have been validated for use during poultry processing (EFSA 2012, 2019b), although some may not be feasible under practical processing conditions. For example, birds in the middle of a transport container may not be visible for assessment, or fast processing speeds (which can exceed 200 birds per minute) may make the detection of some measures very difficult (EFSA 2012, 2019b). Table 1 summarises the animal-based measures relating to the experience of pain which are used to assess animal welfare during euthanasia and slaughter processes.

3 Euthanasia During the Production Period

Sick or injured birds seldom receive individual treatment in commercial poultry production systems. More commonly, they are killed promptly in a painless manner (Leary et al. 2020) to prevent further suffering. This act is properly termed 'euthanasia', a word derived from the Greek terms *eu,* meaning 'good' and *thanatos,* meaning 'death'. The majority of poultry euthanasia methods rely on mechanical methods or inhaled substances. For poultry, the common mechanical methods are cervical dislocation or non-penetrative captive bolt devices, while Inhalational methods involve gases. Inhalational techniques are typically reserved for larger birds, e.g. turkeys and geese, or when large flocks must be killed. Most methods are lethal, but some may require a secondary (terminal) procedure to ensure death, for example, head-only electrical stunning. It should be noted that many studies of poultry euthanasia prioritise lethality as the index of efficacy, rather than the time required eliminating consciousness. Consequently, some recommended euthanasia methods may not produce immediate unconsciousness and, therefore, not be in the best interests of the animals' welfare.

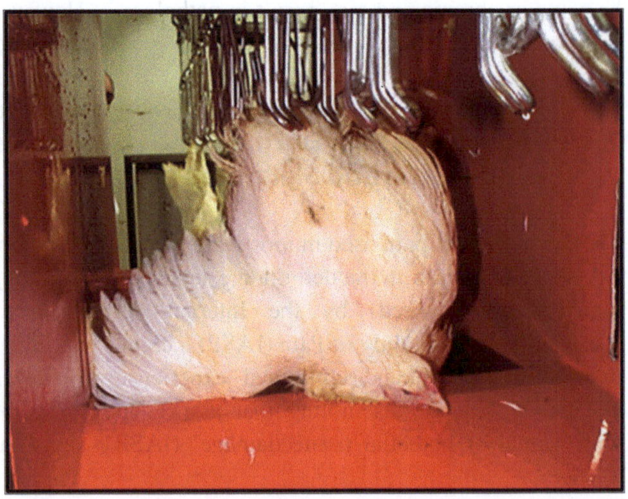

Fig. 3 A chicken restrained on a shackle for electrical stunning (Hewitt 2000)

Table 1 Animal-based measures related to the experience of pain during poultry slaughter

Animal-based measures (ABM) of pain	Definition	References
Escape attempts	Attempt to move, run or fly away from the situation	Graml et al. (2007)
Head shaking	Rapid shaking of the head, usually accompanied by stretching and/or withdrawal movements of the head	Gerritzen et al. (2006)
Injuries	Tissue damage (bruises, scratches, broken bones, dislocations)	EFSA (2012)
Muscle jerks	Muscle contractions similar to spasms, tremors and pedaling movements of the legs	EFSA (2019a)
Vocalisation	Single or repeated short and loud shrieking at higher frequencies	Erasmus et al. (2010c)

Adapted from EFSA (2019b)

3.1 Euthanasia by Cervical Dislocation

Historically, the most common method for the large-scale killing of poultry has been cervical dislocation; there are two different techniques. The first method involves stretching the neck, either manually (Fig. 4) or mechanically, to sever the spinal cord and cause damage to the carotid arteries supplying blood to the brain. This leads to significant blood loss, impaired circulation and the subsequent onset of unconsciousness. The second involves crushing the neck using devices such as the pliers illustrated in Fig. 5. This technique often damages the carotid arteries and jugular veins without consistently severing the spinal cord (Fig. 6) (Hewitt 2000).

The effectiveness of cervical dislocation and its ability to achieve immediate unconsciousness is a matter of scientific debate (FAWC 2009). The prevailing view opposes its use except in exceptional circumstances. The European Food Safety Authority (EFSA 2019a) describes it as a killing, rather than a stunning method and the Humane Slaughter Association (HSA) recommends it only be used in emergencies, when very few birds are involved and when alternative (better) methods are unavailable. Electroencephalographic studies of cervical dislocation reveal it may not always produce immediate unconsciousness (Hernandez et al. 2019). Furthermore, performing effective manual cervical dislocation becomes more challenging with larger, heavier poultry due to the increased force required to properly dislocate the neck and the greater physical strain on the handler. Consequently, European legislation restricts manual cervical dislocation to birds weighing less than 3 kg and limits application to 70 birds per person per day. This is due to concern that an operator's ability to consistently produce effective cervical dislocation diminishes with repetition. Mechanically assisted cervical dislocation has been investigated in the search for more humane methods of dislocation. Mechanical cervical dislocation devices that facilitate neck stretching increase brain trauma, so increase the likelihood of immediate unconsciousness. In contrast, mechanical cervical dislocation devices that crush the neck are not recommended (EFSA 2019a).

3.2 Euthanasia Using Non-penetrative Captive Bolt Devices

Mechanical stunning involves applying a non-penetrating percussive blow to the heads of individual birds (Fig. 7) and has been used for killing poultry and may serve as an alternative to cervical dislocation. Commercially available captive bolt devices for poultry use a non-penetrating concussive force to interrupt neuronal function, disrupt sensory processing, and cause traumatic brain injury, leading to unconsciousness and death. To achieve immediate unconsciousness with mechanical stunning equipment, critical factors include bolt diameter and velocity and application position (EFSA 2004). In poultry, commercial devices produce immediate unconsciousness and can result in the bird's death (Hewitt 2000; EFSA 2004). The use of these

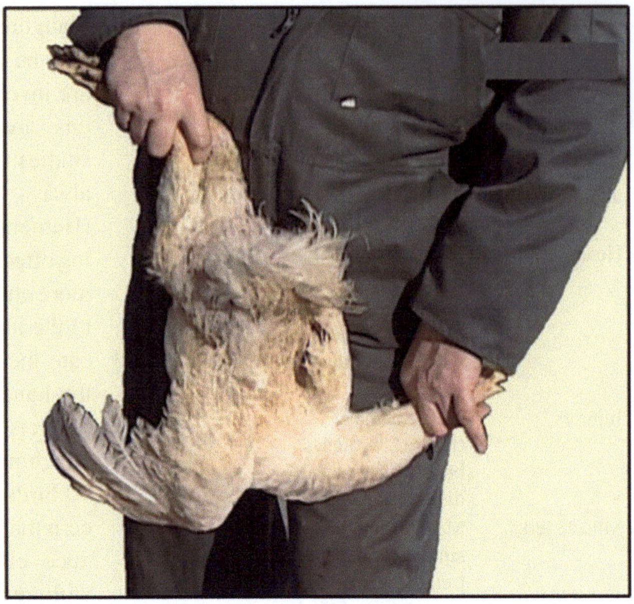

Fig. 4 Manual cervical dislocation of a broiler chicken (HSA 2013)

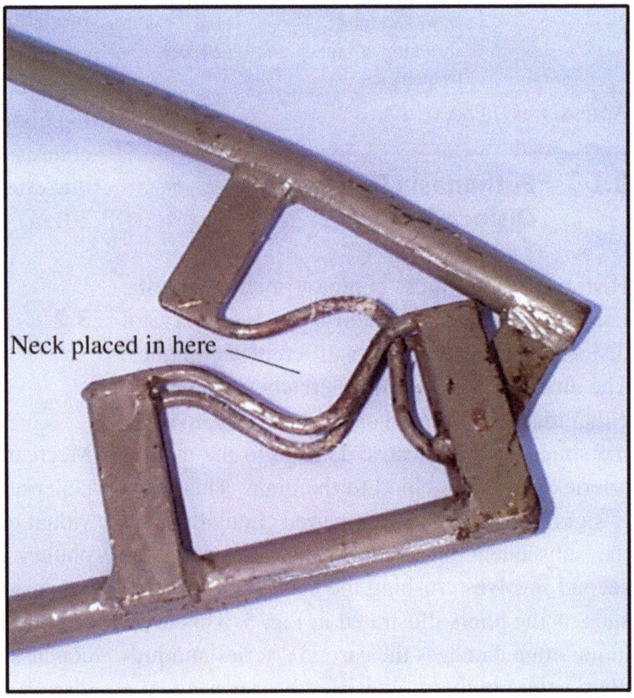

Fig. 5 Pliers used for neck dislocating poultry (Hewitt 2000)

devices has been investigated in chickens, turkeys, ducks and geese, and has been deemed to meet the criteria for a successful euthanasia method (Hewitt 2000; Raj and O'Callaghan 2001; Erasmus et al. 2010a, b; Gibson et al. 2018; Sparrey et al. 2014).

3.3 Euthanasia Using Head-Only Electrical Stunning Devices

Head-only electrical stunning is sometimes used as a euthanasia technique for small numbers of poultry. The equipment is usually manually

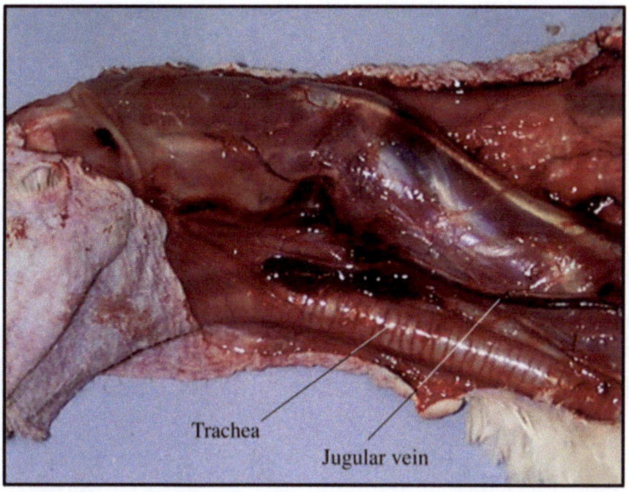

Fig. 6 Dissected turkey neck illustrating jugular vein damage following the application of neck crushing pliers (Hewitt 2000)

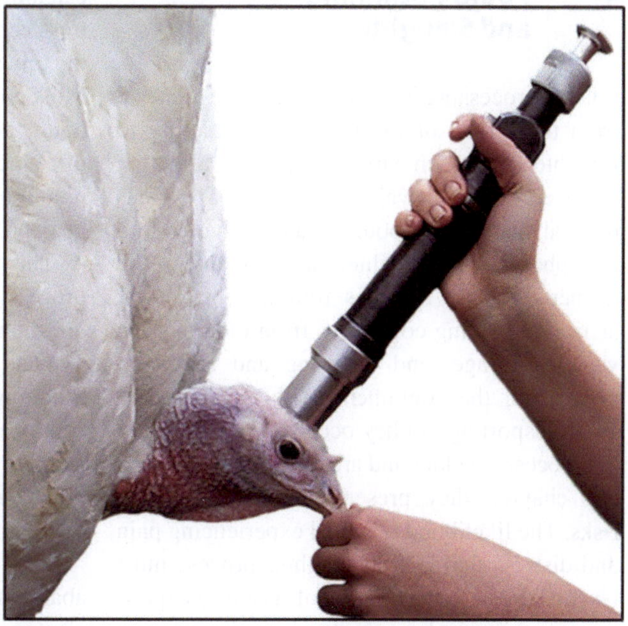

Fig. 7 Non-penetrative captive bolt device for the euthanasia of poultry (Hewitt 2000)

operated, with adjustable electrodes that are applied across the bird's head, spanning the brain (Fig. 8). To facilitate stunning, poultry are restrained in a cone, on a shackle, or held manually by the legs. During head-only electrical stunning, the bird will display a tonic reaction, characterised by stiffening and arching of the neck, rigid extension of the legs, wings folded tightly around the breast, and muscle tremors. After removal of the electrodes, the bird's eyes will remain wide open (with no blinking), and rhythmic breathing will be absent. These signs are indicative of an effective head-only electrical stun, during which the bird is unconscious. To ensure that birds do not regain consciousness after head-only electrical stunning, it is essential to perform neck cutting as soon as unconsciousness has been confirmed.

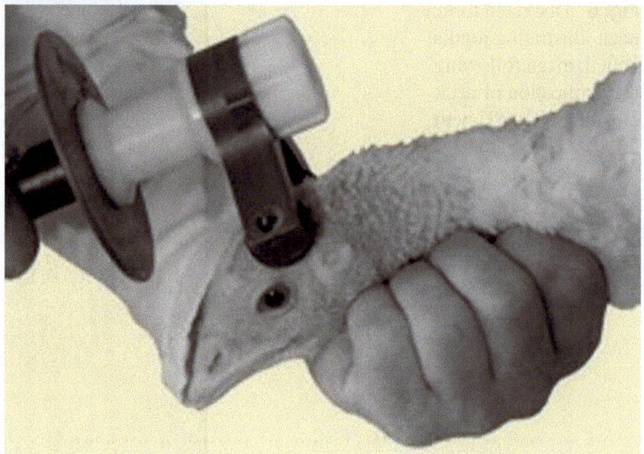

Fig. 8 Hand-held head-only electrical stunning (HSA 2013)

4 Poultry Stunning and Slaughter

Poultry processing involves a sequence of steps, from the arrival of the birds at the facility to slaughter, with each part of the process affecting the overall animal welfare outcome. The commercial slaughter of poultry can be divided into two phases: pre-slaughter and slaughter. The former typically includes four key processes: arrival, unloading containers from the transport vehicle, lairage, and handling and removing birds from the containers. Although catching and transporting poultry occur before arrival at the processing plant and are outside the scope of this chapter, they present significant welfare risks. The likelihood of a bird experiencing pain and distress during the slaughter process must be considered in the context of its entire experience, from arrival on the truck to the point of death.

When poultry arrive at the processing plant, their transport containers are either kept on the truck, placed in lairage or unloaded and moved to designated areas. Placing poultry in well-ventilated locations with shelter and protection from direct sunlight can help mitigate the adverse welfare consequences associated with thermal stress and exposure to inclement weather. Commercial practices typically involve holding birds in their transport containers for several hours, depending on the throughput rate and other operational factors. The interval between catching and slaughter affects the pain and distress experienced by the birds, with longer holding times generally leading to increased mortality. Most poultry are held in lairage for at least a few hours before slaughter, with the holding time often determined by compliance with industry requirements concerning the duration birds can be deprived of feed and water.

The primary goal throughout the slaughter process is to avoid pain and suffering. The key question, however, is how this is achieved? Typically, birds are stunned before slaughter. Stunning induces unconsciousness, which prevents the birds from experiencing pain during neck cutting and bleeding. For stunning to be effective, it must ensure that unconsciousness is maintained until death occurs.

Two methods commonly used in commercial abattoirs for stunning chickens and turkeys are electrical waterbath stunning and controlled atmosphere stunning (CAS), also known as gas stunning. During electrical water bath stunning, an electric current passes from the electrode in the water through the bird and into the shackle line, rendering the bird unconscious. In some waterbath systems, the bird is not fully immersed in water but instead makes direct contact with a wet electrode. Novel stunning techniques, such as Low Atmosphere Pressure Stunning (LAPS), are also being developed and introduced into commercial operations. LAPS operates by gradually removing air from a sealed chamber containing chickens, leading to progressive hypoxia caused by a reduction in oxygen tension (Vizzier et al. 2010; Vizzier 2015; McKeegan et al. 2013).

4.1 Shackling Conscious Birds for Electrical Stunning

When electrical stunning systems are used, poultry are removed from their transport containers and placed onto a shackle line for conveyance to the electrical waterbath (Fig. 3). This is done either by personnel removing birds individually from the crates or by tipping them into a hopper, which transfers them onto conveyor belts leading to a shackling carousel. Shackling involves inverting the birds and inserting their legs into metal shackles. Physiological and behavioural studies have identified acute pain as a consequence of shackling (Gentle 2011). While much research focuses on meat chickens, the welfare risks are probably similar for all birds. The pain results from forces applied to the legs that exceed the mechanical threshold needed to activate nociceptors in the skin (Gentle and Tilston 2000; Shields and Raj 2010). Pain associated with shackling can be exacerbated by rough handling, the design of the shackle, the time between shackling and stunning and features of the processing line that cause birds to flap their wings. The design of individual shackles in relation to the bird's leg size is also crucial, as it affects the degree to which the periosteum and the tarsometatarsal bone are compressed. Poultry processing plants should use shackles with appropriately sized apertures to fit the birds' legs, in order to alleviate pain and reduce the risk of leg injuries. While some modern shackle lines are designed to accommodate birds of different sizes, these are not commonly used under current processing conditions.

Birds that flap their wings after shackling are likely to experience more pain than those that remain calm. It is generally accepted that reducing suspension time may help decrease the amount of wing flapping. The European Food Safety Authority (EFSA 2019b) and the World Organisation for Animal Health (WOAH 2023) recommend a maximum shackling time of 1 min. In a study of broiler chickens (Zulkifli et al. 2019) that evaluated the effect of shackling on plasma levels of adrenaline, noradrenaline and corticosterone, researchers found that shackling for less than 30 s did not elicit a significant physiological stress response. Similarly, a study in turkeys found that while transport increased plasma corticosterone, there was no additional increase in corticosterone levels after shackling (Scanes et al. 2020). Despite these findings, shackling is still considered to diminish animal welfare outcomes during poultry processing. Alternative explanations for the lack of effect on plasma corticosterone such as possible downregulation of the HPA axis due to other stressors, have been suggested.

4.2 Electrical Water Bath Stunning Systems

Globally, electrical water bath stunning is the most widely used technique in commercial poultry processing plants. Birds are inverted and suspended, with their feet restrained in metal shackles, for conveyance through the system. The contact of the head and neck with the water or electrode completes the electric circuit between the water (positive electrode) and shackle (which acts as the earth or negative bar electrode), so that an electric current passes through the bird's head and body. It can result in immediate unconsciousness, which, when followed by neck cutting and bleeding, results in death without recovery.

Water-bath stunning systems vary in design, using either a deep or shallow bath of water (covering an electrode) or a wet electrode (wet plate). In each system, there is a risk of birds receiving a pre-stun shock before they are fully stunned. A pre-stun shock occurs when part of the bird's body, such as the wings or head, comes into contact with an electrified surface such as the water in an electrical water bath, before the bird is properly positioned for stunning. This electrical stimulation can cause pain or distress because the bird is still conscious and able to perceive the shock. The risk of receiving a pre-stun shock increases as the depth of immersion decreases.

This is because birds are typically suspended upside down by their legs and their wings often extend below the head. If the water level is too shallow, the wings or other body parts may touch the electrified water before the head is submerged, causing electrical stimulation. Ensuring good electrical contact in the wet plate system (designed for birds to make direct contact with the electrode) is challenging, as variations in bird size can lead to loss of contact with the metal plate, thereby reducing the effectiveness of the stunning process.

When electrical water bath stunning is used, passing an electrical current through the bird's brain induces a state of generalised epilepsy. The optimal electrical variables for an effective stun are described as a low-frequency sinusoidal AC waveform of 50 Hz, with a minimum current of 120 mA for chickens and 150 mA for turkeys. These settings can abolish brain activity and result in a quiescent EEG, indicating an unequivocal stun, achieving up to 96% effectiveness based on EEG measurements and 100% using non-EEG methods (EFSA 2004, 2012). However, more recent research indicates that specifying a universally effective minimum current for all electrical frequencies and waveforms used in commercial settings is not feasible. The interaction between different variables is complex; for example, higher frequencies require higher currents to achieve an effective stun. There is even debate about whether some higher frequency currents are effective at all (EFSA 2014).

Electrical water bath systems are typically designed to accommodate multiple birds in contact with the water simultaneously. In these systems, all birds passing through the water bath are exposed to a constant voltage. Consequently, the electrical current flowing through each bird depends on their resistance to current flow; birds with lower resistance will receive more current than those with higher resistance.

To summarise, the risk of pain and distress during electrical water bath stunning is associated with:

- Manually removing birds from transport containers and handling them to keep pace with the speed of the shackle line
- Inverting birds and suspending them by their legs in metal shackles
- The size and design of the shackles, which apply pressure to the birds' legs
- Pre-stun shocks at the entrance to the water bath
- Wing flapping and struggling which can cause physical impact and injury
- Variability in the electric current delivered to each bird, potentially leading to inadequate stunning for some individuals

4.3 Controlled Atmosphere Stunning (CAS) Systems

Controlled Atmosphere Stunning (CAS) is achieved by exposing poultry to an atmosphere containing an increased concentration of CO_2 in air (hypercapnic hypoxia), depleted oxygen (anoxia), or a combination of the two (hypercapnic anoxia). Commercial CAS systems are considered to offer welfare advantages over electrical stunning because they do not require conscious birds to be shackled. Instead, birds are either conveyed through the system in their transport containers or by means of a moving conveyor. Investigations into the use of CAS systems under practical conditions have shown excellent stunning effectiveness. However, the time taken to produce unconsciousness varies with gas concentrations. Typically, in commercial CAS systems, birds lose consciousness within a minute (Raj 2006). However, during CAS hypercapnic stunning, birds may experience the unpleasant effects of CO_2 inspiration before they lose consciousness. Several studies have shown that high concentrations of CO_2 are aversive to birds (McKeegan et al. 2007; Sandilands et al. 2011). Additionally, convulsions, which occasionally result in injury, may occur before the birds lose consciousness. These risks are often managed in commercial systems by first exposing birds to low, less aversive CO_2 concentrations (e.g. < than 40% by volume in air) for approximately 60 s, when consciousness is usually lost. Thereafter, birds are exposed to higher concentrations, e.g. approximately 80–90% by volume in air. This ensures that the birds remain stunned until death. Graded exposure as described avoids aversive reactions seen in conscious poultry exposed to high CO_2 concentrations (Gerritzen et al. 2007). Head shaking and gasping at relatively low CO_2 concentrations are commonly observed responses in birds during the initial stages of exposure to controlled atmosphere stunning (CAS) (McKeegan et al. 2007). These behaviours are linked to the bird's natural physiological response to rising CO_2 levels, as the chemoreceptors in the carotid and aortic bodies detect increased CO_2 and trigger an attempt to restore normal oxygen and CO_2 balance. Although these behaviours can be perceived as discomfort or distress, they are not universally regarded as signs of aversion to CO_2 exposure itself (EFSA 2019a, b). During exposure to <40% CO_2, loss of consciousness becomes evident by the loss of posture, where birds become unsteady and either sit down or fall over. Wing flapping may also be observed later in the stunning process, but this is believed to occur after consciousness has been lost.

In general, there are several features of CAS systems that can affect the pain experiences of poultry at slaughter:

- Method of handling in the system; systems that allow poultry to remain in their transport containers seem less aversive
- Aversiveness of the gas used
- Period of exposure and duration of unconsciousness
- State of consciousness at the onset of wing flapping and muscular contractions
- Effective control and monitoring of gas concentrations

4.4 Low Atmospheric Pressure Stunning (LAPS)

Low atmosphere pressure stunning (LAPS) functions by reducing the pressure in a sealed chamber containing poultry and in so doing, lowers O_2 levels, causing gradual unconsciousness through progressive hypoxia (EFSA 2017). Like CAS systems, LAPS systems offer welfare benefits by eliminating the need for shackling and allowing birds to be stunned in their transport containers. However, as with CAS systems, unconsciousness induced by LAPS is not immediate. Therefore, it is crucial to ensure that poultry do not experience pain or distress during the induction of unconsciousness. A study in broiler chickens comparing aversion during LAPS compared with CAS found behavioural indicators of aversion were more pronounced in the CO_2 treatment, with gasping and headshaking occurring earlier and at a greater frequency than with LAPs. The conclusion was that LAPS is likely to be less aversive than CO_2. Research in broiler chickens also indicates that inducing hypoxia with the LAPS system can be achieved without causing injury in the form of physical trauma to the air sacs or infraorbital sinuses (Martin et al. 2020).

A detailed scientific evaluation of the key operational parameters for using LAPS to stun broiler chickens has been completed and aims to ensure that the animal welfare outcomes are at least equivalent to those provided by other permitted stunning methods (EFSA 2017). Any deviations from these conditions could have different consequences for animal welfare. Therefore, the commercial use of the LAPS system has not yet been extended to other types of poultry.

4.5 Neck Cutting and the Bleeding Process

After stunning, poultry are 'bled', usually by severing the carotid arteries supplying oxygenated blood to the brain. Birds die from the loss of circulating blood volume and the resultant cerebral anoxia. For the stunning and slaughter process to be effective, the duration of unconsciousness induced by stunning must exceed the time taken to perform the neck cut and the time required for bleeding to cause neuronal death (EFSA 2004). In poultry, the carotid arteries are superficial compared to mammals, so a deep, complete cut across the front of the ventral neck severs both carotid arteries. In chickens, neck cutting is primarily carried out by automated equipment, which uses one or more rapidly rotating blades. In poultry species that are commonly slaughtered at slower processing speeds, e.g. turkeys, ducks and geese, neck cutting is usually performed manually with a knife. The blood vessels severed during neck cutting (and the resulting loss of blood) will determine the time to permanent loss of brain activity, with the severance of both carotid arteries resulting in a faster decline in EEG activity than when other blood vessels are cut.

4.6 Slaughter Without Stunning

In poultry, slaughter without stunning occurs to fulfill specific market requirements. When conscious birds are slaughtered without stunning, the time taken to lose consciousness is important, as birds will experience pain and distress associated with the neck cut until unconsciousness prevails. A study measuring the time taken for ducks to lose consciousness following slaughter without stunning revealed that it took up to 383 s for some birds to lose consciousness. When the cut was positioned higher on the neck, the onset of consciousness occurred earlier, indicating that the position of the neck cut and the blood vessels severed is an important determinant of the time to loss of brain function (Friedman et al. 2021).

5 Summary

This chapter underscores the importance of ensuring animal welfare during euthanasia and slaughter. It highlights that proper handling techniques and the effective induction of unconsciousness during stunning are essential for preventing pain and distress. It calls for ongoing evaluation and refinement of euthanasia and slaughter practices, emphasising the need to integrate scientific research and technological advancements to enhance animal welfare outcomes.

References

American Veterinary Medical Association (2020) AVMA guidelines for the euthanasia of animals: 2020 edition. American Veterinary Medical Association. https://www.avma.org/sites/default/files/resources/2020-euthanasia-final.pdf

Anastasov MI, Wotton SB (2012) Survey of the incidence of post stun behavioural reflexes in electrically stunned broilers in commercial conditions and the relationship of their incidence with the applied waterbath electrical parameters. Anim Welf 22:247

EFSA (2004) EFSA Scientific Panel on Animal Health and Welfare on a request from the Commission related to welfare aspects of the main systems of stunning and killing the main commercial species of animals. EFSA J 45:1–29

EFSA AHAW Panel (2012) Scientific opinion on the use of animal based measures to assess welfare of broilers. EFSA J 2774:74

EFSA AHAW Panel (Animal Health and Welfare) (2014) Scientific opinion on electrical requirements for poultry waterbath stunning equipment. EFSA J 12(7):3745, 18 pp

EFSA AHAW Panel (EFSA Panel on Animal Health and Animal Welfare), Nielsen SS, Alvarez J, Bicout DJ, Calistri P, Depner K, Drewe JA, Garin-Bastuji B, Gonzales Rojas JL, Gortázar Schmidt C, Miranda Chueca MÁ, Roberts HC, Sihvonen LH, Spoolder H, Stahl K, Velarde Calvo A, Viltrop A, Winckler C, Candiani D, Fabris C, Van der Stede Y, Michel V (2019a) Scientific opinion on the killing for purposes other than slaughter: poultry. EFSA J 17(11):5850, 83 pp

EFSA AHAW Panel (EFSA Panel on Animal Health and Animal Welfare), Nielsen SS, Alvarez J, Bicout DJ, Calistri P, Depner K, Drewe JA, Garin-Bastuji B, Gonzales Rojas JL, Gortázar Schmidt C, Miranda Chueca MÁ, Roberts HC, Sihvonen LH, Spoolder H, Stahl K, Velarde Calvo A, Viltrop A, Winckler C, Candiani D, Fabris C, Van der Stede Y, Michel V (2019b) Scientific opinion on slaughter of animals: poultry. EFSA J 17(11):5849, 91 pp

EFSA Panel on Animal Health and Welfare (2006) Opinion of the Scientific Panel on Animal Health and Welfare (AHAW) on a request from the Commission related with the welfare aspects of the main systems of stunning and killing applied to commercially farmed deer, goats, rabbits, ostriches, ducks, geese. EFSA J 4(3):326, 89 pp

EFSA Panel on Animal Health and Welfare (AHAW) (2012) Scientific opinion on electrical requirements for waterbath equipment applicable for poultry. EFSA J 10(6):2757, 80 pp

EFSA Panel on Animal Health and Welfare (AHAW), More S, Bicout D, Bøtner A, Butterworth A, Calistri P, Depner K, Edwards S, Garin-Bastuji B, Good M, Gortazar Schmidt C, Miranda MA, Nielsen SS,

Sihvonen L, Spoolder H, Willeberg P, Raj M, Thulke HH, Velarde A, Vyssotski A, Winckler C, Cortiñas Abrahantes J, Garcia A, Muñoz Guajardo I, Zancanaro G, Michel V (2017) Low atmospheric pressure system for stunning broiler chickens. EFSA J 15(12)

Erasmus MA, Lawlis P, Duncan IJH, Widowski TM (2010a) Using time to insensibility and estimated time of death to evaluate a nonpenetrating captive bolt, cervical dislocation and blunt trauma for on-farm killing of turkeys. Poult Sci 89:1345–1354

Erasmus MA, Turner PV, Nykamp SG, Widowski TM (2010b) Brain and skull lesions resulting from use of percussive bolt, cervical dislocation by stretching, cervical dislocation by crushing and blunt trauma. Vet Rec 167:850–858

Erasmus MA, Turner PV, Widowski TA (2010c) Measures of insensibility used to determine effective stunning and killing of poultry. J Appl Poult Res 19:288–298

FAWC (2009) Report on the welfare of farmed animals at slaughter or killing part 2: white meat animals. Farm Animal Welfare Council, London

Food and Agriculture Organization of the United Nations (2023) – with major processing by Our World in Data. "Number of poultry birds—FAO" [dataset]. Food and Agriculture Organization of the United Nations, "Production: Crops and livestock products" [original data]. Retrieved January 1, 2025 from https://ourworldindata.org/grapher/poultry-livestock-count

Friedman A, Dalla Costa FA, Dalla Costa OA, Godsell-Ryan A, Gibson TJ (2021) Time to loss of behavioural and brainstem responses of ducks following non-stunned slaughter. Animals 11(12):3531. https://doi.org/10.3390/ani11123531

Gentle MJ (2011) Pain issues in poultry. Appl Anim Behav Sci 135(3):252–258. https://doi.org/10.1016/j.applanim.2011.10.023. ISSN 0168-1591

Gentle MJ, Tilston VL (2000) Nociceptors in the legs of poultry: implications for potential pain in preslaughter shackling. Anim Welf 9:227–236

Gerritzen MA, Raj MAB (2009) Animal welfare and killing for disease control. Food Saf Assurance Vet Public Health 5:191–203

Gerritzen MA, Lambooij E, Stegeman JA, Spruijt BM (2006) Slaughter of poultry during the epidemic of avian influenza in The Netherlands in 2003. Vet Rec 159:39–42

Gerritzen M, Lambooij E, Reimert H, Stegeman A, Spruijt B (2007) A note on behaviour of poultry exposed to increasing carbon dioxide concentrations. Appl Anim Behav Sci 108:179–185

Gibson TJ, Rebelo CB, Gowers TA, Chancellor NM (2018) Electroencephalographic assessment of concussive non-penetrative captive bolt stunning of turkeys. Br Poult Sci 59:13–20

Graml C, Niebuhr K, Waiblinger S (2007) Reaction of laying hens to humans in the home or a novel environment. Appl Anim Behav Sci 113:98–109

Hernandez E, James F, Torrey S et al (2019) Evaluation of brain death in laying hens during on-farm killing by cervical dislocation methods or pentobarbital sodium injection. Front Vet Sci. https://doi.org/10.3389/fvets.2019.00297

Hewitt L (2000) The development of a novel device for humanely dispatching casualty poultry. Veterinary medicine Dissertation University of Bristol, UK

Humane Slaughter Association (HSA) (2013) Practical slaughter of poultry. Available online: https://www.hsa.org.uk/downloads/publications/practicalslaughterofpoultrydownload-cm-updated.pdf. Accessed 10/08/2024

International Association for the Study of Pain (2020). IASP announces revised definition of pain. IASP. Retrieved from https://www.iasp-pain.org. Accessed 08/08/2024

Lambooij E, Korte SM (2011) The use of electroencephalography (EEG) in assessing unconsciousness in poultry. Anim Welf 20(4):557–563

Leary et al (2020) AVMA guidelines for the euthanasia of animals. American Veterinary Medical Association. https://www.avma.org/sites/default/files/2020-02/Guidelines-on-Euthanasia-2020.pdf

Martin JE, McKeegan DEF, Magee DL, Armour N, Pritchard DG (2020) Pathological consequences of low atmospheric pressure stunning in broiler chickens. Animal 14(1):129–137

McKeegan DEF, McIntyre J, Demmers TGM, Lowe JC, Wathes CM, van den Broek PLC, Coenen AML, Gentle MJ (2007) Physiological and behavioural responses of broilers to controlled atmosphere stunning: implications for welfare. Anim Welf 16:409–426

McKeegan DEF, Sandercock DA, Gerritzen MA (2013) Physiological responses to low atmospheric pressure stunning and the implications for welfare. Poult Sci 92:858

Raj ABM (2006) Recent developments in stunning and slaughter of poultry. Worlds Poult Sci J 62(03):467–484

Raj ABM, O'Callaghan M (2001) Evaluation of a pneumatically operated captive bolt for stunning/killing broiler chickens. Br Poult Sci 42:295–299

Sandilands V, Raj ABM, Baker L, Sparks NHC (2011) Aversion of chickens to various lethal gas mixtures. Anim Welf 20:253–262

Scanes CG, Hurst K, Thaxton Y, Archer GS, Johnson A (2020) Effect of transportation and shackling on plasma concentrations of corticosterone and heterophil to lymphocyte ratios in market weight male turkeys in a commercial operation. Poult Sci 99(1):546–554

Shields SJ, Raj AB (2010) A critical review of electrical waterbath stun systems for poultry slaughter and recent developments in alternative technologies. J Appl Anim Welf Sci 14(4):371

Sparrey J, Sandercock DA, Sparks NHC, Sandilands V (2014) Current and novel methods for killing poultry individually on-farm. Worlds Poult Sci J 70:737–758

Vizzier TY (2015) From idea to reality: the development of the low atmosphere stunning system. In: Proceedings of the "Recent advances II"—HSA International Symposium, Zagreb, Croatia, 16–17 July 2015, p 36

Vizzier-Thaxton Y, Christensen KD, Schilling MW, Buhr RJ, Thaxton JP (2010) A new humane method of stunning broilers using low atmospheric pressure. J Appl Poult Res 19:341

Von Holleben K, Von Wenzlawowicz M, Gregory N, Anil H, Velarde A, Rodriguez P (2010) Report on good and adverse practices – Animal welfare concerns in relation to slaughter practices from the viewpoint of veterinary sciences, 2/8/2012; 1.3: 81 p

World Organisation for Animal Health (WOAH) (2023) Terrestrial Animal Health Code

Zulkifli I, Wakiman Z, Sazili AQ, Goh YM, Jalila A, Zunita Z et al (2019) Effect of shackling, electrical stunning and halal slaughtering method on stress-linked hormones in broilers. S Afr J Anim Sci 49(3):598–603

Production Animal Species

Cattle

Alison Small and Danila Marini

Abstract

The lives of humans and cattle have been intertwined for over 9000 years. Cattle have been worshiped as gods, sacrificed to gods, used as beasts of burden and providers of milk, meat and leather. Today, there are approximately 940 million cattle worldwide, the majority being reared for meat or milk production. Cattle are also used in research, as beasts of burden, sporting animals and pets. Both Nietzsche and Frankl acknowledged that life and suffering are entwined, both from philosophical and physical points of view. Parturition engenders pain and suffering and is notable in cattle, while misadventure and injury are a part of life for all animals. Commercial cattle production involves several elective husbandry procedures that are painful. This chapter investigates the domestication and natural history of cattle, the sources of pain and its management.

Keywords

Cows · Calves · Bulls · Welfare · Husbandry procedures · Analgesia

Cows are amongst the gentlest of breathing creatures; none show more passionate tenderness to their young when deprived of them; and, in short, I am not ashamed to profess a deep love for these quiet creatures.—Thomas de Quincey 1785–1859

A. Small
Armidale, Agriculture and Food, CSIRO, FD McMaster Research Laboratory, Chiswick, New England Highway, Australia
e-mail: alison.small@csiro.au

D. Marini (✉)
School of Animal and Veterinary Science, Davies Livestock Research Centre, Adelaide University, Adelaide, SA, Australia
e-mail: danila.marini@adelaide.edu.au

1 Introduction

Modern domestic cattle are descended from aurochs and Highland shorthorns (Isaac 1962). These wild *bovidae* roamed the grasslands of Europe, northern Africa and western Asia, and were hunted by early human beings. Cattle hunts are prominent features in many Paleolithic caves and cattle may have been considered objects of worship (Conrad 1973), symbolising strength, fertility and renewal, as their breeding cycles aligned with the seasons. Throughout the centuries, cattle have continued as objects of reverence: the Minoans, Greeks and Egyptians worshipped a bull god; bulls were sacrificed to Roman gods and presently cattle are sacred to the Hindu religion (Simoons 1968).

R. E. Clutton, G. C. Musk (eds.), *Pain and Suffering in Farmed Animals*,
https://doi.org/10.1007/978-3-032-11407-5_14

2 History of Domestication

Predecessors of modern cattle were domesticated between the 7th and 5th millennia BCE (Toussaint-Samat 1992a, b) and were possibly first gathered into enclosures for sacrificial purposes (Isaac 1962; Simoons 1968) rather than for slaughter or haulage. However, it is possible that sacrificed animals were later consumed ritually, the meat being considered to endow strength and courage on its consumer (Isaac 1962). In time, herding, then breeding of 'domesticated' *bovidae*, and finally milk collection would be practiced (as evidenced by religious texts, e.g. Genesis 18, viii and archaeological finds from around 2500 BCE (Isaac 1962, Toussaint-Samat 1992a, b). With increasing docility would come utility as beasts of labour, cattle being harnessed to ploughs, sleds and then wheeled wagons (Isaac 1962).

Kurgan invaders from Eurasia brought cattle to Europe, India and Iran between around 4400 and 2800 BCE (Gimbutas 1977) in search of fresh pasture (their own being drought-stricken) for their burgeoning herds. Cattle were a symbol of status to the Kurgan and curiously, the Sanskrit word 'gavisti', the word for 'battle', can be translated as 'a desire for cattle' (Lincoln 1981). Similarly, the word 'cattle' has the same origins as 'chattel' and 'capital' in English (Funk 1950), with parallels identifiable in other languages, e.g. Latin and Spanish.

Beef became a popular staple in Europe, particularly among the wealthy. In the fifteenth century, meat consumption rose dramatically after the bubonic plague as the limited labour force demanded greater reward for their efforts (Braudel 1974; Ritchie 1981). This desire to eat beef continued into the mid-sixteenth century, when pasture deterioration through overstocking and the recovery of the human population revised the supply-demand balance.

Before refrigeration, problems with maintaining meat freshness were partly resolved with spices that masked the odours and flavours of putrescent meat. This demand for spices was a major impetus for trade and exploration. During the late fifteenth century, European navigators were motivated to discover and open an Atlantic sea-route to the 'spice islands' of the Indo-Pacific region as a potentially more competitive supply route than the overland 'Silk Road'. Christopher Columbus' expeditions to the Americas revealed huge areas of fresh pasture, prompting him to introduce cattle and horses to the New World—beginning with Haiti in 1494 (Dary 1981). Over the next few centuries, vast numbers of cattle were imported into Spain's colonies in southern and central America, bringing an Iberian cattle culture to the continent from an overstocked Europe (Dary 1981). Similarly, colonisation and clearance (of both indigenous peoples and native bison) of the western plains of North America was driven by the desire to acquire fresh cattle pasture. Cattle ranching received a considerable boost in the mid-nineteenth century with the development of the railways (Webb 1936; Boorstin 1973). These allowed cattle to be rapidly shipped great distances from pastureland to processing facilities. Simultaneously, a similar story was unfolding in Australia, while in French colonial Africa, cattle rearing was shifting from a nomadic herding process (which had been practiced since cattle first entered the continent between 5000 and 2300 BCE) to a European model of enclosed pastures and 'modern' farming practices (Meyn 1970; Lappe 1977).

3 Natural History

Cattle are ruminant, four-legged, bidactyl ungulates of the species *Bovidae*. There are two subspecies: *Bos taurus indicus* (or *Bos indicus*) and *Bos taurus taurus* (or *Bos taurus*) and over 1000 breeds of cattle (Felius 1995). *Bos indicus* cattle include the Zebu, Brahman (Fig. 1) and Nellore types. These are tropically adapted, and are commonly found in hotter regions, e.g. India, Southeast Asia, northern and western Australia, the central and southern Americas, and sub-Saharan Africa. *Bos indicus* may be horned or polled, have drooping ears, a fine, short hair coat and a prominent hump of flesh on the wither, which has a role in the modulation of hydration, similar to the hump of the camel or dromedary. In

Fig. 1 Brahman (*Bos indicus*) young bull. (Source: CSIRO)

contrast, *Bos taurus* are cooler-climate animals, with shorter ears and no hump (Figs. 2 and 3). They, too, may be horned or naturally polled. Through selective breeding, *Bos taurus* can be further classified as dairy-type, beef-type or dual-purpose. The dairy types, e.g. Holstein, Friesian, Jersey or Guernsey breeds, have been selected for milk production (Figs. 4 and 5). They appear almost triangular in body shape with a deep chest and slim hindquarters; the female carries a large and prominent udder. The beef-types, e.g., Angus, Belgian Blue, Limousin and Simmental, are more muscular with a flat back, strong hindquarters and heavy shoulders and neck.

Cattle are social animals and congregate in herds, the size of which is strongly influenced by feed availability. For example, in northern Australia, where feed is sparse, cattle are found in groups of three to six animals. The onset of sexual maturity varies according to breed and body condition, but heifers can show their first oestrus between 9 and 22 months. Bulls may be able to serve cows from the age of 18 months on. When the reproductive cycle is not artificially controlled, cattle are seasonal breeders. Calves are born in the spring, so that the dam can take advantage of fresh herbage during lactation. Gestation is 9 months, and although twin calves do occur, a single calf is the norm. During the breeding period, the female cow cycles every 3 weeks, oestrus behaviour being displayed as increased activity, including mounting of other cattle within the herd. In the absence of human interference, and particularly where feed is scarce, young may be suckled for up to 18 months and cows may carry a calf every other year. However, under modern farming conditions, cattle are usually expected to calve every year, and calves are artificially weaned from 4–6 months to allow the dam to recover from one lactation before beginning the next.

Fig. 2 Belgian Blue (*Bos taurus* beef type) bull. (Source: Sabine Schmoelzl)

Fig. 3 Limousin (*Bos taurus* beef type) bull. (Source: CSIRO)

Fig. 4 Holstein (*Bos taurus* dairy type) heifer. (Source: Alison Small)

Fig. 5 Friesian (*Bos taurus* dairy type) cow and calf. (Source: Sandra Ospina Rios)

4 Current Extent of Utilisation

There are currently approximately 940 million cattle worldwide, with India (304 million) and Brazil (224 million) hosting the greatest populations (FAOSTAT 2023). Cattle are found in various situations as beasts of labour, for biomedical (or animal science) research, as rare breed animals in zoological collections, or as commercially farmed animals. The latter, which may be intensive or extensive, may be used to produce purebred high-value breeding stock, milk or beef production, or a combination enterprise. Commercial farming constitutes most of the cattle population, of which an estimated 264 million are dairy cows.

Considerable diversity of farming practices globally makes it difficult to provide an overall description. Farms range in size from a few to tens to a few hundred thousand animals, which can be intensively or extensively managed.

In stud breeding operations, where the objective is to produce animals, ova and semen of high genetic merit, animals are usually well-cared for being treated as individuals. However, breeding practices are often artificially managed. The oestrus cycle of females can be manipulated using injectable hormone preparations. Superovulation may be followed by flushing of oocytes for in-vitro fertilisation and subsequent embryo transfer, and conception occurs as a result of artificial insemination or embryo implantation. During semen collection, ejaculation may be induced using teaser females and a model 'cow' that the male mounts. Alternatively, electroejaculation may be used, whereby electrical pulses are applied to the prostatic region by a rectal probe.

In the production of beef, breeding cows are usually managed extensively, i.e., at pasture, with bulls being introduced intermittently to the herd for a defined mating period. Consequently, all cows will then calve at a similar time of year, allowing more focused monitoring during that period. Calves remain with their dams until weaning, and often remain on the property of birth in a separate management group. Alternatively, they may be moved elsewhere for the next phase of growth. Some weaner calves are slaughtered for the 'grass-fed veal' trade. Towards the end of the growing period, animals may be placed in feedlots where 'finishing' on a grain-based diet to a defined set of criteria occurs. While some animals enter the feedlot immediately after weaning, others may be 'finished' at pasture. Replacement breeding females may be bought in or selected from the female progeny of the existing breeding herd.

In the process of producing milk for humans, cows are often artificially inseminated so that replacement heifers are of a particular genetic strain. Sexed semen may be used to minimise the number of male calves born (male dairy calves are of little commercial value to the dairy herd as they are insufficiently muscled to be used for beef production). Nevertheless, some male calves (particularly in combined beef-dairy enterprises) will be reared intensively as entire bulls to 12–16 months, after which they will enter the beef herd. In dairy enterprises, calving tends to be seasonal, with two or three calving sessions during each year, to ensure constant milk production from the entire herd; alternatively, calving may occur all year-round. To achieve a consistent milk supply, the oestrus cycle may require artificial manipulation. After calving, calves and dams are usually separated within 48 h, and the former are reared on surplus colostrum and milk from the herd, or with reconstituted powdered milk. If not reared for bull beef, male 'bobby' calves are often sent for slaughter at 1 week of age, while female calves are reared as replacement heifers either for retention in the herd (as followers) or sale. The cow enters the milking herd for her productive lactation, is re-inseminated at the first or second oestrus after calving and is expected to produce a live calf every year. Dairy animals may be housed throughout their lives or live at pasture during the summer months; they rarely live at pasture all year round.

Cattle are favoured in reproductive studies because their reproductive cycle is similar to that of humans (Ménézo and Hérubel 2002). They also feature in animal health and agricultural research, but the numbers involved are relatively small. In 2020, 22,000 cattle were used in Europe for research compared to 4.6 million mice, rats and guinea-pigs

(Hobson 2021; Statistica 2023). Internationally, the care of laboratory animals is covered by ethical codes of practice and/or regulations, and much attention is paid to the prevention or mitigation of pain. For example, animals undergoing surgery or other noxious procedures for research purposes will always be anaesthetized. However, there are situations where anaesthesia and analgesia may be withheld: for example, in agricultural science, when animals act as controls and mimic the on-farm situation when new treatments or analgesics are being validated (Pauly et al. 2012; Petherick et al. 2014a, b; Paull et al. 2015).

Cattle also feature in sport and entertainment (Fig. 6). The bullfights of Spain and Southern America originate from early bull-god religions, wherein the flesh of the sacrificed animal was eaten in the belief that it brought strength and virility. Sacred sacrifice evolved through the stylised man-versus-beast hunting contests popular in Rome, which persists in present-day bullfighting. In this, matadors torment the bull using a cape and lance in a highly ritualized form, leading to the death of the animal. The bulls are specifically bred for aggression, and are minimally handled from birth, leading to an essentially wild animal, as befits a stylised hunting contest. Bull riding is a rodeo sport practiced around the world. The rider attempts to stay on board a bucking animal for 8 seconds, with only one hand holding onto a rope around the bull's chest. The sport may originate from efforts to tame bulls for use on farm, which itself has origins in Minoan culture (Lecompte 1985).

Fig. 6 Bullriding. (Source Troy Kalinowski)

5 Attitudes to Pain in the Species

Identifying pain and suffering in cattle is most feasible when small herds are attended frequently by attentive carers. This arrangement allows the recognition of subtle behavioural and demeanour changes which may not be possible when large free-roaming herds are only examined at intervals of weeks or months. However, the ability to identify the subtle changes that indicate pain in cattle depends on the assessor. While some individuals are sensitive to these changes, others may be unaware of problems until they become severe and prominent (Lehner 1996). An individuals' attitude is also important: some individuals believe that cattle do not experience pain because cattle are not particularly demonstrative (Fajt et al. 2011; Kling-Eveillard et al. 2015). An observational study of dairy farmers in Denmark found that negative sentiments towards cattle and their handling correlated with indicators of poor farm welfare, whereas positive sentiments were associated with benign handling and increased job satisfaction (Andreasen et al. 2020). The attitudes of various stakeholders towards pain in cattle are relatively well researched, with attitudinal surveys across different decades, groups, and husbandry systems that provide useful insight into how attitudes vary and change.

Attitudinal changes to cattle pain over time have occurred in UK veterinarians. Two studies conducted in 2006 (Huxley and Whay (2006) and 2017 (Remnant et al. 2017) gauged the perspectives of buiatricians (cattle veterinarians) towards pain in cattle and the use of pain relief strategies. The former study highlighted significant gender differences, with women being more likely to attribute higher pain scores to various procedures and diseases compared to men. This observation is common within 'attitudes to pain' surveys of this kind, and it is proposed that the likelihood that females score higher pain ratings is related to greater empathy. Empathetic levels towards other humans and animals influence perception of pain in cattle, with higher empathy ratings correlating with the assignment of higher pain scores to perceived painful events in cattle (Wikman et al. 2013; Norring et al. 2014).

The year of qualification also influences attitudes, with recent graduates attributing more pain to diseases such as mastitis and pneumonia than older graduates (Huxley and Whay 2006). That said, those with more extensive experience viewed invasive procedures like claw amputation, dehorning, and disbudding as more painful. Similar results have been obtained from veterinarians in other countries (Kielland et al. 2009) as well as from cattle producers: two Finnish studies found that older producers were more likely to perceive procedures and illnesses as painful than younger and mid-range age producers (Wikman et al. 2013, 2016). There appears to be an asymptote in attitudes—both early-career and late-career livestock carers demonstrate more concern about pain than mid-career carers. The reason for this is unclear: perhaps during mid-career, the pressures of work, family and finances draw attention away from more empathetic consideration; early-career carers could be more idealistic, mid-career more realistic, having lost the idealism; while late-career carers maintain the realism, but develop greater empathy. More research is required to fully understand this phenomenon.

Despite the broadening perception that many invasive procedures are painful, there is often a discrepancy between perceptions and the provision of post-procedural analgesia. For example, while digital amputation and caesarean operations were cited as painful by veterinarians, analgesia was provided by only 61 and 68% of respondents, respectively. Even fewer respondents provided analgesia following castration and disbudding (4.6 and 1.7%, respectively) (Huxley and Whay 2006). The type of analgesia used also varies with the procedure: non-steroidal anti-inflammatory drugs (NSAIDs) are less likely to be used for castration and disbudding compared to similarly scored painful procedures (Remnant et al. 2017). Similar results have been reported in sheep with a discrepancy between perceived procedural pain and the subsequent provision of analgesia (Phillips and Phillips 2010; Larrondo et al. 2018). Several barriers may be responsible for this misalignment in belief and action, including low confidence in pain identification in individual animals and the lack of licensed analgesics.

A follow-up study by Remnant et al. in 2017 intimated that veterinarians were becoming increasingly sensitive towards cattle pain. Pain scores awarded for 12 out of the 27 procedures listed in the initial survey had increased. As with the earlier work, the year of graduation and gender influenced perceptions, with women and graduates from 2010 onwards more likely to assign higher pain scores. Recent graduates displayed a greater tendency to use NSAIDs, in line with current trends. Producer attitudes towards relieving pain were also likely to increase the veterinarians use of analgesics (Remnant et al. 2017).

Producer attitudes appear promising in terms of recognizing that cattle feel pain, which often requires analgesics. Interviews with dairy farmers highlighted the importance they placed in providing pain relief to calves undergoing dehorning and disbudding, citing the need to ensure the calves' comfort and recovery (Saraceni et al. 2022). A survey of Canadian cattle producers found that the majority (> 70%) believed that dehorning, castration and conditions such as dystocia cause pain (Fraser et al. 2012). Interviews with producers found a common theme, i.e., those who had experienced pain, such as burns and cuts, tended to extrapolate the experience to their cattle. Moggy et al. (2017) highlighted the complexity of the decision-making process facing producers when providing pain mitigation to their cattle. Reasons for not providing analgesia included: cost of medication; the labour requirements for administration, the logistics of running large numbers of calves through chutes; a lack of information on the medications available and their effectiveness. Despite these challenges, producers highlighted the importance of trying to reduce pain as much as possible. Proposed options included using alternative or complementary approaches, such as choosing to castrate cattle at a younger age (when they were perceived to experience less pain (Moggy et al. 2017). One producer in the study commented '*...we're just not aware of what we should be doing and if it's going to be beneficial*' (Moggy et al. 2017). The lack of knowledge and resources expressed by producers on the topic of cattle pain management is a common theme in this research field.

Veterinarians have an important role to play in providing information and assisting producers in dispelling unsympathetic attitudes and improving decision-making with respect to pain mitigation in cattle. An interview study with Canadian dairy farmers highlighted the role that veterinarians and education programs had on their uptake of pain relief for dairy calves undergoing disbudding and dehorning (Saraceni et al. 2022). The importance of clear communication between veterinarians and producers for ensuring effective pain relief has been discussed (Tschoner et al. 2021). This study found that producers and veterinarians both rated the pain resulting from invasive husbandry procedures similarly although the use of analgesics was less amongst producers. This discrepancy may have arisen from miscommunication between the two groups. Though veterinarians (66.8%) believed they effectively communicated the need to manage pain in their clients' cattle only 16% of producers agreed. The majority of producers (89.9%) in this study also indicated they were concerned about adverse analgesic drug effects, indicating a gap in knowledge transfer (Tschoner et al. 2021). Producers in Ireland are less likely to provide analgesics to their cattle when they do not understand the outcomes of their use (Browne et al. 2022).

More recent surveys of veterinarians and beef and dairy farmers in the United States found the majority of respondents believed cattle felt pain when experiencing listed procedures and diseases, although some respondents did not believe that cattle felt pain in some procedures, including abdominal surgery, castration and freeze branding (Edwards-Callaway et al. 2023). The US consumer appears, in general, to share the latter view, i.e. that castrating and dehorning with pain mitigation is an impractical way to improve cattle welfare (McKendree et al. 2018). Like Canadian producers, US farmers tend to believe that younger cattle are less likely to experience pain than adults (Edwards-Callaway et al. 2023).

Thus, the provision of analgesia to cattle in the United States becomes more likely as the animal ages (Edwards-Callaway et al. 2023) which mirrors the Canadian pattern (Moggy et al. 2017). When asked about changing beliefs on analgesic use over 10 years, US veterinarians were more likely to provide analgesia (76.4%), although this is not reflected by their clients (32.2%). Barriers to pain mitigation strategies bore some similarities to those identified by Canadian producers, i.e. cost and practicality of application. However, a lack of approved analgesics was regarded as the greatest barrier (Edwards-Callaway et al. 2023).

To summarise, beliefs about pain vary with demographic (age, gender, experience), as well as beliefs and current knowledge. There has been a gradual positive shift over time as research highlights how much pain certain husbandry procedures cause. An increase in the ability to identify and assess pain, along with an enhanced understanding of the benefits of analgesia, means that pain management practices are undoubtedly improving.

6 Anaesthesia and Analgesia

In most countries, the use of analgesics (and other drugs) in food producing animals is controlled in order to ensure the drugs or their metabolites do not enter the human food chain in levels that might affect human health. In food-producing animals, e.g., cattle, it is important to select drugs for which the maximum residue limit (MRL) has been defined, and formulations of those agents that have defined withholding periods (WHP) for meat and/or milk. Unfortunately, such data are not necessarily available for all classes of agent, nor in all regions. In such cases, the supervising veterinarian must make an informed decision, taking into consideration the time between treatment and harvest of animal products for food production and any pharmacokinetic information available. For livestock species, there is a dearth of pharmacokinetic information on many anaesthetics and analgesics, and extrapolation of data from species to species (even between 'similar' species e.g. within the class 'ruminant') is not advisable (Small et al. 2021).

6.1 General Anaesthesia

General anaesthesia is feasible in cattle but seldom conducted on farms or outside research situations. Most surgical procedures are performed with regional (or no) anaesthesia, with or without sedation, often on the standing animal. One of the contraindications of general anaesthesia and deep sedation is ruminal stasis, leading to bloating and respiratory compromise. For sedation, α_2 adrenoreceptor agonists, e.g. xylazine, detomidine or medetomidine, are commonly used. Low doses (e.g. 0.015–0.025 mg/kg xylazine; 40 μg/kg detomidine or 30.0 μg/kg medetomidine) provide standing sedation for up to an hour, while higher doses (0.1–0.2 mg/kg xylazine) can lead to recumbency and deep sedation for an hour or more (Short 1987; Carroll and Hartsfield 1996; Riebold 2001; Clarke et al. 2014). Under laboratory conditions, low doses of α_2 adrenoreceptor agonists provide good post-operative analgesia, but this effect has yet to be confirmed in field conditions. For surgical castration, 0.05 mg/kg xylazine reduced the cortisol response in calves, but had no effect on the behavioural response, while a combination of 0.05 mg/kg xylazine with 0.1 mg/kg ketamine did ameliorate both the behavioural and cortisol responses (Coetzee et al. 2010).

6.2 Local Anaesthesia

Local anaesthesia is commonly used for elective surgeries (e.g. caesarean section, eye enucleation, claw removals) or elective husbandry procedures (e.g. disbudding, dehorning, castration, spaying or tail docking). With administration routes including intravenous regional anaesthesia, nerve block or local infiltration. Lidocaine appeared on the market in the 1940s and is the most commonly assessed local anaesthetic in cattle, because of its ready availability in many jurisdictions. However, its limited duration of action has led to disappointing results when animals are assessed beyond the first hour post-treatment.

Intravenous regional anaesthesia is used for distal limb surgeries, e.g. claw amputation, the agent being injected into the radial vein and/or axial digital vein distal to a tourniquet (Clarke et al. 2014). In controlled studies, a mixture of 2 mg/kg lidocaine with 1.5 mg.kg ketamine provided a faster onset and longer duration of nerve bock in cattle than 2 mg/kg lidocaine alone (Yadav et al. 2023). Previous recommendations of lidocaine dose were 4 mg/kg, which resulted in greater physiological signs of toxicity (reductions in systolic and diastolic blood pressure) than the mixture of 2 mg/kg lidocaine with 1.5 mg.kg ketamine (Yadav et al. 2021) or 2 mg/kg lidocaine alone (Kumar et al. 2020).

The cornual nerve block is commonly used for dehorning or disbudding in *Bos taurus* animals, local anaesthetic being injected onto the nerve as it passes through the supraorbital fossa. The nerve block should be supplemented by local infiltration around the base of the horn, particularly in *Bos indicus* animals, as the horn can be innervated from multiple sources (Graf and Senn 1999; Stafford and Mellor 2005; Winder et al. 2018). Use of the cornual nerve block has been shown to be effective in reducing the cortisol response to caustic paste disbudding (Weyl-Feinstein et al. 2021), and the behavioural response to disbudding using a hot iron (Fig. 22) (Grondahl-Nielsen et al. 1999).

Caudal epidural (at the base of the tail) injection of local anaesthetic can provide good regional anaesthesia of the perineum and results in tail paralysis. This approach is used, e.g. in cases that require vaginal surgery or stitching, for tail docking and for castration. Five to eight minutes after injection of 0.2 mg/kg lidocaine, the tail is paralysed, and the anaesthesia lasts from between 25 to 130 minutes (Lin et al. 2003). In contrast, when clonidine is used (5 μg/kg), the onset of anaesthesia and tail paralysis is delayed (23–44 minutes post-injection), but the duration of effect is extended to 6–11 h (Lin et al. 2003). The addition of dexamethasone to lidocaine can significantly increase the duration of anaesthetic effect of caudal epidural, with an epinephrine-lidocaine mix, intermediate in duration of effect. (Rastabi et al. 2018). Despite the tail paralysis evident when lidocaine (4 mL of 2%) with epinephrine is provided as a caudal epidural, analgesic effects were not detected at 30 mins and days 2 and 6 post rubber ring tail docking of adult dairy cows (Tom et al. 2002a, b). Local anaesthesia by caudal epidural injection has been investigated as a potential approach for burdizzo or surgical castration, with little success. Stilwell et al. (2008) provided 4 mL of 2% lidocaine by caudal epidural, 5 min before castration. In the absence of further analgesia (e.g. with an NSAID), there was no mitigation of the cortisol response, and no improvement in behaviours during castration and 24–48 h post-procedure.

For ring/band castration or burdizzo castration (Fig. 7), local infiltration of local anaesthetic is the most common approach to pain control. In controlled studies effectiveness of pain amelioration varies with castration type and age of animals, and it also appears that the site of infiltration of local anaesthetic is important. Administration of 100 mg of lidocaine into the scrotum and tes-

Fig. 7 Burdizzo castrator. (Photo credit Dr. Joy Crawford)

tes obliterated the cortisol response to ring castration, but not burdizzo castration in 2–4-month-old dairy cross calves (Stafford et al. 2002), while in 3–4-week-old calves, 200 mg of lidocaine injected into the scrotal neck and spermatic cords provided a reduction in cortisol response and pain-related behaviours in both burdizzo and ring-castrated calves (Thüer et al. 2007). Local infiltration of lidocaine into the scrotum and spermatic cords before the procedure does not appear to ameliorate the behavioural responses to pain associated with surgical castration (Stafford et al. 2002; Van der Saag et al. 2019), while application of a lidocaine-bupivacaine mixture (Tri-Solfen®) onto the spermatic cord and wound edges immediately post procedure does mitigate the behavioural responses to surgical castration (McCarthy et al. 2016), but not the cortisol response. A novel bupivacaine liposome preparation injected into the scrotal neck and around each spermatic cord of 4–5-month-old dairy calves prior to surgical castration ameliorates the cortisol response and alterations of weight distribution onto the forelimbs (Martin et al. 2022).

A novel approach to the delivery of local anaesthesia for ring/band castration is lidocaine-impregnated rubber bands. Three sizes of band were developed, to address three different use scenarios: small (suited to very young calves, less than 90 kg bodyweight), middle (calves 90–180 kg) and large (Callicrate® type, for animals 180 kg or more (Fig. 20). The bands each contained between 80 and 210 mg lidocaine, and provided an initial rapid release of the agent, followed by a sustained slow release over the next 48 h (Saville et al. 2020). In field trials, the bands provided a similar analgesic effect in the short term, but superior longer term mitigation of the pain response (measured by nociceptive threshold and infra-red thermography) than lidocaine injected into the scrotal neck and spermatic cords (Saville et al. 2020).

The topical anaesthetic wound dressing, Tri-Solfen®, which contains lidocaine, bupivacaine, adrenaline and cetrimide, has shown promise in improving resolution of Foot and Mouth Disease lesions, and increasing the rate of return to eating and improved lameness scores (Lendzele et al. 2021). It can also help to alleviate pain resulting from foot trimming, when applied to the exposed corium of the hoof, with the added advantage of rapid excretion of metabolites, such that they are undetectable in milk after 6 h post-treatment (Stilwell et al. 2019, 2020), and there may also be an improvement in the resolution of wounds following hot iron disbudding when the wound dressing is applied immediately after the procedure (Sheil et al. 2021).

6.3 Analgesics

6.3.1 Opioids

Although opioids may be used as part of a multimodal anaesthetic-analgesic protocol in cattle (Durand et al. 2021), they are rarely utilised outside the research environment. The use of these drugs is legally restricted due to their potential to be drugs of addiction and abuse. Little is published on their efficacy for routine husbandry procedures. In one study, tramadol (neither 4 mg/kg intravenous nor 100 mg rectal administration) did not ameliorate the behavioural responses to caustic paste disbudding in week-old dairy calves (Braz et al. 2012).

6.3.2 Gabapentinoids

Some interest has been shown in the use of gabapentinoids for pain management in cattle, perhaps in conjunction with local anaesthesia or with an NSAID, e.g. meloxicam or flunixin (Coetzee et al. 2011; Fraccaro et al. 2013; Glynn et al. 2013; Malreddy et al. 2013). The inclusion of gabapentin did not improve lameness scores or step count over the administration of meloxicam alone in a chemical synovitis model (Coetzee et al. 2014), while inconclusive results were also found when co-administering gabapentin with either flunixin or meloxicam in scoop-dehorned calves (Glynn et al. 2013).

6.3.3 Non-steroidal Anti-inflammatory Drugs (NSAIDs)

The NSAIDs exert their influence predominantly through inhibition of the cyclo-oxygenase -prostaglandin synthesis pathway, although other pathways may also be influenced (Voilley et al. 2001; Mamet and Voilley 2002). Several NSAIDs are approved by many jurisdictions for use in cattle, including meloxicam, flunixin meglumine, tolfenamic acid, carprofen and ketoprofen. In cattle, the duration of effect of meloxicam is long compared to other NSAIDs studied, which reflects the differing pharmacokinetics of the individual agents (Erdogan et al. 2019).

Meloxicam is a potent analgesic and has been shown to provide relief in an experimental model of lameness in 4–6-month-old cattle. A dose of 0.5 mg/kg administered orally every 24 h for 4 days resulted in increased steps (activity) and greater weight-bearing on the affected limb (Coetzee et al. 2014). In lactating dairy cows, two injections of 1 mg/kg meloxicam, 24 h apart, significantly reduced the number of lame animals in the group and the overall cortisol response to an induced synovitis (Warner et al. 2021). Meloxicam can also relieve the hyperalgesia associated with mastitis (Milne et al. 2003). A subcutaneous dose of 0.5 mg/kg of meloxicam, given 10 min prior to hot-iron disbudding with cornual nerve block resulted in reductions in ear flicking and head shaking as compared to disbudding with cornual nerve block alone (Heinrich et al. 2010). That study also demonstrated a reduction in pressure sensitivity at the wound site 4 h after disbudding in the meloxicam-treated group, while the same group of authors demonstrated that administration of meloxicam could ameliorate the respiratory and cardiovascular responses to disbudding (Heinrich et al. 2009). Stewart et al. (2009) investigated sympatho-vagal tone and eye temperature for 3 hours post hot-iron disbudding in calves provided with no analgesia, cornual nerve block alone and cornual nerve block with 0.5 mg/kg intravenous meloxicam. They demonstrated that meloxicam provided continued analgesia and suppressed the resurgence of the pain response when the local anaesthetic wore off. In the absence of local anaesthesia, provision of 20 mg meloxicam subcutaneously led to improved growth rates after hot-iron disbudding in 3–6-week-old calves: when cornual nerve block was also used, there was no additional benefit of meloxicam on growth rates (Bates et al. 2015). Oral meloxicam (1 mg/kg) reduced the cortisol, substance P and prostaglandin E_2 (PgE_2) responses to hot-iron disbudding with cornual nerve block (Allen et al. 2013). There was no difference between administration of meloxicam at the time of disbudding, or 12 hours prior to the procedure in terms of cortisol or substance P, but PgE_2remained suppressed for 48 h in the calves administered meloxicam at the time of disbudding, but only for the first 12 h in the calves administered meloxicam 12 h prior to the procedure (Allen et al. 2013). Oral meloxicam also improved activity levels, resting behaviour and time spent feeding in 10-week-old calves dehorned by cautery (Theurer et al. 2012). In 3–4-day-old calves undergoing caustic paste disbudding without local anaesthesia, 45 mg oral meloxicam had little influence on cortisol, substance P, nociceptive thresholds or eye temperatures (Karlen et al. 2021), while in 9-day-old calves undergoing caustic paste disbudding, provision of 0.5 mg/kg meloxicam, subcutaneous, as well as cornual nerve block provided greater amelioration of a range of pain-related responses than cornual nerve block alone (Reedman et al. 2020; Qi et al. 2023). For castration, 0.5 mg/kg meloxicam administered subcutaneously can somewhat mitigate substance P levels, white blood cell parameters and tail flicking after band or surgical castration, but had little effect on other behavioural or physiological parameters (Meléndez et al. 2018). At 1 mg/kg, oral meloxicam provided reductions in cortisol, substance P, heart rate, scrotal inflammation and visual analogue behaviour scores as compared to surgical castration without analgesia (Olson et al. 2016). In that study, meloxicam-treated calves spent less time lying and were more active than the no-analgesic control calves. Gellatly et al. (2021) found little evidence for longer term (up to 62 days post procedure) benefits of meloxicam administration at the time of surgical or band castration. An oral transmucosal formulation of meloxicam, administered into the buccal cavity of animals has been shown to ameliorate the behavioural response to band castration in 3-month-old calves, at a dose rate of 50 mg per calf (Sutherland et al. 2019). Provided at the time of surgical castration, 2 mg/kg oral meloxicam led to improved visual assessment scores, improved categorical attitude scores and improved average daily gain than bulls (all 170–180 kg bodyweight) provided no analgesia (Cull et al. 2022). In feedlots, oral meloxicam provided 24 h prior to surgical castration can reduce the incidence of bovine respiratory disease and pen-level first-pull rate (Coetzee et al. 2012), while oral meloxicam (1 mg/kg) administered prior to transportation also appears to have some beneficial effects on neonatal calves, reducing haptoglobin levels and supporting feed intake and growth (Chibisa et al. 2018).

Flunixin meglumine has potent anti-inflammatory and antipyretic effects and is a useful adjunct to antimicrobial therapy when treating acute endotoxic mastitis in dairy cows (Anderson et al. 1986a, b; Green et al. 1997; Rantala et al. 2002; Wagner and Apley 2004) or bovine respiratory disease (Lockwood et al. 2003). In an experimental lameness model in calves, 1 mg/kg flunixin administered intravenously reduced lameness parameters and reduced the time spent lying as compared with calves given no medication (Schulz et al. 2011). Flunixin is also effective in reducing the cortisol response to surgical castration in calves: in calves of around 176 kg bodyweight, intravenous injection of 2.2 mg/kg flunixin resulted in an improvement in behaviour and attitude, as assessed using a visual analogue scale and a categorical attitude score (Cull et al. 2022); while a transdermal pour-on preparation applied at 3.33 mg/kg significantly reduced the cortisol response (Kleinhenz et al. 2018). In contrast, local administration of flunixin at the scrotal neck and spermatic cords did not ameliorate the behavioural or cortisol responses to band castration in 7–9-week-old calves (Paull et al. 2015). Flunixin also ameliorated the cortisol response to hot-iron dehorning (with cornual nerve block) in 5–9-week-old calves, particularly if the dose (2.2 mg/kg, intravenous) was repeated 3 h after the procedure (Huber et al. 2013), and 2 mg/kg intravenously (with cornual nerve block) reduced head shaking in 7–8-month-old calves dehorned using a scoop dehorner (Fig. 8) followed by cauterisation (Park et al. 2020).

Like flunixin, tolfenamic acid has potent antipyretic and anti-inflammatory effects, and is particularly valued in treatment of respiratory disease in cattle (Deleforge et al. 1994; Bernay et al. 2000). In terms of pain response, tolfenamic acid failed to provide a benefit over application of a plastic shoe in lame dairy cows, as measured by nociceptive threshold or locomotion score (Laven et al. 2008). Efficacy for other painful procedures does not seem to have been assessed.

Carprofen is also effective in ameliorating the clinical signs associated with respiratory disease in calves (Balmer et al. 1996) and fattening cattle (Lockwood et al. 2003; Elitok and Elitok 2004), with 1.2–1.9 mg/kg providing long-acting anti-inflammatory support to antimicrobial treatment, and improving the outcomes as compared to antimicrobial therapy alone. Carprofen has also been shown to reduce the cortisol response (Antanaitis et al. 2018) and rectal temperature (Stilwell et al. 2014) in the first 24 h after calving, and also improves interest in feed post calving (Stilwell et al. 2014). In terms of painful husbandry procedures, carprofen (1.4 mg/kg intravenous) had no effect on inflammatory cytokine production or cortisol response to burdizzo castration without local anaesthesia (Pang et al. 2006, 2011), and negligible effects on the response to hot-iron disbudding (Stock et al. 2016, 2021). In contrast, Stilwell et al. (2008) found that subcutaneous administration of 1.4 mg/kg carprofen in conjunction with caudal epidural anaesthesia did provide significant improvements in the cortisol response of burdizzo-castrated calves, as compared with epidural alone, while intravenous carprofen tended to ameliorate the haptoglobin and fibrinogen responses in band-castrated calves, without local anaesthesia (Pang et al. 2006).

Ketoprofen can also assist in reducing the pyrexia associated with respiratory disease in cattle (Lockwood et al. 2003; De Koster et al. 2022) and has also been demonstrated to reduce the locomotor abnormalities associated with Bovine Ephemeral Fever (Fenwick and Daniel 1996). For lameness, the effect of ketoprofen is at best modest, improvements in locomotion only being seen when 3 mg/kg was given daily over 3 days (Flower et al. 2008), although lameness-

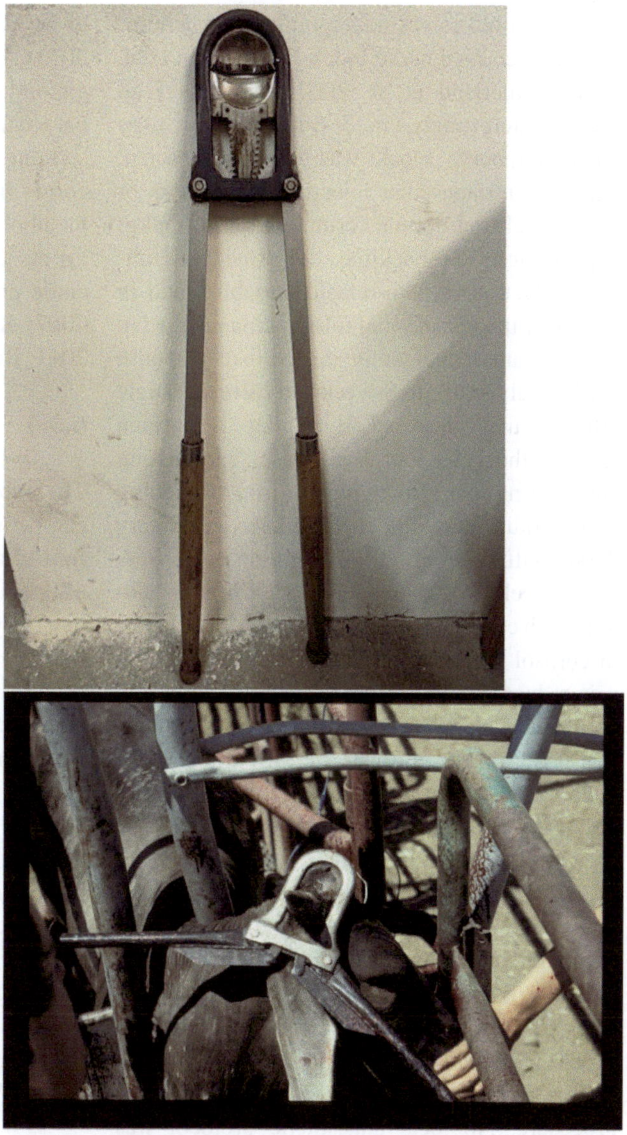

Fig. 8 Scoop dehorner. (Photo credit Dr. Joy Crawford and Dr. Tristan Jubb)

associated hyperalgesia can be reduced using ketoprofen (Whay et al. 2005). However, in cases of digital dermatitis, treated with topical oxytetracycline, 3 mg/kg ketoprofen intramuscular did provide additional benefit, cows having improved locomotion scores 1 week following treatment (Kasiora et al. 2022). Ketoprofen has also been shown to provide some benefit to cows undergoing rumen fistulation surgery or surgical correction of left displacement of the abomasum (Newby et al. 2013, 2014). Ketoprofen is a valuable analgesic in the immediate post-calving period, reducing time spent in lateral lying and increasing time spent in curled sternal recumbency in cows that had required assisted calving (Gladden et al. 2021). Even the calves that had been assisted benefited from a single dose of ketoprofen—showing increased play behaviour compared to calves that received a saline placebo (Gladden et al. 2019). For elective husbandry procedures, intramuscular ketoprofen 3 mg/kg did not provide additional benefit over the use of topical anaesthetic formulation, Tri-Solfen, in calves undergoing scoop dehorning at 2–3 months of age, (Espinoza et al. 2016); however, it did provide a reduction in cortisol response in

3–4-month-old calves undergoing scoop dehorning under cornual nerve block (McMeekan et al. 1998; Sutherland et al. 2002; Stafford et al. 2003). Interestingly, in 3–6-month-old calves where no nerve block was used, ketoprofen, either by intramuscular injection of 3 mg/kg, or via a novel transdermal formulation (10 mg/kg) applied along the backline, did provide significant reduction in pain-related variables, and in particular the transdermal formulation resulted in a significant reduction in the cortisol response (Mills et al. 2020). In 8-week-old calves, a single intramuscular dose of 3 mg/kg ketoprofen reduced the behavioural response (ear flicking and head rubbing) following hot-iron disbudding with cornual nerve block (Faulkner and Weary 2000; Duffield et al. 2010), while in calves less than 2 weeks of age, undergoing hot-iron disbudding with cornual nerve block, the overall change in cortisol levels in the first 3 h post procedure was reduced in the group receiving ketoprofen (Milligan et al. 2004). In several studies, ketoprofen alone did not mitigate the physiological or behavioural responses to surgical or band castration (Moya et al. 2014; Petherick et al. 2014a, b). In contrast, Ting et al. (2003a, b) reported a significant improvement in abnormal standing postures and the cortisol response to surgical castration in 11-month-old bulls provided with 3 mg/kg ketoprofen intravenously, 20 minutes prior to castration, with or without follow-up doses of 1.5 mg/kg ketoprofen at the time of castration, and a further 3 mg/kg at 24 hours post procedure. No local anaesthetic protocol was described in the study. Ketoprofen ameliorated the cortisol response to burdizzo castration, and in conjunction with caudal epidural, pain-related behavioural responses were also reduced (Ting et al. 2003a, b). In conjunction with local infiltration of lidocaine, 3 mg/kg ketoprofen, intravenously, can further mitigate the cortisol response to surgical castration as compared to local anaesthesia alone (Earley and Crowe 2002; Stafford et al. 2002).

Phenylbutazone has been used and investigated in cattle (Sutherland et al. 2002) but is now banned by many regulatory authorities due to toxic metabolites that may remain in the tissues or be excreted into the environment (Bean et al. 2023). There has also been some limited investigation into the efficacy and pharmacokinetics of paracetamol or aspirin (acetylsalicylic acid or sodium salicylate) in calves, but equivocal outcomes in terms of efficacy, combined with rapid metabolism and excretion, suggest that these agents are unlikely to be of value in commercial cattle practice (Janus et al. 2003; Coetzee et al. 2007; Kotschwar et al. 2009; Baldridge et al. 2011; Bergamasco et al. 2011; Onda et al. 2012).

6.3.4 Alternative Analgesic Approaches

The subjective or psychological aspect of pain can be in some manner mitigated by distraction. Immediately following a handling event, e.g. castration, calves will return to their dam (assuming that they are still feeding from the dam) and suckle, whilst play and grooming have been shown to move the affective state towards the positive. Following caesarean section, it is the experience of the author that cows that have a live calf to suckle are brighter in demeanour and appetite returns more readily than cows that have no calf to suckle, who are more listless and inappetant in the days following surgery.

Topical vapocoolant sprays have been shown to ameliorate the pain response to ear tagging or notching (Lomax et al. 2017), but had no effect on the behavioural or ocular thermography responses to surgical castration in 2–4 month-old calves (Van der Saag et al. 2019).

6.3.5 Multimodal Analgesia

Analgesic protocols that utilise combinations of local anaesthesia, sedation and NSAID administration have been shown to provide superior, longer duration pain relief for dehorning (Lepkova et al. 2007; Stafford and Mellor 2011), disbudding (Stilwell et al. 2009, 2012; Herskin and Nielsen 2018; Stock et al. 2021; Weyl-Feinstein et al. 2021), surgical castration (Ting et al. 2003a, b; Currah et al. 2009; Vindevoghel et al. 2019; Martin et al. 2022) and clamp castration (Stilwell et al. 2008). Use of different, complementary agents in combination can allow reductions in the dose of any single agent, provide greater reductions in the physiological and behavioural response to pain, and prolong the duration of analgesia experienced by the animal. Therefore, a multimodal approach to any painful situation is recommended.

7 Causes of Pain in Cattle

Free-ranging aurochs and highland shorthorns would have encountered pain and suffering "naturally", i.e., through "misadventure", tearing muscles and tendons, becoming lame, perhaps breaking limbs. Fighting between animals establishing social structures would be particularly traumatic if the combatants were horned. Reproduction inherently involves pain and suffering resulting from the physical act of mating, the biomechanical adaptations required to carry calves to term, parturition itself, and the discomfort of a full udder being butted and sucked by a calf with developing dentition.

Free-living *bovidae* were, and remain, at risk from predators such as wolves or large cats. Predators hunting singly, pursue animals and leap on their backs before biting the neck and breaking it and, or severing the cervical blood vessels. When hunting in packs, predators separate single animals from the herd and, so as to conserve energy, target the slowest animals. Consequently, prey animals including *bovidae* are adept at disguising weakness, disabilities or pain. This evolutionary imperative poses a challenge to stockmen and researchers concerned with pain and suffering in *bovidae*, as behavioural changes are often subtle and difficult to discern.

Breeding, Conformation, Biomechanics and Health

Humans have manipulated animal breeding to meet their own requirements for centuries. Initially, this would have involved the selective mating of males and females from families with desirable traits, e.g. higher milk yields, larger udders, greater muscling, Over the last 80 years, hereditary principles and early statistical techniques have allowed more systematic methods of genetic selection, e.g. single-trait selection (Hazel 1943). Coupled with advances in reproductive technologies, e.g. artificial control of the reproductive cycle, artificial insemination, etc., the rate of genetic change in commercial cattle populations has accelerated. However, there is evidence that single-trait selection can have adverse effects, and while there have been marked improvements in some desirable traits, increases in undesirable traits have affected animal health and welfare. More recently, genetic improvement has focused on multi-trait selection processes (Sandoe et al. 1999; Lawrence et al. 2004) although it will probably take years to expunge some detrimental consequences of single-trait selection.

For example, in the dairy industry, milk yields have almost doubled in the last 80 years, and over 50% of that increase has been attributed to genetic selection (Oltenacu 2009). However, increased milk yield increases the prevalence of metabolic stress, lameness, mastitis and poor reproductive performance (Pryce et al. 1997; Rauw et al. 1998; Sandoe et al. 1999; Ingvartsen et al. 2003; Pryce et al. 2004).

Lameness is a major contributor to the culling of dairy cows (Esslemont and Kossaibati 1997; Mrode et al. 2000; Whitaker et al. 2000, 2004) and along with mastitis, accounts for much of the cost of running dairy enterprises (Kossaibati and Esslemont 1997; Bennett et al. 1999a, b; Bennett 2003; Orpin and Esslemont 2010). The conformation of modern high-yielding dairy cows, with large and pendulous udders suspended between the pelvic limbs, and the resultant biomechanical stresses elucidate at least four problems faced by these animals. To support the udder's mass and volume, the pelvic limbs are extended backwards, which leads to greater pressure on the front of the toe. In addition, the toes splay widely to support the extra weight. The udder's width pushes the pelvic limbs outwards and forces them to swing around the structure rather than track in a straight line. The greater forces applied to the medial limb and toe predispose to lameness. In dairy cattle, the prevalence of lameness is heavily influenced by environmental and management factors but is moderately heritable and so genetic selection can reduce the problem (Boettcher et al.

1998; Atkins et al. 2002; Pritchard et al. 2013; Weber et al. 2013).

Some breeds of beef cattle, e.g., Belgian White and Blues (BWB), Piedmontese or Limousin, have been selected for 'double muscling' which arises through inactivation of the myostatin gene (Fiems 2012). The carcases of these breeds have exceptional meat yields, but the mismatch between the giant foetus and the dam's narrow pelvic canal means that elective caesarean operation is required in up to 90% of calvings—at least in BWBs (Hanzen et al. 1994; Uystepruyst et al. 2002). The viability of foetuses born via caesarean section is maximized when surgery is delayed until the cervix is fully dilated, because this allows the activation of physiological cascades that optimise respiratory and metabolic adaptation to ex-utero life (Uystepruyst et al. 2002). Consequently, the dam experiences almost the full gamut of sensations and pain associated with parturition, in addition to sensations of surgery and any follow-up therapeutics that may be required. Further, it seems that double-muscled animals are more susceptible to conditions such as lameness, nutritional and metabolic stresses, heat stress, respiratory illness and urolithiasis (Fiems 2012).

The possession of horns in some cattle breeds may confer advantages when competing for mates or when fending off predators (Knierim et al. 2015). When horned animals exist within commercial herds there is an increased prevalence of carcase bruising and consequent economic loss (Meischke et al. 1974; Mendonça et al. 2016). There is also a greater risk of injury for stockmen working with horned cattle (Kling-Eveillard et al. 2015; Knierim et al. 2015). However, some believe horned cattle have better production characteristics than polled animals but this has yet to be proven or refuted (Knierim et al. 2015). One study indicated that polled Charolais bulls could be fatter than horned individuals (Stookey and Goonewardene 1996). Both the increased risk of injury to stockmen and other cattle has prompted significant efforts to develop polled genotypes of cattle (Schafberg and Swalve 2015; Windig et al. 2015; Randhawa et al. 2020; Schuster et al. 2020) although the physical removal of horns (dehorning) or immature horn buds (disbudding) remain commonly practiced husbandry procedures (Hoffsis 1995). Partial failures in breeding for polled animals or in dehorning/disbudding procedures can lead to aberrant horn growth (known as 'scurs'). Abnormally growing horn may penetrate facial tissues including the cheeks, leading to pain and the risk of infection.

Health, Illness and Injury

Cattle are susceptible to numerous microbial and parasitic illnesses; a genetic susceptibility to disease is well documented (Blecha et al. 1984; Muggli et al. 1987; Mrode et al. 1998; Engle et al. 1999; Heringstad et al. 2000; Ødegård et al. 2003; Mirkena et al. 2010; Aleri et al. 2016). Such illnesses may be transmitted from animal to animal, via insect vectors, or via the environment, e.g. soil, dust or feed. Associated pain depends on the body system affected and the inflammatory changes involved. While a detailed discussion of all bovine illnesses is beyond the scope of this chapter a few common conditions are worth reviewing.

An infection of mammary tissue (mastitis) can be intensely painful and contributes significantly to culling rates in dairy cattle (Bennett 2003; Bennett et al., 1999a, b; Esslemont and Kossaibati 1997; Kossaibati and Esslemont 1997; Mrode et al. 2000; Orpin and Esslemont 2010; Whitaker et al. 2000, 2004). Immune responses in infected and inflamed mammary tissue results in increased numbers of white blood cells in milk and constitute the 'somatic cell count' or SCC. A high SCC is indicative of mastitis, and there is a genetic correlation with both SCC and the incidence of mastitis (Mrode et al. 2000; Pritchard et al. 2013). It is unknown whether breeding selection for low SCCs reduces a cow's ability to mount an adequate immune response in the face of challenge, or affects the incidence of the disease and associated morbidities (Oltenacu 2009). Another common affliction more prevalent in dairy cattle is metritis. Infection of the uterus usually results from inadequate involution (the process of the uterus returning to its non-pregnant size and state) after calving and is often associated with retained foetal membranes.

Feedlot cattle are prone to respiratory illnesses, which often spread rapidly from animal to animal. Common underlying predisposing factors include infectious agents (which alone may not cause overt illness) and environmental conditions, e.g. e.g. dust and humidity. Clinical signs range from nasal exudate and catarrh to pneumonia and severe morbidity, and while the subjective experience of the animal is unknown and unmeasurable, its experience of pain and discomfort is likely to be similar to that suffered by humans.

Eye infections, colloquially known as 'pinkeye' (Fig. 9) also spread rapidly in housed or feedlot animals. While this is thought to be largely the result of their 'cheek-by-jowl' existence, it should be acknowledged that few conditions are confined to housed or feedlot animals and may be encountered under extended conditions. Similarly, traumatic injuries can arise under any management system as can others, e.g., 'cancer-eye' (Fig. 10), conditions affecting the gastrointestinal tract, balanoposthitis (inflammation of the penis and prepuce) (Fig. 11), urinary calculi, abscessation of the coronary band (Fig. 12).

The treatment of painful conditions may itself cause stress and pain. Antibiotic injections might involve a large drug volume necessitating the splitting of doses across two sites (Fig. 13). This means two skin punctures with muscular bruising is required—not one. Some drugs, e.g., ketamine, are tissue irritant and cause pain on injection. Some substances, e.g. penicillins, are so irritant that the local anaesthetic procaine is often added to antibiotic formulations.

Reproductive Problems

The physical act of coitus incurs risk of injury to both participants. The physical size and weight of a mature bull mounting a smaller female can cause lumbo-sacral damage, e.g., pelvis and lower back injury in the female (Fig. 14). Bulls may injure their forefeet or forelimbs on dismounting as the heavy forequarters make impact with hard ground. For these and other reasons, many enterprises now favour artificial insemination over natural mating. This approach accelerates the introduction of superior genetic material into the herd, while the risk of physical injuries is

Fig. 9 'Pinkeye' infection. (Source: Alison Small)

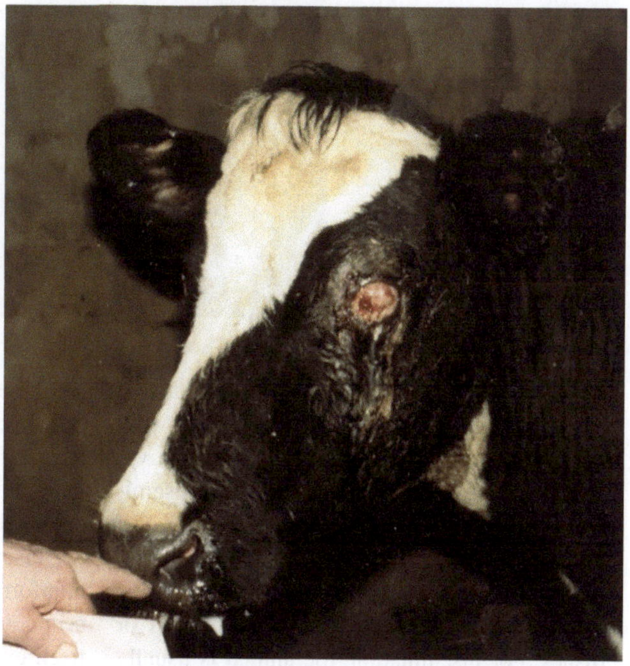

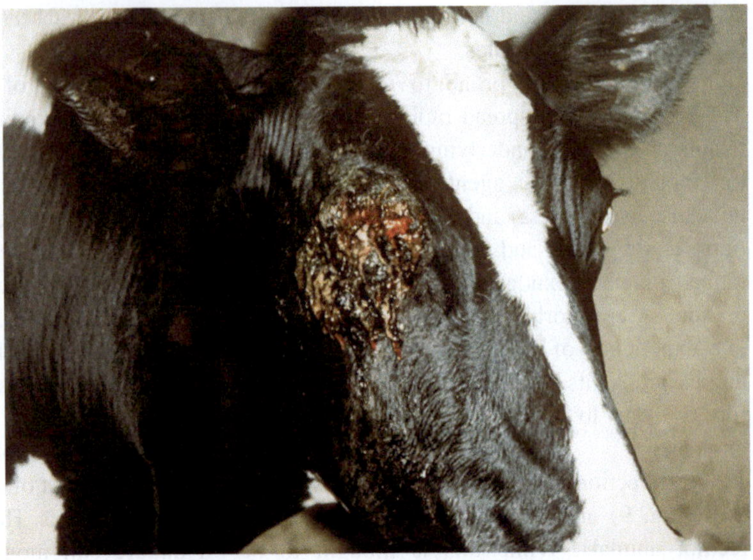

Fig. 10 'Cancer-eye': Squamous Cell Carcinoma of the eye. (Source: Alison Small)

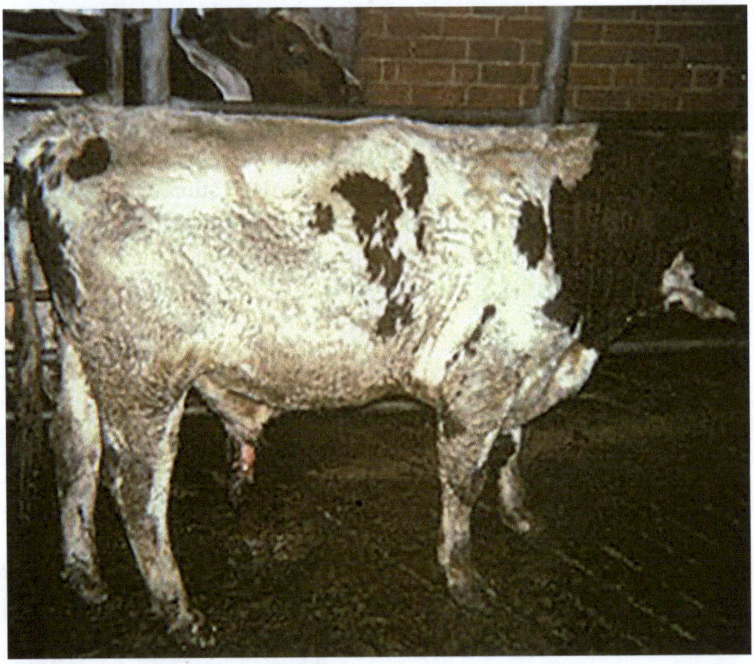

Fig. 11 Balanoposthitis (inflamed penis). (Source: Alison Small)

reduced. The risk to attendants is also reduced: bulls can be unpredictable and belligerent, particularly when cycling (sexually receptive) females are present. That said, artificial insemination brings its own challenges (Fig. 15). Oestrus is often induced using a series of intramuscular injections, each of which may involve stressful handling in restraint chutes, stocks or crushes. For insemination, the animal is usually restrained in a crush or stocks, creating a risk of bruising if struggling occurs and, or the devices are poorly designed and, or ill-fitting. Rectal palpation of the cervix is required for accurate semen deposition into the uterus. Lubrication of the inseminator's arm is required to reduce discomfort during the introduction of the hand and arm into the rectum. The insemination device itself can penetrate the vaginal wall if the animal

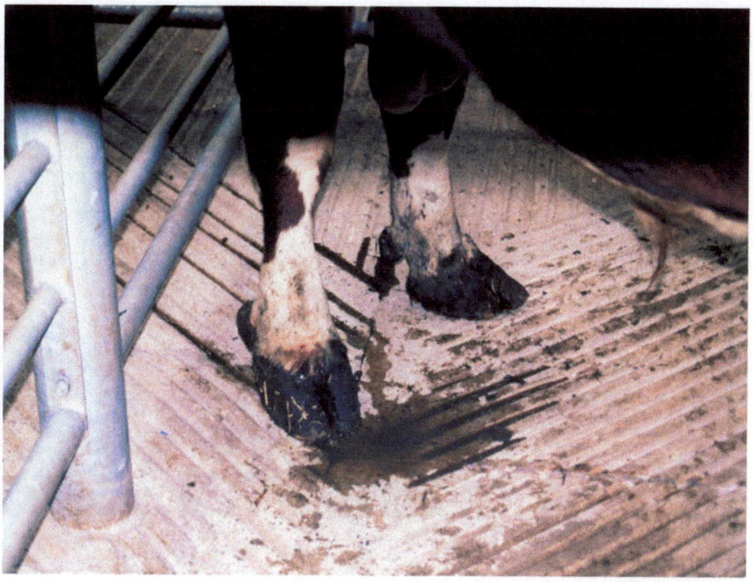

Fig. 12 Coronary band abscess. (Source: Alison Small)

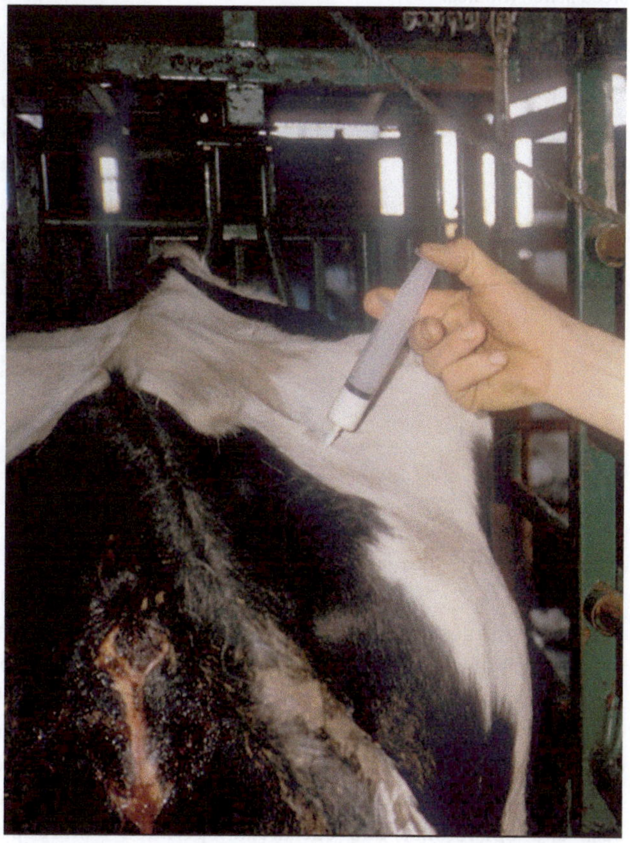

Fig. 13 Administering an injection of antibiotic to a cow suffering from metritis after a difficult calving. (Source: Alison Small)

Fig. 14 A cow with a broken hip after being mated by a heavy bull. (Source: Alison Small)

Fig. 15 Artificial insemination of a beef breed heifer. (Source: CSIRO)

moves violently in response to an inattentive inseminator.

The bovine gestation period lasts 9 months. In the terminal third (trimester), the udder prepares for lactation. Parturition is inherently painful (Gladden et al. 2021) as the head and body of the calf moves through and dilates the cervix. Human assistance may be required in this process and episiotomy and caesarean operation may also be required. Following parturition, the ligaments supporting the reproductive tract in the pelvic canal are stretched and lax, increasing the risk of vaginal and, or uterine prolapse (Fig. 16).

The normally engorged udder appears to be uncomfortable for cattle although mastitis is inescapably more painful. Calves instinctively nuzzle and butt the udder to stimulate milk let-down, before suckling on the teat. As the calf grows larger and stronger head-butting become more powerful (and uncomfortable). As teeth develop in the calf's lower jaw in preparation for grazing, the risk of udder and teat injuries increase as the calf hones its mastication skills.

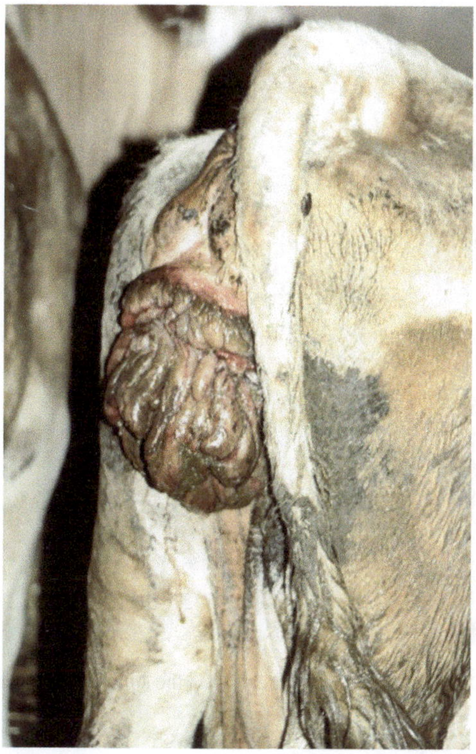

Fig. 16 Vaginal prolapse in a dairy cow. (Source Alison Small)

Metabolic Conditions

Metabolic disturbances are reasonably common in female cattle, particularly in late gestation when the increasing energy demands of the growing foetus drains maternal reserves (Ingvartsen and Andersen 2000; Ingvartsen et al. 2003; Ingvartsen 2006). At parturition, calcium, phosphorus and magnesium imbalances are common as these elements are diverted into lactation. When severe, these imbalances can lead to neurological signs, including weakness, recumbency, and if untreated, death (Littledike et al. 1981; Horst et al. 1997; Kossaibati and Esslemont 1997). The experience of cattle suffering from metabolic imbalances is difficult to discern, although human patients with similar conditions describe muscle cramping or 'burning'. Prolonged recumbency caused by altered nerve conduction and, or myasthenia will lead to reduced muscle blood flow—which is itself painful—and is more likely when the animal lies on unforgiving surfaces such as concrete. While the overt signs of these imbalances are noticeable as weakness, ataxia or recumbency, all body systems are affected. Uterine contractions themselves may be weakened, leading to prolonged labour (and exacerbating further imbalances) necessitating assisted birth. Poor uterine involution post-calving can result in retained foetal membranes and metritis.

The negative energy states some female cows experience in late gestation arise because late-term cows cannot consume sufficient energy to meet maternal and foetal requirements. However, positive energy balances can also occur—and cause pain and suffering. Excessive dietary carbohydrate leads to ruminal acidosis (a metabolic condition in which rumen pH drops below 5.8, which has negative consequences on digestion, appetite and health) and to laminitis. The latter is a painful condition of the foot in which the delicate laminae supporting the hoof capsule degenerate and collapse (Boosman et al. 1991; Vermunt and Greenough 1994; Ossent and Lischer 1998; Naert et al. 1999; Singh et al. 2005; Lean et al. 2013; Passos et al. 2023). Laminitis occurs in both grain-fed fattening cattle and in high-yielding dairy cows (Fig. 17).

Elective Husbandry Procedures

Husbandry procedures visited on cattle as a part of normal production and which are likely to be painful are: ear tagging and notching; branding; castration; disbudding and de-horning, tail docking, spaying and rerouting of the penis.

In many jurisdictions worldwide, animals must be individually identified. This can take the form of radio frequency identification (RFID) technology (electronic chips implanted under the skin, in rumen boluses or embedded in ear tags), ear tags or ear notching. RFID has the advantage that the chips are unique identifiers and can be read electronically, allowing automated alignment of animal identification and other measures (e.g. body weight), and is integral to many in-country traceability programs. Application of ear tags, implantation of subcutaneous RFID chips and ear notching cause brief but clear pain as the instrument punctures the skin, and in the case of ear tags and ear notching, the aural cartilage.

Fig. 17 Grain fed bulls with overgrown and laminitic claws. Discomfort became evident when the animals were transferred onto concrete after being reared on deep straw bedding. (Source: Alison Small)

Although this can be mitigated by the application of a vapocoolant (Lomax et al. 2017), e.g. an 'ice spray' whereby a chemical is applied to the area, and evaporation of that chemical produces a chilling or freezing sensation, reducing the pain sensation at the site. Hot-iron branding and freeze-branding are also widely used and both are acutely painful, both at the time of branding and during the subsequent healing process (that can take 70 days or more after hot-iron branding (Lay Jr et al. 1992, Schwartzkopf Genswein et al. 1997a, b, Watts and Stookey 1999, Lindegaard et al. 2009, Tucker et al. 2014a, b). Freeze branding appears to be less acutely painful than hot-iron branding, with freeze-branded calves showing lower plasma epinephrine concentration, lower heart rate, reduced vocalisation and reduced escape attempts than hot-iron branded calves (Lay Jr et al. 1992). No differences were found between the behaviours of freeze-branded or hot-iron-branded calves over a 10-day post-procedure observation period (Schwartzkopf Genswein et al. 1997a, b).

Male calves are often castrated to reduce their 'bullish' characteristics, e.g., aggression towards other cattle, humans and dogs, and to prevent misalliance. They will be left entire (not castrated) if they are destined to becoming breeding bulls, or if rapid growth is desired. Castrated male cattle are known as 'steers' or 'bullocks'. Castration may be performed surgically (Fig. 18), or by using a rubber ring or band applied around the neck of the scrotum (Figs. 20 and 21). The latter method relies on the physical constriction of blood vessels supplying the testes (ischaemia) such that the testicles shrink and become dysfunctional (atrophy). The Burdizzo™ or 'bloodless castrator' (Fig. 7) also causes testicular atrophy but by crushing the nerves and blood vessels, rather than cutting them. All castration methods are painful but elicit differing behavioural and physiological responses; because of this, claims that one method may be superior to others have been made. However, technique comparisons have not been like-for-like, and irrespective of any objective measurements that have been taken, the animal's subjective experience remains unknowable. One key difference between surgical and ischaemic castration methods is the slower resolution of the wounds following band

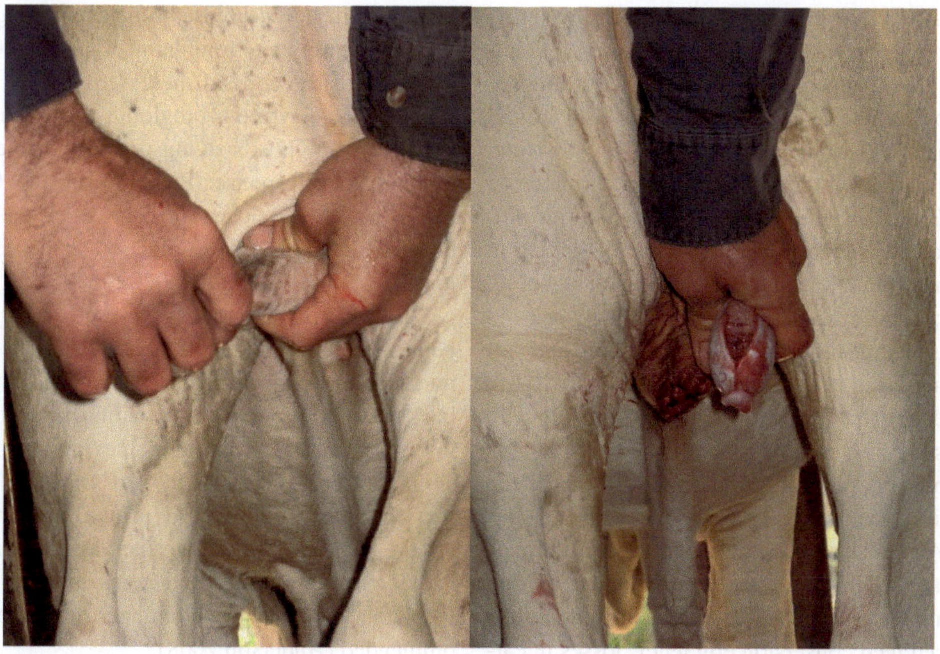

Fig. 18 Surgical castration. A scalpel is used to open the scrotum (left) before extracting the testes (right). (Source CSIRO)

or ring castration: unhealed lesions, physiological and behavioural changes have been observed for some 2–4 weeks after band or ring castration in cattle (Warnock et al. 2012; Petherick et al. 2014a, b, 2015).

Horns are often removed from unpolled breeds either early in life, before the horn bud becomes fixed to the skull (disbudding) or later, by removal of the horn at its base (Hoffsis 1995). Disbudding involves removing or destroying the cells that produce the horn, before the horn attaches to the skull, and can be carried out by scoop knife, hot-iron cautery (Fig. 8) or application of caustic paste; while dehorning involves cutting off the horn at its base, using specially designed scoop dehorners or wire saw. Both processes are considered to be extremely painful, resulting in alterations to behaviour, appetite and physiological measures (Stafford and Mellor 2005; Huxley and Whay 2006; Stilwell et al. 2010; Fajt et al. 2011; Stafford and Mellor 2011; Sutherland et al. 2013; Wikman et al. 2013; Wikman et al. 2016; Adcock et al. 2020; Kongara et al. 2020). In many jurisdictions, local anaesthesia must be provided, but the enervation of the horn is complex, particularly in the *Bos indicus* cattle, and local anaesthesia alone is only partially effective at ameliorating the pain (Doherty et al. 2007; Herskin and Nielsen 2018; Winder et al. 2018). Incorporation of an NSAID, e.g. meloxicam, ketoprofen, carprofen, flunixin or firocoxib, has been shown to provide greater reductions in pain related behavioural and physiological responses (Allen et al. 2013; Ballou et al. 2013; Bates et al. 2015; Duffield et al. 2010; Faulkner and Weary 2000; Heinrich et al. 2007, 2009, 2010; Huber et al. 2013; Karlen et al. 2021; Milligan et al. 2004; Mintline et al. 2013; Qi et al. 2023; Reedman et al. 2020; Stewart et al. 2009, 2008a, b; Stilwell et al. 2009, 2012; Stock et al. 2015; Weyl-Feinstein et al. 2021).

Tail docking was once commonly performed in dairy cows, particularly in New Zealand and the United States (Aubry 2005) in the belief that the removal of the hairy tail 'switch' would lead to cleaner udders, improved milk hygiene, reductions in mastitis incidence: and perhaps could reduce the risk of leptospirosis in farm workers. However, large-scale studies have failed to confirm this (Tucker et al. 2001; Schreiner and Ruegg

2002a, b; Stull et al. 2002) and the practice is becoming less common. Several European countries have banned the procedure. Tail docking was performed by applying a rubber ring around one hand's breadth below the tip of the vulva and allowing the posterior end of the tail to necrose and fall off. There seems to be minimal physiological or behavioural response to tail docking in pregnant heifers or lactating cows (Eicher et al. 2000; Tom et al. 2002a, b), but 3–10-week-old calves demonstrate increased agitation and tail flicking (Tom et al. 2002a, b; Mayer et al. 2020), suggesting that the animals are experiencing pain or discomfort.

Spaying, removal of the ovaries, has been practiced for over 150 years in the United States and Australia to prevent pregnancy in animals that are destined for slaughter. Pregnancy and calving leads to greater decreases in value due to reduced feed conversion efficiency, losses in carcase weight and increased risk of mortality related to dystocia (inability to successfully expel the foetus). (Jubb et al. 2003) Spaying is therefore regarded as necessary in wide range systems where feral bulls are roaming, and the cows are destined for slaughter rather than breeding (Restle and Grassi 1993; Plascencia et al. 2008). Spaying may be carried out *per vaginam* using the Willis Dropped ovary technique (Jubb et al. 2003) or via a flank laparotomy (Petherick et al. 2011) (Fig. 19). It is often performed without anaesthesia or analgesia, and spayed cattle demonstrate clear pain-related behavioural responses in the first 24 hours after surgery: stiff, arched posture; stiff tail; decreased time spent lying and decreased time spent ruminating (Yu et al. 2020). Physiologically, cortisol increased sharply after spaying, and elevated haptoglobin concentrations over the 4-day post-spaying period indicated an inflammatory response (Petherick et al. 2011). Provision of meloxicam at 0.5 mg/kg at the time of spaying was shown to ameliorate the behavioural responses (Yu et al. 2020).

The penis is anatomically 're-routed' in young bulls which are to become 'teaser' animals, i.e., intact males which can incite cows to peak 'standing' oestrus before natural mating or artificial insemination, but without achieving intromission. The penis is diverted towards the flank, such that the teaser animal can urinate and mount cows—but cannot penetrate the vagina. For any surgery to the prepuce and penis, sedation and local anaesthesia or general anaesthesia are recommended (Anderson 2008).

8 Pain Assessment

Recognizing pain in cattle—as in other predated taxa—can be difficult. Furthermore, the resources available to assist recognition are not always available in the environments where cattle are to be found. Pain causes species- and age-specific behavioural and biological changes which can be estimated objectively and subjectively. Physiological measures include heart rate, which increases and stays elevated for nearly 4 hours after dehorning in cattle (Grondahl-Nielsen et al.

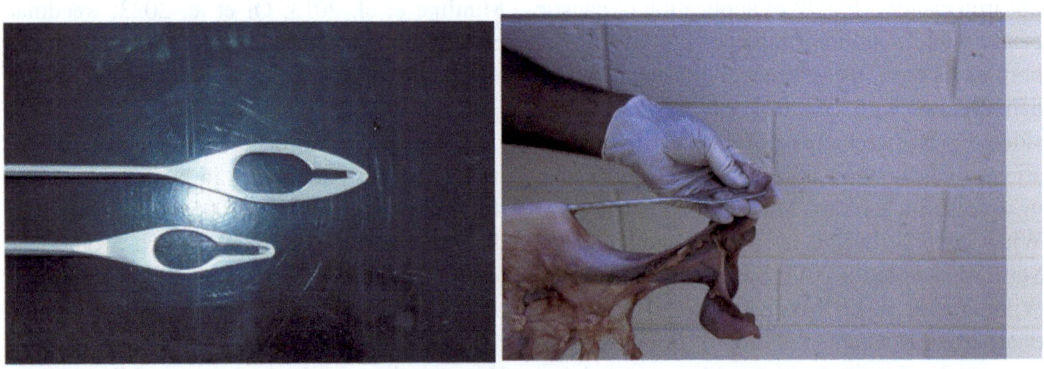

Fig. 19 Willis technique. (Photo credit Dr. Tristan Jubb)

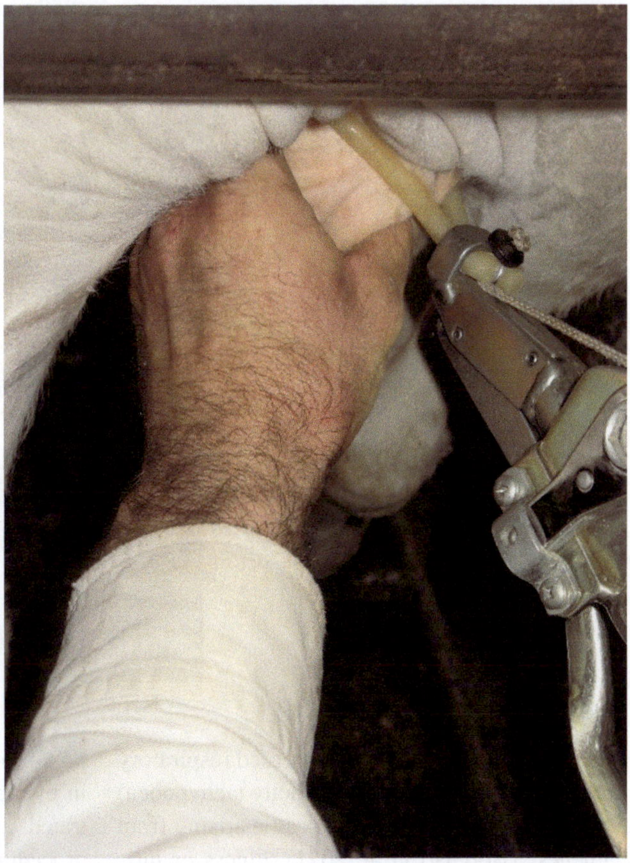

Fig. 20 Applying a Callicrate® band for ischaemic castration of a young bull. (Source CSIRO)

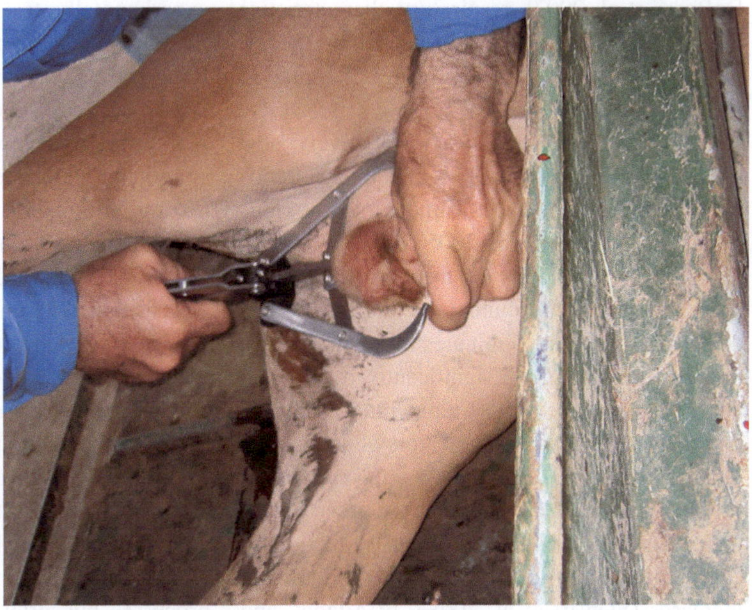

Fig. 21 Applying a TriBander® rubber ring for ischaemic castration of a calf. (Source CSIRO)

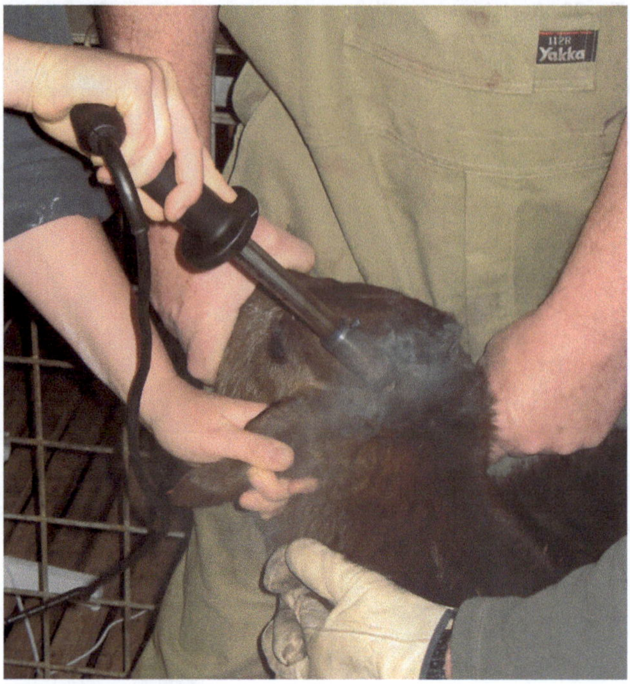

Fig. 22 Disbudding: using a hot iron to cauterise and remove the horn bud. (Source CSIRO)

1999; Heinrich et al. 2009). Increased respiratory rates—known as "blowing" (strictly tachypnoea) is a useful indicator of pain in cattle arising from lameness (Borghart et al. 2021), illness or husbandry practices. Blood pressure (Lin et al. 2003; DeRossi et al. 2007), respiratory characteristics (Heinrich et al. 2009) and increases in body temperature also occur in response to inflammation (Marti et al. 2010a, b).

Infra-red thermography is a relatively new method that has been used in attempts to facilitate pain recognition in cattle undergoing noxious procedures (Stewart et al. 2005, 2008a, b). Using infra-red thermography of the corneal surface, Stewart et al. (2010) found that eye temperatures decreased during the 5 minutes following castration but then increased significantly for at least 20 minutes. These ocular temperature changes corresponded with other indicators of acute pain, including cortisol and heart rate variability. Similar findings have been reported for calves undergoing disbudding: an initial reduction in eye temperature was followed by significant increases (Stewart et al. 2008a, b; Adcock and Tucker 2020). Unfortunately, the sensitivity of this method is limited to the acute (immediate post-procedural) phase, with minimal differences in eye temperature being detected between castrated and handled calves after only 2 post-procedural hours (Kleinhenz et al. 2018).

Electroencephalography (EEG) has been used experimentally to estimate pain perception in livestock (Jongman et al. 2000; Haga and Ranheim 2005; Gibson et al. 2007; Harris et al. 2020). In cattle undergoing 'scoop' dehorning with or without nerve blocks, the EEG revealed significant increases in cortical activity upon noxious stimulation in animals not receiving the nerve block (Gibson et al. 2007). While the EEG can provide an objective estimate of pain-related electrical activity under experimental conditions, its use is limited because of equipment requirements, the need for animals to be almost anesthetised, and the availability of an adept electroencephalographer.

Blood and salivary biomarkers, particularly those related to the hypothalamic-pituitary-adrenal (HPA) axis have been used to gauge the magnitude of the biological response to noxious, processes, but suffer from an inability to discriminate pain from stress and, or anxiety (Fisher et al. 1996; Grondahl-Nielsen et al. 1999; Mellor et al. 2002; Peers et al. 2002; Heinrich et al. 2009; Gonzalez et al. 2010; Petherick et al. 2014a, b; Paull et al. 2015). Catecholamines (Mellor et al. 2002), endorphins (Cooper et al. 1995), haematology (Peers et al. 2002; Petherick et al. 2014a, b; Paull et al. 2015) and neuro-immunological markers (Gonzalez et al. 2010; Johnston et al. 2022) have all been used to gain insights into the presence or absence of pain in cattle.

Some advanced methods have been used in *bovidae* in attempts to quantify responses to noxious stimulation. For example, the gene expression of inflammatory cytokines has been examined after castration (Pang et al. 2011). Morphological signs of neuropathology, e.g. the diameter of cornual and infraorbital nerve axons, or expression of glial fibrillary acidic proteins in response to disbudding, have also been explored (Vidondo et al. 2019). In both studies, no differences between animals that underwent the procedure and those that did not were identified. Some biomarkers, particularly those used as indices of oxidative stress, are reported to alter with pain (Korkmaz et al. 2015). Cattle that have undergone dehorning have increased circulating concentrations of: thiobarbituric acid–malondi aldehyde (TBA-MDA); nitric oxide (NO); plasma antioxidant activity (AOA); cortisol; and significant decreases in glutathione (GSH) and cholesterol. The study revealed MDA, NO, AOA and GSH were the most powerful markers for evaluating the oxidant/antioxidant status in mature cattle (Fidan et al. 2010), but how this translates to pain assessment is yet to be established.

Most methods of gauging physiological pain are suitable for research use and of little value to stockmen observing animals on a day-to-day basis. These rely on changes in behaviour as the principal measure for identifying pain (Table 1). However, as a prey species, cattle's pain-related behaviours can be subdued. Early signs of pain may be limited to reduced activity, reduced appetite and a loss of interest in their surroundings. As

Table 1 Behavioural indicators commonly used to assess pain in calves or cattle

Parameter	Context
Altered time spent performing an activity	
Lying	Lying may be increased if animals have sore feet, or decreased in animals that have been castrated
Play	Play is decreased when animals are in pain. Most commonly seen in calves.
Walking	Lame animals, and animals with laminitis are reluctant to walk. Calves that have been surgically castrated may spend more time standing still and less time walking
Grooming	Animal often lick or groom injuries, so calves that have been castrated may spend more time grooming at their scrotum and flank
Feeding	Animal that are in pain may show limited interest in food. Conversely, very young calves that are suckling form their dam may suckle immediately after a painful procedure as a comfort-seeking behaviour
Ruminating	Cattle that are in pain are likely to show reduced rumination
Altered posture	
Arched back	Associated with abdominal pain; the animal may also tuck its tail
Tail stiffly extended	Associated with tail docking in adult cows
Statue standing	Calves that have been surgically castrated may stand motionless with their hind legs slightly extended backwards
Lateral lying	Cattle that are in pain may spend more time lying flat than in a sternal recumbency
Other abnormal postures	This can include walking with stiff legs; lying in sternal recumbency but in a straight line with the head and neck extended; lying in a twisted posture with forelegs and hind legs pointing in different directions; standing with one foot raised; Standing with head down
Acute pain avoidance behaviours	
Tail flicking	Increases are associated with tail docking and castration in calves
Rolling	Writhing and rolling is associated with band castration in calves

(continued)

Table 1 (continued)

Parameter	Context
Fidgeting or easing quarters	Calves that have been castrated may shift from hind foot to hind foot, or 'squat' slightly with legs apart, attempting to relieve discomfort felt at the scrotum
Foot stamping or kicking	Calves that have been castrated may stamp or kick out with their hind feet
Head shaking	Calves that have been disbudded or dehorned may shake their heads more often than normal
Ear flicking	Calves that have been disbudded or dehorned may flick their ears more often than normal.
Gait abnormalities	Animals that have pain in one foot will show an uneven gait (limping). This is less obvious when multiple feet are affected: a shortened or 'popping' stride may be evident in laminitic animals.

pain becomes more severe, visible changes in respiration and rumination may be observed, and localised pain may become evident, e.g., in the form of lameness. The behavioural responses to acute pain also differ according to the location of the pain. During dehorning, disbudding, branding or castration, the animal may jump, kick out or vocalise (Schwartzkopf Genswein et al. 1997b; Graf and Senn 1999; Grondahl-Nielsen et al. 1999; Watts and Stookey 2001), but after the initial insult, behavioural changes become less ostentatious. For example, after dehorning, calves may shake their heads, or appear to be scratching or grooming, and some authors have described increased lying and kicking (McMeekan et al. 1999; Sylvester et al. 2004; Stafford and Mellor 2005). Cows that have been tail-docked will shake and hold their tails in different postures compared to those that are not tail-docked (Tom et al. 2002a, b). Following castration, the post-surgical response may be reduced activity, altered posture, or weight-shifting, which is typically slow and occurs from one hind foot to the other (Molony et al. 1995; Gonzalez et al. 2010; Devant et al. 2012; Paull et al. 2015). The subtlety of pain signs, occasionally imperceptible may have previously created the impression that cattle do not feel pain, and that analgesic therapy is not required (Howard and Beattie 2018).

In comparison to other cattle rearing systems, the process of modern milk production offers considerable opportunities for the early and accurate detection pain. The high value of dairy cattle coupled with the necessity for production efficiency, has seen dairy farmers employ various technologies, e.g., the MooMoniters⁺, SenseHub™ and automated milking systems, that allow subtle deviations in the behaviours and production of individual animals to be detected, allowing direct examination and early intervention. Using data collected with such technologies, Lemmens et al. (2023) created a system allowing the early detection of lameness. After castration, young beef animals spend less time lying but are less active in the 10 days following the procedure—all of which can be recorded by suitable monitoring equipment (Fan et al. 2022). There are numerous devices currently available for use in the research and commercial settings that can record behavioural time budgets for rumination, lying, standing and walking (Marti et al. 2010a; Devant et al. 2012).

Qualitative Behavioural Assessment (QBA) and facial grimace scales are being used to quantify pain in *bovidae*. QBA is an approach whereby observers view an animal's posture and overall demeanour before assigning descriptive terms, e.g. calm, withdrawn, curious, anxious, etc., to an individual or group of animals. The observer is often given a list of described terms i.e., "withdrawn" animals appear to not be interacting with others or the environment, which they then assign to a visual analogue scale indicating their agreement on viewing the behaviour. This information is then used to define an animal's welfare state (Wemelsfelder et al. 2001). QBAs have been extensively used in experimental settings and have been sensitive enough to differentiate between castrated cattle that did or did not receive pain relief (Kleinhenz et al. 2018). The use of QBAs often requires observer training to reduce inter-observer variation (Bokkers et al. 2012). The use of QBAs in studying cattle behaviour—either directly or through video recordings—has shown strong agreement in assignment of terms, making it a potentially useful tool for remote monitoring where live observation may be impractical (Cooke et al. 2022).

Facial grimace scales (FGSs) were originally developed to assist pain quantification in laboratory mice, and have since been developed for horses, sheep and cattle. Facial expressions convey emotion, and it has been observed in animals that certain facial action units, e.g., eyes, cheeks, nose, etc., change when animals experience pain. A few studies have investigated grimace scales in cattle. For example, Muller et al. (2019) assessed cattle FGSs during hot iron branding, while Gleerup et al. (2015) assessed them in animals with lameness. Attempts to identify which facial changes identify pain in cattle indicate that backwards ear position, dilated nostrils, open mouth, raised brows and a low-carried head position were the most useful.

More complex behavioural techniques have been used to gauge the emotional effect of pain on animals. For example, both conditioned place preference (CPP) and judgement bias testing can be used to determine the duration of pain, the effectiveness of analgesics and the effect of pain on individual animals. That cattle can associate their environment with affective experiences is being used with CPP testing, where cattle are conditioned to two locations with distinct visual cues—one linked to a positive experience, e.g. non-addictive drug treatment, and the other to a neutral control (Ede et al. 2019). After a painful procedure, the time spent in each location reflects their pain experience. While CPP is impractical for day-to-day on-farm application, it is a useful method for identifying pain duration and the efficacy of analgesics.

Judgement bias follows a similar premise: animals are trained to approach a cue or location of a positive reward, e.g., feed reward and avoid the cue or location of a negative reinforcer, e.g., time out, air puffed in the face (Baciadonna and McElligott 2015). They are then offered an ambiguous cue or location between the positive and negative. Their willingness to approach the ambiguous cue/location that is close to the negative reinforcer is seen to represent an optimistic state, i.e., the animals expect to be rewarded in the ambiguous location. Judgement bias requires extensive training in which animals are first habituated to the test setup and approaching a cue (normally the cue associated with the positive reward) to determine willingness to participate and teaching animals to 'go' prior to then introducing the negative and training the 'no-go' (Kremer et al. 2022). The use of judgement bias in dairy calves has involved training calves to either approach or touch a screen; they were then trained to associate a colour cue with a go/no go task (Daros et al. 2014). In the study by Daros et al. 2014, calves were trained to a white screen indicating a reward and a red screen indicating punishment (a time out), following training they were tested with ambiguous cues (colours in-between white and red). When separated from their dam calves were less likely to display a go response when given an ambiguous cue indicating a negative affective state (pessimistic judgement in likelihood of receiving a reward). Judgement bias has been tested on dairy calves following disbudding and revealed similar results, with disbudded calves more likely to exhibit pessimistic biases, indicating negative affect (Neave et al. 2013).

9 Conclusion

Intensively managed animals found in dairy enterprises, feedlots or veal calf rearing units can be monitored daily, which provides the opportunity for those responsible to identify pain-related behavioural changes. This exposure contrasts with the situation with free-living rangeland animals, e.g., in northern Australia, or the rangelands of Canada and the United States. Furthermore, access to therapeutic agents varies widely and impacts the frequency of their use. Local anaesthetics and NSAIDs are generally widely available and registered or licensed for cattle use in several jurisdictions—but not in every cattle-producing country. Where analgesics are available, there may be restrictions in supply: veterinary prescription may be required; indeed, access to veterinary support may be restricted in some areas. The cost of therapeutic agents may be prohibitive for some cattle owners.

Furthermore, when 'natural' causes of pain and suffering arise, the question of treatment thresholds arises. Some 'natural' pain processes are required physiologically: parturition causes a cascade of changes in both the dam and offspring that are necessary for optimal dam-offspring bonding and survival of the neonate (Uystepruyst et al. 2002; Surlis et al. 2017).

Increasing the number of cattle receiving appropriate analgesic therapy for both elective procedures and for 'natural' pain requires further education and extension work. Further research into the financial cost of pain in cattle is needed as a prerequisite to cost-effective pain management in the commercial environment.

References

Adcock SJJ, Tucker CB (2020) The effect of early burn injury on sensitivity to future painful stimuli in dairy heifers. PLoS One 15(6):e0233711

Adcock SJJ, Cruz DM, Tucker CB (2020) Behavioral changes in calves 11 days after cautery disbudding: effect of local anesthesia. J Dairy Sci 103(9):8518–8525

Aleri JW, Hine BC, Pyman MF, Mansell PD, Wales WJ, Mallard B, Fisher AD (2016) Periparturient immunosuppression and strategies to improve dairy cow health during the periparturient period. Res Vet Sci 108:8–17

Allen KA, Coetzee JF, Edwards-Callaway LN, Glynn H, Dockweiler J, KuKanich B, Lin H, Wang C, Fraccaro E, Jones M, Bergamasco L (2013) The effect of timing of oral meloxicam administration on physiological responses in calves after cautery dehorning with local anesthesia. J Dairy Sci 96(8):5194–5205

Anderson DE (2008) Surgery of the prepuce and penis. Vet Clin Food Anim Pract 24(2):245–251

Anderson KL, Smith AR, Shanks RD, Davis LE, Gustafsson BK (1986a) Efficacy of flunixin meglumine for the treatment of endotoxin-induced bovine mastitis. Am J Vet Res 47(6):1366–1372

Anderson KL, Smith AR, Shanks RD, Whitmore HL, Davis LE, Gustafsson BK (1986b) Endotoxin-induced bovine mastitis – immunoglobulins, phagocytosis, and effect of flunixin meglumine. Am J Vet Res 47(11):2405–2410

Andreasen SN, Sandoe P, Waiblinger S, Forkman B (2020) Negative attitudes of Danish dairy farmers to their livestock correlates negatively with animal welfare. Anim Welf 29(1):89–98

Antanaitis R, Januskauskas A, Rutkauskas A, Zilinskas H (2018) The effect of post-calving pain management with carprofen on stress, reproduction, and milk yield in cattle. Acta Vet Brno 87(2):115–118

Atkins G, Shannon J, University A (2002) Minimizing lameness through genetic selection. Western Canadian dairy seminar (WCDS), Red Deer, CANADA

Aubry P (2005) Routine surgical procedures in dairy cattle under field conditions: Abornasal surgery, dehorning, and tail docking. Vet Clin North Am Food Anim Pract 21(1):55

Baciadonna L, McElligott AG (2015) The use of judgement bias to assess welfare in farm livestock. Anim Welf 24(1):81–91

Baldridge SL, Coetzee JE, Dritz SS, Reinbold JB, Gehring R, Havel J, Kukanich B (2011) Pharmacokinetics and physiologic effects of intramuscularly administered xylazine hydrochloride-ketamine hydrochloride-butorphanol tartrate alone or in combination with orally administered sodium salicylate on biomarkers of pain in Holstein calves following castration and dehorning. Am J Vet Res 72(10):1305–1317

Ballou MA, Sutherland MA, Brooks TA, Hulbert LE, Davis BL, Cobb CJ (2013) Administration of anesthetic and analgesic prevent the suppression of many leukocyte responses following surgical castration and physical dehorning. Vet Immunol Immunopathol 151(3–4):285–293

Balmer TV, Williams P, Selman IE, Vet ABC (1996) Clinical efficacy of carprofen as adjunctive therapy in calf pneumonia. XIX World Buiatrics Congress, Edinburgh

Bates AJ, Eder P, Laven RA (2015) Effect of analgesia and anti-inflammatory treatment on weight gain and milk intake of dairy calves after disbudding. N Z Vet J 63(3):153–157

Bean TG, Chadwick EA, Herrero-Villar M, Mateo R, Naidoo V, Ratter BA (2023) Do pharmaceuticals in the environment pose a risk to wildlife? Environ Toxicol Chem. https://doi.org/10.1002/etc.5528

Bennett R (2003) The 'direct costs' of livestock disease: the development of a system of models for the analysis of 30 endemic livestock diseases in Great Britain. J Agric Econ 54(1):55–71

Bennett RM, Christiansen K, Clifton-Hadley RS (1999a) Direct costs of endemic diseases of farm animals in Great Britain. Vet Rec 145(13):376–377

Bennett RM, Christiansen K, Clifton-Hadley RS (1999b) Estimating the costs associated with endemic diseases of dairy cattle. J Dairy Res 66(3):455–459

Bergamasco L, Coetzee JF, Gehring R, Murray L, Song T, Mosher RA (2011) Effect of intravenous sodium salicylate administration prior to castration on plasma cortisol and electroencephalography parameters in calves. J Vet Pharmacol Ther 34(6):565–576

Bernay F, Davot JL, Boisrame B (2000) Tolfenamic acid compared with meloxicam and Carprofen in combined treatment of bovine respiratory disease in beef cattle. Ir Vet J 53(9):468

Blecha F, Boyles SL, Riley JG (1984) Shipping suppresses lymphocyte blastogenic responses in Angus and Brahman x Angus feeder calves. J Anim Sci 59(3):576–583

Boettcher PJ, Dekkers JCM, Warnick LD, Wells SJ (1998) Genetic analysis of clinical lameness in dairy cattle. J Dairy Sci 81(4):1148–1156

Bokkers EAM, de Vries M, Antonissen I, de Boer IJM (2012) Inter- and intra-observer reliability of experienced and inexperienced observers for the qualitative behaviour assessment in dairy cattle. Anim Welf 21(3):307–318

Boorstin DJ (1973) The Americans, the democratic experience. Random House, New York

Boosman R, Nemeth F, Gruys E (1991) Bovine laminitis—clinical aspects, pathology and pathogenesis with reference to acute equine laminitis. Vet Q 13(3):163–171

Borghart GM, O'Grady LE, Somers JR (2021) Prediction of lameness using automatically recorded activity, behavior and production data in post-parturient Irish dairy cows. Ir Vet J 74(1):4

Braudel F (1974) Capitalism and material life, 1400–1800. Fontana, London, p 462

Braz M, Carreira M, Carolino N, Rodrigues T, Stilwell G (2012) Effect of rectal or intravenous tramadol on the incidence of pain-related behaviour after disbudding calves with caustic paste. Appl Anim Behav Sci 136(1):20–25

Browne N, Conneely M, Hudson C (2022) Use of non-steroidal anti-inflammatory drugs and attitudes to pain in pasture-based dairy cows: a comparative study of farmers and veterinarians. Front Vet Sci 9:912564

Carroll GL, Hartsfield SM (1996) General anesthetic techniques in ruminants. Vet Clin North Am Food Anim Pract 12(3):627

Chibisa GE, Vinyard JR, Laarman AH (2018) Effects of meloxicam administration on protein metabolism and growth performance in transported Jersey calves. J Dairy Sci 101(12):11435–11440

Clarke KW, Trim CM, Hall LW (2014) Veterinary Anaesthesia, 11th edn. W.B. Saunders, Oxford

Coetzee JF, Gehring R, Bettenhausen AC, Lubbers BV, Toerber SE, Thomson DU, Kukanich B, Apley MD (2007) Attenuation of acute plasma cortisol response in calves following intravenous sodium salicylate administration prior to castration. J Vet Pharmacol Ther 30(4):305–313

Coetzee JF, Gehring R, Tarus-Sang J, Anderson DE (2010) Effect of sub-anesthetic xylazine and ketamine ('ketamine stun') administered to calves immediately prior to castration. Vet Anaesth Analg 37(6):566–578

Coetzee JF, Mosher RA, Kohake LE, Cull CA, Kelly LL, Mueting SL, KuKanich B (2011) Pharmacokinetics of oral gabapentin alone or co-administered with meloxicam in ruminant beef calves. Vet J 190(1):98–102

Coetzee JF, Edwards LN, Mosher RA, Bello NM, O'Connor AM, Wang B, KuKanich B, Blasi DA (2012) Effect of oral meloxicam on health and performance of beef steers relative to bulls castrated on arrival at the feedlot. J Anim Sci 90(3):1026–1039

Coetzee JF, Mosher RA, Anderson DE, Robert B, Kohake LE, Gehring R, White BJ, KuKanich B, Wang C (2014) Impact of oral meloxicam administered alone or in combination with gabapentin on experimentally induced lameness in beef calves. J Anim Sci 92(2):816–829

Conrad JR (1973) The horn and the sword: the history of the bull as symbol of power and fertility. Greenwood Press, Westport

Cooke AS, Mullan SM, Morten C, Hockenhull J, Lee MRF, Cardenas LM, Rivero MJ (2022) V-QBA vs. QBA-how do video and live analysis compare for qualitative behaviour assessment? Front Vet Sci 9:832239

Cooper C, Evans ACO, Cook S, Rawlings NC (1995) Cortisol, progesterone and beta-endorphin response to stress in calves. Can J Anim Sci 75(2):197–201

Cull CA, Rezac DJ, DeDonder KD, Seagren JE, Cull BJ, Singu VK, Theurer ME, Martin M, Amachawadi RG, Kleinhenz MD, Lechtenberg KF (2022) Behavioral and performance response associated with administration of intravenous flunixin meglumine or oral meloxicam immediately prior to surgical castration in bull calves. J Anim Sci 100(5):skac049

Currah JM, Hendrick SH, Stookey JM (2009) The behavioral assessment and alleviation of pain associated with castration in beef calves treated with flunixin meglumine and caudal lidocaine epidural anesthesia with epinephrine. Can Vet J/Revue Veterinaire Canadienne 50(4):375–382

Daros RR, Costa JHC, von Keyserlingk MAG, Hötzel MJ, Weary DM (2014) Separation from the dam causes negative judgement bias in dairy calves. PLoS One 9(5):e98429

Dary D (1981) Cowboy culture: a saga of five centuries. Knopf, New York

De Koster J, Tena JK, Stegemann MR (2022) Treatment of bovine respiratory disease with a single administration of tulathromycin and ketoprofen. Vet Rec 190(4):e834

Deleforge J, Thomas E, Davot JL, Boisrame B (1994) A field-evaluation of the efficacy of tolfenamic acid and oxytetracycline in the treatment of bovine respiratory-disease. J Vet Pharmacol Ther 17(1):43–47

DeRossi R, Almeida RG, Medeiros U, Righetto FR, Frazilio FO (2007) Subarachnoid butorphanol augments lidocaine sensory anaesthesia in calves. Vet J 173(3):658–663

Devant M, Marti S, Bach A (2012) Effects of castration on eating pattern and physical activity of Holstein bulls fed high-concentrate rations under commercial conditions. J Anim Sci 90(12):4505–4513

Doherty TJ, Kattesh HG, Adcock RJ, Welborn MG, Saxton AM, Morrow JL, Dailey JW (2007) Effects of a concentrated lidocaine solution on the acute phase stress response to dehorning in dairy calves. J Dairy Sci 90(9):4232–4239

Duffield TF, Heinrich A, Millman ST, DeHaan A, James S, Lissemore K (2010) Reduction in pain response by combined use of local lidocaine anesthesia and systemic ketoprofen in dairy calves dehorned by heat cauterization. Can Vet J/Revue Veterinaire Canadienne 51(3):283–288

Durand D, Faure M, Lamberton P, Lemosquet S, des Roches AD (2021) A multiparametric approach to assessing residual pain experienced by dairy cows undergoing digestive tract surgery under multimodal analgesia. Animal 15(9):100338

Earley B, Crowe MA (2002) Effects of ketoprofen alone or in combination with local anesthesia during the castration of bull calves on plasma cortisol, immunological, and inflammatory responses. J Anim Sci 80(4):1044–1052

Ede T, von Keyserlingk MAG, Weary DM (2019) Assessing the affective component of pain, and the efficacy of pain control, using conditioned place aversion in calves. Biol Lett 15:20190642. https://doi.org/10.1098/rsbl.2019.0642

Edwards-Callaway LN, Keller KP, Oselinsky K, Johnstone E, Cramer C, Román-Muñiz N, Stallones L, Coetzee JF (2023) A nationwide survey on producer and veterinarian perceptions of the painfulness of procedures and disease states in dairy and beef cattle. Front Pain Res 4:1059224

Eicher SD, Morrow-Tesch JL, Albright JL, Dailey JW, Young CR, Stanker LH (2000) Tail-docking influences on behavioral, immunological, and endocrine responses in dairy heifers. J Dairy Sci 83(7):1456–1462

Elitok B, Elitok Ö (2004) Clinical efficacy of carprofen as an adjunct to the antibacterial treatment of bovine respiratory disease. J Vet Pharmacol Ther 27(5):317–320

Engle TE, Spears JW, Brown TT, Lloyd KE (1999) Effect and of breed (Angus vs Simmental) on immune function and response to a disease challenge in stressed steers and preweaned calves. J Anim Sci 77(3):516–521

Erdogan H, Akin I, Ural K, Ulutas PA (2019) Effects of selected nonsteroidal anti-inflammatory drugs on acute-phase protein levels after dehorning in Holstein heifers. Medycyna Weterynaryjna-Vet Med Sci Pract 75(7):438–441

Espinoza CA, McCarthy D, White PJ, Windsor PA, Lomax SH (2016) Evaluating the efficacy of a topical anaesthetic formulation and ketoprofen, alone and in combination, on the pain sensitivity of dehorning wounds in Holstein-Friesian calves. Anim Prod Sci 56(9):1512–1519

Esslemont RJ, Kossaibati MA (1997) Culling in 50 dairy herds in England. Vet Rec 140(2):36–39

Fajt VR, Wagner SA, Norby B (2011) Analgesic drug administration and attitudes about analgesia in cattle among bovine practitioners in the United States. J Am Vet Med Assoc 238(6):755–767

Fan B, Bryant R, Greer A (2022) Behavioral fingerprinting: acceleration sensors for identifying changes in livestock health. J 5(4):435–454

FAOSTAT (2023) Retrieved 20 January 2024, from https://www.fao.org/faostat/en/#data/domains_table

Faulkner PM, Weary DM (2000) Reducing pain after dehorning in dairy calves. J Dairy Sci 83(9):2037–2041

Felius M (1995) Cattle breeds: an encyclopedia. C Misset bv, Doetinchem

Fenwick DC, Daniel RCW (1996) Evaluation of the effect of ketoprofen on experimentally induced ephemeral fever in dairy heifers. Aust Vet J 74(1):37–41

Fidan AF, Pamuk K, Ozdemir A, Saritas ZK, Tarakci U (2010) Effects of dehorning by amputation on oxidant-antioxidant status in mature cattle. Revue De Medecine Veterinaire 161(11):502–508

Fiems LO (2012) Double muscling in cattle: genes, husbandry, carcasses and meat. Animals (Basel) 2(3):472–506

Fisher AD, Crowe MA, delaVarga MEA, Enright WJ (1996) Effect of castration method and the provision of local anesthesia on plasma cortisol, scrotal circumference, growth, and feed intake of bull calves. J Anim Sci 74(10):2336–2343

Flower FC, Sedlbauer M, Carter E, von Keyserlingk MAG, Sanderson DJ, Weary DM (2008) Analgesics improve the gait of lame dairy cattle. J Dairy Sci 91(8):3010–3014

Fraccaro E, Coetzee JF, Odore R, Edwards-Callaway LN, KuKanich B, Badino P, Bertolotti L, Glynn H, Dockweiler J, Allen K, Bergamasco L (2013) A study to compare circulating flunixin, meloxicam and gabapentin concentrations with prostaglandin E-2 levels in calves undergoing dehorning. Res Vet Sci 95(1):204–211

Fraser D, Schuppli CA, Spooner JM (2012) Attitudes of Canadian beef producers toward animal welfare. Anim Welf 21(2):273–283

Funk WJ (1950) Word origins and their romantic stories. Wilfred Funk Inc., New York

Gellatly D, Marti S, Pajor EA, Meléndez DM, Moya D, Janzen ED, Yang X, Milani MRM, Schwartzkopf-Genswein KS (2021) Effect of a single subcutaneous injection of meloxicam on chronic indicators of pain and inflammatory responses in 2-month-old knife and band-castrated beef calves housed on pasture. Livest Sci 244:104305

Gibson TJ, Johnson CB, Stafford KJ, Mitchinson SL, Mellor DJ (2007) Validation of the acute electroencephalographic responses of calves to noxious stimulus with scoop dehorning. N Z Vet J 55(4):152–157

Gimbutas M (1977) The 1st-wave of Eurasian steppe pastoralists into copper age Europe. J Indo-Eur Stud 5(4):277–338

Gladden N, Ellis K, Martin J, Viora L, McKeegan D (2019) A single dose of ketoprofen in the immediate postpartum period has the potential to improve dairy calf welfare in the first 48 h of life. Appl Anim Behav Sci 212:19–29

Gladden N, Ellis K, Martin J, McKeegan D (2021) Administration of ketoprofen affects post-partum lying behaviours of Holstein dairy cows regardless of whether parturition is assisted. Vet Rec 189(6):245

Gleerup KB, Andersen PH, Munksgaard L, Forkman B (2015) Pain evaluation in dairy cattle. Appl Anim Behav Sci 171:25–32

Glynn HD, Coetzee JF, Edwards-Callaway LN, Dockweiler JC, Allen KA, Lubbers B, Jones M, Fraccaro E, Bergamasco LL, Kukanich B (2013) The pharmacokinetics and effects of meloxicam, gabapentin, and flunixin in postweaning dairy calves following dehorning with local anesthesia. J Vet Pharmacol Ther 36(6):550–561

Gonzalez LA, Schwartzkopf-Genswein KS, Caulkett NA, Janzen E, McAllister TA, Fierheller E, Schaefer AL, Haley DB, Stookey JM, Hendrick S (2010) Pain mitigation after band castration of beef calves and its effects on performance, behavior, Escherichia coli, and salivary cortisol. J Anim Sci 88(2):802–810

Graf B, Senn M (1999) Behavioural and physiological responses of calves to dehorning by heat cauterization with or without local anaesthesia. Appl Anim Behav Sci 62(2–3):153–171

Green MJ, Green LE, Cripps PJ (1997) Comparison of fluid and flunixin meglumine therapy in combination and individually in the treatment of toxic mastitis. Vet Rec 140(6):149–152

Grondahl-Nielsen C, Simonsen HB, Lund JD, Hesselholt M (1999) Behavioural, endocrine and cardiac responses in young calves undergoing dehorning without and with use of sedation and analgesia. Vet J 158(1):14–20

Haga HA, Ranheim B (2005) Castration of piglets: the analgesic effects of intratesticular and intrafunicular lidocaine injection. Vet Anaesth Analg 32(1):1–9

Hanzen C, Laurent Y, Ward WR (1994) Comparison of reproductive-performance in Belgian dairy and beef-cattle. Theriogenology 41(5):1099–1114

Harris C, White PJ, Mohler VL, Lomax S (2020) Electroencephalography can distinguish between pain and anaesthetic intervention in conscious lambs undergoing castration. Animals 10(3):1498

Hazel LN (1943) The genetic basis for constructing selection indexes. Genetics 28(6):476–490

Heinrich A, Duffield T, Lissemore K, Squires EJ, Millman ST (2007) The efficacy of meloxicam at relieving the pain response to dehorning in dairy calves. J Anim Sci 85:127–127

Heinrich A, Duffield TF, Lissemore KD, Squires EJ, Millman ST (2009) The impact of meloxicam on post-surgical stress associated with cautery dehorning. J Dairy Sci 92(2):540–547

Heinrich A, Duffield TF, Lissemore KD, Millman ST (2010) The effect of meloxicam on behavior and pain sensitivity of dairy calves following cautery dehorning with a local anesthetic. J Dairy Sci 93(6):2450–2457

Heringstad B, Klemetsdal G, Ruane J (2000) Selection for mastitis resistance in dairy cattle: a review with focus on the situation in the Nordic countries. Livest Prod Sci 64(2–3):95–106

Herskin MS, Nielsen BH (2018) welfare effects of the use of a combination of local anesthesia and NSAID for disbudding analgesia in dairy calves-reviewed across different welfare concerns. Front Vet Sci 5:1

Hobson H (2021) Animal research numbers, Great Britain 2020. Understanding Animal Research Retrieved 04 January 2023, from https://www.understandinganimalresearch.org.uk/news/animal-research-numbers-2020

Hoffsis G (1995) Surgical (cosmetic) dehorning in cattle. Vet Clin N Am Food Anim Pract 11(1):159–169

Horst RL, Goff JP, Reinhardt TA, Buxton DR (1997) Strategies for preventing milk fever in dairy cattle. J Dairy Sci 80(7):1269–1280

Howard K, Beattie L (2018) A national producer survey of sheep and cattle husbandry practices. Sydney, Meat & Livestock Australia, 135

Huber J, Arnholdt T, Mostl E, Gelfert CC, Drillich M (2013) Pain management with flunixin meglumine at dehorning of calves. J Dairy Sci 96(1):132–140

Huxley JN, Whay HR (2006) Current attitudes of cattle practitioners to pain and the use of analgesics in cattle. Vet Rec 159(20):662–668

Ingvartsen KL (2006) Feeding- and management-related diseases in the transition cow – physiological adaptations around calving and strategies to reduce feeding-related diseases. Anim Feed Sci Technol 126(3–4):175–213

Ingvartsen KL, Andersen JB (2000) Integration of metabolism and intake regulation: a review focusing on periparturient animals. J Dairy Sci 83(7):1573–1597

Ingvartsen KL, Dewhurst RJ, Friggens NC (2003) On the relationship between lactational performance and health: is it yield or metabolic imbalance that cause production diseases in dairy cattle? A position paper. Livest Prod Sci 83(2–3):277–308

Isaac E (1962) On domestication of cattle. Science 137(3525):195–204

Janus K, Grochowina B, Antoszek J, Suszycki S, Muszczynski Z (2003) The effect of food or water deprivation on paracetamol pharmacokinetics in calves. J Vet Pharmacol Ther 26(4):291–296

Johnston CH, Whittaker AL, Franklin SH, Hutchinson MR (2022) The Neuroimmune Interface and chronic pain through the lens of production animals. Front Neurosci 16:887042

Jongman EC, Morris JP, Barnett JL, Hemsworth PH (2000) EEG changes in 4-week-old lambs in response to castration, tail docking and mulesing. Aust Vet J 78(5):339–343

Jubb TF, Fordyce G, Bolam MJ, Hadden DJ, Cooper NJ, Whyte TR, Fitzpatrick LA, Hill F, D'Occhio MJ (2003) Trial introduction of the Willis dropped ovary technique for spaying cattle in northern Australia. Aust Vet J 81(1–2):66–70

Karlen KJ, Baier FS, Odegard SL, Baumann RM, Coetzee JF, Kehoe SI, Vogel KD (2021) Efficacy of oral meloxicam as primary pain mitigation following caustic paste disbudding of three day old Holstein calves. Transl Anim Sci 5(2):txz151

Kasiora K, Anagnostopoulos A, Bedford C, Menka T, Barden M, Griffiths BE, Achard D, Timms K, Machado VS, Coates A, Oikonomou G (2022) Evaluation of the use of ketoprofen for the treatment of digital dermatitis in dairy cattle: a randomised, positive controlled, clinical trial. Vet Rec 190(6):e977

Kielland C, Skjerve E, Zanella AJ (2009) Attitudes of veterinary students to pain in cattle. Vet Rec 165(9):254–258

Kleinhenz MD, Van Engen NK, Smith JS, Gorden PJ, Ji J, Wang C, Perkins SCB, Coetzee JF (2018) The impact of transdermal flunixin meglumine on biomarkers of pain in calves when administered at the time of surgical castration without local anesthesia. Livest Sci 212:1–6

Kling-Eveillard F, Knierim U, Irrgang N, Gottardo F, Ricci R, Dockes AC (2015) Attitudes of farmers towards cattle dehorning. Livest Sci 179:12–21

Knierim U, Irrgang N, Roth BA (2015) To be or not to be horned-consequences in cattle. Livest Sci 179:29–37

Kongara K, Dukkipati VSR, Tai HM, Heiser A, Murray A, Webster J, Johnson CB (2020) Differential transcription of selected cytokine and neuroactive ligand-receptor genes in peripheral leukocytes from calves in response to cautery disbudding. Animals 10(7):1187

Korkmaz M, Saritas ZK, Bulbul A, Demirkan I (2015) Effect of pre-Emptive Dexketoprofen trometamol on acute cortisol, inflammatory response and oxidative stress to hot-iron disbudding in calves. Kafkas Universitesi Veteriner Fakultesi Dergisi 21(4):563–568

Kossaibati MA, Esslemont RJ (1997) The costs of production diseases in dairy herds in England. Vet J 154(1):41–51

Kotschwar JL, Coetzee JF, Anderson DE, Gehring R, KuKanich B, Apley MD (2009) Analgesic efficacy of sodium salicylate in an amphotericin B-induced bovine synovitis-arthritis model. J Dairy Sci 92(8):3731–3743

Kremer L, van Reenen CG, Engel B, Bokkers EAM, Schnabel SK, van der Werf JTN, Webb LE (2022) Developing a feasible and sensitive judgement bias task in dairy cows. Anim Cogn 25(2):425–445

Kumar S, Khangembam SD, Gangwar AK, Yadav VK, Verma RK (2020) Distal intravenous regional Anaesthesia for claw diseases in bovines. Indian J Anim Res 54(5):614–618

Lappe FM (1977) Food first: beyond the myth of scarcity. Houghton-Mifflin, Boston

Larrondo C, Bustamante H, Gallo C (2018) Sheep farmers' perception of welfare and pain associated with routine husbandry practices in Chile. Animals 8(12):225

Laven RA, Lawrence KE, Weston JF, Dowson KR, Stafford KJ (2008) Assessment of the duration of the pain response associated with lameness in dairy cows, and the influence of treatment. N Z Vet J 56(5):210–217

Lawrence AB, Conington J, Simm G (2004) Breeding and animal welfare: practical and theoretical advantages of multi-trait selection. Anim Welf 13(S1):S191–S196

Lay DC Jr, Friend TH, Grissom KK, Bowers CL, Mal ME (1992) Effects of freeze or hot-iron branding of angus calves on some physiological and behavioral indicators of stress. Appl Anim Behav Sci 33(2–3):137–147

Lean IJ, Westwood CT, Golder HM, Vermunt JJ (2013) Impact of nutrition on lameness and claw health in cattle. Livest Sci 156(1–3):71–87

Lecompte ML (1985) The Hispanic influence on the history of rodeo, 1823-1922. J Sport Hist 12(1):21–38

Lehner PN (1996) Handbook of ethological methods. Cambridge University Press, Cambridge

Lemmens L, Schodl K, Fuerst-Waltl B, Schwarzenbacher H, Egger-Danner C, Linke K, Suntinger M, Phelan M, Mayerhofer M, Steininger F, Papst F, Maurer L, Kofler J (2023) The combined use of automated milking system and sensor data to improve detection of mild lameness in dairy cattle. Animals 13(7):1180

Lendzele SS, Mavoungou JF, Burinyuy KA, Armel KA, Dickmu SJ, Young JR, Thomson PC, Windsor PA (2021) Efficacy and application of a novel topical anaesthetic wound formulation for treating cattle with foot-and-mouth disease: a field trial in Cameroon. Transbound Emerg Dis 68(4):2531–2542

Lepkova R, Sterc J, Vecerek V, Doubek J, Kruzikova K, Bedanova I (2007) Stress responses in adult cattle due to surgical dehorning using three different types of anaesthesia. Berliner Und Munchener Tierarztliche Wochenschrift 120(11–12):465–469

Lin HC, Heath AM, Pugh DG, Trachte EA (2003) Comparison of the analgesic effect of clonidine and lidocaine when administered epidurally in cattle. Bovine Pract 37(1):83–88

Lincoln B (1981) Priests, warriors, and cattle: a study in the ecology of religions. University of California Press, Berkeley, p 242

Lindegaard C, Vaabengaard D, Christophersen MT, Ekstom CT, Fjeldborg J (2009) Evaluation of pain and inflammation associated with hot iron branding and microchip transponder injection in horses. Am J Vet Res 70(7):840–847

Littledike ET, Young JW, Beitz DC (1981) Common metabolic diseases of cattle – ketosis, milk fever, grass tetany, and downer cow complex. J Dairy Sci 64(6):1465–1482

Lockwood PW, Johnson JC, Katz IL (2003) Clinical efficacy of flunixin, carprofen and ketoprofen as adjuncts to the antibacterial treatment of bovine respiratory disease. Vet Rec 152(13):392–394

Lomax S, Witenden E, Windsor P, White P (2017) Effect of topical vapocoolant spray on perioperative pain response of unweaned calves to ear tagging and ear notching. Vet Anaesth Analg 44(1):163–172

Malreddy PR, Coetzee JF, KuKanich B, Gehring R (2013) Pharmacokinetics and milk secretion of gabapentin and meloxicam co-administered orally in Holstein-Friesian cows. J Vet Pharmacol Ther 36(1):14–20

Mamet J, Voilley N (2002) ASIC pH-sensitive ion channels and inflammatory pain. MS-Med Sci 18(8–9):889–895

Marti S, Devant M, Bach A (2010a) Lying time and animal activity after surgical castration of Holstein bulls recorded with pedometers. J Dairy Sci 93:16–17

Marti S, Velarde A, de la Torre JL, Bach A, Aris A, Serrano A, Manteca X, Devant M (2010b) Effects of ring castration with local anesthesia and analgesia in

Holstein calves at 3 months of age on welfare indicators. J Anim Sci 88(8):2789–2796
Martin MS, Kleinhenz MD, Viscardi AV, Curtis AK, Johnson BT, Montgomery SR, Lou ME, Coetzee JF (2022) Effect of bupivacaine liposome suspension administered as a local anesthetic block on indicators of pain and distress during and after surgical castration in dairy calves. J Anim Sci 100(1):skab378
Mayer C, Zablotski Y, Rieger A, Knubben-Schweizer G, Feist M (2020) Behavioural and physiological indicators of pain and distress in calves following tail docking. Tierarztl Prax Ausg Grosstiere Nutztiere 48(05):328–337
McCarthy D, Lomax S, Windsor PA, White PJ (2016) Effect of a topical anaesthetic formulation on the cortisol response to surgical castration of unweaned beef calves. Animal 10(1):150–156
McKendree MGS, Tonsor GT, Wolf CA (2018) Animal welfare perceptions of the U.S. public and cow-calf producers. J Agric Appl Econ 50(4):544–578
McMeekan CM, Stafford KJ, Mellor DJ, Bruce RA, Ward RN, Gregory NG (1998) Effects of regional analgesia and/or a non-steroidal anti-inflammatory analgesic on the acute cortisol response to dehorning in calves. Res Vet Sci 64(2):147–150
McMeekan C, Stafford KJ, Mellor DJ, Bruce RA, Ward RN, Gregory N (1999) Effects of a local anaesthetic and a non-steroidal anti-inflammatory analgesic on the behavioural responses of calves to dehorning. N Z Vet J 47(3):92–96
Meischke HRC, Ramsay WR, Shaw FD (1974) The effect of horns on bruising in cattle. Aust Vet J 50:432–434
Meléndez DM, Marti S, Pajor EA, Moya D, Gellatly D, Janzen ED, Schwartzkopf-Genswein KS (2018) Effect of a single dose of meloxicam prior to band or knife castration in 1-wk-old beef calves: I. Acute pain. J Anim Sci 96(4):1268–1280
Mellor DJ, Stafford KJ, Todd SE, Lowe TE, Gregory NG, Bruce RA, Ward RN (2002) A comparison of catecholamine and cortisol responses of young lambs and calves to painful husbandry procedures. Aust Vet J 80(4):228–233
Mendonça FS, Vaz RZ, Leal WS, Restle J, Pascoal LL, Vaz MB, Farias GD (2016) Genetic group and horns presence in bruises and economic losses in cattle carcasses. Semina-Ciencias Agrarias 37(6):4265–4273
Ménézo YJR, Hérubel F (2002) Mouse and bovine models for human IVF*. Reprod Biomed Online 4(2):170–175
Meyn K (1970) Beef production in East Africa: with a comprehensive bibliography on cattle breeding in East Africa. Munich, IFO Institut für Wirtschaftsforschung
Milligan BN, Duffield T, Lissemore K (2004) The utility of ketoprofen for alleviating pain following dehorning in young dairy calves. Can Vet J-Revue Veterinaire Canadienne 45(2):140–143
Mills PC, Ghodasara P, Satake N, Alawneh J, Fraser B, Kopp S, McGowan M (2020) A novel transdermal Ketoprofen formulation provides effective analgesia to calves undergoing amputation dehorning. Animals 10(12):2442
Milne MH, Nolan AM, Cripps PJ, Fitzpatrick JL (2003) Assessment and alleviation of pain in dairy cows with clinical mastitis. Cattle Pract 11:289–293
Mintline EM, Stewart M, Rogers AR, Cox NR, Verkerk GA, Stookey JM, Webster JR, Tucker CB (2013) Play behavior as an indicator of animal welfare: disbudding in dairy calves. Appl Anim Behav Sci 144(1–2):22–30
Mirkena T, Duguma G, Haile A, Tibbo M, Okeyo AM, Wurzinger M, Sölkner J (2010) Genetics of adaptation in domestic farm animals: a review. Livest Sci 132(1–3):1–12
Moggy MA, Pajor EA, Thurston WE, Parker S, Greter AM, Schwartzkopf-Genswein KS, Campbell JR, Windeyer MC (2017) Management practices associated with pain in cattle on western Canadian cow–calf operations: a mixed methods study1. J Anim Sci 95(2):958–969
Molony V, Kent JE, Robertson IS (1995) Assessment of acute and chronic pain after different methods of castration of calves. Appl Anim Behav Sci 46(1–2):33–48
Moya D, Gonzalez LA, Janzen E, Caulkett NA, Fireheller E, Schwartzkopf-Genswein KS (2014) Effects of castration method and frequency of intramuscular injections of ketoprofen on behavioral and physiological indicators of pain in beef cattle. J Anim Sci 92(4):1684–1695
Mrode RA, Swanson GJT, Winters MS (1998) Genetic parameters and evaluations for somatic cell counts and its relationship with production and type traits in some dairy breeds in the United Kingdom. Anim Sci 66:569–576
Mrode RA, Swanson GJT, Lindberg CM (2000) Genetic correlations of somatic cell count and conformation traits with herd life in dairy breeds, with an application to national genetic evaluations for herd life in the United Kingdom. Livest Prod Sci 65(1–2):119–130
Muggli NE, Hohenboken WD, Cundiff LV, Mattson DE (1987) Inheritance and interaction of immune traits in beef-calves. J Anim Sci 64(2):385–393
Muller BR, Soriano VS, Bellio JCB, Molento CFM (2019) Facial expression of pain in Nellore and crossbred beef cattle. J Vet Behav Clin Appl Res 34:60–65
Naert G, Beeckman D, de Kruif A (1999) Effect of rations differing in fibre on the incidence of subclinical laminitis in beef bulls. Vlaams Diergeneeskd Tijdschr 68(6):282–285
Neave HW, Daros RR, Costa JHC, von Keyserlingk MAG, Weary DM (2013) Pain and pessimism: dairy calves exhibit negative judgement bias following hot-iron disbudding. PLoS One 8(12):e80556
Newby NC, Pearl DL, LeBlanc SJ, Leslie KE, von Keyserlingk MAG, Duffield TF (2013) The effect of administering ketoprofen on the physiology and behavior of dairy cows following surgery to correct a left displaced abomasum. J Dairy Sci 96(3):1511–1520
Newby NC, Tucker CB, Pearl DL, LeBlanc SJ, Leslie KE, von Keyserlingk MAG, Duffield TF (2014) An investigation of the effects of ketoprofen following rumen

fistulation surgery in lactating dairy cows. Can Vet J-Revue Veterinaire Canadienne 55(5):442–448

Norring M, Wikman I, Hokkanen A-H, Kujala MV, Hänninen L (2014) Empathic veterinarians score cattle pain higher. Vet J 200(1):186–190

Ødegård J, Klemetsdal G, Heringstad B (2003) Genetic improvement of mastitis resistance: validation of somatic cell score and clinical mastitis as selection criteria. J Dairy Sci 86(12):4129–4136

Olson ME, Ralston B, Burwash L, Matheson-Bird H, Allan ND (2016) Efficacy of oral meloxicam suspension for prevention of pain and inflammation following band and surgical castration in calves. BMC Vet Res 12(1):102

Oltenacu PA (2009) Genetic improvements for production and animal welfare. In: Smulders FJM, Algers B (eds) Welfare of production animals: assessment and management of risks. Wageningen Academic Publishers, Wageningen, pp 417–441

Onda K, Emura Y, Saito A, Ikeyama T, Sato R, Ochiai H, Kaneko K, Iriki T, Wada Y (2012) Sodium salicylate attenuates acute stress responses to castration in calves. J Anim Vet Adv 11(10):1574–1577

Orpin PG, Esslemont RJ (2010) Culling and wastage in dairy herds: an update on incidence and economic impact in dairy herds in the UK. Cattle Pract 18:163–172

Ossent P, Lischer C (1998) Bovine laminitis: the lesions and their pathogenesis. In Pract 20(8):415

Pang WY, Earley B, Sweeney T, Crowe MA (2006) Effect of carprofen administration during banding or burdizzo castration of bulls on plasma cortisol, in vitro interferon-gamma production, acute-phase proteins, feed intake, and growth. J Anim Sci 84(2):351–359

Pang WY, Earley B, Murray M, Sweeney T, Gath V, Crowe MA (2011) Banding or Burdizzo castration and carprofen administration on peripheral leukocyte inflammatory cytokine transcripts. Res Vet Sci 90(1):127–132

Park SJ, Piao M, Kim H, Kang HJ, Fassah DM, Jung DJS, Kim SY, Na SW, Kim M, Baik M (2020) Effects of dehorning and lidocaine-plus-flunixin treatment on indicators of stress and acute inflammation, behaviors, and their association in Korean cattle bull calves. Livest Sci 241:104198

Passos LT, Bettencourt AF, Ritt LA, Canozzi MEA, Fischer V (2023) Systematic review of the relationship between rumen acidosis and laminitis in cattle. Res Vet Sci 161:110–117

Paull DR, Small AH, Lee C, Labeur L, Colditz IG (2015) Effect of local infusion of NSAID analgesics administered alone or in combination on the pain associated with band castration in calves. Aust Vet J 93(8):271–277

Pauly C, White BJ, Coetzee JF, Robert B, Baldridge S, Renter DG (2012) Evaluation of analgesic protocol effect on calf behavior after concurrent castration and Dehoming. Int J Appl Res Vet Med 10(1):54–61

Peers A, Mellor DJ, Wintour EM, Dodic M (2002) Blood pressure, heart rate, hormonal and other acute responses to rubber-ring castration and tail docking of lambs. N Z Vet J 50(2):56–62

Petherick JC, McCosker K, Mayer DG, Letchford P, McGowan M (2011) Preliminary investigation of some physiological responses of Bos indicus heifers to surgical spaying. Aust Vet J 89(4):131–137

Petherick JC, Small AH, Mayer DG, Colditz IG, Ferguson DM, Stafford KJ (2014a) A comparison of welfare outcomes for weaner and mature Bos indicus bulls surgically or tension band castrated with or without analgesia: 1. Behavioural responses. Appl Anim Behav Sci 157:23–34

Petherick JC, Small AH, Mayer DG, Colditz IG, Ferguson DM, Stafford KJ (2014b) A comparison of welfare outcomes for weaner and mature Bos indicus bulls surgically or tension band castrated with or without analgesia: 2. Responses related to stress, health and productivity. Appl Anim Behav Sci 157:35–47

Petherick JC, Small AH, Reid DJ, Colditz IG, Ferguson DM (2015) Welfare outcomes for 3-and 6-month-old beef calves in a tropical environment castrated surgically or by applying rubber rings. Appl Anim Behav Sci 171:47–57

Phillips CJC, Phillips AP (2010) Attitudes of Australian sheep farmers to animal welfare. J Int Farm Manag 5(2):162–167

Plascencia A, Valenzuela RM, Barreras A, Cervantes M, Gonzalez TM, Zinn RA (2008) Influence of Ovariectomization (spaying) and feeding system on performance and carcass characteristics of implanted beef heifers. J Anim Vet Adv 7(3):354–358

Pritchard T, Coffey M, Mrode R, Wall E (2013) Genetic parameters for production, health, fertility and longevity traits in dairy cows. Animal 7(1):34–46

Pryce JE, Veerkamp RF, Thompson R, Hill WG, Simm G (1997) Genetic aspects of common health disorders and measures of fertility in Holstein Friesian dairy cattle. Anim Sci 65:353–360

Pryce JE, Royal MD, Garnsworthy PC, Mao IL (2004) Fertility in the high-producing dairy cow. Livest Prod Sci 86(1–3):125–135

Qi JC, Fang J, Huang FY, Li ZQ, Kumbhar MA, Cui HM, Guo HR, Shen LH, Ren ZH, Geng Y, Hu YC, Deng JL, Yu SM, Shah AM, Zuo ZC (2023) A clinical trial on the welfare effects of administering meloxicam to 10 to 21 day dairy calves following caustic paste disbudding. J Anim Sci 101

Randhawa IAS, Burns BM, McGowan MR, Porto-Neto LR, Hayes BJ, Ferretti R, Schutt KM, Lyons RE (2020) Optimized genetic testing for Polledness in multiple breeds of cattle. G3 Genes|Genomes|Genetics 10(2):539–544

Rantala M, Kaartinen L, Välimäki E, Stryrman M, Hiekkaranta M, Niemi A, Saari L, Pyörälä S (2002) Efficacy and pharmacokinetics of enrofloxacin and flunixin meglumine for treatment of cows with experimentally induced Escherichia coli mastitis. J Vet Pharmacol Ther 25(4):251–258

Rastabi HI, Guraninejad S, Naddaf H, Hasani A (2018) Comparison of the application of lidocaine, lidocaine-

dexamethasone and lidocaine-epinephrine for caudal epidural anesthesia in cows. Iran J Vet Res 19(3):172–177
Rauw WM, Kanis E, Noordhuizen-Stassen EN, Grommers FJ (1998) Undesirable side effects of selection for high production efficiency in farm animals: a review. Livest Prod Sci 56(1):15–33
Reedman CN, Duffield TF, DeVries TJ, Lissemore KD, Karrow NA, Li ZW, Winder CB (2020) Randomized control trial assessing the efficacy of pain control strategies for caustic paste disbudding in dai calves younger than 9 days of age. J Dairy Sci 103(8):7339–7350
Remnant JG, Tremlett A, Huxley JN, Hudson CD (2017) Clinician attitudes to pain and use of analgesia in cattle: where are we 10 years on? Vet Rec 181(15):400–400
Restle J, Grassi C (1993) Spaying cull cows and its effect on the weight-gain of the cow and calf. Pesquisa Agropecuaria Brasileira 28(12):1437–1441
Riebold TW (2001) Anaesthetic management of cattle. In: Steffey EP (ed) Recent advances in anesthetic management of large domestic animals. International Veterinary Information Service
Ritchie CA (1981) Food in civilization: how history has been affected by human tastes. Methuen Australia, Sydney
Sandoe P, Nielsen BL, Christensen LG, Sorensen P (1999) Staying good while playing god—the ethics of breeding farm animals. Anim Welf 8(4):313–328
Saraceni J, Renaud DL, Nelson E, Van Os JMC, Miltenburg C, Winder CB (2022) Ontario dairy producers' perceived barriers and motivations to the use of pain control for disbudding and dehorning calves: a qualitative study. Animals 12(8):973
Saville JW, Ross JA, Trefz T, Schatz C, Matheson-Bird H, Ralston B, Granot O, Schmid K, Terry R, Allan ND, Wulff JE, Olson M (2020) Development and field validation of lidocaine-loaded castration bands for bovine pain mitigation. Animals 10(12):2363
Schafberg R, Swalve HH (2015) The history of breeding for polled cattle. Livest Sci 179:54–70
Schreiner DA, Ruegg PL (2002a) Effects of tail docking on milk quality and cow cleanliness. J Dairy Sci 85(10):2503–2511
Schreiner DA, Ruegg PL (2002b) Responses to tail docking in calves and heifers. J Dairy Sci 85(12):3287–3296
Schulz KL, Anderson DE, Coetzee JF, White BJ, Miesner MD (2011) Effect of flunixin meglumine on the amelioration of lameness in dairy steers with amphotericin B-induced transient synovitis-arthritis. Am J Vet Res 72(11):1431–1438
Schuster F, Aldag P, Frenzel A, Hadeler KG, Lucas-Hahn A, Niemann H, Petersen B (2020) CRISPR/Cas12a mediated knock-in of the polled Celtic variant to produce a polled genotype in dairy cattle. Sci Rep 10(1):13570
Schwartzkopf Genswein KS, Stookey JM, de Passille AM, Rushen J (1997a) Comparison of hot-iron and freeze branding on cortisol levels and pain sensitivity in beef cattle. Can J Anim Sci 77(3):369–374
Schwartzkopf Genswein KS, Stookey JM, Welford R (1997b) Behavior of cattle during hot-iron and freeze branding and the effects on subsequent handling ease. J Anim Sci 75(8):2064–2072
Sheil M, Chambers M, Polkinghorne A, Sharpe B (2021) Topical application of lidocaine and bupivacaine to disbudding wounds in dairy calves: safety, toxicology and wound healing. Animals 11(3):869
Short CE (1987) Principles and practice of veterinary Anaesthesia. Williams and Wilkins
Simoons FJ (1968) A ceremonial ox of India; the mithan in nature, culture, and history, with notes on the domestication of common cattle, by Frederick J. Simoons, with the assistance of Elizabeth S. Simoons. Illus. by Gene M. Christman. Madison, University of Wisconsin Press
Singh GR, Amarpal, Aithal HP, Kinjavdekar P (2005) Lameness in cattle—a review. Indian J Anim Sci 75(6):723–740
Small A, Fisher AD, Lee C, Colditz I (2021) Analgesia for sheep in commercial production: where to next? Animals 11(4):1127
Stafford KJ, Mellor DJ (2005) Dehorning and disbudding distress and its alleviation in calves. Vet J 169(3):337–349
Stafford KJ, Mellor DJ (2011) Addressing the pain associated with disbudding and dehorning in cattle. Appl Anim Behav Sci 135(3):226–231
Stafford KJ, Mellor DJ, Todd SE, Bruce RA, Ward RN (2002) Effects of local anaesthesia or local anaesthesia plus a non-steroidal anti-inflammatory drug on the acute cortisol response of calves to five different methods of castration. Res Vet Sci 73(1):61–70
Stafford KJ, Mellor DJ, Todd SE, Ward RN, McMeekan CM (2003) The effect of different combinations of lignocaine, ketoprofen, xylazine and tolazoline on the acute cortisol response to dehorning in calves. N Z Vet J 51(5):219–226
Statistica (2023) Retrieved 20 January 2024, from https://www.statista.com/statistics/634374/animals-used-for-experimentation-by-type-european-union-eu/
Stewart M, Webster JR, Schaefer AL, Cook NJ, Scott SL (2005) Infrared thermography as a non-invasive tool to study animal welfare. Anim Welf 14(4):319–325
Stewart M, Schaefer A, Haley DB, Colyn J, Cook NJ, Stafford KJ, Webster JR (2008a) Infrared thermography as a non-invasive method for detecting fear-related responses of cattle to handling procedures. Anim Welf 17(4):387–393
Stewart M, Stafford KJ, Dowling SK, Schaefer AL, Webster JR (2008b) Eye temperature and heart rate variability of calves disbudded with or without local anaesthetic. Physiol Behav 93(4–5):789–797
Stewart M, Stookey JM, Stafford KJ, Tucker CB, Rogers AR, Dowling SK, Verkerk GA, Schaefer AL, Webster JR (2009) Effects of local anesthetic and a non-steroidal antiinflammatory drug on pain responses of dairy calves to hot-iron dehorning. J Dairy Sci 92(4):1512–1519

Stewart M, Verkerk GA, Stafford KJ, Schaefer AL, Webster JR (2010) Noninvasive assessment of autonomic activity for evaluation of pain in calves, using surgical castration as a model. J Dairy Sci 93(8):3602–3609

Stilwell G, Lima MS, Broom DM (2008) Effects of nonsteroidal anti-inflammatory drugs on long-term pain in calves castrated by use of an external clamping technique following epidural anesthesia. Am J Vet Res 69(6):744–750

Stilwell G, de Carvalho RC, Lima MS, Broom DM (2009) Effect of caustic paste disbudding, using local anaesthesia with and without analgesia, on behaviour and cortisol of calves. Appl Anim Behav Sci 116(1):35–44

Stilwell G, Carvalho RC, Carolino N, Lima MS, Broom DM (2010) Effect of hot-iron disbudding on behaviour and plasma cortisol of calves sedated with xylazine. Res Vet Sci 88(1):188–193

Stilwell G, Lima MS, Carvalho RC, Broom DM (2012) Effects of hot-iron disbudding, using regional anaesthesia with and without carprofen, on cortisol and behaviour of calves. Res Vet Sci 92(2):338–341

Stilwell G, Schubert H, Broom DM (2014) Effects of analgesic use postcalving on cow welfare and production. J Dairy Sci 97(2):888–891

Stilwell GT, Ferrador AM, Santos MS, Domingues JM, Carolino N (2019) Use of topical local anesthetics to control pain during treatment of hoof lesions in dairy cows. J Dairy Sci 102(7):6383–6390

Stilwell G, Ferrador AM, Santos MS, Domingues JM (2020) Anesthetic residues in milk after topical application during treatment of hoof lesions in dairy cows. J Dairy Sci 103(1):898–901

Stock ML, Millman ST, Barth LA, Van Engen NK, Hsu WH, Wang C, Gehring R, Parsons RL, Coetzee JF (2015) The effects of firocoxib on cautery disbudding pain and stress responses in preweaned dairy calves. J Dairy Sci 98(9):6058–6069

Stock ML, Barth LA, Van Engen NK, Millman ST, Gehring R, Wang C, Voris EA, Wulf LW, Labeur L, Hsu WH, Coetzee JF (2016) Impact of carprofen administration on stress and nociception responses of calves to cautery dehorning. J Anim Sci 94(2):542–555

Stock ML, Kleinhenz MD, Mazloom R, Jaberi-Douraki M, Barth LA, Van Engen NK, Voris EA, Wang C, Coetzee JF (2021) Field trial comparing four oral nonsteroidal anti-inflammatory drugs on controlling cautery dehorning pain and stress in calves. Transl Anim Sci 5(2):txab041

Stookey JM, Goonewardene LA (1996) A comparison of production traits and welfare implications between horned and polled beef bulls. Can J Anim Sci 76(1):1–5

Stull CL, Payne MA, Berry SL, Hullinger PJ (2002) Evaluation of the scientific justification for tail docking in dairy cattle. J Am Vet Med Assoc 220(9):1298–1303

Surlis C, McNamara K, O'Hara E, Waters S, Beltman M, Cassidy J, Kenny D (2017) Birth delivery method affects expression of immune genes in lung and jejunum tissue of neonatal beef calves. BMC Vet Res 13:391

Sutherland MA, Mellow DJ, Stafford KJ, Gregory NG, Bruce RA, Ward RN (2002) Cortisol responses to dehorning of calves given a 5-h local anaesthetic regimen plus phenylbutazone, ketoprofen, or adrenocorticotropic hormone prior to dehorning. Res Vet Sci 73(2):115–123

Sutherland MA, Ballou MA, Davis BL, Brooks TA (2013) Effect of castration and dehorning singularly or combined on the behavior and physiology of Holstein calves. J Anim Sci 91(2):935–942

Sutherland MA, Bright AL, Schütz KE (2019) Effect of a buccal meloxicam formulation on the behavioural response to ring castration of calves. Anim Prod Sci 59(3):564–569

Sylvester SP, Stafford KJ, Mellor DJ, Bruce RA, Ward RN (2004) Behavioural responses of calves to amputation dehorning with and without local anaesthesia. Aust Vet J 82(11):697–700

Theurer ME, White BJ, Coetzee JF, Edwards LN, Mosher RA, Cull CA (2012) Assessment of behavioral changes associated with oral meloxicam administration at time of dehorning in calves using a remote triangulation device and accelerometers. BMC Vet Res 8:48

Thüer S, Mellema S, Doherr MG, Wechsler B, Nuss K, Steiner A (2007) Effect of local anaesthesia on short- and long-term pain induced by two bloodless castration methods in calves. Vet J 173(2):333–342

Ting STL, Earley B, Crowe MA (2003a) Effect of repeated ketoprofen administration during surgical castration of bulls on cortisol, immunological function, feed intake, growth, and behavior. J Anim Sci 81(5):1253–1264

Ting STL, Earley B, Hughes JML, Crowe MA (2003b) Effect of ketoprofen, lidocaine local anesthesia, and combined xylazine and lidocaine caudal epidural anesthesia during castration of beef cattle on stress responses, immunity, growth, and behavior. J Anim Sci 81(5):1281–1293

Tom EM, Duncan IJH, Widowski TM, Bateman KG, Leslie KE (2002a) Effects of tail docking using a rubber ring with or without anesthetic on behavior and production of lactating cows. J Dairy Sci 85(9):2257–2265

Tom EM, Rushen J, Duncan IJH, de Passille AM (2002b) Behavioural, health and cortisol responses of young calves to tail docking using a rubber ring or docking iron. Can J Anim Sci 82(1):1–9

Toussaint-Samat M (1992a) The history of dairy produce. In: A History of Food. Blackwell Publishing, Oxford, pp 113–124

Toussaint-Samat M (1992b) The history of meat. In: A history of food. Blackwell Publishing, Oxford, pp 85–102

Tschoner T, Sauter-Louis C, Peinhofer V, Feist M (2021) Exploring the attitudes of Bavarian farmers towards pain in cattle and how they differ from the attitudes of bovine veterinarians. Vet Rec 189(8):e515

Tucker CB, Fraser D, Weary DM (2001) Tail docking dairy cattle: effects on cow cleanliness and udder health. J Dairy Sci 84(1):84–87

Tucker CB, Mintline EM, Banuelos J, Walker KA, Hoar B, Drake D, Weary DM (2014a) Effect of a cooling gel on pain sensitivity and healing of hot-iron cattle brands. J Anim Sci 92(12):5666–5673

Tucker CB, Mintline EM, Banuelos J, Walker KA, Hoar B, Varga A, Drake D, Weary DM (2014b) Pain sensitivity and healing of hot-iron cattle brands. J Anim Sci 92(12):5674–5682

Uystepruyst C, Coghe J, Dorts T, Harmegnies N, Delsemme MH, Art T, Lekeux P (2002) Optimal timing of elective caesarean section in Belgian White and blue breed of cattle: the calf's point of view. Vet J 163(3):267–282

Van der Saag D, Lomax S, Windsor PA, Hall E, White PJ (2019) Effect of lignocaine and a topical Vapocoolant spray on pain response during surgical castration of beef calves. Animals 9(4):126

Vermunt JJ, Greenough PR (1994) Predisposing factors of laminitis in cattle. Br Vet J 150(2):151–164

Vidondo B, Stettler S, Stojiljkovic A, Mogel H, Gaschen V, Spadavecchia C, Casoni D, Stoffel MH (2019) Assessment of potential neuropathic changes in cattle after cautery disbudding. Res Vet Sci 126:9–16

Vindevoghel TV, Fleming PA, Hyndman TH, Musk GC, Laurence M, Collins T (2019) Qualitative behavioural assessment of Bos indicus cattle after surgical castration. Appl Anim Behav Sci 211:95–102

Voilley N, de Weille J, Mamet J, Lazdunski M (2001) Nonsteroid anti-inflammatory drugs inhibit both the activity and the inflammation-induced expression of acid-sensing ion channels in nociceptors. J Neurosci 21(20):8026–8033

Wagner SA, Apley MD (2004) Effects of two anti-inflammatory drugs on physiologic variables and milk production in cows with endotoxin-induced mastitis. Am J Vet Res 65(1):64–68

Warner R, Kleinhenz MD, Ydstie JA, Schleining JA, Wulf LW, Coetzee JF, Gorden PJ (2021) Randomized controlled trial comparison of analgesic drugs for control of pain associated with induced lameness in lactating dairy cattle. J Dairy Sci 104(2):2040–2055

Warnock TM, Thrift TA, Irsik M, Hersom MJ, Yelich JV, Maddock TD, Lamb GC, Arthington JD (2012) Effect of castration technique on beef calf performance, feed efficiency, and inflammatory response. J Anim Sci 90(7):2345–2352

Watts JM, Stookey JM (1999) Effects of restraint and branding on rates and acoustic parameters of vocalization in beef cattle. Appl Anim Behav Sci 62(2–3):125–135

Watts JM, Stookey JM (2001) The propensity of cattle to vocalise during handling and isolation is affected by phenotype. Appl Anim Behav Sci 74(2):81–95

Webb WP (1936) The great plains. Houghton Mifflin, Boston

Weber A, Stamer E, Junge W, Thaller G (2013) Genetic parameters for lameness and claw and leg diseases in dairy cows. J Dairy Sci 96(5):3310–3318

Wemelsfelder F, Hunter TEA, Mendl MT, Lawrence AB (2001) Assessing the 'whole animal': a free choice profiling approach. Anim Behav 62:209–220

Weyl-Feinstein S, Shabtay A, Malka H, Adin G, Peleg G, Faktor G, Honig H (2021) Assessing the effect of pain relieving methods during caustic disbudding procedure on dairy calf welfare. Isr J Vet Med 76(3):96–107

Whay HR, Webster AJF, Waterman-Pearson AE (2005) Role of ketoprofen in the modulation of hyperalgesia associated with lameness in dairy cattle. Vet Rec 157(23):729–U737

Whitaker DA, Kelly JM, Smith S (2000) Disposal and disease rates in 340 British dairy herds. Vet Rec 146(13):363–367

Whitaker DA, Macrae AI, Burrough E (2004) Disposal and disease rates in British dairy herds between April 1998 and March 2002. Vet Rec 155(2):43

Wikman I, Hokkanen AH, Pastell M, Kauppinen T, Valros A, Hänninen L (2013) Dairy producer attitudes to pain in cattle in relation to disbudding calves. J Dairy Sci 96(11):6894–6903

Wikman I, Hokkanen AH, Pastell M, Kauppinen T, Valros A, Hänninen L (2016) Attitudes of beef producers to disbudding and perception of pain in cattle. Anim Welf 25(4):429–438

Winder CB, Miltenburg CL, Sargeant JM, LeBlanc SJ, Haley DB, Lissemore KD, Godkin MA, Duffield TF (2018) Effects of local anesthetic or systemic analgesia on pain associated with cautery disbudding in calves: a systematic review and meta-analysis. J Dairy Sci 101(6):5411–5427

Windig JJ, Hoving-Bolink RA, Veerkamp RF (2015) Breeding for polledness in Holstein cattle. Livest Sci 179:96–101

Yadav VK, Gangwar AK, Khangembam SD, Goyal RP, Kumar R (2021) Development of novel intravenous regional anaesthetic technique for desensitization of hoof in cattle. Indian J Anim Sci 91(4):280–284

Yadav VK, Gangwar AK, Khangembam SD, Yadav Y, Goyal RP (2023) Comparison of lidocaine and its combination with ketamine for distal intravenous regional anesthesia (DIVRA) in bovines. Vet Arh 93(4):407–416

Yu A, Van der Saag D, Letchford P, Windsor P, White P (2020) Preliminary investigation to address pain and Haemorrhage following the spaying of female cattle. Animals 10(2):249

Sheep

Danila Marini and Alison Small

Abstract

Sheep have a rich history, with their domestication and utilisation being an integral part of human civilisation's expansion across the world. Sheep are hardy creatures capable of living in some of the most remote and harshest locations and are extensively utilised for their meat and wool, with over 1.2 billion head worldwide in 2023. Beyond agricultural use, sheep are kept as companions and used as models for biomedical research. There is extensive research into the efficacy of various anaesthetics and analgesics and their impact on sheep, as well as investigations into methods to identify pain in sheep. Even with the long history of being raised and utilised by humans, sheep are often misrepresented as being an unintelligent species that has limited experience of pain. It can also be difficult to identify individuals experiencing pain due to their stoic nature. As such, there is still a gap in understanding of the level of pain sheep experience throughout their life in a production system and the availability of analgesic and anaesthesia resources for those that care for them.

Keywords

Ovis aries · Pain in sheep · Analgesia in sheep · Local anaesthesia in sheep · Husbandry procedures in sheep · Post-operative pain in sheep

"Many thanks, too, do we owe to the sheep, both for appeasing the gods and for giving us the use of its fleece"—Pliny the Elder
(Ref) Delphi Complete Works of Pliny the Elder (Illustrated) By Pliny the Elder

D. Marini (✉)
School of Animal and Veterinary Science, Davies Livestock Research Centre, Adelaide University, Adelaide, SA, Australia
e-mail: danila.marini@adelaide.edu.au

A. Small
Armidale, Agriculture and Food, CSIRO, FD McMaster Research Laboratory, Chiswick, New England Highway, Armidale, NSW, Australia

1 History of Domestication

Sheep are amongst some of the earliest animals to have been domesticated, originating from the Fertile Crescent. Sheep were an integral part of the human history of expansion and early settlement. Domestication of modern sheep (*Ovis aries*) occurred around 9000–11,000 years ago, with modern sheep being a descendant of the Asiatic Mouflon (*Ovis orientalis*) (Chessa et al. 2009). The use of mitochondrial DNA to determine sheep phylogeny has led to the suggestion that there were three geographically separate domestication events that led to our modern-day

R. E. Clutton, G. C. Musk (eds.), *Pain and Suffering in Farmed Animals*,
https://doi.org/10.1007/978-3-032-11407-5_15

sheep (Chen et al. 2006; Meadows et al. 2007). The Mediterranean Mouflon (*Ovis aries musimon*), a wild sheep originating from Sardinia, is believed to be an original domesticated line of sheep that was rereleased into the wild (Cassinello 2022; Chessa et al. 2009). Detailed mitochondrial DNA analysis shows that the third common ancestor for sheep is unknown and no longer present in the wild (Hiendleder et al. 1998; Meadows et al. 2007).

A study of a Neolithic site in Central Anatolia has given insight into the gradual shift of the hunting of wild sheep to their eventual management within human settlements (Stiner et al. 2014, 2022). It is likely that the hunting of sheep moved towards a system where wild-caught lambs would have been reared in settlements prior to slaughter, before more permanent herding and raising of sheep within settlements occurred (Stiner et al. 2022). This change of hunting to husbandry is supported by evidence of a shift in diet from a variety of meat sources to a majority caprine meat source, and an increase in the presence of caprine bones (mostly sheep) in the settlements over time (Stiner et al. 2014).

Throughout their domestication, as humans started living in closer quarters with sheep, certain experiences causing pain would have decreased as shepherds started more proactively managing their herds, ensuring the animals' safety and access to new pastures. For one, exposure to predators would have decreased as reliance on shepherds increased. Alongside the domestication of the sheep was the domestication of the dog. With livestock guardians such as the Anatolian shepherd being one of the oldest breeds used to protect the flocks under human care from wolves and other predators (Harned 1982). However, with domestication also came the exposure to new environments, new intensified husbandry conditions and new potential pain experiences. Travelling to new countries required movement over large areas of land and transportation by boat, exposing sheep to new terrain, new pests and diseases, and new vegetation.

The importance of sheep in human history cannot be understated. Sheep travelled with humans across continents to be utilized for their meat, milk, and skins. It was not long after domestication occurred, that sheep were specifically bred and utilised for their fleece, with evidence of what is argued to be a woolly sheep depicted in a 6000-year-old statue found in Iran (Smith 1971). Sheep were a valuable source of wealth, often used as a trade resource, particularly for fleece. It wasn't until approximately 5000 years ago, when a method for spinning wool was created, that sheep were more intensively selected for their wool colour and quality (Vigne and Helmer 2007, 2014). The versatility of sheep's wool has seen it used for clothing and footwear, tents and rugs and even as armour by the Romans.

Beyond being a source of food and clothing, the importance of the sheep flowed into other aspects of human culture. Sheep were highly regarded by the Sumerians and the Egyptians, with sheep depicted as Gods and celebrated in sculptures and art. One such sculpture by the Sumerians, dated 3300–2900 BC, *Head of a Ewe*, was likely a depiction of the Goddess Duttur (Sumerian c. 3300–2900 B.C.). The Egyptians' own creator deity, Khnum, was often depicted as a ram, or human with a ram's head (Britannica 2022; Pinch 2004). There are even records of Egyptians bearing tattoos of sheep (Friedman et al. 2018). Sheep are regarded highly in many religions, often being used as a creature for sacrifice, as a replacement for firstborn sons and as an offer for protection. Within the Bible, sheep and lambs are mentioned several hundred times, making it one of the most referred to animals in the Bible. Lasting impressions of sheep are seen within present-day language, with reference to sheep in idioms, sayings and with nursery rhymes such as 'Mary had a little lamb', 'Little Bo-Peep' and 'Baa Baa Black sheep'.

Over the millennia of domestication, there are now nearly 400 breeds of sheep, specialised and utilised for their meat, milk, wool or a combination, that can be found across the world (Rasali et al. 2006). Through specialised breeding, unique animals have been created, some of which have specific issues, such as the over-muscling of Australian whites and Texels, to the excessive wrinkling of Merinos that makes them susceptible to flystrike. The diversity of landscapes and management styles of sheep husbandry all need to be considered when looking into how to mitigate and prevent pain and suffering.

2 Natural History

Sheep are an even-toed ungulate and comprise some of the smallest ruminants within the Bovidae family. Their subfamily is the caprine, which also consists of goats. Sheep are hardy animals that can adapt to a variety of environmental conditions, such as extreme heat and cold, and can make use of poor-quality forage. It is no surprise that sheep were able to be moved to new environments, with several hundred domesticated breed varieties now found on nearly every continent of the world. This adaptability to harsh environments is observed in the wild species of sheep, with many of them living in arid and remote mountainous regions. The number of wild sheep species has been debated over the years with many classifications based on the morphology and geographical location of the animals (Rezaei et al. 2010). Further research using DNA and mitochondrial information suggests speciation occurred as the animals migrated across continents, leading to six separate species of wild sheep (Table 1) (Rezaei et al. 2010). Sheep species have variation in the number of chromosomes that they carry; for example, Snow sheep carry 52 chromosomes, compared to the 58 of Urial.

Table 1 Species, subspecies and locations of wild sheep, based on (Bunch et al. 2005; Murdoch et al. 2017; Rezaei et al. 2010; Santos et al. 2021; Worley et al. 2004)

Species	Subspecies	Location
Argali (*Ovis ammon*)	Altai argali (*O. a. ammon*) Gobi argali (*O. a. darwin*)	Kazakhstan, Mongolia, West China, Russia
Bighorn sheep (*O. canadensis*)	California (*O. c. californiana*) Rocky Mountain (*O. c. canadensis*) Sierra Nevada (*O. c. sierrae*)	North America
Mouflon (*Ovis orientalis*) and (*Ovis aries*)	Asiatic Mouflon (*O. orientalis*) Mediterranean Mouflon (*O. a. musimon*)	Armenia, Austria, Azerbaijan, Croatia, Czech Republic, France, Germany, Hungary, India, Iran, Italy, Kazakhstan, Pakistan, Slovenia, Slovakia, Turkey
Snow sheep (*O. nivicola*)		North East Asia, Northern Russia
Thinhorn sheep (*O. dalli*)	Dall (*O. d. dalli*) Stone sheep (*O. d. stonei*)	Canada, North America
Urial (*Ovis vignei*)	Ladakh urial (*O. v. vignei*) Punjab urial (O. v. punjabiensis) Baluchistan urial (*O. v. blanfordi*)	Afghanistan, Iran, Kazakhstan, Kashmir, Oman, Pakistan, Tajikistan, Turkmenistan, Uzbekistan

Wild sheep are diverse in their morphology and are often split into two visually identifiable groups based on their body and horn size: bighorn and thinhorn sheep (Fig. 1). Sheep are a sexually dimorphic species, with males growing larger and often carrying bigger horns than the females. Generally, sheep are stocky animals with short legs and coats that tend to be brown or grey in colour with some patches of white (excluding Dall sheep and domesticated breeds selected for wool). Wild sheep naturally shed their hair like fleece yearly, as do some domesticated breeds of sheep such as Damara, Dorpers and Wiltshire Horns. Like many ungulates, their hooves have two digits, which is particularly advantageous for traversing uneven, rocky terrain (McCann 1956). One of the largest and most sexually dimorphic of the wild sheep is the Argali. These rams can reach lengths of 177–200 cm, with a shoulder height of 135 cm. Argali rams have been reported to weigh well over 200 kg and are described as being the size of a pony (Fedosenko and Blank 2005; Reading et al. 1997). The Argali rams also carry some of the biggest horns of any wild sheep, with horn lengths of over 130 cm. The Argali ewe in comparison ranges in weight of 40–100 kg (Fedosenko and Blank 2005; Reading et al. 1997). Bighorn

Fig. 1 Bighorn sheep ram and ewe, NPS/Diane Renkin. (Licence Public domain https://www.flickr.com/photos/yellowstonenps/37991290334/in/album-72157647227252459/)

sheep also reach large sizes with rams weighing between 90 and 200 kg and ewes weighing 40 and 90 kg, depending on the subspecies (Buchalski et al. 2016). The Urial is the least described of the wild sheep, they are reported to be one of the smallest species of sheep (Siraj-ud-Din et al. 2016) with rams weighing 40 kg and ewes only 25 kg (Awan 2006). Mouflon are the most similar in morphology to domesticated sheep; both rams and ewes stand at 70–80 cm at the shoulder and weigh 30–60 kg, unlike the other wild species, the ewes rarely grow horns (Cassinello 2022). Longevity of sheep ranges from 13–20 years depending on the species; however, males often have shorter lifespans of 13–15 years (Buchalski et al. 2016; Cassinello 2022), partially due to their increased risk of predation, use as a hunting trophy and yearly fights during the rut.

Like many ruminants sheep have specialised mouths and teeth to enable grazing. At the front of their mouth sheep have teeth on their lower dental pad but none on the upper dental pad (Cutress 1972). In addition, their upper lip is bisected with a small groove, called the philtrum, which allows sheep to selectively graze their food. Sheep spend the majority of their time foraging throughout the year (Bourgoin et al. 2008; Rosenbaum et al. 2020), particularly in winter as feed availability reduces. Some sheep display seasonal migratory behaviour in seek of quality forage (Enns et al. 2023; Poole et al. 2016; Singh et al. 2010b). Sheep are not strictly grazers and commonly consume forbs (herbaceous flowering plants), shrubs and trees as part of their diet (Marchand et al. 2013).

Sheep are gregarious animals, and this behaviour is seen in both the wild and domestic species. However, the size of flocks and their makeup can vary. The Argali form flock sizes of approximately six sheep (Reading et al. 1997), Snow sheep have an average flock of four ewes, with rams grouping in twos (Bazhenov 2022), whereas flocks of Bighorn ewes have been reported to contain 15 individuals (Buchalski et al. 2016; Mooring et al. 2003). It is also common to see sexual segregation with sheep, for most of the year, ewes stay in groups with their lambs, whereas the rams tend to roam together (Singh et al. 2010a). Segregation based on sex is a common occurrence in sexually dimorphic species and has been hypothesized to have a variety of advantages including, as a reproductive strategy (Main and Coblentz 1990; Main et al. 1996), variation in forage selection (Barboza and Bowyer 2001), variations in activity (Ruckstuhl. 1999) and as a social preference (Bon and Campan 1996). Indeed, sheep are highly social creatures forming close bonds within their flock (Veissier et al. 1998). One of the initial most stable relationships observed is that between an ewe and her female lamb, with these bonds maintained over years (Festa-Bianchet 1991). The biggest preference for social grouping, however, is peer-based (Doyle et al. 2016; Lawrence 1990; Ozella et al. 2020), for example, young Mouflon males have been observed to prefer companionship of other young adult males (Le Pendu et al. 1995). Sexual segregation can also be seen as a survival strategy. It has been observed that the ewes with lambs tend to stay within a home range, with rams more likely to roam over vast areas (Poole et al. 2016). As a result, the ewes don't always have access to the most nutritious feed but tend to be more sheltered and protected from predation (Berger 1991; Bourgoin et al. 2018; Bowyer and Rachlow 1998), whereas the rams end up being more vulnerable to predation (Singh et al. 2010a).

Domestic and wild ewes have seasonal breeding cycles determined by the photoperiod, with the breeding season commencing as the days get shorter in autumn and winter (Lindsay 1988; Rosa and Bryant 2003; Heinzen et al. 2023). Rams also exhibit reproductive seasonality, a trend more defined in wild rams (Lincoln 1998; Lincoln et al. 1990). The seasonality of sheep reproduction has been reviewed by Rosa et al. (2003). Like other ungulates such as deer, the reproductive season of wild sheep is known as the rut. During this time, male and female groups flock together, and the males compete for access to the ewes. Rams engage in these conflicts, performing a run up where they lift up off their front legs, running forward on just the hindlimbs before they come down and clash together with their horns (Bazyan et al. 2016; Hogg and Forbes 1997). Though horns are blunt weapons, the force of butting can lead to injury and death (Hogg et al. 1997). Successful rams will collect their ewes in a small harem that they will defend and mate with (Hogg 1987).

Gestation length in sheep varies between 130 and 170 days depending on the species (Nichols 1978). During partition, the ewe will remove themselves from the flock to give birth, returning once they have had the lamb (Langbein et al. 1998a, b). It is common for sheep to have litter sizes of one or two lambs (Fig. 2) (Awan et al. 2008; Cugnasse et al. 2005); however, domestic sheep production has pushed this to three and even four lambs (Corner-Thomas et al. n.d.). Increases in litter size can have detrimental effects on the ewe and the lamb. As litter sizes increases the nutritional requirements for the ewe changes and if not met can cause metabolic disease. Nutritional requirements during pregnancy, have been extensively discussed in (Edwards et al. 2011a; Ferguson et al. 2017; Young et al. 2016). Lambs that are born in multiple litters generally have lower birth weights compared to singletons and are likely to undergo fetal programming leading to alterations in their response to stress, metabolic changes and have impacted postnatal animal performance (Kenyon et al. 2019; Vautier and Cadaret 2022). In domestic sheep, dystocia is a common birthing issue that is a painful experience for both ewe and lamb and significantly impacts survival of both. Dystocia has been reviewed in depth by (Bruce et al. 2021; Jacobson et al. 2020).

There is much to talk about in regard to the natural history of sheep and the topic could only be briefly touched upon in this chapter. The biggest threat to wild sheep is predation, hunting and resource competition with other domestic livestock, there is also evidence of climate change impacting populations (van de Kerk et al. 2018). Currently there are two wild species listed as threatened on the IUCN Red List (Michel and Ghoddousi 2020; Reading et al. 2020).

Fig. 2 Birthing

3 Current Extent of Utilisation

Sheep are utilised in all aspects of the six environmental domains that are considered in this book, wild-free roaming, zoological collections, pets, laboratories, small scale and commercial farms.

There are several species of wild sheep as discussed in the previous section still free roaming from Asia and the Middle East to Europe and the Americas. The free-roaming populations are often hunted, especially the rams, as trophies for their large horns. The Argali is hunted in China, Mongolia and in neighbouring countries (Frisina and Frisina 2012; Schaller 2012), even as its status creeps toward 'threatened', with the practice being banned in India. Bighorn sheep are also hunted for trophies, as a management practice or method of conservation. Hunting is a heavily debated issue, it is often argued that trophy hunting can generate wealth within regional communities where they occur, and the cost of a licence to hunt an animal can be used for further conservation (Parker et al. 2023). However, there is also risk of negatively influencing breeding outcomes of the hunted species (Coltman et al. 2003), reducing biodiversity (Rashid et al. 2020), impact on predator species (Jumabay-Uulu et al. 2014), as well as corruption issues (Adhikari et al. 2021). The sentiments of local communities living near wild species indicate that they believe the negative aspects of hunting wild sheep outweigh the positives that the sport brings (Harris and Pletscher 2002; Nordbø et al. 2018).

Few of the wild sheep species are kept in zoological collections. As most species are low risk in terms of conservation status, there is no requirement to educate and improve numbers in captivity. You can commonly find domestic sheep as part of 'petting zoos' or in immersion and education areas within zoos, due to their smaller size, trainability and docility. For these reasons sheep are also kept as pets with many website blogs and even books (Weaver 2013) dedicated to sharing information on how to look after your pet sheep. When bottle raised, a lamb becomes imprinted onto the person that raises them and do not stray far, they quickly become sociable and friendly animals often with behaviour akin to a pet dog (Fig. 3). Due to the seasonality of the breeding cycle in sheep its recommended that either wethers (castrated males) or ewes be kept as pets due to the proclivity of rams to be aggressive especially during rut season where testosterone levels increase. Rams will charge and butt without warning and can cause severe injuries to their handlers. As with all pets especially, those that are uncommon, there is always risk of no prior research or poor education from those that keep pet sheep, leading to improper care of the animals and low understanding of the behavioural and physical needs the sheep require.

Sheep can also be found in laboratories, again owing to their size, docility, trainability and ease of keeping compared to other livestock.

Fig. 3 A pet lamb enjoying the comforts of indoor living much like a dog would

Table 2 Areas of biomedical research that use sheep as models, list is not exhaustive

Research area	Study
Bone fracture	Dias et al. (2018), Egermann et al. (2008) Fischer et al. (2023)
Dental fracture and regeneration	Altaii et al. (2016) Andreasen et al. (2006)
Neonatal research	Hunter et al. (2015) Klingenberg et al. (2013) Mani et al. (2022)
Neurodegenerative disease	Morton et al. (2014) Skene et al. (2017)
Traumatic brain injury	Finnie (2001)

Throughout the years sheep have been invaluable for biomedical research, with one of the most recognisable medical marvels being Dolly the sheep. In 1996, Dolly was the first animal cloned through cells collected from an adult animal, the feat caused quite a stir about the potential issues of cloning humans but not on the impact to sheep (Griffin 1999). Beyond the ease of care and adaptability to a laboratory setting, sheep are well-suited to act as models for many biomedical research applications, their lifespan also allows for tracking of disease progression, with measurements able to be taken over a longer period, compared to rat and mice models. They also have many aspects that are similar to humans, such as comparable brains (Jacobsen et al. 2010), and similar foetal development, the scope of their applicability to human health research is reviewed by Scheerlinck et al. (2008). Sheep are currently used in a variety of biomedical research areas (Table 2), their use in research must adhere to ethical standards and follow the strict guidelines involved in the use of animals for research in many countries (Council 2011; NHMRC 2013; Pereira et al. 2004; Union 2010). Part of the compliance for use in research includes assuming that animals feel pain in situations humans would, regardless of their responses and ensuring that pain and distress is minimised and alleviated. This falls under the 3 R's (Replacement, Reduction and Refinement) of animal use, with reducing the impacts of pain and distress associated with the category of refinement. As such sheep used in biomedical science are more closely monitored and cared for on an individual level and are more likely to receive pre- and post-surgical analgesics following invasive procedures. Unlike animals in the production system, there are no restrictions to the types of drugs that can be given to sheep in biomedical research, especially those such as opioids (Ahern et al. 2009; Christou et al. 2015) for use in relieving pain.

When it comes to their use in agriculture sheep are kept in a variety of different systems due to their adaptability to harsh climates and ability to be productive on land unfavourable for cropping or other livestock. They are also a favourable species for low-socioeconomic regions due to their value in producing a variety of products whilst requiring relatively low-input (Kosgey et al. 2008). How sheep are managed ultimately depends on several factors such as the system, location, purpose of the animal (milk, meat, wool or dual purpose) and breed. China, India and Australia are the world's largest sheep producing countries, however China over doubles India and Australia's numbers with 173 million sheep compared to the latter's 69 and 68 million head (IWTO 2022). Currently Australia is the biggest exporter of sheep meat and wool, exporting 562,000 tonnes of sheep and lamb meat and providing 22.1% of the world's clean wool (IWTO 2022; MLA 2023).

Domestic sheep can be roughly split into various groupings depending on the aspect of classification used. For wool there are shedding sheep, those which drop their fleece yearly, a trait common in meat breeds. Long wool breeds, such as Border Leicester and Romney, are often used as dual purpose (meat and wool) with the wool used for rugs and carpets. Medium wool breeds, such as Corriedale, have versatile uses, and then there is fine wool. Fine and superfine wool, like that produced by Merinos has a smaller fibre diameter, making it extremely soft and is used for clothing. Sheep can also be categorised by tail, with thin-tailed sheep (both long and short), fat-rumped tail sheep and fat tail sheep (both long and short) (Kalds et al. 2021). The variety of the breeds leads to a variety of issues that arise specifically to those types of animals. Fat-tailed sheep are unique breeds that accumulate fatty deposits around their rump and tail. This trait is beneficial in semi-arid environments where animals encounter nutritional stress; the fat in the tail mobilises to provide energy during periods of low feed availability (Alves et al. 2013; Atti et al. 2004). However, intensive genetic selection for this trait in production has led to some breeds having issues with movement and breeding (Kalds et al. 2021; Moharrery 2007). Wool breeds whether course or fine, all require yearly shearing as they do not drop their wool annually like the shedding breeds. The practice of shearing is an extremely stressful experience for sheep, studies measuring physiological and behavioural indicators of stress report increased cortisol (Aslan et al. 2013; Mears et al. 1999) and heart rate (Hargreaves and Hutson 1990), altered respiration rate (Aleksiev 2009), and changes in behaviour indicative of negative affect (Hefnawy et al. 2018; Sanger et al. 2011), as well as shearing being shown to induce oxidative stress (Fidan et al. 2009; Piccione et al. 2011). However, shearing is required to maintain animal health, and as wool sheep can produce approximately 3–5 kg of fleece a year (Blackwell and Henderson 1955; Ferguson et al. 2011; Saboulard et al. 1995), which accumulates without shearing, this excessive weight quickly becomes a welfare issue. Yearly shearing is also beneficial for reducing heat stress in wool sheep in regions with high summer temperatures (Pehlivan et al. 2020; Pennisi et al. 2004).

When it comes to production systems, in the past, pastoralism was the main system used for sheep, particularly in the Anatolia region (Fuks and Marom 2021; Makarewicz 2013). Though still practiced in some regions such as the Middle East, China and South America (Morris 2017), today this traditional system is in decline as civilizations settled in cities and competition for land use increased and started encroaching into droving pathways. Now sheep are managed in a variety of systems, such as extensive systems, like those in Australia and New Zealand to more intensive and semi-intensive systems such as feed lotting, or barn and shed systems, like those used in the Northern hemisphere or in the dairy industry. The sizes of domestic flocks can also vary greatly, with sheep kept in groups of hundreds or less, to larger flocks up into the tens of thousands, a stark contrast to their wild counterparts (Dwyer 2008). Each system has its own pros and cons in terms of sheep management and welfare. As there is so much history and diversity in sheep production systems, the following paragraphs will only briefly mention where sheep may encounter pain and distress in some of the main systems.

Extensive systems can occur with large flock sizes on large acreage farms, in remote hills, arid areas or even with small flock sizes held by smallholder producers. In extensive systems, predation is a common issue with domestic sheep, much like with wild sheep. Adult sheep face attacks from bears (Knarrum et al. 2006), wolves (Mattiello et al. 2012), big cats (Odden et al. 2002; Stahl et al. 2002) and wild dogs (Fleming and Korn 1989; Paas et al. 2021). Young lambs are also at risk of attacks from birds of prey and foxes (Alexander et al. 1967; Greentree et al. 2000; Hewson 1984; Saunders et al. 2010). Risk of predation increases with management style, such as coinciding lambing with predator breeding seasons, lambing close to forested habitats vs indoors, use of pest control (baiting, shooting or none) and with the size of farms (Mattiello et al. 2012; Moberly et al. 2003; Stahl et al. 2002; van Bommel and Johnson 2017). In some regions, such as South Africa, where smallholder extensive farms occur, sheep encounter increased health issues, poor nutrition, and low reproduction rates (Fourie et al. 2018; Kosgey et al. 2008).

Intensive systems can refer to both confined housing of animals as well as those kept on pasture. Intensive systems require more input than extensive systems not just with animal management, such as supplemental feeding and out of season breeding, but also in terms of environmental management, such as sowing and rotation of pastures and use of fertilisers (Grace et al. 2018; Morris and Kenyon 2014; Riedel et al. 2007). In countries with harsh winter conditions, sheep are often kept in barns during the winter months to protect them from the cold (Hristov et al. 2012; Jørgensen and Bøe 2011; Simensen et al. 2014), feedlot systems as well as dairy systems often implement similar intensive housing situations (Gelasakis et al. 2012; Longworth et al. 2010). In these systems, sheep can be more closely monitored and managed compared to extensive systems, where in some cases sheep are not observed regularly and potential illnesses or injuries may not be identified (Dwyer 2009; Goddard et al. 2006). If over stocked or poorly managed, intensive systems have an increased risk of high worm burdens (Uriarte and Valderrábano 1990), risk of diseases such as coccidiosis, cryptosporidiosis and ovine respiratory complex (González et al. 2014; González et al. 2016; Ridler 2008), heat stress (Nicolás-López et al. 2021; Stockman et al. 2006), as well as impacts on the animals ability to display normal behaviour (Dwyer 2009).

Live export of sheep overseas can be considered a type of intensive system. The European Union is one of the largest exporters of live sheep, followed by Romania and Spain (WITS 2022). Australia is the seventh largest exporter of live sheep (WITS 2022); however, due to long voyages and high temperatures at sea, Australia's industry has undergone extensive scrutiny due to sheep welfare concerns (McCarthy 2018). As a result, in 2023, the Australian Government announced the phasing out of the practice; however, an end date is yet to be set (DAFF 2023). Sheep on live export ships are kept at high stocking densities and are potentially exposed to high ammonia levels (Pines and Phillips 2013) and much like when kept in feedlots, are at increased risk of heat stress (Carnovale and Phillips 2020; Caulfield et al. 2014) and disease (Higgs et al. 1996; Higgs et al. 1993). Poor conditions on ships can also lead to high mortality rates (Carnovale et al. 2020; Richards et al. 1989).

4 Attitude Towards Pain

For a long time, sheep have been perceived to be, even by those who care for them closely, an animal that doesn't feel pain due to their stoic nature and perceived lack of intelligence. The nature of sheep as prey animals leaves them vulnerable and, therefore, unwilling to give any indication that they may be easily hunted (Gougoulis et al. 2010). Biologically, sheep have all the required mechanisms and pathways to feel pain as well as the emotionality required to internalize pain and its emotional subjective components. Sheep have displayed the ability to remember areas or situations where they have experienced stress or pain, such as yards, and will be hesitant to run through them again, and can also be more difficult to move and handle as a result (Rushen 1986). Therefore, the most ethical stance would be to automatically assume sheep feel pain in situations where humans would. Over time, the public has become increasingly invested and concerned about the welfare of livestock species, especially in relation to the animals experiencing stressful and painful events (Coleman 2018). Looking at different stakeholders' perceptions on the biggest concerns for sheep welfare, shows variations between groups as to what is seen as the most important welfare issue. Consumers often indicate that invasive and painful husbandry procedures are the biggest impact on welfare, compared to producers who cite health (nutrition, parasites, etc.) and environment (heat stress, predation) as the biggest welfare concerns (Cross et al. 2009; Doughty et al. 2017). The disconnect between consumer and producer, what is perceived to be better welfare, such as views that sheep raised extensively have a more 'natural' lifestyle and therefore have better welfare (Dwyer 2009; Goddard et al. 2006), as well as lack of communication can have severe consequences for the producers and their social licence to farm (Coleman 2017).

An example of this is the 2004 targeting of the Australian wool industry by the animal rights groups PETA and Animals Australia over the practice of mulesing sheep, a surgical procedure that involves the removal of skin from the backside of sheep whilst conscious with no pain relief (at the time). Mulesing was targeted as it was seen as a highly painful and cruel practice (Sneddon and Rollin 2010), with research confirming the procedure causes considerable pain and distress that negatively impacts lambs that have undergone mulesing (Hemsworth et al. 2009; Lee and Fisher 2007). The protests instigated by the animal rights organisations, the public and celebrities caused some clothing manufacturers to boycott the Australian wool industry (Coleman 2018; Sneddon et al. 2010). In response, the Australian wool industry declared a purported voluntary 'banning' of the procedure by the industry by 2010. In the lead up to 2010, alternatives to the mulesing procedure for reducing flystrike risk were investigated and taken up by some producers, such as breeding plainer sheep, increasing frequency and changing timing of shearing and crutching (Fisher 2011), and research into the creation of a flystrike vaccine (Kotze and James 2022). There have also been some incentives to move away from the procedure with voluntary reporting of mulesed and unmulesed operations through the national wool declaration, giving access to markets that want unmulesed wool and provide premiums for such. Since 2010, there has been no ban and no end to mulesing, recent surveys of wool producers across Australia show that more than 50% still mules their lambs, with this figure being higher in states more at risk of flystrike such as South Australia and Western Australia (Colvin 2022; Munoz et al. 2022). However, the use of pain relief has increased, and in 2022 it was reported that over 90% of producers that still mules now use some form of pain relief during the procedure (Colvin 2022).

In the United Kingdom, lameness in sheep as a result of footrot, a disease known to cause pain in sheep (McLennan et al. 2016; Zanolari et al. 2021), is a significant welfare issue (Bishop and Nieuwhof 2005; Dickins et al. 2016). There have been considerable efforts to increase producer knowledge on best practice methods for reducing lameness within their flocks (Best et al. 2020) to reach the Farm Animal Welfare Committee (FAWC) target of under 2% prevalence of lameness within flocks that was set for 2021 (Best et al. 2021). Attitudinal surveys of producers have shown that uptake and compliance with best practice methods to reduce lameness is complex and influenced by a variety of factors such as perceived effectiveness of procedures, poor farmer-veterinarian engagement, limited resources and the producers' emotions towards the disease (Best et al. 2020, 2021). Agreeance with negative emotions towards footrot, such as '*Having this disease in my flock makes me feel miserable*', (O'Kane et al. 2017) was linked with an inability to act on treating animals and higher levels of lameness within the producer's flock. The non-compliance of these producers is most likely linked to feelings of hopelessness rather than them being non-caring (O'Kane et al. 2017), highlighting the negative impact poor health of their animals can have on their carers (Best et al. 2021; O'Kane et al. 2017).

When considering producers' attitudes towards livestock welfare, it is a common sentiment that 'happy' healthy animals are those with good welfare. For Australian producers, many claim they wouldn't be in the industry if they didn't care for their animals (Buddle et al. 2021). Further studies assessing producer attitudes to their sheep has shown that positive attitudes towards management of their animals was correlated with better sheep welfare outcomes (Munoz et al. 2019). Australian (Buddle et al. 2021), Brazilian (Tamioso et al. 2017), Chilian (Larrondo et al. 2018) and United Kingdom (Dwyer 2009) producers acknowledge that certain procedures are painful for their animals (such as castration and tail-docking), but they are conducted to prevent future negative welfare impacts (such as flystrike) (Buddle et al. 2021). Although producers see procedures such as castration and tail-docking as painful, there is a disconnect between their beliefs and their behaviour, with these painful husbandry procedures still being undertaken without the provision of analgesia in many cases (Larrondo et al. 2018; Phillips and Phillips 2010). In contrast 30–40% of New Zealand farmers surveyed do not perceive castration and tail docking to be painful and less than 2% used pain relief (Kongara et al. 2023). Identified barriers to the provision of pain relief by producers include, lack of producer knowledge, lack of registered analgesic agents, cost and time (ease of administration) (Colditz et al. 2011; Kongara et al. 2023; Lizarraga and Chambers 2012). Interviews with Irish farmers on the use of pain relief highlight the lack of ability to identify sheep in pain, with many producer's first line of action when confronted with a listless sheep being to administer antibiotics, thinking the animal is sick (Crawford et al. 2023). Uptake of pain relief, such as non-steroidal anti-inflammatory drugs (NSAIDs), was more likely to occur when producers were initially encouraged by their veterinarian and when they witnessed the improved behaviour in their sheep after the administration of analgesia (Crawford et al. 2023). Where pain relief is readily available, such as improved access to registered and approved analgesics, producers are more likely to use it on their sheep (Buddle et al. 2021; Crawford et al. 2023; Howard and Beattie 2018).

A variety of stakeholders acknowledge that sheep feel pain, and addressing consumer concerns about painful husbandry procedures is important. The lack of pain relief use by producers in most countries is not due to the assumption that sheep do not experience pain, but rather due to the difficulty in observing pain, and the lack of registered and approved analgesic drugs available for use in sheep.

5 Anaesthesia, Analgesia and Pain

There are regulations and requirements for those caring for sheep to be able to identify pain and to provide analgesia to ensure good animal welfare. The knowledge and accessibility of pain relief is not equal between the six environmental domains discussed in this chapter. Sheep in zoological collections, those kept as pets and those used in biomedical research can be closely monitored on individual levels, and subtle changes in behaviour are more likely to be noticed when they are in pain. Sheep in these domains are also more likely to undergo invasive surgical procedures under general anaesthesia and be provided with peri-operative analgesia comparable to clinical practice standards. For animals in food, fibre, and dairy production, there are further regulations and guidelines regarding the use of veterinary drugs to ensure food safety for human consumption. These regulations vary on an international level (APVMA 2014), with some countries having access to registered anaesthetics and analgesics that other countries do not. Off-label and unregistered medicines can only be applied under veterinary direction and supervision and require long withholding periods for the animal before it can be sold for consumption. Due to these requirements, many producers that raise sheep do not have access to or the ability to provide pain relief to their sheep.

Research into the efficacy of anaesthetics and analgesics reflects the different requirements for sheep in the different domains. Efficacy of analgesics in biomedical research has previously been reviewed by Stillman and Whittaker (2019), as well as for production animals by Lizarraga et al. (2012) and Small et al. (2021b). Due to the extensive research into the anaesthesia and analgesia in sheep and the detail that is already included in previous reviews (Galatos 2011) and in veterinary textbooks (Pugh et al. 2020; Riviere and Papich 2018; Weil and Baird 2021), this chapter will try to capture and condense the essential knowledge of analgesics and anaesthetics and how that applies to sheep in their current utilization and how that impacts their pain experience and management.

6 Analgesia

There is extensive research into the efficacy of analgesics for pain management in sheep, with the type of analgesic investigated varying on the utilisation of the sheep. Within a biomedical and veterinary aspect, efficacy studies have been conducted on the use of opioids, α_2-adrenorecpetor agonists and non-steroidal anti-inflammatories (NSAIDs). In a production setting, most of the research is focused on NSAIDs and local anaesthetic agents. As previously stated, the efficacy of various analgesics in biomedical research, including agent, dose, method of administration and summary of effect, has been reviewed in Stillman et al. (2019). Similarly, a recent review for sheep in a production setting has been published by Small et al. (2021b), including the pharmacokinetic details of NSAIDs in sheep, alternative analgesic strategies and implication for their use in production settings. Due to the level of detail included in these reviews this section will summarise common analgesics used in the different environmental domains and the impact on sheep.

There are a limited number of pain-alleviating substances registered for use in food and fibre-producing sheep. For sheep in Australia, there are products such as Tri-Solfen® and Numnuts® which contain local anaesthetics registered for use during mulesing, and ring castration and tail-docking, respectively. The only other analgesics available for use in food-producing sheep in Australia include the NSAID, meloxicam, which is available as *Buccalgesic®* for oral transmucoal administration and Metacam injection for lambs undergoing mulesing, castration, and tail-docking. In other countries, the options are even more limited (Weil et al. 2021).

NSAIDs that have been investigated for their efficacy at alleviating pain associated with painful husbandry procedures in production animals include meloxicam, carprofen, flunixin and ketoprofen. Meloxicam is the most extensively studied NSAID in sheep. Pharmacokinetic studies of meloxicam administered through various routes in sheep report a 10–15 h half-life (Shukla et al. 2007; Woodland et al. 2019), with oral administration having the longest half-life at 15 h (Stock et al. 2013). Meloxicam has been shown to ameliorate the response of lambs to the pain associated with mulesing (Hancock et al. 2019), castration, and tail-docking (Small et al. 2014, 2018a, b). Though dose rates of 0.5 mg/kg are insufficient to provide analgesia in sheep (Colditz et al. 2019; Paull et al. 2008a). Use of meloxicam or flunixin for soft tissue surgery (laparotomy) in sheep was shown to provide some analgesia post-surgery but did not completely remove the pain associated with the procedure. (Viscardi et al. 2021). Sheep that had undergone surgery still had signs indicative of pain and inflammation around the surgical site (Viscardi et al. 2021). This is a common outcome for sheep undergoing surgery and painful husbandry procedures when only postoperative NSAID are administered.

Flunixin meglumine is a commonly used NSAID in veterinary practice and is registered for use in other food-producing animals such as cattle in the USA and Europe. A range of pharmacokinetic and efficacy studies have been conducted in sheep in a production setting, but flunixin is still currently unregistered for use in sheep in Australia. Flunixin administered intravenously has a reported elimination half-life of 2.4–3.8 h (Cheng et al. 1998; Welsh et al. 1993), compared to 7.9 h when administered orally through feed (Marini et al. 2016). Administration of flunixin through feed and intramuscular injection is effective at ameliorating the inflammation and pain associated with surgical castration and hot iron docking (Marini et al. 2017). In cases of footrot, flunixin has proven effective in increasing thresholds to noxious mechanical stimulation in sheep (Welsh and Nolan 1995) as well as reducing other indicators of pain, but does not improve recovery time (Kaler et al. 2010).

Carprofen, known for its extended half-life of over 30 h in sheep (Welsh et al. 1992), and demonstrated extended bioavailability when administered orally in sheep (Marini et al. 2015), can serve as a long-lasting analgesic. Furthermore, Carprofen has been found to lower the cortisol response to surgical castration and alleviate behavioural pain responses to ring castration (Paull et al. 2009). For more invasive procedures such as mulesing, carprofen has demonstrated efficacy in mitigating pain-related behaviours following mulesing in lambs, but doesn't show the same impact on physiological pain measures (Colditz et al. 2009; Paull et al. 2009). In contrast, neonatal lambs administered carprofen prior to ring castration and tail-docking were unable to ameliorate the pain experience (Price and Nolan 2001). Similar has been observed in older lambs (12 weeks) administered flunixin when undergoing ring castration and tail-docking (Marini et al. 2021) or application of meloxicam in 4–6 week old lambs undergoing ring castration and hot-iron tail-docking (Kells et al. 2020). However, partial analgesia has been reported in 4-week-old lambs administered flunixin or meloxicam subcutaneously to the site of ring castration (Paull et al. 2012). Studies in cattle have also shown limited efficacy of NSAIDs towards ameliorating ischemic pain (Paull et al. 2015; Petherick et al. 2014).

Even with the knowledge and extensive use of pain relief in the biomedical research context, there is little crossover to the animal production context. This is most likely due to a larger availability of analgesics for research and requirements on the use of registered and approved drugs for animals in food and fibre production. Comparison of the efficacy of different analgesic agents is often difficult due to the different factors in each study, such as age of sheep, type of pain, measures taken, and dosage of the agent used. As interest in animal pain increases, the development of pain assessment tools and pain management strategies is a high priority, especially for the production animal context.

6.1 Local Anaesthesia

The use of local anaesthetics for sheep occurs in both biomedical and production settings. In biomedical research these drugs are regularly administered with general anaesthesia, whereas in production settings, they may be used to provide local anaesthesia and analgesia at the time of procedures conducted on conscious animals. There is less research into the efficacy of various local anaesthetics compared to general anaesthesia. However, within a production and biomedical setting lidocaine (lignocaine) has been studied extensively, with some work conducted on bupivacaine and other local anaesthetics for their efficacy in sheep.

Lidocaine is a quick—acting, short duration local anaesthetic, however the durations and efficacy in providing anaesthesia or analgesia to sheep varies in the literature. Lizarraga et al. (2013) reported sensation blocking within 5 minutes in rams injected with lidocaine into the metacarpal, with sensory blocking to a nociceptive threshold test occurring for up to 40 min. Comparative studies that have looked at the efficacy of lidocaine when applied to animals experiencing pain from castration and tail-docking procedures, the duration of action vary from 10 min (Small et al. 2021a) up to an hour or more (Dinniss et al. 1997a, b). Lidocaine is typically injected subcutaneously in lambs prior to or at the time of the painful procedure to provide analgesia, studies that have been conducted vary in the sex and age of lambs, type of procedure, location, timing, and method of lidocaine application, which has led to the variation in reported efficacy (Table 3). Though timing and length of action varies, lidocaine has consistently been shown to provide some pain amelioration and indication of reducing hyperalgesia in lambs undergoing painful husbandry procedures.

Other application methods for local anaesthetics for producers have been explored due to the health and safety risks of handling needles, the difficulty of handling and variation in procedures conducted on lambs. In Australia one of the most successful and commonly used tools for providing pain relief to lambs is Tri-Solfen®, a topical

Table 3 Detail of studies that applied local anaesthetics to lambs undergoing castration and tail-docking and its efficacy at alleviating pain

Reference	Lamb details	Husbandry procedure	Lidocaine dose and application method	Pain assessment	Duration of effect
Ferrer et al. (2020)	Female 25–35 days	Knife tail–docking	Lidocaine 40.6 g/L, bupivacaine 4.5 g/L, adrenaline 24.8 mg/L, cetrimide 5.0 g/L mix at 0.5 mL/kg Topical after procedure	Cortisol Observation of pain related behaviour	Up to 2.5 h based on behaviour
Graham et al. (1997)	Not reported 3 weeks	Rubber ring, ring and clamp and hot iron tail-docking	Bupivacaine (0.25%) 1 mL injected subcutaneously 1–2 min prior 0.5 mL epidural injection 1–2 min prior	Cortisol Observation of pain related behaviour	Up to 84 min for subcutaneous Up to 36 min for epidural
Kells et al. (2020)	Male 4–6 weeks	Rubber ring castration and hot iron tail-docking	Lidocaine 2%, 5 mL diluted in 5 mL saline Subcutaneous injection to anteromedial and lateral surface of scrotum, each spermatic cord and either side at base of tail immediately prior	Observation of pain related behaviour	Minor amelioration between 15–45 min
Lomax et al. (2010)	Male 6–12 weeks	Mix of knife and hot iron tail docking and castration	Lidocaine 40.6 g/L, bupivacaine 4.5 g/L, adrenaline 24.8 mg/L, cetrimide 5.0 g/L mix at 0.5 mL/kg Topical after procedure	Light touch and pain stimulation Observation of pain related behaviour	Observed between 1 min and 4 h
Mellema et al. (2006)	Male 2–7 days	Castration with rubber rings and clamp	Lidocaine 4 mg/kg Injected at each spermatic cord and subcutaneously around the scrotal neck 5 min prior	Cortisol Observation of pain related behaviour Local palpation	Up to 2 h for rubber ring animals
Paull et al. (2009)	Male 5 weeks	Mix of knife and rubber ring tail docking and castration	Lidocaine 40.6 g/L, bupivacaine 4.5 g/L, adrenaline 24.8 mg/L, cetrimide 5.0 g/L mix at 0.5 mL/kg Topical after procedure	Cortisol Observation of pain related behaviour	Up to 6 h (based on cortisol) to knife procedure animals only
Small et al. (2020)	Male and Female 4–10 weeks	Rubber ring castration and tail-docking	Lidocaine 1.5 mL of 20 mg/mL Injected midline of the testes and tail at time of procedure	Observation of pain related behaviour Mothering up behaviour	Observed between 5 and 20 min
Small et al. (2021c)	Female 2–4 weeks	Ring tail-docking	Lidocaine 1.5 mL of 20 mg/mL Injected midline of the tail at time of procedure	Observation of pain related behaviour Mothering up behaviour	Up to 1 h
Sutherland et al. (1999)	Male 6 weeks	Ring castration and tail-docking	Lidocaine 2% Injected into three sites around the scrotal neck (2 mL total), 5–10 s prior Injected as above 5–10 s after	Cortisol	Observed between 20 and 60 min Note: Lidocaine was not applied to the tail

The variation in methods such as application method, age and sex of lambs, husbandry procedure, pain assessment and length of action is reported

application that contains lidocaine, bupivacaine, adrenaline and cetrimide. Tri-Solfen® is a registered product that does not require a veterinary prescription and is utilised by 96% of Australian sheep producers (Colvin 2022). Originally formulated for application on lambs undergoing mulesing (Lomax et al. 2008) the product has been tested for tail-docking and castration (Lomax et al. 2010) and there is interest in its use for shearing cuts and other painful experiences. Application of lidocaine (40.6 g/L), bupivacaine (4.5 g/L), adrenaline (24.8 mg/L) and cetrimide (5.0 g/L) mix at 0.5 mL/kg, has been shown to rapidly induce analgesia (1 min) (Lomax et al. 2013) for at least a period of 8 h based on wound sensitivity to touch, cortisol levels and behavioural indicators of pain (Lomax et al. 2008, 2013; Paull et al. 2007) and improves wound healing (Lomax et al. 2008) in lambs undergoing mulesing.

In recent years a dual action ring and local anaesthetic injection applicator known as the Numnuts® device (Fig. 4) has been developed and made available for commercial use in Australia and New Zealand. The device allows the producer to conduct ring tail-docking and castration whilst applying a single measured dose (1.5 mL) of lidocaine (2% concentration) adjacent to the ring, removing the requirement for multiple pieces of equipment (Small et al. 2020, 2021c). Other methods that have been explored to allow for simpler anaesthetic application include lidocaine-impregnated rubber rings. In a study conducted by Stewart et al. (2014), lambs castrated with rubber rings coated in 100 mg/mL of lidocaine experienced some pain amelioration but not to the same effect as injected lidocaine. The known dosage absorbed, diffusion and bioavailability of lidocaine to lambs in this study was unknown (Stewart et al. 2014), though it is known that absorption of lidocaine through intact skin can be limited. In cattle, lidocaine impregnated rubber rings appear to have more success and have been reported to contain 80–200 mg of lidocaine (depending on the size of the band) and can provide 0.5–3.5 mg lidocaine per g of surrounding tissue for up to 48 h (Saville et al. 2020).

During surgeries on sheep, local anaesthetics are often administered as an epidural or a nerve block. Similar to research conducted on production animals there is variation in local anaesthetic dose and application method, with reported onset of analgesia varying from 4 to 11 min (Ghadirian and Vesal 2013; Habibian et al. 2011) and duration of action from 30 min to 7 h (Lucky et al. 2007; Shafford et al. 2004). Stillman et al. (2019) have previously summarised biomedical research that implemented lidocaine and bupivacaine in sheep, including the dosage and duration of action.

Fig. 4 Using the Numnuts® device to provide local anaesthesia for ring castration and tail docking

As discussed, the reported duration of action of lidocaine and bupivacaine varies between studies, but bupivacaine is longer acting than lidocaine, with an average duration of action of 110 min (Lizarraga et al. 2013). Due to the relatively short duration of action of lidocaine, to avoid multiple applications, methods such as combining lidocaine with other drugs to increase the duration of effect have been explored. The addition of epinephrine (adrenaline) when administering local anaesthetics has been reported to improve the onset and duration of the anaesthesia (Sinnott et al. 2003). Sheep administered lidocaine (5 mg kg^{-1}) with epinephrine (5 μg mL^{-1}) can experience up to 30 min more analgesia compared to sheep receiving lidocaine alone (Ghadirian et al. 2013). Ghadirian et al. (2013) also evaluated the application of lidocaine with xylazine (0.05 mg kg^{-1}) and found this combination to be even more effective than lidocaine-epinephrine, providing a 3 min faster onset and a further additional 53 min of analgesia. Pharmacokinetics of ropivacaine (50 mg) and bupivacaine (30 mg) with and without epinephrine in sheep plasma have shown that epinephrine increased the elimination half-life of the local anaesthetics (Ratajczak-Enselme et al. 2007). In the study by Ratajczak-Enselme et al. (2007) the local anaesthetics were injected as an epidural, and the pharmacokinetics of the epidural and intrathecal space were also recorded. They found that the addition of epinephrine led to increased concentrations of the local anaesthetics in the epidural space and increased uptake in the intrathecal space.

A novel method of extending the duration of action of lidocaine in sheep has been explored by Militana et al. (2020) using lidocaine (200 mg mL^{-1}) encapsulated in co-polymer poly (lactic-co-glycolic) acid to provide a 7-day sustained release of the anaesthetic. The encapsulated lidocaine was injected to provide a peroneal nerve block in the animals, and provided blockage for 8 h and analgesia (based on nociception) for 90 h. The presence of the encapsulated lidocaine caused no histological abnormalities in the sheep (Militana et al. 2020).

Few studies have looked at a variety of local anaesthetics such as bupivacaine or procaine for application during painful procedures. One study by Small et al. (2021a) evaluated the efficacy of several local anaesthetics delivered at the site of ring application at relieving pain associated with ring castration and tail-docking in 3–5 week old lambs. The bupivacaine (1.5 mL at 10.75 mg) and procaine (1.5 mL with adrenaline, 75 mg) were assessed through observation of pain-related behaviours. Male lambs receiving lidocaine had shown an ameliorated response to the procedure for up to 10 min post procedure, lambs receiving bupivacaine showed no indication of analgesia in the first 20 min with pain avoidance behaviours showing a reduction in the 20–30 min post procedure. Lastly, procaine treated males had ameliorated response 10–20 min post procedure (Small et al. 2021a). In this study, female lambs only received a benefit from the application of procaine for up to 40 min post-tail docking.

Comparison of the efficacy (as a paravertebral thoracolumbar block and analgesia) of lidocaine (2%), levobupivacaine (0.5%) and ropivacaine (0.5%) has been tested in ewes undergoing ruminal fistulation (Oliveira et al. 2016). Lidocaine had a faster onset time of 1.5 min compared to ropivacaine (2.8 min) and levobupivacaine (3.1 min). However, the duration of effect was significantly longer for ropivacaine and levobupivacaine, at 649 min and 590 min, respectively, compared to 80 min for lidocaine (Oliveira et al. 2016). Although analgesic effects through nociceptive testing were investigated in this study, the results were not reported.

When administering local anaesthetic drugs, a combination of different agents can lead to lower doses of each drug, to achieve the desired effects. DeRossi et al. (2015) studied the combination of bupivacaine with an opioid. In the study, the authors compared bupivacaine (0.5 mg/kg 0.25%) alone to bupivacaine (0.25 mg/kg) with methadone (0.15 mg/kg) and methadone alone (0.3 mg/kg) administered by epidural injection. Analgesia was tested by the application of a deep muscle prick to various areas on the body from the tail up to the ribs. The onset of analgesia for bupivacaine and bupivacaine—methadone in this study was 5–10 min. However, the analgesic effects for bupivacaine lasted longer (240 min) compared to bupivacaine—methadone (180 min) (DeRossi et al. 2015). The inclusion of bupivacaine to the treatments also led to motor blockade for up to 85 min, resulting in ataxia and recumbency in animals, whereas methadone alone only led to mild blockade for 20 min. Application of methadone in treatments also leads to mild sedation (DeRossi et al. 2015). The authors report analgesic efficacy in this study, but it could be argued that the ewes were more likely experiencing loss of sensation through anaesthesia.

Due to the non-selective action of local anaesthetics, they can cause potentially lethal cardiovascular and central nervous system effects. Sheep are often used as a model to test the acute toxicity effects of local anaesthetics in biomedical sciences (Mather 2010). Work conducted by Copeland et al. (2008) demonstrated that the toxicity effects of local anaesthetics is affected by whether the animal is conscious or under general anaesthesia. Sheep were injected with the recommended highest dosages of bupivacaine, levobupivacaine, ropivacaine, mepivacaine, prilocaine or lidocaine, during separate test events. Application of either bupivacaine, levobupivacaine, ropivacaine and procaine to conscious sheep induced cardiovascular issues and central nervous system toxicity that resulted in the death of eight sheep (Copeland et al. 2008). Comparatively, when under general anaesthesia, application of toxic levels of local anaesthesia did not cause arrhythmias and resulted in no deaths (Copeland et al. 2008). The dosage that can cause toxicity varies between anaesthetics, in pregnant ewes, toxic effects were induced at doses of 5.0 mg/kg^{-1} for bupivacaine and 7.5 mg/kg^{-1} for ropivacaine with system collapse occurring at 8.5 and 12.9 mg/kg^{-1}, respectively (Santos et al. 1995). Although local anaesthetic toxicity is not likely to be a risk in controlled biomedical settings, there is potential for sheep to experience negative effects of local anaesthetics if they receive multiple doses whilst conscious during husbandry procedures.

Due to the short duration of action and variation in analgesic effects of local anaesthetics reported in research, it is suggested that they be combined with other longer acting analgesics to alleviate pain associated with painful husbandry procedures in a production setting (Durand et al. 2019).

6.2 Other Agents

Few agents that are used for general anaesthesia have analgesic properties and are used as part of pain management in laboratory and research settings. Prior to general anaesthesia, agents such as acepromazine and opioids are often administered to provide pre-emptive analgesia, sedation, and anaesthetic sparing effects.

Acepromazine provides tranquilization and is often used in combination with an opioid prior to general anaesthesia (Aarnes and Muir 2011). The use of acepromazine (0.03 mg/kg) in combination with buprenorphine (0.02 mg/kg) or morphine (0.3 mg/kg) in sheep has been shown to provide adequate sedation, with sheep able to be restrained with minimal resistance (Musk and Wilkes 2018) to a known stressful hold (shearing position) (Niezgoda et al. 1993).

Opioids are commonly used in veterinary medicine and biomedical science during cardiovascular and orthopaedic surgery and for postoperative analgesia. Comparison of intravenously administered methadone (0.3 mg/kg), fentanyl (2 μg/kg) and buprenorphine (0.01 mg/kg) has been conducted on anaesthetised sheep to assist in maintaining analgesia during spinal surgery. The study by Bellini and De Benedictis (2019) found that all three opioids were effective at reducing the impact of spinal surgery through maintenance of cardiovascular measures within clinical levels.

Fentanyl has been shown to increase thermal and mechanical nociceptive thresholds in sheep at a dose rate of 10 μg/kg (Waterman et al. 1990), and reduces pain in sheep following orthopaedic surgery (Ahern et al. 2009). When implemented as a single dose of 1.5 μg/kg^{-1}, fentanyl was found to provide no analgesia to ewes under anaesthesia as determined by anti-nociception testing (Aminkov et al. 2002). However, the study found that fentanyl was able to increase the onset and level of analgesia of the α-agonist xylazine (Aminkov et al. 2002). To induce analgesic effects of fentanyl, it can be administered as a constant rate infusion during surgery. Sedation and adequate analgesia during and post-surgery have also been obtained with fentanyl patches, which provide a transdermal delivery of 2 μg/kg/h (Ahern et al. 2009; Christou et al. 2015). To for fentanyl concentrations in plasma to be adequate to provide analgesia during surgery, the patches need to be applied at least 24 h beforehand (Christou et al. 2015). It is often ideal to use fentanyl patches for transdermal administration as part of the analgesic regime in sheep, as it reduces the requirements of handling and the use of needles.

Although ketamine is used as part of anaesthetic and analgesic protocols in biomedical and veterinary settings (Weil et al. 2021), there are limited studies investigating its efficacy as an analgesic in sheep (Stillman et al. 2019). One study by Guedes et al. (2006) administered ketamine (1 mg/kg) through epidural injection, pre-operatively in sheep undergoing orthopaedic surgery, and found there was no difference in pain scores following surgery when compared to animals not treated with ketamine. However, ketamine treated animals were less likely to require a rescue analgesic and had lower lameness scores in the weeks following surgery (Guedes et al. 2006). Alone, ketamine delivered as a subarachnoid injection at doses of 0.75 mg/kg and 1.5 mg/kg can provide effective short-term analgesia and sedation of up to 30 min for procedures such as transcervical artificial insemination in ewes (Carneiro et al. 2019). When testing the effect of ketamine on the mechanical nociceptive thresholds of healthy sheep, it did not have an effect. However, when sheep were injected with *N*-methyl-d-aspartate (of which ketamine is an antagonist to) to induce mechanical hypersensitivity in the sheep, administration of ketamine increased the sheep's mechanical thresholds (Lizarraga et al. 2008).

In sheep α_2 adrenoreceptor-agonists cause both sedative and analgesic effects and are used (in combination with other drugs) to induce and maintain anaesthesia (Coulson et al. 1989; Kästner 2006). The α_2-adrenoreceptor agonists medetomidine and xylazine have been reported to provide some analgesic effect in sheep, particularly in response to acute pain as demonstrated through increased thermal nociceptive thresholds for medetomidine (Musk et al. 2014) and reduced response to an electrical stimulus for xylazine (Grant et al. 2001; Grant & Upton 2004). At low doses (1.95 mM) the effectiveness of xylazine to provide analgesia can vary between individual sheep (Lizarraga et al. 2008). Medetomidine delivered intraperitoneal through an osmotic pump has also been shown to provide adequate post-surgical analgesia in ewes (Murdoch et al. 2013). In Merino sheep undergoing hoof trimming the use of xylazine as a sedative reduced the animals' stress response to the procedure, with xylazine-treated animals having lower heart rate, respiration rate and cortisol response compared with controls (Amin et al. 2022). Sheep have been reported to be sensitive to the effects of xylazine (Lizarraga et al. 2012), particularly when used for anaesthesia (Stillman et al. 2019), but not when used at analgesic doses (Grant and Upton 2001).

Though xylazine and ketamine individually have been reported to ameliorate acute pain, combined they were found to be less effective at reducing longer-term pain associated with surgery (Hughan et al. 2001). Similar results were reported for xylazine alone by Ley et al. (1991), where application of xylazine to sheep experiencing chronic pain from footrot did not improve their threshold to a mechanical threshold test. Xylazine administered to lambs undergoing mulesing has been shown to reduce impact at the time of the surgery but was unable to effectively ameliorate the pain associated procedure in the hours following (Paull et al. 2008b). When compared with buprenorphine, methadone or flunixin meglumine, again to an electrical stimulus, xylazine was the only agent to provide analgesia in sheep (Grant et al. 1996). Considering the modes of action of the other agents it is unsurprising that they had little impact on the stimulus used. For example, the NSAID flunixin meglumine inhibits cyclooxygenase and is highly effective at reducing inflammatory pain in sheep (Marini et al. 2017; Paull et al. 2008b). It is possible that xylazine provides sedation or anaesthesia, leading to limited reactions in sheep rather than providing analgesia. Indeed, xylazine has been reported to have a local anaesthetic effect (Aziz and Martin 1978) which has been observed in other species such as cattle (Grubb et al. 2002) and horses (Grubb et al. 1992).

Benzodiazepines such as diazepam and midazolam are commonly used for inducing anaesthesia in sheep. Midazolam has been shown to be an effective short duration sedative when administered intravenously and intramuscularly (Kyles et al. 1995; Simon et al. 2017) at doses of 0.5–0.7 mg/kg (Simon et al. 2017; Yavuz et al. 2021). When applied as a sedative, midazolam has a dose-dependent effect; administration leads to decreased respiration and cardiovascular depression (Upton et al. 2009). Beyond sedation, midazolam has been shown to increase nociceptive thresholds in response to thermal and mechanical stimulation (Kyles et al. 1995), indicating an ability to provide analgesia to sheep.

7 Pain in Sheep

Pain is an unavoidable life experience that sheep will encounter in each of the environmental domains. In their natural environment, sheep are likely to encounter situations where they will experience pain and suffering. However, pain is an integral experience of living and pain itself is not necessarily a bad thing. The experience of pain is essential for survival, to learn which situations to avoid and what food should not be consumed.

Injuries incurred through misadventure or fighting between rams, or the pain and stress experienced through birthing, predation, and hunting by humans, are experiences wild sheep are likely to encounter. In zoological settings sheep are likely to experience similar injuries; however, predation and hunting is avoided. These animals are also closely monitored with prompt veterinary intervention if it is determined that an individual is experiencing pain or distress. Pet sheep can also experience similar situations to wild and sheep kept in a zoo environment, depending on the level of care provided.

Through their domestication sheep are now exposed to sources of pain that otherwise would not occur in wild settings. Sheep in laboratory settings undergo invasive surgical procedures and experience pain as a result of surgery or through exposure to certain models of disease.

In a production setting, as humans' intensification of sheep production increased, so did selective breeding sheep for specific traits associated with meat and wool production. This has led to the introduction of certain husbandry procedures required to manage sheep, but also increased the likelihood of sheep experiencing pain in their lifetime. It's common practice in parts of the world for lambs to undergo invasive husbandry procedures. Procedures that are part of yearly routine for many sheep farmers include ear-tagging, castration, tail docking and in some cases mulesing, which have all been shown to cause pain (Dinniss et al. 1997a, b; Fisher 2011; Lay et al. 1992; Stafford and Mellor 2005). These husbandry procedures are carried out to improve long-term health and welfare of the animal. For example, tail-docking to improve carcass traits (Moharrery 2007), tail-docking and mulesing to reduce the incidence of flystrike (Lee et al. 2007; Trentini et al. 2013) or castration to prevent unwanted indiscriminate breeding and reduce aggression in animals (Needham et al. 2017). Ear tagging is often required for identification and for tracking and biosecurity. Husbandry procedures can be conducted using a variety of methods (Table 4), each with varying levels of pain and pain-related responses. Conduction of castration, tail-docking and mulesing varies between countries, systems and sheep type. This includes variation in the age of the sheep when the husbandry procedure occurs, and the methods used.

Table 4 Common husbandry procedures conducted, and the various methods implemented on sheep

Procedures	Method	Description
Castration	Rubber ring	Rubber rings are applied to the neck of the scrotum which causes the scrotal sac and testes to atrophy and eventually drop off due to the prevention of blood flow.
	Bloodless castrator	A clamp is applied to the scrotal neck and crushes the spermatic cord, this also causes the scrotal sac and testes to atrophy
	Surgical	This involves cutting the scrotum and removing the testes via cutting or twisting the spermatic cord, with or without cauterising or clamping.
Identification	Ear marking/notching	Using specialised pliers, the ear is cut in a specific pattern
	Tagging	A tag applicator is used to puncture a plastic tag through the ear
Mulesing	Clips	Clips are applied the skin adjacent to the anus/vulva and tail, causes atrophy and necrosis before skin section drops off
	Cobbett skin traction treatment	Sodium laurel sulphate is injected intradermally around the anus/ vulva area, treated skin becomes necrotic and falls off
	Steinfort liquid nitrogen treatment	The skin around the anus/vulva area is pinched with a liquid nitrogen cooled clamp, the treated skin becomes necrotic and falls off
	Surgical	Using mulesing shears the skin from around the anus/vulva area and/or tail is removed
Tail docking	Rubber ring	Rubber rings are applied at the site of attachment of the caudal folds to the tail
	Bloodless docking	While clamped the tail is cut off distal to the clamp
	Surgical	Tail is cut with a sharp knife
	Hot iron	A clamp device is heated using gas and cauterises the tail as it cuts

Ear tagging is a painful experience for lambs causing immediate behavioural responses to the procedure including struggling and vocalisation which varies slightly with tag and application type (Edwards et al. 2001). There are long healing periods of up to 3 weeks associated with tagging as well as high infection rates experienced by the animals (Edwards et al. 2001). Reported infection rates following tagging are as high as 50–80% in lambs, depending on if the type of tag used (metal tip vs plastic) (Karakuş et al. 2015). When done incorrectly tagging can cause scarring and tearing in the ears. There is minimal intervention for such a procedure and minimal if any pain relief used.

Extensive research has documented the pain associated with castration and tail-docking (Fig. 5), regardless of the method implemented (Lester et al. 1996; Melches et al. 2007; Mellor and Stafford 2000), however, the experience varies due to the amount of tissue involved and the different mechanisms (surgical vs ischemic) (Grant 2004). There are also longer-term consequences of the procedures for example lambs that have their tails docked shorter than recommended (length of 1–2 coccygeal vertebrae vs the recommended practice of 3) are more at risk of developing arthritis (Lloyd et al. 2016). There is also evidence that procedures such as tail-docking in sheep could lead to longer term or chronic pain (French and Morgan 1992; Johnston et al. 2022).

Outside of Australia, mulesing is rarely, if ever, conducted and is a procedure unique to sheep that produce fine wool such as Merino's and often dictated by the climate. Sheep in areas with summer rainfall or sheep with breech areas contaminated with urine and faeces are at higher risk of flystrike, where the *Lucilia cuprina* fly lays eggs in damp wool of the sheep, leading to maggots that start to eat the wool and flesh of the sheep. Flystrike is a painful condition that leads to poor welfare of the animal (Colditz et al. 2005) and can eventually lead to death. Breech modification methods, such as mulesing and tail docking, are done within Australia to reduce the prevalence of the issue (Lee et al. 2007).

Fig. 5 Lamb in a cradle undergoing hot knife tail docking

Conventional mulesing, which involves the surgical removal of skin around the breech area is a known painful experience (Edwards et al. 2011b; Hemsworth et al. 2009; Jongman et al. 2000). Multiple mulesing alternatives have their own advantages and disadvantages in terms of pain and wound healing in sheep. One study, which compared clips (polypropylene clips that's compress blood vessels in the fold of skin) to conventional mulesing, it was found that lambs mulesed with clips displayed significantly less pain-related behaviours and were more likely to walk and interact with their environment compared to conventionally mulesed lambs (Hemsworth et al. 2012). Looking at the systemic effects of mulesing Lepherd et al. (2011a) found that the use of clips for mulesing had significantly reduced acute-phase protein concentrations and haematological, biochemical changes as well as no body weight effects when compared to conventional mulesing. The use of clips has also shown to create significantly less myofibroblast expression at the wound site compared to mulesing (Lepherd et al. 2011b).

Ewes often undergo reproductive surgeries such as laparotomy and multiple ovulation and embryo transfer for improved genetic gain and due to difficulties of artificial insemination in sheep. These procedures are known to cause pain (Teixeira et al. 2011; Viscardi et al. 2021). The birthing process is known to be painful, although of course in the wild there are no interventions and pain relief for animals. As discussed under 'natural history' sheep are likely to experience for pain during lambing, particularly if lambs and ewes undergo traumatic birth experiences, known as dystocia. Currently, the long-term consequences are unknown for either. As we continue to push breeding for higher production (multiple births and higher weights) there could be a detrimental impact to the animal and this process.

There are several factors that can affect a sheep's experience of pain, including the age of the animal, sex (Guesgen et al. 2011, 2014; Marini et al. 2023), breed and presence of stressors (Clark et al. 2011; Pinto-Ribeiro et al. 2004). Female lambs have been shown to have increased pain response to tail-docking compared to male lambs (Marini et al. 2023). Provision of comfortable material and as well as the presence of lambs not experiencing pain can also reduce the animal undergoing tail-docking's experience of pain (Larrondo et al. 2021). Presences of a conspecific not experiencing pain when undergoing castration does not influence the experience of pain however (Cho et al. 2020).

Sheep undergo invasive husbandry practices at a relatively young age, but it ranges from neonatal to several months old. Studies report that experiencing pain at young ages can have impacts on pain experience in later life, leading to subsequent hyperalgesia (Pattinson and Fitzgerald 2004). Castration of lambs at 1 day of age had increased responses (as measured by electroencephalography) to tail-docking at 23 days of age compared to lambs that were only handled at 1 day of age (Impey 2015). Similar has been observed using behaviour measure of pain in lambs, with lambs castrated at one day of age having an increased response to tail-docking at 21 days but not at 42 days (Pifelet 2011). Additionally, a study conducted by McCracken et al. (2010) observed that tail-docking lambs separately from castration revealed an age-dependent effect. Lambs castrated at 1 day old exhibited a more pronounced reaction to subsequent tail-docking compared to lambs castrated at 10 days of age. In lambs undergoing ring castration at two days of age, administration of the local anaesthetic lidocaine impacted the development of lesion and reduced the experience of chronic pain, with hyperalgesia observed in lambs that didn't receive pain relief (Kent et al. 2000). Experience of stressors in early life has also been shown to cause hyperalgesia in lambs undergoing painful procedures (Clark et al. 2011). These studies suggest that the timing of these procedures in lambs, as well as administration of pain relief, can influence how lambs respond to subsequent painful events, highlighting the need to consider age when conducting painful husbandry procedures.

8 Pain Assessment

There are various methods employed to determine the pain state of sheep, current pain assessment is limited to mostly subjective or indirect measures of pain. One of the most common methods of determining if sheep are in pain is the animal's behaviour. Obvious behaviours include those actively displayed by animals at the time of the painful procedure such as struggling, kicking and vocalisation. Following procedures though behaviours can become more subtle and difficult to observe. After castration and tail-docking, lambs exhibit new 'abnormal' behaviours (Grant 2004) often referred to as active pain or avoidance behaviours. These active behaviours include rolling, touching the wound site and kicking (Mellor and Murray 1989; Molony 1993; Kent et al. 2000) and are often displayed by lambs within the first hour following the procedures, after which their observation significantly decreases (Molony et al. 2002). Observation of behaviour during and beyond the first hour to determine pain state, include determining postural behaviours, aberrations in normal movements such as standing and walking, displayed by the lamb. Postures that have been validated to occur and be indicative of lambs in pain include statue standing (standing stiffly akin to a saw-horse), lateral lying (lying with all limbs extended) and abnormal walking (walking stiffly or with an abnormal gait).

Assessment of behaviour requires familiarity with the animal in a specific environment, and careful observation for the subtle changes that occur when a sheep is in pain. For example, following painful husbandry procedures, changes in activity are observed, in the case of mulesing, lambs will often decrease their normal activity, stand stiffly and avoid moving (Edwards et al. 2011b; Hemsworth et al. 2009; Lomax et al. 2008). It's also common to see behaviours classed as 'abnormal' to be present, such as repeated movements of getting up and down when trying to sit, or lambs holding their rumps up off the ground (Grant 2004; Marini et al. 2017; Molony et al. 1993; Small et al. 2018b). However, the behaviours displayed, or the behaviour changes seen vary with the state of pain experienced. For example, when lambs are castrated and tail docked with rings there is an increase in lying behaviour observed (Lester et al. 1996; Molony et al. 1993; Small et al. 2020), however the reverse is true for surgical castration and tail docking (Small et al. 2014). This variation in what could be classed as observable pain, can make it difficult to determine if the animal is in fact experiencing pain and at what point we should intervene.

Alternative non-invasive methods that look to determine the emotional effects of pain states in lambs include grimace scales and qualitative behavioural assessment. Grimace scales have been developed for a variety of animals including rats, rabbits and horses, grimace scales are a non-invasive method that have been validated to identify and assess pain. In lambs and sheep there are five facial action units that change when experiencing pain, they include the nose, mouth, cheek, eyes and ear posture (Guesgen et al. 2016; Häger et al. 2017; McLennan et al. 2016). Sheep grimace scales have been successful at identifying sheep in pain following surgery (Zentrich et al.

2023) and in sheep experiencing footrot and mastitis (McLennan et al. 2016). However the use of grimace scales have reported inconstancies for identifying pain in sheep (Viscardi et al. 2021), something that has been observed through use by the author as well, particularly with Merino lambs, where factors such as wool coverage on the face and wrinkle affected the ability to confi-

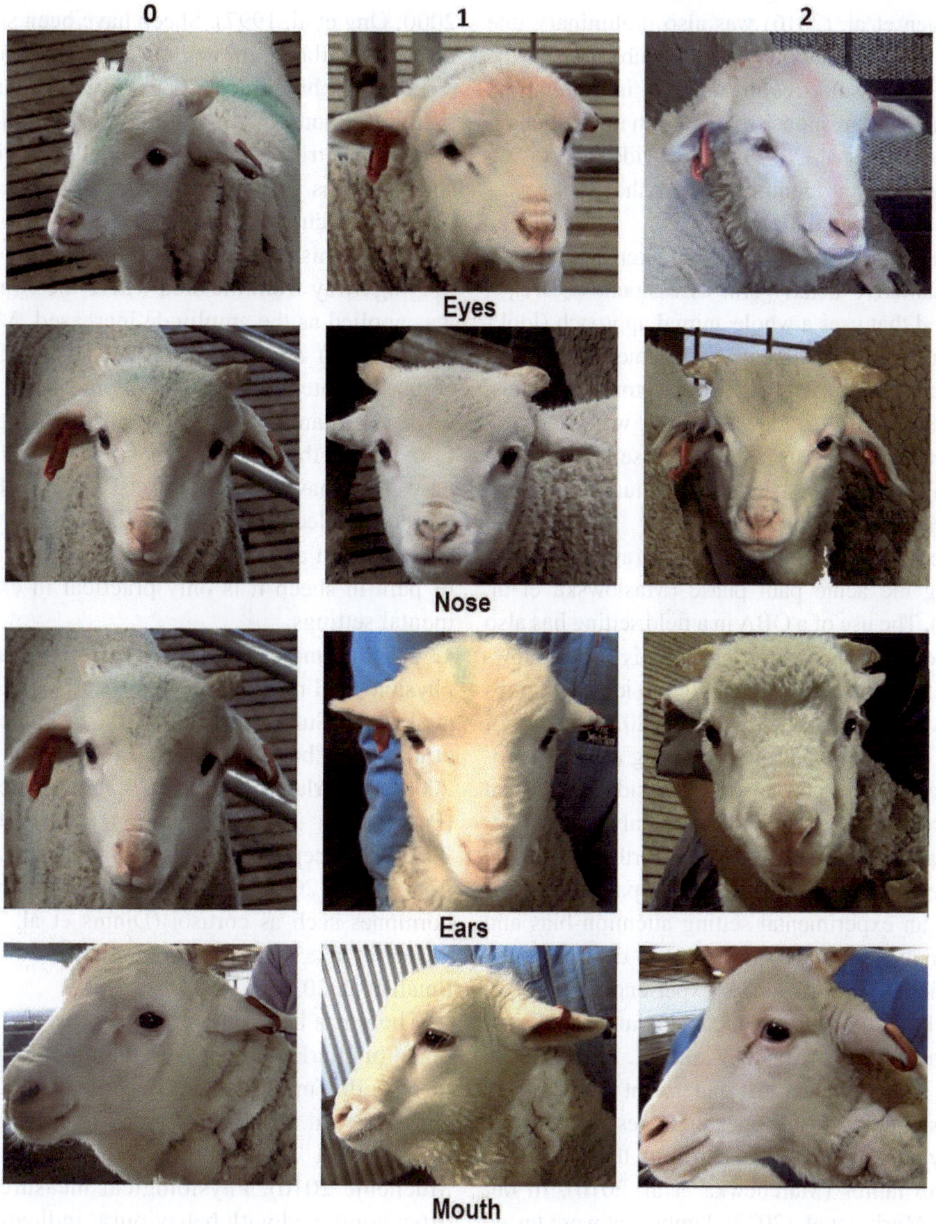

Fig. 6 Examples of the changes and scores given to Merino lambs using the 3 point scale grimace scale, 0 (not present), 1 (moderate), 2 (obvious). When in pain, eyes become more closed, nostrils become tighter and flatter, the mouths normal appearance of a curved 'smile' becomes a flattened line, cheeks appear hollow, ears are rotated backwards and faced downwards. Cheek images are not highlighted as they were difficult to discern in Merino lambs

dently see and score aspects of the grimace scale, such as cheek hollowing and orbital tightening (Fig. 6). A review of sheep and lamb grimace scales has reported that they have a moderate to unsatisfactory validity compared to other animal grimace scales (Evangelista 2022). The study by Guesgen et al. (2016) was also preliminary one and due to the low number of lambs used in the study the results require careful interpretation. Grimacing in response to pain is an innate behaviour (Hicks et al. 2001) that provides promising results for pain identification in sheep and warrants further validation.

There has also been growing interest in the use of qualitative behavioural assessment (QBA), a method that uses a whole animal approach (looking at behaviour, activity and demeanour of the animal) and assigning descriptors (nervous, excited, dull) to identify its welfare state (Wemelsfelder et al. 2001). The use of QBA following painful husbandry procedures has been successfully implemented to discriminate between different methods of castration in lambs during the acute pain phase (Masłowska et al. 2020). The use of a QBA in a field setting has also been effective at identifying animals in pain, however it was not sensitive enough to identify lambs receiving analgesia (Grant et al. 2020). Many of these novel methods are promising and may provide alternative non-invasive pen side assessment to determine if an individual animal is experiencing pain. However, they require further validation between sheep breeds and type of pain.

In an experimental setting attention bias and reactivity tests have been used to determine the emotional state of lambs experiencing pain, so far it has only been conducted on lambs undergoing tail-docking (Marchewka et al. 2016; Marini et al. 2023). Lambs that underwent tail -docking and subjected to an open field test were more likely to stand and abruptly climb the arena than control lambs (Marchewka et al. 2016). In our study Marini et al. (2023), lambs that were tested individual to an emotional reactivity test were more likely to exhibit fear-related behaviour following tail-docking compare to lambs that were handled, indicating that the lambs experiencing pain were likely in a negative affective state.

Brain activity has been used to indicate the pain experience by sheep following husbandry procedures (Harris et al. 2020; Jongman et al. 2000; Ong et al. 1997). Sheep have been shown to have similar brain activity as humans using electroencephalography (EEG) in response to pain. One study by Ong et al. (1997) applied a human electrical stimulus model of pain to sheep. It was found that as the amplitude of a painful electrical stimulus increased, EEG increased, this correlated with the behaviour of moving away from the area where the stimulus was applied as the amplitude increased. Morris et al. (1997) also used EEG and event-related cerebral potential (ERP) to measure pain response to an electrical stimulus in sheep and found that there was a positive correlation between behaviour response, EEG, and ERP to increasing electrical stimulus. Though the measure of brain activity highlights the experience of pain in sheep it is only practical in experimental settings.

Other common indicators of pain include physiological measures such as heart rate (Peers et al. 2002; Stubsjøen et al. 2010a, b; Sutherland et al. 2020), body temperature (Stubsjøen et al. 2009; Sutherland et al. 2020), blood pressure (Peers et al. 2002), respiration (Orsetti et al. 2022) nociceptive thresholds (Fisher 2011; Lomax et al. 2009; Stubsjøen et al. 2010a, b) and hormones such as cortisol (Dinnis et al. 1997; Lester et al. 1996; Mellor and Murray 1989; Small et al. 2018b).

There has been an increase in research into novel biomarkers to measure pain with a focus on specific microRNAs (Lecchi et al. 2018; Qureshi et al. 2016). This area is still new in sheep, with limited research (Fayemi and Muchenje 2016). Physiological measures are often combined with behavioural indicators of pain in order to get a more complete picture. However, many of these methods are time-intensive and costly and though applicable to a research scenario, are unrealistic for use in a production system. Currently, chronic pain isn't something that can be reliably recognized or 'measured'. This has impacts for the lifetime welfare assessment of sheep, especially those kept for the long term.

9 Summary

Sheep have been a consistent part of human civilisation for thousands of years; their domestication and utilisation have seen them spread across the world to a variety of environments where they have been extensively bred for unique uses. Throughout the different environmental domains, sheep will experience pain in their lifetime; however, animals in production settings are more likely to experience pain as a result of the implementation of painful husbandry procedures. Although domesticated, the drive for survival leaves sheep displaying minimal signs of pain, that in turn leaves their guardians less equipped to assist when they may need it most. Determining pain in humans is difficult despite verbal expression of pain, and in sheep it is especially difficult given their propensity to hide signs of pain. Novel techniques such as use of qualitative behavioural assessment and grimace scales provide promising methods to identify individual animals in pain that could be implemented in zoos, on pets, in the laboratory and in production settings. However, even as pain is more reliably identified and quantified, the treatment of pain remains a challenge in this species as the availability of safe and efficacious analgesic agents is limited. Society, veterinarians, animal carers and all those involved in the care and husbandry of sheep have a responsibility to ensure that any pain and suffering inflicted on animals is minimised and treated. Having brought sheep on the journey into dependency on domestication a good life and protection from pain and suffering is required.

References

Aarnes TK, Muir WW (2011) Chapter 26—pain assessment and management. In: Peterson ME, Kutzler MA (eds) Small animal pediatrics. W.B. Saunders, Saint Louis, pp 220–232

Adhikari L, Khan B, Joshi S, Ruijun L, Ali G, Shah GM, Ismail M, Bano K, Ali R, Khan G (2021) Community-based trophy hunting programs secure biodiversity and livelihoods: learnings from Asia's high mountain communities and landscapes. Environ Chall 4:100175

Ahern BJ, Soma LR, Boston RC, Schaer TP (2009) Comparison of the analgesic properties of transdermally administered fentanyl and intramuscularly administered buprenorphine during and following experimental orthopedic surgery in sheep. Am J Vet Res 70:418–422

Aleksiev Y (2009) The effect of shearing on the behaviour of some physiological responses in lactating Pleven blackhead ewes. Bulg J Agric Sci 15:446–452

Alexander G, Mann T, Mulhearn C, Rowley I, Williams D, Winn D (1967) Activities of foxes and crows in a flock of lambing ewes. Aust J Exp Agric 7:329–336

Altaii M, Broberg M, Cathro P, Richards L (2016) Standardisation of sheep model for endodontic regeneration/revitalisation research. Arch Oral Biol 65:87–94

Alves SP, Bessa RJB, Quaresma MAG, Kilminster T, Scanlon T, Oldham C, Milton J, Greeff J, Almeida AM (2013) Does the fat tailed Damara Ovine breed have a distinct Lipid metabolism leading to a high concentration of branched chain fatty acids in tissues? PLoS One 8:e77313

Amin S, Wagner R, Fieseler H, Rizk A, Schmicke M, Spilke J, Mielenz N, Baumgartner W, Schären-Bannert M, Starke A (2022) Sedation with xylazine hydrochloride decreases the stress response in merino meat sheep during routine hoof trimming in a tilt table. Front Anim Sci 2

Aminkov BY, Dinev D, Pascalev M (2002) The antinociceptive and cardiopulmonary effects of extradural fentanyl–xylazine in sheep. Vet Anaesth Analg 29:126–132

Andreasen JO, Munksgaard EC, Bakland LK (2006) Comparison of fracture resistance in root canals of immature sheep teeth after filling with calcium hydroxide or MTA. Dent Traumatol 22:154–156

APVMA. (2014). Food safety studies for veterinary drugs used in food-producing animals. In: A.P.a.V.M. Authority (Ed.): Australian Governments

Aslan R, Bayram I, Cetingul IS, Sahin EH, Sengor E, Yardimci M (2013) Stress responses to comparative handling procedures in sheep. Animal 7:143–150

Atti N, Bocquier F, Khaldi G (2004) Performance of the fat-tailed Barbarine sheep in its environment: adaptive capacity to alternation of underfeeding and re-feeding periods. A review. Anim Res 53:165–176

Awan GA (2006) Conservation of Punjab urial (Ovis vignei punjabiensis) through long-term monitoring of marked individuals. The Rufford Maurice Laing Foundation, London, pp 1–22
Awan GA, Festa-Bianchet M, Gaillard J-M (2008) Early survival of Punjab urial. Can J Zool 86:394–399
Aziz MA, Martin RJ (1978) α Agonist and Local anaesthetic properties of Xylazine. Zentralbl Veterinarmed A 25:181–188
Barboza PS, Bowyer RT (2001) Seasonality of sexual segregation in dimorphic deer: extending the gastrocentric model. Alces 37:275–292
Bazhenov YA (2022) Population structure of the snow sheep (Ovis nivicola, Artiodactyla, Bovidae) in the Kodar mountain ridge, Transbaikalia, Russia. Biol Bull 49:1638–1646
Bazyan S, Asadi H, Rezaei HR, Mesdaghi M (2016) Mating behaviour of wild sheep in captivity (Case study: Laristan Mouflon, Ovis orientalis laristanica). Der Zoologische Garten 85:137–151
Bellini L, De Benedictis GM (2019) Effect of three opioid-based analgesic protocols on the perioperative autonomic-mediated cardiovascular response in sheep. Lab Anim 53:491–499
Berger J (1991) Pregnancy incentives, predation constraints and habitat shifts: experimental and field evidence for wild bighorn sheep. Anim Behav 41:61–77
Best CM, Roden J, Pyatt AZ, Behnke M, Phillips K (2020) Uptake of the lameness Five-Point Plan and its association with farmer-reported lameness prevalence: A cross-sectional study of 532 UK sheep farmers. Prev Vet Med 181:105064
Best CM, Pyatt AZ, Roden J, Behnke M, Phillips K (2021) Sheep farmers' attitudes towards lameness control: Qualitative exploration of factors affecting adoption of the lameness Five-Point Plan. PLoS One 16:e0246798
Bishop SC, Nieuwhof GJ (2005) Costs of the major endemic diseases of sheep in Great Britain and the potential benefits of reduction in disease impact. Anim Sci 81:23–29
Blackwell RL, Henderson CR (1955) Variation in fleece weight, weaning weight and birth weight of sheep under farm conditions 1. J Anim Sci 14:831–843
Bon R, Campan R (1996) Unexplained sexual segregation in polygamous ungulates: a defense of an ontogenetic approach. Behav Process 38:131–154
Bourgoin G, Garel M, Van Moorter B, D. D, Maillard D, Marty E, Gaillard JM (2008) Determinants of seasonal variation in activity patterns of mouflon. Can J Zool 86:1410–1418
Bourgoin G, Marchand P, Hewison AJM, Ruckstuhl KE, Garel M (2018) Social behaviour as a predominant driver of sexual, age-dependent and reproductive segregation in Mediterranean mouflon. Anim Behav 136:87–100
Bowyer RT, Rachlow JL (1998) Habitat selection by Dall's sheep (Ovis dalli): maternal trade-offs. J Zool 245:457–465
Britannica (2022) Khnum. In: T.E.o. Encyclopaedia (Ed.): Encyclopedia Britannica
Bruce M, Young JM, Masters DG, Refshauge G, Thompson AN, Kenyon PR, Behrendt R, Lockwood A, Miller DW, Jacobson C (2021) The impact of lamb and ewe mortality associated with dystocia on Australian and New Zealand sheep farms: a systematic review, meta-analysis and bio-economic model. Prev Vet Med 196:105478
Buchalski MR, Sacks BN, Gille DA, Penedo MCT, Ernest HB, Morrison SA, Boyce WM (2016) Phylogeographic and population genetic structure of bighorn sheep (Ovis canadensis) in North American deserts. J Mammal 97:823–838
Buddle EA, Bray HJ, Ankeny RA (2021) "Of course we care!": a qualitative exploration of Australian livestock producers' understandings of farm animal welfare issues. J Rural Stud 83:50–59
Bunch TD, Wu C, Zhang Y-P, Wang S (2005) Phylogenetic analysis of snow sheep (Ovis nivicola) and closely related taxa. J Hered 97:21–30
Carneiro RPB, Macedo GG, DeRossi R, Costa-e-Silva EV, Souza MIL (2019) Welfare and pregnancy rate of ewes undergoing transcervical artificial insemination with ketamine subarachnoid anesthesia. Trop Anim Health Prod 51:1179–1186
Carnovale F, Phillips CJC (2020) The effects of heat stress on sheep welfare during live export voyages from Australia to the Middle East. Animals 10:694
Cassinello, J. (2022). Ovis aries musimon (European mouflon). CABI
Caulfield MP, Cambridge H, Foster SF, McGreevy PD (2014) Heat stress: a major contributor to poor animal welfare associated with long-haul live export voyages. Vet J 199:223–228
Chen S-Y, Duan Z-Y, Sha T, Xiangyu J, Wu S-F, Zhang Y-P (2006) Origin, genetic diversity, and population structure of Chinese domestic sheep. Gene 376:216–223
Cheng Z, McKeller Q, Nolan A (1998) Pharmacokinetic studies of flunixin meglumine and phenylbutazone in plasma, exudate and transudate in sheep. J Vet Pharmacol Ther 21:315–321
Chessa B, Pereira F, Arnaud F, Amorim A, Goyache F, Mainland I, Kao RR, Pemberton JM, Beraldi D, Stear MJ, Alberti A, Pittau M, Iannuzzi L, Banabazi MH, Kazwala RR, Zhang YP, Arranz JJ, Ali BA, Wang Z, Uzun M, Dione MM, Olsaker I, Holm LE, Saarma U, Ahmad S, Marzanov N, Eythorsdottir E, Holland MJ, Ajmone-Marsan P, Bruford MW, Kantanen J, Spencer TE, Palmarini M (2009) Revealing the history of sheep domestication using retrovirus integrations. Science 324:532–536
Cho AI, Lee C, Small A (2020) Attention behaviours but not pain-related behaviours are modified by the presence of a twin in lambs undergoing castration by rubber ring. Peer J 8:e10081
Christou C, Oliver RA, Rawlinson J, Walsh WR (2015) Transdermal fentanyl and its use in ovine surgery. Res Vet Sci 100:252–256
Clark C, Mendl M, Jamieson J, Arnone A, Waterman-Pearson A, Murrell J (2011) Do psychological and

physiological stressors alter the acute pain response to castration and tail docking in lambs? Vet Anaesth Analg 38:134–145

Colditz I, Walkden-Brown S, Daly B, Crook B (2005) Some physiological responses associated with reduced wool growth during blowfly strike in Merino sheep. Aust Vet J 83:695–699

Colditz I, Lloyd J, Paull D, Lee C, Giraudo A, Pizzato C, Fisher A (2009) Effect of the non-steroidal anti-inflammatory drug, carprofen, on weaned sheep following non-surgical mulesing by intradermal injection of cetrimide. Aust Vet J 87:19–26

Colditz I, Paull D, Hervault G, Aubriot D, Lee C (2011) Development of a lameness model in sheep for assessing efficacy of analgesics. Aust Vet J 89:297–304

Colditz IG, Paull DR, Lloyd JB, Johnston L, Small AH (2019) Efficacy of meloxicam in a pain model in sheep. Aust Vet J 97:23–32

Coleman G (2017) 3—Consumer and societal expectations for sheep products. In: Ferguson DM, Lee C, Fisher A (eds) Advances in sheep welfare. Woodhead Publishing, pp 37–51

Coleman G (2018) Public animal welfare discussions and outlooks in Australia. Anim Front 8:14–19

Coltman DW, O'Donoghue P, Jorgenson JT, Hogg JT, Strobeck C, Festa-Bianchet M (2003) Undesirable evolutionary consequences of trophy hunting. Nature 426:655–658

Colvin A (2022) Trends in mulesing, tail docking and castration practices of Australian woolgrowers: Results of the 2021 AWI Merino Husbandry Practices Survey, Sydney

Copeland SE, Ladd LA, Gu X-Q, Mather LE (2008) The effects of general anesthesia on the central nervous and cardiovascular system toxicity of local anesthetics. Anesth Analg 106:1429–1439

Corner-Thomas RA, Shanks R, Beausoleil NJ (n.d.) The influence of the number of lambs present on the suckling behaviour of triplet-rearing ewes at pasture. N Z J Agric Res:1–22

Coulson NM, Januszkiewicz AJ, Dodd KT, Ripple GR (1989) The cardiorespiratory effects of diazepam-ketamine and xylazine-ketamine anesthetic combinations in sheep. Lab Anim Sci 39:591–597

Council NR (2011) Guide for the care and use of laboratory animals, 8th edn. The National Academies Press, Washington, DC

Crawford PE, Hamer K, Lovatt F, Behnke MC, Robinson PA (2023) Improving analgesia provision for sheep: an analysis of farm medicine records and attitudes towards pain relief on sheep farms in Northern Ireland. Vet Rec Open 10:e75

Cross N, Meng J, Phillips CJC, Wojciechowska J (2009) Perceptions of the importance of different welfare issues in livestock production. Animal 3:1152–1166

Cugnasse J-M, Douvre P, Dubray D, Gaillard J-M, Garel M, Gibert P, Loison A (2005) Reproductive output of female mouflon (Ovis gmelini musimon × Ovis sp.): a comparative analysis. J Zool 266:65–71

Cutress TW (1972) The Incisive Apparatus of the Sheep. Res Vet Sci 13:74–78

DAFF (2023) Phase out of live sheep exports by sea. Department of Agriculture, Fisheries and Forestry

DeRossi R, Jardim P, Hermeto L, Pagliosa R (2015) Comparison of analgesic and systemic effects of bupivacaine, methadone, or bupivacaine/methadone administered epidurally in conscious sheep. Aust Vet J 93:164–169

Dias IR, Camassa JA, Bordelo JA, Babo PS, Viegas CA, Dourado N, Reis RL, Gomes ME (2018) Preclinical and translational studies in small ruminants (sheep and goat) as models for osteoporosis research. Curr Osteoporos Rep 16:182–197

Dickins A, Clark CCA, Kaler J, Ferguson E, O'Kane H, Green LE (2016) Factors associated with the presence and prevalence of contagious ovine digital dermatitis: A 2013 study of 1136 random English sheep flocks. Prev Vet Med 130:86–93

Dinnis A, Stafford K, Mellor D, Bruce R, Ward R (1997) Acute cortisol responses of lambs castrated and docked using rubber rings with or without a castration clamp. Aust Vet J 75:494–496

Dinniss AS, Mellor DJ, Stafford KJ, Bruce RA, Ward RN (1997a) Acute cortisol responses of lambs to castration using a rubber ring and/or a castration clamp with or without local anaesthetic. N Z Vet J 45:114–121

Dinniss AS, Stafford KJ, Mellor DJ, Bruce RA, Ward RN (1997b) Acute cortisol responses of lambs castrated and docked using rubber rings with or without a castration clamp. Aust Vet J 75:494–496

Doughty AK, Coleman GJ, Hinch GN, Doyle RE (2017) Stakeholder Perceptions of Welfare Issues and Indicators for Extensively Managed Sheep in Australia. Animals 7:28

Doyle RE, Broster JC, Barnes K, Browne WJ (2016) Temperament, age and weather predict social interaction in the sheep flock. Behav Process 131:53–58

Durand D, Faure M, de Boyer des Roches A, de la Foye A (2019) Benefits of a multimodal analgesia compared to local anesthesia alone to alleviate pain following castration in sheep: a multiparametric approach. Animal 13:2034–2043

Dwyer CM (2008) Environment and the sheep. In: The Welfare of Sheep. Springer Netherlands, Dordrecht, pp 41–79

Dwyer CM (2009) Welfare of sheep: providing for welfare in an extensive environment. Small Rumin Res 86:14–21

Edwards DS, Johnston AM, Pfeiffer DU (2001) A comparison of commonly used ear tags on the ear damage of sheep. Anim Welf 10:141–151

Edwards JEH, Copping KJ, Thompson AN (2011a) Managing the nutrition of twin-bearing ewes during pregnancy using Lifetimewool recommendations increases production of twin lambs. Anim Prod Sci 51:813–820

Edwards LE, Arnold NA, Butler KL, Hemsworth PH (2011b) Acute effects of mulesing and alternative pro-

cedures to mulesing on lamb behaviour. Appl Anim Behav Sci 133:169–174
Egermann M, Goldhahn J, Holz R, Schneider E, Lill CA (2008) A sheep model for fracture treatment in osteoporosis: benefits of the model versus animal welfare. Lab Anim 42:453–464
Evangelista MC, Monteiro BP, Steagall PV (2022) Measurement properties of grimace scales for pain assessment in nonhuman mammals: a systematic review. PAIN 163(6):e697–e714 https://doi.org/10.1097/j.pain.0000000000002474
Enns GE, Jex B, Boyce MS (2023) Diverse migration patterns and seasonal habitat use of Stone's sheep (Ovis dalli stonei). PeerJ 11:e15215
Fayemi PO, Muchenje V (2016) Molecular marker for detecting slaughter pain in electro-stunned sheep. (pp. 42–46). San Pedro Zacatenco: Cinvestav
Fedosenko AK, Blank DA (2005) Ovis ammon. Mamm Species 2005:1–15
Ferguson MB, Thompson AN, Gordon DJ, Hyder MW, Kearney GA, Oldham CM, Paganoni BL (2011) The wool production and reproduction of Merino ewes can be predicted from changes in liveweight during pregnancy and lactation. Anim Prod Sci 51:763–775
Ferguson D, Lee C, Fisher A (2017) Advances in sheep welfare. Woodhead Publishing
Ferrer LM, Lacasta D, Ortín A, Ramos JJ, Tejedor MT, Borobia M, Pérez M, Castells E, Ruiz de Arcaute M, Ruiz H, Windsor PA (2020) Impact of a Topical Anaesthesia Wound Management Formulation on Pain, Inflammation and Reduction of Secondary Infections after Tail Docking in Lambs. Animals 10:1255
Festa-Bianchet M (1991) The social system of bighorn sheep: grouping patterns, kinship and female dominance rank. Anim Behav 42:71–82
Fidan A, Kucukkurt I, Eryavuz A, Cigerci I, Yardimci M, Ozdemir A (2009) Effects of shearing procedures on oxidant-antioxidant status in Chios sheep. Revue Méd Vét 160:55
Finnie J (2001) Animal models of traumatic brain injury: a review. Aust Vet J 79:628–633
Fischer H, Schmidt-Bleek O, Orassi V, Wulsten D, Schmidt-Bleek K, Heiland M, Steffen C, Rendenbach C (2023) Biomechanical comparison of WE43-based magnesium vs titanium miniplates in a mandible fracture model in sheep. Materials 16:102
Fisher AD (2011) Addressing pain caused by mulesing in sheep. Appl Anim Behav Sci 135:232–240
Fleming P, Korn T (1989) Predation of livestock by wild dogs in eastern New South Wales. Rangeland J 11:61–66
Fourie P, Mahlako S, Van der Westhuizen C (2018) Assessment of the management practices of emerging sheep production systems in the Eastern Free State: can the extensionist make a difference? S Afr J Agric Ext 46:57–68
French NP, Morgan KL (1992) Neuromata in docked lambs' tails. Res Vet Sci 52:389–390
Friedman R, Antoine D, Talamo S, Reimer PJ, Taylor JH, Wills B, Mannino MA (2018) Natural mummies from predynastic Egypt reveal the world's earliest figural tattoos. J Archaeol Sci 92:116–125
Frisina MR, Frisina RM (2012) Influence of trophy harvest on the population age structure of argali (Ovis ammon) in Mongolia. J Bombay Nat Hist Soc 109:173–176
Fuks D, Marom N (2021) Sheep and wheat domestication in southwest Asia: a meta-trajectory of intensification and loss. Anim Front 11:20–29
Galatos AD (2011) Anesthesia and Analgesia in Sheep and Goats. Vet Clin N Am Food Anim Pract 27:47–59
Gelasakis AI, Valergakis GE, Arsenos G, Banos G (2012) Description and typology of intensive Chios dairy sheep farms in Greece. J Dairy Sci 95:3070–3079
Ghadirian S, Vesal N (2013) Brachial plexus block using lidocaine/epinephrine or lidocaine/xylazine in fat-tailed sheep. Vet Res Forum 4:161–167
Goddard P, Waterhouse T, Dwyer C, Stott A (2006) The perception of the welfare of sheep in extensive systems. Small Rumin Res 62:215–225
González J, Lacasta D, Ferrer L, Figueras L, Ramos J, De las Heras M (2014) Natural border disease virus infection in feedlot lambs. Vet Rec 174:69–69
González JM, Bello JM, Rodríguez M, Navarro T, Lacasta D, Fernández A, De las Heras M (2016) Lamb feedlot production in Spain: Most relevant health issues. Small Rumin Res 142:83–87
Gougoulis DA, Kyriazakis I, Fthenakis GC (2010) Diagnostic significance of behaviour changes of sheep: a selected review. Small Rumin Res 92:52–56
Grace C, Boland TM, Sheridan H, Lott S, Brennan E, Fritch R, Lynch MB (2018) The effect of increasing pasture species on herbage production, chemical composition and utilization under intensive sheep grazing. Grass Forage Sci 73:852–864
Graham MJ, Kent JE, Molony V (1997) Effects of four analgesic treatments on the behaviouraland cortisol responses of 3-week-old lambs to tail docking. Vet J 153:87–97
Grant C (2004) Behavioural responses of lambs to common painful husbandry procedures. Appl Anim Behav Sci 87:255–273
Grant C, Upton R (2001) Cardiovascular and haemodynamic effects of intramuscular doses of xylazine in conscious sheep. Aust Vet J 79:58–60
Grant C, Upton R, Kuchel T (1996) Efficacy of intramuscular analgesics for acute pain in sheep. Aust Vet J 73:129–132
Grant EP, Wickham SL, Anderson F, Barnes AL, Fleming PA, Miller DW (2020) Preliminary findings on a novel behavioural approach for the assessment of pain and analgesia in lambs subject to routine husbandry procedures. Animals 10:1148
Greentree C, Saunders G, Mcleod L, Hone J (2000) Lamb predation and fox control in south-eastern Australia. J Appl Ecol 37:935–943

Griffin H (1999) Dolly: the science behind the world's most famous sheep. Cern
Grubb TL, Riebold TW, Huber MJ (1992) Comparison of lidocaine, xylazine, and xylazine/lidocaine for caudal epidural analgesia in horses. J Am Vet Med Assoc 201:1187–1190
Grubb TL, Riebold TW, Crisman RO, Lamb LD (2002) Comparison of lidocaine, xylazine, and lidocaine–xylazine for caudal epidural analgesia in cattle. Vet Anaesth Analg 29:64–68
Guedes AGP, Pluhar GE, Daubs BM, Rudé EP (2006) Effects of preoperative epidural administration of racemic ketamine for analgesia in sheep undergoing surgery. Am J Vet Res 67:222–229
Guesgen MJ, Beausoleil NJ, Minot EO, Stewart M, Jones G, Stafford KJ (2011) The effects of age and sex on pain sensitivity in young lambs. Appl Anim Behav Sci 135:51–56
Guesgen MJ, Beausoleil NJ, Minot EO, Stewart M, Stafford KJ (2014) Social context and other factors influence the behavioural expression of pain by lambs. Appl Anim Behav Sci 159:41–49
Guesgen MJ, Beausoleil NJ, Leach M, Minot EO, Stewart M, Stafford KJ (2016) Coding and quantification of a facial expression for pain in lambs. Behav Process 132:49–56
Habibian S, Bigham AS, Aali E (2011) Comparison of lidocaine, tramadol, and lidocaine–tramadol for epidural analgesia in lambs. Res Vet Sci 91:434–438
Häger C, Biernot S, Buettner M, Glage S, Keubler LM, Held N, Bleich EM, Otto K, Müller CW, Decker S, Talbot SR, Bleich A (2017) The Sheep Grimace Scale as an indicator of post-operative distress and pain in laboratory sheep. PLoS One 12:e0175839
Hancock S, Inglis L, Laurence M, Thompson A (2019) Behavioural measures reflect pain-mitigating effects of meloxicam in combination with Tri-Solfen® in mulesed Merino lambs. Animal 13:2586–2593
Hargreaves AL, Hutson GD (1990) Changes in heart rate, plasma cortisol and haematocrit of sheep during a shearing procedure. Appl Anim Behav Sci 26:91–101
Harned M (1982) Anatolian shepherd dog: an ancient breed. Rangelands Archives 4:63–65
Harris RB, Pletscher DH (2002) Incentives toward conservation of argali Ovis ammon: a case study of trophy hunting in western China. Oryx 36:373–381
Harris C, White PJ, Mohler VL, Lomax S (2020) Electroencephalography Can Distinguish between Pain and Anaesthetic Intervention in Conscious Lambs Undergoing Castration. Animals 10:428
Hefnawy A, Helal MAY, Sabek A, Shousha S (2018) Clinical, behavioral and biochemical alterations due to shearing stress in Ossimi sheep. J Vet Med Sci 80:1281–1286
Heinzen BC, Weber SH, Milczewski V, Maia D, Kozicki LE, Sotomaior CS (2023) Reproductive performance of European-breed ewes in different seasons of the year under mid-latitude. Reprod Domest Anim 58:740–745
Hemsworth PH, Barnett JL, Karlen GM, Fisher AD, Butler KL, Arnold NA (2009) Effects of mulesing and alternative procedures to mulesing on the behaviour and physiology of lambs. Appl Anim Behav Sci 117:20–27
Hemsworth PH, Cronin GM, Barnett JL, Butler KL, Jongman EC, Karlen GA, Coffey A, Arnold NA (2012) Behavioural responses of lambs to plastic clips as an alternative procedure to mulesing. Aust Vet J 90:373–380
Hewson R (1984) Scavenging and Predation Upon Sheep and Lambs in West Scotland. J Appl Ecol 21:843–868
Hicks CL, von Baeyer CL, Spafford PA, van Korlaar I, Goodenough B (2001) The Faces Pain Scale—Revised: toward a common metric in pediatric pain measurement. Pain 93:173–183
Hiendleder S, Mainz K, Plante Y, Lewalski H (1998) Analysis of mitochondrial DNA indicates that domestic sheep are derived from two different ancestral maternal sources: no evidence for contributions from urial and argali sheep. J Hered 89:113–120
Higgs ARE, Norris RT, Richards RB (1993) Epidemiology of salmonellosis in the live sheep export industry. Aust Vet J 70:330–335
Higgs ARB, Norris RT, Baldock FC, Campbell NJ, Koh S, Richards RB (1996) Contagious ecthyma in the live sheep export industry. Aust Vet J 74:215–220
Hogg JT (1987) Intrasexual competition and mate choice in rocky mountain bighorn sheep. Ethology 75:119–144
Hogg JT, Forbes SH (1997) Mating in bighorn sheep: frequent male reproduction via a high-risk "unconventional" tactic. Behav Ecol Sociobiol 41:33–48
Howard, K., & Beattie, L. (2018). A national producer survey of sheep and cattle husbandry practices. Sydney: Meat & Livestock Australia
Hristov S, Maksimović N, Stanković B, Žujović M, Pantelić V, Stanišić N, Zlatanović Z (2012) The most significant stressors in intensive sheep production. Biotechnol Animal Husbandry 28:649–658
Hughan S, Loose J, Caddy D, Canny B, Tilbrook A, Young I (2001) Combined xylazine and ketamine as an analgesic regimen in sheep. Aust Vet J 79:207–211
Hunter DS, Hazel SJ, Kind KL, Liu H, Marini D, Giles LC, De Blasio MJ, Owens JA, Pitcher JB, Gatford KL (2015) Placental and fetal growth restriction, size at birth and neonatal growth alter cognitive function and behaviour in sheep in an age- and sex-specific manner. Physiol Behav 152:1–10
Impey S (2015) The effect of early post-natal castration on subsequent electroencephalogram response to tail docking in lambs. Veterinary Science. Massey University, Palmerston North
IWTO. (2022). World Sheep Numbers & Wool Production. IWTO Market Information Ed.17: IWTO
Jacobsen JC, Bawden CS, Rudiger SR, McLaughlan CJ, Reid SJ, Waldvogel HJ, MacDonald ME, Gusella JF, Walker SK, Kelly JM, Webb GC, Faull RLM, Rees MI, Snell RG (2010) An ovine transgenic Huntington's disease model. Hum Mol Genet 19:1873–1882

Jacobson C, Bruce M, Kenyon PR, Lockwood A, Miller D, Refshauge G, Masters DG (2020) A review of dystocia in sheep. Small Rumin Res 192:106209

Johnston CH, Whittaker AL, Franklin SH, Hutchinson MR (2022) The neuroimmune interface and chronic pain through the lens of production animals. Front Neurosci 16:887042

Jongman E, Morris J, Barnett J, Hemsworth P (2000) EEG changes in 4-week-old lambs in response to castration, tail docking and mulesing. Aust Vet J 78:339–343

Jørgensen GHM, Bøe KE (2011) Outdoor yards for sheep during winter—Effects of feed location, roof and weather factors on resting and activity. Can J Anim Sci 91:213–220

Jumabay-Uulu K, Mishra C, Sharma K, Wegge P (2014) Large carnivores and low diversity of optimal prey: a comparison of the diets of snow leopards Panthera uncia and wolves Canis lupus in Sarychat-Ertash Reserve in Kyrgyzstan. Oryx 48:529–535

Kalds P, Luo Q, Sun K, Zhou S, Chen Y, Wang X (2021) Trends towards revealing the genetic architecture of sheep tail patterning: Promising genes and investigatory pathways. Anim Genet 52:799–812

Kaler J, Daniels SLS, Wright JL, Green LE (2010) Randomized clinical trial of long-acting oxytetracycline, foot trimming, and flunixine meglumine on time to recovery in sheep with footrot. J Vet Intern Med 24:420–425

Karakuş F, Düzgün A, Karakuş M, Aslan L (2015) Can infrared thermography be used to predict ear tags infections in lambs? Scientific Papers, Series D. Anim Sci 58:205–208

Kästner SBR (2006) A2-agonists in sheep: a review. Vet Anaesth Analg 33:79–96

Kells NJ, Beausoleil NJ, R Godfrey AJ, E Littlewood K, Ward RN, Johnson CB (2020) Effect of analgesic strategies on pain behaviour associated with combined ring castration and hot iron tail docking in Merino lambs. Appl Anim Behav Sci 222:104914

Kent JE, Jackson RE, Molony V, Hosie BD (2000) Effects of acute pain reduction methods on the chronic inflammatory lesions and behaviour of lambs castrated and tail docked with rubber rings at less than two days of age. Vet J 160:33–41

Kenyon PR, Roca Fraga FJ, Blumer S, Thompson AN (2019) Triplet lambs and their dams—a review of current knowledge and management systems. N Z J Agric Res 62:399–437

Klingenberg C, Sobotka KS, Ong T, Allison BJ, Schmölzer GM, Moss TJM, Polglase GR, Dawson JA, Davis PG, Hooper SB (2013) Effect of sustained inflation duration; resuscitation of near-term asphyxiated lambs. Arch Dis Child Fetal Neonatal Ed 98:F222

Knarrum V, Sørensen OJ, Eggen T, Kvam T, Opseth O, Overskaug K, Eidsmo A (2006) Brown bear predation on domestic sheep in central Norway. Ursus 17:67–74

Kongara K, Corner-Thomas R, Bruere S, Lawrence K, Gates MC (2023) Practices and opinions of New Zealand sheep farmers towards pain management in lambs during castration and/or tail docking. N Z Vet J 71:8–17

Kosgey IS, Rowlands GJ, van Arendonk JAM, Baker RL (2008) Small ruminant production in smallholder and pastoral/extensive farming systems in Kenya. Small Rumin Res 77:11–24

Kotze A, James P (2022) Control of sheep flystrike: what's been tried in the past and where to from here. Aust Vet J 100:1–19

Kyles AE, Waterman AE, Livingston A (1995) Antinociceptive activity of midazolam in sheep. J Vet Pharmacol Ther 18:54–60

Langbein J, Scheibe K-M, Eichhorn K (1998a) Investigations on periparturient behaviour in free-ranging mouflon sheep (Ovis orientalis musimon). J Zool 244:553–561

Langbein J, Streich J, Scheibe KM (1998b) Characteristic activity patterns of female mouflons (Ovis orientalis musimon) in the lambing period. Appl Anim Behav Sci 58:281–292

Larrondo C, Bustamante H, Gallo C (2018) Sheep farmers' perception of welfare and pain associated with routine husbandry practices in Chile. Animals 8:225

Larrondo C, Orihuela A, Strappini A, Acosta-Jamett G, Mota-Rojas D, Gallo C (2021) Provision of straw and the presence of undocked lambs reduce the behavioural and physiological expressions of pain and stress associated with tail docking in lambs: a preliminary study. Anim Prod Sci 61:423–431

Lawrence AB (1990) Mother-daughter and peer relationships of Scottish hill sheep. Anim Behav 39:481–486

Lay DC, Friend TH, Bowers CL, Grissom KK, Jenkins OC (1992) A comparative physiological and behavioral study of freeze and hot-iron branding using dairy cows. J Anim Sci 70:1121–1125

Le Pendu Y, Briedermann L, Gerard JF, Maublanc ML (1995) Inter-individual associations and social structure of a mouflon population (Ovis orientalis musimon). Behav Process 34:67–80

Lecchi C, Dalla Costa E, Lebelt D, Ferrante V, Canali E, Ceciliani F, Stucke D, Minero M (2018) Circulating miR-23b-3p, miR-145-5p and miR-200b-3p are potential biomarkers to monitor acute pain associated with laminitis in horses. Animal 12:366–375

Lee C, Fisher A (2007) Welfare consequences of mulesing of sheep. Aust Vet J 85:89–93

Lepherd ML, Canfield PJ, Hunt GB, Thomson PC, Bosward KL (2011a) Assessment of the short-term systemic effect of and acute phase response to mulesing and other options for controlling breech flystrike in Merino lambs. Aust Vet J 89:19–26

Lepherd ML, Canfield PJ, Hunt GB, Thomson PC, Bosward KL (2011b) Wound healing after mulesing and other options for controlling breech flystrike in Merino lambs: quantitative and semiquantitative analysis of wound healing and wound bed contraction. Aust Vet J 89:61–69

Lester SJ, Mellor DJ, Holmes RJ, Ward RN, Stafford KJ (1996) Behavioural and cortisol responses of lambs to castration and tailing using different methods. N Z Vet J 44:45–54

Ley S, Waterman A, Livingston A (1991) The influence of chronic pain on the analgesic effects of the α2-adrenoceptor agonist, xylazine, in sheep. J Vet Pharmacol Ther 14:141–144

Lincoln GA (1998) Reproductive seasonality and maturation throughout the complete life-cycle in the mouflon ram (Ovis musimon). Anim Reprod Sci 53:87–105

Lincoln GA, Lincoln CE, McNeilly AS (1990) Seasonal cycles in the blood plasma concentration of FSH, inhibin and testosterone, and testicular size in rams of wild, feral and domesticated breeds of sheep. Reproduction 88:623–633

Lindsay D (1988) Reproductive Behaviour in Survival: A Comparison between Wild and Domestic Sheep. Aust J Biol Sci 41:97–102

Lizarraga I, Chambers JP (2012) Use of analgesic drugs for pain management in sheep. N Z Vet J 60:87–94

Lizarraga I, Chambers JP, Johnson CB (2008) Prevention of N-methyl-d-aspartate-induced mechanical nociception by intrathecal administration of ketoprofen and ketamine in sheep. Anesth Analg 107:2061–2067

Lizarraga I, Janovyak E, Beths T (2013) Comparing lidocaine, bupivacaine and a lidocaine–bupivacaine mixture as a metacarpal block in sheep. Vet J 197:515–518

Lloyd J, Kessell A, Barchia I, Schröder J, Rutley D (2016) Docked tail length is a risk factor for bacterial arthritis in lambs. Small Rumin Res 144:17–22

Lomax S, Sheil M, Windsor PA (2008) Impact of topical anaesthesia on pain alleviation and wound healing in lambs after mulesing. Aust Vet J 86:159–168

Lomax S, Sheil M, Windsor PA (2009) Use of local anaesthesia for pain management during husbandry procedures in Australian sheep flocks. Small Rumin Res 86:56–58

Lomax S, Dickson H, Sheil M, Windsor P (2010) Topical anaesthesia alleviates short-term pain of castration and tail docking in lambs. Aust Vet J 88:67–74

Lomax S, Sheil M, Windsor P (2013) Duration of action of a topical anaesthetic formulation for pain management of mulesing in sheep. Aust Vet J 91:160–167

Longworth J, Brown C, Waldron S (2010) Chinese sheep and wool industries. International Sheep and Wool Handbook, Nottingham University Press, Nottingham

Lucky N, Hashim M, Ahmed J, Sarker K, Gazi N, Ahmed S (2007) Caudal epidural analgesia in sheep by using lignocaine hydrochloride and bupivacaine hydrochloride. Bangladesh J Vet Med:77–80

Main MB, Coblentz BE (1990) Sexual segregation among ungulates: a critique. Wildl Soc Bull (1973–2006) 18:204–210

Main MB, Weckerly FW, Bleich VC (1996) Sexual segregation in ungulates: new directions for research. J Mammal 77:449–461

Makarewicz CA (2013) A pastoralist manifesto: breaking stereotypes and re-conceptualizing pastoralism in the Near Eastern Neolithic. Levant 45:159–174

Mani S, Gugino S, Helman J, Bawa M, Nair J, Chandrasekharan P, Rawat M (2022) Laryngeal mask ventilation with chest compression during neonatal resuscitation: randomized, non-inferiority trial in lambs. Pediatr Res 92:671–677

Marchand P, Redjadj C, Garel M, Cugnasse J-M, Maillard D, Loison A (2013) Are mouflon Ovis gmelini musimon really grazers? A review of variation in diet composition. Mammal Rev 43:275–291

Marchewka J, de Heredia IB, Averós X, Ruiz R, Zanella AJ, Díaz JAC, Estevez I (2016) Behaviour of tail-docked lambs tested in isolation. Ir J Agric Food Res 55:192–199

Marini D, Pippia J, Colditz I, Hinch G, Petherick J, Lee C (2015) Randomised trial of the bioavailability and efficacy of orally administered flunixin, carprofen and ketoprofen in a pain model in sheep. Aust Vet J 93:265–270

Marini D, Pippia J, Colditz IG, Hinch GN, Petherick CJ, Lee C (2016) Palatability and pharmacokinetics of flunixin when administered to sheep through feed. PeerJ 4:e1800

Marini D, Colditz IG, Hinch G, Petherick JC, Lee C (2017) Self-administration by consumption of flunixin in feed alleviates the pain and inflammation associated with castration and tail docking of lambs. Appl Anim Behav Sci 188:26–33

Marini D, Colditz IG, Lee C (2021) Can Lambs in Pain Identify Medicated Feed? Front Anim Sci 2

Marini D, Monk JE, Campbell DLM, Lee C, Belson S, Small A (2023) Sex impacts pain behaviour but not emotional reactivity of lambs following ring tail docking. PeerJ 11:e15092

Masłowska K, Mizzoni F, Dwyer CM, Wemelsfelder F (2020) Qualitative behavioural assessment of pain in castrated lambs. Appl Anim Behav Sci 233:105143

Mather LE (2010) The acute toxicity of local anesthetics. Expert Opin Drug Metab Toxicol 6:1313–1332

Mattiello S, Bresciani T, Gaggero S, Russo C, Mazzarone V (2012) Sheep predation: characteristics and risk factors. Small Rumin Res 105:315–320

McCann LJ (1956) Ecology of the Mountain Sheep. Am Midl Nat 56:297–324

McCarthy M (2018) Independent review of conditions for the export of sheep to the Middle East during the northern hemisphere summer. Department of Agriculture and Water Resources (Australia)

McCracken L, Waran N, Mitchinson S, Johnson CB (2010) Effect of age at castration on behavioural response to subsequent tail docking in lambs. Vet Anaesth Analg 37:375–381

McLennan KM, Rebelo CJB, Corke MJ, Holmes MA, Leach MC, Constantino-Casas F (2016) Development of a facial expression scale using footrot and mastitis as models of pain in sheep. Appl Anim Behav Sci 176:19–26

Meadows JRS, Cemal I, Karaca O, Gootwine E, Kijas JW (2007) Five ovine mitochondrial lineages iden-

tified from sheep breeds of the near East. Genetics 175:1371–1379
Mears GJ, Brown FA, Redmond LR (1999) Effects of handling, shearing and previous exposure to shearing on cortisol and β-endorphin responses in ewes. Can J Anim Sci 79:35–38
Melches S, Mellema SC, Doherr MG, Wechsler B, Steiner A (2007) Castration of lambs: a welfare comparison of different castration techniques in lambs over 10 weeks of age. Vet J 173:554–563
Mellema SC, Doherr MG, Wechsler B, Thueer S, Steiner A (2006) Influence of local anaesthesia on pain and distress induced by two bloodless castration methods in young lambs. Vet J 172:274–283
Mellor DJ, Murray L (1989) Effects of tail docking and castration on behaviour and plasma cortisol concentrations in young lambs. Res Vet Sci 46:387–391
Mellor DJ, Stafford KJ (2000) Acute castration and/or tailing distress and its alleviation in lambs. N Z Vet J 48:33–43
Michel S, Ghoddousi A (2020) Ovis vignei (errata version published in 2021). The IUCN Red List of Threatened Species 2020:e.T54940655A195296049
Militana EA, Campoy L, Martin-Flores M, Gleed RD (2020) Evaluation of a regional nerve block with an experimental formulation of encapsulated lidocaine in sheep. Vet Anaesth Analg 47:819–825
MLA (2023) Industry projections 2023: Australian sheep—July update. Meat and Livestock Australia
Moberly RL, White PCL, Webbon CC, Baker PJ, Harris S (2003) Factors associated with fox (Vulpes vulpes) predation of lambs in Britain. Wildl Res 30:219–227
Moharrery A (2007) Effect of docking and energy of diet on carcass fat characteristics in fat-tailed Badghisian sheep. Small Rumin Res 69:208–216
Molony V, Kent JE, Robertson IS (1993) Behavioural responses of lambs of three ages in the first three hours after three methods of castration and tail docking. Res Vet Sci 55:236–245
Molony V, Kent JE, McKendrick IJ (2002) Validation of a method for assessment of an acute pain in lambs. Appl Anim Behav Sci 76:215–238
Mooring MS, Fitzpatrick TA, Benjamin JE, Fraser IC, Nishihira TT, Reisig DD, Rominger EM (2003) Sexual segregation in desert bighorn sheep (Ovis canadensis mexicana). Behaviour 140:183–207
Morris ST (2017) 2—Overview of sheep production systems. In: Ferguson DM, Lee C, Fisher A (eds) Advances in sheep welfare. Woodhead Publishing, pp 19–35
Morris ST, Kenyon PR (2014) Intensive sheep and beef production from pasture—a New Zealand perspective of concerns, opportunities and challenges. Meat Sci 98:330–335
Morris JP, Ong RM, O'Dwyer JK, Barnett JL, Hemsworth PH, Clarke IJ, Jongman EC (1997) Pain-related cerebral potentials in response to acute painful electrical stimulation in sheep. Aust Vet J 75:883–886
Morton AJ, Rudiger SR, Wood NI, Sawiak SJ, Brown GC, Mclaughlan CJ, Kuchel TR, Snell RG, Faull RLM, Bawden CS (2014) Early and progressive circadian abnormalities in Huntington's disease sheep are unmasked by social environment. Hum Mol Genet 23:3375–3383
Munoz CA, Coleman GJ, Hemsworth PH, Campbell AJD, Doyle RE (2019) Positive attitudes, positive outcomes: the relationship between farmer attitudes, management behaviour and sheep welfare. PLoS One 14:e0220455
Munoz C, Glanville E, Linn B, Kirk B, Tyrell L, Bowles V, Fisher A (2022) Final Report: Phasing out of mulesing: cost, benefits and opportunities, Sydney
Murdoch FR, Maker GL, Nitsos I, Polglase GR, Musk GC (2013) Intraperitoneal medetomidine: a novel analgesic strategy for postoperative pain management in pregnant sheep. Lab Anim 47:66–70
Murdoch JD, Reading RP, Amgalanbaatar S, Wingard G, Lkhagvasuren B (2017) Ecological interactions shape the distribution of a cultural ecosystem service: Argali sheep (Ovis ammon) in the Gobi-Steppe of Mongolia. Biol Conserv 209:315–322
Musk GC, Wilkes GJ (2018) Sedation of sheep following the administration of acepromazine with buprenorphine or morphine is similar. Res Vet Sci 118:41–42
Musk GC, Murdoch FR, Tuke J, Kemp MW, Dixon MJ, Taylor PM (2014) Thermal and mechanical nociceptive threshold testing in pregnant sheep. Vet Anaesth Analg 41:305–311
Needham T, Lambrechts H, Hoffman L (2017) Castration of male livestock and the potential of immunocastration to improve animal welfare and production traits: invited review. SA J Anim Sci 47:731–742
NHMRC (2013) Australian code for the care and use of animals for scientific purposes. National Health and Medical Research Council, Canberra
Nichols L (1978) Dall Sheep Reproduction. J Wildl Manag 42:570–580
Nicolás-López P, Macías-Cruz U, Mellado M, Correa-Calderón A, Meza-Herrera CA, Avendaño-Reyes L (2021) Growth performance and changes in physiological, metabolic and hematological parameters due to outdoor heat stress in hair breed male lambs finished in feedlot. Int J Biometeorol 65:1451–1459
Niezgoda J, Bobek S, Wronska-Fortuna D, Wierzchos E (1993) Response of Sympatho-Adrenal Axis and Adrenal Cortex to Short-Term Restraint Stress in Sheep. J Veterinary Med Ser A 40:631–638
Nordbø I, Turdumambetov B, Gulcan B (2018) Local opinions on trophy hunting in Kyrgyzstan. J Sustain Tour 26:68–84
O'Kane H, Ferguson E, Kaler J, Green L (2017) Associations between sheep farmer attitudes, beliefs, emotions and personality, and their barriers to uptake of best practice: The example of footrot. Prev Vet Med 139:123–133

Odden J, John DCL, Moa PF, Herfindal I, Kvam T, Andersen R (2002) Lynx depredation on domestic sheep in Norway. J Wildl Manag 66:98–105

Oliveira AR, Araújo MA, Jardim PHA, Lima SC, Leal PV, Frazílio FO (2016) Comparison of lidocaine, levobupivacaine or ropivacaine for distal paravertebral thoracolumbar anesthesia in ewes. Vet Anaesth Analg 43:670–674

Ong RM, Morris JP, O'Dwyer JK, Barnett JL, Hemsworth PH, Clarke IJ (1997) Behavioural and EEG changes in sheep in response to painful acute electrical stimuli. Aust Vet J 75:189–193

Orsetti C, Vitale V, Mortola J, Sgorbini M, Bonelli F (2022) Respiratory sinus arrhythmia magnitude quantification as a potential marker of stress and pain in cows and sheep. Vet Res Commun

Ozella L, Langford J, Gauvin L, Price E, Cattuto C, Croft DP (2020) The effect of age, environment and management on social contact patterns in sheep. Appl Anim Behav Sci 225:104964

Paas W, San Martín C, Soriano B, van Ittersum MK, Meuwissen MPM, Reidsma P (2021) Assessing future sustainability and resilience of farming systems with a participatory method: a case study on extensive sheep farming in Huesca, Spain. Ecol Indic 132:108236

Parker BG, Khanyari M, Ambarlı H, Buuveibaatar B, Kabir M, Khanal G, Mirzadeh HR, Onon Y, Farhadinia MS (2023) A review of the ecological and socioeconomic characteristics of trophy hunting across Asia. Anim Conserv 26:609–624

Pattinson D, Fitzgerald M (2004) The neurobiology of infant pain: development of excitatory and inhibitory neurotransmission in the spinal dorsal horn. Reg Anesth Pain Med 29:36–44

Paull D, Lee C, Colditz I, Atkinson S, Fisher A (2007) The effect of a topical anaesthetic formulation, systemic flunixin and carprofen, singly or in combination, on cortisol and behavioural responses of Merino lambs to mulesing. Aust Vet J 85:98–106

Paull D, Lee C, Atkinson S, Fisher A (2008a) Effects of meloxicam or tolfenamic acid administration on the pain and stress responses of Merino lambs to mulesing. Aust Vet J 86:303–311

Paull DR, Colditz IG, Lee C, Atkinson SJ, Fisher AD (2008b) Effectiveness of non-steroidal anti-inflammatory drugs and epidural anaesthesia in reducing the pain and stress responses to a surgical husbandry procedure (mulesing) in sheep. Aust J Exp Agric 48:1034–1039

Paull D, Lee C, Colditz I, Fisher A (2009) Effects of a topical anaesthetic formulation and systemic carprofen, given singly or in combination, on the cortisol and behavioural responses of Merino lambs to castration. Aust Vet J 87:230–237

Paull DR, Small AH, Lee C, Palladin P, Colditz IG (2012) Evaluating a novel analgesic strategy for ring castration of ram lambs. Vet Anaesth Analg 39:539–549

Paull DR, Small AH, Lee C, Labeur L, Colditz IG (2015) Effect of local infusion of NSAID analgesics administered alone or in combination on the pain associated with band castration in calves. Aust Vet J 93:271–277

Peers A, Mellor DJ, Wintour EM, Dodic M (2002) Blood pressure, heart rate, hormonal and other acute responses to rubber-ring castration and tail docking of lambs. N Z Vet J 50:56–62

Pehlivan E, Kaliber M, Konca Y, Dellal G (2020) Effect of shearing on some physiological and hormonal parameters in Akkaraman sheep. Asian Australas J Anim Sci 33:848–855

Pennisi P, Costa A, Biondi L, Avondo M, Piccione G (2004) Influence of the fleece on thermal homeostasis and on body condition in Comisana ewe lambs. Anim Res 53:13–19

Pereira S, Veeraraghavan P, Ghosh S, Gandhi M (2004) Animal experimentation and ethics in India: the CPCSEA makes a difference. Altern Lab Anim 32:411–415

Petherick JC, Small AH, Mayer DG, Colditz IG, Ferguson DM, Stafford KJ (2014) A comparison of welfare outcomes for weaner and mature Bos indicus bulls surgically or tension band castrated with or without analgesia: 1. Behavioural responses. Appl Anim Behav Sci 157:23–34

Phillips C, Phillips A (2010) Attitudes of Australian sheep farmers to animal welfare. J Int Farm Manag 5:1–26

Piccione G, Fazio F, Casella S, Pennisi P, Caola G (2011) Influence of shearing on oxidative stress and some physiological parameters in ewes. Anim Sci J 82:481–485

Pifelet S (2011) The effect of early painful experiences on subsequent pain sensitivity in lambs. Physiology. Massey University, Manawatu

Pinch G (2004) Egyptian mythology: a guide to the gods, goddesses, and traditions of ancient Egypt. Oxford University Press, Oxford

Pines MK, Phillips CJC (2013) Microclimatic conditions and their effects on sheep behavior during a live export shipment from Australia to the Middle East 1. J Anim Sci 91:4406–4416

Pinto-Ribeiro F, Almeida A, Pêgo JM, Cerqueira J, Sousa N (2004) Chronic unpredictable stress inhibits nociception in male rats. Neurosci Lett 359:73–76

Poole KG, Serrouya R, Teske IE, Podrasky K (2016) Rocky mountain bighorn sheep (Ovis canadensis canadensis) winter habitat selection and seasonal movements in an area of active coal mining. Can J Zool 94:733–745

Price J, Nolan AM (2001) Analgesia of newborn lambs before castration and tail docking with rubber rings. Vet Rec 149:321–324

Pugh DG, Baird NN, Edmondson M, Passler T (2020) Sheep, goat, and cervid medicine-E-Book. Elsevier Health Sciences

Qureshi RA, Tian Y, McDonald MK, Capasso KE, Douglas SR, Gao R, Orlova IA, Barrett JE, Ajit SK, Sacan A (2016) Circulating microRNA signatures in rodent models of pain. Mol Neurobiol 53:3416–3427

Rasali DP, Shrestha JNB, Crow GH (2006) Development of composite sheep breeds in the world: a review. Can J Anim Sci 86:1–24

Rashid W, Shi J, Rahim Iu, Dong S, Sultan H (2020) Issues and opportunities associated with trophy hunting and tourism in Khunjerab National Park, Northern Pakistan. Animals 10:597

Ratajczak-Enselme M, Estebe J-P, Rose F-X, Wodey E, Malinovsky J-M, Chevanne F, Dollo G, Ecoffey C, Le Corre P (2007) Effect of epinephrine on epidural, intrathecal, and plasma pharmacokinetics of ropivacaine and bupivacaine in sheep. Br J Anaesth 99:881–890

Reading RP, Amgalanbaatar S, Mix H, Lhagvasuren B (1997) Argali Ovis ammon surveys in Mongolia's South Gobi. Oryx 31:285–294

Reading R, Michel S, Amgalanbaatar S (2020) Ovis ammon. The IUCN Red List of Threatened Species 2020:e.T15733A22146397

Rezaei HR, Naderi S, Chintauan-Marquier IC, Taberlet P, Virk AT, Naghash HR, Rioux D, Kaboli M, Pompanon F (2010) Evolution and taxonomy of the wild species of the genus Ovis (Mammalia, Artiodactyla, Bovidae). Mol Phylogenet Evol 54:315–326

Richards RB, Norris RT, Dunlop RH, McQuade NC (1989) Causes of death in sheep exported live by sea. Aust Vet J 66:33–38

Ridler AL (2008) Disease threats to sheep associated with intensification of pastoral farming. N Z Vet J 56:270–273

Riedel JL, Casasús I, Bernués A (2007) Sheep farming intensification and utilization of natural resources in a Mediterranean pastoral agro-ecosystem. Livest Sci 111:153–163

Riviere JE, Papich MG (2018) Veterinary pharmacology and therapeutics. John Wiley & Sons

Rosa HJD, Bryant MJ (2003) Seasonality of reproduction in sheep. Small Rumin Res 48:155–171

Rosenbaum B, Amgalanbaatar S, Reading PR (2020) Seasonal daytime activity budget of Mongolian Gobi Argali Sheep (Ovis ammon). Mongol J Biol Sci 19:3–15

Ruckstuhl. (1999) To synchronise or not to synchronise: a dilemma for young bighorn males? Behaviour 136:805–818

Rushen J (1986) Aversion of sheep to electro-immobilization and physical restraint. Appl Anim Behav Sci 15:315–324

Saboulard M, Russell WC, Riley ML (1995) Selection for lambing rate and clean fleece weight in sheep. J Anim Sci 73:3195–3198

Sanger ME, Doyle RE, Hinch GN, Lee C (2011) Sheep exhibit a positive judgement bias and stress-induced hyperthermia following shearing. Appl Anim Behav Sci 131:94–103

Santos AC, Arthur RG, Wlody D, De Armas P, Morishima HO, Finster M (1995) Comparative systemic toxicity of ropivacaine and bupivacaine in nonpregnant and pregnant Ewes. Anesthesiology 82:734–740

Santos SHD, Peery RM, Miller JM, Dao A, Lyu F-H, Li X, Li M-H, Coltman DW (2021) Ancient hybridization patterns between bighorn and thinhorn sheep. Mol Ecol 30:6273–6288

Saunders GR, Gentle MN, Dickman CR (2010) The impacts and management of foxes Vulpes vulpes in Australia. Mammal Rev 40:181–211

Saville JW, Ross JA, Trefz T, Schatz C, Matheson-Bird H, Ralston B, Granot O, Schmid K, Terry R, Allan ND, Wulff JE, Olson M (2020) Development and field validation of lidocaine-loaded castration bands for bovine pain mitigation. Animals 10:2363

Schaller GB (2012) Tibetan Wild Sheep Scandal. In: Schaller GB (ed) Tibet wild: a naturalist's journeys on the roof of the world. Island Press/Center for Resource Economics, Washington, DC, pp 259–272

Scheerlinck J-PY, Snibson KJ, Bowles VM, Sutton P (2008) Biomedical applications of sheep models: from asthma to vaccines. Trends Biotechnol 26:259–266

Shafford HL, Hellyer PW, Turner AS (2004) Intra-articular lidocaine plus bupivacaine in sheep undergoing stifle arthrotomy. Vet Anaesth Analg 31:20–26

Shukla M, Singh G, Sindhura BG, Telang AG, Rao GS, Malik JK (2007) Comparative plasma pharmacokinetics of meloxicam in sheep and goats following intravenous administration. Comparative Biochemistry and Physiology Part C: Toxicology & Pharmacology 145:528–532

Simensen E, Kielland C, Hardeng F, Bøe KE (2014) Associations between housing and management factors and reproductive performance in 327 Norwegian sheep flocks. Acta Vet Scand 56:26

Simon BT, Scallan EM, O O, Ebner LS, Cerullo MN, Follette C, Cox SK, Doherty TJ, Lizarraga I (2017) Pharmacokinetics and pharmacodynamics of midazolam following intravenous and intramuscular administration to sheep. Am J Vet Res 78:539–549

Singh NJ, Bonenfant C, Yoccoz NG, Côté SD (2010a) Sexual segregation in Eurasian wild sheep. Behav Ecol 21:410–418

Singh NJ, Yoccoz NG, Lecomte N, Côté SD, Fox JL (2010b) Scale and selection of habitat and resources: Tibetan argali (Ovis ammon hodgsoni) in high-altitude rangelands. Can J Zool 88:436–447

Sinnott CJ, Cogswell LP, Johnson A, Strichartz GR (2003) On the mechanism by which Epinephrine Potentiates Lidocaine's peripheral nerve block. Anesthesiology 98:181–188

Siraj-ud-Din M, Minhas RA, Khan M, Ali U, Bibi SS, Ahmed B, Awan MS (2016) Conservation Status of Ladakh Urial (Ovis vignei vignei Blyth, 1841) in Gilgit Baltistan, Pakistan. Pak J Zool 48

Skene DJ, Middleton B, Fraser CK, Pennings JLA, Kuchel TR, Rudiger SR, Bawden CS, Morton AJ (2017) Metabolic profiling of presymptomatic Huntington's disease sheep reveals novel biomarkers. Sci Rep 7:43030

Small A, Belson S, Holm M, Colditz I (2014) Efficacy of a buccal meloxicam formulation for pain relief in Merino lambs undergoing knife castration and tail docking in a randomised field trial. Aust Vet J 92:381–388

Small AH, Marini D, Dyall T, Paull D, Lee C (2018a) A randomised field study evaluating the effectiveness of buccal meloxicam and topical local anaesthetic formulations administered singly or in combination at improving welfare of female Merino lambs undergoing surgical mulesing and hot knife tail docking. Res Vet Sci 118:305–311

Small AH, Marini D, le Floch M, Paull D, Lee C (2018b) A pen study evaluation of buccal meloxicam and topical anaesthetic at improving welfare of lambs undergoing surgical mulesing and hot knife tail docking. Res Vet Sci 118:270–277

Small AH, Jongman EC, Niemeyer D, Lee C, Colditz IG (2020) Efficacy of precisely injected single local bolus of lignocaine for alleviation of behavioural responses to pain during tail docking and castration of lambs with rubber rings. Res Vet Sci 133:210–218

Small A, Fetiveau M, Smith R, Colditz I (2021a) Three studies evaluating the potential for lidocaine, bupivacaine or procaine to reduce pain-related behaviors following ring castration and/or tail docking in lambs. Animals 11:3583

Small A, Fisher AD, Lee C, Colditz I (2021b) Analgesia for sheep in commercial production: where to next? Animals 11:1127

Small A, Marini D, Colditz I (2021c) Local anesthetic delivered with a dual action ring and injection applicator reduces the acute pain response of lambs during tail docking. Animals 11:2242

Smith PEL. (1971). Iran, 9000–4000 B.C.: Expedition Magazine

Sneddon J, Rollin B (2010) Mulesing and animal ethics. J Agric Environ Ethics 23:371–386

Stafford KJ, Mellor DJ (2005) The welfare significance of the castration of cattle: a review. N Z Vet J 53:271–278

Stahl P, Vandel JM, Ruette S, Coat L, Coat Y, Balestra L (2002) Factors affecting lynx predation on sheep in the French Jura. J Appl Ecol 39:204–216

Stewart M, Beausoleil NJ, Johnson CB, Webster JR, Schütz KE, Cox N, Stafford KJ (2014) Do rubber rings coated with lignocaine reduce the pain associated with ring castration of lambs? Appl Anim Behav Sci 160:56–63

Stillman MW, Whittaker AL (2019) Use and efficacy of analgesic agents in sheep (Ovis aries) used in biomedical research. J Am Assoc Lab Anim Sci 58:755–766

Stiner MC, Buitenhuis H, Duru G, Kuhn SL, Mentzer SM, Munro ND, Pöllath N, Quade J, Tsartsidou G, Özbaşaran M (2014) A forager–herder trade-off, from broad-spectrum hunting to sheep management at Aşıklı Höyük, Turkey. Proc Natl Acad Sci 111:8404–8409

Stiner MC, Munro ND, Buitenhuis H, Duru G, Özbaşaran M (2022) An endemic pathway to sheep and goat domestication at Aşıklı Höyük (Central Anatolia, Turkey). Proc Natl Acad Sci U S A 119:e2110930119

Stock ML, Coetzee JF, KuKanich B, Smith BI (2013) Pharmacokinetics of intravenously and orally administered meloxicam in sheep. Am J Vet Res 74:779–783

Stockman C, Barnes A, Pethick D (2006) The heat is on in Western Australian summer feedlots. In: Cronje P, Maxwell D (eds) Wool meets meat. Sheep CRC, pp 259–260

Stubsjøen SM, Flø AS, Moe RO, Janczak AM, Skjerve E, Valle PS, Zanella AJ (2009) Exploring non-invasive methods to assess pain in sheep. Physiol Behav 98:640–648

Stubsjøen SM, Bohlin J, Skjerve E, Valle PS, Zanella AJ (2010a) Applying fractal analysis to heart rate time series of sheep experiencing pain. Physiol Behav 101:74–80

Stubsjøen SM, Valle PS, Zanella AJ (2010b) The use of a hand-held algometer as a method to measure mechanical nociceptive thresholds in sheep. Anim Welf 19:31–36

Sumerian (c. 3300-2900 B.C.) (n.d.). Head of a Ewe. (p. Terracotta Sculpture)

Sutherland M, Mellor D, Stafford K, Gregory N, Bruce R, Ward R, Todd S (1999) Acute cortisol responses of lambs to ring castration and docking after the injection of lignocaine into the scrotal neck or testes at the time of ring application. Aust Vet J 77:738–741

Sutherland MA, Worth GM, Dowling SK, Lowe GL, Cave VM, Stewart M (2020) Evaluation of infrared thermography as a non-invasive method of measuring the autonomic nervous response in sheep. PLoS One 15:e0233558

Tamioso PR, Guimarães PRB, Molento CFM (2017) Attitudes of South Brazilian sheep farmers to animal welfare and sentience. Ciência Rural 47

Teixeira PPM, Padilha LC, Motheo TF, Silva MAM, Oliveira MEF, da Silva ASL, Barros FFPC, Coutinho LN, Flôres FN, Lopes MCS, Rodrigues LFS, Vicente WRR (2011) Ovariectomy by laparotomy, a video-assisted approach or a complete laparoscopic technique in Santa Ines sheep. Small Rumin Res 99:199–202

Trentini R, Maitino AG, Fede Ed, Iannetti L, Villa Pd (2013) Tail docking of sheep and animal welfare: a review. Large Animal Rev 19:21–31

Union E (2010) Directive 2010/63/EU of the European Parliament and of the Council of 22 September 2010 on the protection of animals used for scientific purposes. Off J Eur Union. L276/33

Upton RN, Martinez AM, Grant C (2009) Comparison of the sedative properties of CNS 7056, midazolam, and propofol in sheep. BJA 103:848–857

Uriarte J, Valderrábano J (1990) Grazing management strategies for the control of parasitic diseases in intensive sheep production systems. Vet Parasitol 37:243–255

van Bommel L, Johnson C (2017) Predation control. In: Advances in Sheep Welfare. Elsevier, pp 177–196

van de Kerk M, Verbyla D, Nolin AW, Sivy KJ, Prugh LR (2018) Range-wide variation in the effect of spring snow phenology on Dall sheep population dynamics. Environ Res Lett 13:075008

Vautier AN, Cadaret CN (2022) Long-term consequences of adaptive fetal programming in ruminant livestock. Front Anim Sci 3

Veissier I, Boissy A, Nowak R, Orgeur P, Poindron P (1998) Ontogeny of social awareness in domestic herbivores. Appl Anim Behav Sci 57:233–245

Vigne J-D, Helmer D (2007) Was milk a "secondary product" in the Old World Neolithisation process? Its role in the domestication of cattle, sheep and goats. Anthropozoologica 42:9–40

Vila E, Helmer D (2014) 2. The expansion of sheep herding and the development of wool production in the ancient near East: an archaeozoological and iconographical approach. Wool Economy in the Ancient Near East 17:22

Viscardi AV, Reppert EJ, Kleinhenz MD, Wise P, Lin Z, Montgomery S, Daniell H, Curtis A, Martin M, Coetzee JF (2021) Analgesic comparison of flunixin meglumine or meloxicam for soft-tissue surgery in sheep: a pilot study. Animals 11:423

Waterman AE, Livingston A, Amin A (1990) The antinociceptive activity and respiratory effects of fentanyl in sheep. J Assoc Vet Anaesth GB IRL 17:20–23

Weaver S (2013) The Backyard Sheep—an introductory guide to keeping productive pet sheep. Storey, New York

Weil AB, Baird AN (2021) 18—anesthetic and pain management. In: Pugh DG, Baird AN, Edmondson MA, Passler T (eds) Sheep, goat, and cervid medicine, 3rd. Elsevier, St. Louis, pp 461–478

Welsh EM, Nolan AM (1995) Effect of flunixin meglumine on the thresholds to mechanical stimulation in healthy and lame sheep. Res Vet Sci 58:61–66

Welsh EM, Baxter P, Nolan AM (1992) Pharmacokinetics of carprofen administered intravenously to sheep. Res Vet Sci 53:264–266

Welsh EM, McKellar QA, Nolan AM (1993) The pharmacokinetics of flunixin meglumine in the sheep. J Vet Pharmacol Ther 16:181–188

Wemelsfelder F, Hunter TEA, Mendl MT, Lawrence AB (2001) Assessing the 'whole animal': a free choice profiling approach. Anim Behav 62:209–220

WITS (2022) Sheep; live exports by country in 2022. World Intergrated Trade Solution

Woodland AN, Van der Saag D, Kimble B, White PJ, Govendir M, Lomax S (2019) Plasma pharmacokinetic profile and efficacy of meloxicam administered subcutaneously and intramuscularly to sheep. PLoS One 14:e0215842

Worley K, Strobeck C, Arthur S, Carey J, Schwantje H, Veitch A, Coltman DW (2004) Population genetic structure of North American thinhorn sheep (Ovis dalli). Mol Ecol 13:2545–2556

Yavuz Ü, Yener K, Şahan A (2021) The effect of midazolam and its reversal flumazenil on sedative and cardiopulmonary variables in sheep. Kafkas Univ Vet Fak Derg 27

Young JM, Behrendt R, Curnow M, Oldham CM, Thompson AN (2016) Economic value of pregnancy scanning and optimum nutritional management of dry, single- and twin-bearing Merino ewes. Anim Prod Sci 56:669–678

Zanolari P, Dürr S, Jores J, Steiner A, Kuhnert P (2021) Ovine footrot: a review of current knowledge. Vet J 271:105647

Zentrich E, Wassermann L, Struve B, Selke K, Buettner M, Keubler LM, Reifenrath J, Angrisani N, Kempfert M, Krause A, Bellmann O, Kopaczka M, Merhof D, Bankstahl M, Bleich A, Häger C (2023) Postoperative severity assessment in sheep. Eur Surg Res 64:27–36

Pigs

Kenneth M. D. Rutherford

Abstract

Pigs (animals of the Genus *Sus*), of which there are nine extant species within the wider Family *Suidae*, are adaptable polytocous (giving birth to multiple offspring at one time) omnivores, that occupy a variety of ecological niches in temperate zones largely across Europe and Asian. Pigs have a long history of use by humans, culminating in the process of domestication, which resulted in the variety of breeds of domesticated pigs (*Sus scrofa domesticus*) existing currently. Today, pigs have economic, scientific, and cultural significance around the world. The domestic pig is farmed in large numbers, and increasingly, pigs are used as animal models in biomedical research because of their biological similarity to humans. Pigs are also kept as pets, and in some areas feral pig populations or wild boar are considered an invasive pest species.

In a farming context, pigs may experience pain during necessary parts of the production cycle, e.g. parturition, through deliberate management interventions, e.g. castration, tail docking, teeth resection, or from various diseases or disorders that may occur at increased prevalence in intensive farming systems. In the research context, pigs may experience induced pain in studies where pain is an outcome measure or, more commonly, as the result of ineffective pain management after noxious scientific procedures. A variety of approaches have been used for pain assessment in the pig, and approaches to pain mitigation have been extensively researched to increase the ethical justification for using the species in research. Yet despite this work, pain remains a central welfare problem for large numbers of pigs across the world in both the farmed and research contexts.

Keywords

Pigs · Pain · Procedures · Pain assessment

1 Introduction

1.1 History of the Domestic Pig

The domestic pig (*Sus scrofa domesticus*) is a member of the Suidae family within the Artiodactyla (even-toed ungulates) order of mammals. In phylogenetic terms, the nearest species is the Eurasian Wild Boar from which the

K. M. D. Rutherford (✉)
Animal Behaviour and Welfare, SRUC, Edinburgh, UK
e-mail: Kenny.Rutherford@sruc.ac.uk

R. E. Clutton, G. C. Musk (eds.), *Pain and Suffering in Farmed Animals*,
https://doi.org/10.1007/978-3-032-11407-5_16

domestic pig originated (on multiple independent occasions from various wild boar subspecies, across Europe, Western Asia, and China (Larson et al. 2005). Estimates for the timing of domestication vary but the process is thought to have been underway at least 9000 years ago and to have followed a similar path to dog domestication i.e. beginning with local Wild boar populations adapting to human settlements, foraging for food close to humans, before eventually humans started to take them captive, controlling their breeding and adapting them to a human presence from early life.

Within the species, many breeds have been produced via artificial selection after initial domestication. These breeds differ in various anatomical, physiological, reproductive, and behavioural characteristics. Domestication in pigs has produced—amongst other physical changes—a clear reduction in brain size (by around one quarter: Balcarcel et al. 2022), yet domesticated breeds are known to retain most behavioural traits of their wild progenitors. One major distinction is between European and Chinese pig breeds, which are thought to have diverged at least 2000 years ago (Giuffra et al. 2000). That said, planned breeding in the 18th and 19th centuries used Chinese pig genetics to improve European breeds, leading to the most common breed variants (Large White/Landrace) used in intensive pig production today. Initial selection by humans focused on physical traits, but from the mid-twentieth century attention increasingly turned to traits increasing the production output and value of pigs, e.g. lean carcase proportion, growth rate, litter size. Modern genetic selection methods continue to focus on improvements in these features, although focus has shifted to selection methods that balance pure production traits with outcomes relating to pig health and welfare, or more generically, to 'robustness' (Knap 2005). One notable development in pig breeding within the last 100 years has been the development of 'minipig' breeds (such as the Yucatan and Göttingen minipigs), which were specifically bred to increase the utility of pigs in biomedical laboratory research (Rozkot et al. 2015). These breeding lines retain many of the biological benefits of pigs as a model species, but being smaller, are less costly to maintain and have been adapted via selection to a laboratory environment where they offer other advantages when size-relevant models for adult humans are not required.

1.2 General Biology

Species within the suidae are polytocous, and the domestic pig is polyoestrous, so unlike the Wild Boar, which breeds seasonally, domestic pigs reproduce year-round. Pigs are omnivores, with a great capacity to adapt to differing conditions due to their behavioural plasticity and reproductive flexibility. Their monogastric stomach enables a variety of diets to be consumed, further enhancing their capability to adapt to differing environmental niches, including—in early domestication and continuing small-scale production—their beneficial ability to consume waste food products considered unfit for human consumption. Pig species have a particularly well-developed sense of smell, and the olfactory bulb occupies around 7% of brain tissue, cf ~0.01% in humans (Schild and Rørvang 2023). Ancestral wild boar populations show a matriarchal social structure where groups of females of differing ages form, whilst reproductively mature males are generally semi-solitary and only join social groups for breeding. Many of these ancestral traits are still present to a varying degree in domestic pigs, and many of the health and welfare issues faced by domestic pigs in captivity can be traced back to their wild origins.

1.3 Cultural Significance

In general, pigs are considered intelligent and emotionally expressive animals. Culturally, pigs have an important historical role in many countries, particularly China, where, in addition to being the main meat used in cuisine, they are a common symbol of fertility and prosperity, and across many European countries, where they have a notable role in national or regional speciality foods and seasonal feast celebrations. Pigs also have a positive place in the public consciousness as reflected by fictional accounts in literature such as *Animal Farm, Winnie the Pooh*, and *Charlotte's Web*, as well as film depictions such as *Babe*, and more recently *Okja*. Conversely, consumption of pork products is forbidden on religious grounds, most notably within Jewish and Muslim faiths, most likely due to perceptions that omnivorous dietary habits led to pigs being unclean (compared with ruminant species) (Essig 2015). Such perceptions persist in the broader public mind, and the terms 'pig' or 'swine' are sometimes used in a derogatory fashion, implying a dirty or lazy characteristic of the target.

1.4 Farming of Pigs

Pigs were suitable subjects for domestication because of their adaptability and prolificacy, as well as their capacity to survive on waste food from human populations. For human populations in the transition to agriculture, through to the modern day, pigs represented an effective way of storing food resources over winter or other lean periods. As with other species, pig production underwent a transition from local subsistence agriculture to large-scale intensive 'factory' farming methods in Western countries in the period following the Second World War. This led to major changes in traditional pig-keeping methods and large increases in farm size. These developments led to a major increase in the overall numbers of pigs being raised in commercial production (Fig. 1)—initially in Europe and North America, and then, on an even larger scale, in China. China has been the largest producer of pork for many decades, with around half of the world's commercial pig population (Lander et al. 2020).

The pig industry faces various endemic or emerging disease threats, most notably infectious diseases, such as African Swine Fever (ASF). Some diseases are zoonotic and of major concern to human health authorities. The high level of prophylactic antimicrobial use in intensive pig production has also caused major concerns about antimicrobial resistance in various disease agents, including those threatening public health. Concerns about environmental impact are also being expressed, and although the relative contribution of pig meat production to gas emissions is low compared to other products, the scale of global pig production means it is a significant contributor to adverse environmental effects. Animal welfare remains a source of controversy from an industry perspective, and engagement with welfare varies considerably around the globe. Alongside behavioural restriction, issues relating to pain (and the lack of its alleviation) rank highly with respect to pig welfare (Rioja-Lang et al. 2020).

1.5 Pigs in Research

Pigs have a long history in research, which dates to Greek and Roman times (Rozkot et al. 2015). However, suidae currently represent a minor portion of the world's research animal population. Within the European Union 73.5 K pigs were used in 2020, increasing to 87 K in 2022 (European Commission 2024), for studies relating to pork production, pig health and welfare and general biomedicine. However, numbers of both commercial animals and laboratory minipigs in research are increasing (Gutierrez et al. 2015) and, as the use of non-human primates is increasingly condemned, pigs are being promoted as the better model species for translational research into a wide range of the biosciences including cardiovascular, neurological, dermatological and oncological disease research, studies of early life development, organ transplantation, toxicology, pharmaceuticals and metabolism. More recently, genetic modification has been used to create specific model lines of pigs in areas such as cystic fibrosis research. The use of pigs as a laboratory animal brings with it several species-specific welfare concerns, as well as a variety of issues that apply across laboratory animals generally (Herskin and Marchant 2023).

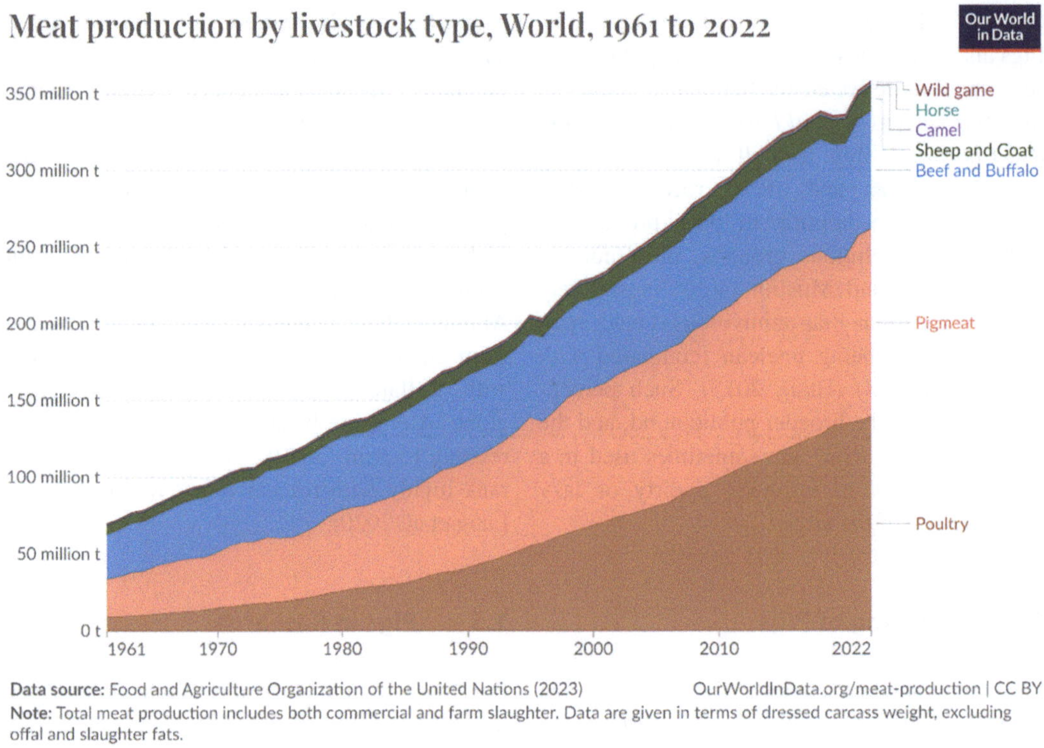

Fig. 1 Global meat production levels since the 1960s (https://ourworldindata.org/meat-production)

1.6 Other Pig Populations

In comparison with farmed animals, the number of pigs being kept as pets (generally minipigs or Vietnamese potbellied pigs) is small. Pet pig may experience various welfare issues relating to the extent to which their intrinsic needs can be met in the household environment. They may also experience health disorders (e.g. locomotor, gastrointestinal, respiratory disorders and various skin problems: Sipos et al. 2007, Carr 2004), which cause pain (Mozzachio and Tynes 2013).

Feral populations of pigs (domestic and wild boar species, and their hybrids) are a problem in many areas around the world, as they can act as a reservoir for infectious disease and have an impact on the natural environment and on crops and pastures used by humans. In these conditions, they may be viewed as an invasive pest species, and various approaches to population management are used to limit the damage feral pigs do.

2 Sources of Pain in Pigs

Pain is a welfare concern in pigs in a variety of contexts relating to human use of the species. Pain occurs when deliberate noxious interventions are conducted by humans for farming or scientific purposes, or various disease or injury states, which may occur at an increased prevalence due to management practices in a production context.

2.1 Breeding Females

Parturition Dystocia in pigs is thought to be of low prevalence (e.g. ~5%: Ison et al. 2016a, 2016b, 2016c) and in commercial production, most sows give birth without human intervention (though uncertainties remain regarding the recognition of dystocia in pigs: Walls et al. 2022). However, increases in average pig litter sizes brought about by genetic selection (Rutherford et al. 2013) are likely to have increased the overall painfulness of the farrowing experience, and various studies have examined parturition in pigs from the perspective of pain (Mainau and Manteca 2011; Ison et al. 2016b, 2017; Navarro et al. 2020).

Shoulder ulcers While feeding is seldom restricted during the lactation period, sows may still fail to eat enough to match the energy requirements of nursing a large litter of piglets. Consequently, loss of body condition is common during lactation, particularly when sows are restrained in farrowing crates. This can lead to the development of shoulder ulcers: open sores on the bony prominences of the shoulder resulting from a reduced blood supply (due to the sow lying for prolonged periods in the same position) to the tissues overlying these prominences (Herskin et al. 2010). It is presumed that these lesions are painful for the animals bearing them, and behavioural differences between sows with and without ulcers have been seen in some (Larsen et al. 2015), though not all (Gaab et al. 2022) studies.

Nose ringing Pigs in outdoor production settings may have a ring inserted into their nasal septum to prevent damage to paddocks caused by natural rooting behaviour. Rings are effective in reducing rooting activities on pasture but also impair feeding in a way that suggests a trade-off between pain and feeding (Horrell et al. 2000, 2001).

2.2 Painful Neonatal Procedures

Neonatal pigs commonly experience pain during and after the various management procedures inflicted upon them, often in the first few days of life. Such procedures may be for production benefit (castration, identification) or to prevent animals from causing injuries to conspecifics (teeth resection, tail docking). The pain caused by these procedures is well established (Marchant-Forde et al. 2009, 2014; Herskin and Di Giminiani 2024) as is its persistence for several days after the procedure (Hay et al. 2003).

Castration Castration is a practice used in commercial pig production to mitigate against 'boar taint' (an unpleasant taste, perceived by some people, in pig meat products caused by male sex hormones such as androstenone). Castration also limits unhelpful behavioural traits in male pigs, including aggression and the possibility of unplanned breeding in animals that reach reproductive maturity before slaughter. (Male pigs may become sexually active at 3 months of age.) A large body of work over many decades has clearly established the painfulness of piglet castration (von Borell et al. 2009; Rault and Lay 2011; Dzikamunhenga et al. 2014; O'Connor et al. 2014, 2016). Acute pain is caused by both the initial incision and by the severing of the spermatic cord, prior to removal of the testes (with the latter being considered the more significant cause of distress for the animals concerned (Garcia et al. 2023). Subsequently, piglets experience inflammatory post-surgical pain for several days (Hay et al. 2003). While the routine castration of commercial pigs is normally conducted at the start of their lives, a limited number of adult animals may be castrated (Romeo et al. 2023). For example, in Italy, breeding boars may be castrated before being slaughtered for meat.

Tail docking Tail docking—the removal of part of the tail—is widely conducted in intensive pig production (often despite legislative controls on its use: D'Eath et al. 2016) to limit the later problem of tail biting. Docking may be conducted using clippers or a heated cautery iron that cauterises the remaining tail tip (Marchant-Forde et al. 2009) and is widely recognised as a painful procedure (D'Eath et al. 2016).

Teeth resection Teeth resection involves clipping or grinding the 'needle' (canines and third incisors) teeth and is conducted in the first few days of life to minimise piglets' capacity to physically damage their littermates (during agonistic teat competition interactions) or their dam's udder or teats. Performed ineptly, the procedures may also cause mouth lesions and pulpitis—which are likely to be painful (Marchant-Forde et al. 2009).

Identification and Injections Commercial pigs are often identified using ear tags, which are inserted (e.g. at weaning, upon entry to the breeding herd, or before leaving the holding) using methods involving puncturing the ear lobe. Alternatively, pigs may be identified by ear notching, which involves cutting distinctive shapes into the outer ear margins of piglets. Commonly, both ears are cut with a system used to identify litter of origin on one ear and the number within the litter on the other. Notching is preferred by some producers as it provides a permanent and inexpensive method of animal identification. Both methods involve pain (Marchant-Forde et al. 2009; Leslie et al. 2010; Lomax et al. 2018).

Piglets are frequently injected with medications for the benefit of their health and development. For example, the administration of iron salts is a common management procedure conducted early in life to treat 'physiological anaemia'. Similarly, pigs of various ages may be given prophylactic treatments as a matter of course, while therapeutics may be given as required. Whilst some of these substances may be delivered orally (in water or feed), other treatments require direct—usually intramuscular—injection. Traditionally, this has been achieved using hypodermic needles of a size more suited to the physical properties of the injectate, rather than the pain engendered in the recipient. More recently, methods for needle-free injection have been developed and are currently in commercial use for pigs (Dalmau et al. 2021).

2.3 Older Animals

Lameness While estimates of the prevalence of lameness in pigs vary considerably (4.5% in sows: Willgert et al. 2014; 17% in sows, 20% in finisher pigs: Kilbride et al. 2009b), there is authoritative consensus that lameness is a major cause of poor welfare within pig farming. Pathological alterations to gait, i.e. lameness, may result from injuries, overgrown claws, and joint infections. While lameness can also arise from the non-painful mechanical disruption of normal ambulation, it is generally regarded as a painful condition. Unfortunately, there is substantial variability amongst farmers and vets in grading the painfulness of different causes of lameness (Ison and Rutherford 2014; Jensen et al. 2012). Research in the painfulness of lameness and its relief has either used naturally occurring lameness (Mustonen et al. 2011; Conte et al. 2014; Meijer et al. 2014, 2015; Macfadyen et al. 2019; Forseth et al. 2023) or artificially induced models (e.g. injection of monosodium-iodacetate into the intercarpal joint to create an osteoarthritis model: Uilenreef et al. 2019; injection of Amphotericin into claw joints to create a transient synovitis: Karriker et al. 2013). The use of naturally occurring pain states is often perceived as being more ethically acceptable when studying pain management and has the scientific value that all aspects of the condition are present (Uilenreef et al. 2019). However, this approach may compromise data quality given uncertainties about onset, variable expression of the condition and lack of control in study subjects. Translational pain research may also require deliberate infliction of the injury state if the target problem rarely spontaneously occurs in the model species (e.g. most translational/basic neuropathic pain research).

Tail biting Tail biting usually occurs in captive pigs when various risk factors are met (Taylor et al. 2010). Key amongst these is the failure to provide appropriate levels of environmental stimulation that allow an outlet for the pig's natural oral exploratory behaviour. As a result, pigs with intact tails may end up biting each other, with the main outlet for that being the tail (or occasionally ears, or the flank). Once tail biting has started within a group of pigs, it can spread across individuals and—in the absence of intervention—individual pigs may suffer severe, and occasionally fatal injuries. Consequently, tail-bitten pigs may suffer anything from mild acute pain to severe pain associated with the removal of most of the tail and injuries to the surrounding areas of the rump. Even in the absence of such severe wounds, being the victim of tail biting creates a chronic stress state in affected animals (Munsterhjelm et al. 2013).

Disease Commercial pig health is challenged by numerous bacterial and/or viral diseases, many of which arise as a direct result of intensive management and several of which are painful. Gastric ulceration is a highly prevalent disorder in commercial pig production (Swaby and Gregory 2012), which is caused by erosions of the *pars oesphagea* area of the stomach associated with the consumption of finely ground food. Behavioural differences between pigs with or without ulceration indicate the former are experiencing pain (Rutherford et al. 2018). Respiratory disease is also common in pigs and is probably painful, because the demeanour of affected animals improves with analgesic therapy (Hälli et al. 2018).

2.4 Experimental Studies

Pigs involved in scientific research may suffer pain in three general ways: (i) for reasons discussed previously, environmental or managemental, and unrelated to their 'laboratory animal' status; (ii) because the pain generated by noxious scientific procedures, e.g. surgery; and (iii) as recipients of painful stimuli in pain studies. Almost certainly, the second category represents the most painful pig studies, where, it must be emphasised, the infliction of pain is incidental, but not deliberate. Many biomedical studies, i.e. those intended to either elucidate basic biological mechanisms or to yield data that are relevant in solving human medical problems, involve major procedures, e.g. extensive telemetric or surgical operations, that are painful in people. Being experimental, however, means that methods of pain management may not be fully developed or as effective in pigs. As such, further progress in the refinement principle (Russell and Burch 1959) continues to be required to overcome this deficiency.

Most pain-focused studies (in which pain is intentionally inflicted, or deliberately unmanaged) in pigs intend to improve the recognition and mitigation of pain found in commercial pigs in the real world and are ethically justified because the benefits to large pig numbers outweigh the adverse experiences of the few. This is arguably less contentious than the use of pigs as pain models for humans, which is small but increasing (Meijs et al. 2021). In reviewing translational pain studies, Meijs et al. (2021) identified 23 distinct porcine pain models, including many classically involving laboratory rodents, e.g. post-surgical pain, skin incisions (Shiff et al. 2023), inflammatory challenge, and nerve damage (e.g. Castel et al. 2016; Hellman et al. 2021a, b; Hilgart et al. 2023). They also included studies where a putative pain 'model' was produced in the context of researching an ongoing welfare, e.g. forms of lameness, castration or tail docking, parturition, and shoulder ulcers.

3 Overview of Pain Assessment Methods

Pain assessment in animals is a critical prerequisite to achieving high animal welfare standards. Understanding the biological variables that can be used to provide a valid and reliable inference of an animal's pain experience is necessary for the effective mitigation of that pain as well as improving the scientific methods required to develop and improve pain relieving products, i.e. analgesic drugs.

Verbal self-report is the gold standard for pain assessment in most human beings. However, in animals, other variables must be assessed to draw inferences about underlying pain experiences (as is the case in non-verbal human infants, and others). Various measures have been used as indicators of pain in pigs (Ison et al. 2016a; Sheil and Polkinghorne 2020; Herskin and Di Giminiani 2024). These range from methods which are only of use in a research context to approaches for assessing pain in a clinical setting (e.g. on farm). The latter methods are subject to requirements additional to repeatability and validity, e.g. provision of an immediate metric, low cost, rapid application, not requiring repeated or invasive sampling.

3.1 Pain Scoring

Although, pain scoring systems are often developed for use in clinical situations where more detailed research methods are rendered impractical due to key limitations (e.g. expense, time-consuming, retrospective outcomes), they may also be applied in research studies. Scoring schemes vary widely in nature from simple Visual-Analogue Scales (VAS; e.g. Lagoda et al. 2021) or Numerical Rating Scales (NRS; e.g. Castel et al. 2014), to scales that involve something closer to behavioural quantification and include several different assessed variables. (See chronic pain chapter.)

More recently, efforts have been made to create a composite pain scale that incorporates knowledge relating to behaviour change into a scale application in a more clinical context (the Unesp-Botucatu Pig composite Acute Pain Scale (UPAPS): Luna et al. 2020; da Silva et al. 2023; Robles et al. 2023). This scale involves assessment of various behavioural categories (such as posture, interest in surroundings, activity), each scored on a four-point scale (0–3), indicating the degree to which the trait is present during the observation. This tool combines some of the benefits of a quantitative behavioural approach (focus on certain clearly defined behaviours) but avoids the need for formal behavioural quantification, meaning that the scale can be readily and efficiently applied with immediate outcome.

3.2 Quantitative Sensory Testing

Quantitative sensory testing (QST) includes various methods that aim to measure the responses of animals to standardised mechanical (applied pressure) or thermal (heat or cold) stimulation. The threshold at which animals are seen to respond to the stimulation is used as a measure of sensitivity. Testing of the mechanical nociceptive threshold (MNT) may involve the use of handheld von Frey hairs or similar, or application of an algometer device. Testing thermal nociceptive thresholds (TNT) may either involve the application of a device that can apply standardised levels of heat or assess alterations in response to a single challenge, e.g. cold tests. These methods can be used to characterise alterations to nociceptive function, representing either reduced (hypoalgesic) or increased (hyperalgesic/allodynic) sensitivity. QST is primarily applied in biomedical research, although it can be used clinically to assess, for example, areas of increased sensitivity after injury and the effects of analgesics. QST methods (both MNT and TNT) have been used in pigs of different ages, in commercial situations as well as in minipigs, to investigate nociceptive changes in various contexts (Table 1). Others have investigated basic response variables for QST in pigs of varying ages to test reliability and identify extraneous sources of variability, e.g. degree of habituation, social context, animal age/weight, in response to tests (MNT: Janczak et al. 2012; Nalon et al. 2013; Di Giminiani et al.

Table 1 Summary of studies applying QST in pigs

Context	QST method	References
Naturally occurring lameness	MNT	Fosse et al. (2011a, b); Forseth et al. (2023)
Induced lameness	MNT, TNT	Tapper et al. (2013); Mohling et al. (2014); Pairis-Garcia et al. (2014)
Surgical incision	MNT	Castel et al. (2014, 2017); Cui et al. (2023); Wojtalewicz et al. (2025); Yang et al. (2024)
Tail docking	MNT	Di Giminiani et al. (2017a)
Parturition	TNT	Jarvis et al. (1997)
Inflammatory challenge	MNT	Sandercock et al. (2009, 2011)
	MNT, TNT	Di Giminiani et al. (2014a, b)
Neurofibromatosis model	MNT, TNT	Khanna et al. (2019)
Peroneal nerve injury	MNT, TNT	Hellman et al. (2021a, b)
Peripheral neuritis trauma	MNT	Castel et al. (2016)
High Frequency Stimulation of ulnar nerve	MNT, TNT	Meijs et al. (2025)
Tourniquet	MNT, TNT	Petrucci et al. (2024)

2016b, TNT: Herskin et al. 2009). These studies have generally found high levels of test reliability (e.g. Rettore Andreis et al. 2023) but do draw attention towards the need for high-quality experimental designs to avoid or control potential confounding factors.

Additionally, whilst analgesics, i.e. substances reducing pain, often alter nociceptive response sensitivity, this does not necessarily reflect the ongoing pain experience of the animal. Therefore, whilst QST methods have an important place in identifying mechanistic alterations and altered sensitivities which may have implications for an animal's welfare experience, these methods do not allow direct inferences about the animal's experience to be made.

3.3 Quantification of Behaviour

Historically, the focus on pain assessment in animal welfare research has focused on behavioural quantification. High-quality studies in this area apply detailed and specific ethograms, with planned observation methods, and clear attention to limiting inter- or intra-observer variability. Such methods have usually involved video-based, i.e. retrospective, and time-consuming evaluation, but when conducted properly, can provide a repeatable assessment of behavioural responses. The focus is on a combination of presumed, or validated, 'pain behaviours' as well as alterations in the performance of normal behaviours and postures (Ison et al. 2016a; Table 2). Detailed quantification of behaviour can provide a sound inference about the presence of pain, although gaining a clear understanding of how differing levels and types of behavioural change relate to the underlying experience remains challenging. There are problems with specificity, because behaviour changes may reflect other features of the animal's experience beyond pain alone, e.g. fear, sickness, hunger, etc. Depending on the study context, this uncertainty may add or detract from the validity with which research questions are answered. That said, approaches relying on quantification alone may still raise questions about the relationship between statistical and biological relevance, i.e. the uncertainty about the extent to which a certain degree of behavioural change (induced by an experimental treatment or disease) or the partial return to normality (resulting from an intervention) really matters to the animal. Other behavioural approaches, such as motivational or affective state testing, may provide a higher level of inference in that regard (Weary et al. 2017).

Grimace Scales The seminal work of Langford et al. (2010) resulted in the development and testing of facial grimace scaling in mice, which involved quantifying changes in facial action units putatively in response to pain. Since then, many studies have investigated the use of such 'Grimace Scales' for pain assessment in a variety of species (McLennan et al. 2019; Fischer-Tenhagen et al. 2022), including pigs (Table 2).

Vocalisation Studies conducted in the 1990s (Noonan et al. 1994; Weary et al. 1998; Horn et al. 1999; Taylor and Weary 2000) have revealed the usefulness of vocalisations in porcine pain assessment, since they have become one of the most commonly applied outcome measures (Dzikamunhenga et al. 2014; O'Connor et al. 2014, 2016; Ison et al. 2016a; Sheil and Polkinghorne 2020), particularly in relation to acute procedures (Table 2). These studies either make use of a human classification to crudely categorise vocalisation or use more detailed biophysical analysis of vocalisations in terms of structure (i.e. frequency, amplitude, intensity etc.; Coutant et al. 2023). More recently, there has been a growing interest in the application of machine learning approaches to the assessment of pig vocalisations (da Silva et al. 2019).

Motivational-Affective testing The definition of pain (Raja et al. 2020) notes both the sensory-discriminative and the motivational-affective (emotional) dimensions of the experience. Whilst these two elements tend to co-vary, it has become increasingly clear that the more mechanistic aspects of the sensory dimension do not necessarily predict the effect that any given pain state has on individual animals. In plain English, alter-

Table 2 Recent studies applying assessment of behaviour, grimace scales or vocalisations in studies of pig pain

Method	Context	References
Behavioural quantification	Tail docking	Herskin et al. (2016); Di Giminiani et al. (2017b); Viscardi and Turner (2019b); Nagel et al. (2024)
	Castration	Gottardo et al. (2016); Viscardi and Turner (2018a, b, 2019a); Burkemper et al. (2019); Yun et al. (2019); Lou et al. (2022b); Nagel et al. (2024)
	Teeth resection	Timphaus et al. (2024)
	Injections	McGlone et al. (2016); Sutherland et al. (2017); Temple et al. (2017, 2020); Dalmau et al. (2021); Lewis et al. (2024)
	Parturition	Ison et al. (2016b, 2017, 2018); Nowland et al. (2019); Ward et al. (2022)
	Lameness	Whalin et al. (2016); Ala-Kurikka et al. (2017)
	Gastric ulcers	Rutherford et al. (2018)
Grimace Scale	Tail docking	Di Giminiani et al. (2016a); Viscardi et al. (2017); Viscardi and Turner (2019b)
	Castration	Di Giminiani et al. (2016a); Viscardi et al. (2017) Viscardi and Turner (2018a, b, 2019a); Lou et al. (2022a); Neary et al. (2024)
	Cryptorchidectomy surgery	Vullo et al. (2020); Carbonari et al. (2022)
	Parturition	Navarro et al. (2020)
Vocalisation	Castration	Sutherland et al. (2017); Hancock et al. (2018); Viscardi and Turner (2018a, b, 2019a); Valenzuela et al. (2019); Sheil et al. (2020, 2021); Coutant et al. (2022, 2023)
	Tail docking	Di Giminiani et al. (2017b); Tallet et al. (2019); Viscardi and Turner (2019b); Morrison and Hemsworth (2020a, b)
	Injections	Scollo et al. (2020); Temple et al. (2020); Dalmau et al. (2021); Lewis et al. (2024)

ations in sensory thresholds (sensory-discriminative) in arthritic sows do not necessarily reflect how miserable (affective-motivational) they feel. Recent developments in translational pain research have focused on developing methods that allow inferences to be made about the motivational-affective experience of the animals under study. This has arisen because it is increasingly felt that prioritizing the affective-motivational consequences of pain will lead to the development of more effective analgesics for human use—after all, it is the 'hurt' of pain and its emotional effects that people want mitigated (Mogil 2022). This argument also pertains to animals from their own perspective, i.e. it could be presumed that the degree to which the pain state wields motivational power over behavioural decisions provides the strongest inference of what matters to that animal (Weary et al. 2017). Such methodologies either directly test the motivational imperative associated with putative pain, or its alleviation, by: (i) placing the animal in a situation requiring a 'behavioural trade-off' (Fuchs and McNabb 2012), (ii) assessing the effect on naturally motivated spontaneous behaviours (Jirkof et al. 2019), (iii) assessing conditioned responses to induction of pain or its alleviation (Li 2013), or (iv) aiming to directly assess affective changes associated with pain, i.e. 'extended pain emotions' (Price and Barrell 2012). Whilst various approaches to assessing motivational-affective aspects of animal pain have been developed, and these are now being used in animal welfare research, there has been relatively little application of this approach, to date, in pig pain studies.

Various motivational trade-off tests have been applied to investigate trade-offs made by animals between avoiding painful stimulation of a body part and an alternative motivational priority (Fosse et al. 2011a; Bos et al. 2015; Bilsborrow et al. 2016; Coutant et al. 2023), applying the principle that—since pain can be best viewed as a motivational state—alterations to pain state, may increase trade-offs with other motivational priorities. Alternatively, aversion tests may be used to test whether an applied treatment reduces the willingness of the animal to encounter the location where the treatment happened (Dalmau et al. 2021) or humans generally (Tallet et al. 2019; Coutant et al. 2023). Both approaches have been used in young pigs to assess pain associated with induced lameness (Fosse et al. 2011a), castration (Bilsborrow et al. 2016; Davis et al. 2017; Abendschön et al. 2020; Reynolds et al. 2020), tail docking (Tallet et al. 2019), injections (Dalmau et al. 2021), and in older animals, to assess lameness pain (Bos et al. 2015).

Intriguingly, some studies (Castration: Bilsborrow et al. 2016; Davis et al. 2017; Reynolds et al. 2020; Tail docking: Tallet et al. 2019; Injections: Dalmau et al. 2021) have found alterations in the predicted direction (i.e. motivational trade-offs caused by pain, reversal by analgesia). For instance, Bilsborrow et al. (2016) tested whether castration affected the willingness of piglets to traverse a 1.8 m long chute, equipped with 2 small hurdles, to return to their littermates and dam. The time it took piglets to return was found to be sensitive to the castration itself and to the benefits of analgesia (in this study, and subsequently, in relation to traditional (Reynolds et al. 2020) and novel forms of analgesia provision (Davis et al. 2017)). However, some applications of the approach have not helped discriminate between pigs based on differing degrees of pain (Abendschön et al. 2020). Similarly, Coutant et al. (2023) developed a social motivation test, which recorded the time taken for newly-castrated piglets to traverse a 2 m long arena whilst having to negotiate three hurdles of increasing size to re-establish contact with three littermates. Subsequently, the associations between this and more traditional tests such as vocalisation, counting of pain behaviours, home-pen behaviour, salivary cortisol, and responses in human interaction tests were examined and demonstrated a general lack of association between different types of outcome measures (Coutant et al. 2023). Some differences may be explained by differences in timing in relation to the castration procedure, but even variables covering a similar period (such as immediate 'resistance' movements and vocalisations at the time of castration) were poorly correlated. This points to the fact that despite many research studies investigating piglet procedural pain, uncertainties about the best outcome measures remain.

Dalmau and others (2021) used an aversion learning test in 4-week-old piglets to assess responses to different vaccination injection methods. For this, piglets were first trained to proceed along a 4 m long narrow chute, at the end of which they received a milk replacer award. Study animals were then allocated to different injection treatments conducted at the chute's 'reward' end or a handled control. The animals were subsequently re-tested at various times post-injection, with the outcome measure being the time taken to traverse the chute (where an increased latency would be interpreted as an aversion response to the test treatment). Piglets exposed to a standard intramuscular injection took longer to move along the test chute 10 min after treatment compared to either controls or piglets vaccinated with a needle-free device. Tallet et al. (2019) developed a human aversion test and demonstrated that tail-docked piglets were more reluctant to approach an unfamiliar human than control animals.

While most studies in this area have used piglets, Bos et al. (2015) investigated the motivational-affective importance of lameness pain in sows by offering them the chance to walk between two feeder stations, which delivered feed alternately. Sows are kept severely feed-restricted in commercial breeding herds, so are normally highly motivated by food; however, the lamest animals showed a reduction in the number of feed rewards gained during the test period. This outcome gives a clearer indication of the imperative value that avoiding further pain has for these animals.

Machine vision Machine vision is a branch of artificial intelligence where computer algorithms are selected for their ability to discriminate visual images (after previous exposure to training images selected for the desired outcome measure). Machine vision algorithms have demonstrated a capability to be trained to recognise individual pigs based on facial features (Hansen et al. 2018), to discriminate, based on facial expression, between pigs based on their stress status (Hansen et al. 2021), to provide useful early detection of problems such as tail-biting (D'Eath et al. 2021) and to be able to detect various postures and behaviours that could have application in the area of pain assessment (Ojukwu et al. 2020; Chen et al. 2023; de Paula et al. 2024). Such methods hint at future systems which provide around-the-clock monitoring of pain status in animals based on facial/behavioural traits.

3.4 Physiology

Stress hormones The most used physiological metric for pain in acute neonatal studies is cortisol because it is an indicator of adreno-cortical activation via the hypothalamic-pituitary-adrenal axis. Numerous studies have documented the acute stress response triggered by surgical castration procedures in pigs (Table 3). The conviction that cortisol levels reflect the animal's pain experience is often taken to mean that lower adrenocortical activation indicates pain mitigation, e.g. in comparison to alternative methods, or following analgesic administration. Cortisol may be measured from blood samples, although handling and the sampling procedure itself may cause a stress response (Coutant et al. 2022) and potentially confound other aspects of the study—for instance, on-going behavioural assessments. Alternatively, cortisol can be measured in saliva although the very low saliva flow in neonatal animals limits this approach, while collection still requires stressful handling.

Heart Rate/Heart Rate Variation Assessment of heart rate is widely incorporated in numerous pain assessment methods. It is used to monitor anaesthesia in unconscious pigs, and to quantify painful stimulation in conscious animals (Table 3). Most studies use raw heart rate as a clinical response variable although more focussed analyses generate indices of heart rate variability (e.g. Byrd et al. 2020), as an indicator of sympathetic/parasympathetic autonomic nervous balance.

Temperature/Thermal imaging Thermography has been used in various situations assessing animal welfare and/or pain (Whittaker et al. 2023). The technique represents a non-invasive approach to assessing an animal's skin temperature, which is used either to assess the degree of inflammation associated with an injury (Table 3) or to assess the whole-body autonomic nervous response to pain.

Other biomarkers A variety of other neuroendocrinological markers have been used in porcine pain studies (Dzikamunhenga et al. 2014). These include Substance P (Sutherland et al. 2012), salivary microRNAs (Lecchi et al. 2020), chromogranin A (Hofmann et al. 2019; Temple et al. 2017), catecholamines (Rauh et al. 2019), acute phase proteins (Temple et al. 2017), lactate (Lonardi et al. 2015), and beta-endorphin (Marchant-Forde et al. 2009, 2014). These substances reflect activation of nociceptive pathways or other aspects of bodily response to injury, but altering levels are unlikely to be linearly associated with individual pain experiences, which is better reflected in behavioural outcomes.

4 Pain Alleviation in Pigs: The 3Ss

The 3Ss approach represents a valuable schema for considering different approaches to pain alleviation in pigs. First suggested by Guatteo et al. (2012) the concept identifies three separate pathways, (Suppress, Substitute and Sooth), to reduce the overall pain burden of animals in any given situation, although its application is directed towards animals in production and planned procedures. 'Suppress' represents opportunities to avoid the causes of pain entirely. 'Substitution' should be attempted when the source of pain cannot be avoided entirely and involves identifying

Table 3 Recent examples of physiological assessment in pig pain assessment

Method	Context	References
HPA Stress physiology	Castration	Davis et al. (2017); Byrd et al. (2020); Saller et al. (2020); Viscardi et al. (2020); Scollo et al. (2021); Coutant et al. (2022); Merenda et al. (2022); Garcia et al. (2023); Lopez-Soriano et al. (2023)
	Tail docking	Morrison and Hemsworth (2020a, b)
	Multiple procedures	Backus and McGlone (2018); Coetzee et al. (2019); Nixon et al. (2021, 2022)
	Injections	Sutherland et al. (2017); Scollo et al. (2020); Dalmau et al. (2021)
	Ear Tagging	Numberger et al. (2016)
	Lameness	Contreras-Aguilar et al. (2019); Lagoda et al. (2021)
	Farrowing	Plush et al. (2021)
	Shoulder ulcers	Nystén et al. (2018)
	Biopsy	Yun et al. (2017)
Heart rate/HRV	Castration	Byrd et al. (2020); Saller et al. (2020); Söbbeler et al. (2022); Werner et al. (2022)
	Cryptorchidectomy	Carbonari et al. (2022)
	Laparotomy	Goutchtat et al. (2021)
Temp/Thermal imaging	Shoulder ulcers	Staveley et al. (2022)
	Castration	Lonardi et al. (2015); Viscardi et al. (2020)

alternative methods producing the least severe pain for the animal. To 'soothe' implies the application of established pharmacological methods (general or local anaesthesia or analgesia) to prevent and or mitigate pain. Alternatively, traditional and or putative non-pharmaceutical approaches may be applied.

4.1 Suppress

Neonatal procedures The docking of pigs' tails has, in theory, been banned within EU countries since the 1990s. However, most European pigs still undergo tail docking, due to a provision in the legislation allowing tail docking when it is necessary to prevent tail biting in later life (D'Eath et al. 2016; De Briyne et al. 2018). Within the EU, docking—where it is deemed necessary to control tail biting—can be conducted without pain alleviation up to 7 days of age (after which it must be carried out by a veterinarian and with pain relief). Some countries (Finland, Sweden, Norway and Switzerland) exert stricter legislative control of docking, i.e. impose a total ban. Experiences in Finland (Valros 2022) and elsewhere (De Briyne et al. 2018) have revealed the necessary measures available to reduce the risk of tail biting to manageable levels; these include: reduced stocking density, better provision of manipulable materials, optimal provision of feed and water, high health status and stable environmental conditions. Unfortunately, many of these measures have cost or practical implications, and in general, pig farmers are reluctant to experiment with keeping intact pigs (Nalon and De Briyne 2019). Until stronger legal controls are enacted, the strategy with the greatest hope of overcoming the current situation is one based on the development of measures that allow for the detection and rapid intervention of tail biting when it begins. This would probably involve a combination of traditional stock-handling skills with modern technology, such as the use of machine vision to automatically detect early signs of biting (D'Eath et al. 2021). While the beneficiaries of a tail docking ban are the piglets themselves, farms may also benefit from reduced labour and time costs, as well as potential price benefits from a differentiated product. Furthermore, the industry would benefit more generally by improving consumer perceptions about the pig industry (De Briyne et al. 2018).

The 'suppress' option in piglet castration is rearing intact, i.e. uncastrated, animals (von Borell et al. 2020). In 2010, workshop discussions led to the formation of a working group consisting of pig producers and other stakeholders who published a 'European Declaration on alternatives to surgical castration of pigs'. This statement declared that from January 2012 castration should only be conducted 'with prolonged analgesia and/or anaesthesia' and that of January 2018 piglet castration should be abandoned entirely. Largely because of this, the percentage of pigs castrated without pain relief dropped from 79% to 54% between 2010 and 2016 (De Briyne et al. 2016). However, the study also concluded that the declaration's goals (being a voluntary initiative) were far from being met. This has subsequently been proven with surgical castration being far from abandoned within Europe (Backus et al. 2018).

Entire male production is common in Ireland (100%), the UK (98%) Portugal (85%) Spain (80%) and is appreciable in the Netherlands (65%) France (22%) and Germany (20%) (Backus et al. 2018). However, market forces are likely to continue a demand for meat products from older animals, which are nevertheless devoid of boar taint. For instance, in Spain, the production of entire males has declined due to the export market (Borrisser-Pairó et al. 2016). Some studies have investigated possible genetic selection for reduced levels of the responsible chemicals, which are heritable (Zamaratskaia and Squires 2009). However, the theoretical possibility of selecting against boar taint has not been implemented in practice because of concerns with reproductive capability and other traits (Duarte et al. 2021).

The pain of surgical castration can also be avoided by immuno-castration (IC), which works to reduce plasma concentrations of testosterone and androstenone, whilst accelerating the metabolism of skatole (Quiniou et al. 2012). It is effective in pigs, and products are available for use in commercial pig production. However, IC has so far had limited success in the EU, with Belgium enjoying the greatest uptake at 15% (Backus et al. 2018). This is difficult to understand given the welfare benefits and other advantages such as production efficiency and reduced environmental effects (Werner et al. 2021). Greater uptake has been seen in other territories, for instance, the estimated adoption of this practice is 60% of male pigs in Australia (Sinnett et al. 2022). Behavioural challenges associated with intact males can also be managed using single sex groups, altering stocking density, and minimising feed competition and social mixing. However, these measures are costly and reduce production efficiency. A fuller implementation of the final goal of the declaration will require wide and complete engagement across the full pig supply chain, including some element of consumer demand.

Efforts in pig production could also be made to reduce the incidence of lameness, gastric ulcers, and other painful conditions. Production diseases and injuries can often be avoided by limiting (ideally eliminating) known causative factors. For instance, in the case of lameness, by improving flooring conditions (Kilbride et al. 2009a), or for gastric ulcers by feeding lower-risk feed.

4.2 Substitution

Neonatal procedures Marchant-Forde and others (2009) studied piglet behavioural and physiological reactions when different methods for achieving common husbandry ends were applied, e.g. clipping vs the grinding of teeth; ear tagging vs ear notching; intramuscular iron injections vs oral iron administration; castration via cord cutting or cord tearing; cold- vs hot-clip docking. The work revealed the benefits of grinding (vs teeth clipping), cold (vs hot) tail docking, ear tagging (vs notching), although minimal differences were found between the iron administration methods, or between the castration methods. Sutherland and others (2008) found that hot iron docking provoked less of a cortisol response but a similar behavioural response to a cold iron, so the relative benefit may vary depending on the equipment used and the extent to which the operator can avoid accidental burns when using the hot iron method. Studying castration, Garcia et al. (2023) compared pain responses to a one vs a two-incisional approach. Piglets castrated with one incision enjoyed an improved growth rate through to weaning, implying they experienced the less harmful option. However, the study noted that the method was technically more difficult and if done incorrectly, could cause more harm. Whether this method could improve piglet welfare or not, therefore, depends on the quality of staff training. A cost–benefit analysis comparing the cost of increased procedural time against the improvements in production variables is also required.

Some studies have examined the effect of age (as well as method) on the painfulness of neonatal husbandry procedures. This is important because historically there has been a widely held perception that conducting painful procedures earlier may be 'better' based on the assumption that very young animals have undeveloped pain perception pathways. (This unchallenged assumption also forms the basis for legislation restricting the ages at which husbandry procedures can be carried out by non-veterinarians without the need for anaesthesia). However, more recent work challenges this idea, revealing that age-related changes in pain responses tend to be negligible once the fact that conducting procedures in older, and larger animals inevitably involves more tissue damage and therefore pain is factored in. For example, Torrey et al. (2009) found no difference in pain indicators in piglets tail docked or ear notched at 1 or 3 days of age.

4.3 Soothing Pain

Parturition The effects of analgesia administration in the peri-parturient period vary, according to the nature of the study, i.e. experimental vs real-life, the outcome measures (production vs pain indicators) and the timing and nature of intervention (Mainau and Manteca 2011). Some studies have demonstrated a beneficial impact of post-farrowing analgesia on various welfare and production indices (Sabaté et al. 2012; Homedes et al. 2014; Navarro et al. 2021). Beyond pharmaceutical interventions, it seems that environmental conditions may contribute to pain experiences around farrowing time, with both environmental enrichment (Espejo-Beristain et al. 2022) and loose farrowing (Nowland et al. 2019) being beneficial.

Neonatal procedures

A detailed systematic review, and associated expert grading of the literature, conducted a decade ago and funded by the American pig industry (Dzikamunhenga et al. 2014; O'Connor et al. 2014, 2016) was limited in the recommendations it could provide to pig producers regarding pain mitigation for neonatal procedures due to various deficiencies in the available scientific data at the time (despite there being an appreciable volume of work to review). At the time of review, most studies focused on castration pain, with some examining docking and even fewer examining other procedures. Mitigation approaches investigated were almost evenly split between general or local anaesthesia, or the use of non-steroidal anti-inflammatory drugs (NSAIDs) (Dzikamunhenga et al. 2014). Despite limitations, the expert group strongly advised against the use of CO_2/O_2 anaesthesia (largely based on the poor risk/reward ratio, especially when used in field). They less strongly advised against lidocaine as a local anaesthetic (based on practical limitations for field use, and its limited efficacy against post-procedural pain) and made a weak recommendation for the use of NSAIDs to alleviate post-procedural pain (based on some evidence of efficacy and the very limited potential for harm). However, the general picture emerging from that work is that findings related to piglet procedures are particularly variable in terms of outcome measures and their interpretation (notably for the assessment of vocalisations and pain behaviours). The literature also often fails to provide sufficient detail in order that findings can be adequately appraised to reach consensual conclusions about efficacy (O'Connor et al. 2014). A secondary problem is that there is still some uncertainty about the best outcome measures to use in attempting to demonstrate pain mitigation (Coutant et al. 2023). Many studies also fail to apply high-quality designs, for instance which account for the possible effects of social contact and emotional contagion (Yun et al. 2019).

Various approaches to the mitigation of castration pain have been investigated. These have included local or general anaesthesia to limit incisional pain, and numerous NSAIDs to counter the subsequent inflammatory pain. Unfortunately, effective methods have major disadvantages 'in the field' chiefly cost and/or impracticality. Consequently, the piglet castration declaration is far from being met. In Europe, the situation is highly variable, with some countries (Iceland, Norway, and Sweden) having 99%–100% of piglets castrated with analgesia and anaesthesia, while others (Austria, Finland, Germany) rely predominantly on analgesia. Other countries have 90–100% of male piglets castrated without any form of pain mitigation (Backus et al. 2018). Several European countries have banned piglet castration without pain control (Switzerland in 2010; Denmark in 2019; Germany in 2021; France in 2022). In Germany, two approaches are legal: injectable or inhalation anaesthesia (Winner et al. 2022). Injectable anaesthesia must be carried out by a veterinarian, but farm staff can undergo training and competence assessment to be certified to conduct inhalation anaesthesia. The German state also introduced a subsidy scheme to help farmers purchase anaesthetic equipment (Winner et al. 2022). However, legitimate questions remain about the additional costs of these strategies. A study of pig farmers in Belgium found that only a minority (6%) thought that the extra cost of pain relief should be borne by the farmer, whilst the majority (68%) thought that the consumer should pay (Tuyttens et al. 2012).

Other neonatal procedures have not been studied to the same extent as castration. But studies have, for instance, identified pain alleviation approaches for procedures such as ear notching (Lomax et al. 2018).

While much of the 'soothe' related work has focussed on traditional methods of pain relief, recent studies have investigated newer approaches including topical local anaesthesia (Burkemper et al. 2019; Sheil et al. 2021; Garcia et al. 2023), intra-nasal administration of analgesia (Lopez-Soriano et al. 2023), and trans-mammary analgesia for piglets undergoing husbandry procedures (Coetzee et al. 2019; Lopez-Soriano et al. 2022). While the primary method of pain mitigation is often pharmacological, a few studies have investigated non-pharmaceutical interventions. For example, the efficacy of sucrose, which has demonstrable analgesic properties in other species, has produced mixed results in piglets. Rand et al. (2002) found no benefit of sucrose before docking, teeth clipping or ear notching piglets, while Davis et al. (2017) found benefit before castration. No benefit of enrichment provision in farrowing accommodation was found on pain responses to tail docking, castration, or the combination of the two (Backus and McGlone 2018), even though enrichment did have a general beneficial effect on welfare.

Only the most severe cases of lameness are likely to be treated with pain relief under production conditions, where the use of analgesics is limited for reasons including, human attitudes, cost and time constraints and concerns over side-effects (see Clutton 2018 for details of major known side-effects of analgesic drugs in pigs). Additional pain relief for lame pigs has been found to have minimal side effects (Zoric et al. 2016; Gorissen et al. 2017). In a production setting, the drugs of choice are NSAIDs with or without paracetamol. Careful product choice should improve pain relief alongside interventions that resolve the inciting causes of the lameness.

4.3.1 Biomedical Studies

The management of pain in laboratory pigs is not limited by the same factors, or to the same degree, as those affecting their use in the commercial herd. For example, concerns about drugs entering the human food chain are less important because the bodies of many laboratory animals will undergo post-mortem examination before incineration, as part of research data collection. Similarly, laboratory pig herds are of limited size, so problems with large-scale medication are less important. In contrast, the types of pain encountered in new experiments may be poorly understood, meaning it may take some time to identify the most effective therapies. A common but unique consideration when dealing with pain in laboratory pigs compared with commercial animals is that the unconsidered choice of analgesic agents may undermine research objectives, so drugs and methods must be chosen with care. Previously, many researchers used this concern to avoid the provision of analgesia, arguing that pain relief limits the comparison of resulting data with earlier work. However, it is increasingly being recognised that unmitigated pain itself represents a far more potent confounding factor in experimental studies, particularly where the work is aimed at translation to human populations who might be expected to be receiving treatment for their pain. The ethical review of animal studies is based on the 3Rs principle of Russell and Burch (1959), who considered anaesthesia and analgesia to be critical elements of refinement, one of the three pillars of ethical and humane experimental technique.

Application and reporting of pain management Coulter and others (2009) reviewed and compared the analgesics used during experimental surgical procedures conducted on pigs and other species during the years 2000/01 and 2005/06. They found that the use of systemic analgesia was reported in about two-thirds of studies during both periods, but whilst analgesic use in rabbits, sheep, dogs, and primates had increased over this time, their use in pigs has not. And while most studies did involve some form of pain mitigation, uncertainties were expressed whether a failure to describe pain mitigation reflects failure to use them, or simply not to report their use. Irrespective of this, the work suggested that pain management—at that time—was inadequate in studies involving major surgical procedures, e.g. laparotomy and thoracotomy, in pigs. Subsequently, Bradbury et al. (2016) examined the literature (from 2012 to early 2014) describing surgical pig models and found that whilst the majority (83%) of studies reported using drugs with analgesic properties, only 37% specifically described the use of post-operative analgesia. The authors suggest that this apparently large fall in pain relief provision may be a consequence of their search strategy leading to the inclusion of a broader range of relevant papers rather than an actual decline in practice. They concluded, however, that substantial improvement was required both in the management of pain in pigs undergoing experimental surgical procedures, and in the details provided when reporting anaesthesia and analgesia practices. Covering a similar period (2012–2018) Alemán-Laporte et al. (2019) also found a general paucity of reporting of information—as required by the ARRIVE guidelines (Percie du Sert et al. 2020)—in surgical studies involving experimental pigs. They found this particularly worrying in relation to the use of analgesics, where only 41% of reviewed studies reported the use of analgesia, with many missing key details about dose and route of administration. As well as practices relating to direct pain mitigation, Liedel et al. (2023) reviewed reports of *Streptococcus suis* infections in pigs for details of practices relating to monitoring the disease and endpoints. They identified numerous studies where monitoring was limited and consisted of prolonged intervals between assessments where pigs were left open to unnecessary suffering. Application of formal scoring schemes with clearly defined endpoints was also uncommon. Similarly, a broader survey of preclinical study reporting in Denmark found a major deficit in content relating to the 3Rs (Kousholt et al. 2023), suggesting that advances are often not effectively shared across the animal research community.

These reviews help identify current deficiencies in the conduct and reporting of surgical studies using pigs. Other types of experimental intervention where pain might be expected have not been subject to similar scrutiny, but the situation is probably similar. The onus is on researchers, as well as funders, local ethical committees, and journal editors, to improve practice and apply existing knowledge regarding pig pain management (Fleischmann et al. 2023) as well as investigating potential improvements.

Despite the efforts of reductionists and replacers, the use of pigs in biomedical research is likely to increase as they represent, in the eyes of most, an ethical and more acceptable alternative to primates as the optimum animal model for humans. The onus lies on the legislations in those parts of the world where this increase is most likely, and where animal protection laws are poorly developed, if they exist at all. Despite major advances in the understanding of what laboratory pigs require so that they have a life worth living, the implementation of this knowledge has far to go. Funding in the UK and the EU in favour of animal replacement studies probably means that laboratory pigs are more likely to be replaced by in vitro, ex situ or in silico techniques before the holy grail of abolishing laboratory pig pain is finally discovered.

5 Conclusions

Global pig numbers are unlikely to fall and may increase given their pivotal role in feeding humans and in translational research. In both these situations, active efforts are ongoing to better identify and manage pain and its associated suffering when it occurs. Whilst these efforts have had some success, large numbers of pigs still experience pain because of their use by humans. In human medicine, the differential between desired and received pain relief is termed 'oligoanalgesia'. Narrowing the oligoanalgesic gap for pigs requires meeting and overcoming two key challenges. The first (which is also the case in people: Green 2012) is better understanding the perspective of animals under care. Scientific efforts are needed to better understand the 'desire' side, i.e. to accurately gauge the extent of undesirability a pig may place on certain pain states it experiences, and conversely, their requirements to have those states alleviated. In humans, that desire can be measured via verbal self-report in most subjects. Although animal studies provide no absolute certainty about the states being experienced, thoughtful and rigorous scientific studies can be used to provide a high strength of inference regarding the effect of putative pain states in pigs, and the motivational importance they might place on having those states mitigated. In humans, the term 'pain' is used to describe situations ranging from the trivial and transient to the overwhelming. The ongoing challenge is to better understand where different animal pain states occur on that continuum: to ensure attention is focused on situations where intervention will have the most benefit to the animals concerned. A first critical step towards this is improving the ability to make strong inferences about the underlying pain experiences of animals—based on behavioural decisions animals make whilst putatively in pain or having that pain relieved. In parallel with methods improving the understanding of pig pain, efforts are also needed to produce validated approaches to quantifying pain.

However, better understanding the pain experiences of pigs in different contexts does not guarantee that action will be taken to mitigate this pain. Consequently, a second necessary step is understanding the perspectives of key stakeholders (farm managers and workers, veterinarians, researchers, government bodies and business owners) who are involved in decision making about how pigs are cared for, as well as understanding views expressed by the public (as consumers of pigs products or funders of animal research). Understanding the underlying attitudes, knowledge base, social context and practical barriers that lead to failures to treat pain adequately is an important step towards identification of strategies to improve matters.

There is evidence that the veterinary profession is shifting towards a more welfare-driven, pain-oriented perspective on animals. Numerous studies reveal that younger and female vets have a more open and proactive approach to pain avoidance and mitigation (Raekallio et al. 2003; Huxley and Whay 2006; Ison and Rutherford 2014). However, in the case of pigs, that professional trend is partially counteracted by a general shift away from large animal practice, with the result that veterinarians with specialist pig expertise are increasingly rare.

Farmers—at least in the UK—have a generally positive attitude to pain alleviation in their herd, though this does not always translate to behaviour. Pain management is just one of many factors that animal managers will be attending to fulfil their role. Research studies could do a better job of identifying the additional benefits of analgesia provision.

Historically, researchers have been cautious about implementing new protocols for standard models involving pain (Fenwick et al. 2011). Societal perspectives on the use of animals in research are built on the implicit idea of a social contract between researchers and the public, who feel that some use of animals in research is justified as long as animals are only used when necessary, that the right number of animals are used, and that the harms imposed on animals is at the minimal level required to achieve legitimate and important scientific benefits. A wider community focus on pain mitigation in research studies would allow researchers to better meet their side of the social contract, as well as address personal, institutional, and legal responsibilities.

Production and consumption of pig meat products is likely to rise in the future, given the growing worldwide demands for meat protein and increasing concerns about alternative products that are deemed to be more environmentally harmful, e.g. beef. The immediate likelihood for improvements in pig welfare relating to pain mitigation is low, both within and between countries and regions of the globe. Market forces may not be a strong driver of change because many members of the public may be largely unaware of current practices (Neary et al. 2023).

References

Abendschön N, Senf S, Deffner P, Miller R, Grott A, Werner J, Saller AM, Reiser J, Weiß C, Zablotski Y, Fischer J, Bergmann S, Erhard MH, Baumgartner C, Ritzmann M, Zöls S (2020) Local anesthesia in piglets undergoing castration-A comparative study to investigate the analgesic effects of four local anesthetics based on defensive behavior and side effects. Animals 10:1752. https://doi.org/10.3390/ani10101752

Ala-Kurikka E, Heinonen M, Mustonen K, Peltoniemi O, Raekallio M, Vainio O, Valros A (2017) Behavior changes associated with lameness in sows. Appl Anim Behav Sci 193:15–20. https://doi.org/10.1016/j.applanim.2017.03.017

Alemán-Laporte J, Alvarado G, Sa Garcia-Gomes M, Fonseca Brasil Antiorio AT, Zúñiga-Montero M, Cabrera Mori CM (2019) Quality of adherence to the ARRIVE guidelines in the material and methods section in studies where swine were used as surgical biomodels: a systematic review (2013-2018). Animals 9:947. https://doi.org/10.3390/ani9110947

Backus BL, McGlone JJ (2018) Evaluating environmental enrichment as a method to alleviate pain after castration and tail docking in pigs. Appl Anim Behav Sci 204:37–42. https://doi.org/10.1016/j.applanim.2018.04.009

Backus G, Higuera M, Juul N, Nalon E, De Briyne N (2018) Second progress report 2015–2017 on the European declaration on alternatives to surgical castration of pigs. Available online: https://www.vitalevarkenshouderij.nl/download/387/Downloads/boars-on-the-way/Second-progress-report-2015-2017-final-1.pdf

Balcarcel AM, Geiger M, Clauss M, Sánchez-Villagra MR (2022) The mammalian brain under domestication: discovering patterns after a century of old and new analyses. J Exp Zool B 338:460–483. https://doi.org/10.1002/jez.b.23105

Bilsborrow K, Seddon YM, Brown J, Waldner C, Stookey JM (2016) An investigation of a novel behavioural test to assess pain in piglets following castration. Can J Anim Sci 96:376–385. https://doi.org/10.1139/cjas-2015-0109

Borrisser-Pairó F, Panella-Riera N, Zammerini D, Olivares A, Garrido MD, Martínez B, Gil M, García-Regueiro JA, Oliver MA (2016) Prevalence of boar taint in commercial pigs from Spanish farms. Meat Sci 111:177–182. https://doi.org/10.1016/j.meatsci.2015.10.001

Bos EJ, Nalon E, Maes D, Ampe B, Buijs S, van Riet MM, Millet S, Janssens GP, Tuyttens FA (2015) Effect of locomotion score on sows' performances in a feed reward collection test. Animal 9:1698–1703. https://doi.org/10.1017/S1751731115001275

Bradbury AG, Eddleston M, Clutton RE (2016) Pain management in pigs undergoing experimental surgery; a literature review (2012-4). Br J Anaesth 116:37–45. https://doi.org/10.1093/bja/aev301

Burkemper MC, Pairis-Garcia MD, Moraes LE, Park RM, Moeller SJ (2019) Effects of oral meloxicam and topical lidocaine on pain associated behaviors of piglets undergoing surgical castration. J Appl Anim Welf Sci 23:209–218. https://doi.org/10.1080/10888705.2019.1590717

Byrd CJ, Radcliffe JS, Craig BA, Eicher SD, Lay DC Jr (2020) Measuring piglet castration pain using linear and non-linear measures of heart rate variability. Anim Welf 29:257–269. https://doi.org/10.7120/09627286.29.3.257

Carbonari A, Lillo E, Cicirelli V, Sciorsci RL, Rizzo A (2022) The use of abdominal ultrasound to improve the cryptorchidectomy of pigs. Animals 12:1763. https://doi.org/10.3390/ani12141763

Carr J (2004) Survey of clinical problems identified in pet pigs in the UK. Vet Rec 155:269–271. https://doi.org/10.1136/vr.155.9.269

Castel D, Sabbag I, Brenner O, Meilin S (2016) Peripheral neuritis trauma in pigs: a neuropathic pain model. J Pain 17:36–49. https://doi.org/10.1016/j.jpain.2015.09.011

Castel D, Sabbag I, Meilin S (2017) The effect of local/topical analgesics on incisional pain in a pig model. J Pain Res 10:2169–2175. https://doi.org/10.2147/JPR.S144949

Castel D, Willentz E, Doron O, Brenner O, Meilin S (2014) Characterization of a porcine model of post-operative pain. Eur J Pain 18:496–505. https://doi.org/10.1002/j.1532-2149.2013.00399.x

Chen Z, Lu J, Wang H (2023) A review of posture detection methods for pigs using deep learning. Appl Sci 13:6997. https://doi.org/10.3390/app13126997

Clutton RE (2018) A review of factors affecting analgesic selection in large animals undergoing translational research. Vet J 236:12–22. https://doi.org/10.1016/j.tvjl.2018.04.006

Coetzee JF, Sidhu PK, Seagen J, Schieber T, Kleinhenz K, Kleinhenz MD, Wulf LW, Cooper VL, Mazloom R, Jaberi-Douraki M, Lechtenberg K (2019) Transmammary delivery of firocoxib to piglets reduces stress and improves average daily gain after castration, tail docking, and teeth clipping. J Anim Sci 97:2750–2768. https://doi.org/10.1093/jas/skz143

Conte S, Bergeron R, Gonyou H, Brown J, Rioja-Lang FC, Connor L, Devillers N (2014) Measure and characterization of lameness in gestating sows using force plate, kinematic, and accelerometer methods. J Anim Sci 92:5693–5703. https://doi.org/10.2527/jas.2014-7865

Contreras-Aguilar MD, Escribano D, Martínez-Miró S, López-Arjona M, Rubio CP, Martínez-Subiela S, Cerón JJ, Tecles F (2019) Application of a score for evaluation of pain, distress and discomfort in pigs with lameness and prolapses: correlation with saliva biomarkers and severity of the disease. Res Vet Sci 126:155–163. https://doi.org/10.1016/j.rvsc.2019.08.004

Coulter CA, Flecknell PA, Richardson CA (2009) Reported analgesic administration to rabbits, pigs, sheep, dogs and non-human primates undergoing experimental surgical procedures. Lab Anim 43:232–238. https://doi.org/10.1258/la.2008.008021

Coutant M, Malmkvist J, Foldager L, Herskin MS (2023) Relationship among indicators of pain and stress in response to piglet surgical castration: an exploratory analysis. J Vet Behav 67:20–32. https://doi.org/10.1016/j.jveb.2023.07.001

Coutant M, Malmkvist J, Kaiser M, Foldager L, Herskin MS (2022) Piglets' acute responses to local anesthetic injection and surgical castration: effects of the injection method and interval between injection and castration. Front Vet Sci 9:1009858. https://doi.org/10.3389/fvets.2022.1009858

Cui H, Lyu L, Bian J, Xu S, Chen R, Cai C, Chen Y, Xu ZR (2023) LC-MS/MS quantification of ropivacaine and local analgesic and adverse effects of Long-acting Ropivacaine Injection based on pharmacokinetic-pharmacodynamic modelling in Bama minipigs. J Chromatogr B Analyt Technol Biomed Life Sci 1223:123716. https://doi.org/10.1016/j.jchromb.2023.123716

Da Silva GV, Pivato GM, Peres BG, Luna SPL, Pairis-Garcia MD, Trindade PHE (2023) Simplified assessment of castration-induced pain in pigs using lower complexity algorithms. Sci Rep 13:21237. https://doi.org/10.1038/s41598-023-48551-1

da Silva JP, de Alencar Nääs I, Abe JM, da Silva Cordeiro AF (2019) Classification of piglet (Sus Scrofa) stress conditions using vocalization pattern and applying paraconsistent logic Eτ. Comput Electron Agr 166:105020. https://doi.org/10.1016/j.compag.2019.105020

Dalmau A, Sánchez-Matamoros A, Molina JM, Xercavins A, Varvaró-Porter A, Muñoz I, Moles X, Baulida B, Fàbrega E, Velarde A, Pallisera J, Puigredon A, Contreras-Jodar A (2021) Intramuscular vs. intradermic needle-free vaccination in piglets: relevance for animal welfare based on an aversion learning test and vocalizations. Front Vet Sci 8:715260. https://doi.org/10.3389/fvets.2021.715260

Davis K, Seddon Y, Creutzinger K, Bouvier M, Brown J (2017) An investigation into the use of sucrose to reduce castration pain in piglets. Can J Anim Sci 97:439–447. https://doi.org/10.1139/cjas-2016-0170

De Briyne N, Berg C, Blaha T, Temple D (2016) Pig castration: will the EU manage to ban pig castration by 2018? Porc Health Manag 2:29. https://doi.org/10.1186/s40813-016-0046-x

De Briyne N, Berg C, Blaha T, Palzer A, Temple D (2018) Phasing out pig tail docking in the EU - present state, challenges and possibilities. Porcine Health Manag 4:27. https://doi.org/10.1186/s40813-018-0103-8

de Paula TMCG, de Sousa RV, Sarmiento MP, Kramer T, de Souza Sardinha EJ, Sabei L, Machado JS, Vilioti M, Zanella AJ (2024) Deep learning pose detection model for sow locomotion. Sci Rep 14:16401. https://doi.org/10.1038/s41598-024-62151-7

D'Eath RB, Niemi JK, Vosough Ahmadi B, Rutherford KMD, Ison SH, Turner SP, Anker HT, Jensen T, Busch ME, Jensen KK, Lawrence AB, Sandøe P (2016) Why are most EU pigs tail docked? Economic and ethical analysis of four pig housing and management scenarios in the light of EU legislation and animal welfare outcomes. Animal 10:687–699. https://doi.org/10.1017/S1751731115002098

D'Eath RB, Foister S, Jack M, Bowers N, Zhu Q, Barclay D, Baxter EM (2021) Changes in tail posture detected by a 3D machine vision system are associated with injury from damaging behaviours and ill health on commercial pig farms. PLoS One 16:e0258895. https://doi.org/10.1371/journal.pone.0258895

Di Giminiani P, Brierley VL, Scollo A, Gottardo F, Malcolm EM, Edwards SA, Leach MC (2016a) The assessment of facial expressions in piglets undergoing tail docking and castration: toward the development of the piglet grimace scale. Front Vet Sci 3:100. https://doi.org/10.3389/fvets.2016.00100

Di Giminiani P, Edwards SA, Malcolm EM, Leach MC, Herskin MS, Sandercock DA (2017a) Characterization of short-and long-term mechanical sensitisation following surgical tail amputation in pigs. Sci Rep 7:4827. https://doi.org/10.1038/s41598-017-05404-y

Di Giminiani P, Nasirahmadi A, Malcolm EM, Leach MC, Edwards SA (2017b) Docking piglet tails: how much does it hurt and for how long? Physiol Behav 182:69–76. https://doi.org/10.1016/j.physbeh.2017.09.028

Di Giminiani P, Petersen LJ, Herskin MS (2014a) Capsaicin-induced neurogenic inflammation in pig skin: a behavioural study. Res Vet Sci 96:447–453. https://doi.org/10.1016/j.rvsc.2014.03.023

Di Giminiani P, Petersen LJ, Herskin MS (2014b) Characterization of nociceptive behavioural responses in the awake pig following UV-B-induced inflammation. Eur J Pain 18:20–28. https://doi.org/10.1002/j.1532-2149.2013.00340.x

Di Giminiani P, Sandercock DA, Malcolm EM, Leach MC, Herskin MS, Edwards SA (2016b) Application of a handheld Pressure Application Measurement device for the characterisation of mechanical nociceptive thresholds in intact pig tails. Physiol Behav 165:119–126. https://doi.org/10.1016/j.physbeh.2016.07.006

Duarte DAS, Schroyen M, Mota RR, Vanderick S, Gengler N (2021) Recent genetic advances on boar taint reduction as an alternative to castration: a review. J Appl Genet 62:137–150. https://doi.org/10.1007/s13353-020-00598-w

Dzikamunhenga RS, Anthony R, Coetzee J, Gould S, Johnson A, Karriker L, McKean J, Millman ST, Niekamp SR, O'Connor a M (2014) Pain management in the neonatal piglet during routine management procedures. Part 1: a systematic review of randomized and non-randomized intervention studies. Anim Health Res Rev 15:14–38. https://doi.org/10.1017/S1466252314000061

Espejo-Beristain G, Ahuja-Aguirre C, Carrasco-García AA, Hernandez-Cruz B, Paredes-Ramos P (2022) Environmental enrichment for primiparous and multiparous pregnant sows and its effect on cortisol and behavior at farrowing and production parameters at weaning. Livest Sci 265:105103. https://doi.org/10.1016/j.livsci.2022.105103

Essig M (2015) Lesser beasts: a snout-to-tail history of the humble pig. Basic Books, New York

European Commission (2024) Summary Report on the statistics on the use of animals for scientific purposes in the Member States of the European Union and Norway in 2022. Brussels

Fenwick N, Danielson P, Griffin G (2011) Survey of Canadian animal-based researchers' views on the Three Rs: replacement, reduction and refinement. PLoS One 6:e22478. https://doi.org/10.1371/journal.pone.0022478

Fischer-Tenhagen C, Meier J, Pohl A (2022) "Do not look at me like that": is the facial expression score reliable and accurate to evaluate pain in large domestic animals? A systematic review. Front Vet Sci 9:1002681. https://doi.org/10.3389/fvets.2022.1002681

Fleischmann T, Clutton RE, Haga HA, van Oostrom H, Weisskopf M (2023) Anesthesia and analgesia in laboratory pigs. In: Anesthesia and analgesia in laboratory animals. 3rd, Academic Press, pp 411–439

Forseth AK, Karriker LA, Millman ST, Stalder KJ, Parsons RL, Azarpajouh S, Johnson AK (2023) Validation of standing and locomotion scoring, behavioral assessments, and mechanical nociceptive threshold testing on naturally occurring sow lameness. Animals 13:1801. https://doi.org/10.3390/ani13111801

Fosse TK, Spadavecchia C, Horsberg TE, Haga HA, Ranheim B (2011a) Pharmacokinetics and pharmacodynamic effects of meloxicam in piglets subjected to a kaolin inflammation model. J Vet Pharmacol Ther 34:367–375. https://doi.org/10.1111/j.1365-2885.2010.01237.x

Fosse TK, Toutain PL, Spadavecchia C, Haga HA, Horsberg TE, Ranheim B (2011b) Ketoprofen in piglets: enantioselective pharmacokinetics, pharmacodynamics and PK/PD modelling. J Vet Pharmacol Ther 34:338–349. https://doi.org/10.1111/j.1365-2885.2010.01236.x

Fuchs PN, McNabb CT (2012) The place escape/avoidance paradigm: a novel method to assess nociceptive processing. J Integr Neurosci 11:61. https://doi.org/10.1142/S0219635212500045

Gaab T, Nogay E, Pierdon M (2022) Development and progression of shoulder lesions and their influence on sow behavior. Animals 12:224. https://doi.org/10.3390/ani12030224

Garcia A, Sutherland M, Vasquez G, Quintana A, Thompson G, Willis J, Chandler S, Niure K, McGlone J (2023) An investigation of the use of ethyl chloride and meloxicam to decrease the pain associated with a single or double incision method of castration in piglets. Front Pain Res 4:1113039. https://doi.org/10.3389/fpain.2023.1113039

Giuffra E, Kijas JM, Amarger V, Carlborg O, Jeon JT, Andersson L (2000) The origin of the domestic pig: independent domestication and subsequent introgression. Genetics 154:1785–1791. https://doi.org/10.1093/genetics/154.4.1785

Gorissen BMC, Uilenreef JJ, Bergmann W, Meijer E, van Rietbergen B, van der Staay FJ, Weeren PRV, Wolschrijn CF (2017) Effects of long-term use of the preferential COX-2 inhibitor meloxicam on growing pigs. Vet Rec 181:564. https://doi.org/10.1136/vr.104175

Gottardo F, Scollo A, Contiero B, Ravagnani A, Tavella G, Bernardini D, De Benedictis GM, Edwards SA (2016) Pain alleviation during castration of piglets: a comparative study of different farm options. J Anim Sci 94:5077–5088. https://doi.org/10.2527/jas.2016-0843

Goutchtat R, Chetboun M, Wiart JF, Gaulier JM, Pattou F, Allorge D, Hubert T (2021) Long-term analgesia following a single application of fentanyl transdermal solution in pigs. Eur Surg Res 62:115–120. https://doi.org/10.1159/000516828

Green SM (2012) There is oligo-evidence for oligoanalgesia. Ann Emerg Med 60:212–214. https://doi.org/10.1016/j.annemergmed.2012.06.006

Guatteo R, Levionnois O, Fournier D, Guémené D, Latouche K, Leterrier C, Mormède P, Prunier A, Servière J, Terlouw C, Le Neindre P (2012) Minimising pain in farm animals: the 3S approach—"Suppress, Substitute, Soothe". Animal 6:1261–1274. https://doi.org/10.1017/S1751731112000262

Gutierrez K, Dicks N, Glanzner WG, Agellon LB, Bordignon V (2015) Efficacy of the porcine species in biomedical research. Front Genet 6:293. https://doi.org/10.3389/fgene.2015.00293

Hälli O, Haimi-Hakala M, Laurila T, Oliviero C, Viitasaari E, Orro T, Peltoniemi O, Scheinin M, Sirén S, Valros A, Heinonen M (2018) Effect of oral KETOPROFEN treatment in acute respiratory disease outbreaks in finishing pigs. Porcine Health Manag 4:7. https://doi.org/10.1186/s40813-018-0081-x

Hancock TM, Caulkett NA, Pajor EA, Grenwich L (2018) An investigation of the effects of intratesticular alfaxalone and lidocaine during castration in piglets. Vet Anaesth Analg 45:858–864. https://doi.org/10.1016/j.vaa.2018.07.002

Hansen MF, Smith ML, Smith LN, Abdul Jabbar K, Forbes D (2018) Automated monitoring of dairy cow body condition, mobility and weight using a single 3D video capture device. Comput Ind 98:14–22. https://doi.org/10.1016/j.compind.2018.02.011

Hansen MF, Baxter EM, Rutherford KMD, Futro A, Smith ML, Smith LN (2021) Towards facial expression recognition for on-farm welfare assessment in pigs. Agriculture 11:847. https://doi.org/10.3390/agriculture11090847

Hay M, Vulin A, Génin S, Sales P, Prunier A (2003) Assessment of pain induced by castration in piglets: behavioral and physiological responses over the subsequent 5 days. Appl Anim Behav Sci 82(3):201–218. https://doi.org/10.1016/S0168-1591(03)00059-5

Hellman A, Maietta T, Clum A, Byraju K, Raviv N, Staudt MD, Jeannotte E, Ghoshal G, Shin D, Neubauer P, Williams E, Heffter T, Burdette C, Qian J, Nalwalk J, Pilitsis JG (2021a) Pilot study on the effects of low intensity focused ultrasound in a swine model of neuropathic pain. J Neurosurg 135:1508–1515. https://doi.org/10.3171/2020.9.JNS202962

Hellman A, Maietta T, Clum A, Byraju K, Raviv N, Staudt MD, Jeannotte E, Nalwalk J, Belin S, Poitelon Y, Pilitsis JG (2021b) Development of a common peroneal nerve injury model in domestic swine for the study of translational neuropathic pain treatments. J Neurosurg 135:1516–1523. https://doi.org/10.3171/2020.9.JNS202961

Herskin MS, Di Giminiani P (2024) Pain in pigs: characterisation and indicators. In: Camerlink I, Baxter EM (eds) Advances in pig welfare. Woodhead Publishing, pp 23–48. https://doi.org/10.1016/B978-0-08-101012-9.00011-3

Herskin MS, Marchant JN (2023) Pigs as laboratory animals. In: Camerlink I, Baxter EM (eds) Advances in pig welfare. Woodhead Publishing, pp 471–493. https://doi.org/10.1016/B978-0-323-85676-8.00021-3

Herskin MS, Di Giminiani P, Thodberg K (2016) Effects of administration of a local anaesthetic and/or an NSAID and of docking length on the behaviour of piglets during 5h after tail docking. Res Vet Sci 108:60–67. https://doi.org/10.1016/j.rvsc.2016.08.001

Herskin MS, Bonde MK, Jørgensen E, Jensen KH (2010) Decubital shoulder ulcers in sows: a review of classification, pain and welfare consequences. Animal 5:757–766. https://doi.org/10.1017/S175173111000203X

Herskin MS, Ladewig J, Arendt-Nielsen L (2009) Measuring cutaneous thermal nociception in group-housed pigs using laser technique—effects of laser power output. Appl Anim Behav Sci 118:144–151. https://doi.org/10.1016/j.applanim.2009.02.016

Hilgart DR, Iversen MM, Peters AY, Zabriskie MS, Hoareau GL, Vapniarsky N, Clark GA, Shah LM, Rieke V (2023) Non-invasive central nervous system assessment of a porcine model of neuropathic pain demonstrates increased latency of somatosensory-evoked potentials. J Neurosci Methods 396:109934. https://doi.org/10.1016/j.jneumeth.2023.109934

Hofmann K, Rauh A, Harlizius J, Weiß C, Scholz T, Schulze-Horsel T, Escribano D, Ritzmann M, Zöls S (2019) Schmerz- und Stressbestimmung bei der Injektion und Kastration von Saugferkeln unter Lokalanästhesie mit Procain und Lidocain [Pain and distress responses of suckling piglets to injection and castration under local anaesthesia with procaine and lidocaine - part 1: cortisol, chromogranin A, wound healing, weights, losses]. Tierarztl Prax Ausg G Grosstiere Nutztiere 47:87–96. https://doi.org/10.1055/a-0861-9640

Homedes J, Salichs M, Sabaté D, Sust M, Fabre R (2014) Effect of ketoprofen on pre-weaning piglet mortality on commercial farms. Vet J 201:435–437. https://doi.org/10.1016/j.tvjl.2014.05.038

Horn T, Marx G, von Borell E (1999) Verhalten von Ferkeln während der Kastration mit und ohne Lokalanästhesie [Behavior of piglets during castration with and without local anesthesia]. Dtsch Tierarztl Wochenschr 106:271–274

Horrell I, A'Ness P, Edwards SA, Riddoch I (2000) Nose-rings influence feeding efficiency in pigs. Anim Sci 71:259–264. https://doi.org/10.1017/S1357729800055090

Horrell RI, A'Ness PJ, Edwards SA, Eddison JC (2001) The use of nose-rings in pigs: consequences for rooting, other functional activities, and welfare. Anim Welf 10:3–22. https://doi.org/10.1017/S0962728600023228

Huxley JN, Whay HR (2006) Current attitudes of cattle practitioners to pain and the use of analgesics in cattle. Vet Rec 159:662–668. https://doi.org/10.1136/vr.159.20.662

Ison SH, Rutherford KMD (2014) UK farmers and veterinarians' use of anti-inflammatories in breeding pigs and attitudes towards pain and the use of pain relief in pigs. Vet J 202:622–627. https://doi.org/10.1016/j.tvjl.2014.10.003

Ison SH, Di Giminiani P, Clutton RE, Rutherford KMD (2016a) A review of pain assessment in pigs. Front Vet Sci 3:108. https://doi.org/10.3389/fvets.2016.00108

Ison SH, Jarvis S, Rutherford KMD (2016b) The identification of potential behavioural indicators of pain in periparturient sows. Res Vet Sci 109:114–120. https://doi.org/10.1016/j.rvsc.2016.10.002

Ison SH, Jarvis S, Rutherford KMD (2016c) A survey of sow management at farrowing in the UK. Anim Welf 25:309–317

Ison SH, Jarvis S, Ashworth CJ, Rutherford KMD (2017) The effect of post-farrowing ketoprofen on sow feed intake, nursing behaviour and piglet performance. Livest Sci 202:115–123. https://doi.org/10.1016/j.livsci.2017.06.001

Ison SH, Jarvis S, Hall SA, Ashworth CJ, Rutherford KMD (2018) Periparturient behavior and physiology: further insight into the farrowing process for primiparous and multiparous sows. Front Vet Sci 5:122. https://doi.org/10.3389/fvets.2018.00122

Janczak AM, Ranheim B, Fosse TK, Hild S, Nordgreen J, Moe RO, Zanella AJ (2012) Factors affecting mechanical (nociceptive) thresholds in piglets. Vet Anaesth Analg 39:628–635. https://doi.org/10.1111/j.1467-2995.2012.00737.x

Jarvis S, McLean KA, Chirnside J, Deans LA, Calvert SK, Molony V, Lawrence AB (1997) Opioid-mediated changes in nociceptive threshold during pregnancy and parturition in the sow. Pain 72:153–159. https://doi.org/10.1016/S0304-3959(97)00027-4

Jensen TB, Kristensen HH, Toft N (2012) Quantifying the impact of lameness on welfare and profitability of finisher pigs using expert opinions. Livest Sci 149(3):209–214. https://doi.org/10.1016/j.livsci.2012.07.013

Jirkof P, Rudeck J, Lewejohann L (2019) Assessing affective state in laboratory rodents to promote animal welfare—what is the progress in applied refinement research? Animals 9:1026. https://doi.org/10.3390/ani9121026

Karriker LA, Abell CE, Pairis-Garcia MD, Holt WA, Sun G, Coetzee JF, Johnson AK, Hoff SJ, Stalder KJ (2013) Validation of a lameness model in sows using physiological and mechanical measurements. J Anim Sci 91:130–136. https://doi.org/10.2527/jas.2011-4994

Khanna R, Moutal A, White KA, Chefdeville A, Negrao de Assis P, Cai S, Swier VJ, Bellampalli SS, Giunta MD, Darbro BW, Quelle DE, Sieren JC, Wallace MR, Rogers CS, Meyerholz DK, Weimer JM (2019) Assessment of nociception and related quality-of-life measures in a porcine model of neurofibromatosis type 1. Pain 160:2473–2486. https://doi.org/10.1097/j.pain.0000000000001648

KilBride A, Gillman C, Green L (2009b) A cross-sectional study of the prevalence of lameness in finishing pigs, gilts and pregnant sows and associations with limb lesions and floor types on commercial farms in England. Anim Welf 18:215–224. https://doi.org/10.1017/S0962728600000464

KilBride A, Gillman C, Ossent P, Green L (2009a) Impact of flooring on the health and welfare of pigs. In Pract 31:390–395. https://doi.org/10.1136/inpract.31.8.390

Knap PW (2005) Breeding robust pigs. Aust J Exp Agric 45:763–773. https://doi.org/10.1071/EA05041

Kousholt BS, Præstegaard KF, Stone JC, Thomsen AF, Johansen TT, Ritskes-Hoitinga M, Wegener G (2023) Reporting of 3Rs approaches in preclinical animal experimental studies-A nationwide study. Animals 3:3005. https://doi.org/10.3390/ani13193005

Lagoda ME, Boyle LA, Marchewka J, O'Driscoll K (2021) Early detection of locomotion disorders in gilts using a novel visual analogue scale; associations with chronic stress and reproduction. Animals 11:2900. https://doi.org/10.3390/ani11102900

Lander B, Schneider M, Brunson K (2020) A history of pigs in China: from curious omnivores to industrial pork. J Asian Stud 79:865–889. https://doi.org/10.1017/S0021911820000054

Langford DJ, Bailey AL, Chanda ML, Clarke SE, Drummond TE, Echols S, Glick S, Ingrao J, Klassen-Ross T, Lacroix-Fralish ML, Matsumiya L, Sorge RE, Sotocinal SG, Tabaka JM, Wong D, van den Maagdenberg AMJM, Ferrari MD, Craig KD, Mogil JS (2010) Coding of facial expressions of pain in the laboratory mouse. Nat Methods 7:447–449. https://doi.org/10.1038/nmeth.1455

Larsen T, Kaiser M, Herskin MS (2015) Does the presence of shoulder ulcers affect the behaviour of sows? Res Vet Sci 98:19–24. https://doi.org/10.1016/j.rvsc.2014.11.001

Larson G, Dobney K, Albarella U, Fang M, Matisoo-Smith E, Robins J, Lowden S, Finlayson H, Brand T, Willerslev E, Rowley-Conwy P, Andersson L, Cooper A (2005) Worldwide phylogeography of wild boar reveals multiple centers of pig domestication. Science 307:1618–1621. https://doi.org/10.1126/science.1106927

Lecchi C, Zamarian V, Gini C, Avanzini C, Polloni A, Rota Nodari S, Ceciliani F (2020) Salivary microRNAs are potential biomarkers for the accurate and precise identification of inflammatory response after tail docking and castration in piglets. J Anim Sci 98:skaa153. https://doi.org/10.1093/jas/skaa153

Leslie E, Hernández-Jover M, Newman R, Holyoake P (2010) Assessment of acute pain experienced by piglets from ear tagging, ear notching and intraperitoneal injectable transponders. Appl Anim

Behav Sci 127:86–95. https://doi.org/10.1016/j.applanim.2010.09.006

Lewis I, Wishart H, Breeze E, Setter P, Amory J (2024) Needle-free intradermal vaccination, an opportunity to improve commercial pig welfare. Anim Welf 233:e48. https://doi.org/10.1017/awf.2024.53

Li JX (2013) The application of conditioning paradigms in the measurement of pain. Eur J Pharmacol 716:158–168. https://doi.org/10.1016/j.ejphar.2013.03.002

Liedel C, Rieckmann K, Baums CG (2023) A critical review on experimental Streptococcus suis infection in pigs with a focus on clinical monitoring and refinement strategies. BMC Vet Res 19:188. https://doi.org/10.1186/s12917-023-03735-9

Lomax S, Hall E, Oehlers L, White P (2018) Topical vapocoolant spray reduces nociceptive response to ear notching in neonatal piglets. Vet Anaesth Analg 45:366–373. https://doi.org/10.1016/j.vaa.2016.08.012

Lonardi C, Scollo A, Normando S, Brscic M, Gottardo F (2015) Can novel methods be useful for pain assessment of castrated piglets? Animal 9:871–877. https://doi.org/10.1017/S1751731114003176

Lopez-Soriano M, Merenda VR, Anderson S, Trindade PHE, Leidig MS, Messenger K, Ferreira JB, Pairis-Garcia MD (2023) Efficacy of inguinal buffered lidocaine and intranasal flunixin meglumine on mitigating physiological and behavioral responses to pain in castrated piglets. Front Pain Res 4:1156873. https://doi.org/10.3389/fpain.2023.1156873

Lopez-Soriano M, Rocha Merenda V, Esteves Trindade PH, Loureiro Luna SP, Pairis-Garcia MD (2022) Efficacy of transdermal flunixin in mitigating castration pain in piglets. Front Pain Res 3:1056492. https://doi.org/10.3389/fpain.2022.1056492

Lou ME, Porter ST, Massey JS, Ventura B, Deen J, Li Y (2022a) The application of 3D landmark-based geometric morphometrics towards refinement of the piglet grimace scale. Animals 12:1944. https://doi.org/10.3390/ani12151944

Lou M, Ventura B, Deen J, Li Y (2022b) Surgical castration changes struggle behavior and vocalizations in male piglets. J Appl Anim Welf Sci 25:410–417. https://doi.org/10.1080/10888705.2021.1916938

Luna SPL, de Araújo AL, da Nóbrega Neto PI, Brondani JT, de Oliveira FA, Azerêdo LMDS, Telles FG, Trindade PHE (2020) Validation of the UNESP-Botucatu pig composite acute pain scale (UPAPS). PLoS One 15:e0233552. https://doi.org/10.1371/journal.pone.0233552

Macfadyen MA, Daniel Z, Kelly S, Parr T, Brameld JM, Murton AJ, Jones SW (2019) The commercial pig as a model of spontaneously-occurring osteoarthritis. BMC Musculoskelet Disord 20:70. https://doi.org/10.1186/s12891-019-2452-0

Mainau E, Manteca X (2011) Pain and discomfort caused by parturition in cows and sows. Appl Anim Behav Sci 135:241–251. https://doi.org/10.1016/j.applanim.2011.10.020

Marchant-Forde JN, Lay DC Jr, McMunn KA, Cheng HW, Pajor EA, Marchant-Forde RM (2014) Postnatal piglet husbandry practices and Well-being: the effects of alternative techniques delivered in combination. J Anim Sci 92:1150–1160. https://doi.org/10.2527/jas.2013-6929

Marchant-Forde JN, Lay DC, McMunn KA, Cheng HW, Pajor EA, Marchant-Forde RM (2009) Postnatal piglet husbandry practices and Well-being: the effects of alternative techniques delivered separately. J Anim Sci 87:1479–1492. https://doi.org/10.2527/jas.2008-1080

McGlone J, Guay K, Garcia A (2016) Comparison of intramuscular or subcutaneous injections vs. castration in pigs—impacts on behavior and welfare. Animals 6:52. https://doi.org/10.3390/ani6090052

McLennan KM, Miller AL, Dalla Costa E, Stucke D, Corke MJ, Broom DM, Leach MC (2019) Conceptual and methodological issues relating to pain assessment in mammals: the development and utilisation of pain facial expression scales. Appl Anim Behav Sci 217:1–15. https://doi.org/10.1016/j.applanim.2019.06.001

Meijer E, Oosterlinck M, van Nes A, Back W, van der Staay FJ (2014) Pressure mat analysis of naturally occurring lameness in young pigs after weaning. BMC Vet Res 10:193. https://doi.org/10.1186/s12917-014-0193-8

Meijer E, van Nes A, Back W, van der Staay FJ (2015) Clinical effects of buprenorphine on open field behaviour and gait symmetry in healthy and lame weaned piglets. Vet J 206:298–303. https://doi.org/10.1016/j.tvjl.2015.10.016

Meijs S, Andreis FR, Janjua TAM, Graven-Nielsen T, Jensen W (2025) High-frequency electrical stimulation increases cortical excitability and mechanical sensitivity in a chronic large animal model. Pain 166:e18–e26. https://doi.org/10.1097/j.pain.0000000000003354

Meijs S, Schmelz M, Meilin S, Jensen W (2021) A systematic review of porcine models in translational pain research. Lab Anim 50:313–326. https://doi.org/10.1038/s41684-021-00862-4

Merenda VR, Wagner BK, Arruda AG, Lopez Soriano M, Montgomery S, Coetzee JF, Pairis-Garcia MD (2022) Impact of transdermal flunixin administration on serum prostaglandin E2 and cortisol concentrations in piglets following castration. Am J Vet Res 83:ajvr.21.12.0201. https://doi.org/10.2460/ajvr.21.12.0201

Mogil JS (2022) The history of pain measurement in humans and animals. Front Pain Res 3:1031058. https://doi.org/10.3389/fpain.2022.1031058

Mohling CM, Johnson AK, Coetzee JF, Karriker LA, Stalder KJ, Abell CE, Tyler HD, Millman ST (2014) Evaluation of mechanical and thermal nociception as objective tools to measure painful and nonpainful lameness phases in multiparous sows. J Anim Sci 92:3073–3081. https://doi.org/10.2527/jas.2013-7521

Morrison R, Hemsworth P (2020a) Tail docking of piglets 1: stress response of piglets to tail docking. Animals 10:1701. https://doi.org/10.3390/ani10091701

Morrison R, Hemsworth P (2020b) Tail docking of piglets 2: effects of meloxicam on the stress response to tail docking. Animals 10:1699. https://doi.org/10.3390/ani10091699

Mozzachio K, Tynes VV (2013) Recognition and treatment of pain in pet pigs. In: Egger CM, Love L, Doherty T (eds) Pain management in veterinary practice. https://doi.org/10.1002/9781118999196.ch34

Munsterhjelm C, Simola O, Keeling L, Valros A, Heinonen M (2013) Health parameters in tail biters and bitten pigs in a case-control study. Animal 7:814–821. https://doi.org/10.1017/S1751731112002194

Mustonen K, Ala-Kurikka E, Orro T, Peltoniemi O, Raekallio M, Vainio O, Heinonen M (2011) Oral ketoprofen is effective in the treatment of non-infectious lameness in sows. Vet J 190:55–59. https://doi.org/10.1016/j.tvjl.2010.09.017

Nagel D, Ralston B, Hanson A, Burwash L, Matheson-Bird H, Olson B, Schatz C, Olson M (2024) Efficacy and safety of Lidocam topical gel (4% lidocaine—0.3% meloxicam) for pain and inflammation management during castration and tail docking in piglets. Animals 14:930. https://doi.org/10.3390/ani14060930

Nalon E, De Briyne N (2019) Efforts to ban the routine tail docking of pigs and to give pigs enrichment materials via EU law: where do we stand a quarter of a century on? Animals 9:132. https://doi.org/10.3390/ani9040132

Nalon E, Maes D, Piepers S, van Riet MM, Janssens GP, Millet S, Tuyttens FA (2013) Mechanical nociception thresholds in lame sows: evidence of hyperalgesia as measured by two different methods. Vet J 198:386–390. https://doi.org/10.1016/j.tvjl.2013.08.016

Navarro E, Mainau E, Manteca X (2020) Development of a facial expression scale using farrowing as a model of pain in sows. Animals 10:2113. https://doi.org/10.3390/ani10112113

Navarro E, Mainau E, De Miguel R, Temple D, Salas M, Manteca X (2021) Oral meloxicam administration in sows at farrowing and its effects on piglet immunity transfer and growth. Front Vet Sci 8:574250. https://doi.org/10.3389/fvets.2021.574250

Neary JM, Ali AB, Jacobs L (2024) Application of an attention bias test after surgical castration in piglets. Livest Sci 290:105612. https://doi.org/10.1016/j.livsci.2024.105612

Neary JM, Guthrie AP, Jacobs L (2023) Public and industry knowledge and perceptions of US swine industry castration practices. Anim Welf 32:e79. https://doi.org/10.1017/awf.2023.99

Nixon E, Carlson AR, Routh PA, Hernandez L, Almond GW, Baynes RE, Messenger KM (2021) Comparative effects of nonsteroidal anti-inflammatory drugs at castration and tail-docking in neonatal piglets. PLoS One 16:e0254409. https://doi.org/10.1371/journal.pone.0254409

Nixon E, Chittenden JT, Baynes RE, Messenger KM (2022) Pharmacokinetic/pharmacodynamic modeling of ketoprofen and flunixin at piglet castration and tail-docking. J Vet Pharmacol Ther 45:450–466. https://doi.org/10.1111/jvp.13083

Noonan G, Rand J, Priest J, Ainscow J, Blackshaw J (1994) Behavioural observations of piglets undergoing tail docking, teeth clipping and ear notching. Appl Anim Behav Sci 39:203–213. https://doi.org/10.1016/0168-1591(94)90156-2

Nowland TL, van Wettere WHEJ, Plush KJ (2019) Allowing sows to farrow unconfined has positive implications for sow and piglet welfare. Appl Anim Behav Sci 221:104872. https://doi.org/10.1016/j.applanim.2019.104872

Numberger J, Ritzmann M, Übel N, Eddicks M, Reese S, Zöls S (2016) Ear tagging in piglets: the cortisol response with and without analgesia in comparison with castration and tail docking. Animal 10:1864–1870. https://doi.org/10.1017/S1751731116000811

Nystén M, Orro, T, Peltoniemi O (2018) Systemic inflammatory response to shoulder ulcers and lack of preventive effect of postpartum pain medication with ketoprofen in sows. Livest Sci 214:9–17

O'Connor A, Anthony R, Bergamasco L, Coetzee JF, Dzikamunhenga RS, Johnson AK, Karriker LA, Marchant-Forde JN, Martineau GP, Millman ST, Pajor EA, Rutherford K, Sprague M, Sutherland MA, von Borell E, Webb SR (2016) Review: an assessment of completeness of reporting in studies evaluating pain management in the neonatal piglet during routine management procedures (review). Animal 10:660–670. https://doi.org/10.1017/S1751731115002323

O'Connor A, Anthony R, Bergamasco L, Coetzee J, Gould S, Johnson AK, Karriker LA, Marchant-Forde JN, Martineau GS, McKean J, Millman ST, Niekamp S, Pajor EA, Rutherford K, Sprague M, Sutherland M, von Borell E, Dzikamunhenga RS (2014) Pain management in the neonatal piglet during routine management procedures. Part 2: grading the quality of evidence and the strength of recommendations. Anim Health Res Rev 15:39–62. https://doi.org/10.1017/S1466252314000073

Ojukwu CC, Feng YZ, Jia GF, Zhao HT, Tan HQ (2020) Development of a computer vision system to detect inactivity in group-housed pigs. Int J Agric Biol Eng 13:42–46

Pairis-Garcia MD, Johnson AK, Stalder KJ, Karriker LA, Coetzee JF, Millman ST (2014) Measuring the efficacy of flunixin meglumine and meloxicam for lame sows using nociceptive threshold tests. Anim Welf 23:219–229. https://doi.org/10.7120/09627286.23.2.219

Percie du Sert N, Ahluwalia A, Alam S, Avey MT, Baker M, Browne WJ, Clark A, Cuthill IC, Dirnagl U, Emerson M, Garner P, Holgate ST, Howells DW, Hurst V, Karp NA, Lazic SE, Lidster K, MacCallum CJ, Macleod M, Pearl EJ, Petersen OH, Rawle F, Reynolds P, Rooney K, Sena ES, Silberberg SD, Steckler T, Würbel H (2020) Reporting animal research: explanation and elaboration for the ARRIVE guidelines 2.0. PLoS Biol 18:e3000411. https://doi.org/10.1371/journal.pbio.300041

Petrucci M, Spadavecchia C, Rieben R, Casoni D (2024) Mechanical and thermal thresholds before and after application of a conditioning stimulus in healthy Göttingen Minipigs. PLoS One 19:e0309604. https://doi.org/10.1371/journal.pone.0309604

Plush KJ, Pluske JR, Lines DS, Ralph CR, Kirkwood RN (2021) Meloxicam and dexamethasone administration as anti-inflammatory compounds to sows prior to farrowing does not improve lactation performance. Animals 11:2414. https://doi.org/10.3390/ani11082414

Price DD, Barrell JJ (2012) Inner experience and neuroscience: merging both perspectives. The MIT Press, Cambridge, MA

Quiniou N, Monziols M, Colin F, Goues T, Courboulay V (2012) Effect of feed restriction on the performance and behaviour of pigs immunologically castrated with Improvac®. Animal 6:1420–1426. https://doi.org/10.1017/S1751731112000444

Raekallio M, Heinonen KM, Kuussaari J, Vainio O (2003) Pain alleviation in animals: attitudes and practices of finnish veterinarians. Vet J 165:131–135. https://doi.org/10.1016/s1090-0233(02)00186-7

Raja SN, Carr DB, Cohen M, Finnerup NB, Flor H, Gibson S, Keefe FJ, Mogil JS, Ringkamp M, Sluka KA, Song XJ, Stevens B, Sullivan MD, Tutelman PR, Ushida T, Vader K (2020) The revised International Association for the Study of Pain definition of pain: concepts, challenges, and compromises. Pain 161:1976–1982. https://doi.org/10.1097/j.pain.0000000000001939

Rand JS, Noonan GJ, Priest J, Ainscow J, Blackshaw JK (2002) Oral administration of a 12% sucrose solution did not decrease behavioural indicators of distress in piglets undergoing tail docking, teeth clipping and ear notching. Anim Welf 11:395–404. https://doi.org/10.1017/S0962728600025124

Rauh A, Hofmann K, Harlizius J, Weiß C, Numberger J, Scholz T, Schulze-Horsel T, Otten W, Ritzmann M, Zöls S (2019) Schmerz- und Stressbestimmung bei der Injektion und Kastration von Saugferkeln unter Lokalanästhesie mit Procain und Lidocain [Pain and distress response of suckling piglets to injection and castration under local anaesthesia with procaine and licocaine - part 2: Defence behaviour, catecholamines, coordination of movements]. Tierarztl Prax Ausg G Grosstiere Nutztiere 47:160–170. https://doi.org/10.1055/a-0866-6694

Rault J-L, Lay DC (2011) Nitrous oxide by itself is insufficient to relieve pain due to castration in piglets. J Anim Sci 89:3318–3325. https://doi.org/10.2527/jas.2011-4104

Rettore Andreis F, Mørch CD, Jensen W, Meijs S (2023) On determining the mechanical nociceptive threshold in pigs: a reliability study. Front Pain Res 4:1191786. https://doi.org/10.3389/fpain.2023.1191786

Reynolds K, Johnson R, Brown J, Friendship R, O'Sullivan TL (2020) Assessing pain control efficacy of meloxicam and ketoprofen when compounded with iron dextran in nursing piglets using a navigation chute. Animals 10:1237. https://doi.org/10.3390/ani10071237

Rioja-Lang FC, Connor M, Bacon HJ, Lawrence AB, Dwyer CM (2020) Prioritization of farm animal welfare issues using expert consensus. Front Vet Sci 6:495. https://doi.org/10.3389/fvets.2019.00495

Robles I, Luna SPL, Trindade PHE, Lopez-Soriano M, Merenda VR, Viscardi AV, Tamminga E, Lou ME, Pairis-Garcia MD (2023) Validation of the Unesp-Botucatu pig composite acute pain scale (UPAPS) in piglets undergoing castration. PLoS One 18:e0284218. https://doi.org/10.1371/journal.pone.0284218

Romeo SMT, Morrone S, Akl T, Scanu A, Columbano N (2023) Sutureless technique for surgical castration in adult boars: a feasibility study. Animals 13:407. https://doi.org/10.3390/ani13030407

Rozkot M, Václavková E, Bělková J (2015) Minipigs as laboratory animals–review. Res Pig Breed 9:10–14

Russell WMS, Burch RL (1959) The principles of humane experimental technique. Methuen and Co Ltd., London, UK

Rutherford KMD, Baxter EM, D'Eath RB, Turner SP, Arnott G, Roehe R, Ask B, Sandøe P, Moustsen VA, Thorup F, Edwards SA, Berg P, Lawrence AB (2013) The welfare implications of large litter size in the domestic pig I: biological factors. Anim Welf 22:199–218. https://doi.org/10.7120/09627286.22.2.199

Rutherford KMD, Thompson CS, Thomson JR, Lawrence AB, Nielsen EO, Busch ME, Haugegaard S, Sandøe P (2018) A study of associations between gastric ulcers and the behaviour of finisher pigs. Livest Sci 212:45–51. https://doi.org/10.1016/j.livsci.2018.03.013

Sabaté D, Salichs M, Bosch J, Ramió P, Homedes J (2012) Efficacy of ketoprofen in the reduction of pre-weaning piglet mortality associated with sub-clinical forms of Post-partum Dysgalactia Syndrome in sows. Pig J 67:19–23

Saller AM, Werner J, Reiser J, Senf S, Deffner P, Abendschön N, Weiß C, Fischer J, Schörwerth A, Miller R, Zablotski Y, Bergmann S, Erhard MH, Ritzmann M, Zöls S, Baumgartner C (2020) Local anesthesia in piglets undergoing castration-A comparative study to investigate the analgesic effects of four local anesthetics on the basis of acute physiological responses and limb movements. PLoS One 15:e0236742. https://doi.org/10.1371/journal.pone.0236742

Sandercock DA, Gibson IF, Brash HM, Rutherford KMD, Scott EM, Nolan AM (2009) Development of a mechanical stimulator and force measurement system for the assessment of nociceptive thresholds in pigs. J Neurosci Methods 182:64–70. https://doi.org/10.1016/j.jneumeth.2009.05.019

Sandercock DA, Gibson IF, Rutherford KMD, Donald RD, Lawrence AB, Brash HM, Scott EM, Nolan AM (2011) The impact of prenatal stress on basal nociception and evoked responses to tail-docking and inflammatory challenge in juvenile pigs. Physiol Behav 104:728–737. https://doi.org/10.1016/j.physbeh.2011.07.018

Schild SLA, Rørvang MV (2023) Pig olfaction: the potential impact and use of odors in commercial pig husbandry. Front Anim Sci 4:1215206. https://doi.org/10.3389/fanim.2023.1215206

Scollo A, Contiero B, De Benedictis GM, Galli MC, Benatti D, Gottardo F (2021) Analgesia and/or anaesthesia during piglet castration—part I: efficacy of farm protocols in pain management. Italian J Anim Sci 20:143–152. https://doi.org/10.1080/1828051X.2021.1873707

Scollo A, Minervini S, Galli MC, Cevidalli A, Bortoletto G, Romano G, Gottardo F (2020) Evaluation of pain and stress in three-week old piglets in relation to route of vaccine administration. Livest Sci 233:103939. https://doi.org/10.1016/j.livsci.2020.103939

Sheil ML, Chambers M, Sharpe B (2020) Topical wound anaesthesia: efficacy to mitigate piglet castration pain. Aust Vet J 98:256–263. https://doi.org/10.1111/avj.12930

Sheil M, Polkinghorne A (2020) Optimal methods of documenting analgesic efficacy in neonatal piglets undergoing castration. Animals 10:1450. https://doi.org/10.3390/ani10091450

Sheil M, De Benedictis GM, Scollo A, Metcalfe S, Innocent G, Polkinghorne A, Gottardo F (2021) Efficacy of intra-operative topical wound anaesthesia to mitigate piglet castration pain-A large, multi-centred field trial. Animals 11:2763. https://doi.org/10.3390/ani11102763

Shiff J, Schwartz K, Hausman B, Seshadri DR, Bogie KM (2023) Development and use of a porcine model with clinically relevant chronic infected wounds. J Tissue Viability 32:527–535. https://doi.org/10.1016/j.jtv.2023.08.004

Sinnett A, Dunshea F, D'Souza D, Malcolm B, Griffith G (2022) A benefit cost analysis of entire male pig immuno-castration. AFBM J 19:79–93

Sipos W, Schmoll F, Stumpf I (2007) Minipigs and pot-bellied pigs as pets in the veterinary practice–a retrospective study. J Vet Med A 54:504–511. https://doi.org/10.1111/j.1439-0442.2007.00968.x

Söbbeler FJ, Wendt S, Briese A, Tünsmeier J, Waldmann KH, Kästner SBR, von Altrock A (2022) Comparative study of pain-related responses of male piglets up to seven days of age to the application of different local anaesthetics and subsequent castration. Animals 1220:2833. https://doi.org/10.3390/ani12202833

Staveley LM, Zemitis JE, Plush KJ, D'Souza DN (2022) Infrared thermography for early identification and treatment of shoulder sores to improve sow and piglet welfare. Animals 12:3136. https://doi.org/10.3390/ani12223136

Sutherland MA, Backus BL, Brooks TA, McGlone JJ (2017) The effect of needle-free administration of local anesthetic on the behavior and physiology of castrated pigs. J Vet Behav 21:71–76. https://doi.org/10.1016/j.jveb.2017.07.003

Sutherland MA, Davis BL, Brooks TA, Coetzee JF (2012) The physiological and behavioral response of pigs castrated with and without anesthesia or analgesia. J Anim Sci 90:2211–2221. https://doi.org/10.2527/jas.2011-4260

Sutherland MA, Bryer PJ, Krebs N, McGlone JJ (2008) Tail docking in pigs: acute physiological and behavioural responses. Animal 2:292–297. https://doi.org/10.1017/S1751731107001450

Swaby H, Gregory NG (2012) A note on the frequency of gastric ulcers detected during post-mortem examination at a pig abattoir. Meat Sci 90:269–271. https://doi.org/10.1016/j.meatsci.2011.06.015

Tallet C, Rakotomahandry M, Herlemont S, Prunier A (2019) Evidence of pain, stress, and fear of humans during tail docking and the next four weeks in piglets (Sus scrofa domesticus). Front Vet Sci 6:462. https://doi.org/10.3389/fvets.2019.00462

Tapper KR, Johnson AK, Karriker LA, Stalder KJ, Parsons RL, Wang C, Millman ST (2013) Pressure algometry and thermal sensitivity for assessing pain sensitivity and effects of flunixin meglumine and sodium salicylate in a transient lameness model in sows. Livest Sci 157:245–253. https://doi.org/10.1016/j.livsci.2013.07.017

Taylor A, Weary D (2000) Vocal responses of piglets to castration: identifying procedural sources of pain. Appl Anim Behav Sci 70:17–26. https://doi.org/10.1016/S0168-1591(00)00143-X

Taylor NR, Main DC, Mendl M, Edwards SA (2010) Tail-biting: a new perspective. Vet J 186:137–147. https://doi.org/10.1016/j.tvjl.2009.08.028

Temple D, Escribano D, Jiménez M, Mainau E, Cerón JJ, Manteca X (2017) Effect of the needle-free "intra dermal application of liquids" vaccination on the welfare of pregnant sows. Porc Health Manag 3:1–7. https://doi.org/10.1186/s40813-017-0056-3

Temple D, Jiménez M, Escribano D, Martín-Valls G, Díaz I, Manteca X (2020) Welfare benefits of intradermal vaccination of piglets. Animals 10:1898. https://doi.org/10.3390/ani10101898

Timphaus CB, kleine Kruthaup FA, Freise F, Herbrandt S, große Beilage E, Fels M (2024) Effects of two different methods of teeth grinding on dental injuries, skin lesions, growth and behaviour of suckling piglets compared to a non-treated control group. Animals 14:1318. https://doi.org/10.3390/ani14091318

Torrey S, Devillers N, Lessard M, Farmer C, Widowski T (2009) Effect of age on the behavioral and physiological responses of piglets to tail docking and ear notching. J Anim Sci 87:1778–1786. https://doi.org/10.2527/jas.2008-1354

Tuyttens FA, Vanhonacker F, Verhille B, De Brabander D, Verbeke W (2012) Pig producer attitude towards surgical castration of piglets without anaesthesia versus alternative strategies. Res Vet Sci 92:524–530. https://doi.org/10.1016/j.rvsc.2011.02.017

Uilenreef J, van der Staay FJ, Meijer E (2019) A monosodium Iodoacetate osteoarthritis lameness model in growing pigs. Animals 9:405. https://doi.org/10.3390/ani9070405

Valenzuela C, Torres C, Muñoz V, Simbaina JC, Sánchez A, Bustamante T, Sepúlveda JM, Piron R, Del Campo

M, Lagos N (2019) Evaluation of Neosaxitoxin as a local anesthetic during piglet castration: a potential alternative for Lidocaine. Toxicon 164:26–30. https://doi.org/10.1016/j.toxicon.2019.03.021

Valros A (2022) Review: the tale of the Finnish pig tail - how to manage non-docked pigs? Animal 16:100353. https://doi.org/10.1016/j.animal.2021.100353

Viscardi AV, Turner PV (2018a) Efficacy of buprenorphine for management of surgical castration pain in piglets. BMC Vet Res 14:1–12. https://doi.org/10.1186/s12917-018-1643-5

Viscardi AV, Turner PV (2018b) Use of meloxicam or ketoprofen for piglet pain control following surgical castration. Front Vet Sci 5:299. https://doi.org/10.3389/fvets.2018.00299

Viscardi AV, Turner PV (2019a) Use of meloxicam, buprenorphine, and Maxilene® to assess a multimodal approach for piglet pain management, part 1: surgical castration. Anim Welf 28:487–498. https://doi.org/10.7120/09627286.28.4.487

Viscardi AV, Turner PV (2019b) Use of meloxicam, buprenorphine, and Maxilene® to assess a multimodal approach for piglet pain management, part 2: tail-docking. Anim Welf 28:499–510. https://doi.org/10.7120/09627286.28.4.499

Viscardi AV, Cull CA, Kleinhenz MD, Montgomery S, Curtis A, Lechtenberg K, Coetzee JF (2020) Evaluating the utility of a CO2 surgical laser for piglet castration to reduce pain and improve wound healing: a pilot study. J Anim Sci 98:skaa320. https://doi.org/10.1093/jas/skaa320

Viscardi AV, Hunniford M, Lawlis P, Leach M, Turner PV (2017) Development of a piglet grimace scale to evaluate piglet pain using facial expressions following castration and tail docking: a pilot study. Front Vet Sci 4:51. https://doi.org/10.3389/fvets.2017.00051

von Borell E, Baumgartner J, Giersing M, Jäggin N, Prunier A, Tuyttens FA, Edwards SA (2009) Animal welfare implications of surgical castration and its alternatives in pigs. Animal 3:1488–1496. https://doi.org/10.1017/S1751731109004728

von Borell E, Bonneau M, Holinger M, Prunier A, Stefanski V, Zöls S, Weiler U (2020) Welfare aspects of raising entire male pigs and immunocastrates. Animals 10:2140. https://doi.org/10.3390/ani10112140

Vullo C, Barbieri S, Catone G, Graïc JM, Magaletti M, Di Rosa A, Motta A, Tremolada C, Canali E, Dalla Costa E (2020) Is the piglet grimace scale (PGS) a useful welfare indicator to assess pain after cryptorchidectomy in growing pigs? Animals 10:412. https://doi.org/10.3390/ani10030412

Walls A, Hatze B, Lomax S, Bathgate R (2022) Defining "Normal" in pig parturition. Animals 12:2754. https://doi.org/10.3390/ani12202754

Ward SA, Kirkwood RN, Song Y, Garg S, Plush KJ (2022) Effect of dexamethasone and route of administration on sow farrowing behaviours, piglet delivery and litter performance. Animals 12:847. https://doi.org/10.3390/ani12070847

Weary DM, Droege P, Braithwaite VA (2017) Behavioral evidence of felt emotions: approaches, inferences, and refinements. In: Advances in the study of behavior, vol 49. Academic Press, pp 27–48. https://doi.org/10.1016/bs.asb.2017.02.002

Weary DM, Braithwaite LA, Fraser D (1998) Vocal response to pain in piglets. Appl Anim Behav Sci 56:161–172. https://doi.org/10.1016/S0168-1591(97)00092-0

Werner D, Baldinger L, Bussemas R, Büttner S, Weißmann F, Ciulu M, Mörlein J, Mörlein D (2021) Early immunocastration of pigs: from farming to meat quality. Animals 11:298. https://doi.org/10.3390/ani11020298

Werner J, Saller AM, Reiser J, Senf S, Deffner P, Abendschön N, Fischer J, Grott A, Miller R, Zablotski Y, Steiger K, Bergmann S, Erhard MH, Ritzmann M, Zöls S, Baumgartner C (2022) Evaluation of two injection techniques in combination with the local anesthetics lidocaine and mepivacaine for piglets undergoing surgical castration. Animals 12:1028. https://doi.org/10.3390/ani12081028

Whalin L, Pairis-Garcia M, Proudfoot K, Stalder K, Johnson A (2016) Validating behavioral sampling techniques for lame sows administered flunixin meglumine and meloxicam. Livest Sci 191:103–107. https://doi.org/10.1016/j.livsci.2016.07.017

Whittaker AL, Muns R, Wang D, Martínez-Burnes J, Hernández-Ávalos I, Casas-Alvarado A, Domínguez-Oliva A, Mota-Rojas D (2023) Assessment of pain and inflammation in domestic animals using infrared thermography: a narrative review. Animals 13:2065. https://doi.org/10.3390/ani13132065

Willgert KJ, Brewster V, Wright AJ, Nevel A (2014) Risk factors of lameness in sows in England. Prev Vet Med 113:268–272. https://doi.org/10.1016/j.prevetmed.2013.10.004

Winner EM, Beisl M, Gumbert S, Härtel H, Kaiser J, Wernecke A, Senf S, Zablotski Y, Ritzmann M, Zöls S (2022) Implementation of piglet castration under inhalation anaesthesia on farrowing farms. Porcine Health Manag 8:20. https://doi.org/10.1186/s40813-022-00263-0

Wojtalewicz S, Shuckra J, Barger K, Erickson S, Vizmeg J, Niederauer S, Simpson A, Davis J, Schauder A, Hifi O, Castel D, Meilin S, Agarwal J, Lade C, Davis B (2025) An injectable oleogel-based bupivacaine formulation for prolonged non-opioid post-operative analgesia. Drug Deliv Transl Res 15:1424–1443. https://doi.org/10.1007/s13346-024-01688-6

Yang P, Yang S, Durham LB, Lester PA, Myers DD (2024) Evaluation of efficacy of 2 extended-release bupivacaine products in a porcine model of incisional pain. J Am Assoc Lab Anim Sci 63:422–427. https://doi.org/10.30802/AALAS-JAALAS-23-000106

Yun J, Björkman S, Pöytäkangas M, Peltoniemi O (2017) The effects of ovarian biopsy and blood sampling methods on salivary cortisol and behaviour in sows. Res Vet Sci 114:80–85

Yun J, Ollila A, Valros A, Larenza-Menzies P, Heinonen M, Oliviero C, Peltoniemi O (2019) Behavioural alterations in piglets after surgical castration: effects of analgesia and anaesthesia. Res Vet Sci 125:36–42. https://doi.org/10.1016/j.rvsc.2019.05.009
Zamaratskaia G, Squires EJ (2009) Biochemical, nutritional and genetic effects on boar taint in entire male pigs. Animal 3:1508–1521. https://doi.org/10.1017/S1751731108003674
Zoric M, Schmidt U, Wallenbeck A, Wallgren P (2016) Lameness in piglets - should pain killers be icluded at treatment? Porc Health Manag 2:8. https://doi.org/10.1186/s40813-016-0022-5

Goats

Preet Singh, Dinakaran Venkatachalam,
Paul Chambers, and Kavitha Kongara

Abstract

Goats are among the oldest domesticated animals and are raised globally for meat, milk, fiber, and hides. Recently, there has been an increased focus on goat welfare, particularly as management practices shift from traditional extensive systems to semi-intensive and intensive settings. Research has identified several useful indicators, related to the animals, their management, and available resources, for assessing the on-farm welfare of goats. While animal welfare is influenced by multiple factors, some reliable indicators include body condition, hair coat cleanliness, udder and skin health, and claw condition. The impact of routine husbandry procedures is also considered in welfare assessments. Studies have highlighted prevalent welfare issues on large commercial goat farms, such as lameness due to claw overgrowth, skin and udder lesions, poor hair coats, and complications from improper disbudding and dehorning. Goat kids are often disbudded to improve their welfare and ensure the safety of other livestock, farm staff, and veterinarians, as well as to facilitate various production and management tasks. However, numerous studies have documented physiological and behavioral changes that indicate pain and distress following disbudding, which can significantly compromise the welfare of goat kids. Recognizing and measuring pain is essential for its management after such procedures. In light of growing concerns about animal welfare, pain mitigation strategies for disbudding have garnered considerable attention. Various methods are now available, with current best practices recommending the use of sedation and analgesia through an alpha-2 agonist, the application of a two-point cornual nerve block, followed by an NSAID for postoperative pain relief.

This chapter is based on the Authors articles https://doi.org/10.3390/ani13243814 and https://doi.org/10.3390/ani14040555 (Kongara et al. 2023, Singh et al. 2024).

P. Singh
School of Veterinary Science, Massey University, Palmerston North, New Zealand

D. Venkatachalam
Independant researcher, Rowan University, Palmerston North, New Zealand

P. Chambers
Associate Professor (Retired) School Of Veterinary Science, Massey University, Rowan University, Palmerston North, New Zealand

K. Kongara (✉)
Anatomy and Physiology, Shreiber School of Veterinary Medicine, Rowan University, Glassboro, NJ, USA
e-mail: Kongarak@rowan.edu

R. E. Clutton, G. C. Musk (eds.), *Pain and Suffering in Farmed Animals*,
https://doi.org/10.1007/978-3-032-11407-5_17

Keywords

Goat welfare · Husbandry procedures · Goat kid disbudding · Pain assessment · Physiological measures · Behavioural methods · Pain mitigation

1 Evolution of Domestic Goats

Goats were one of the first species of animals to be domesticated (Nomura et al. 2013) and as with other species of animals, evolution by natural selection, proposed by Charles Darwin in his origin of species, has been the key phenomenon for evolution of domestic goats (*Capra hircus*) from their wild ancestors (*Capra aegagrus*). Selective breeding to meet human needs and preferences also played an important role in the evolution of the domestic goat. Current-day domestic goats are spread across all continents (except Antarctica) by virtue of their extraordinary adaptability and ability to thrive in a range of environments that include tropical rainforests, dry, hot deserts, and cold, hypoxic high mountains (Amills et al. 2009).

Goats belong to a broad, populous family, 'Bovidae', which also includes other ruminants such as wild cattle and sheep (Savage and Long, 1986). All Bovids share common traits such as hooves, an herbivorous diet and cranial appendages (all males and some females) such as horns (Geist 1984). A few members of the family, including goats, ibex-like species and sheep, had diverged into a subfamily 'Caprinae'. The genus "*Capra*" includes species of both domestic and wild goats, including the most ancient ancestor, Bezoar ibex (*Capra aegagrus*). Many archaeological and molecular genetic studies indicated the origin of the ancient ancestors (Bezoar ibex) in western Asia (Naderi et al. 2007) 10,000 YBP (Years Before the Present; Amills et al. 2017). But the domestication process has been suggested to occur at different places in the world, including Turkey, Iran, Pakistan, Southern Levant and China (Nomura et al. 2013). Domestic goats were dispersed across the globe accompanying human migration, and today, nearly 1.2 billion goats, of more than 200 breeds, are kept in a wide variety of global environments/systems for the production of milk, meat, fiber, and hides (Nomura et al. 2013; Colli et al. 2018). The table below shows the distribution of goat population across continents (Table 1; Mazinani and Rude 2020).

2 Goat Production Systems and Welfare

Goats are raised in a wide range of management systems across the continents. Intensive (zero-grazing), semi-intensive and extensive management systems have been the main farming systems in goat production globally.

Extensive management systems involve a large number of animals in extensive rangelands with low agricultural productivity. Therefore, the stocking density will be low, and inputs of labor/money are small compared to the size of the land. Some countries in the world, e.g., USA, Australia, Canada and New Zealand, use large pastoral systems where meat and/or dairy goats are raised on extensive pastures and forages. The extensive systems are usually more palatable to consumers and the general public as they provide a free, natural environment.

Routine, painful husbandry procedures such as castration and disbudding/dehorning, neonatal mortality and predation are potential welfare issues for animals raised on extensive systems (Temple and Manteca 2020). Due to their extensive movement and consequently remote locations, monitoring of animals for pain, parturition difficulties and neonatal problems, body condition, etc. is challenging in extensive systems.

Table 1 Goat populations around the world

Continent	Population (millions)	%
Asia	556	55.4
Africa	388	38.7
Europe	17	1.7
America	38	3.8
Oceania	4	0.4

In the semi-intensive system, goats are confined during nighttime and allowed to graze pastures for a few hours in a day. Poor hair condition, faecal soiling, overgrown claws and improper disbudding or dehorning have been reported to be the main welfare problems of dairy goats in a semi-intensive system under drought conditions (Larrondo et al. 2021). Studies from Northwestern Italy demonstrated an adequate level of animal well-being both in intensive and semi-intensive farming systems (Tiezzi et al. 2019). Researchers did not observe any difference in animal health indicators such as hair coat condition, integument cleanliness and lameness between the two systems.

Intensive (total confinement with zero grazing) management systems have been used in large commercial goat farming. Lameness has been a prevalent problem in commercial dairy goat farms. It is an important behavioral indicator of pain (Anzuino et al. 2010). Claw overgrowth, causing pain during walking, is one of the major causes of lameness in commercial (intensive) dairy goat farms worldwide. It is assumed to be due to insufficient wear of claws due to soft bedding (Anzuino et al. 2010; Ajuda et al. 2020; Silva Salas et al. 2021). Udder and teat lesions, skin lesions, poor haircoat and pruritis have been reported in welfare assessment studies conducted in commercial dairy goat farms in the UK, Portugal, Northern Italy and Mid-Western United States (Anzuino et al. 2010; Hempstead et al. 2021a; Can et al. 2016; Battini et al. 2016). Rapid milking rates, large herd sizes, high goat: stockperson ratios and minimal hygiene routines at milking have been ascribed to udder and teat lesions. In the majority of the farms, horned goats were grouped with hornless goats. This raises welfare concerns, as the horned goats may exhibit bullying behavior and attack others, leading to injuries.

Scurs (distorted horn regrowth), which develop due to inefficient and incomplete disbudding, have also been reported in these studies. Scurs would break off easily, leaving open wounds that are at increased risk of infection (Hartnack et al. 2018). Scurs and/or horns can pose a threat and can have traumatic interactions with conspecifics.

Cranial appendages such as horns and antlers have evolved in animals to protect against predators and to establish social ranking within a herd (Smith and Sherman 2009). Their presence may be advantageous in the wild, but not desirable in a commercial farming condition. Goats with horns are dangerous to herd mates and handlers. They often engage in fighting to establish dominance, during which severe injuries or death may occur. In addition, horned goats are difficult to handle and transport, cause damage to farm infrastructure such as fences and pen partitions, require more feeding and trough space, and make it difficult to use head bales or automatic feeders (Loretz et al. 2004; Smith and Sherman 2009). To avoid these problems, disbudding is commonly performed in goats, particularly in dairy goat kids.

Disbudding involves the removal or destruction of the germinal epithelial tissue forming the horn bud before it attaches to the skull horn producing cells, to prevent the growth of horns (Winder et al. 2016). The removal of the horn after its attachment to the skull is called dehorning. Dehorning is much more stressful and painful than disbudding of the animal, and has complications such as epistaxis, sinusitis, and delayed wound healing (Hartnack et al. 2018). Therefore, disbudding is more humane and preferred over dehorning. As the horns of goats grow rapidly, it is recommended to disbud goats at a younger age, commonly between 2 and 4 weeks (Hartnack et al. 2018). Several disbudding methods have been reported in goat kids including thermal cauterization (commonly used method), application of caustic paste or liquid nitrogen, and injection of clove oil (Molaei et al. 2015; Hempstead et al. 2018c, 2018d, 2018e, 2018f). Irrespective of the method or age of the animal, disbudding causes behavioural, physiological and neuroendocrine changes which indicate that the procedure is stressful and painful (Hempstead et al. 2017; Alvarez et al. 2015), compromising goat welfare.

The need for disbudding can be circumvented by selectively breeding for polledness. Although this approach is effective in cattle, it tends to trigger severe reproductive complications in certain dairy goats with European lineage, such as

Saanen, Alpine, and Toggenburg breeds (Smith and Sherman 2023). In these breeds, the existence of horns is governed by a recessive gene that leads to infertility. Consequently, female goats with a homozygous polled genotype will mature as infertile intersex individuals, while male goats with the same genotype face a heightened likelihood of developing sperm granulomas. This underscores the necessity for disbudding within these goat breeds (Smith and Sherman 2023).

3 Evidence and Attitudes to Disbudding Pain and Animal Welfare

Application of hot-iron (thermal cautery) is the most commonly used method of disbudding kids on dairy, meat and other purpose goat farms globally, including the USA, Canada, UK, Australia and New Zealand (Fig. 1). Cautery disbudding causes thermal destruction of the horn bud cells to prevent horn growth. It is considered to cause acute pain and distress (Alvarez et al. 2009; Ingvast-Larsson et al. 2011a, b; Hempstead et al. 2017, 2018a). Possible complications associated with the procedure include burning of the normal skin around the cauterised horn tissues, thermal damage to the skull and brain, and meningoencephalitis (Sanford 1989; Thompson et al. 2005), which could be due to prolonged or excessive pressure with the hot iron, i.e., inappropriate technique (Sanford 1989). Disbudding without administration of effective anaesthesia and analgesia, and conducted with inadequate technical skill therefore significantly diminishes the welfare of goat kids.

Global consumer concern regarding the suffering and well-being of animals used for food production has been increasing. The United Kingdom and several European countries mandate that disbudding in goats is exclusively carried out by a veterinarian, using appropriate anesthesia and analgesia (Wagmann et al. 2018). Consequently, in New Zealand and Australia, a significant shift has occurred: pain relief measures have become obligatory for disbudding goat kids (Hempstead et al. 2018c, 2018d, 2018e, 2018f). Nearly 95% of the goat operations in the USA use cautery disbudding: analgesics or anaesthetics were used routinely on only 30.4% of the operations during the disbudding (USDA 2020). The use of pain relief for disbudding has not been mandated by the US Federal law.

Recognition and assessment of pain is complex in small ruminants and cervids as they are timid, prey species. On commercial farm operations, an estimate of the intensity of the pain is made, mostly based on animal behaviour, which would depend on stockpeople's experience, training, perceptions, attitudes, and empathy. Also,

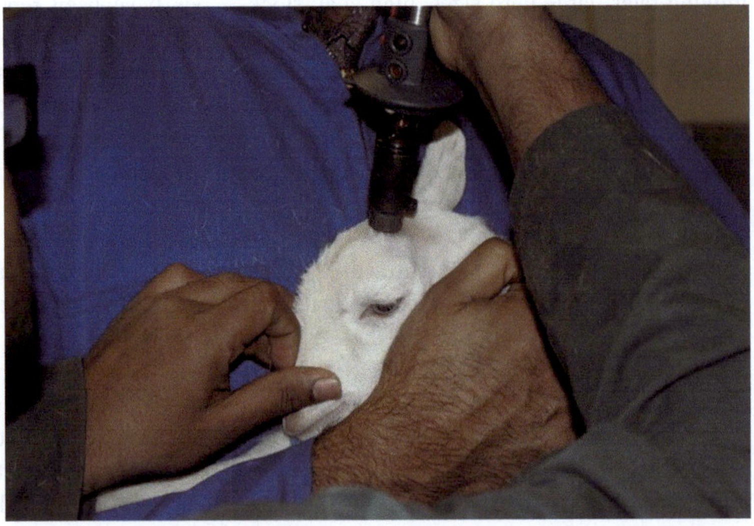

Fig. 1 Goat kid being physically restrained for the disbudding procedure. (Photo: Dr. Preet Singh)

their ability to recognize and respond to pain in their livestock plays a significant role in pain treatment and/or management, which in turn influences animal welfare (Muri and Valle 2012). The interrelationships between dairy goat stockpeoples' pain perception and provision of veterinary attention, and their association with stockpeoples' demographics, attitudes and empathy have been explored in a Norwegian survey (Muri and Valle 2012). Females, older stock people and those who grew up in a rural setting were significantly more likely to consult a veterinarian for painful conditions in their livestock. The researchers suggested training of the stockpeople for better evaluation and management of pain for high welfare standards.

Prey species, including sheep and goat do not often manifest overt signs of pain and it is difficult to accurately assess behavioural signs of pain in these animals (Matthews et al. 2019). In research studies, a combination of physiological and behavioural measures has been developed and used for assessment of pain and the efficacy of analgesic strategies.

4 Methods of Disbudding Pain Assessment

4.1 Physiological Measures

4.1.1 Plasma Cortisol, Glucose and Lactate

The physiological variable that has been historically studied to assess the pain related distress following routine husbandry procedures is the total plasma cortisol (Mellor and Stafford 2000). Cortisol is a hormone, which is released in high concentrations into circulation after activation of hypothalamo-pituitary-adrenal (HPA) axis in response to stressful conditions including pain (Molony and Kent 1997). It thus gives an indication of distress, rather than pain directly. Increases in plasma cortisol concentrations following disbudding and other painful husbandry procedures such as castration and tail docking, and its alleviation by analgesic administration has commonly been used to assess pain in calves, lambs, and goat kids (Alvarez et al. 2009; Alvarez and Gutierrez 2010; Sutherland and Tucker 2011). The degree of plasma cortisol response was also used to compare the pain intensity between different methods of goat kid disbudding (Hempstead et al. 2018c, 2018d, 2018e, 2018f). Disbudding of kids with liquid nitrogen or caustic paste has been found to cause more severe changes in cortisol, indicating intense acute distress, and possibly pain, compared to cautery disbudding (Hempstead et al. 2018c, 2018d, 2018e, 2018f).

To measure plasma cortisol, blood samples are commonly collected from the jugular veins and cortisol concentrations are measured using different assay techniques, including the enzyme-linked immunosorbent assay and radioimmunoassay (Sutherland and Tucker 2011). Although many studies have used plasma cortisol for the assessment of pain-induced distress, there are some caveats to this method. Plasma cortisol levels were also influenced by diurnal rhythms (Andersson et al. 2000) and stressors that are not associated with pain, such as animal handling (Weary et al. 2006), feeding (Eriksson et al. 1989), and sexual excitement (Colborn et al. 1991). Trends in plasma cortisol concentrations measured from blood samples collected before and at multiple time points after a specific procedure would give a more reliable estimate of the stress associated with pain following a noxious procedure in production animals (although blood sampling will increase plasma cortisol). In analgesia studies, comparing the cortisol levels between sham-handled and treatment groups, and correlating cortisol response with other pain assessment measures such as pain behaviours, would make plasma cortisol evaluations a more valid way to estimate the efficacy of analgesic drugs (Tschoner 2021). It should be noted that some drugs which are used as analgesics in ruminants, such as alpha 2 adrenergic agonists, probably have a direct depressant effect on the adrenal glands in goats, as they do in pigs (Jager et al. 1998).

Plasma metabolites, such as glucose and lactate, were measured in conjunction with plasma cortisol following disbudding of goat kids

(Hempstead et al. 2018c, 2018d, 2018e, 2018f, 2020). These blood constituents were measured based on the hypothesis that increased secretion of cortisol in response to pain-induced distress stimulates mobilization of glycogen, which results in increased production of glucose and lactate (Prunier et al. 2005). None of the studies found significant changes in plasma lactate and glucose despite a significant elevation of plasma cortisol after disbudding. The researchers opined that these two metabolites are not as useful as plasma cortisol responses to acute pain associated with disbudding.

4.1.2 Autonomic Responses

Other physiological measures resulting from the activation of sympathetic nervous outflow in response to noxious stimulation, such as heart rate, respiratory rate, blood pressure and body temperature can be monitored to assess pain following disbudding in goat kids (Alvarez et al. 2009). Although changes in these variables are influenced by the type of noxious stimulus based on the method of husbandry procedure, in general, animals in pain often have an increase in these variables (Stubsjoen et al. 2009, 2010). A major limitation to the use of the sympathetically driven variables, similar to cortisol, is that changes are not necessarily specific to pain (Peers et al. 2002) and are usually limited to acute pain responses and hence less useful for the assessment of longer-lasting pain. Also, changes in these autonomic variables have poor correlation with pain perception as these are modulated by centres of the brain stem below the level of the higher centres involved in cognitive pain perception (Holton et al. 1998).

4.1.3 Electroencephalography

The EEG is a record of the spontaneous electrical activity of the cerebral cortex. It is an objective tool for the assessment of neural (sensory) processing of noxious stimuli, which is known as nociception. EEG provides a direct and quick profile of cortical neuronal activity in response to noxious stimulation in animals. The characteristic configuration of neurones (particularly pyramidal type) in the layers of the cerebral cortex leads to formation of electrical vectors that facilitate transmission of electrical currents. The EEG is the far-field potential (electrical field created by inherent neuronal activity and recorded away from them) of the vector currents recorded using electrodes on the surface of the head (Murrell and Johnson 2006). Fast Fourier transformation, a mathematical procedure, converts the raw EEG signal into its component sine waves of different frequencies, characterized by corresponding amplitude. Thus, the power spectrum generated is simply a distribution and derivation of spectral EEG variables from the corresponding frequencies and amplitudes (Murrell and Johnson 2006). There are also several other proprietary methods of analysing raw EEG data, which are used in instruments designed for people.

Changes in the EEG power spectrum have been used for the quantification of nociception in minimally anaesthetised red deer, lambs, and calves following noxious procedures such as velvet antler removal, castration, and slaughter through a ventral neck-cut, respectively (Johnson et al. 2005a, b; Gibson et al. 2009). Changes in frequency spectra of EEG in lightly anesthetized animals (with halothane, which is minimally analgesic) reflect changes in activity of cerebral cortex in response to pain perception (Johnson et al. 2005a, b). This concept is further supported by studies in goats that did not find a difference in EEG spectral frequencies between conscious animals and those that were lightly anaesthetised during slaughter (Sabow et al. 2017). Advantages of the minimal anaesthesia model developed for recording EEG for quantification of pain in animals include the minimization of the impact on the EEG of extraneous electrical activity (such as from muscles) and loss of conscious pain perception by study animals, thereby reducing suffering from pain during noxious stimulation (Murrell and Johnson 2006).

In goats, EEG power spectra have been used to delineate loss of awareness from cognitive perception of pain following the application of various techniques for euthanasia and pre-slaughter stunning (Sutherland et al. 2016). Also, there have been studies that described the EEG frequency band analysis to find the effect of pre-slaughter stress, experimental pain and ontological changes in brain activity in goats (Raghazli et al. 2021; Huozha et al. 2011; Bergamasco et al. 2006). So far, no studies are available on the use of EEG spectral frequency changes to quantify pain associated with husbandry procedures such as disbudding/dehorning in goats.

Although an electroencephalogram can non-invasively capture the sensory neuronal processing of noxious stimuli and indirectly reflect cortical pain perception, it may not represent the motivational states such as aversiveness, fear and anxiety associated with pain perception, all of which are subjective experiences. Another limitation to the use of the EEG is that it can only be used for acute pain assessment, as an animal can be kept minimally anaesthetized for only a limited amount of time. A combined approach such as concurrent evaluation of cognitive pain perception manifested by behavioural changes, will be more reliable for pain assessment using EEG in farm animals.

4.1.4 Mechanical Nociceptive Threshold Testing

Mechanical stimuli, such as pressure, are usually applied to the skin and can be used to produce quantifiable nociception or pain in farm animals, including goats. The degree of mechanical pressure that an animal can tolerate before showing an avoidance (behavioural) response or withdrawal reflex is defined as a mechanical nociceptive (pressure) threshold of that animal (Hempstead et al. 2017, 2018c, 2018d, 2018e, 2018f). Mechanical stimuli are used to induce experimental pain to test or compare analgesics in normal animals. In animals subjected to noxious procedures, these stimuli are used to test the pain sensitivity at the site of tissue damage (wound site) and/or areas surrounding the wound. A variety of commercial, hand-held algometers are available (e.g. Force one FDIX 50, Wagner Instruments, Riverside, CT, USA; ProdPlus algometer, TopCat Metrology Ltd., Ely, UK), which are validated for use in various farm animal species, including goats (Stubsjoen et al. 2010; Mirra et al. 2018; Alvarez et al. 2019) (Fig. 2).

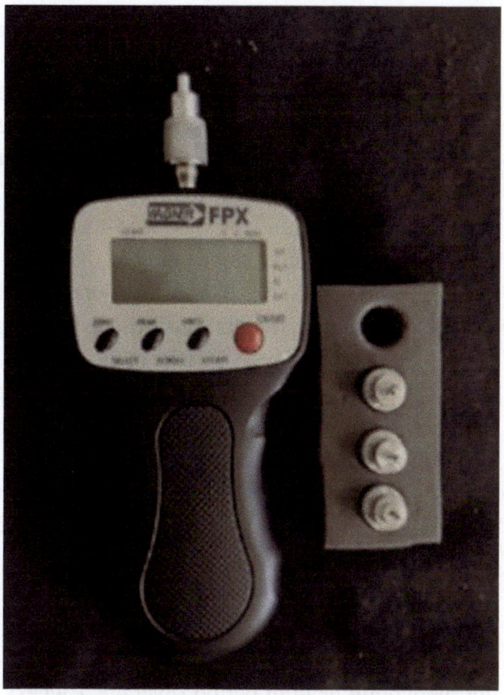

Fig. 2 A modified Wagner hand-held algometer (FPX 20: Wagner instruments, Greenwich, CT, USA.) with 2 mm diameter steel probe

Mechanical nociceptive thresholds are usually measured in Newtons (N) or Kilo-gram Force (Kgf). Force is applied perpendicular to the skin surface at a constant rate and in gradual increments to avoid a sudden, jerky increase in readings without corresponding attainment of threshold manifested by signs of discomfort specific to the species being tested. Head withdrawal with a specific ear flick has been noted as the end response by goat kids being tested at the disbudded site (Hempstead et al. 2018c, 2018d, 2018e, 2018f; Frahm et al. 2020). To attain a good correlation between threshold values and the amount of pressure applied, selecting an algometer probe that matches the size of the animal is also important. A round rubber tip probe of 1 cm^2 has been used on a Force One Wagner algometer (Hempstead et al. 2018c, 2018d, 2018e, 2018f), and a 2 mm diameter metal probe has been used on a ProdPlus algometer in disbudded goat kids (Alvarez et al. 2019; Frahm et al. 2020). Pressure algometry has been used to compare the pain sensitivity between three different methods of disbudding, and cautery disbudding was found to cause less acute pain hypersensitivity than the caustic paste and cryosurgery methods in goat kids (Hempstead et al. 2018c, 2018d, 2018e, 2018f). In other studies, it was used to find the welfare benefits of alternative methods of disbudding (Frahm et al. 2020) and the duration of wound hypersensitivity after hot-iron disbudding in goat kids (Alvarez et al. 2019). In the study by Frahm et al. (2020) development of acute mechanical hypersensitivity around the horn buds following the injection of clove oil or isoeugenol has been reported (Fig. 3). Tissue irritation by the injected substances was suggested as the potential cause of heightened pain sensitivity and, therefore, the welfare benefits of the method over hot-iron disbudding are questionable. Persistent mechanical hypersensitivity throughout the wound healing period (average 7 weeks) has been reported in goat kids by Alvarez et al. (2019).

Animal handling before the application of the stimulus has the potential to influence the threshold readings. A good human–animal relationship, prior habituation of the animal to the device, and testing in home pens close to dam/other pen mates can minimize the confounding effect of handling stress on threshold measurements (Frahm et al. 2020). Other types of nociceptive stimuli that have been used to test the effect of local anaesthetic (LA) administration prior to goat kid disbudding were pin pricks (Venkatachalam et al. 2021). The observer gently pricked the skin around the horn buds with a needle point at approximately 30 s to 1 min intervals until no response is observed to confirm analgesia after infiltration with LA. This qualitative analgesia testing method is rapid and practical to use in field conditions, and reliable to assess the onset of analgesia prior to disbudding.

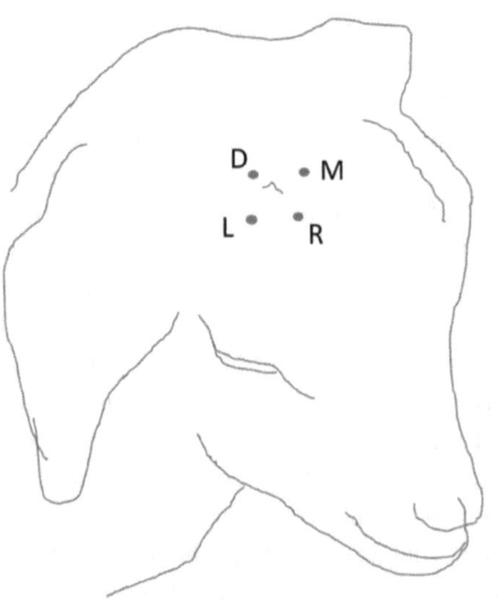

Fig. 3 Schematic illustration of the four locations around the horn buds that were used to measure the mechanical nociceptive thresholds of the goat kids in a study by Frahm et al. (2020)

In summary, all the devices and methods developed to test the mechanical nociceptive thresholds of farm animals let us assess the level of pain hypersensitivity, efficacy of analgesics and/or analgesic approaches, and development of 'plasticity' in the nervous system, following noxious husbandry procedures in field conditions.

4.1.5 Infrared Thermography

Infra-red thermography (IRT) is a non-invasive method of recording infrared radiation emitted by bodies, i.e., heat (Travain and Valsecchi 2021). It can measure changes in surface temperature due to activation of autonomic nervous system in response to stress. A stress-induced secretion of catecholamines causes an increase in internal body temperature called stress-induced hyperthermia, and also causes a reduction in blood flow in the skin around the eyes. A drop in eye temperature, assessed using IRT, has been reported to be associated with the onset of acute pain in calves following hot-iron dehorning (Stewart et al. 2005a, 2008). Sympathetically mediated vasoconstriction in response to acute pain has been proposed to be the cause of the drop in eye temperature. The IRT recording can be taken at a distance from the animal without the need to contact/restrain the animal. This overcomes the effect of animal handling stress on actual data, which is the drawback associated with the measurement of other physiological (and autonomic) variables in response to stress and pain (Stewart et al. 2005a, 2005b) Care must be taken that the hair coat must be free of dirt, grease and foreign material while obtaining IRT images of the area of interest (Stewart et al. 2008).

A study in disbudded goat kids used IRT to take the images of the horn bud (~3 cm diameter around horn bud), and surface skin temperature was calculated from a ~2 cm diameter area around the horn buds (Hempstead et al. 2018c, 2018d, 2018e, 2018f). Disbudded goat kids were found to have a higher skin surface temperature around the horn buds than sham kids. Another study has used IRT to assess the wound inflammation and surface temperature during healing after cautery disbudding of goat kids (Alvarez et al. 2019) and reported that the necrotic tissue formed during the inflammatory phase was hotter than the epithelium. Studies are required to find the use of the IRT to evaluate the effect of analgesic agents and strategies after disbudding in goat kids.

4.1.6 Biomarkers

Immunological, inflammatory and pain biomarkers were measured in tissues and body fluids of young farm animals following husbandry procedures (Pang et al. 2011; Mirra et al. 2018; Sa; Kongara et al. 2020). Acute phase inflammatory proteins such as haptoglobins were measured in goat kids disbudded using different methods (Hempstead et al. 2018c, 2018d, 2018e, 2018f). Clove oil injection under the horn bud caused a significant increase in serum haptoglobin concentrations 24 h after treatment, which indicates marked inflammation associated with the method of disbudding (Hempstead et al. 2018c, 2018d, 2018e, 2018f). Measurement of immunoreactive proteins such as β-endorphin concentration in the blood plasma is a useful method of acute pain assessment following hot-iron disbudding of goat kids (Greenwood and Shutt 1990). With the advent of new molecular biological techniques in animals, it has been possible to demonstrate the efficacy of non-steroidal anti-inflammatory drugs in suppressing the expression of mRNA (in peripheral leucocytes of a blood sample) for inflammatory cytokine and nociceptive marker genes in calves disbudded by thermocautery (Kongara et al. 2020). More advanced techniques, such as plasma proteome analysis, allowed identification of candidate protein biomarkers that are associated with nociceptive and inflammatory processes following surgical dehorning of calves (Ghodasara et al. 2022). Biomarkers such as these may be useful in the future to assess various methods of pain mitigation for husbandry procedures in farm animals, including goats.

4.2 Behavioural Measures

Changes in normal behaviour are commonly observed in animals in response to painful stimuli. Hence, monitoring normal behavioural patterns and deviations from these is a significant component of pain assessment in animals (Morton and Griffiths 1985). Species-specific behavioural changes that can be identified and quantified have been used to evaluate pain and analgesia in goat kids, lambs, piglets, and calves following routine husbandry procedures (Molony and Kent 1997; Alvarez et al. 2009; Leslie et al. 2010). A person assessing pain behaviours, either directly from an animal or indirectly from video recordings, should be familiar with the animal's normal behavioural patterns, have adequate training and experience in recognition of changes in normal behaviours in response to noxious procedures (Sanford et al. 1986; Bufalari et al. 2007).

A multitude of factors including severity of insult to the tissues and neuronal pathways influence the manifestation of pain behaviours by animals. Age, previous pain experience, social hierarchy in the herd, human presence, environmental conditions are among the others (Fitzpatrick et al. 2006). In young farm animals such as goat kids, lambs and calves, a significant alteration in their normal behaviour, body posture, locomotor activity, orientation toward dam, and response to manipulations have been used as a basic template for the behavioural assessment of pain and analgesia following husbandry procedures, including disbudding.

There appears to be two phases to goat kids' behaviour in response to hot iron disbudding: acute avoidance type behaviours in responses to application of the iron and abnormal behaviour in the subsequent hours. Behavioural indicators of acute stress and pain during cautery disbudding in goat kids have been first reported by Alvarez et al. (2009). Struggles (kicks), such as a vigorous movement of the legs and attempts to escape, and vocalisations in the form of bleats with an open or closed mouth were reported. A change in the frequency of behaviours along with plasma cortisol levels was used to find the effect of analgesic treatment on disbudding pain (Alvarez and Gutierrez 2010).

In a sham controlled trial, Hempstead et al. (2017) recorded behavioural changes associated with cautery disbudding of female Saanen dairy goat kids. Video recordings of pre- and post-treatment behaviours were analysed by a trained and an experienced observer to build an ethogram of behaviour patterns. Although the frequency of 11 behaviours in the ethogram was chosen to study, individual behaviours such as head shaking, rubbing, scratching, and body shaking were significantly different between treated and control kids. Due to a large inter-individual variation, self-grooming behaviour did not reach statistical significance despite an apparent in-crease in its frequency in the disbudded group. Individual variation in a behavioural response, within the same species, to a noxious procedure is one important confounding factor in pain assessment studies. This could be due to differences in pain perception and expression between individual animals (Small et al. 2020). The following table (Table 1) provides an overview of behavioural changes in goat kids following different methods of disbudding.

In addition to the assessment of frequency of specific behaviours, some studies used a visual analogue scale (from 1 to 10) to score pain behaviours to evaluate the efficacy of anaesthetic and analgesic treatments administered to goat kids (Sutherland and Tucker 2011; Wagmann et al. 2018). A summary of the pain behaviours was used to assign a score from the scale. Also, a vocalisation score, which is the total number of vocalizations during the disbudding procedure for both horn buds, was used to compare the alternative methods to cautery disbudding (Still Brooks et al. 2021).

From these studies in goat kids, it is evident that disbudding induces specific changes in behaviour, which include increased frequency of struggles, vocalisations, head shaking, scratching, and rubbing, body and tail shaking, lying bouts and duration. Changes in these behaviours have been used to evaluate relative noxiousness of different (alternative) methods of disbudding and efficacy of anaesthetic and analgesic protocols in various studies.

4.3 Facial Grimace Scale

Due to the limitations of behavioural methods (such as laborious and time-consuming video and audio recording and analyses and the need for specialised equipment), and physiological measures (such as expensive laboratory analytical pro-cedures), Lou (2020) developed a goat kid grimace scale for pain assessment following thermal disbudding. It is a pain assessment method that enables instantaneous identification and assessment of changes in facial expression after a noxious procedure (Langford et al. 2010). The basic template for the scale has been drawn from sheep and lamb grimace scales (Guesgen et al. 2016; Häger et al. 2017; Wagmann et al. 2018) and adapted for a goat face. Orbital and lip tightening, nostril dilatation and change in ear position have been scored on a 3-point scale (0 = Not Present, 1 = Moderately Present, and 2 = Obviously Present), from photographic images, to find the effect of analgesic administration and thermal disbudding in goat kids. Orbital tightening and ear position scores were found to be more reliable indicators of treatment effects than lip tightening and nostril dilatation. More research needs to be conducted to further explore the validity of this facial grimace scale in goat kids.

Both the behavioural and physiological systems are involved in the response to stress and pain, and in majority of research studies, a combination of both methods is used in pain assessment. Individual behaviours or an ethogram in conjunction with plasma cortisol measurement has been the basic pain assessment paradigm in a large number of goat disbudding studies. Some studies used other physiological methods such as mechanical nociceptive threshold testing, infrared thermography and plasma inflammatory marker analyses, heart rate and body temperature changes, and body weight gains, etc. Only one study used a facial grimace scale in combination with plasma cortisol evaluations in disbudded goat kids. Using a combination of pain assessment methods allows to offset some of the disadvantages associated with each technique and to gain as much accurate and comprehensive information as possible from an experimental study (Johnson 2016). In clinical practice, a combination of pin prick testing for analgesia before disbudding and behavioural assessment afterwards is probably most practical.

5 Mitigation of Disbudding Pain

The pain from hot iron disbudding has at least two phases, which probably require different approaches to alleviation. The application of a hot iron causes intense acute pain, and then, the release of inflammatory mediators from the burnt tissue will cause a lower-grade but longer-lasting pain. Both phases should be treated to ensure animal welfare. There are a variety of ways of accomplishing this: general anaesthetics and analgesics with or without local anaesthetics and then non-steroidal anti-inflammatory drugs (NSAIDs) for postoperative pain. The efficacy of local anaesthetics and general anaesthetics for disbudding pain in goat kids has not been extensively investigated or reported in the literature compared with calves. Most of the drugs recommended or referenced in this chapter are used off-label, as they are not registered for use in food-producing animals. Additionally, the registration status and availability of these drugs differ across countries and regions.

5.1 General Anesthesia and Systemic Analgesia

Xylazine hydrochloride is probably the most commonly used alpha-2 adrenergic agonist in ruminant veterinary practice despite its association with significant complications like hypoxemia and pulmonary edema. These complications are linked to the activation of pulmonary intravascular macrophages (Celly et al. 1999) and may involve the release of TNF-Alpha. TNF suppression agents, such as choline chloride given before xylazine, have a small beneficial effect (Singh et al. 2017). Other inflammatory mediators are almost certainly involved as well. In adult sheep, the degree of hypoxemia is similar to xylazine, romifidine, detomidine, and medetomidine (Celly et al. 1997). Goats exhibit higher sensitivity to xylazine compared with other ruminants (Taylor 1991). As a result, careful dose calculation and monitoring of the effects of xylazine are particularly important in goats, especially in goat kids, when compared with other ruminants.

Sedatives such as xylazine have proven effective in reducing stress during restraint and offer partial analgesic effects, yet they do not provide comprehensive pain relief during disbudding. Wagmann et al. (2018) investigated the efficacy of a mixture of xylazine (0.05 mg/kg) and ketamine (20 mg/kg) administered prior to disbudding by certified Swiss farmers (Wagmann et al. 2018). The authors concluded that this mixture did not provide adequate anesthesia and analgesia in goat kids and suggested that refinement to this protocol is required. Dexmedetomidine administered intramuscularly may be more effective for analgesia as compared with the administration of lidocaine around the horn bud and intramuscular injections of meloxicam

(Nfor et al. 2016). In the UK, where disbudding is carried out under general anesthesia in the first week of life, xylazine overdose is considered the most common cause of death (Matthews and Dustan 2019). Alphaxalone has been recommended for disbudding performed on the farm at a dose of 6 mg/kg administered intravenously (Buttle et al. 1986). As it provides negligible analgesia (Varga 2014), a pre- and postsurgical analgesic protocol should be followed.

Alpha-2 agonists are easy to administer while performing this procedure on the farm; however, they produce adverse effects such as hypothermia and cardiovascular and respiratory depression and often require a reversal agent such as yohimbine or atipamezole postoperatively (which will also reverse analgesia).

Nonsteroidal anti-inflammatory drugs have demonstrated their ability to alleviate post-disbudding pain; however, they have fallen short in preventing the acute pain induced during the disbudding process itself (Ingvast-Larsson et al. 2011a, b). Another disadvantage is that, as goats are disbudded shortly after birth, drugs requiring liver metabolism may have a variable, but probably long, duration of action. Goat kids, like other neonates, almost certainly have a restricted capacity to metabolize and eliminate drugs. The administration of sedatives such as xylazine may reduce stress and make it easier to administer local anesthetics.

Inhalant general anesthetics, especially isoflurane, have demonstrated effectiveness in mitigating pain during disbudding. However, their practicality in commercial farming scenarios is questionable given the need for anesthetic equipment and the potential for increased disbudding costs (Hempstead et al. 2018c, 2018d, 2018e, 2018f). Using inhalation anesthetics in oxygen delivered by an anesthetic machine via a face mask used to be considered the best practice (Buttle et al. 1986) but can be dangerous in the presence of a hot iron, as oxygen supports combustion.

The United Kingdom and several European countries mandate that disbudding in goats is exclusively carried out by a veterinarian, utilizing appropriate anesthetics and analgesics (Wagmann et al. 2018). Consequently, in New Zealand and Australia, a significant shift has occurred, and pain relief measures have become mandatory for disbudding goat kids (MPI 2018). This represents a noteworthy advancement in animal welfare practices.

5.2 Local Anesthesia

The administration of a local anesthetic to produce a nerve block is a common strategy for alleviating pain during the disbudding process (Smith and Sherman 2023). Usually, two nerves are blocked on each side: the cornual branches of the infratrochlear nerve (a branch of the ophthalmic division of the trigeminal nerve CNV_1) and the zygomaticotemporal nerve (also referred to as the cornual branch of the lacrimal nerve, a branch of the maxillary division of the trigeminal nerve CNV_2) (Fig. 4).

Usually, 0.5 mL of a local anesthetic is injected at each site, but because there may be several branches in the infratrochlear nerve (Vitums 1954), sometimes larger volumes are used to achieve a greater spread.

All local anesthetics have a potential for toxicity (Lemke and Dawson 2000). Goat kids, given their small size, are susceptible to overdosing. Overdose can cause cardiac effects (signs of reduced cardiac output), sedation, convulsions, and death. The need to block four nerves for both horn buds (in contrast to a single nerve per side in calves) increases the amount of local anesthetic required. Additionally, the vascularity at the site of the nerve block (Harwood 2012) exacerbates the risk.

Goat kids are different from small calves (Hempstead et al. 2020) insofar as they undergo disbudding at a younger age, with thinner skulls and much lower body weights (Hempstead et al. 2020). For instance, goat breeds like pygmy and

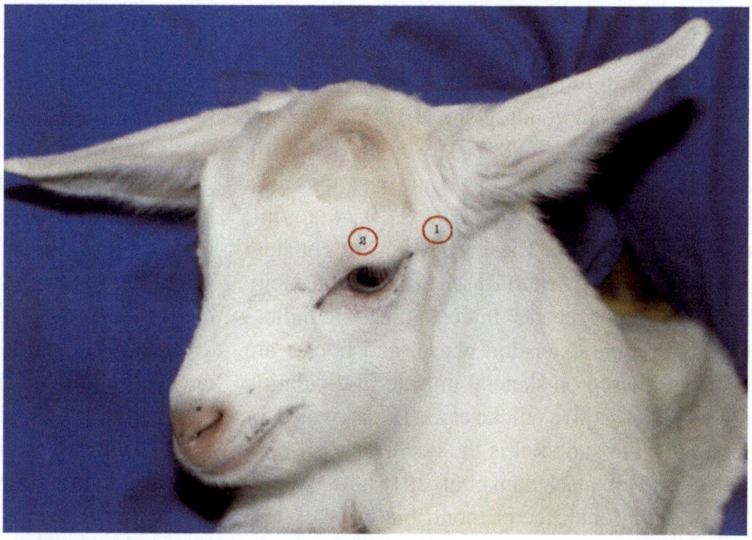

Fig. 4 Each horn bud in the goat kids is innervated by two cornual nerves, one from the zygomaticotemporal nerve and the other from the infratrochlear nerve. Therefore, two injection sites (as shown in the image) on each side are recommended to effectively alleviate pain during the process of disbudding

Nigerian Dwarf, which can weigh under 2 kg, might encounter issues when 2 mL of 2% lidocaine (20 mg/mL), equivalent to 20 mg/kg, is injected for nerve blocks. This dosage could easily lead to plasma concentrations sufficient to cause convulsions (Venkatachalam et al. 2018).

It is crucial to emphasize that the techniques used for disbudding calves should not be directly applied to goat kids. Thinner skulls in goat kids increase the risk of brain damage due to thermal injuries, resulting in convulsions and death. Pathological findings from goat kids disbudded with hot irons have revealed central areas of cavitation in the brain, both in grey and white matter. Histological lesions included extensive hemorrhages and coagulation necrosis (Thompson et al. 2005).

The thermal lesions caused by cautery disbudding can be infected by bacteria, leading to a potential risk of bacterial invasion and the development of meningoencephalitis (Sanford 1989). This can be treated with a course of broad-spectrum antibiotics (Thompson et al. 2005). Surviving goat kids display signs of incoordination, paraplegia, and convulsions even up to 3 weeks. The application time is also crucial, as the placement of a cautery iron for 15–20 s has been shown to cause severe brain injuries (Hempstead et al. 2021b).

5.3 Lidocaine

Lidocaine is the most used local anesthetic in veterinary practice (Taylor 1991) in most places, apart from the EU, and is thus used in goat medicine. It is a cheap, effective, usually safe, and readily available drug. Only one study has reported convulsions in a goat kid following the intramuscular injection of lidocaine at approximately 10 mg/kg (Morishima et al. 1981). In a dose-ranging investigation, a dosage of 7 mg/kg of body weight administered intravenously over a 60 s interval yielded no observable signs of toxicity (Venkatachalam et al. 2018). Consequently, this dosage is presumed safe for cornual nerve blocks in goat kids, apart from possibly accidental intravenous administration. This dose may also be deemed safe for other localized and regional nerve blocks in goat kids, although more extensive safety studies involving more animals are necessary.

The minimum dose necessary to induce seizures in young goats (12.42 mg/kg) is lower than the comparable dose observed in newborn lambs (18.40 mg/kg), and the average plasma level associated with convulsions in young goats (13.59 ± 2.34 μg/mL) (Venkatachalam et al. 2018) is lower than the level observed in newborn lambs (16.6 ± 1.2 μg/mL) (Morishima et al.

1981). Comparatively, in dogs, the lidocaine concentration leading to toxicity averages 8.21 + 1.69 μg/mL, which is notably below the level recorded in young goats (13.59 ± 2.34 μg/mL) (Wilcke et al. 1983). Meyer et al. (2001) found that horses exhibited intoxication at a serum concentration of 3.24 ± 0.74 μg/mL (Meyer et al. 2001). These variations might stem from the diverse criteria used to assess toxicity across different species. In dogs, the toxic indicator is the tonic extension phase, and horses are evaluated based on skeletal muscle fasciculation (Meyer et al. 2001; Wilcke et al. 1983), while for goat kids, the endpoint is the occurrence of convulsions (Venkatachalam et al. 2018).

An additional factor contributing to these observed disparities might be the variation in the rate of drug administration across different species or the fact that young animals display reduced sensitivity to lidocaine toxicity when compared with adult animals, a phenomenon likely attributable to the higher volume of distribution, a characteristic of younger individuals (Morishima et al. 1981).

The subcutaneous administration of 0.5 mL of 1% lidocaine (such as around a superficial nerve) exhibits rapid absorption in goat kids, with an average T_{max} of 0.33 ± 0.11 h [18]. Rapid absorption has also been reported following cornual nerve blocks in goat kids (Venkatachalam et al. 2021). The C_{max} (0.58 ± 0.17 μg mL^{-1}) was around four times less than the concentration (2.55 ± 0.41 μg mL^{-1}) observed at 1 min following intravenous administration of 8 mg kg^{-1} for over 60 s, the maximum dose that did not show any observable toxicity signs. The elimination rate of both lidocaine and its main metabolite in most species, monoethylglycinexylidide (MEGX), is moderate, indicated by mean $t_{1/2\lambda z}$ values of 2.28 h and 3.20 h, respectively. After subcutaneous administration, the mean peak plasma concentration of lidocaine (2.12 ± 0.81 μg/mL) is roughly 6.5 times lower than the mean plasma concentration associated with convulsions (13.59 ± 2.34 μg/mL) (Ingvast-Larsson et al. 2011a, b). Decreasing peak plasma concentrations lowers the likelihood of encountering toxicity (Vnuk et al. 2006). Given that the C_{max} resulting from a 0.5 mL/site injection of 1% lidocaine hydrochloride is significantly below the toxic plasma concentration, this dosage is likely to be safe for cornual nerve blocking in goat kids. No pain-related behavioral signs have been observed during disbudding, indicating that it is also effective. Nevertheless, most of the goat kids begin exhibiting behaviors like head scratching and head shaking around 20 min after nerve blocks, indicating that the anesthetic effect lasts only for about 20 min.

Another major concern of using lidocaine in food-producing animals is its metabolism to dimethylaniline (DMA, 2,6 xylidine). Amide-type local anesthetics like lidocaine undergo hepatic biotransformation through specific cytochrome P450 isoforms (CYP3A4) (Tydén et al. 2004). In the liver, lidocaine is transformed into various metabolites, including MEGX and glycinexylidide (GX), via oxidative N-dealkylation and DMA via hydrolysis (Bill et al. 2004). MEGX further undergoes biotransformation into DMA, although MEGX does not appear to be produced in significant quantities in adult cattle (Hoogenboom et al. 2015). DMA is oxidized in the liver into N-(2,6-dimethylphenyl) hydroxylamine (DMHA) and 4-amino-3,5-dimethylphenol (DMAP). DMHA then goes through phase II biotransformation, specifically acetylation, forming reactive esters that ultimately convert into a reactive nitrenium ion through phase 2 metabolism (Tydén et al. 2004). Both DMHA and the nitrenium ion can bind covalently to DNA, potentially leading to the development of tumors. DMHA can also react with hemoglobin, resulting in the formation of hemoglobin adducts through covalent binding with cysteine residues (Tydén et al. 2004). The other metabolite, DMAP, undergoes oxidation into an iminoquinone, a highly reactive electrophile with genotoxic properties (Gan et al. 2001). DMAP can also be generated from the nitrenium ion or DMHA. DMA is also rapidly converted into 4-hydroxyDMA, which accounts for most urinary excretions in adult cattle (Hoogenboom et al. 2015). DMA, based on toxicology studies in rats, has been categorized as a possible carcinogen (Group B) by the

International Agency for Research on Cancer (Anonymous 1995). The study revealed that the chronic oral administration of DMA at a dosage of 3000 mg/kg resulted in the development of various cancers in rats, including nasal papilloma and carcinoma, rhabdomyosarcoma, subcutaneous fibromas and fibrosarcomas (Program 1990). Furthermore, the mortality rates were higher in animals that received 1000 and 3000 mg/kg of DMA compared with the control group. These findings led to the conclusion that DMA acts as a carcinogen in rats. In dogs, chronic oral administration of DMA causes weight reduction, hyperbilirubinemia, hypoproteinemia, and significant fatty degeneration. In vitro studies have further confirmed the genotoxic and mutagenic properties of DMA (Tydén et al. 2004). Several human studies have also indicated a potential link between DMA and an increased risk of bladder cancer (Duan et al. 2008; Tao et al. 2013). Collectively, these reports strongly suggest that DMA may indeed be a carcinogen (at high doses), making the use of lidocaine in animals undesirable.

5.4 Articaine

Articaine is also an amino-amide class of local anesthetic and has a distinct molecular structure with a thiophene ring instead of a benzene ring, as well as an ester group. The presence of the thiophene ring enhances its lipid solubility, enabling articaine to penetrate nerve fibers more efficiently and rapidly (Yapp et al. 2011). Moreover, its unique molecular composition allows for improved penetration through both bone and soft tissues compared with other local anesthetics (Oertel et al. 1997; Skjevik et al. 2011). The ester group present in articaine makes it noteworthy among amino-amide local anesthetics, as it undergoes rapid hydrolysis via esterases in tissues and plasma (Oertel et al. 1997; Skjevik et al. 2011; Yapp et al. 2011) into inactive articainic acid. Its structure means that DMA is not formed during metabolism. Therefore, articaine hydrochloride could be a safer and better option than lidocaine in goat kids.

An investigation involving goat kids was conducted to assess the toxicity and pharmacokinetics of articaine (Venkatachalam et al. 2021). This study involved the intravenous administration of articaine hydrochloride IV over a 60 s interval, and cornual nerve blocks were administered using 1.5% articaine hydrochloride at a dosage of 0.5 mL per site. The goat kids exhibited no indications of toxicity associated with nerve blocks throughout these procedures. The average IV dosage required to produce convulsions was 16.24 ± 1.79 mg kg^{-1}, and the average plasma concentrations of articaine and articainic acid at convulsions were 9.90 ± 2.38 μg mL^{-1} and 1.52 ± 0.91 μg mL^{-1}, respectively.

A cornual nerve block (0.5 mL/site) using 1.5% articaine hydrochloride took about 4 min to anesthetize the horn buds based on the absence of a withdrawal response and vocalization during disbudding. This confirms the effective analgesic properties of articaine. However, post-procedure observations of the animals indicated the emergence of pain-related behaviors, including head scratching and head shaking, approximately 25 min after administration (Venkatachalam et al. 2021).

The absorption of articaine after cornual nerve block is rapid with a mean C_{max} of 586.58 ± 175.10 ng mL^{-1} at 0.22 ± 0.09 h (T_{max}) (Venkatachalam et al. 2021). The rapid absorption of articaine has also been reported in red deer (Venkatachalam et al. 2017) and humans (Snoeck 2012). This could be because of vasodilatation, similar to most other local anesthetics (Becker and Reed 2012). The short elimination half-lives ($t_{½λz}$) of articaine following intravenous administration (0.66 ± 0.14 h) and subcutaneous administration (1.26 ± 0.34 h) indicate that articaine is rapidly eliminated following systemic absorption (Venkatachalam et al. 2021). Rapid elimination has also been reported in deer (Venkatachalam et al. 2017) and people (Snoeck 2012). The rapid elimination of articaine may be due to rapid hydrolysis via plasma esterases into articainic acid (Oertel et al. 1997). Articainic acid is an inactive metabolite, whereas several primary metabolites of lidocaine are active and can increase the risk of toxicity during accidental intravenous administration or overdosage (Strong et al. 1975; Yapp et al. 2011). The plasma clearance of articaine was rapid in goat kids with a mean CLss of 5.33 ± 0.66 L kg^{-1} (Venkatachalam et al. 2021).

There have been no documented cases of toxicity associated with articaine in humans, except, rarely, paresthesia. Paresthesia is characterized by persistent anesthesia or altered sensations in the form of neuropathy [43]. The underlying causes of paresthesia are still not fully understood, although it is believed to be related to the concentration of local anesthetics used. It has been observed that paresthesia occurs more frequently following the administration of 4% local anesthetic formulations, such as articaine and prilocaine, compared with the use of 2% local anesthetic formulations (Piccinni et al. 2015).

While articaine is generally regarded as a safe local anesthetic, it can potentially cause systemic toxicity (including central nervous system and cardiovascular toxicity) in the same way as other local anesthetics when toxic concentrations are achieved through inadvertent intravenous administration or overdose. However, the risk of systemic toxicity from overdosing is relatively low compared with lidocaine and other amide-type local anesthetics, as articaine undergoes rapid hydrolysis following systemic absorption (Yapp et al. 2011). The clearance of articaine was found to be 10 times greater than that of lidocaine in humans (Simon et al. 1998). Rapid hydrolysis into an inactive metabolite and rapid elimination indicate that articaine may be safer than lidocaine for cornual nerve blocks in goat kids. Because of its wider margin of safety, articaine is clinically used as a 4% solution, whereas lidocaine is used as a 2% solution (Oertel et al. 1997).

5.5 Bupivacaine

Both lidocaine and articaine are effective for less than 30 min. The most commonly used local anesthetic in people, bupivacaine, has a much longer duration in most species. It was used for nerve blocks in adult goats in (Adami et al. 2011) for the surgical examination of the stifle joint conducted under general anesthesia and sciatic–femoral nerve blocks. In that study, 0.5% bupivacaine provided effective analgesia with minimal adverse effects. A group receiving a higher dose of bupivacaine exhibited unilateral motor blockade. There are no reports of its use in goat kids. Extended-release formulations of bupivacaine have been used for disbudding in calves (Venkatachalam et al. 2022). However, long-acting local anesthetics such as bupivacaine may not be a good choice for goat kids, as bupivacaine possesses a greater risk of cardiac toxicity than articaine and lidocaine. This is an area that needs further research.

5.6 Postoperative Analgesia

Since many of the currently available local anesthetics for farm animals are short-acting, additional pain relief, such as systemic NSAIDs, may be required to control post-disbudding pain. Meloxicam has been reported to reduce post-disbudding pain in goat kids (Hempstead et al. 2018c, 2018d, 2018e, 2018f). The subcutaneous administration of meloxicam 1 h prior has been shown to minimize the expression of inflammatory cytokines in calves (Kongara et al. 2020). The administration of both local anesthesia and systemic NSAIDs appears to have the potential to provide better analgesia (compared with their single use) for disbudding pain. However, this did not provide complete postoperative pain relief in goat kids (Hempstead et al. 2018c, 2018d, 2018e, 2018f; Knauer et al. 2023). A combination of meloxicam, lidocaine, and xylazine demonstrated greater effectiveness in the initial hour compared with using only lidocaine (Knauer et al. 2023). Therefore, a multimodal analgesic approach is likely to prove more beneficial than the use of a single class of drugs.

6 Alternative Methods of Disbudding

Various alternative methods to thermal cautery disbudding have been explored in goat kids (Hempstead et al. 2018a). These include cryosurgical and chemical disbudding techniques. Cryosurgery involves the use of liquid nitrogen, while the chemical method employs caustic paste (usually sodium, calcium, or potassium hydroxide paste) to destroy horn buds. However, both cryosurgical and caustic paste methods have been shown to induce more pain than cautery disbudding, as indicated by physiological and behavioral changes (Hempstead et al. 2018a). Caustic paste can also pose the risk of damaging the eyes of goat kids and the udders of does.

There is a need for a simple, safe, cost-effective technique to prevent the growth of horn buds in goat kids. An approach involving clove oil injection into the buds has been explored for horn bud destruction both in goat kids (Molaei et al. 2015) and in calves (Molaei et al. 2014), but the degree of distress experienced during or after the procedure has not been extensively reported. Histopathological changes in the horn buds injected with clove oil have revealed coagulative necrosis of the epidermis and the infiltration of neutrophils. In a pilot study involving 12 calves injected with clove oil and isoeugenol at different volumes, it was found that the injection volume plays a crucial role in successful disbudding in calves (Schoiswohl et al. 2020). Although clove oil or isoeugenol injections cause less tissue damage compared with hot iron disbudding, the success rate is lower, as scur formation was observed 6 months post-procedure (Hempstead et al. 2018b; Schoiswohl et al. 2021). Therefore, clove oil injection may not be entirely effective in preventing horn bud growth, as evidenced by the emergence of scurs in a substantial number of goat kids.

Cloves have been used as an anesthetic for fish (Sladky et al. 2001) and a product (Aqui-S, Aqui-S New Zealand Ltd., Lower Hutt, New Zealand) is currently registered to enable handling. It has traditionally been used as a local anesthetic agent in human dentistry; however, at higher doses, it exhibits cytotoxic effects (Markowitz et al. 1992). As cloves and their oil have been used since antiquity as food flavorings, there are fewer concerns about residues in food-producing animals. Consequently, these characteristics render them a viable alternative for disbudding.

Concerns regarding adverse effects in goat kids, including swelling around the horn buds, have been raised because of hypersensitivity reactions caused by eugenol, the main component of clove oil (Hempstead et al. 2018a). The pain experienced during the administration of clove oil for disbudding in calves is significantly less than that of the cautery disbudding method. However, calves injected with clove oil still exhibited signs of discomfort (Sutherland et al. 2018).

A similar study with calves subjected to clove oil injection showed that, while it might delay horn bud growth, complete elimination is not always achieved. Over a 16-month observation period, calves developed horns or scurs, suggesting that clove oil injection might not offer full prevention of horn bud growth (Sutherland et al. 2019). These findings emphasize the complexities and limitations associated with alternative disbudding methods in terms of both effectiveness and potential adverse reactions.

While clove oil appears promising as an alternative to cautery disbudding, it falls short of the effectiveness demonstrated by traditional methods. Further efforts should be directed toward refining the formulation of clove oil or eugenol to enhance its distribution in the horn bud. Research is essential to understanding the pharmacokinetics and systemic absorption of eugenol/clove oil and determining if a slow-release formulation is necessary. Such advancements could potentially reduce the volume used and mitigate associated side effects.

7 Conclusions

The ideal analgesia for goat kid disbudding is probably a combination of a general anesthetic with postoperative analgesia, but this is not practical or economical on a farm. The current best practice for disbudding on the farm suggests sedation and analgesia with an alpha-2 agonist, the placement of a two-point nerve block, and then an NSAID for postoperative pain. Even this is unlikely to be completely effective in all cases, and more research is needed in this area.

Local anesthetics are an essential part of the analgesic protocol, but goat kids are susceptible to overdose. Articaine is a safe and effective local anesthetic for cornual nerve block in goat kids with some potential advantages over lidocaine. However, comprehensive future studies, involving different doses and concentrations of articaine hydrochloride within a larger population, are essential to definitively establish its safety and efficacy for disbudding. Articaine is not registered for use in livestock and lacks defined minimum residual limits (MRLs). This absence of MRLs is a significant obstacle hindering its use in livestock. Residue studies to allow an MRL to be set for articaine would require substantial financial investment and a collaborative effort from the animal industry.

In addition to local anesthetic nerve blockades, sedation with an alpha-2 agonist will minimize stress and pain during local anesthetic injections. As recovery may be prolonged in neonatal animals, an antagonist should be available. Postoperative pain management using NSAIDs is essential, although long-acting local anesthetics may be available in the future. Evaluating the safety and efficacy of this protocol and its administration via alternative routes, such as transdermal or oral, for disbudding in goat kids should be a focus of future studies.

Looking ahead, further research into drugs that will prevent the growth of horn buds safely and effectively is needed. Addressing the issue of pain associated with disbudding in goat kids necessitates a collaborative effort across multiple disciplines, including clinical veterinarians, pharmacologists, chemists, and animal welfare scientists. By pooling expertise from these diverse fields, a more comprehensive and effective solution may be developed to enhance the well-being of goat kids during the disbudding process. This multidisciplinary approach is crucial for the successful resolution of the problem and the development of humane practices in goat farming.

References

Adami C, Bergadano A, Bruckmaier RM, Stoffel MH, Doherr MG, Spadavecchia C (2011) Sciatic-femoral nerve block with bupivacaine in goats undergoing elective stifle arthrotomy. Vet J 188:53–57. https://doi.org/10.1016/j.tvjl.2010.02.008

Ajuda I, Battini M, Mattiello S, Arcuri C, Stilwell G (2020) Evaluation of pain mitigation strategies in goat kids after cautery disbudding. Animals 10:277

Alvarez L, Gutierrez J (2010) A first description of the physiological and behavioural responses to disbudding in goat kids. Anim Welf 19:55–59

Alvarez L, Nava RA, Ramirez A, Ramirez E, Gutierrez J (2009) Physiological and behavioural alterations in disbudded goat kids with and without local anaesthesia. Appl Anim Behav Sci 117:190–196

Alvarez L, De Luna JB, Gamboa D, Reyes M, Sánchez A, Terrazas A, Rojas S, Galindo F (2015) Cortisol andpain-related behavior in disbudded goat kids with and without cornual nerve block. Physiol Behav, 138:58–61. https://doi.org/10.1016/j.physbeh.2014.10.026

Alvarez L, Adcock SJJ, Tucker CB (2019) Sensitivity and wound healing after hot-iron disbudding in goat kids. J Dairy Sci 102:10152–10162

Amills M, Ramírez O, Tomàs A, Badaoui B, Marm J et al (2009) Mitochondrial DNA diversity and origins of South and Central American goats. Anim Genet 40:315–322

Amills M, Capote J, Tosser-Klopp G (2017) Goat domestication and breeding: a jigsaw of historical, biological and molecular data with missing pieces. Anim Genet 48(6):631–644

Andersson H, Lillpers K, Rydhmer L, Forsberg M (2000) Influence of light environment and photoperiod on plasma melatonin and cortisol profiles in young domestic boars, comparing two commercial melatonin assays. Domest Anim Endocrinol 19:261–274

Anonymous (1995) Occupational exposures of hairdressers and barbers and personal use of hair colourants; some hair dyes, cosmetic colourants, industrial dyestuffs and aromatic amines. IARC Monogr Eval Carcinog Risks Hum 57:43–118. https://doi.org/10.1016/0003-2670(95)90241-4

Anzuino K, Bell N, Bazeley K, Nicol C (2010) Assessment of welfare on 24 commercial UK dairy goat farms based on direct observations. Vet Rec 167(20):774–780. https://doi.org/10.1136/vr.c5892

Battini M, Barbieri S, Vieira A, Stilwell G, Mattiello S (2016) Results of testing the prototype of the AWIN welfareassessment protocol for dairy goats in 30 intensive farms in Northern Italy. Ital J Anim Sci 15:283–293

Becker DE, Reed KL (2012) Local anesthetics: review of pharmacological considerations. Anesth Prog 59:90–101. https://doi.org/10.2344/0003-3006-59.2.90. quiz 102–3

Bergamasco L, Macchi E, Facello C, Badino P, Odore R, Re G, Osella MC (2006) Electroencephalographic power spectral analysis of growing goat kids (Capra hircus). Small Rumin Res 66:265–272

Bill TJ, Clayman MA, Morgan RF, Gampper TJ (2004) Lidocaine metabolism pathophysiology, drug interactions, and surgical implications. Aesthet Surg J 24:307–311. https://doi.org/10.1016/j.asj.2004.05.001

Bufalari A, Adami C, Angeli G, Short CE (2007) Pain assessment in animals. Vet Res Commun 31(suppl. S1)55–58

Buttle H, Mowlem A, Mews A (1986) Disbudding and dehorning of goats. In Pract 8:63. https://doi.org/10.1136/inpract.8.2.63

Can E, Vieira A, Battini M, Mattiello S, Stilwell G (2016) On-farm welfare assessment of dairy goat farms using animal-based indicators: the example of 30 commercial farms in Portugal. Acta Agric Scand Sect A Anim Sci 66:43–55. https://doi.org/10.1080/09064702.2016.1208267

Celly CS, McDonell WN, Young SS, Black WD (1997) The comparative hypoxaemic effect of four alpha 2 adrenoceptor agonists (xylazine, romifidine, detomidine and medetomidine) in sheep. J Vet Pharmacol Ther 20:464–471

Celly CS, Atwal OS, McDonell WN, Black WD (1999) Histopathologic alterations induced in the lungs of sheep by use of alpha2-adrenergic receptor agonists. Am J Vet Res 60(154):161

Colborn DR, Thompson DL, Roth TL, Capehart JS, White KL (1991) Responses of cortisol and prolcatin tosexual excitement and stress in stallions and geldings. J Anim Sci 69:2556–2562

Colli L, Milanesi M, Talenti A et al (2018) Genome-wide SNP profiling of worldwide goat populations reveals strong partitioning of diversity and highlights post-domestication migration routes. Genet Sel Evol 50:58

Duan J-D, Jeffrey AM, Williams GM (2008) Assessment of the medicines lidocaine, prilocaine, and their metabolites, 2,6-dimethylaniline and 2-methylaniline, for DNA adduct formation in rat tissues. Drug Metab Dispos 36:1470–1475. https://doi.org/10.1124/dmd.107.019950

Fitzpatrick J, Scott M, Nolan A (2006) Assessment of pain and welfare in sheep. Small Rumin Res 62:55–61

Eriksson L, Teravainen TL (1989) Circadian rhythm of plasma cortisol and blood glucose in goats. Aust J Appl Sci 2:202–203

Frahm S, Di Giminiani P, Stanitznig A, Schoiswohl J, Krametter-Frötscher R, Wittek T, Waiblinger S (2020) Nociceptive threshold of calves and goat kids undergoing injection of clove oil or isoeugenol for disbudding. Animals 10(7):1228

Gan J, Skipper PL, Tannenbaum SR (2001) Oxidation of 2,6-dimethylaniline by recombinant human cytochrome P450s and human liver microsomes. Chem Res Toxicol 14:672–677. https://doi.org/10.1021/tx000181i

Geist V (1984) In: Macdonald D (ed) The encyclopedia of mammals. Facts on File, New York, pp 584–587. ISBN 0-87196-871-1

Ghodasara P, Satake N, Sadowski P, Kopp S, Mills PC (2022) Investigation of cattle plasma proteome inresponse to pain and inflammation using next generation proteomics technique, SWATH-MS. Mol Omics 18:133–142

Gibson TJ, Johnson CB, Murrell JC, Hulls CM, Mitchinson SL, Stafford KJ, Johnstone AC, Mellor DJ (2009) Electroencephalographic responses of halothane-anaesthetised calves to slaughter by ventral-neck incision without prior stunning. N Z Vet J 57:77–83

Greenwood PL, Shutt DA (1990) Effects of management practices on cortisol, b-endorphin and behavior in young goats. Aust Soc Anim Prod Adelaide Aust 18:224–227

Guesgen MJ, Beausoleil NJ, Leach M, Minot EO, Stewart M, Stafford KJ (2016) Coding and quantification of a facial expression for pain in lambs. Behav Process 132:49–56

Häger C et al (2017) The Sheep Grimace Scale as an indicator of post-operative distress and pain in laboratory sheep. PLoS one 12:0175839

Hartnack AK, Jordan ME, Roussel AJ (2018) Complications associated with surgical dehorning in goats: a retrospective study of 239 cases. Vet Surg 47(2):188–192. https://doi.org/10.1111/vsu.12743. Epub 2017 Dec 14

Harwood D (2012) Disbudding goats. Vet Rec 170:343. https://doi.org/10.1136/vr.e2367

Hempstead MN, Waas JR, Stewart M, Cave VM, Sutherland MA (2017) Behavioural response of dairy goat kids to cautery disbudding. Appl Anim Behav Sci 194:42–47

Hempstead MN, Waas JR, Stewart M, Cave VM, Sutherland MA (2018a) Evaluation of alternatives to cautery disbudding of dairy goat kids using physiological measures of immediate and longer-term pain. J Dairy Sci 101:5374–5387. https://doi.org/10.3168/jds.2017-13814

Hempstead M, Waas JR, Stewart M, Cave VM, Sutherland MA (2018b) Evaluation of alternatives to cautery disbudding of dairy goat kids using behavioural measures of post-treatment pain. Appl Anim Behav Sci 206:32–38. https://doi.org/10.1016/j.applanim.2018.06.009

Hempstead MN, Waas JR, Stewart M, Zobel G, Cave VM, Julian AF, Sutherland MA (2018c) Pain sensitivity and injury associated with three methods of disbudding goat kids: cautery, cryosurgical and caustic paste. Vet J 239:42–47

Hempstead MN, Waas JR, Stewart M, Cave VM, Sutherland MA (2018d) Evaluation of alternatives to cautery disbudding of dairy goat kids using physiological measures of immediate and longer-term pain. J Dairy Sci 101:5374–5387

Hempstead MN, Waas JR, Stewart M, Dowling SK, Cave VM, Lowe GL, Sutherland MA (2018e) Effect of isoflurane alone or in combination with meloxicam on the behavior and physiology of goat kids following cautery disbudding. J Dairy Sci 101:3193–3204. https://doi.org/10.3168/jds.2017-13507

Hempstead MN, Waas JR, Stewart M, Cave VM, Turner AR, Sutherland MA (2018f) The effectiveness of clove oil and two different cautery disbudding methods on preventing horn growth in dairy goat kids. PLoS One 13:e0198229. https://doi.org/10.1371/journal.pone.0198229

Hempstead M, Waas J, Stewart M, Sutherland M (2020) Goat kids are not small calves: species comparisons in relation to disbudding. Anim Welf 29:293–312. https://doi.org/10.7120/09627286.29.3.293

Hempstead MN, Lindquist TM, Shearer JK, Shearer LC, Cave VM, Plummer PJ (2021a) Welfare assessment of 30 dairy goat farms in the Midwestern United States. Front Vet Sci 8:646715. https://doi.org/10.3389/fvets.2021.646715

Hempstead MN, Shearer JK, Sutherland MA, Fowler JL, Smith JS, Smith JD, Lindquist TM, Plummer PJ (2021b) Cautery disbudding iron application time and brain injury in goat kids: a pilot study. Front Vet Sci 7:568750. https://doi.org/10.3389/fvets.2020.568750

Holton LL, Scott EM, Nolan AM, Reid J, Welsh E (1998) Relationship between physiological factors and clinical pain in dogs scored using a numerical rating scale. J Small Anim Pract 39:469–474

Hoogenboom RLAP, Zuidema T, Essers M, van Vuuren AM, van Wikselaar PG, van Eijkeren JCH, Mengelers MJB, Zeilmaker MJ, Bulder AS (2015) Concentrations of dimethylaniline and other metabolites in milk and tissues of dairy cows treated with lidocaine. Food Addit Contam Part A 32:1256–1264. https://doi.org/10.1080/19440049.2015.1064172

Huozha R, Rastogi SK, Korde JP, Madan AK (2011) Electroencephalographic changes during experimental pain induction in goats. Vet arhiv 81(3):359–368

Ingvast-Larsson C, Hogberg M, Mengistu U, Olsen L, Bondesson U, Olsson K (2011a) Pharmacokinetics of meloxicam in adult goats and its analgesic effect in disbudded kids. J Vet Pharmacol Ther 34:64–69

Ingvast-Larsson C, Högberg M, Mengistu U, Olsén L, Bondesson U, Olsson K (2011b) Pharmacokinetics of meloxicam in adult goats and its analgesic effect in disbudded kids. J Vet Pharmacol Ther 34:64–69. https://doi.org/10.1111/j.1365-2885.2010.01195.x

Jager LP, De Graaf GJ, Widjaja-Greefkes H (1998) Effects of atipamezole, detomidine and medetomidine onrelease of steroid hormones by porcine adrenocortical cells in vitro. Eur J Pharmacol 346:71–76. https://doi.org/10.1016/S0014-2999(98)00044-2

Johnson CB (2016) Research tools for the measurement of pain and nociception. Animals 6:71. https://doi.org/10.3390/ani6110071

Johnson CB, Woodbury WM, Caulkett N, Wilson P (2005a) Comparison of lidocaine and antler pedicle compression for analgesia during antler removal in red deer (Cervus elaphus) anaesthetised by halothane in oxygen: EEG effects. J Vet Anaesth Analg 32:16–71

Johnson CB, Stafford KJ, Sylvester SP, Ward RN, Mitchinson S, Mellor DJ (2005b) Effects of age on the electroencephalographic response to castration in lambs anaesthetised using halothane in oxygen. N Z Vet J 53:433–437

Knauer WA, Barrell EA, Guedes AGP, Ventura BA (2023) Effects of multimodal pain management strategies on acute physiological and behavioral response to cautery disbudding in neonatal goat kids. J Dairy Sci 106:2830–2845. https://doi.org/10.3168/jds.2022-22433

Kongara K, Dukkipati VSR, Tai HM, Heiser A, Murray A, Webster J, Johnson CB (2020) Differential transcription of selected cytokine and neuroactive ligand-receptor genes in peripheral leukocytes from calves in response to cautery disbudding. Animals 10:1187. https://doi.org/10.3390/ani10071187

Kongara K, Singh P, Venkatachalam D, Chambers JP (2023) Pain assessment in goat kids: focus on disbudding. Animals 13(24):3814. https://doi.org/10.3390/ani13243814

Langford DJ, Bailey AL, Chanda ML, Clarke SE, Drummond TE (2010) Coding of facial expressions of pain in the laboratory mouse. Nat Methods 7:447–449

Larrondo C, Leiva J, de la Cruz-Cruz L (2021) Dairy goat welfare in semi-intensive production systems and drought conditions. Anim Welf 30:371

Lemke KA, Dawson SD (2000) Local and regional anesthesia. Vet Clin North Am Small Anim Pract 30:839–857. https://doi.org/10.1016/s0195-5616(08)70010-x

Leslie E, Hernández-Jover M, Newman R, Holyoake P (2010) Assessment of acute pain experienced by piglets from ear tagging, ear notching and intraperitoneal injectable transponders. Appl Anim Behav Sci 127:86–95

Loretz C, Wechsler B, Hauser R, Rüsch P (2004) A comparison of space requirements of horned and hornless goats at the feed barrier and in the lying area. Appl Anim Behav Sci 87:275–283

Lou ME (2020) The development of a goat kid grimace scale following thermal disbudding. Thesis chapter; MSc

Markowitz K, Moynihan M, Liu M, Kim S (1992) Biologic properties of eugenol and zinc oxide-eugenol a clinically oriented review. Oral Surg Oral Med Oral Pathol 73:729–737. https://doi.org/10.1016/0030-4220(92)90020-q

Matthews J, Dustan B (2019) Disbudding of goat kids. In Pract 41:433–444. https://doi.org/10.1136/inp.l6028

Mazinani M, Rude B (2020) Population, world production and quality of sheep and goat products. Am J Anim Vet Sci 15(4):291–299

Mellor DJ, Stafford KJ (2000) Acute castration and tailing distress and its alleviation in lambs. N Z Vet J 48:33–43

Meyer GA, Lin HC, Hanson RR, Hayes TL (2001) Effects of intravenous lidocaine overdose on cardiac electrical activity and blood pressure in the horse. Equine Vet J 33:434–437. https://doi.org/10.2746/042516401776254871

Mirra A, Spadavecchia C, Bruckmaier R, Gutzwiller A, Casoni D (2018) Acute pain and peripheral sensitization following cautery disbudding in 1- and 4-week-old calves. Physiol Behav 184:248–260

Molaei MM, Azari O, Esmaeilzadeh S (2014) Study of calves disbudding following injection of clove oil under horn bud. J Vet Res 4:363–369

Molaei MM, Mostafavi A, Kheirandish R, Azari O, Shaddel M (2015) Study of disbudding goat kids following injection of clove oil essence in horn bud region. Vet Res Forum 6(17):22

Molony V, Kent JE (1997) Assessment of acute pain in farm animals using behavioral and physiological measurements. J Anim Sci 75(1997):266–272

Morishima HO, Pedersen H, Finster M, Sakuma K, Bruce SL, Gutsche BB, Stark RI, Covino BG (1981) Toxicity of lidocaine in adult, newborn, and fetal sheep. Anesthesiology 55:57–61. https://doi.org/10.1097/00000542-198107000-00012

Morton DB, Griffiths PH (1985) Guidelines on the recognition of pain, distress and discomfort in experimental animals and hypothesis for assessment. Vet Rec 116:431–436

MPI (2018) Painful husbandry procedures. Animal Welfare Act

Muri K, Valle P (2012) Human-animal relationships in the Norwegian dairy goat industry: assessment of pain and provision of veterinary treatment (Part II). Anim Welf 21:547–558. https://doi.org/10.7120/09627286.21.4.547

Murrell J, Johnson C (2006) Neurophysiological techniques to assess pain in animals. J Vet Pharmacol Ther 295:325–335

Naderi S, Rezaei HR, Taberlet P, Zundel S, Rafat SA et al (2007) Large-scale mitochondrial DNA analysis of the domestic goat reveals six haplogroups with high diversity. PLoS One 2:e1012

Nfor ON, Chan JPW, Kere M, Peh HC (2016) Disbudding pain: the benefits of disbudding goat kids with dexmedetomidine hydrochloride. Small Rumin Res 139:60–66. https://doi.org/10.1016/j.smallrumres.2016.05.005

Nomura K, Yonezawa T, Mano S, Kawakami S, Shedlock AM, Hasegawa M et al (2013) Domestication process of the goat revealed by an analysis of the nearly complete mitochondrial protein-encoding genes. PLoS One 8(8):e67775. https://doi.org/10.1371/journal.pone.0067775

Oertel R, Rahn R, Kirch W (1997) Clinical pharmacokinetics of articaine. Clin Pharmacokinet 33:417–425. https://doi.org/10.2165/00003088-199733060-00002

Pang WY, Earley B, Murray M, Sweeney T, Gath V, Crowe MA (2011) Banding or Burdizzo castration and carprofen administration on peripheral leukocyte inflammatory cytokine transcripts. Res Vet Sci 90:127–132

Peers A, Mellor DJ, Wintour EM, Dodic M (2002) Blood pressure, heart rate, hormonal and other acute responses to rubber-ring castration and tail docking of lambs. N Z Vet J 50:56–62

Piccinni C, Gissi DB, Gabusi A, Montebugnoli L, Poluzzi E (2015) Paraesthesia after local anaesthetics: an analysis of reports to the FDA adverse event reporting system. Basic Clin Pharmacol Toxicol 117:52–56. https://doi.org/10.1111/bcpt.12357

Program NT (1990) NTP toxicology and carcinogenesis studies of 2,6-Xylidine (2,6-dimethylaniline) (CAS no. 87-62-7) in Charles River CD rats (feed studies). Natl Toxicol Program Tech Rep Ser 278:1–138

Prunier A, Mounier AM, Hay M (2005) Effects of castration, tooth resection, or tail docking on plasma metabolites and stress hormones in young pigs. J Anim Sci 83:216–222

Raghazli R, Othman AH, Kaka U, Abubakar AA, Imlan JC, Hamzah H, Sazili AQ, Goh YM (2021) Physiological and electroencephalogram responses in goats subjected to pre-and during slaughter stress. Saudi J Biol Sci 28:6396–6407

Sabow, A.B.; Goh, Y.M.; Zulkifli, I.; Sazili, A.Q.; Kadir, M.Z.A.A.; Kaka, U.; Khadijah, N.; Adeyemi, K.D. (2017) Ebrahimi, M. Electroencephalographic responses to neck cut and exsanguination in minimally anaesthetized goats. S Afr J Anim Sci 47, 34–40

Sanford J, Ewbank R, Melony V,Tavenor WD, Uvarov O (1986) Guidelines for the recognition and assessment ofpain in animals. Vet Rec 118:334–338

Sanford SE (1989) Ontario. Meningoencephalitis caused by thermal disbudding in goat kids. Can Vet J 30:832

Savage RJG, Long MR (1986) Mammal evolution: an illustrated guide. New York: Facts on file. pp 232–235. ISBN978-0-8160-1194-0

Schoiswohl J, Stanitznig A, Waiblinger S, Frahm S, Krametter-Froetscher R, Wittek T (2020) Suppression of horn growth in cattle by clove oil and isoeugenol. J Vet Behav 36:1–3. https://doi.org/10.1016/j.jveb.2020.01.003

Schoiswohl J, Stanitznig A, Sigmund M, Kneissl S, Thaller D, Frahm S, Waiblinger S, Palme R, Tichy A, Wittek T, Krametter-Froetscher R (2021) Comparison of alternative disbudding methods with hot-iron dehorning of goat kids. J Vet Behav 46:31–39. https://doi.org/10.1016/j.jveb.2021.07.002

Silva Salas MÁ, Mondragón-Ancelmo J, Jiménez Badillo MR, Rodríguez Licea G, Napolitano F (2021) Assessing dairy goat welfare in Mexican intensive or semi-intensive farming conditions. J Dairy Sci 104:6175–6184. https://doi.org/10.3168/jds.2020-19557

Simon MAM, Vree TB, Gielen MJM, Booij LHDJ (1998) Comparison of the effects and disposition kinetics of articaine and lidocaine in 20 patients undergoing intravenous regional anaesthesia during day case surgery. Pharm World Sci 20:88–92. https://doi.org/10.1023/a:1008622018161

Singh PM, Reid K, Gaddam R, Bhatia M, Smith S, Jacob A, Chambers P (2017) Effect of choline chloride premedication on xylazine-induced hypoxaemia in sheep. Vet Anaesth Analg 44(5):1149–1155. https://doi.org/10.1016/j.vaa.2017.01.002

Singh P, Venkatachalam D, Kongara K, Chambers P (2024) Pain mitigation strategies for disbudding in goat kids. Animals 14(4):555. https://doi.org/10.3390/ani14040555

Skjevik ÅA, Haug BE, Lygre H, Teigen K (2011) Intramolecular hydrogen bonding in articaine can be related to superior bone tissue penetration: a molecular dynamics study. Biophys Chem 154:18–25. https://doi.org/10.1016/j.bpc.2010.12.002

Sladky KK, Swanson CR, Stoskopf MK, Loomis MR, Lewbart GA (2001) Comparative efficacy of tricaine methanesulfonate and clove oil for use as anesthetics in red pacu (Piaractus brachypomus). Am J Vet Res 62:337–342. https://doi.org/10.2460/ajvr.2001.62.337

Small A, Fisher A, Lee C and Colditz I (2020) Gap evaluation of pain alleviation research. (Final Report to AWI. wool.com/globalassets/wool/sheep/research-publications/welfare/improved-pain-relief/project-final-report-on-gap-evaluation-of-pain-alleviation.pdf)

Smith MC, Sherman DM (2009) Goat medicine, 2nd edn. Wiley-Blackwell, Ames, pp 723–731

Smith MC, Sherman DM (2023) Goat Medicine. Wiley. https://doi.org/10.1002/9780813818825

Snoeck M (2012) Articaine: a review of its use for local and regional anesthesia. Local Reg Anesth 5:23–33. https://doi.org/10.2147/lra.s16682

Stewart M, Webster JR, Schaefer AL, Cook NJ, Scott SL (2005a) Infrared thermography as a non-invasive tool to study animal welfare. Anim Welf 14:319–325

Stewart M, Webster JR, Verkerk GA, Colyn JJ, Schaefer AL (2005b) Infrared thermography as a non-invasive measure of stress in dairy cows. J Anim Sci 83(Suppl 1):633

Stewart M, Stafford K, Dowling S, Schaefer A, Webster J (2008) Eye temperature and heart rate variability of calves disbudded with or without local anaesthetic. Physiol Behav 93:789–797

Still Brooks KM, Hempstead MN, Anderson JL, Parsons RL, Sutherland MA, Plummer PJ, Millman TS (2021) Characterization of efficacy and animal safety across four caprine disbudding methodologies. Animals 11:430

Strong JM, Mayfield DE, Atkinson AJ, Burris BC, Raymon F, Webster LT (1975) Pharmacological activity, metabolism, and pharmacokinetics of glycinexylidide. Clin Pharmacol Ther 17:184–194. https://doi.org/10.1002/cpt1975172184

Stubsjoen SM, Flo AS, Moe RO, Janczak AM, Skjerve E, Valle PS, Zanella AJ (2009) Exploring non-invasive methods to assess pain in sheep. Physiol Behav 98:640–648

Stubsjoen SM, Bohlin J, Skjerve E, Valle PS, Zanella AJ (2010) Applying fractal analysis to heart rate time series of sheep experiencing pain. Physiol Behav 101:74–80

Sutherland MA, Tucker C (2011) The long and short of it; a review of tail docking in farm animals. Appl Anim Behav Sci 135:179–191

Sutherland MA, Watson TJ, Johnson CB, Millman ST (2016) Evaluation of the efficacy of a non-penetrating captive bolt to euthanase neonatal goats up to 48 hours of age. Anim Welf 25:471–479

Sutherland MA, Larive J, Cave V, Zobel G (2018) Behavioural and physiological responses to clove oil injected under the horn bud of calves. Appl Anim Behav Sci 204:29–36. https://doi.org/10.1016/j.applanim.2018.03.004

Sutherland M, Julian A, Huddart F (2019) Clove oil delays rather than prevents scur/horn growth in dairy cattle. Vet Sci 6:102. https://doi.org/10.3390/vetsci6040102

Tao L, Day BW, Hu B, Xiang Y-B, Wang R, Stern MC, Gago-Dominguez M, Cortessis VK, Conti DV, Berg DVD, Pike MC, Gao Y-T, Yu MC, Yuan J-M (2013) Elevated 4-aminobiphenyl and 2,6-dimethylaniline hemoglobin adducts and increased risk of bladder cancer among lifelong nonsmokers—the Shanghai Bladder Cancer Study. Cancer Epidemiol Biomarkers Prev 22:937–945. https://doi.org/10.1158/1055-9965.epi-12-1447

Taylor PM (1991) Anaesthesia in sheep and goats. In Pract 13:31–36. https://doi.org/10.1136/inpract.13.1.31

Temple D, Manteca X (2020) Animal welfare in extensive production systems is still an area of concern. Front Sustain Food Syst 4:545902. https://doi.org/10.3389/fsufs.2020.545902

Thompson K, Bateman R, Morris P (2005) Cerebral infarction and meningoencephalitis following hot-iron disbudding of goat kids. N Z Vet J 53:368–370. https://doi.org/10.1080/00480169.2005.36578

Tiezzi F, Tomassone L, Mancin G, Cornale P, Tarantola M (2019) The assessment of housing conditions, management, and animal-based measures of dairy goats' welfare and its association with productive and reproductive traits. Animals 9:893. https://doi.org/10.3390/ani9110893

Travain T, Valsecchi P (2021) Infrared thermography in the study of animals' emotional responses: a critical review. Animals 11:2510

Tschoner (2021) Methods for pain assessment in calves and their use for the evaluation of pain during different procedures-a review. Animals 11(5):1235

Tydén E, Tjälve H, Larsson P (2004) Metabolic activation of 2,6-Xylidine in the nasal olfactory mucosa and the mucosa of the upper alimentary and respiratory tracts in rats. Toxicol Sci 81:263–272. https://doi.org/10.1093/toxsci/kfh219

USDA (2020) Sheep and goats. National Agricultural Statistics Service (NASS), Washington, DC

Varga M (2014) Textbook of rabbit medicine (Second Edition). Part II: rabbit med, pp 178–202. https://doi.org/10.1016/b978-0-7020-4979-8.00004-2

Venkatachalam D, Chambers JP, Kongara K, Singh P (2017) Pharmacokinetics of articaine hydrochloride and its metabolite articainic acid after subcutaneous administration in red deer (Cervus elaphus). N Z Vet J 66(1):16. https://doi.org/10.1080/00480169.2017.1391141

Venkatachalam D, Chambers P, Kongara K, Singh P (2018) Toxicity and pharmacokinetic studies of lidocaine and its active metabolite, monoethylglycinexylidide, in goat kids. Animals 8:142. https://doi.org/10.3390/ani8080142

Venkatachalam D, Chambers JP, Kongara K, Ward N, Jacob A, Singh P (2021) Pharmacokinetics, efficacy, and convulsive dose of articaine hydrochloride in goat kids. Vet Anaesth Analg 48:264–271. https://doi.org/10.1016/j.vaa.2019.12.012

Venkatachalam D, Kells N, Chambers P, Jacob A, Ward N, Singh P (2022) Pharmacokinetics and efficacy of a novel long-acting bupivacaine formulation for cornual nerve block in calves. Front Vet Sci 9:1060951. https://doi.org/10.3389/fvets.2022.1060951

Vitums A (1954) Nerve and arterial blood supply to the horns of the goat with reference to the sites of anesthesia for dehorning. J Am Vet Méd Assoc 125:284–286

Vnuk D, Lemo N, Radisic B, Nesek-Adam V, Musulin A, Kos J (2006) Serum lidocaine concentration after epidural administration in dogs. Vet Med 51:432–436. https://doi.org/10.17221/5569-vetmed

Wagmann N, Spadavecchia C, Morath-Huss U, Schüpbach-Regula G, Zanolari P (2018) Evaluation of anaesthesia and analgesia quality during disbudding of goat kids by certified Swiss farmers. BMC Vet Res 14:220. https://doi.org/10.1186/s12917-018-1544-7

Weary DM, Niel L, Flower FC, Fraser D (2006) Identifying and preventing pain in animals. Appl Anim Behav Sci 100:64–76

Wilcke JR, Davis LE, Neff-Davis CA (1983) Determination of lidocaine concentrations producing therapeutic and toxic effects in dogs. J Vet Pharmacol Ther 6:105–111. https://doi.org/10.1111/j.1365-2885.1983.tb00386.x

Winder CB, Leblanc SJ, Haley DB, Lissemore KD, Godkin MA, Duffield TF (2016) Practices for the disbudding and dehorning of dairy calves by veterinarians and dairy producers in Ontario, Canada. J Dairy Sci 99:10161–10173. https://doi.org/10.3168/jds.2016-11270

Yapp KE, Hopcraft MS, Parashos P (2011) Articaine: a review of the literature. Br Dent J 210:323–329. https://doi.org/10.1038/sj.bdj.2011.240

Deer

Rob A. Gregory

Abstract

Deer, of the Family Cervidae, are fascinating animals that have captured the imagination of people since the dawn of time. Found on almost every continent on Earth, they are unique among mammals in that, with one exception (the Chinese water deer), they possess annually regenerating tissue in the form of antlers. Despite their popularity in mythology and modern culture, deer are relatively underutilised as a species, having only recently been subject to any real form of domestication. As a result, there has been little work conducted on their experience of pain and suffering, the majority of which has been performed in relation to the practice of velvet removal.

Keywords

Cervidae and pain · Pain in deer · Post-operative pain in deer · Deer and anaesthetics · General anaesthesia in deer · Local anaesthesia in deer · Non-classical wound healing · Atypical wound healing · Wild capture of deer · Deer farming · Velvet removal

R. A. Gregory (✉)
Deer Industry New Zealand,
Wellington, New Zealand
e-mail: info@deernz.org

1 History of Domestication

The relationship between human beings and deer goes back millennia. Primitive art found at the Asphendou Cave, in Crete, depicting a now extinct species of dwarf deer (*Candiacervus*), has been dated to the Upper Palaeolithic period, around 11,000 years ago (Strasser et al. 2018). Evidence of deer being hunted to supplement early agricultural efforts has been found in Palermo, Sicily, dating back six-thousand years (Martin et al. 2023) and in western Iberia, around 3000 BCE (Almeida et al. 2021); while in Northern Europe, red deer (*Cervus elaphus*) remains have been associated with the Pitted Ware Culture of the mid-Neolithic period, also around 3000 BCE (Price et al. 2021). During the same general period (6000–3000 BCE) in north central Idaho, archaeological records suggest that deer (*Odocoileus* spp.) were the preferred source of animal protein for the locals of the time, with other species being consumed on an opportunistic basis (Chadez and Sappington 2017).

Perhaps the earliest recorded attempt at domesticating deer involved Persian fallow deer (*Dama mesopotamica*) being brought into Cyprus around 10,500 BCE (Vigne 2011). Although this endeavour ultimately failed (Vigne 2011), during the period of the Roman Empire (27 BCE—476 CE), fallow deer (*Dama dama*) were successfully introduced to southern and western Europe, including the British Isles (Sykes et al. 2011).

Whether this type of introduction and establishment of a species into a new environment rep-

R. E. Clutton, G. C. Musk (eds.), *Pain and Suffering in Farmed Animals*,
https://doi.org/10.1007/978-3-032-11407-5_18

resents domestication as we would accept it today, i.e. actively adapting and managing an animal for a specific purpose, while also taking responsibility for meeting its basic needs, is debatable. For this reason, it can be argued that true domestication of deer did not begin until approximately two thousand years ago, with the use of reindeer (*Rangifer tarandus*) for draught transport by nomadic peoples in the Arctic Circle, a practice which continues to this day (Losey et al. 2021).

In terms of farming deer, the world's first commercial deer farm was established in Jilin Province, China, in 1733 (Li 1993). It is possible that another major deer farming country, Korea also started commercially farming deer around the same time. However, this is hard to verify, as much of the available information from the Far East is still largely inaccessible to Western audiences, even with modern computer technology.

More recently, at the start of the twentieth century, Russia began farming maral deer (*Cervus elaphus maral*) for their velvet antler on large, extensive, fenced off blocks of mountainous land (FEDFA 2023). Then, in 1966, in New Zealand, Gordon McNutt became the first person to capture a live deer from a helicopter, which ultimately led to the creation of the largest farmed deer industry in the world today (DINZ 2023).

Since then, commercial deer farming has become established in a number of other countries, including Australia, Canada and throughout Europe. The Federation of European Deer Farmers Associations (FEDFA) boasts 16 member countries, the majority of which date the establishment of their deer farming organisations to the 1970s, 1980s or later (FEDFA 2023).

It is quite possibly because of the long historical association between people and deer and their relatively recent domestication, that they are still generally viewed with a sense of majesty and mystique that many other livestock species are not. Whether this will continue as their domestication progresses remains to be seen. One can only hope that it does.

2 Natural History

Depending on the source, there are somewhere between 40 and 55 species of Cervidae in the world today (Fletcher 2013; Smith-Jones 2022). Their range of habitats is enormous and spans almost every continent, with the exception of Antarctica (Smith-Jones 2022). They can be found in the mountains of Tibet, the snowy wastes of arctic Russia, the temperate woodlands of Europe, the arid regions of the Middle East, the tropical rainforests of South-East Asia and across both North and South America (Feldhamer and McShea 2012).

Equally impressive is the range of diversity that exists between the various species. For example, the northern pudu (*Pudu mephistophiles*) of the South American Andes is tiny, standing around 32 centimetres (12.5 inches) high at the shoulder, with vestigial tusks and single spiked antlers measuring around 6 centimetres (2.25 inches) long (Smith-Jones 2022). In contrast, the moose (*Alces alces*), which is common across most of the far northern hemisphere, can stand as high as 2.15 metres (7 feet, 1 inch) and has huge, palmate antlers that can reach over one metre (39 inches) in length (Fletcher 2013; Smith-Jones 2022). In between these two extremes are deer with branched antlers, such as red deer and white-tailed deer; deer with both antlers and enlarged canine teeth, such as muntjac deer (*Muntiacus* spp.) that are used for slashing opponents during the breeding season; and even a species of deer with no antlers at all, *Hydropotes inermis*, the Chinese water deer (Feldhamer and McShea 2012). Some species, such as red deer, fallow deer and chital (*Axis axis*), are social in nature and congregate in single sex groups outside of the breeding season, while others, such as pudu deer, muntjac deer, brocket deer (*Mazama* spp.) and moose, tend towards a more solitary lifestyle (Smith-Jones 2022). Finally, there is the unique and extremely rare Père David's deer (*Elaphrus davidianus*), which bucks the trend completely by growing its antlers during the winter months and having a most unusual combination of features, including a donkey-like tail, cow-like feet and a camel-like neck (Smith-Jones 2022).

In terms of classification, the Cervidae, or true deer, are mammals that belong to the Order Artiodactyla, the even-toed ungulates, Sub-Order Ruminantia, the ruminants and Infra-Order Pecora, the horned ruminants (Fletcher 2013; Goss 1983; Smith-Jones 2022). Within the Family Cervidae are several sub-classifications. The Cervinae (also known as the Plesiometacarpalia or Old World Deer) contains two groups: the Cervini and the Muntiacini, while the Capreolinae (also known as the Telemetacarpalia or New World Deer) contains three groups: the Odocoileini, the Capreolini and the Alceini (Fletcher 2013). Examples of the members of each of these groups are presented in Table 1.

Being ruminants, deer are herbivores, with a four-chambered stomach that allows them to rapidly ingest large amounts of vegetation and digest it later at their relative leisure (Feldhamer and McShea 2012). Browsing and grazing are the predominant forms of obtaining food and while plant matter forms the main part of their diet, deer have also been observed to eat carrion, eggs and small animals, including crabs (Smith-Jones 2022). In addition to this, chewing and eating cast antlers and bones (Osteophagia) has also been reported in several species, including chital, where it has been linked to an increased demand for calcium during antler growth and mineralisation (Barrette 1985).

In common with other ruminants, deer have developed specialised teeth for dealing with their vegetative diet. Molars and pre-molars are similar in design to increase efficiency and while lower incisors are present, upper ones are notably absent (Goss 1983; Smith-Jones 2022). Similarly, while lower canines are present as modified incisors, upper canines also tend to be absent, with the general exception of some members of the genus *Cervus* and those deer possessing rudimentary antlers, in which case the canines have been retained as tusks (Goss 1983; Smith-Jones 2022). Figure 1 shows the upper dentition of a young male red deer.

Most members of the Cervidae only give birth to a single offspring during each breeding cycle; however, as always, there are exceptions to this rule. Chinese water deer typically produce three or more offspring, while mule deer (*Odocoileus hemionus*), white-tailed deer, roe deer and moose tend to produce twins (Smith-Jones 2022). Among those deer that live in highly seasonal climates, birth is generally timed to coincide with spring, the period of greatest food availability and takes place in a secluded, well-covered location to protect the newly born deer from predation.

However, far more fascinating than the number of offspring that are produced or when or where they are born, is the behaviour that precedes it, mating.

The 'rut' or 'roar' is the name usually applied to the mating season of deer; the latter term relating to the noise made by sexually active males to announce their presence and deter potential competitors. While rutting behaviour is most strongly associated with deer in seasonal environments, even those that live in places with little or no seasonal variation, e.g. chital, which are native to the Indian sub-continent, will still show rutting

Table 1 Example members of each of the five sub-groups of the cervidae

Cervinae (Old World Deer)		Capreolinae (New World Deer)		
Cervini	Muntiacini	Odocoileini	Capreolini	Alceini
Red deer (*Cervus elaphus*)	Red muntjac (*Muntiacus muntjak*)	Whitetail deer (*Odocoileus virginianus*)	European roe deer (*Capreolus capreolus*)	Moose (*Alces alces*)
Fallow deer (*Dama dama*)	Tufted deer (*Elaphodus cephalophus*)	Northern Humeul (*Hippocamelus antisensis*)		
Sambar (*Rusa unicolour*)		Reindeer (*Rangifer tarandus*)		

Fig. 1 Young male red deer skull, showing the upper molars and pre-molars. Only five teeth are present on each side, indicating that this is not a mature specimen. Note the round hole at the bottom left of the image, which would normally contain a short upper canine tooth. (Photo credit: R. A. Gregory)

behaviour around a female in oestrus (Smith-Jones 2022).

No discussion about rutting behaviour would be complete without mentioning antlers. Antlers are secondary sexual characteristics, analogous to horns in other species, that develop from the pedicles; elongations of the frontal bone of the skull (Asher 1988; Fennessy 1981). They are unique to the Cervidae, both in terms of their composition and growth (Feldhamer and McShea 2012). Unlike horns, which are made of keratin and grow from the base, antlers are made of bone and grow from the tip (Goss 1983; Smith-Jones 2022). Furthermore, antlers are deciduous and a new pair is grown and shed for each breeding cycle. This means that they are the only mammalian tissue that is capable of fully regenerating (Goss 1983).

In seasonal climates, as day length increases, testosterone levels in male deer decline, which triggers the beginning of the antler growth cycle (Smith-Jones 2022; Suttie et al. 1994). At the cellular level, mesenchymal cells in what will be the antler tip rapidly proliferate, divide, become cartilaginous and subsequently ossify (Kierdorf and Kierdorf 2002; Landete-Castillejos et al. 2019; Price et al. 2005). At the same time, the pedicle skin becomes compressed, then stretches, before new skin is grown to compensate, which subsequently becomes velvet antler, characterised by its soft covering of hair, loss of sweat glands and arrector pili muscles and the addition of large sebaceous glands (Li and Suttie 2000).

While testosterone plays an important part in the timing of velvet antler growth, other compounds such as insulin-like growth factor (IGF 1) play a more important role in actual growth (Suttie and Fennessy 1991, 1992), the rate of which is phenomenal. Velvet antler grows at up to two centimetres (almost one inch) per day, which makes it the fastest-growing mammalian tissue on Earth (Goss 1983). It not only contains cartilage, but is also well supplied with blood vessels and nerves, both of which have to develop at a similar rate, which is equally astounding (Suttie 1990). The blood supply is somewhat unusual in that the arteries are almost entirely peripheral, with venous drainage occurring through the centre of the antler (Adams 1979). As for the nerve supply, this is extensive and while not believed to be necessary for growth per se, at least in first antler development (Li et al. 1993), it is believed to help the deer protect the growing antler from damage (Goss 1983); something that is assisted by the relative docility conferred by low circulating testosterone levels during the spring and summer months.

As summer turns to autumn and day lengths begin to shorten, testosterone levels increase, velvet growth slows and eventually stops and ossification intensifies (Flint 2012). As the antler calcifies and hardens, the blood and nerve supplies, along with the velvet outer layer, die and are sloughed off through a process known as 'stripping' or 'cleaning', which reveals the dead, bony antlers that are used for competition and combat (Landete-Castillejos et al. 2019). At the same time, the increase in testosterone makes the males more aggressive. Consequently, male social groupings, where they exist, disperse and the roar begins.

During the roar and depending upon the species, males may either occupy specific territories or 'leks', where they engage in displays of dominance; seek to obtain a harem of females, which they mate with and defend during the breeding season; or compete over a specific female, as in the case of the chital, described earlier (Smith-Jones 2022).

Perhaps the most detailed descriptions of roaring behaviour have been made on red deer, a harem-holding species. During the breeding season, adult red deer males (stags) will attempt to acquire a harem that they subsequently defend for several weeks (Clutton-Brock and Albon 1979). During this time and in common with many other species of deer, they will be solely focused on mating and as a result, can lose a significant amount of body condition; up to 20% in some cases (Clutton-Brock and Albon 1979; Goss 1983). Vocal roaring not only deters younger or mismatched competitors from challenging a more dominant animal, it also allows more evenly matched individuals to assess each other's likely fighting ability, in a more energy-efficient and less risky manner than engaging in an actual fight. As with many of the larger and more social species of deer, fighting among red deer is usually a last resort, as it carries with it the risk of serious injury or even death, especially if the loser is unable to retreat to safety (Smith-Jones 2022). In wild red deer, fighting injuries were noted in 23% of stags, with around 6% of those being permanent (Clutton-Brock et al. 1979).

When two relatively evenly matched stags encounter each other, a roaring contest usually ensues. This may continue for several minutes, until the challenger either retires or the pair engages in what is known as 'parallel walking'; moving slowly up and down at right angles to the direction that the challenger came from (Clutton-Brock and Albon 1979). Parallel walking gives the opponents another opportunity to assess each other's potential and to flee before fighting occurs, which is initiated by one of the pair lowering their heads in an invitation to challenge; although as noted by Clutton-Brock and Albon (1979), it is rare for a harem holder to give up without a fight, as it is not in their best interest to do so. Following the fight, the loser retreats and the winner assumes control of the harem until the next challenge occurs.

Although fighting is not that common among the more social species of deer, it is among the smaller, solitary species, especially those with tusks. During the breeding season, these deer will establish strongly defended territories and compete vigorously for breeding access, often incurring serious injuries in the process (Smith-Jones 2022). The resulting need for speed and agility may explain why males of black muntjac (*Muntiacus crinifrons*) and Fea's muntjac (*Muntiacus feae*) are smaller than the females, a reversal of the sexual dimorphism seen elsewhere among the Cervidae (Smith-Jones 2022).

At the end of the breeding period, increasing day length reduces circulating testosterone levels and males become less aggressive. Single sex social groupings re-establish among those species that have them, and during this time, it is not unusual to see previous competitors living side by side, still carrying their full set of antlers. Then, at the start of spring, the intermediary layer between the pedicle and antler changes, the old antlers fall off during 'casting' and within a few days, the antler growth cycle begins again in preparation for the next breeding season.

Before moving away from the subject of breeding altogether, a few further points are worth noting. First, although antlers are important for breeding success in the wild, they are less important in a farmed environment. Removing the antlers has been shown to have little or no effect on the dominance hierarchy of red deer or their ability to engage in agonistic encounters—and by implication, on their subsequent breeding success (Suttie 1980). However, anecdotal reports suggest that if yearling deer with their first set of antlers (so-called 'Spikers') are mixed with older, velvetted deer, then the younger deer will dominate the older animals, simply by virtue of having weapons when the others do not. Second, while antlers are primarily the preserve of male deer, there is one species, reindeer, in which the females also habitually grow them. Also, unlike most other deer, male reindeer regrow their antlers during the winter, while females retain theirs, which gives them a competitive advantage when food is at its least plentiful (Smith-Jones 2022). Finally, it should always be remembered that because of the increase in testosterone that occurs during the breeding season, deer with hard antlers are both extremely unpredictable and highly dangerous. Sadly, there have been numerous instances of hand-reared stags injuring or killing their owners during the roar, because this simple fact has been overlooked. Even experienced deer farmers tend to avoid interacting directly with their stags during the roar, regardless of whether they have been velvetted or not.

While this is only the briefest of overviews of the natural history of deer, there are two further aspects worth mentioning: their sensory capabilities and their scent glands. Deer have exceptional hearing and are extremely sensitive to movement, which befits their role as an animal that serves as prey to a wide range of other species (Smith-Jones 2022). However, these senses pale into insignificance compared to their sense of smell, which is many times more sensitive than that of humans (Goss 1983; Smith-Jones 2022). Related to this is the fact that most deer are equipped with a wide range of scent glands, which they use for numerous purposes, including territorial marking, advertising their sexual state and recognition (Smith-Jones 2022). This is in addition to other pheromone signaling outlets, including urine, faeces and skin secretions, which have been postulated to play a role in 'biostimulation'; the ability of animals to influence the reproductive physiology and performance of others of the same species through sensory cues (Landaeta-Hernández et al. 2022). In terms of scent glands, while these may be found at either end of the body and also between the digits, the majority are clustered around the head, with the most obvious being the pre-orbital glands in front of the eyes (Goss 1983; Smith-Jones 2022). Normally closed, they can open wide when deer are stressed (Bartošová-Víchová et al. 2007) and have been used as an indicator of such in at least one habituation study (Ceacero et al. 2014). Anecdotally, deer farmers in New Zealand use the opening of the pre-orbital glands as an indicator of potential aggression when handling deer in close confinement.

3 Current Extent of Utilisation

Within the six environmental domains considered by this book, deer are utilised by humans in three or possibly four of them. These are commercial and small-scale farms, in the wild and as part of zoological collections.

The relatively recent domestication of deer, combined with the costs of providing sufficient space and suitable containment, not to mention the challenges around managing males during the roar, means that they are not commonly chosen as pets. The same factors also mean that deer are not widely found in laboratory environments. That is not to say that they have not been used in research; however, such work has generally been aimed at addressing questions of relevance to deer (see later in this chapter), rather than using them as a model for other species.

Deer form a part of many zoological collections around the world, where they may be held for conservation or education purposes. It is also likely that their herbivorous, browsing nature lends itself to being kept in captivity, especially for the smaller species, which require less space for containment, although this is purely speculation on the part of the author, rather than being something that has been documented in the scientific literature.

With regard to small-scale farming, if this is taken to mean utilisation in enterprises such as smallholdings, lifestyle blocks, farmlets, or subsistence systems, then deer are not common, for the same reasons that they are not generally kept as pets; the costs of containment and their specific management requirements mean that other, more commonly farmed species are a more attractive option. However, if small-scale is considered in relation to the size of the farm, then deer are commonly used. For example, in Germany, the average size of a commercial deer operation is 2.5 hectares, with an average of 20 deer on it (FEDFA 2023).

In terms of utilisation on commercial farms, Table 2 shows the average farm size, main breeds and total number of deer farmed across a range of European countries, as well as in New Zealand, which has the world's largest farmed deer industry.

From Table 2, it can be seen that fallow and red deer are the most commonly farmed species of deer. In Europe, the majority are kept either semi-wild, in park-like enclosures, or more intensively, on purpose-built deer farms (FEDFA 2023). In New Zealand, farmed deer are all kept on deer farms (DINZ 2023). The primary products obtained from farmed deer are venison and velvet antler, although the latter is not utilised for commercial sale in many European countries or the United Kingdom (see next section for discussion).

Table 2 Average farm size, main breeds farmed and number of deer across a range of European countries and New Zealand

Country	Average farm size (ha)	Main breeds used	Number of deer
Austria	6	Fallow	17,900
Czech Republic	10	Fallow, Red deer	18,700
Germany	2.5	Fallow, Red deer, Sika (*Cervus nippon*)	180,000
Latvia	80	Fallow, Red deer	10,800
Netherlands	15	Fallow, Red deer	800
Poland	10	Fallow, Red deer	22,000
Russia	336.4	Fallow, Maral, Red deer, Sika	5793
Slovakia	3.5	Fallow, Red deer	10,180
Switzerland	3	Fallow, Red deer, Reindeer, Sika, Wapiti (*Cervus canadensis*)	11,000
United Kingdom	60	Fallow, Red deer	30,000
New Zealand	625	Fallow, Red deer, Wapiti	833,000

Sources: European data—FEDFA (2023), New Zealand data—DINZ (2023)

Although up-to-date information is not readily available, it is worth mentioning the Chinese deer industry in passing. In a paper by Chunyi Li (1993), the Chinese deer industry comprised 300,000 deer, 85% of which were sika, with the remainder being made up of wapiti and a few other species. Unlike the European or Australasian deer industries, at that time, deer farming in China was focused solely on velvet antler production, with deer being kept either in walled enclosures (pens) for the duration of their lives, or else with a greater or lesser degree of access to pasture (Li 1993).

Finally, hunting is perhaps the most common form of utilisation of deer in the wild and, as noted at the beginning of this chapter, is a practice that has been going on for many thousands of years (Strasser et al. 2018). Wild deer may be hunted for a number of reasons, including for their meat (sustenance), for their antlers (as trophies) or as part of population control/management efforts (as pests), e.g. Flaherty et al. (2019). Estimating the number of deer that are hunted each year across the globe is an impossible task; however, given the continued popularity of the practice, the number is likely to be significant. In support of this, one study, looking at wild-game meat in relation to achieving food security in West Virginia, estimated the annual kill to be around ten million pounds (~4.5 million kilograms), 96% of which came from white-tailed deer (Hall et al. 2020).

4 Attitudes to Pain in the Species

Attitudes towards deer can be highly variable and extremely personal. For example, a deer farmer might consider their animals to be a valuable asset and look upon them with a sense of pride, while viewing the wild deer on the other side of the fence as disease-carrying pests in dire need of extermination. In the eyes of the deer hunter, however, it may be the wild animals that are seen as being of value, while the farmed deer are considered a nuisance. To a park ranger in one country, deer may be seen as worthy of conservation, while to a ranger in another country, they might be the target of control measures, in order to conserve other, more vulnerable species. And to any of those people involved in the estimated 1.5 million road accidents involving deer in the USA each year (Okwueze et al. 2007), deer may simply be considered a menace, regardless of what species they are or where they come from.

Where attitudes towards deer have been assessed, the results have sometimes been surprising. For example, an investigation into attitudes towards white-tailed deer and rare plant management, conducted in the Canaan Valley region of West Virginia (Flaherty et al. 2019), found that while residents who had suffered crop damage as a result of deer activity were unsurprisingly less tolerant of them than non-residents; hunters, who were expected to prioritise deer over plants, were more tolerant of management to control deer populations than those in previous studies, e.g. Holsman and Peyton (2003). This perhaps reflects the fact that attitudes can and do change over time and that while previous emotional experiences play a role in influencing future experiences, they do not necessarily determine them (Hicks 2017).

Valente et al. (2020) studied the attitudes of three groups of Portuguese society (general public, farmers and hunters) towards red deer, to inform appropriate strategies for managing the rapidly growing population of deer in the country. While they expected perceptions to be generally negative, due to the increased potential for disease transmission, crop and wildlife damage and road accidents, the results suggested the opposite. Most respondents, except crop farmers, viewed deer positively and wanted an increase in wild populations, which the authors put down to a lack of underlying public knowledge about wildlife impacts, combined with the romantic notion of deer portrayed in the traditional and contemporary media (Valente et al. 2020). Positive attitudes towards white-tailed deer were also reported by Messner (2011) in relation to semi-urban populations in the USA. However, it was also noted that these could change as a result of damage to property, including through vehicle collisions and as a result of increased awareness of deer as a potential vector for disease (see Böhm et al. 2007 for further discussion).

When looking at deer population management in the South-East of England, Austin et al. (2013) found that, despite financial or ecological reasons being given as the primary reason for taking action, all but one of the landowners surveyed viewed deer as being damaging to the surrounding woodland, i.e. a pest species.

Turning to attitudes towards pain in deer, there is a dearth of published literature available. What little there is relates to the practice of velvetting and is general in nature. 'Velvetting' refers to the surgical removal of the antler above the pedicle, before extensive calcification has occurred, i.e. while the antler is still actively growing. As mentioned previously, velvet antler is extensively innervated, with branches originating from the zygomaticotemporal, infratrochlear and cervical nerves (Adams 1979). These nerves are believed to play a role in nociception, as they have been shown to respond to mild electrical stimulation (Matthews et al. 1992) and, as such, support the idea that they exist to help the stag protect the antlers while they are growing and vulnerable to damage (Goss 1983). When velvet antlers were removed without any analgesia, stags showed significant aversion behaviours, further indicating that the process was noxious (Matthews and Cook 1991). Consequently, removal without some form of anaesthesia is highly likely to cause pain and if the results of the above attitudinal studies are to be believed, that would be unacceptable to the majority of people.

Not all countries that farm deer conduct velvet removal. Across much of Europe, the practice is neither traditional nor necessary, largely because farm sizes are small (see Table 2) and the focus is on raising females, rather than males. In the United Kingdom, the practice was banned on animal welfare grounds in the early 1980s, although it is still permitted to be conducted in an emergency, or to treat injury or disease, but only by a veterinarian, using appropriate anaesthesia (Anon 1980, 1982). It is interesting to speculate whether the same outcome would have been reached had the fledgling British deer industry of the time been further advanced, or the British Deer

Society, which predominantly represented traditional deer stalkers (hunters), been less successful in their lobbying (Fletcher 1990). However, the determination was made and has not been revisited to date, despite the procedure not appearing to adversely affect the welfare of stags when performed with adequate analgesia (Matthews 1995).

In contrast to the United Kingdom, New Zealand has taken a more pragmatic approach and velvetting has long been permitted in law, most recently recognised under the Animal Welfare (Care and Procedures) Regulations 2018. However, it is still considered to be a surgical procedure and as such, is restricted to veterinarians and appropriately trained individuals, with local anaesthesia being a mandatory requirement. This has led to the creation of a rigorous training and certification programme, run by the National Velvetting Standards Body (NVSB), which celebrated its 30th anniversary in 2023 (DINZ 2023). The NVSB programme is a partnership between veterinarians and deer farmers and is predicated on ensuring the welfare of stags during velvet removal, using a system of theoretical and practical training, annual veterinary assessment and independent audit to achieve this goal and meet societal expectations (DINZ 2023).

Velvet antler is removed for several reasons including protecting stags from injury during the roar, noting that on large commercial farms, many stags may be present and opportunities to escape aggressive encounters can be limited; to protect those working with deer from injury; to minimise the risk of pain through antler damage during transport; and for profit, as velvet antler is considered a valuable commodity in some Asian countries (English 1991a; Wilson 1989). Regarding that final point, from a farming and ethical standpoint, the question has to be asked: 'Is it not better for both stag and farmer to remove velvet antler using effective pain relief, at a safe time of the year and derive an income from it, than to wait until a more dangerous time and remove hard antler without pain relief or financial gain?' The answer may not be as simple to resolve as at first glance, as it will depend upon public, market and government attitudes, but either way, it could shape the future of commercial deer industries across the globe.

5 Anaesthesia, Analgesia and Pain

Unlike attitudes towards pain in deer, there is a reasonable amount of published literature looking at mitigating pain and suffering across the species, through the use of analgesic and anaesthetic agents. This work, much of it coming from studies into the management of wild and farmed deer, has made an invaluable contribution to improving our understanding of how to minimise stress, suffering and poor welfare outcomes when engaging with these animals.

5.1 General Anaesthesia

Successful general anaesthesia of deer has been achieved using several methods. In studies looking at the efficacy of different methods of achieving analgesia for velvet removal, a combination of propofol and ketamine was used to induce anaesthesia in male red deer, with subsequent anaesthesia being maintained at a 'light plane', using halothane in oxygen (Johnson et al. 2005; Woodbury et al. 2005). The advantage of this method from an experimental perspective was that the light plane of anaesthesia achieved, allowed a non-analgesic control group to be included, which would otherwise have been precluded on ethical grounds if the animals had been conscious.

In contrast to the light anaesthesia model used above, deep anaesthesia, in order to perform orthopaedic surgery on a fallow deer, was successfully achieved using a combination of butorphanol and medetomidine, followed by induction with ketamine and diazepam, with maintenance achieved using isoflurane in oxygen and fentanyl (Mansour et al. 2022). In this instance, a multimodal or 'balanced anaesthesia' approach was taken, to provide pain relief and sedation prior to induction of anaesthesia, with additional pain relief via continuous infusion of fentanyl, provided throughout the procedure. The inclusion of diazepam at the induction phase, allowed easier intubation of the subject, by providing additional muscle relaxation (Mansour et al. 2022).

While halothane and isoflurane are the two main inhalational agents recommended for maintaining extended anaesthesia in deer, especially when conducting highly invasive procedures (Caulkett and Haigh 2007), when it comes to inducing anaesthesia, immobilisation or sedation, a far wider range of agents are available.

Xylazine, an alpha-2 agonist, has long been used in deer, as it provides predictable, potent sedation, with a rapid onset of action (Walsh et al. 2001). For these reasons and the fact that it is relatively inexpensive compared to other anaesthetics, it has been used extensively as part of the velvet removal protocol (along with local anaesthesia) in New Zealand, for over 40 years (Suttie 1982).

Deer have a tendency to suffer from hypoxaemia when under anaesthesia and xylazine, like other alpha-2 agonists, can exacerbate this (Caulkett et al. 2000; Caulkett and Haigh 2007). Further anaesthetic risk factors include hyperthermia, ruminal tympany, bloat, regurgitation and smothering when animals are anaesthetised in groups (Mackintosh and Waldrup 1994).

While severe adverse reactions to xylazine are rare, sudden deaths have been reported by Walker and Middelberg (1988). In those cases, apparently healthy stags died some hours or days following xylazine administration. The cause of death was believed to be due to a delayed hypersensitivity to the drug and occurred at an estimated rate of less than two in a thousand animals (Walker and Middelberg 1988). Support for the hypersensitivity hypothesis was provided by Mackintosh and Cross (1989), who also noted a significant decline in basophil counts shortly after xylazine administration. The susceptibility of basophils to xylazine was confirmed in studies by Cross et al. (1989, 1992), along with strong evidence for splenic sequestration of red blood cells in xylazine sedated deer, which may play a role in aggravating hypoxaemia during anaesthesia.

Fortunately, the effects of xylazine can be rapidly and reliably reversed using Yohimbine (Bubenik and Brown 1989; Mackintosh 1985a), which has also been proposed as a suitable treatment for humans, following accidental self-injection (Mackintosh 1985b).

Other agents have been investigated in combination with yohimbine to assess their effects on xylazine reversal, with mixed results. The addition of physostigmine showed a small but non-significant improvement in recovery times (Haigh and Alsager 1987), while its use with 4-aminopyridine was cautioned against, on the basis that it could cause severe adverse effects in deer (Mackintosh 1985a).

Although xylazine is generally used as the sole sedative during velvet removal, it lends itself well to combination with a range of other anaesthetics, as part of a balanced anaesthesia approach. Balanced anaesthesia allows sedation to be maintained or enhanced, using lower doses of each drug than would be required if they were administered alone, e.g. English (1991b). This is particularly important when using opioids, due to their extreme potency, low margin of safety and high risk of serious consequences following accidental exposure, not to mention the risk of addiction through misuse (Walsh et al. 2001; Walsh and Wilson 2002a). Balanced anaesthesia also allows the unwanted side effects of particular drugs, e.g. muscle rigidity associated with ketamine and tiletamine, to be mitigated, making them easier to use (Walsh et al. 2001). Consequently, xylazine has been formulated with a range of agents, including ketamine, as 'Xylaket' and fentanyl/azaperone, as 'Fentazin' (Mackintosh and Waldrup 1994; Walsh et al. 2001).

Xylazine has also been suggested as a useful component for assisting in the capture of wild deer, when used in combination with phencyclidine, tiletamine and the benzodiazepine, zolazepam (Caulkett and Haigh 2007). However, it should be noted that special care needs to be taken when using anaesthetics on wild deer, as dose rates generally need to be higher than for captive animals and they are at greater risk of acute stress, self-injury and capture myopathy, the latter of which can often be fatal (Caulkett and Haigh 2007). In a recent study exploring the effect of oxygen supplementation to combat hypoxaemia in white-tailed deer anaesthetised using xylazine–ketamine or xylazine–tiletamine–zolazepam, both combinations were deemed suitable for immobilisation, with supplemental oxygen resulting in higher arterial oxygen levels and lower heart rates than in control animals (Mitchell et al. 2021).

Newer, more potent alpha-2 agonists have also been used to successfully anaesthetise deer. These include detomidine, medetomidine and its active enantiomer, dexmedetomidine (Bouts et al. 2011; Walsh et al. 2001), all of which can be readily reversed using the specific antagonist, atipamezole (Walsh and Wilson 2002a). Medetomidine, which is between 20 and 40 times more potent than xylazine, has found particular popularity as a component of balanced anaesthesia and has been recommended for use on wild deer, when combined with ketamine (Caulkett and Haigh 2007). Bouts et al. (2011) subsequently compared the effects of medetomidine-ketamine and dexmedetomidine-ketamine on wild Chinese water deer and found both treatments to cause rapid and deep immobilisation, with no adverse effects following reversal.

Medetomidine, like xylazine above, has also been used with tiletamine and zolazepam, to provide a more potent combination and faster induction than with ketamine (Walsh and Wilson 2002a). Rauch et al. (2022) used this combination to investigate how seasonal variation and diet affected physiological parameters in red deer. Using remote heart rate sampling and direct arterial blood pressure monitoring, they found that anaesthetised deer had lower heart rates and blood pressures in winter compared to summer and that deer on a restricted diet, or one supplemented with omega-6 pellets, had lower heart rates, but higher blood pressures, than those fed *ad libitum* or supplemented with omega-3 pellets. The key message from this work was that nutritional status and time of year need to be taken into account when administering anaesthetics to highly seasonal mammals, such as deer.

Medetomidine-tiletamine-zolazepam was also used in a study by Einwaller et al. (2020) to investigate the effects of the alpha-2 agonist, vatinoxan, on red deer. Unlike atipamezole, vatinoxan has limited ability to cross the blood–brain barrier and has been shown to have a positive impact on cardiovascular indicators in medetomidine anaesthetised dogs, without affecting the level of sedation (Salla et al. 2022). The study found that while vatinoxan had no significant effect on oxygen saturation, it did reduce mean arterial blood pressure for the duration of the study and temporarily increased heart rate, without altering the level of sedation, suggesting that it could be a valuable component in reducing the undesirable side effects of using medetomidine on deer (Einwaller et al. 2020).

As an alternative to using phencyclidines, which do not have specific reversal agents, the effects of medetomidine, in combination with the neuroleptic, azaperone and neurosteroid, alfaxalone, were investigated on captive white-tailed deer in a study by Pon et al. (2016). The authors reported that 15 minutes after administering medetomidine-azaperone-alfaxalone, all but one animal were laterally recumbent and remained so for the duration of the study (60 minutes). While the drug combination did cause hypoxaemia and hypoventilation in all subjects and temporary arrhythmia in two, all were quickly and successfully reversed with atipamezole, with no subsequent adverse effects. The same drug combination was used to investigate the possibility of using sodium nitroprusside as a vasodilatory agent to improve blood oxygenation in white-tailed deer (Lakhdhir et al. 2021). Unfortunately, while nitroprusside did show a positive effect on arterial oxygen levels, the effect was temporary and was also associated with severe hypotension, so it was not recommended for further consideration.

A further anaesthetic combination involving medetomidine has received interest from researchers. It is a mixture of the synthetic opioid butorphanol, azaperone and medetomidine, commonly abbreviated to BAM. BAM has successfully been used in relation to semen collection in white-tailed deer, without impacting sperm quality (Kirschner and Rodenkirch 2017) and for immobilising wild white-tailed deer, following capture by a variety of methods including darting, drop nets and Clover traps (McDermott et al. 2020). In the latter study, BAM was associated with a reduction in oxygen saturation, heart rate, respiration rate and body temperature, although the effects were not severe and the authors concluded that BAM was safe to use for post-capture restraint of white-tailed deer. Interestingly, in a separate study, conducted some years before those above, the level of anaesthesia provided by BAM and a xylazine–carfentanil mix was examined using electroencephalography (EEG), relative to a baseline of general anaesthesia under isoflurane (Kreeger et al. 2010). By establishing a minimum alveolar concentration (MAC) under isoflurane and measuring the bispectral index (BIS) of the resulting electroencephalograms, the study suggested that neither anaesthetic treatment provided general anaesthesia, i.e. animals were not fully unconscious and that they should be classed as neuroleptanalgesics, rather than true anaesthetics.

Despite their inherent risks, the ultra-potent synthetic opioids, carfentanil and fentanyl and the semi-synthetic opioid, etorphine, have all been and continue to be used as anaesthetics for deer (Mackintosh and Waldrup 1994; Walsh et al. 2001; Walsh and Wilson 2002a). Due to the fact that they can cause unwanted side effects when used alone, they are often combined with xylazine or other anaesthetics (Caulkett and Haigh 2007; English 1991b).

English (1991b) reviewed the use of carfentanil, with particular regard to deer and noted that the excellent synergy between it and xylazine meant that rusa/sambar hybrid stags could be velvetted successfully, when the same procedure was impossible to perform under deep xylazine sedation alone. Waldrup et al. (1992) extended English's work and showed that carfentanil-xylazine, at a lower intramuscular dose than that used by English, provided a suitable level of immobilisation to conduct velvetting of red deer and wapiti. However, a higher dose, administered intravenously, was required for more invasive procedures, e.g. surgical vasectomy. Interestingly, species-specific differences were observed in that study, with red deer taking longer to enter sternal recumbency than wapiti, but being able to be handled sooner once they had reached that point (Waldrup et al. 1992). Carfentanil-xylazine was also used successfully to immobilise a wapiti suffering from obstructive urolithiasis, without additional pain relief being required before surgery (Larsen et al. 2000).

The effects of etorphine on fallow deer, in combination with xylazine, acepromazine, or both drugs together, as well as in the presence or absence of atropine, were investigated by Pearce and Kock (1989). As expected, the anaesthetic agents had a depressive effect on respiratory rates and blood oxygen levels and increased pulmonary arterial blood pressures. However, in those deer given atropine, heart rates were higher than in non-treated animals, reflecting the effect of the drug on inhibiting acetylcholine receptors and allowing sympathetic stimulation to predominate. Unfortunately, several animals suffered cardiac arrests during induction and anaesthesia, which reinforces the challenges and risks associated with an aesthetising semi-wild animals.

One advantage of opioids over some other classes of anaesthetic is that, like alpha-2 agonists, they can be readily reversed. Antagonism has been achieved using nalorphine, diprenorphine, naloxone, nalmefene and naltrexone (Caulkett and Haigh 2007; Waldrup et al. 1992). Although naloxone is popular because it has the shortest half-life of the opioid antagonists, it has been associated with re-narcotisation, so it requires careful post-recovery monitoring of treated animals (Caulkett and Haigh 2007). It is also worth noting that naloxone has successfully been used at a low dose to counter the effects of hypoxaemia and cardiopulmonary depression in wapiti anaesthetised with carfentanil-xylazine, without affecting the level of sedation (Moresco et al. 2001). This has particular implications for mitigating the adverse effects of opioid anaesthesia under field conditions, where supplementary oxygen may not be readily available.

Other compounds that have been used for achieving anaesthesia in deer, examples of which have been mentioned above, include benzodiazepines and neuroleptics. Neuroleptics act as tranquilisers; however, as they lack specific analgesic properties, they are frequently combined with other drugs, primarily alpha-2 agonists, to increase their efficacy (Walsh et al. 2001). Commonly used neuroleptics in deer include acetyl promazine (also known as acepromazine or acetopromazine), azaperone, haloperidol, perphenazine and zuclopenthixol (Caulkett and Haigh 2007; Walsh et al. 2001; Walsh and Wilson 2002a). Depending on the neuroleptic used, their effects can last for hours or even days (Caulkett and Haigh 2007), which can be problematic, as specific reversal agents are not available (Walsh et al. 2001).

In a study looking at the suitability of neuroleptics without additional anaesthesia, to reduce stress in red deer, Diverio et al. (1996) found that a combination of perphenazine and zuclopenthixol had a positive effect on calming deer and that it could potentially form part of handling protocols for farm management and wild capture. A similar study, using azaperone or haloperidol administered singly to wild roe deer captured with drive nets, found them to exert a calming effect on the animals, as evidenced by a faster stabilisation of heart rate, reduction in hypotensive indicators and lower serum cortisol levels than control deer (Mentaberre et al. 2010). In contrast, Ortega et al. (2020) used a combination of the benzodiazepine, midazolam and the neuroleptic, azaperone, to reduce stress and potential injury to mule deer caught by a helicopter net gun. Using a range of physiological and behavioural indicators, as well as faecal glucocorticoids, they concluded that the drug combination in their study did not improve welfare outcomes compared to control animals and was, in fact, associated with additional risks to the animal post-release, due to the ongoing effects of sedation.

Finally, it is worth noting, if only as a warning, that succinylcholine has been investigated as an immobilising agent in white-tailed deer. In a study by Schwartz et al. (1997), the drug was compared to xylazine-ketamine, as part of a capture-euthanasia protocol. While succinylcholine was more effective than xylazine-ketamine for achieving rapid immobilisation, it has no anaesthetic or analgesic properties, so it cannot be used humanely for procedures where animals remain conscious.

5.2 Local Anaesthesia

In comparison to general anaesthetics, far fewer local anaesthetic agents have been used with deer. This is likely because, in most cases, deer are either immobilised or sedated for non-invasive procedures, e.g. handling and tagging during wild-capture, or the general anaesthesia provided is sufficient to enable surgery to be performed without the need for additional anaesthesia. What follows is largely based on studies investigating the use of local anaesthetics for velvetting.

The amino-amide, lignocaine (lidocaine) hydrochloride, is the most commonly used local anaesthetic for velvet removal (Walsh and Wilson 2002b; Wilson and Stafford 2002). It has an onset of action of approximately 30 seconds (Bartels et al. 2001) and a mean duration of between 90 minutes and 2 hours (Bartels et al. 2001; Matthews et al. 1992). Amino-amide local anaesthetics can cause local vasodilation, and while lignocaine has been formulated with adrenaline (epinephrine) in the past, this combination is no longer used for velvetting, as the vasoconstrictive effect of adrenaline has been linked to an increased risk of clostridial infection through localised anoxia of the pedicle (Seifert 1997; Walsh and Wilson 2002b). A metabolite of lignocaine, xylazine and potentially other amino-amide anaesthetics, 2,6-xylidine, has also been identified as being carcinogenic in rats (see Walsh et al. 2001 for discussion). While this is not likely to be a problem when used on non-food-producing animals, it does highlight the need for veterinarians and others using amino-amide drugs on farmed deer to take appropriate steps to manage the risk of residues in the resulting animal products. The same is true for all drugs, including anaesthetics, when they are used on a discretionary or 'off-label' basis, in which case, particular consideration needs to be given to determining an appropriate withholding time, or using the default withholding period if one is available. In the case of managing lignocaine residues in velvet antler, this is achieved in practice through the simple expedient of applying a tourniquet (above arterial pressure) before the administration of local anaesthetic. An alternative to using amino-amide local anaesthetics, is to use amino-ester variants, e.g. procaine and chloroprocaine, which are metabolised differently and do not break down into 2,6-xylidine. However, amino-esters are associated with occasional allergic reactions, due to the production of the metabolite para-aminobenzoic acid (Walsh and Wilson 2002b).

In terms of the administration of lignocaine for velvetting, a ring block around the pedicle was found to be superior to a specific nerve block, as measured by fewer aversive behavioural responses and lower average heart rates in the ring block group, compared to the nerve block group (Matthews et al. 1992). Further studies by Wilson et al. (1999, 2000a, b) showed that a high-dose ring block of lignocaine, equating to approximately one millilitre per centimetre of pedicle circumference, was more effective at providing local anaesthesia than a lower dose ring block or a specific nerve block, regardless of whether the nerve block was applied close to the skull (low position) or further up the pedicle (high position). Interestingly, when a saw cut was made on a piece of wood held against the animals' pedicles, to investigate the potential effect of noise, vibration, or activity around the head as a confounding factor, some animals responded, which suggested that not all reactions to subsequent velvet removal may have been due solely to pain (Wilson et al. 1999).

Other local anaesthetics have also been investigated for use on deer. Bartels et al. (2001) compared the onset and respective durations of action of lignocaine, bupivacaine and mepivacaine, administered alone and in a variety of combinations, as a high-dose ring block (1 ml/cm pedicle circumference for all treatments). While lignocaine and mepivacaine had a similar onset of action (approximately 30 seconds), mepivacaine had a significantly longer mean duration of action (271 minutes versus 88 minutes for lignocaine). Bupivacaine had the longest onset of action (48/86 seconds, depending on the formulation used) and the longest duration of action (273/461 minutes). Combinations tended to produce intermediate results and, in some cases, sped up the onset of action, without substantially decreasing the duration of anaesthesia. While the findings suggested that combinations of local anaesthetics could improve welfare outcomes following velvet removal, it should be noted that bupivacaine has a greater cardio-depressant effect than other local anaesthetics, so additional care may be needed when using it (Walsh and Wilson 2002b).

More recently, articaine, an amino-amide local anaesthetic possessing an additional ester group, has been investigated in relation to velvetting (Venkatachalam et al. 2018, 2019; Sahebjam et al. 2020). When applied as a standard high-dose ring block, articaine had a longer onset of action than lignocaine, but did provide sufficient analgesia to permit velvet removal (Venkatachalam et al. 2019). It also had a longer duration of action than lignocaine (Sahebjam et al. 2020). Investigation of the pharmacokinetics of articaine showed that it was quickly metabolised into articainic acid and eliminated from the body (Venkatachalam et al. 2018). The main advantage of articaine over lignocaine was viewed as the fact that it did not break down into 2,6-xylidine (P. Singh, personal communication). However, the need for further investigation into the safety of articaine and its metabolites was stressed by the authors before it could be recommended for use with deer (Venkatachalam et al. 2018, 2019).

Finally, Mansfield et al. (2006) reviewed the anaesthetic agents used for conducting tooth removal for ageing purposes in conscious deer. Only two published studies were available. In the first, a low dose of lignocaine (0.5 ml in total) was used; however, it was not clear how it was administered or how effective it was. In the second, xylazine-ketamine was used to provide general anaesthesia, without supplementary local anaesthesia being provided.

5.3 Analgesia

Analgesia differs from anaesthesia; in that it refers to pain relief without complete loss of consciousness or sensation (USDANAL 2023). From an examination of the available literature, it appears to be an area that has received little attention from deer researchers to date.

Acetyl salicylate, a non-steroidal anti-inflammatory drug (NSAID), was proposed by Mackintosh and Waldrup (1994) as a means of extending the window of analgesia by several hours beyond that provided by lignocaine. This was as a result of a study by Pollard et al. (1992), which, despite finding no main effects of acetyl salicylate on the behaviour of deer following velvetting, did note several interactive effects of treatment and velvetting status that persisted for the 4-hour duration of the experiment. These included more grooming attempts, more aggression and differences in standing posture in velvetted deer receiving saline, than those receiving acetyl salicylate, or non-velvetted control animals. However, a later study failed to find a consistent effect of acetyl salicylate on modifying the behaviour of velvetted deer, with the additional observation that there were no significant differences in behaviour between velvetted and non-velvetted stags at the end of the study (24 hours) and only a significant difference in ear-flicking and head shaking, 7 hours after velvetting (Pollard et al. 1994). Subsequently, interest in acetyl salicylate waned, and the drug appears not to have found widespread use as a pain-relieving agent for deer.

Flunixin is another NSAID that has been briefly looked at in deer. Walsh and Wilson (2002b) noted that it had been used in a pilot study by Bartels, along with azaperone, as an option for reducing the effect of handling stress on plasma cortisol. Unfortunately, it did not appear to have the desired effect, as there was no significant difference in cortisol levels between NSAID and control animals during the study (Walsh and Wilson 2002b). Flunixin was also reported as being given immediately prior to tooth extraction in xylazine-sedated deer, by Mansfield et al. (2006); however, no specific details about the subsequent effect of the NSAID were given, apart from the general observation that the procedure resulted in no apparent adverse effects.

While NSAIDs are commonly used in both veterinary and human medicine to provide analgesia, care needs to be taken when administering them to deer that may be dehydrated or in shock, especially if multiple NSAIDs are administered over a period of days, due to the risk of causing NSAID-induced renal papillary necrosis (Stern et al. 2010).

Non-chemical analgesia, in the form of Transcutaneous Electrical Nerve Stimulation (TENS), a technique used in human dentistry, was investigated by Matthews et al. (1999) and Woodbury et al. (2001), as a potential means of achieving temporary pain relief in deer during velvetting, without the risk of introducing drug residues into the antler. However, in both studies, not only was the application of so-called 'electro-analgesia' aversive to stags, it was also less effective and less reliable than a high-dose ring block of lignocaine (Matthews et al. 1999; Woodbury et al. 2001). Consequently, the technique was not recommended for use on deer.

A further non-chemical analgesic technique, involving the use of a specially designed, high-pressure band, the 'NaturO ring™', has successfully been used for velvetting yearling deer (Matthews and Suttie 2001). However, it is not discussed further here, as its specificity renders it unsuitable for wider application to other surgical procedures.

5.4 Pain (and Its Assessment)

Causes of pain in deer are likely to vary with the particular environment the animal is in. For example, in a zoo environment, pain is likely to be limited to the result of aggression, misadventure, or illness, due to the relatively small number of animals held in containment and the relatively high degree of human oversight that is provided. In the wild, there are additional risks associated with injury, especially self-injury during capture and handling (Diverio et al. 1996; Ortega et al.

2020). In a farmed environment, deer face a wider range of potentially painful interventions in addition to those mentioned above. These are intended to make handling, productivity and management more efficient and include electro-ejaculation, laparoscopic artificial insemination, velvetting and minor surgery (Mackintosh and Waldrup 1994). In almost all cases, pain relief, through the use of general or local anaesthetics, is used to mitigate the negative effects of these procedures on the animals.

When it comes to the assessment of pain, basic principles are covered elsewhere in this book. However, it is worth noting that because deer are a prey species, pain might be outwardly hard to detect, as it does not pay to draw attention to oneself if the act of doing so will attract potential predators (Mansfield et al. 2006). Nonetheless, there have been several studies that have provided useful insights into the assessment of pain in deer. These are discussed below.

Matthews et al. (1992) pioneered the use of mild electrical stimulation (as opposed to electro-analgesia) to assess antler sensitivity in relation to the onset and duration of anaesthesia before velvetting. By varying the voltage and current delivered to the antlers, they were able to establish a reproducible, nine-point scale of aversive behaviours, ranging from minor head movements at the lower end, through to whole body displacement (escape response) at the upper end of stimulation. This was subsequently used as a reference to assess the effectiveness of different local anaesthetic applications, as well as the duration of analgesia. Using this method, a ring block was shown to be more effective than a specific nerve block at providing local anaesthesia, with a duration of action lasting somewhere between 30 and 120 minutes.

Electrical stimulation and behavioural responses according to a three-point scale were subsequently used to confirm the superiority of a high-dose ring block over a regional nerve block by Wilson et al. (1999, 2000a, b). In a later study, advances in the technology allowed Bartels et al. (2001) to initially stimulate antlers remotely following application of different local anaesthetics, although manual stimulation had to be used later on, due to the increasing number of observations required. Despite this, the methodology allowed for a more sensitive insight to be gained into the duration of action of the anaesthetics than was permitted by Matthews et al. (1992) and showed that lignocaine had a mean duration of around 90 minutes (Bartels et al. 2001).

More recently, a different technique, pressure algometry, has been used to assess the effectiveness of articaine in deer, when given as a local anaesthetic during velvetting (Sahebjam et al. 2020). Pressure algometry involves using a probe to apply incremental pressure to a tissue, with resulting movement or avoidance indicating noxiousness. Although pressure algometry has been widely used for assessing pain in some species (see Sahebjam et al. 2020 for discussion), there was considerable variation in responses during threshold testing and a very short acclimation period used in the study, which suggests that the animals were not fully habituated to the process, before anaesthetic testing. That said, the technique did show statistically significant differences in responses associated with the return of sensation following velvetting, so it is worthy of further investigation as a potential pain assessment tool for use in deer.

Other studies have used direct observation of behaviour, with or without accompanying physiological measures, to assess pain in deer. Pollard et al. (1991) measured changes in the behaviour and heart rate of restrained stags during velvetting, compared to those that were restrained without being velvetted. Measurements were taken both during treatment and up to 9 hours afterward. Changes in heart rate suggested an anticipatory response (increased heart rate) from stags immediately before the removal of their second antler, with a habituation response (decreased heart rate) from non-velvetted stags at the same time point. However, the difference was only significant for the non-velvetted control group. The anticipatory response was supported by more stags struggling immediately prior to velvet removal than those that were only restrained, but this difference was not evident during removal itself. For 3 hours following treatment, velvetted stags shook their heads, flicked their ears, groomed themselves and sat down more than control stags, as well as eating less, which the authors took to indicate discomfort following the procedure. Beyond this point, the only difference in behaviour was that velvetted stags ate more than control stags, which was put down to a compensatory response, as a result of reduced feed intake earlier in the day (Pollard et al. 1991).

As previously reported, stags velvetted with lignocaine and saline, showed a number of behavioural differences, compared to those velvetted with lignocaine and acetyl salicylate, or non-velvetted control animals (Pollard et al. 1992). Head shaking, ear flicking and nosing the ground were all increased immediately following velvetting in stags given lignocaine and saline; however, in contrast to their earlier study (Pollard et al. 1991), ear flicking was non-significant after 10 minutes and remained that way for the rest of the study. After 10 minutes, grooming attempts were higher in the velvetted lignocaine/saline group and were still significant after 4 hours, along with nosing the ground, time spent feeding, standing with the head down or at shoulder level and aggression. Interestingly, head shaking and ear flicking were significantly higher in velvetted than non-velvetted stags, regardless of whether saline or analgesic was used, which suggests that they may have been more closely linked to the novelty of losing weight from the head than to pain. This potential confounding factor does not appear to have been addressed in subsequent studies, so is worth investigating further.

A later study by Pollard et al. (1994), used behavioural observations to further assess the effectiveness of acetyl salicylate over an extended period of time. Using a similar approach to Pollard et al. (1992), with additional intravenous administration of either acetyl salicylate or saline at either 7 or 24 hours, the study found more head shaking and ear flicking in velvetted than non-velvetted stags, which persisted for 7 hours, but had diminished by 24 hours. Other than that, there were few consistencies in behavioural responses to the treatments and little evidence that the analgesic had any significant effect. The discrepancy between these findings and their earlier work (Pollard et al. 1992) was attributed to potential sub-optimal administration of the local anaesthetic ring block, but also highlights the inherent variability in behavioural responses between individual animals and the need to interpret such observations with care. That last point was reinforced by a highly detailed video analysis of factors influencing successful velvetting, which suggested that relatively subtle behaviours, 'flinch', 'head arch', 'head down' and 'head rotate', were more sensitive indicators of acute pain than gross movements of the head or body used in previous studies (Flint 2012).

Concerning the question of any pain or discomfort experienced by velvetted stags once local anaesthesia has worn off, based on the studies described above (Pollard et al. 1991, 1992, 1994), it would appear that it is short-lived, with a likely duration of not more than 24 hours. While further work is required to confirm this and better understand the mechanism of antler regeneration following removal, it is possible that the unique nature of antler tissue means that it does not follow the classical model of wound healing common to other somatic tissues, with the result that the associated pain/inflammatory response is also altered or ameliorated.

Finally, as mentioned at the start of this section, electroencephalography (EEG) has been used, in conjunction with light general anaesthesia, to assess pain in deer (Johnson et al. 2005; Woodbury et al. 2005). While the technique is useful in that it allows for non-analgesic control groups to be included, it is only suitable for assessing acute noxious insults, due to the effect of background bio-electrical 'noise' masking the measurement and interpretation of chronic pain signals (C. Johnson, personal communication).

To date, research into the assessment of pain in deer appears to have focused on acute pain, particularly that associated with velvetting. This is likely due to the challenges associated with assessing chronic pain in a species that outwardly hides displays of discomfort, along with the confounding effect of stress on potential physiological indicators of pain (see Matthews et al. 1994; Sahebjam et al. 2020 for discussion). However, in the future, it may be possible to use techniques from other areas of scientific endeavour, such as remote sensing, self-medication, transcriptomics, proteomics and DNA methylation, to investigate this more challenging area of the pain spectrum with success.

6 Summary

Despite deer having been part of the collective human consciousness for thousands of years and the continued interest in hunting and farming them for their meat and antlers, very little is known about their subjective experience of pain and suffering. As a result of increasing interest in minimising harm during capture, handling and management, this is starting to change; however, the area is still in its infancy and is ripe for further investigation. What little we do know has largely arisen over the last 50 years, as the result of large-scale farming operations in the Southern Hemisphere, motivated by a genuine desire to maximise good welfare outcomes, in order to meet societal expectations.

References

Adams JL (1979) Innervation and blood supply of the antler pedicle of the red deer. N Z Vet J 27:200–201

Almeida NJ, Cerrillo-Cuenca E, Saladié P (2021) Framing agricultural intensification in western Iberia during the late Neolithic: a new insight through the faunal record from Los Barruecos site. J Archaeol Sci Rep 36:102815

Anon (1980) The welfare of livestock (deer) order 1980. Her Majesty's Stationery Office

Anon (1982) The welfare of livestock (prohibited operations) regulations 1982. Her Majesty's Stationery Office

Asher GW (1988) Antlers. In: Allen PL, Asher GW (eds) Progressive fallow farming: proceedings of a course on fallow deer farming. Ruakura Agricultural Centre

Austin Z, Raffaelli DG, White PC (2013) Interactions between ecological and social drivers in determining and managing biodiversity impacts of deer. Biol Conserv 158:214–222

Barrette C (1985) Antler eating and antler growth in wild Axis deer. Mamm 49(4):491–500. https://doi.org/10.1515/mamm.1985.49.4.491

Bartels MM, Wilson PR, Caulkett N et al (2001) Dynamics of local anaesthetics for velvet antler analgesia. In: Proceedings of a deer course for veterinarians, vol 18, pp 100–108

Bartošová-Víchová J, Bartos L, Sˇvecova L (2007) Preorbital gland opening in red deer (*Cervus elaphus*) calves as an indicator of stress. J Anim Sci 85(2):494–496

Böhm M, White PC, Chambers J et al (2007) Wild deer as a source of infection for livestock and humans in the UK. Vet J 174(2):260–276

Bouts T, Taylor P, Berry K et al (2011) Evaluation of medetomidine-ketamine and dexmedetomidine-ketamine in Chinese water deer (*Hydropotes inermis*). Vet Anaesth Analg 38(2):106–112

Bubenik GA, Brown RD (1989) The effect of yohimbine on plasma levels of T3, T4 and cortisol in xylazine-immobilised white-tailed deer. Comp Biochem Physiol C: Comp Pharm 92(2):315–318

Caulkett N, Haigh JC (2007) Deer (cervids). In: West G, Heard D, Caulkett N (eds) Zoo animal and wildlife immobilization and anesthesia, 1st edn. Blackwell, Iowa, pp 607–612

Caulkett NA, Cribb PH, Haigh JC (2000) Comparative cardiopulmonary effects of carfentanil-xylazine and medetomidine-ketamine used for immobilization of mule deer and mule deer/white-tailed deer hybrids. Can J Vet Res 64(1):64–68

Ceacero F, Landete-Castillejos T, Bartošová J et al (2014) Habituating to handling: factors affecting preorbital gland opening in red deer calves. J Anim Sci 92(9):4130–4136

Chadez J, Sappington RL (2017) The Holocene exploitation of mammals in the Clearwater and lower Snake River regions of north-central Idaho. J Arch Sci: Rep 16:258–265

Clutton-Brock TH, Albon SD (1979) The roaring of red deer and the evolution of honest advertisement. Behav 69(3–4):145–170

Clutton-Brock TH, Albon SD, Gibson RM et al (1979) The logical stag: adaptive aspects of fighting in red deer (*Cervus elaphus L.*). Anim Behav 27:211–225

Cross JP, Mackintosh CG, Griffin JFT (1989) Further observations on xylazine and haematological parameters in red deer (*Cervus elaphus*): the effect on reference values and on splenctomised animals. Proc NZ Vet Assoc Deer Branch 6:146–150

Cross JP, Griffin JFT, Mackintosh CG (1992) Influence of xylazine on hematology values in farmed red deer. In: Brown RD (ed) The biology of deer. Springer, New York, pp 136–140

DINZ (2023) Deer Industry New Zealand. https://www.deernz.org/. Accessed 31 Jan 2023

Diverio S, Goddard PJ, Gordon IJ (1996) Use of long-acting neuroleptics to reduce the stress response to management practices in red deer. App Anim Behav Sci 49(1):83–88

Einwaller J, Painer J, Raekallio M et al (2020) Cardiovascular effects of intravenous vatinoxan (MK-467) in medetomidine–tiletamine–zolazepam anaesthetised red deer (*Cervus elaphus*). Vet Anaesth Analg 47(4):518–527

English AW (1991a) Velvet antler harvest from farmed deer - animal welfare and drug supply issues in Australia in 1991. In: Proceedings of a deer refresher course for veterinarians. NZVA Deer Branch, Sydney

English AW (1991b) The use carfentanil in deer. In: Proceedings of a deer course for veterinarians, vol 8, pp 191–199

FEDFA (2023) Federation of European Deer Farmers Associations. https://www.fedfa.com/fedfa-members/. Accessed 31 Jan 2023

Feldhamer GA, McShea WJ (2012) Deer: the animal answer guide. JHU Press, Baltimore

Fennessy PF (1981) Antler growth in deer. In: Proceedings of a deer seminar for veterinarians, Queenstown. Veterinary Association Deer Advisory Panel

Flaherty KL, Turk PJ, Anderson JT (2019) Comparing stakeholder attitudes toward white-tailed deer and rare plant management in Canaan Valley, West Virginia. Glob Ecol Conserv 17:e00519

Fletcher J (1990) The welfare of farmed deer: the UK perspective. Deer branch course (New Zealand)

Fletcher J (2013) Deer. Reaktion Books, London

Flint P (2012) Velvet antler removal from red deer. Doctoral dissertation, Massey University

Goss RJ (1983) Deer antlers: regeneration, function and evolution. Academic Press, New York

Haigh JC, Alsager R (1987) Yohimbine and physostigmine as antidotes to xylazine in elk (*Cervus elaphus*). J Zoo Anim Med 18(2/3):70–72

Hall JC, Gum H, Pietkoski K (2020) Wild alternatives: accounting for and rethinking the relationship between wild game and food security in Appalachian food systems. Appl Geogr 124:102329

Hicks JR (2017) Mixed emotions: a qualitative exploration of northern Illinois whitetail deer-related decision making. J Out Recreat Tour 18:113–121

Holsman RH, Peyton RB (2003) Stakeholder attitudes toward ecosystem management in southern Michigan. Wild Soc Bull 31:349–361

Johnson CB, Wilson PR, Woodbury MR et al (2005) Comparison of analgesic techniques for antler removal in halothane-anaesthetised red deer (*Cervus elaphus*): electroencephalographic responses. Vet Anaesth Analg 32(2):61–71

Kierdorf U, Kierdorf H (2002) Pedicle and first antler formation in deer: anatomical, histological and developmental aspects. Zeit Für Jagd 48(1):22–34

Kirschner SM, Rodenkirch R (2017) Assessment of butorphanol-azaperone-medetomidine combination as anesthesia for semen collection and evaluation of semen quality in white-tailed deer (*Odocoileus virginianus*). Anim Reprod Sci 184:196–203

Kreeger TJ, Mama K, Huizenga M et al (2010) Bispectral index analysis of opioid immobilization of Rocky Mountain elk. J Wildl Manage 74(4):902–905

Lakhdhir S, Caulkett N, Duke-Novakovski T (2021) Evaluation of intramuscular sodium nitroprusside injection to improve oxygenation in white-tailed deer (*Odocoileus virginianus*) anesthetised with medetomidine–alfaxalone–azaperone. Vet Anaesth Analg 48(1):65–73

Landaeta-Hernández AJ, Ungerfeld R, Chenoweth PJ (2022) Biostimulation and pheromones in livestock: a review. Anim Reprod Sci 248:107154

Landete-Castillejos T, Kierdorf H, Gomez S et al (2019) Antlers-evolution, development, structure, composition and biomechanics of an outstanding type of bone. Bone 128:115046

Larsen RS, Cebra CK, Wild MA (2000) Diagnosis and treatment of obstructive urolithiasis in a captive rocky mountain wapiti (*Cervus elaphus nelsoni*). J Zoo Wild Med 31(2):236–239

Li C (1993) Chinese deer farm industry. In: Woodhouse I (ed) Proceedings of the first world deer congress, Christchurch

Li C, Suttie JM (2000) Histological studies of pedicle skin formation and its transformation to antler velvet in red deer (*Cervus elaphus*). Anat Rec 260(1):62–71

Li C, Sheard PW, Corson ID et al (1993) Pedicle and antler development following sectioning of the sensory nerves to the antlerogenic region of red deer (*Cervus elaphus*). J Exp Zool 267(2):188–197

Losey RJ, Nomokonova T, Arzyutov DV et al (2021) Domestication as enskilment: harnessing reindeer in arctic Siberia. J Archaeol Method Theo 28:197–231. https://doi.org/10.1007/s10816-020-09455-w

Mackintosh CG (1985a) Reversal of xylazine in red deer. Vet Rec 117(19):507

Mackintosh CG (1985b) Potential antidote for Rompun (xylazine) in humans. N Z Med J 89(785):714–715

Mackintosh CG, Cross JP (1989) Xylazine study report. Proceedings of a deer course for veterinarians 6:136–143

Mackintosh CG, Waldrup K (1994) Practical anaesthesia and analgesia in farmed deer. In: Milne JA (ed) Recent developments in deer biology. Proceedings of the 3rd international congress on the biology of deer. MLURI, Aberdeen, p 318

Mansfield KG, Verstraete FJ, Pascoe PJ (2006) From the field: mitigating pain during tooth extraction from conscious deer. Wild Soc Bull 34(1):201–202

Mansour C, Hout NE, Hassoun R (2022) Anaesthetic management in adult fallow deer undergoing orthopaedic surgery. Vet Rec Case Rep 10(2):e344

Martín P, Messana C, Di Simone G et al (2023) Animal husbandry in Sicilian prehistory: the zooarchaeological perspective from Vallone Inferno (Scillato, Palermo). J Archaeol Sci Rep 48:103813

Matthews LR (1995) Animal welfare in the deer industry: a provocative view. In: Proceedings of a deer course for veterinarians, vol 12, pp 97–100

Matthews LR, Cook CJ (1991) Deer welfare research—Ruakura findings. In: Proceedings of a deer course for veterinarians, vol 8, pp 120–127

Matthews LR, Suttie JM (2001) Research progress in non-chemical techniques for inducing analgesia prior to velvet antler removal. In: Sim JS, Sunnwoo HH, Hudson RJ, Jeong BT (eds) Antler science and product

technology. Antler Science and Product Technology Research Centre, Edmonton, p 411
Matthews LR, Ingram J, Cook C et al (1992) Induction and assessment of velvet analgesia. In: Proceedings of a deer course for veterinarians, vol 9, pp 69–76
Matthews LR, Carragher JF, Ingram JR (1994) Post-velveting stress in free-ranging red deer. In: Proceedings of a deer course for veterinarians, vol 11, pp 138–146
Matthews L, Pollard J, Ingram J et al (1999) Non-chemical techniques for inducing analgesia prior to velveting: I. Electronic-analgesia. In: Proceedings of a deer course for veterinarians, vol 16, pp 189–192
McDermott JR, Leuenberger W, Haymes CA et al (2020) Safe use of butorphanol–azaperone–medetomidine to immobilise free-ranging white-tailed deer. Wild Soc Bull 44(2):281–291
Mentaberre G, López-Olvera JR, Casas-Díaz E et al (2010) Use of haloperidol and azaperone for stress control in roe deer (*Capreolus capreolus*) captured by means of drive-nets. Res Vet Sci 88(3):531–535
Messner TC (2011) White-tailed deer management strategies and domestication processes. Hum Ecol 39:165–178
Mitchell K, Barletta M, Giguère S et al (2021) Physiologic and blood gas effects of xylazine–ketamine versus xylazine–tiletamine–zolazepam immobilization of white-tailed deer before and after oxygen supplementation: a preliminary study. Vet Anaesth Analg 48(3):356–363
Moresco A, Larsen RS, Sleeman JM et al (2001) Use of naloxone to reverse carfentanil citrate-induced hypoxemia and cardiopulmonary depression in Rocky Mountain wapiti (*Cervus elaphus nelsoni*). J Zoo Wildl Med 32(1):81–89
Okwueze M, Thompson T, Spear M et al (2007) Motor vehicle vs deer: an antler injury to the ring finger. Wilderness Environ Med 18(1):67–69
Ortega AC, Dwinnell SP, Lasharr TN et al (2020) Effectiveness of partial sedation to reduce stress in captured mule deer. J Wildlife Manag 84(8):1445–1456
Pearce PC, Kock RA (1989) Physiological effects of etorphine, acepromazine and xylazine on the black fallow deer (*Dama dama*). Res Vet Sci 46(3):380–386
Pollard JC, Suttie JM, Littlejohn RP et al (1991) Measurement of behaviour and heart rate to assess the aversiveness of handling treatments used for red deer. In: Proceedings of a deer course for veterinarians, vol 8, pp 109–119
Pollard JC, Littlejohn RP, Johnstone P et al (1992) The effects of post-treatment analgesia on behavioural responses to velvet antler removal in red deer. In: Proceedings of a deer course for veterinarians, vol 9, pp 49–55
Pollard JC, Littlejohn R, Waldrup KA et al (1994) The effects of analgesia on the behaviour of stags at 0, 7 and 24 hours following velvet antler removal. In: Proceedings of a deer course for veterinarians, vol 11, pp 156–161
Pon K, Caulkett N, Woodbury M (2016) Efficacy and safety of a medetomidine–azaperone–alfaxalone combination in captive white-tailed deer (*Odocoileus virginianus*). J Zoo Wild Med 47(1):29–37
Price JC, Allen S, Faucheux C et al (2005) Deer antlers: a zoological curiosity or the key to understanding organ regeneration in mammals? J Anat 207:603–618
Price TD, Klassen L, Sjögren KG (2021) Pitted ware culture: isotopic evidence for contact between Sweden and Denmark across the Kattegat in the middle Neolithic, ca. 3000 BC. J Anthropol Archaeol 61:101254
Rauch H, Pohlin F, Einwaller J et al (2022) Effect of season and diet on heart rate and blood pressure in female red deer (*Cervus elaphus*) anaesthetised with medetomidine-tiletamine-zolazepam. PLoS One 17(6):e0268811
Sahebjam F, Kongara K, Chambers JP et al (2020) pressure algometry validation and determination of efficacy of articaine hydrochloride ring block in antler removal in red deer (*Cervus elaphus*). Animals 10(11):2023
Salla KM, Turunen HA, Kallio-Kujala IJ et al (2022) Effects of vatinoxan in dogs premedicated with medetomidine and butorphanol followed by sevoflurane anaesthesia: a randomised clinical study. Vet Anaesth Analg 49(6):563–571
Schwartz JA, Warren RJ, Henderson DW (1997) Captive and field tests of a method for immobilization and euthanasia of urban deer. Wild Soc Bull 25(2):532–541
Seifert D (1997) Post-velvetting infections. In: Proceedings of a deer course for veterinarians, vol 14, pp 229–238
Smith-Jones C (2022) A guide to the deer of the world. Quiller, Stroud
Stern AW, Ritchey JW, Hall B et al (2010) Nonsteroidal anti-inflammatory drug—associated renal papillary necrosis in a white-tailed deer (*Odocoileus virginianus*). J Vet Diagn Invest 22(3):476–478
Strasser TF, Murray SC, Geer AVD et al (2018) Palaeolithic cave art from Crete, Greece. J Arch Sci Rep 18:100–108
Suttie JM (1980) The effect of antler removal on dominance and fighting behaviour in farmed red deer stags. J Zool 190(2):217–224
Suttie JM (1982) Velvet and velveting. British Deer Farmers Association Newsletter 10:11–14
Suttie JM (1990) Experimental manipulation of the neural control of antler growth. In: Bubenik GA, Bubenik AB (eds) Horns, pronghorns and antlers. Springer, New York, pp 359–370
Suttie JM, Fennessy PF (1991) Growth promoting hormones and antler development. In: Bobek B, Perzanowski K, Regelin W (eds) Global trends in wildlife management. Trans. 18th IUGB Congress, Krakow 1987. Swiat Press, Krakow-Warszawa
Suttie JM, Fennessy PF (1992) Recent advances in the physiological control of velvet antler growth. In: Brown RD (ed) The biology of deer. Springer, New York, pp 471–486
Suttie JM, Li C, Sadighi M et al (1994) Physiological control of antler growth. In: Milne JA (ed) Recent developments in deer biology. Macaulay Land Use

Research Institute, Aberdeen and Moredun Research Institute
Sykes NJ, Baker KH, Carden RF et al (2011) New evidence for the establishment and management of the European fallow deer (*Dama dama dama*) in Roman Britain. J Arch Sci 38(1):156–165
USDANAL (2023) United States Department of Agriculture National Agricultural Library. https://www.nal.usda.gov/animal-health-and-welfare/animal-use-alternatives/reducing-animals-pain-and-distress#:~:text=Analgesia%20is%20the%20relief%20of,%2C%20Isoflurane%2C%20etc. Accessed 31 Jan 2023
Valente AM, Acevedo P, Figueiredo AM et al (2020) Dear deer? Maybe for now. People's perception on red deer (*Cervus elaphus*) populations in Portugal. Sci Total Environ 748:141400
Venkatachalam D, Chambers JP, Kongara K et al (2018) Pharmacokinetics of articaine hydrochloride and its metabolite articainic acid after subcutaneous administration in red deer (*Cervus elaphus*). N Z Vet J 66(1):16–20. https://doi.org/10.1080/00480169.2017.1391141
Venkatachalam D, Chambers JP, Kongara K et al (2019) Analgesic efficacy of articaine hydrochloride for velvet antler removal in red deer (*Cervus elaphus*) and analysis of drug residues in the harvested velvet antlers. N Z Vet J 67(5):228–233. https://doi.org/10.1080/00480169.2019.1611503
Vigne JD (2011) The origins of animal domestication and husbandry: a major change in the history of humanity and the biosphere. Com Ren Biol 334(3):171–181
Waldrup KA, Mackintosh CG, Key E (1992) The use of carfentanil in red deer and elk. In: Proceedings of the deer branch of the NZVA, vol 9, pp 175–180
Walker IH, Middelberg A (1988) Post-velveting deaths in stags. In: Proceedings of a deer course for veterinarians, vol 5, pp 128–134
Walsh VP, Wilson PR (2002a) Sedation and chemical restraint of deer. N Z Vet J 50(6):228–236
Walsh VP, Wilson PR (2002b) Chemical analgesia for velvet antler removal in deer. N Z Vet J 50(6):237–243
Walsh V, Wilson P, Chambers P et al (2001) Current concerns, considerations and consequences of velvet antler harvest in New Zealand. In: Proceedings of a deer course for veterinarians, vol 18, pp 79–87
Wilson PR (1989) Velvet harvesting moral, ethical and legal aspects. In: Proceedings of a deer course for veterinarians, vol 6, pp 104–117
Wilson PR, Stafford KJ (2002) Welfare of farmed deer in New Zealand. 2. Velvet antler removal. N Z Vet J 50(6):221–227
Wilson PR, Thomas DG, Stafford KJ (1999) Routes and doses of lignocaine hydrochloride for analgesia of the velvet antler of stags. N Z Vet J 47(5):167–174
Wilson PR, Stafford KJ, Mellor DJM et al (2000a) Optimum local anaesthetic administration techniques for analgesia of antlers. In: Proceedings of a deer course for veterinarians, vol 17, pp 155–160
Wilson PR, Stafford KJ, Thomas DG et al (2000b) Evaluation of techniques for lignocaine hydrochloride analgesia of the velvet antler of adult stags. N Z Vet J 48(6):182–187
Woodbury MR, Caulkett NA, Baumann D et al (2001) Comparison of analgesic techniques for antler removal in wapiti. Can Vet J 42(12):929
Woodbury MR, Caulkett NA, Johnson CB et al (2005) Comparison of analgesic techniques for antler removal in halothane-anaesthetised red deer (*Cervus elaphus*): cardiovascular and somatic responses. Vet Anaesth Analg 32(2):72–82

Birds

Thomas Dutton

Abstract

Birds, within the class Aves, are a hugely diverse group of warm-blooded vertebrates. There are over 11,000 living species of birds with a huge variation in their size. The Class Aves are characterised by having feathers, toothless beaked jaws and oviparous reproduction (produce hard-shelled eggs). All bird species have wingsalthough some have evolved over time to be flightless, with wings having vestigial function such as Ratites (i.e., Ostriches) and Sphenisciformes (i.e., penguins). Other island species of birds include the Kakapo, which is the world's only flightless parrot. All birds that are commonly kept for food production are classified as Poultry. Poultry birds include domestic fowl, turkeys, waterfowl, guinea fowl, quail, pigeons, ratites and game birds (pheasant/partridge) kept in captivity to produce meat, eggs or restocking game. The most common poultry bird is the chicken, who is the ancestor of the red jungle fowl—with domestication thought to have occurred around 8000 years ago in Southeast Asia. Despite thousands of years of captivity only recently has significant research begun into the mechanisms of pain, effective analgesia and minimising suffering in captivity.

T. Dutton (✉)
The Avian Vet, Fairford, UK
e-mail: tom@theavianvet.com

Keywords

Anaesthesia · Analgesia · Avian · Pododermatitis · Dermatitis

1 History of Domestication

Chickens (*Gallus* spp) are the most numerous and ubiquitous domesticated bird, with significant importance for food production and worldwide food security. While it is widely accepted that domestication of chickens started with domestication of red jungle fowl in Southeast Asia the time and location where this occurred is unclear (Peters et al. 2016). One hypothesis holds that middle Holocene village economies in Northern China practiced poultry farming by at least ~8.0 ka calBP (a measure of radiocarbon dating) which corresponds to the early Neolithic age. Due to a favourable climate in South China, this site is considered the most likely location of early domestication, with flocks being transported north where human assistance and shelter would have been required to allow survival (West and Zhou 1989). While the Iron Age was the main period for dispersion of chickens throughout Europe, they were already present in some areas during the late Neolithic and early Bronze Age. It is proposed that the earliest European birds arrived in Russia from China. West and Zhou (1989) although an alternate hypothesis is that birds moved to Europe through South Asia.

R. E. Clutton, G. C. Musk (eds.), *Pain and Suffering in Farmed Animals*,
https://doi.org/10.1007/978-3-032-11407-5_19

Examination of early archaeological specimens showed the size of domesticated fowl started to increase compared to their wild ancestors—likely due to regular and sufficient feeding and protection against predators, which reduced the need to fly. Energy intake was then focused on optimising food conversion into meat and fat (Von den Driesch 1987).

There is much archaeological debate about the origins of some bird specimens and confusion is created because morphological comparison of specimens identified as *Gallus* and featured in Chinese archaeological literature revealed that they are not derived from chickens, but from pheasants (Eda et al. 2016). This observation leads to an alternate theory that domestication of chickens occurred much later around 1650 BC and 1250 BC at a site called Ban Non Wat in central Thailand (Peters et al. 2022). Cereal farming is suggested to be a trigger for Chicken domestication—as this industry attracted wild birds to a more open and less forested environment, which encouraged occupation by junglefowl.

There is little evidence that the first domesticated hens were butchered for meat. Other proposed reasons for domestication include use as early alarm calls and utilising their egg production. The first depiction of chicken as meat was thought to be on Babylonian carvings from around 600 BC.

Domestication of Pigeons (*Columba livia*) was thought to have happened around 4000–5000 years ago. Far less work has been done on determining the precise time through radio dating and forensic archaeology compared to exploring the domestication of chickens. Pigeons are kept for a myriad of purposes—messaging, racing, tumbling (acrobatic flight) and meat production. Rock pigeons are native to Europe, North Africa, the Middle East and South Asia, and they were probably domesticated at several times and places (Shapiro and Domyan 2013).

More recently, 'exotic' species such as the majestic Ostrich have been kept and farmed. Commercial farming of ratites is a relatively new and evolving form of agriculture. The first commercial Ostrich farm was thought to have been founded in South Africa in 1860 (Moseki et al. 2021). Ratite farming spread into other parts of Africa, into Australasia and to the USA. Aside from the small Kiwi, the other ratites are the largest birds in the world. China, South Africa and Brazil are the biggest farmers of Ratites for meat with emerging markets in the Middle East and Europe (Kistner 2019).

2 Natural History

Birds are some of the most studied animals on the planet. They are generally easy to identify and count, are considered engaging to watch, and their populations are directly affected by climate and the environment, thus are a good indication of global ecological health. Most species are identified by a combination of sight and sound without specialist equipment and as a result have become a model group for understanding species-environment relationships. Consequently, their taxonomic, functional and phylogenetic diversity, geographic distributions, ecology and conservation status is better understood than any other comparable group of organisms (Lees et al. 2022).

Depending on the source the current estimate of living species of bird is between 9700 and 11,000, with almost 50% of these being passerine (perching) birds (Callaghan et al. 2021; Lees et al. 2022). Most birds are flighted—giving them an incredible capacity for dispersal. Birds can be found virtually anywhere on the Earth's surface from pole to pole and at least seasonally from the remotest ocean basins to the most barren desert and highest mountains (Lees et al. 2022). Birds live on all continents and all countries in the world with the lowest number in Antarctica and the greatest diversity of avian species in the Tropics. Unlike most species being restricted to living on habitable land masses, birds can live up to 11.5 km in the sky with a Rüppell's Vulture (*Gyps rueppelli*) recorded colliding with an aircraft at an altitude of 11,300 m (Laybourne 1974). While many pelagic species may remain on the high seas thousands of kilometres from land for much of the year, an Emperor Penguin (*Aptenodytes forsteri*) has been recorded diving to 564 m below sea level (Wienecke et al. 2007). These data illustrate the range of heights and depths at which birds are physiologically capable of operating. For some species nearly all function occurs 'on the wing' with Swifts and Frigatebirds eating, sleeping and copulating on the wing (Weimerskirch et al. 2016). The greatest proportion of birds are invertivores, closely followed by omnivores, contrasting with far smaller numbers of scavengers and nectarivores. There is also incredible diversity in the size of birds—from the Bee Hummingbird weighing only a few grams up to the Ostrich weighing up to 130 kg and standing over 2.5 m in height.

Virtually all metrics of avian biodiversity are exhibiting globally consistent negative trends, with the International Union for Conservation of Nature's Red List Index showing a steady deterioration in the conservation status of the global avifauna over the past three decades (Lees et al. 2022). The resultant loss of avian diversity is non-random and has, so far, disproportionately affected large, flightless and insular species, with the median mass of extinct species seven times larger than that of extant ones (Fromm and Meiri 2021).

There are multiple opinions on the evolutionary history of the class Aves, with much controversy. Most theories include different Classes within the Dinosauria (Clarke and Middleton 2006) with various extinction events driving biodiversity and evolution.

More than half of all bird species are restricted to the tropics, but a remarkable 91% of all birds have geographic ranges that intersect at least seasonally with the tropics via migration (Barlow et al. 2018). Avian species richness is unevenly distributed across biogeographic realms, with the Neotropical realm hosting ~36% of all known landbird species, followed by the Afrotropical (~21%), Indomalayan (~18%), Australasian (~17%), Palearctic (~10%), Nearctic (~8%) and Oceanic (~2%) realms (Lees et al. 2022).

Avian species and taxonomy were studied extensively between 1700 and 1900 when the majority of avian species were described. However, many species have also been discovered/identified in recent years, with approximately 250 in the last 75 years (Sangster and Luksenburg 2015). Recent changes in the way avian vocabulary and molecular analysis have been used to describe species have led to new species being described that had previously been placed together due to similarities in plumage but have significantly different evolution (Sangster 2014).

Wild birds and products derived from them are also economically important as food (meat, eggs and, in some cases, nests) or guano as fertilizer. The chicken (*Gallus gallus domesticus*), which is the domesticated Red Junglefowl (*Gallus gallus*) is now the most abundant bird on Earth with current estimates of approximately 20 million alive at any time (Robinson et al. 2014). Including all the other species of captive poultry (waterfowl, ratites, pigeon, etc), the biomass is approximately threefold higher than that of wild birds (Bar-On et al. 2018).

Approximately half of all extant bird species are undergoing population decline, 40% are stable, 6% increasing and the remaining are unknown (BirdLife International 2020). Migratory bird species are most affected by population declines (Sanderson et al. 2006) and grassland and farmland species are most negatively affected largely due to the intensification of agriculture (Gregory et al. 2019).

3 Current Extent of Utilisation

Almost half of all bird species (4561 bird species; 46%) have been recorded as used for one purpose or another by humans. It is estimated that 37% of bird species are utilised as pets, 14% are hunted for food and 4% are hunted for sport (Butchart 2008).

Unsustainable extraction for food, leading to population declines, sometimes coupled with sport hunting, remains an issue for wild avian species. For example, 11–36 million birds are estimated to be killed/taken illegally in the Mediterranean region, including two million in

Italy (Brochet et al. 2016). Migratory birds are at particular risk of overharvesting. For example, Ortolan Buntings (*Emberiza hortulana*) trapped in Southwest France come from declining northern and western European populations—a finding that supports a ban on the harvest of the species (Jiguet et al. 2019). There is also legal and illegal wildlife trade for pet, or wild animal products. Species like the Helmeted hornbill (*Rhinoplax vigil*) are desirable in Chinese medicine, leading to significant population declines, principally in Sumatra and Borneo (Beastall et al. 2016). Of the approximately 10,000 bird species, approximately 45% have been recorded in the wildlife trade, and traded species are more threatened than nontraded ones (Scheffers et al. 2019).

Broiler chickens (meat) and laying hens (eggs) provide most of the poultry products for human consumption globally. Other poultry species include waterfowl, ratites, pigeons and gamebirds. The poultry industry is the leading supplier of efficient, high-quality animal proteins to the world (Korver 2023). Poultry meat provides a favourable balance of proteins, amino acids, energy and micronutrients.

Poultry meat compares favourably with other animal products in terms of protein content and amino acid balance, energy and micronutrients (Bohrer 2017). Poultry meat is lean, with low amounts of intramuscular fat when compared to mammalian meat (Pereira and Vicente 2013). Most poultry meat species show rapid and efficient growth and there is a very short generation interval compared to commercially reared mammals. Eggs provide an excellent amino acid profile, a good source of energy, essential fatty acids and high levels of many vitamins and minerals (Bohrer 2017).

Poultry utilisation is increasing, particularly in the developing world, and less economically developed countries, as poultry rearing has relatively lower housing and start-up costs compared to most mammalian farming systems—making it an excellent option for improving nutrition in impoverished humans (Korver 2023).

Poultry housing systems vary widely. Broilers can be in barns or free-range. Laying hens can be in cages, enriched cages, barns and free-range systems. Space and welfare requirements vary from country to country. Most poultry meat production is realized with intensive production systems, and extensive rearing systems (ERS) of poultry (organic, free-range, and low-input) represent only a small portion of poultry production in the EU (about 5%) (Dal Bosco et al. 2021).

Pigeon as a source of meat is growing in popularity, with significant growth in the industry expected in the next 15 years. Pigeon rearing systems are typically free-range, allowing birds to express natural behaviour, which provides significant ethical advantages over other traditional poultry species. Pigeons, as a meat animal, mature early, have very short incubation times, are considered desirable to eat, and production systems have low input costs (Kalita and Borah 2024).

Domestic ducks trace back to two species of the mallard duck (*Anas platyrhynchos*) and the Muscovy duck (*Cairina moschata*), while the origins of geese are the Greylag goose (*Anser anser*) and the Swan goose (*Anser cygnoides*) (El Sabry and Almasri 2023). Waterfowl have been identified as an area of agriculture which could be upscaled to provide improved global food security (Pingel and Germany 2011). Several farming systems are used with waterfowl—from intensive housed systems to extensive lower input free-range systems. Some integrated systems utilise waterfowl for pest/weed control, where they consume pest species such as rasses and broadleaved weed plants in many crops, where pesticides cannot be used (Patil et al. 2021). The total size of the global duck population is about 1,099,332,000 head, and most of them are distributed throughout Eurasia. The largest duck populations are found in Asian countries. Goose numbers are smaller with 363,132,000 geese worldwide, with about 89.8% of the goose population found in Asia, 7.3% in Africa and 2.6% in Europe. China produces approximately 80% of the global duck meat and egg production, with a global production estimated at around six million tonnes (El Sabry and Almasri 2023). The contribution of duck meat to global poultry meat production is about 4.1%, which is significantly lower compared to the contribution of chicken (Patil et al. 2021).

Ratites are still a very niche part of avian agriculture. Ostriches (Struthio camelus) are the largest living birds originating from Africa and the Middle East. With the males being larger than the females, they are sexually dimorphic; males have distinctive black feathers and wing plumes. The common Rhea (*Rhea americana*) and Darwin's or lesser Rhea (*Rhea pterocnemia*) are found in South America. The emu (*Dromaius novaehollandiae*) originates from Australia. Emus are second in size to the ostrich, showing reverse sexual dimorphism with the females slightly larger than males. Ostriches are the most commonly farmed Ratite, with Emu also considered commercially important. They are reared commercially for their meat, hide and feathers (Ostrich only). There is also a small amount of consumption of Ratite eggs. In comparison to mammalian livestock, Ostrich farming can compare favourably commercially because of the higher and faster financial returns of ostrich projects. Whereas a cow produces a calf that reaches market weight after 654 days from conception, yielding 250 kg of meat, an ostrich produces approximately 40 chicks annually that reach market age after only 407 days from conception (42 days incubation +365 days of age) and yield 1800 kg of meat, 50 m^2 of leather and 36 kg of feathers each year. The net weight of meat represents 50% of live weight in ostriches, a percentage much higher than that of other farm animals such as cattle, sheep and most other poultry (Shanawany 2013). Ostrich are considered to have lower fat and cholesterol but higher protein then both beef and chicken. Like other commercial poultry production, Ratites can be farmed in free range, semi-intensive and intensive systems. In extensive production, there is a choice of natural and artificial incubation.

The other ratites have no role in agriculture and very limited appearance in zoological collections. There are three cassowary species, the Double-wattled (*Casuarius casuarius*), Single-wattled (*Casuarius unappendiculatus*) and the Dwarf cassowary (*Casuarius bennetti*), who are rainforest species from Australasia. Kiwi are very different from the other ratite species—being somewhat small and confined to New Zealand; the Great Spotted Kiwi (*Apteryx haastii*), the Little Spotted Kiwi (*Apteryx owenii*) and the Brown Kiwi (*Apteryx australis*). Tinamous are contentious, with some classifying them as ratites, others considering them a very close relative. They are small flighted ground-dwelling birds from the forests of South America.

Outside of commercial farming, there has been a significant increase in the number of small-scale, backyard/hobby flocks. One US Department of Agriculture study found that 0.8% of all households in the United States own chickens and 4% of households without chickens were planning to obtain them within 5 years (USDA 2021). Typically (but not always) these are provided with more extensive accommodation, are largely free range, and can be treated more as pets by some owners than as farm animals. Owners are prepared to provide advanced veterinary treatments that would not be viable financially on a commercial farm. Birds are kept into later stages of life—and chronic older age health problems are seen far more commonly. Although their ability to display natural behaviours is enhanced, and general welfare is considered higher than intensive commercially farmed birds, they are often kept to lower biosecurity standards, which exposes birds to the risk of infectious disease with associated pain, morbidity and mortality.

Pet bird keeping trends vary from country to country. Due to a lack of registration in many countries accurate data is not available. In the USA, pet birds are the fourth most commonly kept species after dogs, cats and fish (Kidd and Kidd 1998). While in North America and Europe, there has been a move away from caged birds as companions, in some parts due to welfare concerns, this trend is not seen in all countries. India, for example, is a rapidly growing market for pet bird keeping. An emerging middle class in developing countries, falling human birth rates, and in certain countries like China a reduction is legislation has led to an increase in pet keeping.

Birds are commonly kept species in zoological collections, both private and licensed. The range of species kept can be vast. Birds can be kept for flying displays—particularly raptors—and psittacines in aviaries. Zoo licensing legislation varies from country to country, but there is a significant move in many European countries to reduce practices such as tethering and pinioning in licensed zoological collections.

4 Attitudes to Pain

There has been significant improvement into the attitudes towards pain in birds in the last 10–20 years, with an increased body of research (Prunier et al. 2013). Improved research and understanding of the mechanisms of pain in birds shows that from transduction to transmission, modulation, projection and perception, birds possess the neurologic components necessary to respond to painful stimuli and they likely perceive pain like mammals (Douglas et al. 2018).

In farmed birds the vast majority of poultry are still kept in intensive rearing systems. This approach to husbandry can lead to a number of conditions that cause pain, such as pododermatitis, keel sores, feather picking, vent picking and cannibalism (Duncan 2001). While efforts to improve the welfare of farmed birds are ongoing—some of the problems seen and sometimes tolerated in birds would be considered completely unacceptable in mammals.

Painful procedures such as pinioning, beak trimming and devoicing (in cockerels) are generally either becomingly increasingly restricted or completely banned depending on the country or system utilised. Beak trimming is particularly controversial—as it may prevent other painful problems including feather pecking, vent picking and cannibalism later in life. This benefit is traded against the pain—acute, chronic, or both associated with the procedure and complications such as neuroma formation. A bird's beak is a complex functional organ with an extensive nerve supply and various sensory receptors. The complexity and plasticity of the nervous system and poultry's inability to communicate verbally make pain difficult to measure directly. Alternative methods, such as utilising genetic selection, has been shown to eliminate the incidence of feather picking and cannibalism. Initially, these strategies are complex and more costly but prevent unnecessary pain and welfare compromise in the long term (Cheng 2005). When comparisons in beak-trimmed (T) and non-beak-trimmed (NT) barn laying flocks have been performed, mortality tended to be higher in NT flocks compared with T flocks (14.2 vs. 8.6%; $P = 0.06$). The prevalence of keel bone damage was higher among hens with poor plumage condition than hens with moderate/good plumage condition (31.5 vs. 22.2%; $P < 0.001$). Thus, it was concluded that omitting beak trimming had negative consequences for the condition of plumage, skin, and keel bone, and tended to increase mortality, highlighting the risk of reduced welfare when keeping layers with intact beaks (Riber and Hinrichsen 2017).

As well as the discussion above on beak trimming, physiological and behavioural experiments have identified the problem of acute pain following shackling (suspended from legs prior to slaughter), feather pecking and environmental pollution. Chronic pain is often considered an even greater welfare problem because it can last for extended time periods—and potentially throughout a significant proportion of an animal's life. Evidence for possible chronic pain in poultry has been evidenced in beak trimming in older birds (typically >10 days age), orthopaedic disease in broiler and bone breakage due to metabolic bone disease in laying hens. Experiments on pain in chickens have not only identified acute and chronically painful conditions but also have provided information on qualitative differences in the pain experienced, as well as identifying a cognitive component providing evidence of conscious pain perception (Gentle 2011).

The evidence behind increased pain associated with beak trimming in birds >10 days of age is controversial. Jongman et al. 2008 showed with beak sensitivity studies that the beak of birds trimmed at hatch and re-trimmed at 14 weeks may be more sensitive than in birds whose beaks were trimmed at the time of hatching only. There is limited evidence that the beaks of trimmed birds have an altered threshold to potentially painful stimuli. The risk of neuroma formation is reported to be higher in older birds and as low as zero in young birds <10 days age (Gentle 1986).

Several different orthopaedic conditions have been reported in broiler chickens. Soft tissue (tendon or ligament) injuries are commonly reported along with femoral head avulsion fractures due to disturbed endochondral ossification (Duff and Hocking 1986). The most common skeletal pathologies of young broilers affect the growth plate. In adult (breeding stock) progressive degeneration of the articular cartilage results in osteoarthrosis and degenerative joint disease, lameness and a consequential loss of reproductive performance (Thorp 1994). In laying hens, osteoporosis leads to fragile bones and a risk of pathological fractures, caused by bone loss during folliculogenesis prior to egg laying. There are relationships between genetics, environment and nutrition in influencing the onset of osteoporosis in laying hens (Fleming et al. 2006). Genetic selection, freedom to exercise, and feeding particulate limestone all have a positive effect on layers bone strength. Increased freedom to exercise, however, can increase the risk of keel damage (Fleming et al. 2006). The European ban on battery changes has forced change towards barn and furnished cage systems, which has increased the prevalence of keel bone fractures (Nasr et al. 2012).

All farmed avian species have evolved from wild birds that are natural prey species. Even the largest living bird, the Ostrich, has wild predators. Prey species have evolved to hide signs of pain as a means of survival and as such, these animals do not demonstrate obvious pain behaviours. Wild birds showing pain will be targeted by predators and picked out of a flock or group.

5 Anaesthesia, Analgesia and Pain

In the last 10–20 years the amount of published research on general anaesthesia, analgesia and treating pain in birds has grown. Most of the research however has been conducted in a very small number of species and care must be taken when extrapolating between species—even those who seem evolutionarily close. Species differences are most apparent where digestive function is markedly different between species—such as carnivorous vs seed eating birds. While birds don't show the same gastrointestinal variability as a carnivorous vs ruminant mammal, there is still significant anatomic variation between the anatomy of the proventriculus and ventriculus in birds depending on the digestive evolution. While the variation between species is likely to limit the value of extrapolation, extrapolation between chicken breeds has also been shown to be unreliable (Souza et al. 2021). Compared to studies in White Leghorn hens, meloxicam in Wyandotte hens has a longer terminal half-life, greater area under the plasma concentration-versus-time curve from time 0 to infinity, a smaller elimination rate constant, and a longer mean residence time-versus-time curve from time 0 to infinity, and longer drug persistence in egg yolks (Souza et al. 2021). While not an analgesic or anaesthetic drug the coccidiostat lasalocid has also been investigated in chickens and the breed (Isa Brown and Green-legged Partridge laying hens) significantly influenced not only absorption but also the elimination of lasalocid. This variability in pharmacokinetics may result in breed-related differences in drug residue accumulation in eggs, and ultimately, the risk associated with consumer exposure to drug residues may also vary (Bello et al. 2024).

6 General Anaesthesia

Due to economics and scale general anaesthesia is rarely utilised in most farmed birds. The cost of an anaesthetic procedure would far outweigh the financial value of an individual bird in most cases. With poultry, back-yard and small-scale flock birds are much more common anaesthetic patients than their farmed counterparts, as are some economically important breeding birds. Companion pet and zoological birds are also far more common anaesthetic patients than wild birds—and these species are where much of the research has been conducted.

Reported mortality rates within 48 h of anaesthesia are relatively high in avian species

(1.76–3.95%) compared to feline (0.24%), canine (0.17%), and human (0.005–0.02%) patients (Brodbelt et al. 2008; Carter and Story 2013). One retrospective case series assessing outcomes following inhalation anaesthesia in birds demonstrated an anaesthesia-associated mortality rate of 7.7% (Seamon et al. 2017). A more recent study looking at how The American Society of Anesthesiologists (ASA) physical status grading system related to anaesthetic mortality reported that bird patients assigned an ASA grade of 1 had a mortality rate of 0% (95% confidence interval [CI] 0.0–0.2), and patients assigned an ASA grade of 2 had a mortality rate of 0.6% (CI 0.2–1.3%). Patients assigned ASA grades 3, 4, and 5 had mortality rates of 5.9% (CI 4.3–8.0), 18.8% (CI 13.4–25.7) and 50.0% (CI 29.9–70.1) respectively. Patients assigned an ASA grade of 3 were found to have 11.5 times the odds of a mortality event (95% CI 5.0–31.8) compared to ASA grades 1 and 2 (Hollwarth et al. 2022).

The majority of general anaesthesia in birds will be conducted in a dedicated clinic or hospital with access to anaesthetic machines, and electrical monitoring equipment. However, field anaesthesia is also possible—and may be utilised on farms, in remote areas and in wild birds (such as surgically fitting radiotrackers). Specialist mobile anaesthetic machines have been developed for us in field conditions (Small et al. 2004).

7 Avian Anaesthesia Anatomy

Birds have evolved a very different respiratory anatomy to mammals. This anatomical variation has significant implications on gaseous exchange and anaesthesia. In short, the avian respiratory system consists of nares, nasopharynx, choana, larynx, glottis, trachea, syrinx, lungs and air sacs. The main difference to the mammalian system is a far simpler glottis, which makes tracheal intubation more straightforward than in mammals, a longer but wider trachea, a syrinx at the tracheal bifurcation that acts as a birds 'voice box', rigid but very thin-walled and efficient lung structure and a system of airsacs, which act as bellows.

Some hornbill cranes and waterfowl have a cartilaginous 'spike' ventral to the glottis (crista ventralis) that can affect the size of the endotracheal tube that can be placed, and increases the risk of volatile agent leaking into the environment (Lawton 2000). The trachea consists of complete tracheal rings—this means the ability of the tracheal lumen diameter to increase is severely limited. As a result, endotracheal tubes should be non-cuffed to prevent the risk of pressure necrosis on the tracheal mucosa and the development of post-intubation tracheal structures. The trachea can have significant variation in anatomy between species. There is a medial tracheal septum in some penguins and petrels (bifurcation of trachea), an inflatable saclike diverticulum in emus (*Dromaius novaehollandiae*) and Ruddy ducks (*Oxyura jamaicensis*), and finally a complex system of tracheal loops and coils located in the caudal neck, keel or thorax in some cranes, swans and shovellers (Orosz 1997; McLelland 1989). In most species, the tracheal diameter narrows towards the bifurcation into the two mainstem bronchi. As such, an endotracheal tube that fits well in the distal trachea could be excessively tight and traumatising the wall in the proximal trachea. A solution for this is using stepped ET tubes called 'Coles' tubes that keep the tip of the ET tube off the tracheal mucosa. Reducing the risk of mechanical trauma.

Another unique anatomical structure in the avian respiratory system is the syrinx. This structure is within thorax of the bird (in comparison to the larynx in the throat region of mammals). The syrinx consists of consists of ossified syringeal cartilages (tympanum, pessulus, tracheal and bronchial syringeal cartilages). In male ducks, there is a tracheal bulmous expansion at the syrinx. The main function of the syrinx is production of voice/sounds.

Most birds have a system of nine air sacs. The cranial air sacs are the clavicular, cervical (paired) and cranial thoracic (paired). The caudal air sacs are the paired caudal thoracic and abdominal. Movement of the respiratory muscles—intercostals, superficial and deep pectorals lead to a chance in the potential volume of the air sacs so they act as bellows to move air across the rigid lung structure. They are very thin-walled, poorly vascularised, typically transparent and have limited function in gaseous exchange.

Birds can have bones that link to the respiratory air sacs. This feature is species specific—in most birds the humerus, keel and femur are pneumatised. However, in pelicans, the ulna is also pneumatised. This is particularly important to remember when treating fractures or if placing intraosseous catheters (where pneumatised bones must not be used).

The lung structure of birds is very different to mammalian species. The lungs are rigid, and the airsacs are required to work as bellows to drive the unidirectional flow of air through the lungs. Being rigid, the gaseous exchange surfaces can be thinner and more efficient than mammalian alveoli. The continuous unidirectional flow across the exchange surfaces with a cross current blood flow creates a respiratory system, which is considered to be up to 10× more efficient than mammals. This leads to rapid induction and change in respiratory depth when using volatile anaesthetic agents.

8 Local Anaesthesia

There has recently been a significant increase in the use of local anaesthetic and regional anaesthetic techniques in avian species. Early reports of toxicity appear to be unfounded—and if species specific dosing and careful allometric scaling is followed birds do not appear to be any more sensitive to local anaesthetics than mammalian species. Brandão et al. (2014) demonstrated that 6 mg/kg IV lidocaine did not produce any adverse cardiac affects in chickens. Lidocaine has been used in cutaneous anaesthesia in several avian species (Clyde et al. 1999). Brandão et al. (2014) demonstrated that 6 mg/kg IV lidocaine did not produce any adverse cardiac effects in chickens.

Bupivacaine typically has a longer duration of action and has been less frequently used in birds due to a relative lack of pharmacokinetic activity and concerns about the potential for prolonged toxicity.

Machin et al. demonstrated plasma half-lives are a single dose of bupivacaine of 6–12 h. But unreliable efficacy was reported with bupivacaine when developing a brachial plexus block technique in mallard ducks (Anas platyrhynchos) (Brenner et al. 2010). In broiler chickens, there is a 50% probability of clinically significant change in mean arterial pressure at doses of 1.96 mg/kg bupivacaine (DiGeronimo et al. 2017).

Standing sedation and a lidocaine line block has been advocated as an alternative to general anaesthesia while performing surgery to remove crop foreign bodies and impactions in chickens. The proposed benefits are reduced risk of regurgitation and aspiration. This is likely a procedure only performed in backyard or valuable breeding birds.

9 Analgesia

As with the anaesthetic medications, species differences in pharmacokinetics and pharmacodynamics in birds require very careful use of extrapolation of analgesic drug doses. There has been a significant shift in the last 10 years—particularly in relation to the use of opioid medications in birds. Early studies of opioid receptors in birds showed overrepresentation of kappa receptors (up to 76%) in the forebrain of pigeons (Mansour et al. 1988). This finding led to the routine use of kappa agonist opioids like butorphanol for analgesia in birds. More recent studies have demonstrated the presence of μ-opioid receptors in the central nervous system (CNS) of birds including chickens and it is expected that the analgesic effect of those drugs is like that reported in mammals (Khurshid et al. 2009; Csillag et al. 1990). This has been reinforced with recent pharmacokinetic and dynamic studies into alternative opioids such as buprenorphine, hydromorphone and fentanyl, which have shown promising efficacy in certain bird species. As well as having different representations of opioid receptors than mammals, there is also variation in the structure of these receptors compared to humans. When analysed the u, k and delta receptors in the peregrine falcon (*Falco peregri nus*), snowy owl (*Bubo scandiacus*), and blue-fronted Amazon parrot (*Amazona aestiva*) had nucleotide homologies with a similarity compared with humans ranging from 77% to 82%, which might account for some of the differences seen in the efficacy of the opiate drugs (Duhamelle et al. 2018).

One of the problems with some of the modern potent synthetic opioid drugs is that the duration of action in birds is very short. The stress associated with repeated handling and restraint may become problematic as some drugs may need to be administered multiple times a day. Several sustained-release and depot formulations are starting to be investigated in birds to see if an increased dosing interval can be achieved.

Carprofen administered at a dose of 1 mg/kg substantially increased the speed of lame birds, providing evidence that birds with moderate lameness suffer pain when they walk (McGeown et al. 1999). Carprofen has also been evaluated for use as an in-feed medication. When administered in feed at 40 mg/kg BW/day, a plasma concentration of 8.3 ug/ml was produced, which would be considered therapeutic based on known values in dogs and cats (Danbury et al. 2000). Carprofen administration in feed at doses from 25–100 ppm has been investigated in broiler chicks. At 100 ppm 12% mortality was seen and at 5 ppm 4% mortality was seen. Carprofen was considered nephrotoxic and hepatotoxic at these doses, causing visceral gout in broiler chicks (Dabhi et al. 2015).

In laying hens, butorphanol provided analgesia in birds with keel bone fractures compared to a saline control. The marker of effectiveness was birds with keel fractures having improved mobility scores a decreased latency to land when treated with butorphanol.

Non-steroidal anti-inflammatories (NSAIDs) are very commonly used in avian species and are an excellent choice for pain management, especially in a farm context. The expression of COX-1 and COX-2 enzymes varies between species but both enzymes have important roles in pain and inflammation (Guzman & Hawkins 2023). Broad tissue distribution of cyclo-oxygenase (COX) has been demonstrated in chickens (Mathonnet et al. 2001). Very few substantiated reports of toxicity following NSAID administration in birds exist.

Fallen (i.e., dead) stock containing pharmaceutical drugs can lead to the risk of toxicosis in birds who feed on carrion, particularly if fallen stock are not promptly removed from the farm. There has been decimation of the wild vulture population due to the widespread use of diclofenac in farmed animals, such as cattle, who are fed on after death by the carrion-eating vultures. This behaviour has led to renal failure and visceral gout in these birds. Another anti-inflammatory, nimesulide, is also known to be toxic at the concentrations found in the muscle of treated cattle (Cuthbert et al. 2016; Galligan et al. 2022).

Meloxicam is the most widely used NSAID in avian medicine, with liquid, tablet and injectable formulations being available. Pharmacokinetic studies have been performed in a several avian species, across varying orders including Galliformes, Anseriformes, Psittacines, Falconiformes, and Ratites. Differences in dosage, bioavailability, clearance, and elimination half-lives most likely contribute to the clinical differences seen with the use of this NSAID. For example, the elimination half-life of meloxicam in chickens and pigeons is 3 times as long as that in ducks, ostriches, and turkeys (Baert and Backer 2003). Significant differences have also been seen in breeds of the same species, such as the example earlier comparing meloxicam administered to Wyandotte and White Leghorn hens. This difference led to the conclusion that the dosing interval of meloxicam administered orally should be 24 h in the Wyandotte as opposed to 12 h in White Leghorn chickens (Guzman and Hawkins 2023). It has been suggested that the highly variable bioavailability in avian species (in comparison to mammals, where it is usually 85% or greater) could be due to the variability of pH in the proventriculus in avian species.

There has been interest in sustained-release formulation of meloxicam (Meloxicam SR) in avian species. However, this formulation is not currently recommended for use in parrots as the pharmacokinetic variability is high, and in flamingos, for example, there was little difference in effect compared to the standard formulation (Guzman and Hawkins 2023).

10 Pain (and Its Assessment)

There are several methods for assessing pain in avian species. Withdrawal reflexes have been demonstrated in poultry (Evrard and Balthazart 2002). A demonstrated automatic withdrawal of the foot occurred with thermal stimuli when the foot of a Japanese quail was plunged into hot water. Evasive and defensive behaviours during painful interventions involving attempts to escape (jumps, wing flapping) have been observed in chickens subjected to feather removal. Furthermore, when birds could not escape from painful stimuli, they ceased defensive, flight behaviours, and crouched immobile (Gentle and Hunter 1991). Immobility may also be seen in response to stress and fear, making these emotions difficult to differentiate in birds. For this reason, remote observation is important to understand bird behaviour without the influence of observers.

Vocalisations are associated with painful stimuli in poultry during repeated manual removal of feathers from different anatomical regions (leg, cape, etc.) and during pecking (Gentle and Hunter 1991; Collias 1987). The vocalisations during plucking were different than those recorded as distress calls during non-painful procedures such as capture. In chicken flocks where feather pecking was a persistent problem, the flocks recorded the emission of more vocalisations, particularly squawks (Bright 2008).

Behaviours to avoid stimulation of a painful region or anatomical area have been recorded in birds. Birds that have had beak trimming demonstrated failure to peck as frequently as control groups that had not undergone beak trimming (Gentle and Hunter 1991). With pelvic limb orthopaedic disease, limping, and weight shifting posture/positioning is described. Lame chickens spent more time lying and walked less (Weeks and Kestin 1997; Weeks et al. 2000), and turkeys with musculoskeletal disorders were less generally active than controls (Duncan et al. 1991). Lame chickens also display less normal grooming behaviours such as dust bathing (Vestergaard and Sanotra 1999). When induced orthopaedic pain was created with an intra-articular injection of urate, hens reduced their grooming, feeding and drinking activities (Hocking et al. 1997, 2001), and chickens with musculoskeletal disorders dust-bathed less than healthy chickens.

While not a direct measurement of pain—standardised welfare assessments can be made for clinical conditions known to cause pain in birds. Hock burn is a condition that occurs in broiler chickens due to prolonged contact with litter material, which leads to skin damage, inflammation, and discomfort in the hock region, and is correlated with footpad dermatitis. Standardised scoring systems such as the Welfare Quality® scoring system have been developed to assess bird objectively, so systems where high scores (increased severity) are identified, management changes can be implemented (Kwon et al. 2024).

Gait scoring is another method in broilers to assess pain associated with musculoskeletal problems in heavy intensively reared birds. Tibial dyschondroplasia, femoral head necrosis, and osteomyelitis are the most common painful orthopaedic conditions that lead to poor gait scores. For gait scoring birds must walk 3–5 meters on flat ground—birds are scored 0–5 with 0 representing a normal gait and 5 representing complete immobility (Kwon et al. 2024).

Pododermatitis or footpad dermatitis can affect many avian species. Pododermatitis is a cause of significant welfare concern—particularly in the poultry industry but also in zoological and pet birds. There are various scoring systems for this condition: 1–5 is used in most pet and zoo birds while in poultry a 0–2 or 0–4 system is utilised (Berg 2004).

11 Summary

Despite birds being the most numerous vertebrate animal used in food production and the fourth most commonly kept pet species, there is still a relative lack of information about their subjective experience of pain and suffering. Recent research efforts and a focus on animal welfare in kept and farmed birds is leading to improvements in welfare. Much change is consumer-driven—in the UK the move away from caged laying hens was driven more by supermarkets rejecting the product than legislation preventing the use of enriched cages. Unnecessary pain and suffering are becoming socially unacceptable in captive birds in large parts of the world, which is driving change and improvements in the sector.

References

Baert K, De Backer P (2003) Comparative pharmacokinetics of three non-steroidal anti-inflammatory drugs in five bird species. Comp Biochem Physiol C Toxicol Pharmacol 134:25–33

Barlow J, França F, Gardner TA, Hicks CC, Lennox GD, Berenguer E et al (2018) The future of hyperdiverse tropical ecosystems. Nature 559(7715):517–526

Bar-On YM, Phillips R, Milo R (2018) The biomass distribution on earth. Proc Natl Acad Sci 115(25):6506–6511

Beastall C, Shepherd CR, Hadiprakarsa Y, Martyr D (2016) Trade in the Helmeted Hornbill *Rhinoplax vigil*: the 'ivory hornbill'. Bird Conserv Int 26:137–146

Bello A, Mochel JP, Jajor P, Bobrek K, Świtała M, Poźniak B (2024) Population pharmacokinetics and absolute oral bioavailability of Lasalocid after single intravenous and Intracrop Administration in Laying Hens. J Agric Food Chem 72(23):13348–13359. doi

Berg C (2004) Pododermatitis and hock burn in broiler chickens. In: Measuring and auditing broiler welfare. CABI Publishing, Wallingford, pp 37–49

BirdLife International (2020) IUCN Red List for birds

Bohrer BM (2017) Nutrient density and nutritional value of meat products and non-meat foods high in protein. Trends Food Sci Technol 65:103–112

Brandão J, da Cunha A, Pypendop B et al (2014) Cardiovascular tolerance of intravenous lidocaine in broiler chickens (*Gallus gallus domesticus*) anesthetized with isoflurane. Vet Anaesth Analg 41(5)

Brenner DJ, Larsen RS, Dickinson PJ, Wack RF, Williams DC, Pascoe PJ (2010) Development of an avian brachial plexus nerve block technique for perioperative analgesia in mallard ducks (Anas platyrhynchos). J Avian Med Surg 24(1):24–34

Bright A (2008) Vocalisations and acoustic parameters of flock noise from feather pecking and non-feather pecking laying flocks. Br Poult Sci 49(3):241–249

Brochet AL, Van den Bossche W, Jbour S, Ndang'Ang'A PK, Jones VR et al (2016) Preliminary assessment of the scope and scale of illegal killing and taking of birds in the Mediterranean. Bird Conserv Int 26:1–28

Brodbelt DC, Blissitt KJ, Hammond RA, Neath PJ, Young LE, Pfeiffer DU, Wood JL (2008) The risk of death: the confidential enquiry into perioperative small animal fatalities. Vet Anaesth Analg 35(5):365–373

Butchart SHM (2008) Red List Indices to measure the sustainability of species use and impacts of invasive alien species. Bird Conserv Int 18(suppl):245–262

Callaghan CT, Nakagawa S, Cornwell WK (2021) Cornwell, global abundance estimates for 9,700 bird species. Proc Natl Acad Sci USA 118:20210525

Carter J, Story DA (2013) Veterinary and human anaesthesia: an overview of some parallels and contrasts. Anaesth Intensive Care 41(6):710–718

Cheng H (2005) Morphopathological changes and pain in beak trimmed laying hens. Worlds Poult Sci J 62(1):41–52

Clarke J, Middleton K (2006) Bird evolution. Curr Biol 16(10):R350–R354

Clyde VL, Paul-Murphy J (1999) Avian analgesia. In: ME Fowler, RE Miller (Ed). Zoo and wild animal medicine: current therapy 4. Philidelphia: W.B. Saunders, pp 309–314

Collias NE (1987) The vocal repertoire of the red junglefowl: a spectrographic classification and the code of communication. Condor 89(3):510–524

Csillag A, Bourne RC, Stewart MG (1990) Distribution of mu, delta, and kappa opioid receptor binding sites in the brain of the one-day-old domestic chick (Gallus do- mesticus): an in vitro quantitative autoradiographic study. J Comp Neurol 302(3):543–551

Cuthbert RJ, Taggart MA, Mohini S et al (2016) Continuing mortality of vultures in India associated with illegal veterinary use of diclofenac and a potential threat from nimesulide. Oryx 50:104–112

Dabhi PB, Ghodasara DJ, Pandey S, Ghodasara PD, Patil VM, Kapadiya K, Joshi BP (2015) Toxicopathological studies of Carprofen in broiler chicks. Indian J Vet Pathol 39:181

Dal Bosco A, Mattioli S, Cartoni Mancinelli A, Cotozzolo E, Castellini C (2021) Extensive rearing systems in poultry production: the right chicken for the right farming system. A review of twenty years of scientific research in Perugia University, Italy. Animals (Basel) 11(5):1281. Published 2021 Apr 29. https://doi.org/10.3390/ani11051281

Danbury TC et al (2000) Self-selection of the analgesic drug carprofen by lame broiler chickens. Vet Rec [Online] 146(11):307–311

DiGeronimo PM, da Cunha AF, Pypendop B, Brandão J, Stout R, Rinaldi M, Tully TN Jr (2017) Cardiovascular tolerance of intravenous bupivacaine in broiler chickens (Gallus gallus domesticus) anesthetized with isoflurane. Vet Anaesth Analg 44(2):287–294

Douglas JM, Sanchez-Migallon Guzman D, Paul-Murphy JR (2018) Pain in birds: the anatomical and physiological basis. Vet Clin North Am Exot Anim Pract 21(1):17–31. https://doi.org/10.1016/j.cvex.2017.08.008. PMID: 29146030

Duff SRI, Hocking PM (1986) Chronic orthopaedic disease in adult male broiler breeding fowls. Res Vet Sci 41(3):340–348

Duhamelle A, Raiwet DL, Langlois I et al (2018) Preliminary findings of structure and expression of opioid receptor genes in a Peregrine Falcon (Falco peregrinus), a Snowy Owl (Bubo scandiacus), and a Blue-fronted Amazon Parrot (Amazona aestiva). J Avian Med Surg 32:173–184

Duncan IJ (2001) Animal welfare issues in the poultry industry: is there a lesson to be learned? J Appl Anim Welf Sci 4(3):207–221

Duncan IJH, Beatty ER, Hocking PM, Duff SRI (1991) Assessment of pain associated with degenerative hip disorders in adult male turkeys. Res Vet Sci 50(2):200–203

Eda M, Lv P, Kikuchi H, Li Z, Li F, Yuan J (2016) Reevaluation of early Holocene chicken domestication in northern China. J Archaeol Sci 67:25–31

El Sabry MI, Almasri O (2023) Global waterfowl production: stocking rate is a key factor for improving productivity and well-being-a review. Trop Anim Health Prod 55(6):419. Published 2023 Nov 24. https://doi.org/10.1007/s11250-023-03835-6

Evrard HC, Balthazart J (2002) The assessment of nociceptive and non-nociceptive skin sensitivity in the Japanese quail (Coturnix japonica). J Neurosci Methods 116(2):135–146

Fleming RH, McCormack HA, McTeir L, Whitehead CC (2006) Relationships between genetic, environmental and nutritional factors influencing osteoporosis in laying hens. Br Poult Sci 47(6):742–755

Fromm A, Meiri S (2021) Big, flightless, insular and dead: characterising the extinct birds of the quaternary. J Biogeogr 48(9):2350–2359

Galligan TH, Green RE, Wolter K et al (2022) The nonsteroidal anti-inflammatory drug nimesulide kills Gyps vultures at concentrations found in the muscle of treated cattle. Sci Total Environ 807:150788

Gentle MJ (1986) Beak trimming in poultry. Worlds Poult Sci J 42(3):268–275

Gentle MJ (2011) Pain issues in poultry. Appl Anim Behav Sci 135(3):252–258

Gentle MJ, Hunter LN (1991) Physiological and behavioural responses associated with feather removal in Gallus gallus var domesticus. Res Vet Sci 50(1):95–101

Gregory RD, Skorpilova J, Vorisek P, Butler S (2019) An analysis of trends, uncertainty and species selection shows contrasting trends of widespread forest and farmland birds in Europe. Ecol Indic 103:676–687

Guzman DSM, Hawkins MG (2023) Treatment of pain in birds. Vet Clin North Am Exotic Anim Pract 26(1):83–120

Hocking PM, Gentle MJ, Bernard R, Dunn LN (1997) Evaluation of a protocol for determining the effectiveness of pretreatment with local analgesics for reducing experimentally induced articular pain in domestic fowl. Res Vet Sci 63(3):263–267

Hocking PM, Robertson GW, Gentle MJ (2001) Effects of anti-inflammatory steroid drugs on pain coping behaviours in a model of articular pain in the domestic fowl. Res Vet Sci 71(3):161–166

Hollwarth AJ, Pestell ST, Byron-Chance DH, Dutton TA (2022) Mortality outcomes based on ASA grade in avian patients undergoing general anesthesia. J Exot Pet Med 41:14–19

Jiguet F, Robert A, Lorrillière R, Hobson KA, Kardynal KJ et al (2019) Unravelling migration connectivity reveals unsustainable hunting of the declining ortolan bunting. Sci Adv 5:eaau2642

Jongman EC, Glatz PC, Barnett JL (2008) Changes in behaviour of laying hens following beak trimming at hatch and re-trimming at 14 weeks. Asian Australas J Anim Sci 21(2):291–298

Kalita J, Borah D (2024) Comprehensive review on pigeon farming: a sustainable livelihood option for rural communities in Assam. The Pharma Innovation 12:2411–2413

Khurshid N, Agarwal V, Iyengar S (2009) Expression of μ-and δ-opioid receptors in song control regions of adult male zebra finches (Taenopygia guttata). J Chem Neuroanat 37(3):158–169

Kidd AH, Kidd RM (1998) Problems and benefits of bird ownership. Psychol Rep 83(1):131–138

Kistner C (2019) 20203376874, English, Journal article, Germany, 1617–2906, 53, (1), Cuxhaven, Lohmann Information, Lohmann & Co. AG, Ostrich production today: the (eco)logical way to economic success, (2019), Article 3

Korver DR (2023) Review: current challenges in poultry nutrition, health, and welfare. Animal 17(Suppl 2):100755. https://doi.org/10.1016/j.animal.2023.100755. Epub 2023 Mar 7. PMID: 36966027

Kwon BY, Park J, Kim DH, Lee KW (2024) Assessment of welfare problems in broilers: focus on musculoskeletal problems associated with their rapid growth. Animals 14(7):1116

Lawton MPC (2000) Anaestesia. In: Samour J (ed) Avian medicine. Harcourt Publishers Limited, London, pp 80–97

Laybourne RC (1974) Collision between a vulture and an aircraft at an altitude of 37,000 feet. Wilson Bull 86(4):461–462

Lees AC, Haskell L, Allinson T, Bezeng SB, Burfield IJ, Renjifo LM, Rosenberg KV, Viswanathan A, Butchart SHM (2022) State of the worlds birds. Annu Rev Environ Resour 47:231–260

Mansour A, Khachaturian H, Lewis ME et al (1988) Anatomy of CNS opioid receptors. Trends Neurosci 11:308–314

Mathonnet M, Lalloue F, Danty E et al (2001) Cyclooxygenase 2 tissue distribution and developmental pattern of expression in the chicken. Clin Exp Pharmacol Physiol 28:425–432

McGeown D, Danbury TC, Waterman-Pearson AE, Kestin SC (1999) Effect of carprofen on lameness in broiler chickens. Vet Rec 144(24):668–671

McLelland J (1989) Larynx and trachea. In: King AS, McLelland J (eds) Form and function in birds, vol 4. Academic, London, pp 69–103

Moseki M, Manyeula F, Moreki J (2021) History and present status of ostrich farming in Botswana. J World's Poult Res 11:481–486

Nasr MA, Nicol CJ, Murrell JC (2012) Do laying hens with keel bone fractures experience pain? PLoS One 7:e42420

Orosz SE (1997) Avian anatomy and physiology of the respiratory system: a review. Proc Annu Conf Assoc Avian Vet:3–7

Patil SS, Shinduja R, Suresh KP, Phukan S, Kumar S, Sengupta PP et al (2021) A systematic review and meta-analysis on the prevalence of infectious diseases of duck: a world perspective. Saudi J Biol Sci 28(9):5131–5144

Pereira PMDCC, Vicente AFDRB (2013) Meat nutritional composition and nutritive role in the human diet. Meat Sci 93(3):586–592

Peters J, Lebrasseur O, Deng H, Larson G (2016) Holocene cultural history of red jungle fowl (*Gallus gallus*) and its domestic descendant in East Asia. Quat Sci Rev 142:102–119

Peters J, Lebrasseur O, Irving EK, Pease PD, Paxinos JB, Smallman R, Callou C, Gardeisen A, Trixl S, Frantz L, Sykes N, Fuller DQ, Larson G (2022) The biocultural origins and dispersal of domestic chickens. Proc Natl Acad Sci U S A 119(24):e2121978119

Pingel H, Germany L (2011) Waterfowl production for food security. Lohmann Inf 46(2):32–42

Prunier A et al (2013) Identifying and monitoring pain in farm animals: a review. Animal 7(6):998–1010

Riber AB, Hinrichsen LK (2017) Welfare consequences of omitting beak trimming in barn layers. Front Vet Sci 4:222. https://doi.org/10.3389/fvets.2017.00222

Robinson TP, Wint GW, Conchedda G, Van Boeckel TP, Ercoli V et al (2014) Mapping the global distribution of livestock. PLoS One 9:e96084

Sanderson FJ, Donald PF, Pain DJ, Burfield IJ, Van Bommel FP (2006) Long-term population declines in Afro-Palearctic migrant birds. Biol Conserv 131:93–105

Sangster G (2014) The application of species criteria in avian taxonomy and its implications for the debate over species concepts. Biol Rev 89:199–214

Sangster G, Luksenburg JA (2015) Declining rates of species described per taxonomist: slowdown of progress or a side-effect of improved quality in taxonomy? Syst Biol 64:144–151

Scheffers BR, Oliveira BF, Lamb I, Edwards DP (2019) Global wildlife trade across the tree of life. Science 366:71–76

Seamon AB, Hofmeister EH, Divers SJ (2017) Outcome following inhalation anesthesia in birds at a veterinary referral hospital: 352 cases (2004–2014). J Am Vet Med Assoc 251(7):814–817

Shanawany MM (2013) Recent developments in ostrich farming. Agric Food Sci

Shapiro MD, Domyan ET (2013) Domestic pigeons. Curr Biol 23(8):R302–R303

Small MF, Baccus JT, Waggerman GL (2004) Mobile anesthesia unit for implanting radiotransmitters in birds in the field. Southwest Nat 49(2):279–282

Souza MJ, Gerhardt LE, Shannon L, Fortner C, Davis R, Condon M, Bergman JB, Cox SK (2021) Breed differences in the pharmacokinetics of orally administered meloxicam in domestic chickens (*Gallus domesticus*). J Am Vet Med Assoc 259(1):84–87. https://doi.org/10.2460/javma.259.1.84. PMID: 34125605

Thorp BH (1994) Skeletal disorders in the fowl: a review. Avian Pathol 23(2):203–236

United States Department of Agriculture (2021) Urban chicken ownership in four U.S. cities. Available at: https://www.nal.usda.gov/exhibits/ipd/frostonchickens/exhibits/show/backyardchickens/usdaresearch-backyard. Accessed 6 Feb 2021

Vestergaard S, Sanotra GS (1999) Relationships between leg disorders and changes in the behaviour of broiler chickens. Vet Rec 144(8):205–209

von den Driesch A (1987) Evolution und Haustiere – Evolution. Aufl. G. Fischer, Stuttgart/New York, pp 393–414

Weeks CA, Kestin SC (1997) The effect of leg weakness on the behviour of broilers. In: Unknown. pp 117–118

Weeks CA, Danbury TD, Davies HC, Hunt P, Kestin SC (2000) The behaviour of broiler chickens and its modification by lameness. Appl Anim Behav Sci 67(1–2):111–125

Weimerskirch H, Bishop C, Jeanniard-du-Dot T, Prudor A, Sachs G (2016) Frigate birds track atmospheric conditions over months-long transoceanic flights. Science 353(6294):74–78

West B, Zhou B-X (1989) Did chickens go north? New evidence for domestication. Worlds Poult Sci J 45(3):205–218

Wienecke B, Robertson G, Kirkwood R, Lawton K (2007) Extreme dives by free-ranging emperor penguins. Polar Biol 30:133–142

Fish

Izabella Maree Norris

Abstract

The exponentiating use of fish for food, biomedical research, or as ornaments has prompted an increased interest in their welfare, with one question in particular—their capacity to experience pain—receiving major attention. Those dismissing this possibility argue that fish neither possess the neuroanatomical and physiological processes necessary for the full (presumably mammalian) capacity to feel pain nor the behavioural repertoire necessary to convince human observers that they do. However, using the sentience framework developed by Birch et al. (Review of the evidence of sentience in cephalopod molluscs and decapod crustaceans. LSE, The London School of Economics and Political Science, 2021) this chapter provides evidence that fish experience pain and that optimising fish welfare depends on accurate pain recognition and the subsequent delivery of effective analgesic strategies. The chapter reveals the various signs that are used to assess pain in fish, e.g. physical appearance, production-related, physiological and behavioural changes, and discusses the role of anaesthetics, e.g. MS-222 and isoeugenol, and analgesics, e.g. morphine and lidocaine, in pain management.

I. M. Norris (✉)
Animal Welfare Science and Bioethics Centre,
Massey University, Palmerston North, New Zealand

Keywords

Fish pain · Fish welfare · Fish farming · Pain relief in fish · Fish and anaesthetics · Fish and analgesics · Atlantic salmon · Nile tilapia · Zebrafish · Rainbow trout

1 History of Domestication

The cultivation of fish is suspected to have first started 8000 years ago in China with the Common carp (*Cyprinus carpio*) (Teletchea 2021). Evidence of farming Nile tilapia (*Oreochromis niloticus*) has been found in Egyptian tombs, dating back 3500 years ago (Teletchea 2019, 2021; Popma and Masser 1999). The first written word of fish breeding came from China, 475 BC (Teletchea 2019). The farming of Milkfish (*Chanos chanos*), a popular fish eaten in Asia, reportedly began in the fifteenth century (Teletchea 2019). This was the first report of fish in brackish water, i.e. water with a salinity that is between that of freshwater and seawater, similar to what is measured in estuaries. Farming of fish is an old practice; however, the development of aquaculture outside of Asia is relatively more recent, e.g. about 100 years ago in North America, about 80 years ago in Africa and mid-1900s in Australia and New Zealand (Teletchea 2019, 2021).

R. E. Clutton, G. C. Musk (eds.), *Pain and Suffering in Farmed Animals*,
https://doi.org/10.1007/978-3-032-11407-5_20

Over the past 50 years, the farming of fish has expanded rapidly into a global practice producing protein for humans and animals. The farming of salmonid species, particularly the Atlantic salmon (*Salmo salar*), has increased enormously in global production from the early 1900s to the present day and is a staple of the aquaculture industry (FAO 2022; Pandey et al. 2023) (Fig. 1). Outside of cultivation, or farming and food production, fish can be found in other environments fulfilling different purposes, across the globe. For example, Rainbow trout (*Oncorhynchus mykiss*) are cultured and released into lakes and river systems to provide stock for game fishing (Haworth 2010), and Zebrafish (*Danio rerio*) are cultured for experimental use in laboratories and as pets in the ornamental industry (Avdesh et al. 2012) (Fig. 2). It is estimated that there are over 30,000 different species of fish, with new species being discovered till this day (Hastings et al. 2015). With only a small proportion of the total number of species studied, we still have much to uncover about these finned animals underwater.

2 Potential Sources of Pain

Of the six environments of interest in this book, fish are utilised by humans, to some extent, in all six. However, of particular interest in this chapter, is the cultivation of fish for human consumption in commercial farms and the use of fish in laboratory experiments. In these environments, fish are used for a specific purpose and rely on humans to meet their basic needs. This section will briefly highlight sources of potential pain (noxious stimuli) for fish used in commercial farms and laboratories.

2.1 Water Temperature and Oxygen Saturation

As poikilotherms, fish body temperatures fluctuate as water temperatures change (Huntingford and Kadri 2014). Due to this, water temperatures can influence fish metabolic rate and oxygen requirements (Stien et al. 2013; Noble et al. 2018a). Increases in water temperature increase the metabolic rate of fish and thus their oxygen

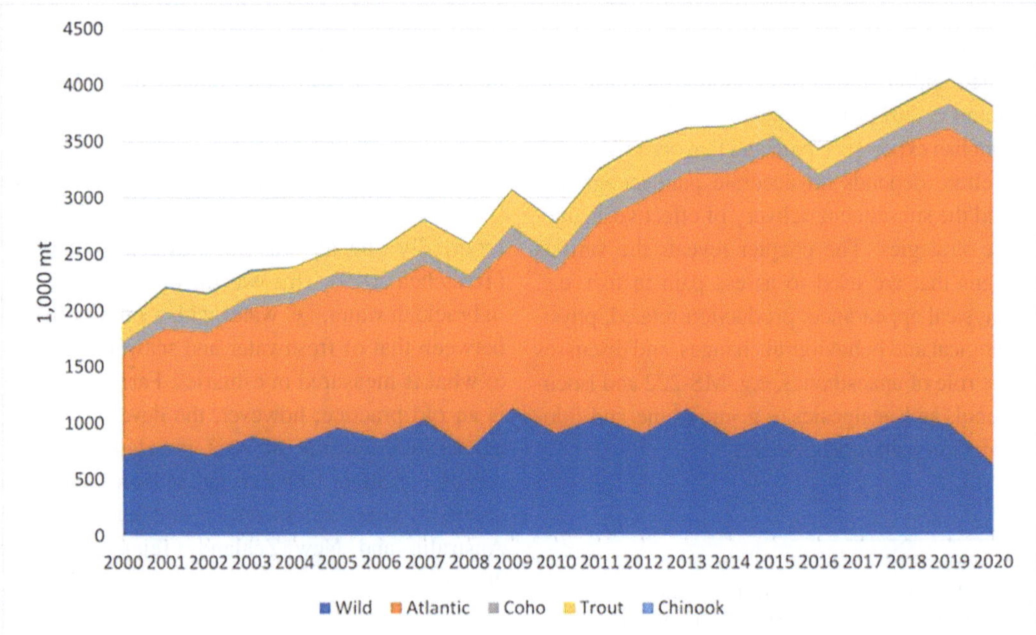

Fig. 1 Trends in global salmon production by farmed species and wild from 2000–2020. (Source: Pandey et al. (2023) https://doi.org/10.1016/j.aquaculture.2023.739972)

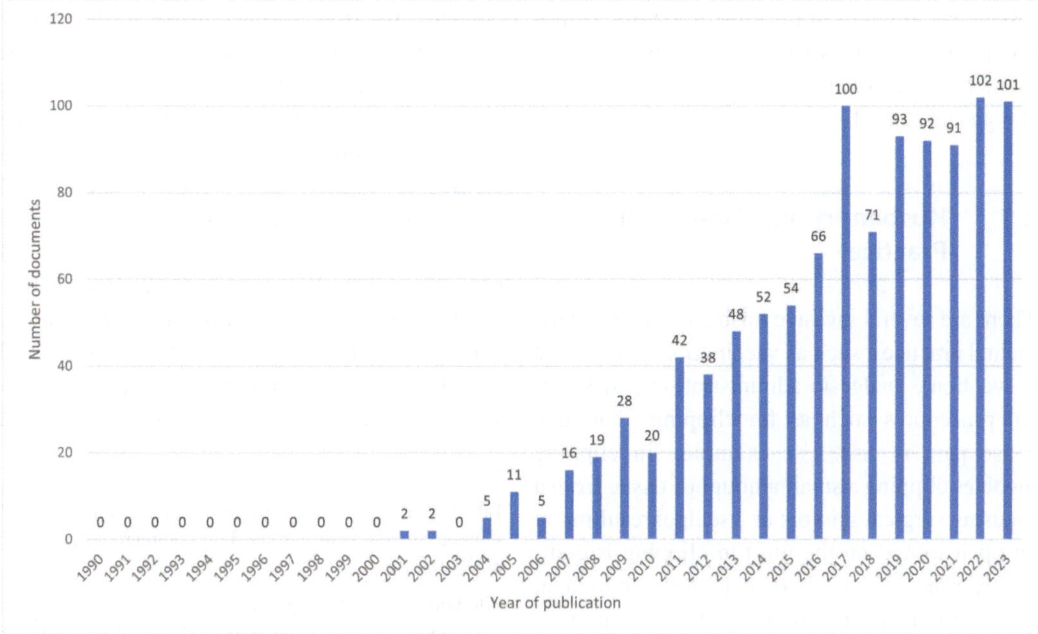

Fig. 2 The number of publications in Scopus per year in response to keywords 'zebrafish' and 'biomedical'

requirements. If oxygen levels in the water significantly decline but temperatures remain high, fish will be unable to meet their increased oxygen requirements which can induce a physiological and behavioural stress response (Noble et al. 2018a; Sundh et al. 2010). For example, sudden exposure of Atlantic salmon to noxiously high water temperatures resulted in abnormal behaviours such breaking the water surface and bouts of rapid swimming, behavioural indicators of pain (Nilsson et al. 2019).

2.2 Carbon Dioxide and Ammonia Concentration

Carbon dioxide (CO_2) and ammonia are normal waste products of respiration and digestion. Build-up of these compounds can occur in systems with low water flows, limiting waste removal. At high concentrations, both compounds are toxic to fish, impairing osmoregulation (Noble et al. 2018a; van de Vis et al. 2020). Osmoregulation is necessary for the maintenance of cellular fluid composition and volume. The relative osmotic concentrations of body fluids and environmental water influence the movement of water and ions across the gills, where solutes move down concentration gradients from areas of high concentration to areas of low concentration (Evans 2011). Inadequate osmoregulation can lead to the build-up of CO_2 and ions such as chloride, sodium and ammonia, which can be toxic to fish, and also lead to dehydration if physiological changes in the gut fail to occur (Evans 2010).

Ammonia is excreted by the gills and can react with water to form ammonium ions. If ammonium is not cleared from the water efficiently, it can impair ammonia excretion from the gills, leading to build-up in the blood and ammonia intoxication. Ammonia intoxication is associated with increased gill membrane permeability, impairing osmoregulation (Eddy 2005). Carbon dioxide can dissolve in water-forming carbonic acid. Carbonic acid decreases water pH. If water CO_2 concentrations increase and water pH decreases, blood concentrations follow, i.e. blood CO_2 concentrations will increase and blood pH will decrease (Noble et al. 2018a). Elevated concentrations of blood CO_2 can impair haemoglobin oxygen-carrying capability and subsequently respiration. Low blood pH can also result in a loss of calcium ions in the gills, impairing osmoregulation by increasing the permeability of gill membranes to water and other ions (Wendelaar Bonga and Lock 1992).

2.3 Husbandry and Experimental Practices

There are several invasive husbandry and experimental practices, such as vaccination (for disease prevention), analgesic administration and surgical procedures such as fin clipping, that may cause pain in fish. For example, fin clipping involves clipping a small amount of tissue from a fin using surgical scissors or a scalpel resulting in an open wound on the fin. Fin clipping is commonly used as a method to identify individual fish or groups of fish used in fish research or on farm (Eriksen et al. 2011). It is also used as a non-lethal method to collect fish tissue samples for stable isotope analysis (Sanderson et al. 2009) (a method used in studies of fish migration) monitoring harmful substances in the environment, such as mercury concentrations (Rolfhus et al. 2008), and to extract genetic material for genotyping (Tilley et al. 2020). Clipping induces behavioural responses, e.g. increased anxiety-like behaviours such as increased thigmotaxis and decreased feeding, and physiological changes, e.g. increased mucus production and increased plasma cortisol levels, associated with the experience of pain following traumatic injury in fish (Tilley et al. 2020; Roques et al. 2010). Other surgical procedures, such as implantation of electronic tags, may cause pain in fish. For more detailed information regarding surgical procedures in fish, see Harms (2005) and Vergneau-Grosset and Weber III (2021).

2.4 Stunning and Slaughter

The lifecycle of farmed and laboratory fish ends at harvest, where fish are ideally stunned (rendered unconscious) and then slaughtered (killed). Stunning and slaughter methods should result in rapid loss of consciousness and/or death. This prevents fish from being sensible to the experiences of pain and distress associated with the slaughter process (van de Vis et al. 2014). The stunning and slaughter methods known to be used on commercial farms and in laboratories include anaesthetic and carbon dioxide (CO_2) immersion, percussion, electrical stunning, and brain spiking (Table 1). Each of these methods, if not done properly, can result in significantly painful experiences themselves, and subsequently may leave fish sensible to the painful experience of exsanguination (bleeding out via slitting the throat arteries—the final step of fish

Table 1 Brief description of common stunning and slaughter methods of fish used on commercial farms and in laboratories

Method	Description
Anaesthetic and CO_2 immersion	The combination of anaesthetic and carbon dioxide is used to adequately stun fish before manual slitting of the throat arteries to bleed out fish. Anaesthetic immersion produces depression of the central nervous system—This depression is reversible. CO_2 immersion prolongs the stunning effect as the acidification of blood plasma by CO_2 impacts neurological and physiological systems, leading to narcosis and, potentially, death (Hayashi et al. 2004)
Percussion	Percussion requires fish to be removed from the water and a blow to be delivered to the head to produce rapid insensibility (van de Vis et al. 2020). If a sufficient force is applied, the gross trauma to the brain can result in rapid unconsciousness and death (Roth et al. 2007)
Electrical stunning	A current is passed through the fish's brain (van de Vis et al. 2020). The electrical current depolarises neurons and triggers action potentials, which disrupt the normal neural functioning of the brain, inducing a seizure-like state (Grimsbø et al. 2016)
Brain spiking	A spike is inserted, either by hand or pneumatic gun, into the brain and rotated (Robb et al. 2000; Robb and Kestin 2002; Poli et al. 2005). The rotation of the spike aims to physically destroy the cerebellum and/or medulla oblongata to result in immediate brain death

slaughter). For a detailed review of stunning and slaughter methods and fish welfare, see Lines and Spence (2011) and van de Vis et al. (2003)

3 Attitudes to Pain in Fish

The expansion and intensification of aquaculture and the use of fish in laboratory experiments have prompted an increased interest amongst researchers, governments, and the public in farmed fish welfare. Understanding the Piscean capacity to perceive pain has become particularly important and relevant. Pain is defined by Molony and Kent (1997) as 'an aversive sensory and emotional experience representing an awareness by the animal of damage or threat to the integrity of its tissues… it changes the animal's physiology and behaviour to reduce or avoid damage, to reduce the likelihood of recurrence and to promote recovery'. Pain has both a sensory, e.g. detection of noxious stimuli, and emotional, e.g. negatively valenced mental experience, components.

3.1 What Is Animal Welfare?

Contemporarily viewed, animal welfare is characterised as a dynamic state within an animal, which reflects the summation of all its mental experiences at a certain point in time. Mental experiences are generated through the processing of sensory information regarding the animal's internal functional state, e.g. biological mechanisms, and external environment, e.g. species-specific behaviours (Mellor and Beausoleil 2019). Sensory inputs are cognitively processed by integrative brain regions, generating mental experiences that may be experienced as positive or negative, i.e. have affective quality or valence attached to them (Harvey et al. 2020). The valence assigned to sensory inputs is subjective, influenced by the animal's past experience and genetic traits (Beausoleil et al. 2022). In general terms, animal welfare reflects the animal's perception of its internal state and external environment, i.e. animal welfare reflects how an animal experiences its own life.

An animal's welfare state is considered to range on a continuum from very good to very bad, i.e. an animal's welfare state is dynamic (Mellor et al. 2020). Negative experiences, such as pain and fear, adversely affect an animal's welfare and push its welfare state towards the poor side of the continuum. In contrast, the provision of positive experiences, such as playful interactions with conspecifics or the opportunity to explore, acts to push an animal's welfare state in the opposite direction. However, rectifying or preventing negative experiences is not enough to achieve 'good' welfare. It can only ever, at best, achieve neutral welfare. Without first minimising an animal's negative experiences, the provision of opportunities for positive experiences does not always lead to a positive welfare state (Mellor et al. 2020). Opportunities for positive experiences may be provided to the animal, but without first ensuring that negative experiences are avoided, or alleviated, the animal may not be able, or want, to utilise the provided opportunities.

3.2 Can Fish Feel Pain?

Sentience in fish, particularly their capacity to experience pain, has been the subject of debate for several years. The school of thought opposed to the subjective experience of pain in fish is dominated by arguments of insufficient neuroanatomy and behavioural flexibility. Rose et al. (2014) and Key (2016) assign a singular neural substrate, the neocortex, as the core of consciousness. Key (2016) claims that conscious neural processing requires signal amplification and global integration. Such neural processing is thought to only occur in the human neocortex due to its cellular structure, specifically the layering (or lamination) of the cortex and columnar organisation. Here, lamination is required to separate discrete features of sensory information, while columnar organisation allows for their integration (Key 2016). Among the school of thought against pain in fish, behavioural studies are often scrutinised in isolation of other pain-relevant evidence, and thus claims are made that such behavioural studies only provide evidence of reflexive rather than flexible behaviour (Rose et al. 2014).

Proving sentience, or the capacity for subjective experience, in non-human animals is rightfully termed the 'Hard Problem' (Feinberg and Mallat 2016). The colour red that you see may not be the same colour that your neighbour sees, and either of these red colours may or may not be the same colour red that a fish sees. In other words, it is not possible to objectively know and describe *what* or *how* exactly an animal, or even another human being, may perceive colour, and the same is true for affective experiences, i.e. emotions. Although it is possible to describe affective experiences in terms of their physical properties such as chemical and neural pathways, there will always be an 'explanatory gap' in which the personal, subjective experience itself will remain truly unknown (Feinberg and Mallat 2016; Dawkins 2001). In isolation, each argument against 'fish pain' described above rightly do not 'prove' the capacity for subjective experience in fish. However, when brought together, individual pieces of evidence relevant to the experience of pain in fish help to build a bridge across the 'explanatory gap' (Dawkins 2001).

Dawkins (2001) uses a bridge analogy to explain the possibility of using scientific evidence to infer what is not directly, scientifically accessible—affective experiences. The use of such analogy does not claim to undoubtably prove affective experiences in non-human animals, but instead, serves as a reminder that measurable and observable evidence can only be used to *infer* such experiences. Therefore, the strength of any inference made is dependent on how the evidentiary bridge is built, i.e. dependent on the quality and quantity of evidence used. Dawkins' bridge analogy is reflected in the criteria used by Birch et al. (2021) to evidence the experience of pain in cephalopods and decapods:

1. 'The animal possesses receptors sensitive to noxious stimuli (nociceptors);
2. The animal possesses integrative brain regions capable of integrating information from different sensory sources;
3. The animal possesses neural pathways connecting the nociceptors to the integrative brain regions;
4. The animal's behavioural response to a noxious stimulus is modulated by chemical compounds affecting the nervous system through endogenous neurotransmitters and/or putative local anaesthetics, analgesics, anxiolytics, or anti-depressants;
5. The animal shows motivational trade-offs in which the disvalue of a noxious or threatening stimulus is weighed against the value of an opportunity for reward, leading to flexible decision making;
6. The animal shows flexible self-protective behaviour of a type likely to involve representing the bodily location of a noxious stimulus;
7. The animal shows associative learning in which noxious stimuli become associated with neural stimuli, and/or in which novel ways of avoiding noxious stimuli are learned through reinforcement;
8. The animal shows that it values putative analgesics or anaesthetics when injured'.

Each criterion on its own does not prove a capacity to experience pain. However, together, the eight criteria build a bridge across the 'explanatory gap' and provide a strong evidence base to infer the experience of pain. As such, the remainder of this section will present evidence, relevant to the criteria of Birch et al. (2021), to support the capacity for the subjective experience of pain in fish. A brief comparison between the putative pain pathways of fish and mammals is provided Table 2.

3.2.1 Criterion 1: Possession of Nociceptors

The detection of a noxious stimulus, i.e. a potentially painful stimulus, requires nociceptors. Nociceptors are high-threshold sensory receptors found at the nerve endings of peripheral somatosensory fibres (Birch et al. 2021) (see Chapter "Physiology of Acute Pain" by Andy Bell). Nociceptors transduce and encode information relating to actual or potential threats of tissue damage, i.e. noxious stimuli.

Table 2 Major similarities and differences between putative pain pathways in fish and mammals with reference to Birch et al. (2021) sentience criteria

Criterion	Fish	Mammals (ref Andy Bell Chapter "Physiology of Acute Pain")
1	Evidence of nociceptors that respond to mechanical, thermal and chemical stimulation (Sneddon 2003a; Mettam et al. 2012) Immunohistochemical evidence of pain pathway activation: increased expression of COX-2 gene in response to noxious stimulation (Steenbergen and Bardine 2014)	Peripheral nociceptors are activated by a noxious stimulus. Noxious stimuli may include thermal, mechanical or chemical stimuli Inflammatory mediators (e.g. prostaglandins) increase in response to noxious stimulation
2	Integrative brain structures that are associated with pain include optic tectum and pallium Optic tectum—lamination of grey and white matter similar to mammalian neocortex associated with multisensory mapping capacity (Feinberg and Mallat 2016; Woodruff 2017; Kotrschal and Kotrschal 2020) Pallium—receives sensory information and associated with learning and memory capability (Woodruff 2017; Kotrschal and Kotrschal 2020) Reciprocal communication between optic tectum and pallium alters the motor output given by the brainstem, similar to the communication of the mammalian neocortex, thalamus and brainstem (Woodruff 2017; Kotrschal and Kotrschal 2020)	Brain structures associated with pain include the thalamus, somatosensory cortex (specialised region of the neocortex), amygdala and hypothalamus
3	A-delta fibres and C-fibres identified in nociceptive pathways of fish. Identified in both the tail of the Common carp (Roques et al. 2010) and head of Rainbow trout (Sneddon et al. 2003). Although Rainbow trout have a lower density of C-fibres than mammals Evidence of activation of A-delta and C-fibres in spinal cords, cerebellum, tectum, and pallial structures in response to noxious stimuli (Nordgreen et al. 2007)	Specific sensory neurons that connect nociceptors to integrative brain regions. These are myelinated A-delta (fast, acute pain) and unmyelinated C-fibres (slow, burning pain).
4	Behavioural and physiological indicators of fish pain reduced by morphine/buprenorphine (Steenbergen and Bardine 2014; Sneddon 2003b; Costa et al. 2019) (opioids) and diclofenac (Costa et al. 2019) (NSAID) administration. The effects of opioids are reversed by opioid antagonist, naloxone (Steenbergen and Bardine 2014; Costa et al. 2019) Opioid receptor genes (OPRM1, OPRK1 and OPRD1) are conserved in fish (Gräns et al. 2014). These genes code for the mu, kappa, and delta opioid receptors Zebrafish exhibit drug-seeking behaviour reflective of morphine (opioid) addiction—a behaviour reliant on the presence of mu receptors (Bossé and Peterson 2017).	Pain relief achieved through release of endogenous and exogenous opioids via inhibitory opioid receptor action. NSAIDs provide an analgesic effect by blocking the prostaglandin synthesis activity of COX enzymes

As such, nociceptors have high thresholds for activation, which require extreme stimuli for excitation (Birch et al. 2021). Different types of nociceptors are activated by different modalities of stimuli and at varying intensities or durations. Sneddon (2003a) was the first to identify and characterise trigeminal somatosensory nerves on the head of Rainbow trout as bona fide nociceptors. Electrophysiological recordings provided evidence of five different receptor types that responded to mechanical, thermal and chemical stimulation (Sneddon 2003a). Mettam et al. (2012) confirmed the nociceptive function of trigeminally innervated receptors in Rainbow trout through electrophysiological recordings that showed only irritant chemicals, e.g. acetic acid and carbon dioxide, evoked responses in the identified receptors. Immunohistochemical evidence of nociceptive pathway activation has also been reported.

Steenbergen and Bardine (2014) observed increased expression of the COX-2 gene in Zebrafish immersed in acidic water. The COX-2 enzyme is involved in the synthesis of inflammatory prostaglandins at the site of inflamed tissue. These inflammatory molecules are associated with nociceptor sensitisation as they reduce the threshold for nociceptor activation (Birch et al. 2021). Thus, providing electrophysiological and molecular evidence of the presence and activation of nociceptive pathways.

3.2.2 Criterion 2: Possession of Integrative Brain Regions

Higher processing and integration of sensory information is required for the awareness and perception of nociceptive input as an unpleasant experience. The fish brain possesses neural structures, e.g. the fish optic tectum and pallium, that are capable of integrating multiple sources of sensory information (Woodruff 2017). The optic tectum is the main centre of visual input processing in fish and also receives input from other sensory modalities, e.g. touch, auditory, taste, mechanical and electroception (Feinberg and Mallat 2016; Kotrschal and Kotrschal 2020). The cortex-like layering, or lamination, of grey and white matter, indicates a multisensory mapping capacity of the tectum. Different layers of the tectum are proposed to receive different modalities of sensory information, similar to that of the mammalian neocortex (Feinberg and Mallat 2016). The different sensory modalities are integrated in the tectum, which allows for the generation of a unified representation of the fish's sensory 'space'. This information can then be used to direct motor output through communication with the brainstem (Feinberg and Mallat 2016; Kotrschal and Kotrschal 2020).

The Piscean optic tectum also has reciprocal neural connections with the pallium—a structure considered homologous to the mammalian cerebral-cortex. The dorsal pallium receives sensory information of all modalities, while the medial pallium is associated with learning and memory (Woodruff 2017; Kotrschal and Kotrschal 2020). Reciprocal communication between the tectum and pallium allows for memories and affective experiences to alter tectal output to the brainstem. The optic tectum is responsible for fast, crude decisions, while the pallium is responsible for more processed and nuanced decisions. It has been shown that fish with damaged pallia retain the ability to catch prey and behave relatively normally but show reduced rates of declarative learning and reduced engagement in complex social behaviours (Feinberg and Mallat 2016; Woodruff 2017; Kotrschal and Kotrschal 2020). It is likely that retaining a functional optic tectum allows for the selective attention required to catch prey (Feinberg and Mallat 2016), however, without a functional pallium learning and memory are impaired (Woodruff 2017).

3.2.3 Criterion 3: Possession of Neural Pathways Connecting Nociceptors to Integrative Brain Regions

Neural connections between nociceptors and integrative brain regions are required for the ability to consciously interpret and experience nociceptive input. Sneddon et al. (2003) identified A-delta fibres (fast, sharp pain) and C-fibres (slow, burning pain) in the trigeminal somatosensory nerves of Rainbow trout. However, Rainbow trout were found to have a lower density of C-fibres than that found in mammals. Electrophysiological recordings have shown that a noxious stimulus applied to the tail of Rainbow trout will activate A-delta and C-fibres in the spinal cord, cerebellum, tectum, and telencephalon (the location of the piscean pallium) (Dunlop and Laming 2005). An electric shock applied to the tail of Atlantic salmon has also been shown to elicit somatosensory evoked potentials in the dorsal telencephalon. The somatosensory evoked potentials provided evidence of pallial processing of input from A-delta and C-fibres (Nordgreen et al. 2007). Both studies demonstrate connectivity between nociceptors and integrative brain regions in fish.

3.2.4 Criterion 4: Chemical Modulation of Behavioural Responses to Noxious Stimuli

The behavioural response of fish to noxious stimuli can be modulated by exogenous pain relief. Rainbow trout displayed aversive behaviour following an injection with acetic acid (a noxious chemical stimulus) and were compared to those receiving acetic acid plus the opioid analgesic, morphine (Sneddon 2003b). Trout injected with both acid and morphine displayed significantly fewer aversive behaviours than trout injected with acid alone. The expression of adverse behaviour occurred alongside physiological changes such as opercular beat rate. Trout injected with acid alone had higher opercular beat rates compared to trout injected with morphine (Sneddon 2003b). Similarly in Zebrafish, exposure to acetic acid via immersion or injection produces aversive behaviours that can be reduced by opioid and nonsteroidal anti-inflammatory drug administration. Morphine and diclofenac reduced abnormal changes in Zebrafish body curvature induced by acetic acid exposure (Costa et al. 2019) and pretreatment with buprenorphine prevented acetic acid-induced swimming activity changes (Steenbergen and Bardine 2014). All effects of opioid drugs on Zebrafish behaviour were blocked or reversed by naloxone, an opioid antagonist (Steenbergen and Bardine 2014; Costa et al. 2019). Thus, it is possible to modify fish responses to noxious stimuli in ways that are consistent with the hypothesis that these pain-relieving drugs reduce the experience of pain, distress, or harm in fish.

3.2.5 Criterion 5: Motivational Trade-Offs

Motivational trade-offs that indicate the experience of pain require an animal to weigh the disvalue of a noxious stimulus against the value of a potential rewarding opportunity. Motivational trade-offs involve flexible decision-making and reveal a capacity for integrative processing and evaluation of different sensory information (Birch et al. 2021). Several studies have demonstrated the capacity of fish to make motivational trade-offs. Ashley et al. (2009) showed that shock-avoidance behaviour of Rainbow trout significantly changed in the presence of a conspecific. When isolated, trout avoided locations associated previously with electric shock stimuli. However, in the presence of a conspecific, trout are willing to remain close to the conspecific despite receiving electric shocks (Ashley et al. 2009). This suggests shock-avoidance is not merely a reflex, as Rainbow trout made a motivational trade-off between remaining with a conspecific and receiving an aversive shock.

Ashley et al. (2009) also investigated the importance of pain compared to predation and social status in Rainbow trout. Noxiously stimulated fish did not show the expected anti-predator responses, suggesting pain was the priority affective experience in this situation. However, in the presence of an unfamiliar group, noxiously stimulated dominant-type fish maintained normal aggression levels, suggesting that maintenance of the social hierarchy was imperative. Similarly, Rosa et al. (2022) gave Zebrafish a choice between remaining near conspecifics in an aquarium of either phosphate-buffered solution or acetic acid or an empty aquarium with untreated water. Acetic acid is known to increase inter-fish distance; however, Zebrafish will choose to be close to acetic acid-treated fish over an empty aquarium.

3.2.6 Criterion 6: Flexible Protective Behaviours

Flexible self-protective behaviours provide stronger evidence of the experience of pain than reflex behaviours. Flexibility, according to situational context, requires higher cognitive processing than simple, reflex responses to noxious stimulation. Flexible self-protective behaviours include the expression of targeted behaviours that demonstrate an animal's awareness of the area of noxious stimulation or behaviours that decrease subsequent stimulation of that area. In response to lip injections of bee venom and acetic acid, Rainbow trout performed rocking and lip rubbing behaviours (Sneddon et al. 2003; Reilly et al. 2008). Lip rubbing behaviour indicates an awareness of the site of damage, while rocking behaviour has been interpreted as a coping behaviour similar to that seen in injured mammals (Sneddon et al. 2003). Reductions in swimming activity and increased latency to feed have been observed in Rainbow trout and Zebrafish following acetic acid injections (Reilly et al. 2008) and in Atlantic salmon after intraperitoneal vaccination (Bjørge et al. 2011). Social behaviours of Zebrafish (Rosa et al. 2022) and Atlantic salmon (Bjørge et al. 2011) are also reduced directly after injection. These behaviours serve a protective function in which injured areas are protected from further stimulation and energy is reserved for healing. In contrast, Nilsson et al. (2019) observed more erratic behaviour, including an immediate increase in swimming speed, collisions, head shaking and surface breaks, in Atlantic salmon exposed suddenly to high-temperature water. These erratic behaviours reflect protective escape behaviours relevant to the type of noxious stimulation, e.g. full body immersion in high temperature water. Therefore, there is evidence of fish expressing flexible self-protective behaviour according to the type of noxious stimulation.

3.2.7 Criterion 7: Associative Learning

Associative learning requires the animal to be able to form mental representations of two stimuli and link them together. For example, Wong et al. (2014) used a conditioned place avoidance paradigm (a motivational test with a learning and memory component) to understand Zebrafish aversion to various anaesthetics. Zebrafish were conditioned to associate anaesthetic exposure (a noxious stimuli) with a preferred side of a light/dark box. Fish spent less time in their 'preferred side' when an anaesthetic was present, suggesting aversion to the anaesthetic. After conditioning, anaesthetic was removed from the 'preferred side' and Zebrafish demonstrated complete avoidance of their preferred side despite the absence of the noxious stimulus. Wong et al. (2014) therefore demonstrated that Zebrafish are not only willing to trade off residing in their preferred environment to avoid anaesthetic exposure (Criterion 5) but also that they are able to form a learned association between a noxious stimulus and physical location (Criterion 7). Associative learning tests, such as conditioned place aversion, test the animal's response to a cue that is associated with a past experience. In this example, the conditioned place aversion test reflects the fish's aversion to a noxious stimulus that it had learnt to expect. The expression of place aversion without direct exposure to the anaesthetic suggests previous exposure events were significant enough to influence the animal's future behaviour. These studies are demonstrative of the integrative processing capacity of fish and reflect the importance of affective experience. If a fish remembers the experience of pain and avoids it later, this indicates the possession of high cognitive processing.

3.2.8 Criterion 8: No Current Evidence

Taken individually, each criterion does not unequivocally signify sentience or the experience of pain. However, the coherence of several lines of evidence provides greater confidence in the assumption of the capacity for sentience and pain experience in fish. To the author's knowledge, research has yet to be conducted demonstrating evidence of criterion 8—animal demonstrating value of putative pain relief when injured. However, there exists, as demonstrated above, scientific evidence of the ability of fish to fulfil a majority of Birch et al.'s (2021) criteria. Therefore, it can be assumed that fish are sentient beings that have the capacity to experience some forms of pain and thus, pain assessment strategies in fish should be explored and applied.

If fish are capable of experiencing pain, similar to mammals, then attention should perhaps be focussed on: farmed fish during vaccination and slaughter; laboratory species undergoing noxious procedures, including surgery; and any fish exposed to sub-optimal environmental conditions, e.g. poor water quality. Optimising fish welfare requires an ability to identify (and ideally avoid) those factors causing piscean pain, to recognise pain when it is present and to control it effectively.

4 Pain and Its Assessment in Fish

Identifying when a fish may be experiencing pain is a prerequisite to applying mitigation strategies. It also allows farmers, or the carers of fish to implement prevention strategies where appropriate. There is no single measurement or observation that indicates the affective experience of pain in fish. Animal pain assessments, similar to holistic animal welfare assessment (see Mellor et al. 2020), rely on the use of measurable and observable signs or indices, i.e. welfare indicators, to identify the experience of pain. The following section introduces welfare indicators and examples that provide evidence of pain in fish. It also emphasises important considerations when using welfare indicators in pain assessment.

4.1 Indicators of Pain in Fish

What it *feels* like for a fish to be in pain is subjective and internal, i.e. the experience of pain is not directly measurable. However, the affective experiences of animals can be indirectly assessed using measurable and observable signs. In animal welfare science, these tools are termed 'welfare indicators' (Mellor et al. 2020). Welfare indicators are measurements and observations of an animal's biological function, behaviour and external environment that can be used to cautiously infer the affective experiences of animals (Mellor et al. 2020). For example, increases in escape behaviour following immersion in a noxious substance is indicative of nociceptive pathway activation, providing evidence of the potential experience of pain. For the purposes of this chapter, 'welfare indicators' will be referred to as 'pain indicators'. This chapter identifies four different categories of pain indicators: physical appearance, production-related, behavioural and physiological (Table 3).

4.2 Physical Appearance and Production-Related Indicators of Pain

Changes in the physical appearance and production-related indicators may reflect changes in the biological functioning of fish in response to painful or stressful situations. For example, Nile tilapia injected with formalin display a change in body colour and develop lateral black stripes, which are characteristic of exposure to noxious stimuli in this species (Ide and Hoffmann 2002). Exposure to noxious levels of dissolved oxygen (130–220%) results in the appearance of subcutaneous bubbles along the bodies of Atlantic salmon smolts (Espmark et al. 2010). Subcutaneous bubbles are a sign of 'gas bubble disease'—a disease in which gas emboli in the blood vessels and major organs lead to blood flow occlusion, tissue damage and mortality (Espmark et al. 2010). Indicators of other diseases infecting Atlantic salmon and Rainbow trout are detailed in Noble et al. (2018b, 2020) respectively.

Lesions (or wounds) are a common sign of disease and are likely to be painful. Tissue damage in fish activates nociceptive neural pathways and elicits physiological and behavioural responses reminiscent of a painful experience (Birch et al. 2021). Case definitions for external skin lesion syndromes have been described for Chinook salmon in Aotearoa New Zealand (Johnston et al. 2021). Bjørge et al. (2011) have also described a system for evaluating the severity of lesions following intraperitoneal vaccination of Atlantic salmon according to peritoneal thickness and opaqueness, adhesions between organs, and the extensiveness of melanin pigmentation. Tissue inflammation, ulceration, haemorrhage and healing can also be examined at a histological level. At both gross and histological levels, focal diffuse thickening of the peritoneum indicates severe tissue damage, which is likely to be associated with the experience of pain (Bjørge et al. 2011).

Table 3 Pain indicators in various fish species adapted from Sneddon and Roques (2023)

Indicator type and pain indicator	Species and reference number			
	S. salar	*O. niloticus*	*O. mykiss*	*D. rerio*
Physical appearance				
Presence and appearance of lesions or wounds, e.g. signs of inflammation, ulceration or haemorrhage	Bjørge et al. (2011)			
Subcutaneous gas bubbles or blisters	Espmark et al. (2010)			
Changes in body and/or fin colouration		Ide and Hoffmann (2002)		
Production-related				
Bodyweight			Rizzo et al. (2017)	
Mortality rates	Espmark et al. (2010)			
Physiological				
Blood parameters	Nordgreen et al. (2013)	Roques et al. (2010)	Ashley et al. (2009), Rizzo et al. (2017), Mettam et al. (2011), and Jones et al. (2012)	
Changes in inflammation gene expression				Steenbergen and Bardine (2014), Zhao et al. (2021)
Changes in anti-oxidant gene expression				Zhao et al. (2021)
Heart rate		Ide and Hoffmann (2002)	Gräns et al. (2014)	
Mucus production		Roques et al. (2010)		
Na^+/K^+-ATPase activity		Roques et al. (2010)		
Respiration rate			Sneddon et al. (2003), Sneddon (2003b), Ashley et al. (2009), Reilly et al. (2008), Gräns et al. (2014), Rizzo et al. (2017), and Mettam et al. (2011)	Costa et al. (2019), Reilly et al. (2008)
Behavioural				
Aggressive behaviour, e.g. striking, biting	Bjørge et al. (2011), Jarvis et al. (2021)		Ashley et al. (2009)	
Body curvature index/bend	Nilsson et al. (2019)			Costa et al. (2019)

(continued)

Table 3 (continued)

Indicator type and pain indicator		Species and reference number *S. salar*	*O. niloticus*	*O. mykiss*	*D. rerio*
Changes in cover use/hiding behaviour		Nordgreen et al. (2013)	Roques et al. (2010)	Ashley et al. (2009), Reilly et al. (2008)	
Erratic behaviour or spontaneous movements:	C-turns/fast direction changes	Nilsson et al. (2019), Espmark et al. (2010)		Wagner and Stevens (2000)	Rosa et al. (2022)
	Circling	Nilsson et al. (2019)			
	Collisions	Nilsson et al. (2019)			
	Head shakes	Nilsson et al. (2019)			
	Panic episodes/ startled behaviour	Espmark et al. (2010), Jarvis et al. (2021)			
	Surface breaks	Nilsson et al. (2019)			
	Tail flick/ tail waft/fin movement			Jones et al. (2012)	Schroeder and Sneddon (2017)
Immobility/freezing				Ide and Hoffmann (2002)	Costa et al. (2019)
Lip rubbing				Sneddon et al. (2003), Sneddon (2003b), and Reilly et al. (2008)	
Reduced feed intake				Mettam et al. (2011)	
Reduced shoaling behaviour/ social interaction		Bjørge et al. (2011)			Rosa et al. (2022)
Response to food pellets		Bjørge et al. (2011), Nordgreen et al. (2013)		Sneddon (2003b), Rizzo et al. (2017), and Mettam et al. (2011)	
Rocking behaviour				Sneddon et al. (2003), Sneddon (2003b), and Reilly et al. (2008)	
Swimming activity, e.g. swimming rate, amount of time spent active, total distance travelled		Nilsson et al. (2019), Bjørge et al. (2011), and Jarvis et al. (2021)	Roques et al. (2010)	Sneddon (2003b), Ashley et al. (2009), Reilly et al. (2008), and Mettam et al. (2011)	Steenbergen and Bardine (2014), Costa et al. (2019), Reilly et al. (2008), Zhao et al. (2021), and Lopez-Luna et al. (2017a, b, c)
Thigmotaxis					Lopez-Luna et al. (2017a)
Vertical position in tank		Bjørge et al. (2011)		Rizzo et al. (2017)	Costa et al. (2019)

Note: Thigmotaxis can be described as an animal's avoidance the interior area of a new environment and preference for physical contact of a solid surface on the perimeter of the new environment. For example, when placed into a new tank, Zebrafish will remain close to the tank wall and avoid the centre of the tank (Lopez-Luna et al. 2017a)

Body condition is a useful long-term indicator of fish welfare. For example, an emaciated body condition may indicate reduced feeding activity, which often follows exposure to noxious stimuli. Mettam et al. (2011) reported a reduction in feeding behaviour following acetic acid treatment in Rainbow trout. Declining body condition and feeding behaviour may be confirmed by monitoring, i.e. repeatedly measuring body weight. Whilst often insidious, changes in bodyweight or body condition are useful indicators of chronic stress and/or exposure to noxious stimuli. These types of indicators are valuable because obvious changes in overall fish behaviour may not be apparent in some forms of chronic pain. Physical appearance and changes in production-related variables, e.g. growth rate, can reveal chronic changes that might reflect underlying pain.

4.3 Physiological Indicators of Pain

While physical appearance and production-related indicators are useful indicators of long-term welfare, short-term welfare changes are best revealed by physiological signs. For example, the respiration rate in Rainbow trout significantly increases 2 h following an abdominal incision but return to baseline within 24 h (Gräns et al. 2014). Similar physiological changes were found in Rainbow trout injected with bee venom and acetic acid (Reilly et al. 2008). The return of respiration rates to baseline rates coincided with reductions in the expression of anomalous pain-related behaviours such as rocking side-to-side and rubbing the injection site (Reilly et al. 2008), indicating some association between respiration rate and known behavioural indicators of pain.

Care must be taken when selecting physiological indicators as indicators of acute pain because the method of measuring indicators may influence assessment results. Gräns et al. (2014) found that the respiratory and heart rates of noxiously stimulated Rainbow trout were lower when non-invasive rather than invasive methods were used. Non-invasive methods of measuring respiration and heart rates rely on electrodes placed in water tanks to detect signals originating from cardio-ventilatory activity (Gräns et al. 2014). The non-invasive method does not involve direct handling of fish. However, they are only practical for measuring the respiratory and heart rates of individually housed fish. More invasive methods involve direct handling and insertion of intramuscular electrodes. The stress of handling and insertion of electrodes into the body may increase cardio-ventilatory activity per se and obfuscate measurements made when test noxious stimuli are applied.

Interpreting physiological indicators requires consideration of context. Changes in some variables commonly linked with nociception can also result from handling stress, individual housing, and exposure to novel testing environments. Rizzo et al. (2017) observed unexpected decreases (rather than stress-induced increases) in plasma cortisol levels following acetic acid injection in Rainbow trout. This was attributed to adrenal exhaustion resulting from the cumulative stress of individual housing, handling, and surgery. Consequently, physiological indicators should be used in conjunction with other methods, e.g. behavioural indicators, and with consideration for context, whenever the presence of pain is being critically evaluated in fish.

4.4 Behavioural Indicators of Pain

Behavioural indicators are highly valuable in the assessment of fish pain as they provide evidence of a mental pain experience at the time of observation. Behaviour has been described as the outcome of an animal's decision-making in relation to integrated stimuli from their internal and external environment (Dawkins 2001). Atlantic salmon exposed to highly oxygenated water displayed escape behaviours such as jumping away in an erratic and spontaneous way, as well as an increase in c-turns (which are fast directional changes) (Espmark et al. 2010). These escape behaviours provide direct evidence that fish experience pain when immersed in super oxygenated water. Consequently, behavioural signs may be used as measures of an animal's 'real-time' perception of their internal and external environments.

Behavioural indicators also allow for the contextualisation of other types of indicators. As previously mentioned, many physiological indicators arise during stressful mental experiences such as fear and anxiety, or excitement and anticipation—as well as pain. Following fin clipping, Nile tilapia have demonstrated escape behaviour via increased swimming activity (Roques et al. 2010), which coincided with increased plasma cortisol levels. Together, increased swimming activity and plasma cortisol levels indicate that tail fin clipping is probably a painful experience for Nile tilapia (Roques et al. 2010). Similarly, Ashley et al. (2009) observed elevated respiration rates in Rainbow trout exposed to predator pheromone. This threat coincided with the increased use of cover and probably indicates the affective experience of fear (of potential tissue damage) and pain in the event of a physical attack. In short, behavioural changes are a near necessity for the contextualisation and accurate interpretation of physiological pain indicators.

Behavioural indicators possess some disadvantages. First, recognising fish behaviours in an aquatic setting is impeded by the relative inaccessibility of individuals. In terrestrial agriculture settings, e.g. farming of cattle and sheep, farmers and researchers can build relationships with individual animals because the relative ease of access to land animals allows for regular, close interactions between farmers/researchers and land animals. Regular interactions facilitate the recognition of abnormal behaviour or changes in behaviour patterns (Macaulay et al. 2021). The inability to form such relationships with fish may impair an observer's ability to recognise behavioural expressions in the first instance. It will also impair the recognition of variations in the behaviour of individuals; this is important because behavioural responses to the same stimuli will differ between fish. Secondly, the qualitative analysis of fish behaviour is subjective (e.g. individual differences in behavioural definition interpretation) and requires the observer to have knowledge of appropriate behaviours for a given lifecycle stage, production system and species (Noble et al. 2018b; Stien et al. 2020). Thirdly, assessing absolute or quantitative changes in fish behaviour is laborious as it requires the analysis of collected data (Noble et al. 2018b). Such limitations must be considered when planning experiments or making on-farm assessments.

4.5 Overall Pain Assessment: Important Considerations

When assessing pain in fish, contextual information must be considered when interpreting potential pain indicators. No single measurement represents an unequivocal pain behaviour. Therefore, pain assessments should examine multiple variables across multiple modalities, i.e. using indicators from two or more categories listed previously. For example, to experimentally assess the pain response of Nile tilapia to the injection of noxious chemicals, observers should assess physical appearance and production-related indicators, e.g. changes in body colour, physiological indicators, e.g. plasma cortisol and glucose levels, cardio-ventilatory activity and expression of inflammatory gene marker, and behavioural signs, e.g. 'freezing' behaviour and a reduced response to food pellets (Ide and Hoffmann 2002). The use of indicators across several modalities creates a more holistic understanding of a fish's mental experience at a certain point in time (Beausoleil and Mellor 2017). However, confounding factors such as species and individual variation in response to noxious stimulation should also be considered during assessment.

Pain indicators vary amongst species and amongst individuals of the same species. Lip rubbing and rocking behaviours are displayed by Rainbow trout but not by Zebrafish in response to labial acetic acid injection. Behavioural and physiological responses also vary according to individual coping styles, e.g. plasma cortisol levels are typically higher in reactive, compared to proactive[1] fish (Jones et al. 2012; Butler et al.

[1]Much like other animals, fish are known to form social hierarchies. Within their social groups, fish generally

2018; Vargas et al. 2018). Hierarchical positions are also important: subordinate fish show a longer latency to eat following noxious stimulation compared to dominant fish (Mettam et al. 2011; Butler et al. 2018; Vargas et al. 2018). Not unexpectedly, behaviour is also influenced by the type of noxious stimulus experienced. Exposure to acetic acid is associated with reduced swimming activity in Zebrafish (Lopez-Luna et al. 2017a), although citric acid exposure or direct transfer into water at 28 °C results in increased swimming activity in Zebrafish (Lopez-Luna et al. 2017a) and Atlantic salmon (Nilsson et al. 2019), respectively. Therefore, behavioural responses may be species, personality, and stimulus-specific.

Pain assessment in fish must also consider environmental factors within which assessments are being made: the external environment may influence physiological and/or behavioural signs. For example, grouping: plasma cortisol levels in experimental fish depend on whether they are individually or grouped housed. Following acetic acid injection, individually housed Rainbow trout had significantly higher plasma cortisol levels than control fish (Rizzo et al. 2017). However, in the same study, no significant difference was found in plasma cortisol levels when Rainbow trout were housed in social groups. Environmental context may also influence the usefulness of some pain indicators. In farmed fish, for example, there is difficulty in monitoring the welfare of individual fish due to the sheer number of animals involved. Costs, resources, and the time taken to measure some pain indicators may also be limiting factors. Such is the case when monitoring inflammatory marker genes expression, which requires capturing individual fish to obtain blood samples, which in turn are submitted for laboratory analysis (Noble et al. 2018b). Group-based indicators, such as shoaling behaviour, are more appropriate for on-farm use. In contrast, observing fish under laboratory conditions allows more meticulous monitoring of individual fish especially when the more laborious indicators of pain may be used. In both the farm and laboratory environments, the awareness of a human (observer) during fish evaluation may trigger stress responses in the latter or skew results based on learned behaviours (Rizzo et al. 2017).

5 Pain Mitigation in Fish

Fish in the care of humans are subjected to invasive experimental and husbandry procedures that may result in pain. To maintain fish welfare, anaesthetics and analgesics can be used to ameliorate or minimise associated pain or discomfort. This section reviews the scientific literature describing analgesics and anaesthetics for use on farmed or laboratory fish.

5.1 Fish Anaesthesia

Anaesthesia is a controlled and reversible state of consciousness induced by drugs that renders subjects less rousable to all conscious modalities, including pain (Musk et al. 2020). Anaesthetics are used in farmed fish, and during veterinary or experimental procedures to minimise the stress caused by handling, to limit pain experiences and to facilitate handling. Ideally, anaesthetics should produce unconsciousness with minimal stress, produce immobilisation and suppress arousal by noxious stimuli and produce a prompt recovery free of acute and chronic adverse effects (Musk et al. 2020). The most common method of inducing anaesthesia in fish is by immersion, which is achieved by placing the subject in a bath of dissolved anaesthetic. Most of the drug is absorbed across the gill epithelium in the ventilating fish, although some absorption can occur across the skin, diffusing into the bloodstream to eventually cross the blood–brain barrier (Zahran et al. 2021). Immersion anaesthesia enables the simultaneous induction of anaesthesia in multiple fish, reducing the stress and practical constraints related to

adopt one of the two different coping styles to stressors: proactive or reactive. Dominant fish are typically display proactive coping styles. These fish are described as aggressive or bold individuals (Beausoleil and Mellor 2017; Butler et al. 2018). Whereas subordinates typically display a reactive coping style. These fish are described as passive or timid individuals (Beausoleil and Mellor 2017; Butler et al. 2018).

restraining and individually injecting each fish. The efficacy of the anaesthesia induced by an agent can be assessed by considering induction and recovery times, along with behavioural and physiological indicators. The depths of anaesthesia in fish during its induction have been categorised into distinct stages of anaesthetic planes (Table 4) (Sneddon 2012, Zahl et al. 2012). Stages range from 0 (normal; no sedation) to 4 (impending death). Stage 3 (anaesthesia) is divided into three planes: 1—light anaesthesia, 2—surgical anaesthesia and 3—deep narcosis. For further explanation of anaesthetic stages in fish, see Sneddon (2012) and Zahl et al (2012)

5.2 Anaesthetic Agents for Fish

Currently, a diverse array of agents is employed for fish anaesthesia. Among the commonly used ones are tricaine (MS-222), benzocaine and eugenol-based drugs like clove oil, eugenol and iso-eugenol. Metomidate, etomidate, 2-phenoxyethanol and propofol are also used. The effectiveness of each drug depends on factors such as dosage, biological characteristics, e.g. fish species, life-cycle stage, and body weight and environmental conditions (including water pH and temperature). These also influence the fish's metabolic rate, so affect drug uptake rate across the gills, as well as the pharmacokinetic behaviour of the agent (Iversen et al. 2003). For detailed information regarding dose ranges and the effects of agents on specific species, please refer to Table 5.

5.2.1 Tricaine (MS-222) and Benzocaine

Tricaine and benzocaine are two of the most commonly used anaesthetics in fish. Both are local anaesthetics that interfere with voltage-gated sodium channels on neuronal cell membranes (Nordgreen et al. 2014). Inhibition of these channels prevents sodium ion influx required for cell membrane depolarisation and the generation and propagation of action potentials (Musk et al. 2020). When administered by injection these agents produce localised anaesthesia, including analgesia. However, when administered via immersion, they enter the circulation through the gills and produce general anaesthesia by blocking neural transmission in peripheral and central pathways (Zahl et al. 2012). MS-222 and benzocaine are similar in chemical composition, although agent preparation varies. MS-222 is water-soluble and has an acidic pH that requires buffering, i.e. sodium bicarbonate must be added to achieve an appropriate pH (Sneddon 2012). In contrast, benzocaine is only slightly water-soluble and requires dissolving in organic solvents, e.g. ethanol, before fish immersion (Sneddon 2012).

Doses and effects of MS-222 vary between species. At 168 mg L^{-1} MS-222 produces deep anaesthesia in Zebrafish and allows caudal fin clipping with no behavioural responses (Musk et al. 2020). Nordgreen et al. (2014) demonstrated that acute behavioural responses to experimental noxious stimulation returned as soon as Zebrafish regained consciousness from MS-222 anaesthesia with little residual effect of anaesthesia on behaviour. Tranquilisation with MS-222 during transport reduced mortality of Nile tilapia (Sherif et al. 2022). MS-222 anaesthesia in Rainbow trout produced lower plasma cortisol levels during recovery compared to control fish (Wagner et al. 2003). While Musk et al. (2020) reported no obvious aversion to MS-222 in Zebrafish, Pounder et al. (2018) observed erratic swimming and surface breaks in Rainbow trout upon exposure to 100 mg L^{-1} MS-222 solutions. This coincided with increased respiration rates and plasma cortisol levels, indicating a physiological stress response (Pounder et al. 2018). MS-222 produces deep anaesthesia and reduces transportation stress in fish, although it may cause a stress response per se.

Benzocaine doses and anaesthetic effects are also species dependent. At concentrations of 33 mg L^{-1} benzocaine did not cause responses in Rainbow trout (Pounder et al. 2018). However, induction was prolonged and at 108 mg L^{-1}, trout showed greater heart rate variability and lower absolute heart rates as exposure time increased (Cotter and Rodnick 2006). Benzocaine is an effective anaesthetic

Table 4 Anaesthesia stages in fish sourced from Sneddon (2012)

Stage		Level of anaesthesia	General demeanour	Activity	Equilibrium	Gill ventilation rate	Reactivity	Heart rate	Muscle tone	Examples of procedures
0		Conscious	Normal	Normal	Normal	Normal	Normal	Normal	Normal	
I		Lightly sedated	Disoriented	Reduced	Normal	Normal	Reduced	Normal	Normal	
II		Excitation	Agitated	Increased	Difficulty	Increased	Increased	Increased	Normal	
III	1	Light anaesthesia	Anaesthetised	None	Loss	Decreased	Reflex responses	Regular	Decreased	Weighing; close examination; external non-invasive tagging; gill scrape
	2	Surgical	Anaesthetised	None	Loss	Shallow	None	Reduced	Decreased	Invasive tagging; tissue removal; injection; blood sampling; gill biopsy; lesion dressing; recovery surgery
	3	Deep	Anaesthetised	None	Loss	Rare movements	None	Reduced	Relaxed	Non-recovery surgery
IV		Overdose	Apparently dead	None	Loss	None	None	Cardiac failure	None	

Table 5 Summary of anaesthetic agents used in fish including the type of analgesic, dose range, species investigated, and reported pros and cons of each agent

Anaesthetic agent	Dose range (mg L^{-1})	Species	Reported pros	Reported cons
Benzocaine	25–150	*O. mykiss* (Pounder et al. 2018; Cotter and Rodnick 2006; Saunders et al. 2020)	Local anaesthetic; likely to provide pain relief Not aversive in terms of behavioural responses Induction slow compared with other drugs	Increased respiration rate Increased heart rate variability Potential oxidative stress on haemoglobin
	25–100	*S. salar* (Iversen et al. 2003; Saunders et al. 2020; Zahl et al. 2010)	No stress-related increase in plasma glucose	Increase in plasma cortisol Potential oxidative stress on haemoglobin
	80–600	*O. niloticus* (de Miranda Cabral Gontijo et al. 2003)	No genotoxicity	Doses above 160 mg L^{-1} may cause cardiac arrest
Clove oil (eugenol)	100	*O. mykiss* (Pounder et al. 2018)	Relative low plasma cortisol levels Faster induction time compared with other agents	Increased respiration rate compared to sham handling
	30–100	*S. salar* (Iversen et al. 2003)	Prevents increases in plasma cortisol above resting levels	Increased plasma lactate Reduced respiration rate
Iso-eugenol (Aqui-S™)	5.4–100	*S. salar* (Iversen et al. 2003; Zahl et al. 2010; Chance et al. 2018)	Repeat use does not result in tolerance or habituation Does not induce inflammatory gill changes Low increases in plasma cortisol above resting levels	Significant upregulation of osmotic control marker genes Epithelial lifting Increased plasma lactate Reduced respiration rate
Etomidate	2–10	*O. mykiss* (Pounder et al. 2018)	Low plasma cortisol increase No significant increase in respiration rate	Relatively prolonged induction time Induces aversive behaviour
Metomidate	3–12	*S. salar* (Iversen et al. 2003; Zahl et al. 2010; Chance et al. 2018)	Low cortisol release No stress-related increase in plasma glucose Repeat use does not result in tolerance or habituation Does not induce inflammatory gill changes	Increase in plasma cortisol levels
MS-222	168	*D. rerio* (adult) (Musk et al. 2020; Nordgreen et al. 2014)	Brief recovery time compared to other agents No response to caudal fin clipping during anaesthesia No obvious aversion observed	Preparation of buffered MS-222 is laborious and poses a safety risk to humans

(continued)

Table 5 (continued)

Anaesthetic agent	Dose range (mg L^{-1})	Species	Reported pros	Reported cons
	60–150	*O. mykiss* (Pounder et al. 2018; Wagner et al. 2003; Pirhonen and Schreck 2003)	No significant changes in growth hormone and thyroid hormones	Elicits aversive behaviour Elevated plasma cortisol levels and respiration rates (100 mg L^{-1}) Depressive effect on feeding behaviour
	40	*O. niloticus* (Sherif et al. 2022)	Reduced mortality after transport	Profuse mucus production
	55–80	*S. salar* (Zahl et al. 2010; Chance et al. 2018)	Repeated use does not result in habituation	Induced a physiological stress response
2-phenoxyethanol	0.5–6	*O. mykiss* (Pounder et al. 2018)	Rapid induction compared to other drugs Low plasma cortisol Low behavioural stress response	Increased respiration rate at low doses
Propofol	0.8–12.5	*O. niloticus* (Zahran et al. 2021; Luís Félix et al. 2021; Obirikorang et al. 2020)	No significant changes in liver function Fillet quality unaffected No histological gill changes No stress-related behaviour Minimal effects on blood constituents	Transient increases in plasma cortisol and glucose Decreased ventilation rate
		D. rerio (Gressler et al. 2021; Valentim et al. 2016)	Anti-anxiety during novel dive test Response to food regained within an hour post-exposure Return to normal behaviour 5 h exposure	Higher doses decrease ventilation rate Induces shrinkage of gill structure

at 33 mg L^{-1} in Atlantic salmon with induction achieved between 3 and 4 min after exposure. Benzocaine, however, may oxidise the haemoglobin of Atlantic salmon because methaemoglobin (haemoglobin with the iron oxidised to its ferric state) percentages increase by 49% following exposure. Gontijo et al. (2003) reported no genotoxicity, i.e. no damage to genetic information, of benzocaine in Nile tilapia, although doses above 160 mg L^{-1} may cause cardiac arrest. Benzocaine has benefits as an anaesthetic in fish at varying doses but should perhaps be avoided in experiments in which haematological variables must not change.

5.2.2 Eugenol-Based Anaesthetics

Clove oil derivatives are commonly used as fish anaesthetics. Clove oil actually contains eugenol, iso-eugenol, and methyl-eugenol, although the primary anaesthetic is eugenol (85–95%) (Cotter and Rodnick 2006). Eugenol 'inhibits sodium, potassium, and calcium channels, inhibits N-methyl-D-aspartate (NMDA) receptors and potentiates inhibitory gamma-aminobutyric acid type A ($GABA_A$) receptors' (Zahl et al. 2010). The isomer of eugenol, iso-eugenol, is the active ingredient of Aqui-S a drug approved for use as an anaesthetic in farmed fish in New Zealand and Australia (Chance et al. 2018). The manufacturer recommends fish are immersed in

15–20 mg L^{-1} aqueous Aqui-S solutions (https://www.aqui-s.com/products/aqui-s-aquatic-anaesthetic). Aqui-S contains 540 g L^{-1} of the active ingredient iso-eugenol (Zahl et al. 2012). Iso-eugenol acts in a similar manner to eugenol as an anaesthetic, i.e. by decreasing action potential generation in nerve axons it results in general depression of the central nervous system (Zahl et al. 2010).

Eugenol-based anaesthetics are considered to have greater benefits as anaesthetics than others such as MS-222. Clove oil and iso-eugenol elicit lower plasma cortisol increases than MS-222 and produced a more rapid induction (Pounder et al. 2018; Zahl et al. 2010). Compared with MS-222, iso-eugenol also results in lower upregulation of inflammatory and stress genes in the head kidney of Atlantic salmon (Chance et al. 2018). Eugenol and iso-eugenol can be effective at doses as low as 30 mg mL^{-1} with little effect on the stress response (Iversen et al. 2003). However, repeated exposure to iso-eugenol causes 'lifting' of the gill epithelium (Chance et al. 2018). Separation of the epithelial layer from gill capillaries is described as a 'defence mechanism to reduce surface contact with the external environment' (Chance et al. 2018). Severe epithelial lifting may be followed by epithelial cell necrosis leading to respiratory and osmotic regulatory disturbances (Alize et al. 2020). Despite some well-recognised disadvantages, eugenol-based anaesthetics have some major advantages over MS-222.

5.2.3 Etomidate and Metomidate

Etomidate and its methyl analogue, metomidate, modulate and activate $GABA_A$ receptors, which increase chloride conductance of neuronal cell membranes and prevent the generation and propagation of action potentials. Similar central actions result in sedation. Metomidate is an effective anaesthetic for Rainbow trout at 2 mg L^{-1} with no increases in plasma cortisol or stress-related plasma glucose levels being recorded (Iversen et al. 2003). However, both etomidate and metomidate suppress inter-renal production of cortisol (Zahl et al. 2012). It is, therefore, possible that fish anaesthetised with these drugs may be experiencing stress when blood cortisol levels would indicate otherwise. Metomidate also depresses ventilation, unlike etomidate at 2 mg L^{-1} which does not. However, etomidate does have a relatively prolonged induction time (Pounder et al. 2018) which may be useful for monitoring anaesthesia stages.

5.2.4 2-Phenoxyethanol

2-phenoxyethanol is found in anti-bacterial vaccines, skin care products, and perfumes. Unsurprisingly, its exact mode of action is not fully known, although it has been suggested that the compound may expand neuronal cell membranes, causing general CNS depression (Zahl et al. 2012; Sneddon 2012). In comparison with other anaesthetics, 2-phenoxyethanol at concentrations of 0.5 mg L^{-1} produces brief induction times, low behavioural stress during induction, and lower plasma cortisol levels—at least in Rainbow trout (Pounder et al. 2018). However, side effects have been reported in this species, in particular reductions in immune function and impaired cardioventilatory function (Sneddon 2012).

5.2.5 Propofol

Propofol is a strong modulator of the $GABA_A$ receptor whose activation increases chloride ion permeability and prevents the propagation of new action potentials. In mammals, propofol is known for its rapid recovery rates. Zahran et al. (2021) described propofol (2.5 mg L^{-1}) induction of Nile tilapia as smooth and stated that observers were readily able to distinguish different stages of anaesthesia, including 'deep' anaesthesia. No mortalities along with minimal, reversible increases in blood constituent measures, and no adverse effects on liver function enzymes were reported (Zahran et al. 2021). Propofol causes a transient physiological stress responses in Nila tilapia following transfer into a stable tank (Zahran et al. 2021) and a vibrating tank to simulate transportion (Luís Félix et al. 2021). Depressed respiratory activity occurs in both Nile tilapia (Obirikorang et al. 2020) and Zebrafish (Valentim et al. 2016). In Zebrafish, anti-anxiety behaviour was observed after exposure to propofol, suggesting a residue effect. However, further research needs to be done across different fish species to confirm this.

5.3 Analgesics for Fish

Less is known about the pharmacology of analgesics in fish compared with anaesthetics (Chatigny et al. 2018a) although it is assumed they offer the same advantages of anaesthetics in fish as they do in mammals. Not all anaesthetics are analgesic insofar that neuroendocrinological responses to noxious stimuli may not be entirely suppressed, even in unconscious animals. In these circumstances, analgesic drugs and techniques may be required to optimise post-interventional welfare.

While many analgesics can be given by immersion, the intramuscular (IM), intravenous (IV), or intraperitoneal (IP) routes are also popular (Chatigny et al. 2018a). As with anaesthetics, doses, biological factors, and environmental conditions influence the pharmacokinetic behaviour of analgesic drugs. The major classes of analgesic drugs discussed here are shown in Table 6.

5.4 Opioids

In mammals, opioid drugs mimic the action of endogenous opioid peptides through interacting with the mu, kappa, and delta opioid receptors located on neuronal cell membranes. Opioid drugs act by inhibiting nociceptive neurotransmitter release through facilitation of postsynaptic K^+ efflux, i.e. membrane hyperpolarisation, or by reduction of presynaptic Ca^{2+} influx, i.e. by prevention of vesicle release (Chatigny et al. 2018a). Inhibition of neurotransmitter release blocks nociceptor activity and nociceptive signal transmission (Sneddon 2012). Studies have identified the presence of an endogenous opioid system, similar to that of mammals, in fish (Sneddon 2020). The possession of opioid receptors indicates the potential suitability of opioid drugs to provide pain relief in fish.

5.4.1 Morphine

Morphine is the most well-studied opioid in fish. As a full mu (μ) agonist, morphine evokes a maximal response from opioid receptors to produce profound analgesia (Chatigny et al. 2018a). Morphine is effective in preventing antinociceptive behaviours induced by acetic acid in Rainbow trout (Sneddon 2003b) and Zebrafish (Lopez-Luna et al. 2017a). In Zebrafish larvae, morphine had beneficial effects on the 'time spent active' during exposure to (noxiously high) water temperatures (Lopez-Luna et al. 2017c) and carbon dioxide (Lopez-Luna et al. 2017b). Morphine also confers beneficial physiological effects in noxiously stimulated fish. For example, it reduces elevated opercular beat rates in Rainbow trout induced by acetic acid (Sneddon 2003b). Morphine also has potential anxiolytic effects in fish. Nordgreen et al. (2013) observed morphine treated Atlantic salmon spent significantly more time in the open during a novel object test compared to control salmon. Similarly, Wong et al. (2010) observed reduced neophobia in Zebrafish during novel tank tests. A novel tank test is a behavioural assay that uses vertical distribution in a novel environment as an indicator of 'anxiety' in fish. Morphine is a potent analgesic in all fish species studied and also provides beneficial anxiolytic effects. Interestingly, pain-free fish have demonstrated drug seeking behaviour when taught to self-administer morphine (Bossé and Peterson 2017). This behaviour is reflective of addictive behaviour demonstrated in mammals (Bossé and Peterson 2017; Horký et al. 2021).

5.4.2 Buprenorphine

Buprenorphine is a synthetic opioid analgesic that acts on three target genes that are conserved in fish: OPRM1, OPRK1 and OPRD1 (Gräns et al. 2014). These genes provide instructions for synthesising the mu, kappa, and delta receptors, respectively. Behavioural changes induced by acetic acid (Steenbergen and Bardine 2014) and acute thermal aversion (Curtright et al. 2015) are prevented when Zebrafish larvae are immersed in buprenorphine solutions. However, buprenorphine was unable to prevent both increased latency to feed and reduced swimming activity after expo-

Table 6 Summary of analgesic agents used in fish including the type of analgesic, dosage range, route of administration, species investigated, and reported pros and cons of each agent

Analgesic agent		Dose range (route)	Species	Reported pros	Reported cons
Opioids	Buprenorphine	0.1 mg L^{-1} (immersion)	*D. rerio* (larvae) (Steenbergen and Bardine 2014)	Prevents behavioural changes induced by acetic acid and high temperatures Does not cause sedation	Doses <0.05 mg L^{-1} are ineffective at preventing behavioural changes induced by acetic acid
		0.01–0.1 mg kg^{-1} (IM)	*O. mykiss* (Gräns et al. 2014; Mettam et al. 2011)	Efficacy as an analgesic requires further study	Unable to prevent the increased latency to feeding and reduced swimming activity induced by acetic acid Produces sedation and reduces cardioventilatory function
	Butorphanol	0.2–0.5 mg L^{-1} (immersion)	*D. rerio* (adult) (Schroeder and Sneddon 2017)		Does not reduce behavioural responses to fin clipping
	Morphine	1–48 mg L^{-1} (immersion)	*D. rerio* (larvae) (Lopez-Luna et al. 2017a, b, c)	Prevents reductions in activity induced by acetic acid, carbon dioxide and high water temperatures	Relative slow rate of uptake when delivered by immersion
		2 mg L^{-1} (immersion)	*D. rerio* (adult) (Wong et al. 2010)	Anxiolytic effects during novel tank test	
		1–100 mg kg^{-1} (IP) 0.3 g L^{-1} (IM)	*O. mykiss* (Sneddon 2003b; Jones et al. 2012)	Reduces behavioural and physiological changes induced by acetic acid and electrical stimulation	
		300 mg kg^{-1} (IM)	*S. salar* (Nordgreen et al. 2013)		Analgesic effect following vaccination not apparent
NSAIDs	Aspirin	1–2.5 mg L^{-1} (immersion)	*D. rerio* (larvae) (Lopez-Luna et al. 2017a, b, c)	Beneficial effects on swimming activity in response to 0.1% acetic acid	Ineffective at providing demonstrable analgesia in response to elevated carbon dioxide and water temperatures levels
		1–2.5 mg L^{-1} (immersion)	*D. rerio* (adult) (Schroeder and Sneddon 2017)	Reduces abnormal fin wafting behaviour following fin clipping Maintains normal activity levels and respiration rate following fin clipping	

(continued)

Table 6 (continued)

Analgesic agent		Dose range (route)	Species	Reported pros	Reported cons
	Carprofen	1–5 mg kg^{-1} (IM) 2.5 mg kg^{-1} (IM, IV, oral)	*O. mykiss* (Mettam et al. 2011; Uney et al. 2021)	Beneficial effects on resumption of eating following acetic acid injection Favourable pharmacokinetics such as good bioavailability and long terminal elimination half-life Does not induce stress response when administered alone	Unable to prevent reduction in swimming activity induced by acetic acid
	Flunixin	8–20 mg L^{-1} (immersion)	*D. rerio* (larvae) (Lopez-Luna et al. 2017a, b, c)	Able to prevent pain induced reductions in swimming activity following exposure to high water temperatures Further research required to understand analgesic efficacy	Unable to prevent pain induced reductions in swimming activity following CO_2 and acetic acid exposure
		0.5 mg kg^{-1} (IM)	*O. mykiss* (Rizzo et al. 2017)	No evidence of gastrointestinal or renal toxicity	
	Ketoprofen	8 mg kg^{-1} (IM)	*O. niloticus* (Greene et al. 2020)	No evidence of gastrointestinal or renal toxicity	Further research on analgesic efficacy required
		2 mg kg^{-1} (IM) 3 mg kg^{-1} (IM, IV)	*O. mykiss* (Rizzo et al. 2017; Greene et al. 2020)	No evidence of gastrointestinal or renal toxicity	Further research on analgesic efficacy required
	Ketorolac	0.5 mg kg^{-1} (IM)	*O. mykiss* (Rizzo et al. 2017)	No evidence of gastrointestinal or renal toxicity	Further research on analgesic efficacy required
Local anaesthetic	Lidocaine	5 mg L^{-1} (immersion)	*D. rerio* (larvae) (Lopez-Luna et al. 2017a, b, c)	Prevented behavioural changes induced by acetic acid, and elevated carbon dioxide and water temperature levels. No apparent adverse side effects	
		2–5 mg L^{-1} (immersion)	*D. rerio* (adult) (Schroeder and Sneddon 2017)	Reduced behavioural and physiological responses following tail fin clipping	
		05–2 mg per fish (IM) 10–20 mg kg^{-1} (IM)	*O. mykiss* (Mettam et al. 2011; Chatigny et al. 2018b, c)	Does not cause a stress response when used alone Beneficial effects across behavioural and physiological variables following acetic acid injection	Repeated IM injections may cause muscle damage

IM intramuscular, *IV* intravenous, *IP* intraperitoneal

sure to acetic acid in Rainbow trout (Mettam et al. 2011). Mettam et al. (2011) suggests that buprenorphine may be more suited for pain relief in cases of long-term injury, such as post-surgical procedures, as opposed to use in acute, somatic pain tests. However, assessment of buprenorphine as a post-surgical analgesic in Rainbow trout suggests otherwise (Gräns et al. 2014). Buprenorphine produces physiological changes in both painful and pain-free Rainbow trout, in which respiration rate and heart rate remained below baseline values for 7 days following treatment (Gräns et al. 2014). The reduction of both respiration and heart rate, alongside reduced swimming activity, of both painful and pain-free Rainbow trout suggest a potential sedative effect of buprenorphine as opposed to an analgesic effect in this species (Gräns et al. 2014; Mettam et al. 2011). The analgesic value of buprenorphine has been demonstrated in Zebrafish larvae, but not Rainbow trout. The analgesic effects of this drug needs further evaluation before use in other species of fish.

5.4.3 Butorphanol

Butorphanol is a partial opioid agonist with agonist effects at the kappa receptor and antagonist effects at the mu receptor. Therefore, the drug is able to provide analgesia with minimal depression of the CNS as seen with full opioid agonists (Chatigny et al. 2018a). As butorphanol is soluble in water, it is delivered by immersion for amphibians with success in providing an analgesic effect. However, Schroeder and Sneddon (2017) found little evidence to support the possibility that butorphanol was an effective analgesic when administered via immersion in Zebrafish. Butorphanol was unable to reduce behavioural changes, e.g. fin wafting, induced by tail fin clipping in Zebrafish (Schroeder and Sneddon 2017). The analgesic effect of butorphanol in fish is not yet clear.

5.5 Nonsteroidal Anti-inflammatory Drugs

The analgesic effects NSAIDs arise primarily from their ability to inhibit COX enzymes. As mentioned in a previous Sect. 3.2.1 (Criterion 1: Possession of nociceptors), COZ enzymes are involved in the pain and inflammation process via facilitation of the production of thromboxanes and prostaglandins, i.e. mediators of inflammatory responses (Lopez-Luna et al. 2017a). Inhibition of COX enzymes reduces prostaglandin synthesis and impairs pain fibre transmission. NSAIDs can be used to alleviate both somatic (acute and sharp) and visceral (chronic and dull) pain (Rizzo et al. 2017).

Aspirin (acetylsalicylic acid) at 2.5 mg kg^{-1} prevents pain-related behaviours induced by 0.1% acetic acid in Zebrafish larvae (Lopez-Luna et al. 2017a). Aspirin also reduces fin wafting and maintains normal activity levels including opercular beat rate after fin clipping in adult Zebrafish (Schroeder and Sneddon 2017). However, pre-exposure to aspirin, i.e. pre-emptive analgesia, did not produce significant analgesic effects in Zebrafish larvae exposed to high CO_2 concentrations (Lopez-Luna et al. 2017b) and water at 40 °C (Lopez-Luna et al. 2017c). Similarly, flunixin was unable to prevent reduction in the swimming activity of noxiously stimulated Zebrafish larvae Lopez-Luna et al. 2017a, b, c). However, Lopez-Luna et al. (2017a) noted that flunixin is perhaps more effective at higher doses (>20 mg L^{-1}) as the noxious effects of acetic acid were more pronounced at doses of 8 mg L^{-1} compared to the higher dose of 20 mg L^{-1}. Ketorolac (0.5 mg kg^{-1} IM) and ketoprofen (2 mg kg^{-1} IM) in Rainbow trout (Rizzo et al. 2017) and Nile tilapia (8 mg kg^{-1}) do not cause adverse gastrointestinal (ulceration) and renal (acute failure) side effects that are seen in mammals. However, more research is needed to determine their analgesic efficacy in fish. Administration of carprofen (5 mg kg^{-1} IM) in Rainbow trout produced beneficial effects on the resumption of eating but was unable to prevent reduction in swimming activity induced by acetic acid (Mettam et al. 2011). Carprofen (2.5 mg kg^{-1}) administered IM and orally to Rainbow trout demonstrates good bioavailability and a prolonged half-life (Uney et al. 2021) suggesting it may be of benefit when wide dose intervals are required. In comparison to IM administration, oral administration of carprofen, may involve less labour for farmers/researchers and less stress to fish because it can be non-invasively administered through medicated feed (Uney et al. 2021). However, the exact dose ingested by fish is less controlled. More research is required regarding the analgesic efficacy and pharmacokinetics of NSAIDs in fish; this will allow the development of more effective and practical guidelines for their use on farms and in laboratories.

5.6 Local Anaesthetics

Local anaesthetics alleviate pain by blocking all nerve activity at sites where adequate tissue concentrations are achieved. The mode of action by which local anaesthetics provide their analgesic effects has been of great interest to the pharmacology field since the channel blocking action of tetrodotoxin was discovered in the early 1960s (Narahashi 2008). Tetrodotoxin is a potent neurotoxin first found in pufferfish that works by inactivating voltage-gated sodium channels (Lago et al. 2015). Blockade of sodium channels prevents the influx of sodium ions into nerve membranes that is required to produce action potentials (Chatigny et al. 2018b; Lago et al. 2015). Inhibition of nerve conduction prevents the transmission of noxious stimuli input to higher brain centres (Chatigny et al. 2018c). Local anaesthetics used in fish work via a similar mode of action.

Local anaesthetics provide localised analgesia in several fish species. Simultaneous injection of lidocaine and acetic acid suppresses all pain indicators measured by Mettam et al (2011) For example, 1 and 2 mg kg^{-1} of lidocaine both reduced latency to feed and prevented changes in swimming activity and respiration rates following acetic acid exposure. Lidocaine (20 mg kg^{-1}) also reduced escape behaviour in Rainbow trout (Chatigny et al. 2018c) following placement in a novel environment, indicating a potential anxiolytic or sedative effect. There have been no significant adverse side effects reported at 10–20 mg kg^{-1} injected either side of the dorsal fin, but care is advised if repeated injections are necessary: it has been recommended that a waiting period of 30 days should be observed before re-injection to prevent muscle damage (Chatigny et al. 2018b, c). Pre-treatment via immersion has proven effective in preventing pain-related behavioural changes in Zebrafish larvae induced by acetic acid, elevated carbon dioxide and water temperature levels. Similarly, immersing adult Zebrafish in dissolved lidocaine reduces behavioural and physiological responses associated with tail fin clipping. In summary, lidocaine appears to be an effective and safe analgesic in fish although concerns about metabolite genotoxicity and meat levels in farmed fish (as have been raised in mammalia) might limit its use on a commercial scale, despite its cheap cost.

6 Conclusion

Fish have been cultivated by humans for thousands of years. However, only recently has the notion that fish may feel pain—and suffer—gained traction. This is in line with the increasing body of scientific evidence supporting the possibility. Interest in the capacity of fish to feel pain, and fish welfare in general increases as the cultivation of fish for human consumption and use in laboratory experiments increases. Changes in the physical appearance, production measures, physiology and behaviour of fish provide evidence of piscine pain. Indicators of this depend on the species and source of noxious stimulation. Identification of pain in fish allows those in charge of their welfare to implement mitigation strategies such as the use of anaesthetics, e.g. MS-222 and isoeugenol, and analgesics, e.g. morphine and lidocaine. Work must continue to understand the capacity for fish to feel pain in order to optimise their welfare whilst under human care.

References

Alize D, Sutriana A, Nazaruddin N, Armansyah T, Etriwati E, Hanafiah M, Hafizuddin H, Hasan DI, Awaluddin A, Ulfa B (2020) Histopathological changes in the gills of *Oreachromis mossambicus* exposed to mercury chloride ($Hgcl_2$). In: International conference on veterinary, animal, and environmental sciences. Atlantis Press, pp 74–80

Aqui-S™ Aqui-S: aquatic anaesthetic https://www.aqui-s.com/products/aqui-s-aquatic-anaesthetic

Ashley PJ, Ringrose S, Edwards KL, Wallington E, McCrohan CR, Sneddon LU (2009) Effect of noxious stimulation upon antipredator responses and dominance status in Rainbow trout. Anim Behav 77:403–410

Avdesh A, Chen M, Martin-Iverson MT, Mondal A, Ong D, Rainey-Smith S, Taddei K, Lardelli M, Groth DM, Verdile G (2012) Regular care and maintenance of a zebrafish (Danio rerio) laboratory: an introduction. J Vis Exp 69:e4196

Beausoleil NJ, Mellor DJ (2017) Validating indicators of sheep welfare. In: Achieving sustainable production of sheep. Burleigh Dodds Science Publishing, pp 349–370

Beausoleil NJ, McKeegan DE, Martin JE (2022) Avian welfare: fundamental concepts and scientific assessment. In: Colin GS, Dridi S (eds) Sturkies avian physiology. Academic, pp 1099–1109

Birch J, Burn C, Schnell A, Browning H, Crump A (2021) Review of the evidence of sentience in cephalopod molluscs and decapod crustaceans. LSE: The London School of Economics and Political Science

Bjørge MH, Nordgreen J, Janczak AM, Poppe T, Ranheim B, Horsberg TE (2011) Behavioural changes following intraperitoneal vaccination in Atlantic salmon (Salmo salar). Appl Anim Behav Sci 133:127–135

Bossé GD, Peterson RT (2017) Development of an opioid self-adminstration assay to study drug seeking in zebrafish. Behav Brain Res 335:158–166

Butler JM, Whitlow SM, Roberts DA, Maruska KP (2018) Neural and behavioural correlates of repeated social defeat. Sci Rep 8:6818

Chance RJ, Cameron GA, Fordyce M, Noguera P, Wang T, Collins C, Secombes CJ, Collet B (2018) Effects of repeated anaesthesia on gill and general health of Atlantic salmon, Salmo salar. J Fish Biol 93:1069–1081

Chatigny F, Creighton CM, Stevens ED (2018a) Updated review of fish analgesia. J Am Assoc Lab Anim Sci 57:5–12

Chatigny F, Groman DB, Martinson SA, Stevens ED (2018b) Evaluation of tissue changes following intramuscular infiltration of lidocaine in Rainbow trout Oncorhynchus mykiss. J Fish Biol 92:888–900

Chatigny F, Creighton CM, Stevens ED (2018c) Intramuscular infiltration of a local anesthetic, lidocaine, does not result in adverse behavioural side effects in Rainbow trout. Sci Rep 8:1–9

Costa FV, Rosa LV, Quadros VA, Santos ARS, Kalueff AV, Rosemberg DB (2019) Understanding nociception-related phenotypes in adult zebrafish: behavioral and pharmacological characterization using a new acetic acid model. Behav Brain Res 359:570–578

Cotter PA, Rodnick KJ (2006) Differential effects of anesthetics on electrical properties of the Rainbow trout (Oncorhynchus mykiss) heart. Comp Biochem Physiol A Mol Integr Physiol 145:158–165

Curtright A, Rosser M, Goh S, Keown B, Wagner E, Sharifi J, Raible DW, Dhaka A (2015) Modeling nociception in zebrafish: a way forward for unbiased analgesic discovery. PLoS One 10:e0116766

Dawkins MS (2001) Who needs consciousness? Anim Welf 10:S19–S29

de Miranda Cabral Gontijo ÁM, Barreto RE, Speit G, Reyes VAV, Volpato GL, Salvadori DMF (2003) Anesthesia of fish with benzocaine does not interfere with comet assay results. Mutat Res 534:165–172

Dunlop R, Laming P (2005) Mechanoreceptive and nociceptive responses in the central nervous system of goldfish (Carassius auratus) and trout (Oncorhynchus mykiss). J Pain 6:561–568

Eddy FB (2005) Ammonia in estuaries and effects on fish. J Fish Biol 67:1495–1513

Eriksen TB, Fraser T, Gregersen H, Kristiansen M, Olufsen M, Rajan B, Ramirez R, Ruangsri J, Røn Ø, Sarmiento MG (2011) Should fin clipping be used as a method for identification of fish. Norecopa

Espmark ÅM, Hjelde K, Baeverfjord G (2010) Development of gas bubble disease in juvenile Atlantic salmon exposed to water supersaturated with oxygen. Aquaculture 306:198–204

Evans DH (2010) A brief history of fish osmoregulation: the central role of the Mt. Desert Island biological laboratory. Front Physiol 1:13

Evans DH (2011) Osmotic, ionic and nitrogenous-waste balance. In: Farrell AP (ed) Encyclopedia of fish physiology from genome to environment. Elsevier, pp 1348–1353

FAO (2022) The state of world fisheries and aquaculture 2022: towards blue transformation. Food and Agriculture Organization of the United Nations, Rome

Feinberg TE, Mallat JM (2016) The ancient origins of consciousness: how the brain created experience. The MIT Press, Cambridge, MA

Gräns A, Sandblom E, Kiessling A, Axelsson M (2014) Post-surgical analgesia in Rainbow trout: is reduced Cardioventilatory activity a sign of improved animal welfare or the adverse effects of an opioid drug? PLoS One 9:e95283

Greene W, Mylniczenko ND, Storms T, Burns CM, Lewbart GA, Byrd L, Papich MG (2020) Pharmacokinetics of Ketoprofen in Nile tilapia (Oreochromis niloticus) and Rainbow trout (Oncorhynchus mykiss). Front Vet Sci 7:585324

Gressler LT, Xavier MLG, Junior GB, Loebens L, Barboza VDS, da Costa ST, Baldisserotto B (2021) Behavioral and histological features of zebrafish following sedation with eugenol or propofol. Appl Anim Behav Sci 244:105482

Grimsbø E, Nortvedt R, Hjertaker BT, Hammer E, Roth B (2016) Optimal AC frequency range for electro-stunning of Atlantic salmon (Salmo salar). Aquaculture 451:283–288

Harms CA (2005) Surgery in fish research: common procedures and postoperative care. Lab Anim (NY) 34:28–34
Harvey AM, Beausoleil NJ, Ramp D, Mellor DJ (2020) A ten-stage protocol for assessing the welfare of individual non-captive wild animals: free-roaming horses (Equus ferus caballus) as an example. Animals 10(1):148
Hastings PA, Walker HJ, Galland GR (2015) Fishes: a guide to their diversity. University of California Press, Oakland
Haworth J, Swimming upstream: how salmon farming developed in New Zealand (Wily Publications Ltd., 2010)
Hayashi M, Kita J, Ishimatsu A (2004) Acid-base responses to lethal aquatic hypercapnia in three marine fishes. Mar Biol 144:153–160
Horký P, Grabic R, Grabicová K, Brooks BW, Douda K, Slavik O, Hubená P, Sancho Santos EM, Randák T (2021) Methamphetamine pollution elicits addiction in wild fish. J Exp Biol 224(13):jeb242145
Huntingford FA, Kadri S (2014) Defining, assessing and promoting the welfare of farmed fish. Rev Sci Tech 33(1):233–244
Ide LM, Hoffmann A (2002) Stressful and behavioral conditions that affect reversible cardiac arrest in the Nile tilapia, Oreochromis niloticus (Teleostei). Physiol Behav 75:119–126
Iversen M, Finstad B, McKinley RS, Eliassen RA (2003) The efficacy of metomidate, clove oil, Aqui-S™ and Benzoak® as anaesthetics in Atlantic salmon (Salmo salar L.) smolts, and their potential stress-reducing capacity. Aquaculture 221:549–566
Jarvis S, Ellis MA, Turnbull JF, Planellas SR, Wemelsfelder F (2021) Qualitative behavioral assessment in juvenile farmed Atlantic salmon (Salmo salar): potential for on-farm welfare assessment. Front Vet Sci 8:702783
Johnston H, Symonds J, Walker S, Preece M, Lopez C, Nowak B (2021) Case definitions for skin lesion syndromes in chinook salmon farmed in Marlborough Sounds, New Zealand. J Fish Dis 44:141–147
Jones SG, Kamunde C, Lemke K, Stevens ED (2012) The dose–response relation for the antinociceptive effect of morphine in a fish, Rainbow trout. J Vet Pharmacol Ther 35:563–570
Key B (2016) Why fish do not feel pain. Anim Sentience 1:1
Kotrschal A, Kotrschal K (2020) Fish brains: anatomy, functionality, and evolutionary relationships. In: The welfare of fish. Springer, pp 129–148
Lago J, Rodriguez LP, Blanco L, Vieites JM, Cabado AG (2015) Tetrodotoxin, an extremely potent marine neurotoxin: distribution, toxicity, origin and therapeutical uses. Mar Drugs 13:6384–6406
Lines JA, Spence J (2011) Safeguarding the welfare of farmed fish at harvest. Fish Physiol Biochem 38:153–162
Lopez-Luna J, Al-Jubouri Q, Al-Nuaimy W, Sneddon LU (2017a) Reduction in activity by noxious chemical stimulation is ameliorated by immersion in analgesic drugs in zebrafish. J Exp Biol 220:1451–1458
Lopez-Luna J, Canty MN, Al-Jubouri Q, Al-Nuaimy W, Sneddon LU (2017b) Behavioural responses of fish larvae modulated by analgesic drugs after a stress exposure. Appl Anim Behav Sci 195:115–120
Lopez-Luna J, Al-Jubouri Q, Al-Nuaimy W, Sneddon LU (2017c) Impact of analgesic drugs on the behavioural responses of larval zebrafish to potentially noxious temperatures. Appl Anim Behav Sci 188:97–105
Luís Félix RC, Rita Sequeira CR, Sandra Monteiro LA, José Silva CV, Valentim A (2021) Ms-222 and propofol sedation during and after the simulated transport of Nile tilapia (Oreochromis niloticus). Biology 10:1309
Macaulay G, Bui S, Oppedal F, Dempster T (2021) Challenges and benefits of applying fish behaviour to improve production and welfare in industrial aquaculture. Rev Aquac 13:934–948
Mellor DJ, Beausoleil NJ (2019) Moving beyond a problem-based focus on poor animal welfare toward creating opportunities to have positive welfare experiences. In: McMillan FD (ed) Mental health and well-being in animals. CAB International
Mellor DJ, Beausoleil NJ, Littlewood KE, McLean AN, McGreevy PD, Jones B, Wilkins C (2020) The 2020 five domains model: including human–animal interactions in assessments of animal welfare. Animals 10(10):1870
Mettam JJ, Oulton LJ, McCrohan CR, Sneddon LU (2011) The efficacy of three types of analgesic drugs in reducing pain in the Rainbow trout, Oncorhynchus mykiss. Appl Anim Behav Sci 133:265–274
Mettam JJ, McCrohan CR, Sneddon LU (2012) Characterisation of chemosensory trigeminal receptors in the Rainbow trout, Oncorhynchus mykiss: responses to chemical irritants and carbon dioxide. J Exp Biol 215:685–693
Molony V, Kent JE (1997) Assessment of acute pain in farm animals using behavioral and physiological measurements. J Anim Sci 75:266–272
Musk GC, Ezzy BJ, Kenchington LM, Hopper WA, Callahan LM (2020) A comparison of buffered tricaine methanesulfonate (Ms-222) and isoeugenol anesthesia for caudal fin clipping in zebrafish (Danio rerio). J Am Assoc Lab Anim Sci 59:732–736
Narahashi T (2008) Tetrodotoxin - a brief history. Proc Jpn Acad Ser B Phys Biol Sci 84(5):147–154
Nilsson J, Moltumyr L, Madaro A, Kristiansen TS, Gåsnes SK, Mejdell CM, Gismervik K, Stien LH (2019) Sudden exposure to warm water causes instant behavioural responses indicative of nociception or pain in Atlantic salmon. Vet Anim Sci 8:100076
Noble C, Gismervik K, Iversen MH, Kolarevic J, Nilsson J, Stien LH, Turnbull JF (2018a) Welfare indicators for farmed Atlantic salmon: tools for assessing fish welfare. Nofima
Noble C, Gismervik K, Iversen MH, Kolarevic J, Nilsson J, Stien LH, Turnbull JF, NOFIMA AS (2018b) Welfare indicators for farmed Atlantic salmon: tools for assessing fish welfare. Nofima

Noble C, Gismervik K, Iversen MH, Kolarevic J, Nilsson J, Stien LH, Turnbull JF (2020) Welfare indicators for farmed Rainbow trout: tools for assessing fish welfare. Fiskeri- og havbruksnæringens forskningsfinansiering

Nordgreen J, Horsberg TE, Ranheim B, Chen ACN (2007) Somatosensory evoked potentials in the telencephalon of Atlantic salmon (Salmo salar) following galvanic stimulation of the tail. J Comp Physiol A 193:1235–1242

Nordgreen J, Bjørge MH, Janczak AM, Poppe T, Koppang EO, Ranheim B, Horsberg TE (2013) The effect of morphine on changes in behaviour and physiology in intraperitoneally vaccinated Atlantic salmon (Salmo salar). Appl Anim Behav Sci 145:129–137

Nordgreen J, Tahamtani FM, Janczak AM, Horsberg TE (2014) Behavioural effects of the commonly used fish anaesthetic tricaine methanesulfonate (Ms-222) on zebrafish (Danio rerio) and its relevance for the acetic acid pain test. PLoS One 9:e92116

Obirikorang KA, Asante-Tuoh DT, Agbo NW, Amponsah AK, Skov PV (2020) Anaesthetic potential of propofol for Nile tilapia (Oreochromis niloticus): effect of anaesthetic concentration and body weight. Sci Afr 10:e00595

Pandey R, Asche F, Misund B, Nygaard R, Adewumi OM, Straume HM, Zhang D (2023) Production growth, company size, and concentration: the case of salmon. Aquaculture 577:739972

Pirhonen J, Schreck CB (2003) Effects of anaesthesia with Ms-222, clove oil and CO_2 on feed intake and plasma cortisol in steelhead trout (Oncorhynchus mykiss). Aquaculture 220:507–514

Poli BM, Parisi G, Scappini F, Zampacavallo G (2005) Fish welfare and quality as affected by pre-slaughter and slaughter management. Aquacult Int 13:29–49

Popma T, Masser M (1999) Tilapia life history and biology

Pounder KC, Mitchell JL, Thomson JS, Pottinger TG, Sneddon LU (2018) Physiological and behavioural evaluation of common anaesthesia practices in the Rainbow trout. Appl Anim Behav Sci 199:94–102

Reilly SC, Quinn JP, Cossins AR, Sneddon LU (2008) Behavioural analysis of a nociceptive event in fish: comparisons between three species demonstrate specific responses. Appl Anim Behav Sci 114:248–259

Rizzo AL, Wooster GA, Guanzini LE, Peterson CM, Fenderson K, Erb HN, Bowser PR, Martin ME (2017) Biochemical, histopathologic, physiologic, and behavioral effects of nonsteroidal antiinflammatory drugs in Rainbow trout (Oncorhynchus mykiss). Comp Med 67:106–111

Robb DHF, Kestin SC (2002) Methods used to kill fish: field observations and literature reviewed. Anim Welf 11:269–282

Robb DHF, Wotton SB, McKinstry JL, Sørensen NK, Kestin SC, Sørensen NK (2000) Commercial slaughter methods used on Atlantic salmon: determination of the onset of brain failure by electroencephalography. Vet Rec 147:298–303

Rolfhus KR, Sandheinrich MB, Wiener JG, Bailey SW, Thorenson KA, Hammerschmidt CR (2008) Analysis of fin clips as a nonlethal method for monitoring mercury in fish. Environ Sci Technol 42:871–877

Roques JAC, Abbink W, Geurds F, van de Vis H, Flik G (2010) Tailfin clipping, a painful procedure: studies on Nile tilapia and common carp. Physiol Behav 101:533–540

Rosa LV, Costa FV, Gonçalves FL, Rosemberg DB (2022) Acetic acid-induced nociception modulates sociability in adult zebrafish: influence on shoaling behavior in heterogeneous groups and social preference. Behav Brain Res 434:114029

Rose JD, Arlinghaus R, Cooke SJ, Diggles BK, Sawynok W, Stevens ED, Wynne CDL (2014) Can fish really feel pain? Fish Fish 15:97–133

Roth B, Slinde E, Robb DHF (2007) Percussive stunning of Atlantic salmon (Salmo salar) and the relation between force and stunning. Aquac Eng 36:192–197

Sanderson BL, Tran CD, Coe HJ, Pelekis V, Steel A, Reichert WL (2009) Nonlethal sampling of fish caudal fins yields valuable stable isotope data for threatened and endangered fishes. Trans Am Fish Soc 138:1166–1177

Saunders J, Speare DJ, McConkey S (2020) Methemoglobin concentrations in three salmonid species following exposure to benzocaine or tricaine methanesulfonate. Fish Physiol Biochem 46:2257–2263

Schroeder PG, Sneddon LU (2017) Exploring the efficacy of immersion analgesics in zebrafish using an integrative approach. Appl Anim Behav Sci 187:93–102

Sherif AH, Eldessouki EA, Sabry NM, Ali NG (2022) The protective role of iodine and Ms-222 against stress response and bacterial infections during Nile tilapia (Oreochromis niloticus) transportation. Aquac Int 31(1):401–416

Sneddon LU (2003a) Trigeminal somatosensory innervation of the head of a teleost fish with particular reference to nociception. Brain Res 972:44–52

Sneddon LU (2003b) The evidence for pain in fish: the use of morphine as an analgesic. Appl Anim Behav Sci 83:153–162

Sneddon LU (2012) Clinical anesthesia and analgesia in fish. J Exot Pet Med 21:32–43

Sneddon LU (2020) Can fish experience pain? In: The welfare of fish. Springer, pp 229–249

Sneddon LU, Braithwaite VA, Gentle MJ (2003) Do fishes have nociceptors? Evidence for the evolution of a vertebrate sensory system. Proc R Soc Lond Ser B Biol Sci 270:1115–1121

Sneddon LU, Roques JA (2023) Painrecognition in fish. Veterinary Clinics: Exot Anim Pract 26(1):1–10

Steenbergen PJ, Bardine N (2014) Antinociceptive effects of buprenorphine in zebrafish larvae: an alternative for rodent models to study pain and nociception? Appl Anim Behav Sci 152:92–99

Stien LH, Bracke MBM, Folkedal O, Nilsson J, Oppedal F, Torgersen T, Kittilsen S, Midtlyng PJ, Vindas MA, Øverli Ø (2013) Salmon Welfare Index Model (SWIM 1.0): a semantic model for overall welfare assessment of caged Atlantic salmon: review of the selected wel-

fare indicators and model presentation. Rev Aquac 5:33–57
Stien LH, Bracke M, Noble C, Kristiansen TS (2020) Assessing fish welfare in aquaculture. In: The welfare of fish. Springer, pp 303–321
Sundh H, Kvamme BO, Fridell F, Olsen RE, Ellis T, Taranger GL, Sundell K (2010) Intestinal barrier function of Atlantic salmon (Salmo salar L.) post smolts is reduced by common sea cage environments and suggested as a possible physiological welfare indicator. BMC Physiol 10:22
Teletchea F (2019) Fish domestication: an overview. In: Teletchea F (ed) Animal domestication. IntechOpen
Teletchea F (2021) Fish domestication in aquaculture: 10 unanswered questions. Anim Front 11:87–91
Tilley CA, Carreño Gutierrez H, Sebire M, Obasaju O, Reichmann F, Katsiadaki I, Barber I, Norton WH (2020) Skin swabbing is a refined technique to collect DNA from model fish species. Sci Rep 10:18212
Uney K, Corum DD, Terzi E, Corum O (2021) Pharmacokinetics and bioavailability of carprofen in Rainbow trout (Oncorhynchus mykiss) broodstock. Pharmaceutics 13:990
Valentim AM, Félix LM, Carvalho L, Diniz E, Antunes LM (2016) A new anaesthetic protocol for adult zebrafish (Danio rerio): propofol combined with lidocaine. PLoS One 11:e0147747
van de Vis H, Kestin S, Robb D, Oehlenschlager J, Lambooij B, Munker W, Kuhlmann H, Kloosterboer K, Tejada M, Huidobro A, Ottera H, Roth B, Sorensen NK, Akse L, Bryne H, Nesvadba P (2003) Is humane slaughter of fish possible for industry? Aquac Res 34:211–220
van de Vis JW, Abbink W, Lambooij B, Bracke MBM (2014) Stunning and killing of farmed fish: how to put it into pratice? In: Encyclopedia of meat sciences, 2nd edn. Academic, pp 421–426
van de Vis H, Kolarevic J, Stein LH, Kristiansen TS, Gerritzen M, van de Braak K, Abbink W, Sæther B-S, Noble C (2020) Welfare of farmed fish in different production systems and operations. In: Kristiansen TS, Fernö A, Pavlidis MA, van de Vis H (eds) The welfare of fish. Springer, Switzerland, pp 323–361
Vargas R, Balasch JC, Brandts I, Reyes-Lopez F, Tort L, Teles M (2018) Variations in the immune and metabolic response of proactive and reactive *Sparus aurata* under stimulation with *Vibrio anguillarum* vaccine. Sci Rep 8:17352
Vergneau-Grosset C, Weber ES III (2021) Fish surgery. In: Bennett RA, Pye GW (eds) Surgery of exotic animals. Wiley Blackwell, pp 46–62
Wagner GN, Stevens ED (2000) Effects of different surgical techniques: suture material and location of incision site on the behaviour of Rainbow trout (Oncorhynchus mykiss). Mar Freshw Behav Physiol 33:103–114
Wagner GN, Singer TD, Scott McKinley R (2003) The ability of clove oil and Ms-222 to minimize handling stress in Rainbow trout (Oncorhynchus mykiss Walbaum). Aquac Res 34:1139–1146
Wendelaar Bonga SE, Lock RAC (1992) Toxicants and osmoregulation in fish. Neth J Zool 42:478–493
Wong K, Elegante M, Bartels B, Elkhayat S, Tien D, Roy S, Goodspeed J, Suciu C, Tan J, Grimes C (2010) Analyzing habituation responses to novelty in zebrafish (Danio rerio). Behav Brain Res 208:450–457
Wong D, von Keyserlingk MAG, Richards JG, Weary DM (2014) Conditioned place avoidance of zebrafish (Danio rerio) to three chemicals used for euthanasia and anaesthesia. PLoS One 9:e88030
Woodruff ML (2017) Consciousness in teleosts: there is something it feels like to be a fish. Anim Sentience 2:1
Zahl IH, Kiessling A, Samuelsen OB, Olsen RE (2010) Anesthesia induces stress in Atlantic salmon (Salmo salar), Atlantic cod (Gadus Morhua) and Atlantic halibut (Hippoglossus hippoglossus). Fish Physiol Biochem 36:719–730
Zahl IH, Samuelsen O, Kiessling A (2012) Anaesthesia of farmed fish: implications for welfare. Fish Physiol Biochem 38:201–218
Zahran E, Risha E, Rizk A (2021) Comparison propofol and eugenol anesthetics efficacy and effects on general health in Nile tilapia. Aquaculture 534:736251
Zhao Z, Xiao Q, Tchivelekete GM, Reilly J, Jiang H, Shu X (2021) Quantification of computational fluid dynamics simulation assists the evaluation of protection by Gypenosides in a zebrafish pain model. Physiol Behav 229:113223

Cephalopods

Robyn J. Crook

Abstract

Cephalopod molluscs are a large and complex taxonomic group, acting as keystone species in multiple global food webs. Their value in fisheries and aquaculture continues to grow as finfish stocks decline, and there is growing interest in expanding aquaculture to new species of cephalopods. While welfare concerns for cephalopods are now at the forefront in the realms of scientific research and in educational settings, welfare of fished or aquacultured cephalopods remains largely overlooked. Here, welfare risks to fished and aquacultured cephalopods are considered with a primary focus on pain and suffering at points of human contact, including slaughter. Critical gaps in knowledge are highlighted, which, if addressed, could lead to measurable improvements in welfare, and suggest a set of key welfare indicators that can be used by veterinary and other professionals charged with assessing well-being and health status of aquacultured cephalopods.

Keywords

Octopus · Squid · Cuttlefish · Husbandry · Analgesia · Anaesthesia

R. J. Crook (✉)
Department of Biology, San Francisco State University, San Francisco, CA, USA
e-mail: rcrook@sfsu.edu

1 Natural History

Cephalopods are molluscs, and first appear in the fossil record about 530 million years ago, having evolved from a monoplacophoran-like mollusc through conversion of their external shell into a buoyancy apparatus (Kröger et al. 2011). There are approximately 800 described, extant species of cephalopods spread across two classes, Nautiloidea and Coleoidea. Nautiloidea are easily recognisable among cephalopods as the only group that retained their external shells. Though fossil evidence suggests nautiluses were once prolific in the oceans, they are now a tiny minority of all extant cephalopods and are rarely seen in the wild. Nautiluses are occasionally displayed in public aquaria and are fished for the ornamental shell trade, but are not fished commercially for food. (Ward and Saunders 1997).

Coleoidea, the 'modern' cephalopods, includes all extant cuttlefish, squid and octopuses in two superorders, Decapodiformes and Octopodiformes. Decapodiformes refers to the 10 appendages present in cuttlefish and squid, which all have eight arms and two tentacles. Octopuses have eight arms and no tentacles. Arms and tentacles are distinguished by anatomy and by use; arms are lined in suckers from the base to the tip and are always visible, whereas tentacles have long, smooth stalks and a bulbous, sucker covered tip and are held coiled in two pouches beneath the buccal mass. Tentacles are shot

R. E. Clutton, G. C. Musk (eds.), *Pain and Suffering in Farmed Animals*,
https://doi.org/10.1007/978-3-032-11407-5_21

toward prey in a high-speed striking movement but are otherwise kept internally.

Cephalopods share certain basic features of their body organisation with other molluscs. The typical molluscan shell is greatly reduced or lost in coleoid cephalopods. Instead of relying on the protection that an external shell might offer, coleoid cephalopods instead defend against predation by being exceptionally fast; their soft body allows for rapid jet-propulsion mediated by giant motor neurons that contract both sides of the main body cavity (the mantle) within a fraction of second. This produces extremely fast backward escape movement, which is often combined with ink release. Additionally, cephalopods have a unique ability to change colour on millisecond timescales, using pigment-filled organs called chromatophores. Chromatophore patterns are used in camouflage, but also in intra-specific signalling during mating and aggressive interactions. Some deep-water species and many squid also have photophores and other bioluminescent skin elements for camouflage and signalling (Bush et al. 2009; Johnsen et al. 1999; Robison et al. 2003).

The nervous system of cephalopods is among the largest and most complex of all the invertebrates, both in terms of raw number of cells and in the number of different discrete lobes within the brain. The brain mass/body mass ratio surpasses that of the primitive or lower vertebrates (Hanlon and Messenger 2018; Nixon and Young 2003). In general, the nervous system consists of a large central brain located between the eyes, along with multiple peripheral ganglia with varying degrees of elaboration within the mantle and arms. Octopuses have around 500 million neurons, considerably more that many vertebrate animals, although approximately 2/3 are outside the central brain mass (Nixon and Young 2003; Young 1963), mainly in the arms. The arm ganglia govern local movements, including complex, coordinated movements involved in grasping, prey handling and defensive withdrawal (Gutfreund et al. 2006; Hague et al. 2013; Matzner 2000; Nesher et al. 2014; Sumbre et al. 2001), which can occur without input from the central brain (Fig. 1).

Wild cephalopods are obligate carnivores, primarily active hunters of fish, crustaceans and other molluscs. Hunting and other behaviour in the wild are quite different among the different taxa. Excluding the pelagic clades where individuals spend their entire lives in the water column, most octopuses settle to the substrate after a brief paralarval period, and then tend to occupy a home den where they spend much of their time, leaving to patrol a surrounding area to catch food. Denning octopuses are non-social, and although some tolerate close conspecifics, or engage in cooperative hunting with fish, in general they are solitary. Squid and cuttlefish have habits that vary according to their life stage and ecology. Many squid are open ocean, pelagic animals throughout their entire lives, only coming close to shore for their terminal reproductive event. Some reef squid are quasi-social, forming loose shoals that improve predator detection and provide safety in numbers; however, there is minimal evidence of true sociality and instead congregation is opportunistic, for the exploitation of a shared resource. Cuttlefish behaviour varies, with many species moving from offshore nektobenthic habitats to shallow inshore waters to feed and reproduce.

Cephalopod reproduction is semelparous (having a single reproductive period, usually at the end of the life), with either a single egg-laying event or multiple broods laid over a relatively short period. Cephalopods are gonochoristic (dioecious) and oviparous, and fertilisation is achieved by direct mating (Arnold and Williams-Arnold 1977; Rocha et al. 2001; Wells and Wells 1977). Sexual maturity is controlled by secretion of hormones from the optic gland (Di Cosmo et al. 2001; Di Cosmo and Di Cristo 1998; Polese et al. 2015). To some degree, sexual maturity is controlled by body size and water temperature, giving it a seasonal aspect (Hernández-García et al. 2002; Lipiński and Underhill 1995; Martins 1982; Otero et al. 2007). Mating in squid and cuttlefish is often preceded by elaborate courtship displays and sometimes-violent male contests for access to females. In octopuses, more typically males and females approach each other with caution before mating occurs, but the duration of courtship and the behavioural sequence

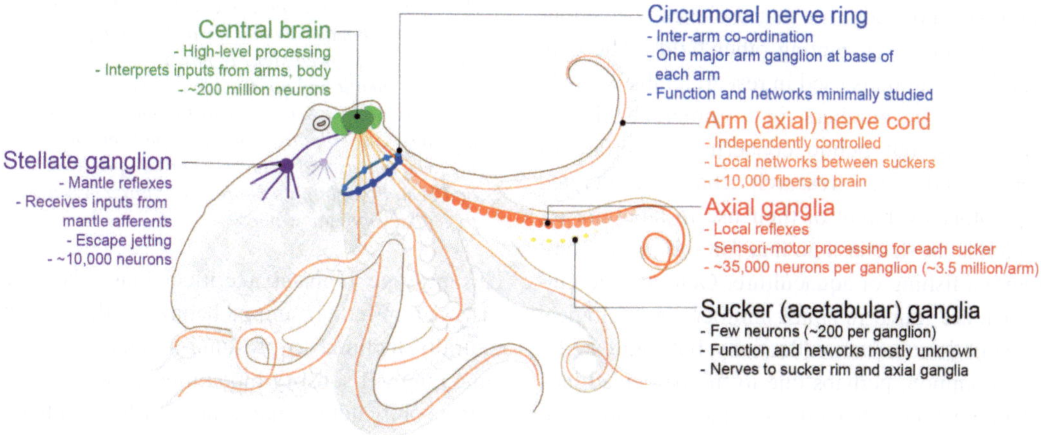

Fig. 1 The primary nervous system structures of the octopus

leading to mating are more cryptic. During mating, the male transfers a spermatophore to the female along a muscular groove in one hectocotlysed arm, with the specific arm being variable among the taxa. The tip of the arm is guided presumably by olfactory cues to the oviduct inside the mantle cavity of the female, where spermatophores can be stored for months prior to egg laying. Females often store sperm from multiple males and multiple paternity is common (Huffard and Godfrey-Smith 2010; Thompson and Voight 2003; Wodinsky 2008).

Care of fertilised eggs is variable. Octopus females lay their eggs inside a den, then remain with the eggs more or less uninterrupted until they hatch, using their siphon to aerate the egg masses during development. Less commonly, females may hold egg masses in a protective posture in their arms while they rest on the substrate (Boletzky 1998; Drazen et al. 2003; Rocha et al. 2001). Incubation periods vary from a typical 1–3 months up to more than 2 years in some deep-sea species (Robison et al. 2014). During incubation, females cease feeding and lose body weight and condition, and this period is associated with a drastic decline in the health of skin, muscle and the nervous system. Females either die spontaneously after incubation or are predated. Males take no part in egg care and enter a period of terminal decline shortly after maturation of their spermatophores, whether or not they mate successfully.

Squid and cuttlefish lay their eggs attached to a substrate, where they incubate without parental care from either sex. Some exceptions to these general rules are known; the deep-sea squid, *Gonatus onyx*, for example, broods its eggs by holding them in its arms as it swims (Seibel et al. 2000), and the octopus Argonauta secretes a shell-like encasement to protect its eggs until hatching (Finn and Norman 2010). There is no evidence of post-hatching parental care in any cephalopod, and hatchlings are capable of hunting shortly after emergence. Hatchling growth rates are exceptionally high, which is unsurprising given the very short life spans of most cephalopods. Demand for food among early-stage hatchlings remains the primary barrier to successful aquaculture of many cephalopod species.

2 Current Extent of Utilisation

Cephalopod fisheries date to ancient times, with near-shore fisheries present throughout the Mediterranean and around much of Asia. More recently, modern commercial exploitation of cephalopod stocks has grown to represent some 4.3% of the total global seafood catch by volume (Global Fisheries Watch), indicating a strong market for cephalopods for human consumption.

Cephalopods are also important in research laboratories worldwide and have produced foundational knowledge in neuroscience, behaviour and microbiology, among other fields. The total number of animals used in research labs is vastly smaller than numbers in commercial fishing operations, but public concern for welfare of cephalopods in research has accelerated regulatory oversight of their use in laboratories, which is far stricter than that for animals in commercial fishing or aquaculture. Octopus and cuttlefish are found in small numbers in zoological and public aquarium collections, but squid are far less common, perhaps due to the perceived lack of interesting behaviours produced (many sepiolid squid that are amenable to aquarium culture spend most of their daylight hours buried in sand), or due to the challenges of housing and maintaining pelagic squid. Octopus and cuttlefish are occasionally seen in the retail aquarium trade, but numbers are small compared to ornamental marine fishes.

3 Attitudes to, and Evidence for, Pain in Cephalopods

Modern public discourse around cephalopods emphasises their intelligence and uniqueness among invertebrates, which has promoted long-running debates over their capacity to experience pain or suffering. Historical views of cephalopods are less complimentary; among the most famous descriptions of octopus behaviour is Aristotle, in his History of Animals (350 BC) stating 'the octopus is a stupid creature' a conclusion at which he arrived by observing *Octopus vulgaris* appearing unafraid of a human hand waved in front of it in the water. More recent reference to octopus behaviour (and pertinent to discussion on pain physiology) was made by the famous marine documentarian Jacques Cousteau in his book 'Octopuses and Squid, the soft intelligence' (1973), where he writes

> *One day at Octopus City, in the Bay of Alicaster, Dumas dived with an underwater rocket and began waving it in front of an octopus' house. Nothing happened. The animal reacted not at all. He did not try to hide, or to escape. Dumas then turned the beam directly onto the octopus, which did not even draw [sic] its arms. The game was called off, however, when Dumas saw that it was becoming cruel. The octopus showed signs of having been burned. But even then it had not tried to escape from it. … This surprising insensitivity to fire has been confirmed by Guy Hilpatric, one of the pioneers of diving, who told us that he has seen an octopus, which had been brought onto shore, cross through a fire to get back into the water.*

This passage hints at a critical bias when we observe non-human animal behaviour that we are naturally inclined to presume animals have the same range of sensory modalities as humans do. In fact, obligately aquatic animals like cephalopods have no specific selection pressures driving evolution of receptors tuned to extreme heat, but this does not necessarily imply they lack other relevant neural apparatus for the perception of pain.

Recently, there has been sustained public interest in the question of sentience, pain and suffering in invertebrate animals (Birch et al. 2021). Given that cephalopods are by far the most neurologically complex among them, much of the debate has focused on cephalopods (along with other comparably large-brained invertebrates, including decapod crustaceans and beneficial insects like bees). In parallel with a renewed interest in cephalopods as research models, growing concern from animal welfare organisations and lay people has brought into focus the ethical issues that come with using a sentient animal in invasive research (Crook 2023). While this has promoted a renewed focus (and some progress) on welfare-enhancing refinements for cephalopods, most large funding agencies remain uninterested in funding studies aimed primarily at improving the welfare of research animals.

3.1 Regulation and Law as Applied to Cephalopods

Concern for the welfare of invertebrate animals has accelerated government efforts to provide oversight of their use, a movement most noticeable in the growing regulatory frameworks for cephalopods in zoos, public aquaria and research institutions (Crook 2023). Public awareness of cephalopod intelligence and potential sentience is much higher than for other invertebrates, but so far there is minimal evidence that welfare concerns drive consumer choice for consumption of cephalopod products. However, the recent public backlash to efforts to develop commercial-scale octopus aquaculture (Jacquet et al. 2019) shows that public perceptions of animal suffering may be a significant, market-driving force in future.

An interesting case study of differing public and governmental opinions relating to cephalopod welfare in different contexts can be seen in the contrasting views of aquacultured vs fished cephalopods in two US states. In California and Washington (both coastal states with strong fisheries, active aquaculture sectors, strong regulatory oversight of animal welfare generally, and left-leaning political cultures), two recent bills have been passed focusing on octopus aquaculture (Washington HB 1153 stating 'a person may not participate in octopus aquaculture in Washington', and California Bill AB.3162 m banning 'aquaculture of any species of octopus for the purpose of human consumption'). These measures can be viewed as direct responses to public pressure and lobbying from animal rights organisations, given that no commercially-farmed cephalopods destined for human consumption are presently grown in these states. Notably, neither state enacts regulations on the welfare of cephalopods (although there are firm regulations over catch effort, closures and gear types) in their fisheries.

Regulations governing cephalopods in research are more common and more widespread. Nearly all western nations regulate research using cephalopods in some capacity (e.g. EU: Directive 2010/63/EU, Canada, Canadian Council on Animal Care (CCAC, 1994), UK: Animals (Scientific Procedures) Act 1984, and UK Animal Welfare (Sentience) Act 2022, Australia (Code for the care and use of animals in scientific procedures, 2013), New Zealand (Animal Welfare Act, 1999). A notable exception is the United States, where cephalopods (along with all other invertebrates) are excluded from the Public Health Service definition of an animal. Although there has been some recent interest in changing the definition of an animal to include cephalopods (see NIH NOT-OD-23-176), at the time of writing, there is still no federal or state regulation of research animal welfare that includes cephalopods in the United States.

3.2 Evidence for Pain and Suffering in Cephalopods

Cephalopods have the largest and most vertebrate-like brain of any of the invertebrate clades (Shigeno et al. 2018), however, the scientific evidence for similar capacities for subjective experience is still patchy (Birch et al. 2021, and see also New York Declaration on Animal Consciousness, 18 April 2024). In general, there has been a marked shift in attitudes toward potential cephalopod suffering over the past 10–15 years, as evidenced by increasing attention paid to cephalopod welfare by regulatory agencies and oversight bodies (Crook 2023; Ponte et al. 2019).

Pain experience is a two-step process that requires both sensory circuits tuned to the detection of noxious stimuli (containing in some cases a dedicated class of sensory neurons termed nociceptors), and a higher processing step that produces a subjective experience for the animal as a result of that sensory input (Tracey 2017). Nociception is the sensory capacity, and pain is the subjective state. There is evidence that cephalopods are capable of both nociception and experiencing pain.

Evidence for the presence of nociceptors in cephalopods is strong. Multiple taxa across decapods and octopods have been shown to have nociceptive sensory neurons, which undergo changes to their excitability after injury that are highly similar to those studied in great detail in mammalian models (Crook et al. 2013; Kuo et al. 2022). Nociceptors in cephalopods are probably polymodal (a single neuron responds to different kinds of stimuli such as mechanical and chemical), but this remains to be confirmed by detailed molecular and genetic profiling of individual cells, which has been limited in cephalopods. Cephalopod nociceptors certainly respond to noxious mechanical force, chemical stimuli, and pH, but there is an apparent lack of thermosnociceptors tuned to either damaging heat or cold (Crook et al. 2013). The local circuits controlling reflex nocifensive behaviours in cephalopods are readily sensitised by noxious stimuli, producing a range of behavioural effects on defensive and antipredator behaviours (Crook et al. 2011, 2014), foraging (Howard et al. 2019; Zepeda et al. 2017), general activity (Alupay et al. 2014; Crook et al. 2013) and social interactions (Crook et al. 2011; Oshima et al. 2016).

Evidence for pain and suffering in cephalopods comes from a number of recent studies that show complex, non-reflexive behavioural responses to stimulation that would be perceived as painful in mammals, and from neural recordings showing central processing of noxious stimuli (Crook 2021; Kuo et al. 2022). This second point is critical, because cephalopods have an extensive and complex peripheral nervous system that provides the arms with high degrees of autonomy (Hochner et al. 2006). The question of what information is sent to the brain has remained largely unaddressed. The anatomical features of the arm nervous system suggest very large volumes of information are processed entirely in the periphery; in an octopus, each arm contains around 40 million neurons, but the centrally projecting axon tracts that carry information to the central brain from the base of each arm contain only around 10,000 fibres (Young 1963). This suggests that only a tiny portion of what the arm senses is sent to the brain for further processing. If we assume that the arm itself is not sentient (although the concept of embodied cognition (Hochner 2012) may include some capacity for the arm to be a cognitive entity in its own right), then for the animal to experience pain or suffering as a result of peripheral sensory input, that information must be prioritised over almost all others to be sent via the highly bottlenecked pathway to the brain. In recent studies of neural signal conduction in the arm, relatively little activity was recorded in locations in proximal (i.e. closer to the brain) sites when stimulation was presented to progressively more distal arm regions (Alupay et al. 2014; Fouke and Rhodes 2020), suggesting that non-noxious sensory inputs are processed locally or at least regionally within the arm. However, newer evidence now suggests that, almost certainly, pain-specific sensory signals are conveyed with high fidelity to the central brain (Crook 2021), thus refuting the notion that the brain of a cephalopod may not 'know' about injuries or noxious sensory input in the periphery.

Behavioural studies also provide strong evidence that cephalopods are capable of pain and suffering. After injury or noxious stimulation, squid (*Doryteuthis pealei*) change their behaviour to become cautious and more vigilant in schools, and these changes persist for multiple days (Oshima et al. 2016). Similarly, in bobtail squids (*Euprymna* spp.), injury induces changes to antipredator defences that differ if injuries were experienced acutely as adults or in early life (Howard et al. 2019). Cuttlefish with injuries to the tentacle clubs (used in high-speed prey strikes) more rapidly learn to inhibit their hunting behaviour if the clubs impact onto non-food items, suggesting that the impact is painful (Butler Struben and Crook 2018). While all these examples are suggestive of central processing that goes beyond simple nociceptive reflexes, the strongest evidence for affective pain experience comes from behavioural assays adapted directly from those used in mammals; conditioned place preference (CPP) and conditioned place avoidance (CPA). When octopuses received an injection of acetic acid into the skin of one arm, they learned to avoid a previously preferred location within which they were confined after injection (Crook 2021). This simple choice experiment demonstrates the capacity for cephalopods to evaluate their internal state and decide whether it is positive or negative; in this case, the injection caused the animals to feel more negative about a location they previously preferred, which suggests strongly that noxious injection causes negative affect—pain. In a follow-up study, octopuses were given a local anaesthetic injection over the acetic acid site, and these animals developed a strong preference for a location they previously did not prefer, suggesting that relief from pain was rewarding, just as it is in mammals (Crook 2021). The two experiments described briefly here are the strongest current evidence for pain experience in cephalopods. More study is needed to validate these findings in cuttlefish and squid.

Of course, poor welfare and negative affect can be caused by non-painful experiences as well. In a recent study using cuttlefish, chronic stress produced by poor housing conditions and once-daily, brief handling resulted in cuttlefish displaying negative judgement bias in a visually-cued food choice experiment, the first evidence for negative affect in cuttlefish and non-pain-related negative affect in any cephalopod (Giancola-Detmering and Crook 2024). Additional study is needed, particularly in squid, which tend to be overlooked for behavioural studies due to the challenges of housing larger groups in research settings.

4 Anaesthesia and Analgesia

Rendering animals unconscious or immobile is a standard practice in research and in aquaculture. Unlike vertebrates, cephalopods have received relatively little study with respect to modern anaesthetic or analgesic drugs, and even less research on humane methods of killing, particularly at scale. The standard anaesthetic drugs that are validated for cephalopods are combinations of magnesium salts in high concentrations, ethanol and isoflurane, all of which are added to seawater and used as anaesthetic baths (Abbo et al. 2021; Butler-Struben et al. 2018; Di Cosmo et al. 2023; Gleadall 2013; Polese et al. 2014; Roumbedakis et al. 2020). There are no validated injectable general anaesthetic agents for cephalopods at the time of writing. Anaesthesia efficacy and depth in cephalopods is assessed mainly by behavioural indicators such as chromatophore relaxation (body paling), loss of righting reflex, loss of reflex responses to noxious stimulation such as skin pinch, and measurement of pupillary light reflex (Deutsch et al. 2023). In some cases, neural recordings combined with behavioural indicators have been used, and these show general concordance (although not perfect temporal correlation) between loss of behavioural reactivity and decline of afferent and efferent neural signals (Abbo et al. 2021; Butler-Struben et al. 2018).

The study of analgesia is likewise in its infancy; only one recent study has tested analgesic agents, and in only one species, a bobtail squid (Deutsch et al. 2023). The findings from this study suggested limited efficacy for buprenorphine, ketorolac and dexmedetomidine, but considerable further work is needed to optimise dosage and test efficacy in other species and other conditions. The use of systemic analgesics in cephalopods in currently rare and not standardised in any way, and thus is unlikely to be common practice in research laboratories, aquaculture or fisheries in the near future. Local anaesthesia using lidocaine has been studied in octopus and cuttlefish, and it appears to be effective at blocking sensation and preventing or relieving pain-related behaviours (Crook et al. 2021; Butler-Struben et al. 2018; Kuo et al. 2022).

5 Likely Causes of Pain and Suffering in Research, Fishing and Aquaculture

Cephalopods are appealing for aquaculture due to their high growth rates and good food conversion ratios (octopus: (Cerezo Valverde and García 2017; Iraba et al. 2023; Navarro et al. 2014; Pham and Isidro 2009) cuttlefish: (Capaz et al. 2020; Domingues et al. 2001; Sykes et al. 2003); squid: (Forsythe and Hanlon 1989; Lee et al. 1998; Walsh et al. 2002), making them economically valuable for intensive culture systems. However, their reliance on live feeds, poor tolerance for substandard water quality and territorial and aggressive natures are significant barriers to their successful culture.

Overwhelmingly, cephalopods are fished rather than aquacultured, with cephalopod fisheries making up 4.3% of the total global fisheries by volume (Global Fishing Watch 2023). Welfare of fished cephalopods is poorly regulated, given that the majority of the catch comes from distant water fisheries in poorly regulated fishing zones. Squid are caught either by trawl netting or by jigging (Arkhipkin et al. 2015), each of which likely produces significant suffering in captured animals prior to death. Trawl netting results in death by suffocation due to crushing during winding in of the nets, or asphyxiation for animals emersed and sorted on board. Jigging, even with barbless hooks, produces traumatic puncture and tearing injuries to the arms and tentacles prior to death from suffocation as animals are hauled onto decks. There are no established guidelines for humane killing of cephalopods in fisheries.

In aquaculture, welfare risks extend beyond slaughter across the lifetime of the animal. Cephalopods are prone to injuries in holding tanks due to their thin, delicate skin and jet-propelled swimming, which can result in injuries from collisions with tank structures (Hanlon et al. 1984) Intraspecific aggression is also common in many species and is exacerbated by high-density stocking (Zheng et al. 2023), thus injuries and death inflicted through fighting are also a challenge. In cuttlefish, maintaining balanced or only slightly biased sex ratios is important for limiting male-male fighting and excessive harassment of females as they approach sexual maturity, while maximising fecundity (Guo et al. 2022). In small-scale settings such as research laboratories, it is common practice to move males into female housing for short periods and otherwise keep the sexes separate, but this approach is likely not feasible for large-scale aquaculture.

Research into methods of humane slaughter for cephalopods raised for human consumption is extremely limited. Typical euthanasia in research and education settings involves immersion of whole animals into a chemical anaesthetic bath (Butler-Struben et al. 2018), which is impractical both for scale and food quality in aquaculture. Killing of cephalopods in artisanal fisheries may involve inversion of the mantle in octopuses, leading to asphyxiation on an uncertain time scale, percussive blows or cuts over the brain, or rapid butchery aimed at preserving meat quality (Andrews et al. 2013). The degrees of possible suffering encompassed in each of these methods remained unquantified, but a basic understanding of cephalopod nervous systems suggests that each of these methods is likely to result in considerable suffering of uncertain duration prior to death. There is an urgent need for more research into effective stunning and killing of cephalopods both in fisheries and aquaculture.

Octopus aquaculture presents additional welfare challenges due to the obligately solitary nature of the littoral octopus species most likely to be viable for culture. Octopuses in the wild are central place foragers that occupy dens that they defend from conspecifics and predators, and close proximity to other octopuses is generally not preferred (Forsythe and Hanlon 1997; Mather 1991). Octopuses crowded into small spaces may engage in increased aggressive interactions and potential cannibalism, which is likely to damage product quality as well as creating significant welfare issues.

6 Disease, Stress and Injuries in Aquacultured Cephalopods

Control of infection and diseases in cephalopods is most readily achieved by maintaining good water quality and proactive health monitoring. Interventions and treatments for illnesses are poorly researched, and as such, there is minimal information on the type, dosing or efficacy of veterinary drugs for cephalopods. Wild-caught cephalopods are often infested with protozoan and other parasites, and control of their spread is imperative in captive settings. Most often control is achieved by maintenance of pristine water quality and short-term treatments of animals or eggs with low-salinity dips, although these should be used with caution given the low tolerance for salinity or pH variation in cephalopods. Injured and exposed tissue is also rapidly colonised by bacterial and protozoal species, which can lead to poor recovery or infection leading to death (reviewed in (Sykes and Gestal 2014). As is typical for molluscs, there is only an innate immune system and no evidence for immunological memory mechanisms, so immune responses are limited to haemocyte infiltration.

As animals age, there is reduced capacity to heal wounds and respond to other challenges, especially after the onset of the terminal reproductive phase (Anderson et al. 2002). For animals grown for harvest, senescence is less likely to present welfare risks, presuming animals are killed prior to reproductive onset. For broodstock, management of senescence and determinations of humane endpoints as animals enter a terminal decline phase is an important factor in limiting potential suffering at end of life, but currently there are no robust standards in place supporting such determinations in aquaculture settings (although there are some guidelines available for animals in education and display, these are specific to a single species which is not suited to aquaculture (Holst and Miller-Morgan 2020; Kirby et al. 2023).

7 Assessment of Pain and Welfare in Cephalopods

There are currently no validated rubrics or guides for evaluating the presence or severity of pain in cephalopods. Here, a set of simple metrics for evaluating welfare (including potential pain) of captive cephalopods is outlined, focused on easily observable indicators requiring minimal handling.

7.1 External Appearance

External appearance can be divided into external physical condition and behavioural condition, each of which provides an initial point observation of any welfare concern. Evaluation of the integument should reveal injuries or areas of diseased or infected tissues. Healthy cephalopod skin is smooth and covered in active chromatophores that should express changes in response to visual and other cues sensed by the animal. Cephalopods with areas of white skin should be evaluated further to determine whether this is exposed muscle tissue resulting from injury to the epithelium (Hanlon et al. 1984) or areas of dysfunctional chromatophores, which may indicate infection or other health concerns. Animals with small areas of broken skin or exposed muscle should be monitored closely, whereas those with extensive areas affected (more than 5% of the body surface) might be most appropriately euthanised. Other issues of concern of the outer appearance are missing or damaged limbs, indicating conspecific aggression, injuries from live food handling or autophagy. Autophagy is particularly concerning given that it is often a response to extremes of stress or of terminal neurological decline, and there is some evidence that it releases a chemical cue from affected animals that promotes the onset of autophagy in other cephalopods sharing the same water volume (Budelmann 1998). While less concerning than autophagy, injuries from aggressive interactions and prey handling should be addressed by isolation of the injured animal, and interventions that reduce crowding and food competition. While some cephalopods are obligate consumers of live prey, it is better from a welfare perspective (and possibly also a nutritional perspective) to avoid feeding of live prey if possible.

Behavioural observation can also reveal emerging health and welfare issues. Most cephalopods are reclusive by nature and will naturally avoid unfamiliar objects, including human observers. Animals showing reduced responsiveness to visual cues and attempted capture are likely to be unhealthy or injured and should be evaluated further. In general, a healthy octopus will be inside a shelter and largely stationary (unless actively hunting), whereas healthy squid and cuttlefish may be either in shelters or swimming actively in the water column. Animals that appear unable to regulate their buoyancy and are floating, sinking or swimming at an angle (with either an unchanging head-up or head-down position) should be evaluated further (Sykes and Gestal 2014).

7.2 Internal Measures

There is ongoing debate about measurement of stress hormones in cephalopods, with differing results on the presence of corticosteroids and other metabolic stress indicators (Chancellor et al. 2021; Hu et al. 2024; Maskrey et al. 2024). In general, there is little consensus on evaluation metrics for health and disease in cephalopods. While haemolymph sampling is relatively easy in cephalopods, its relevance in routine health maintenance is less clear.

Cephalopods are host to multiple endo- and ectoparasites in the wild, including bacteria, coccidia, cestodes, trematodes, nematodes and small crustaceans such as copepods (Castellanos-Martínez and Gestal 2013). While the presence of such organisms likely represents normal findings in wild animals, their overgrowth or explosion may lead to impaired health and welfare in aquacultured settings. Cephalopods also host normal endosymbionts in some genera, including the bioluminescent bacterium *Vibrio fisheri*, and a highly unusual endoparasite, dicyemida, which lives inside the renal sacs and some other organs of some cephalopods. Whether these normal symbionts have beneficial or adverse effects on growth, health and welfare in aquacultured settings is poorly known.

7.3 Holistic Evaluation of Pain and Welfare

In general, cephalopods behaving abnormally (i.e. not displaying species-typical behaviour) or with evidence of wounds or injuries, should be evaluated as possibly experiencing pain. If closer evaluation reveals wound-tending, wound guarding, lack of normal visual responsiveness, reduced hunting behaviour or abnormal social behaviours, or elevated and sustained defensive behaviours such as jetting and inking, pain should be considered likely. The most appropriate intervention—given the very limited range of validated treatments for pain and its causes—depends on the life stage and habit of the animal. Older animals or those with significant injuries or abnormalities may be best euthanised, given that there are no well-validated analgesics for cephalopods (Deutsch et al. 2023). Isolation and close ongoing monitoring may be acceptable for others. There is an urgent need for research into veterinary care and appropriate treatments for injuries and diseases of cephalopods that may cause pain.

8 Summary and Future Directions

In general, the impression a reader may get from this chapter is that there is little concrete knowledge on the potential for pain and suffering in farmed and fished cephalopods. Largely this is correct; the science of pain and suffering, and its alleviations in cephalopods is in its infancy, with much unknown about the affective state and subjective experience of cephalopods in any context. There are also significant gaps in knowledge about best practices for large-scale growing and harvesting of cephalopods for commercial use. The majority of cephalopods on the global market are fished by distant ocean fleets with minimal regulation or oversight relating to capture and processing. Smaller, near-shore fisheries are also variable in the target species, the method of capture and the method of killing. The use of welfare-enhancing methods of slaughter is minimal and poorly researched. Cephalopod aquaculture is an attractive prospect given the rising global demand for cephalopods, their fast growth rates, short life cycles and high food conversion ratios; however, there is still very limited understanding of best practices for rearing, growing, breeding and humane killing of farmed cephalopods. There is an urgent need both for further research in welfare-enhancing methods and for the development and validation of operational welfare indicators. Additionally, implementation and enforcement of basic welfare standards will be essential as the potential for large-scale cephalopod aquaculture is realised.

References

Abbo LA, Himebaugh NE, DeMelo LM, Hanlon RT, Crook RJ (2021) Anesthetic efficacy of magnesium chloride and ethyl alcohol in temperate octopus and cuttlefish species. J Am Assoc Lab Anim Sci 60(5):556–567. https://doi.org/10.30802/AALAS-JAALAS-20-000076

Alupay JS, Hadjisolomou SP, Crook RJ (2014) Arm injury produces long-term behavioral and neural hypersensitivity in octopus. Neurosci Lett 558:137–142. https://doi.org/10.1016/j.neulet.2013.11.002

Anderson RC, Wood JB, Byrne RA (2002) Octopus senescence: the beginning of the end. J Appl Anim Welf Sci 5(4):275–283. https://doi.org/10.1207/S15327604JAWS0504_02

Andrews PLR, Darmaillacq A-S, Dennison N, Gleadall IG, Hawkins P, Messenger JB, Osorio D, Smith VJ, Smith JA (2013) The identification and management of pain, suffering and distress in cephalopods, including anaesthesia, analgesia and humane killing. J Exp Mar Biol Ecol 447(null):46–64. https://doi.org/10.1016/j.jembe.2013.02.010

Arkhipkin AI, Rodhouse PGK, Pierce GJ, Sauer W, Sakai M, Allcock L, Arguelles J, Bower JR, Castillo G, Ceriola L, Chen C-S, Chen X, Diaz-Santana M, Downey N, González AF, Granados Amores J, Green CP, Guerra A, Hendrickson LC et al (2015) World squid fisheries. Rev Fish Sci Aquacult 23(2):92–252. https://doi.org/10.1080/23308249.2015.1026226

Arnold J, Williams-Arnold L (1977) Cephalopoda: decapoda. In: Giese A (ed) Reproduction of marine invertebrates, pp 243–290

Birch J, Burn C, Schnell A, Browning H, Crump A (2021) Review of the evidence of sentience in cephalopod molluscs and decapod crustaceans. The London School of Economics and Political Science. https://www.researchgate.net/profile/Charlotte-Burn/publication/356459802_Review_of_the_Evidence_of_Sentience_in_Cephalopod_Molluscs_and_Decapod_Crustaceans/links/619cc06ad7d1af224b1a513d/Review-of-the-Evidence-of-Sentience-in-Cephalopod-Molluscs-and-Deca

Boletzky SV (1998) Cephalopod eggs and egg masses. Oceanogr Mar Biol Annu Rev 36:341–371. https://www.researchgate.net/profile/Sigurd_Von_Boletzky/publication/279192323_Boletzky_1998/links/558d874108ae47a3490bcbc6.pdf

Budelmann BU (1998) Autophagy in octopus. S Afr J Mar Sci 20(1):101–108. https://doi.org/10.2989/025776198784126502

Bush SL, Robison BH, Caldwell RL (2009) Behaving in the dark: locomotor, chromatic, postural, and bioluminescent behaviors of the deep-sea squid *Octopoteuthis deletron* Young 1972. Biol Bull 216(1):7–22. https://doi.org/10.1086/BBLv216n1p7

Butler Struben H, Crook RJ (2018) Injury enhances learning but does not affect spontaneous exploratory behaviors in cuttlefish. SICB Abstracts 58:E27

Butler-Struben HM, Brophy SM, Johnson NA, Crook RJ (2018) In vivo recording of neural and behavioral correlates of anesthesia induction, reversal, and euthanasia in cephalopod molluscs. Front Physiol 9:109. https://doi.org/10.3389/fphys.2018.00109

Capaz JC, Hernández-Brooke D, Balvet S, Couto AT, Alves AC, Gonçalves RA, Frias PA, Andrade JP, Sykes AV (2020) Control of Zootechnology leads to improved cuttlefish (Sepia officinalis, L.) reproduction performance up to pre-industrial levels. Front Mar Sci 7:112. https://doi.org/10.3389/fmars.2020.00112

Castellanos-Martínez S, Gestal C (2013) Pathogens and immune response of cephalopods. J Exp Mar Biol Ecol 447:14–22. https://doi.org/10.1016/j.jembe.2013.02.007

Cerezo Valverde J, García BG (2017) High feeding and growth rates in common octopus (*Octopus vulgaris*) fed formulated feeds with an improved amino acid profile and mixture of binders. Aquac Res 48(7):3308–3319. https://doi.org/10.1111/are.13157

Chancellor S, Abbo L, Grasse B, Sakmar T, Brown J, Scheel D, Santymire R (2021) Determining the effectiveness of using dermal swabs to evaluate the stress physiology of laboratory cephalopods: a preliminary investigation. Gen Comp Endocrinol 314:113903. https://www.sciencedirect.com/science/article/pii/S0016648021001969?casa_token=RR_eKmDQDsYAAAAA:IkJ4-6JePnoor0a9QefPM520NZ_vy_sk9yZ9ZiThq81T-2z7Oav_YT4bOkW1hE8Kj0rjMAp0m

Crook RJ (2021) Behavioral and neurophysiological evidence suggests affective pain experience in octopus. iScience 24(3):102229. https://doi.org/10.1016/j.isci.2021.102229

Crook RJ (2023) Perspective: social license as a lens for improving ethical and welfare standards in cephalopod research. Integr Comp Biol 63(6):1307–1315. https://doi.org/10.1093/icb/icad099

Crook RJ, Lewis T, Hanlon RT, Walters ET (2011) Peripheral injury induces long-term sensitization of defensive responses to visual and tactile stimuli in the squid *Loligo pealeii*, Lesueur 1821. J Exp Biol 214(Pt 19):3173–3185. https://doi.org/10.1242/jeb.058131

Crook RJ, Hanlon RT, Walters ET (2013) Squid have nociceptors that display widespread long-term sensitization and spontaneous activity after bodily injury. J Neurosci 33(24):10021–10026. https://doi.org/10.1523/JNEUROSCI.0646-13.2013

Crook RJ, Dickson K, Hanlon RT, Walters ET (2014) Nociceptive sensitization reduces predation risk. Curr Biol 24(10):1121–1125. https://doi.org/10.1016/j.cub.2014.03.043

Deutsch S, Parsons R, Shia J, Detmering S, Seng C, Ng A, Uribe J, Manahan M, Friedman A, Winters-Bostwick G, Crook RJ (2023) Evaluation of candidates for systemic analgesia and general anesthesia in the emerging model cephalopod, Euprymna berryi. Biology 12(2):201. https://doi.org/10.3390/BIOLOGY12020201

Di Cosmo A, Di Cristo C (1998) Neuropeptidergic control of the optic gland of Octopus vulgaris: FMRF-amide and GnRH immunoreactivity. J Comp Neurol 398(1):1–12. https://doi.org/10.1002/(SICI)1096-9861(19980817)398:1<1::AID-CNE1>3.0.CO;2-5

Di Cosmo A, Di Cristo C, Paolucci M (2001) Sex steroid hormone fluctuations and morphological changes of the reproductive system of the female of Octopus vulgaris throughout the annual cycle. J Exp Zool 289(1):33–47. https://doi.org/10.1002/1097-010X(20010101/31)289:1<33::AID-JEZ4>3.0.CO;2-A

Di Cosmo A, Maselli V, Cirillo E, Norcia M, de Zoysa HKS, Polese G, Winlow W (2023) The use of isoflurane and adjunctive magnesium chloride provides fast, effective anaesthetization of Octopus vulgaris. Animals 13(22):3579. https://doi.org/10.3390/ani13223579

Domingues PM, Kingston T, Sykes A, Andrade JP (2001) Growth of young cuttlefish, *Sepia officinalis* (Linnaeus 1758) at the upper end of the biological distribution temperature range. Aquac Res 32(11):923–930. https://doi.org/10.1046/j.1365-2109.2001.00631.x

Drazen JC, Goffredi SK, Schlining B, Stakes DS (2003) Aggregations of egg-brooding deep-sea fish and cephalopods on the gorda escarpment: a reproductive hot spot. Biol Bull 205(1):1–7. https://doi.org/10.2307/1543439

Finn JK, Norman MD (2010) The argonaut shell: gas-mediated buoyancy control in a pelagic octopus. Proc R Soc B Biol Sci 277(1696):2967–2971. https://doi.org/10.1098/rspb.2010.0155

Forsythe JW, Hanlon RT (1989) Growth of the Eastern Atlantic squid, Loligo forbesi Steenstrup (Mollusca: Cephalopoda). Aquac Res 20(1):1–14. https://doi.org/10.1111/j.1365-2109.1989.tb00437.x

Forsythe JW, Hanlon RT (1997) Foraging and associated behavior by Octopus cyanea Gray, 1849 on a coral atoll, French Polynesia. J Exp Mar Biol Ecol 209(1):15–31. http://www.sciencedirect.com/science/article/pii/S0022098196000573

Fouke KE, Rhodes HJ (2020) Electrophysiological and motor responses to chemosensory stimuli in isolated cephalopod arms. Biol Bull 238(1):1–11. https://doi.org/10.1086/707837

Giancola-Detmering SE, Crook RJ (2024) Stress produces negative judgement bias in cuttlefish. Biol Lett 20(10):20240228. https://doi.org/10.1098/rsbl.2024.0228

Gleadall IG (2013) The effects of prospective anaesthetic substances on cephalopods: summary of original data and a brief review of studies over the last two decades. Clin Psychol Rev 33(7):23–30. https://doi.org/10.1016/j.cpr.2013.04.004

Guo H, Zhang D, Näslund J, Wang L, Zhang X (2022) Effects of spawning group sex ratio and stocking density on the outcome of captive reproduction in golden cuttlefish Sepia esculenta. Aquaculture 559:738416. https://doi.org/10.1016/j.aquaculture.2022.738416

Gutfreund Y, Matzner H, Flash T, Hochner B (2006) Patterns of motor activity in the isolated nerve cord of the octopus arm. Biol Bull 211(3):212–222. http://www.ncbi.nlm.nih.gov/pubmed/17179381

Hague T, Florini M, Andrews PLR (2013) Preliminary in vitro functional evidence for reflex responses to noxious stimuli in the arms of Octopus vulgaris. J Exp Mar Biol Ecol 447:100–105. https://doi.org/10.1016/j.jembe.2013.02.016

Hanlon R, Messenger J (2018) Cephalopod behaviour. https://books.google.com/books?hl=en&lr=&id=oppPDwAAQBAJ&oi=fnd&pg=PR11&dq=cephalopod+behaviour&ots=C-09O1h0JB&sig=6iEMzJXubnvAcFrzlilfGJtjxh8

Hanlon RT, Forsythe JW, Cooper KM, Dinuzzo AR, Folse DS, Kelly MT (1984) Fatal penetrating skin ulcers in laboratory-reared octopuses. J Invertebr Pathol 44(1):67–83. https://doi.org/10.1016/0022-2011(84)90047-8

Hernández-García V, Hernández-López JL, Castro-Hdez JJ (2002) On the reproduction of Octopus vulgaris off the coast of the Canary Islands. Fish Res 57(2):197–203. https://doi.org/10.1016/S0165-7836(01)00341-1

Hochner B (2012) An embodied view of octopus neurobiology. Curr Biol 22(20):R887–R892. https://doi.org/10.1016/j.cub.2012.09.001

Hochner B, Shomrat T, Fiorito G (2006) The octopus: a model for a comparative analysis of the evolution of learning and memory mechanisms. Biol Bull 210(3):308–317. http://www.jstor.org/stable/10.2307/4134567

Holst M, Miller-Morgan T (2020) The use of a species specific health and welfare assessment tool to monitor health, stress and welfare of the giant Pacific octopus, Enterocotpus dofleini. J Appl Anim Welf Sci 24(3):272–291

Howard RB, Lopes LN, Lardie CR, Perez PP, Crook RJ (2019) Early-life injury produces lifelong neural hyperexcitability, cognitive deficit and altered defensive behaviour in the squid Euprymna scolopes. Philos Trans R Soc Lond B Biol Sci 374(1785):20190281. https://doi.org/10.1098/RSTB.2019.0281

Hu Y, Zheng J, Han F, Fan L, Liu Y, Qian Y, Zhang Q, Zheng X (2024) Effects of shelter on the welfare of juvenile Amphioctopus fangsiao: survival, growth, behavior and physiology. Aquaculture 578:740040. https://doi.org/10.1016/j.aquaculture.2023.740040

Huffard CL, Godfrey-Smith P (2010) Field observations of mating in Octopus tetricus Gould, 1852 and Amphioctopus marginatus (Taki, 1964) (Cephalopoda: Octopodidae). Molluscan Res 30(2):81–86. http://www.academia.edu/download/34061570/HuffardGodfreySmith2010Tetricus_Marginatus_Mating_OPT.pdf

Iraba N, Yahya S, Mang'ena J, Malesa F (2023) Comparison of growth and survival rates of big blue octopus (Octopus cyanea, 1849) fed on natural and formulated diets in captivity. West Indian Ocean J Mar Sci 22(1):47–55. https://doi.org/10.4314/wiojms.v22i1.5

Jacquet J, Franks B, Godfrey-Smith P, Sancez-Suarez W (2019) The case against octopus farming. Issues Sci Technol 35(2):37–44. https://www.jstor.org/stable/26948988?casa_token=wqWOMxaKlI8AAAAA:VPBXgzgJbdm4Kjr0O_yPERiEOImAPSGJmosbwbrLSDh1V2MMm5JdxXaECLOm6JQYXCnGX-h8Dz0CpX3_fUfLyfgS-HPq0R8p_fq5OH6FWRrH-d0Yot

Johnsen S, Balser EJ, Fisher EC, Widder EA (1999) Bioluminescence in the Deep-Sea Cirrate Octopod Stauroteuthis syrtensis Verrill (Mollusca: Cephalopoda). Biol Bull 197(1):26–39. https://doi.org/10.2307/1542994

Kirby AJ, Balko JA, Goertz CEC, Lewbart GA (2023) Characterization of current husbandry and veterinary care practices of the Giant Pacific octopus (Enteroctopus dofleini) using an online survey. Vet Sci 10(7):448. https://doi.org/10.3390/vetsci10070448

Kröger B, Vinther J, Fuchs D (2011) Cephalopod origin and evolution: a congruent picture emerging from fossils, development and molecules. BioEssays 33(8):602–613. https://doi.org/10.1002/bies.201100001

Kuo T, Sneddon L, Spencer J, Chiao CC (2022) Impact of lidocaine on pain-related grooming in cuttlefish. Biology 11(11):1560. https://doi.org/10.3390/biology11111560

Lee P, Turk P, Forsythe J, DiMarco F (1998) Cephalopod culture:physiological, behavioral and environmental requirements. Aquacult Sci 46(3):417–422. https://doi.org/10.11233/aquaculturesci1953.46.417

Lipiński MR, Underhill LG (1995) Sexual maturation in squid: quantum or continuum? S Afr J Mar Sci 15(1):207–223. https://doi.org/10.2989/02577619509504844

Martins HR (1982) Biological studies of the exploited stock of loligo forbesi (mollusca: Cephalopoda) in the azores. J Mar Biol Assoc U K 62(4):799–808. https://doi.org/10.1017/S002531540007034X

Maskrey BH, Costas C, Méndez-Martínez L, Guerrero-Peña L, Tur R, García P, Touriñan P, Chavarrias D, Canario AV, Scott AP, Rotllant J (2024) Studies on cortisol, corticosterone and 17β-estradiol indicate these steroids have no role in stress or reproduction in the common octopus (*Octopus vulgaris*). Am J Physiol Endocrinol Metab 328(1):E105–E115. https://doi.org/10.1152/ajpendo.00251.2024

Mather JA (1991) Navigation by spatial memory and use of visual landmarks in octopuses. J Comp Physiol A Neuroethol Sens Neural Behav Physiol 168(4):491–497. https://doi.org/10.1007/BF00199609

Matzner H (2000) Neuromuscular system of the flexible arm of the octopus: physiological characterization. J Neurophysiol 83(3):1315–1328. http://jn.physiology.org/content/83/3/1315.short

Navarro JC, Monroig Ó, Sykes AV (2014) Nutrition as a key factor for cephalopod aquaculture. In: Cephalopod culture. Springer Netherlands, pp 77–95. https://doi.org/10.1007/978-94-017-8648-5_5

Nesher N, Levy G, Grasso FW, Hochner B (2014) Self-recognition mechanism between skin and suckers prevents octopus arms from interfering with each other. Curr Biol 24(11):1271–1275. https://doi.org/10.1016/j.cub.2014.04.024

Nixon M, Young JZ (2003) The brains and lives of cephalopods. Oxford University Press, Oxford. http://www.lavoisier.fr/livre/notice.asp?id=OR6WSAA3LXROWN

Oshima M, di Pauli von Treuheim T, Carroll J, Hanlon RT, Walters ET, Crook RJ (2016) Peripheral injury alters schooling behavior in squid, Doryteuthis pealeii. Behav Processes 128:89–95. https://doi.org/10.1016/j.beproc.2016.04.008

Otero J, González ÁF, Sieiro MP, Guerra Á (2007) Reproductive cycle and energy allocation of Octopus vulgaris in Galician waters, NE Atlantic. Fish Res 85(1–2):122–129. https://doi.org/10.1016/j.fishres.2007.01.007

Pham CK, Isidro E (2009) Growth and mortality of common octopus (*Octopus vulgaris*) fed a monospecific fish diet. J Shellfish Res 28(3):617–623. https://doi.org/10.2983/035.028.0326

Polese G, Winlow W, Di Cosmo A (2014) Dose-dependent effects of the clinical anesthetic isoflurane on Octopus vulgaris: a contribution to cephalopod welfare. J Aquat Anim Health 26(4):285–294. https://doi.org/10.1080/08997659.2014.945047

Polese G, Bertapelle C, Di Cosmo A (2015) Role of olfaction in Octopus vulgaris reproduction. Gen Comp Endocrinol 210:55–62. https://doi.org/10.1016/j.ygcen.2014.10.006

Ponte G, Andrews P, Galligioni V, Pereira J, Fiorito G (2019) Cephalopod welfare, biological and regulatory aspects: an EU experience. Springer, Cham, pp 209–228. https://doi.org/10.1007/978-3-030-13947-6_9

Robison BH, Reisenbichler KR, Hunt JC, Haddock SHD (2003) Light production by the arm tips of the deep-sea cephalopod *Vampyroteuthis infernalis*. Biol Bull 205(2):102–109. https://doi.org/10.2307/1543231

Robison B, Seibel B, Drazen J (2014) Deep-sea Octopus (Graneledone boreopacifica) conducts the longest-known egg-brooding period of any animal. PLoS One 9(7):e103437. https://doi.org/10.1371/journal.pone.0103437

Rocha F, Guerra A, González AF (2001) A review of reproductive strategies in cephalopods. Biol Rev Camb Philos Soc 76(3):291–304. https://doi.org/10.1017/S1464793101005681

Roumbedakis K, Alexandre MN, Puch JA, Martins ML, Pascual C, Rosas C (2020) Short and long-term effects of anesthesia in Octopus maya (Cephalopoda, Octopodidae) juveniles. Front Physiol 11:697. https://doi.org/10.3389/FPHYS.2020.00697/BIBTEX

Seibel BA, Hochberg FG, Carlini DB (2000) Life history of Gonatus onyx (Cephalopoda: Teuthoidea): deep-sea spawning and post-spawning egg care. Mar Biol 137(3):519–526. https://doi.org/10.1007/s002270000359

Shigeno S, Andrews PLR, Ponte G, Fiorito G (2018) Cephalopod brains: an overview of current knowledge to facilitate comparison with vertebrates. In: Frontiers in physiology (Vol. 9, Issue JUL).

Frontiers Media S.A. https://doi.org/10.3389/fphys.2018.00952

Sumbre G, Gutfreund Y, Fiorito G, Flash T, Hochner B (2001) Control of Octopus arm extension by a peripheral motor program. Science 5536:1845–1848. http://www.sciencemag.org/content/293/5536/1845.short

Sykes AV, Gestal C (2014) Welfare and diseases under culture conditions. In: Cephalopod Culture. Springer Netherlands, pp 97–112. https://doi.org/10.1007/978-94-017-8648-5_6

Sykes AV, Domingues PM, Loyd M, Sommerfield A, Andrade JP (2003) The influence of culture density and enriched environments on the first stage culture of young cuttlefish, Sepia officinalis (Linnaeus, 1758). Aquac Int 11(6):531–544. https://doi.org/10.1023/B:AQUI.0000013262.15437.e4

Thompson JT, Voight JR (2003) Erectile tissue in an invertebrate animal: the Octopus copulatory organ. J Zool 261(1):101–108. https://doi.org/10.1017/S0952836903003996

Tracey WD (2017) Nociception. Curr Biol 27(4):R129–R133. https://doi.org/10.1016/j.cub.2017.01.037

Walsh LS, Turk PE, Forsythe JW, Lee PG (2002) Mariculture of the loliginid squid Sepioteuthis lessoniana through seven successive generations. Aquaculture 212(1–4):245–262. https://doi.org/10.1016/S0044-8486(02)00126-6

Ward PD, Saunders WB (1997) Allonautilus: a new genus of living nautiloid cephalopod and its bearing on phylogeny of the Nautilida. J Paleontol 71(6):1054–1064. http://www.jstor.org/stable/10.2307/1306604

Wells M, Wells J (1977) Cephalopoda: Octopoda. In: Giese A (ed) Reproduction of marine invertebrates, pp 291–330

Wodinsky J (2008) Reversal and transfer of spermatophores by Octopus vulgaris and O. hummelincki. Mar Biol 155(1):91–103. https://doi.org/10.1007/s00227-008-1010-3

Young JZ (1963) The number and sizes of nerve cells in Octopus. In: Proceedings of the Zoological Society of London. Blackwell Publishing Ltd, Oxford, pp 229–254. http://onlinelibrary.wiley.com/doi/10.1111/j.1469-7998.1963.tb01862.x/abstract

Zepeda EA, Veline RJ, Crook RJ (2017) Rapid associative learning and stable long-term memory in the squid Euprymna scolopes. Biol Bull 232(3):212–218. https://doi.org/10.1086/693461

Zheng J, Qian Y, Zheng X (2023) Effects of stocking density on juvenile Amphioctopus fangsiao (Mollusca: Cephalopodasca: Cephalopoda): survival, growth, behavior, stress tolerance and biochemical response. Aquaculture 567:739243. https://doi.org/10.1016/j.aquaculture.2023.739243

Crustaceans

Andrew Crump, Amaya Albalat,
Christopher J. Coates, and Adam Powell

Abstract

Decapod crustaceans are a diverse group of invertebrates, which underpin global aquaculture and fishing industries. Yet they are routinely underrepresented in animal welfare debates, science, and law. In this chapter, we discuss pain and welfare considerations for decapod crustaceans. We summarise the scientific evidence for pain in decapods and advocate a precautionary approach, which means that we should consider their welfare. We outline how humans use decapods and highlight potential welfare challenges in aquaculture and fisheries, emphasising the many knowledge gaps. For vets, we suggest a health inspection procedure and introduce key welfare indicators. We finish by describing relevant legislation and industry standards. We hope this chapter advances the emerging science of decapod pain and welfare.

Keywords

Aquaculture · Decapod · Fisheries · *Penaeus vannamei* · Shellfish

A. Crump (✉)
Animal Welfare Science & Ethics, Pathobiology & Population Science, Royal Veterinary College, London, UK

Centre for Philosophy of Natural & Social Science, London School of Economics & Political Science, London, UK

A. Albalat
Institute of Aquaculture, University of Stirling, Stirling, UK

C. J. Coates
Zoology, Ryan Institute, School of Natural Sciences, University of Galway, Galway, Ireland

A. Powell
Institute of Biological, Environmental & Rural Sciences (IBERS), Aberystwyth University, Gogerddan, Aberystwyth, Ceredigion, UK

1 Introduction

Crustaceans are a diverse invertebrate group of around 67,000 species, mostly aquatic and distinguished by an external calcified segmented exoskeleton with biramous appendages. This chapter focuses on decapod ('ten-footed') crustaceans: over 17,000 species of familiar and often commercially exploited species, including crabs, lobsters, crayfish, prawns and shrimp (De Grave et al. 2009, 2023).

Preventative and therapeutic veterinary care is limited for crustaceans. This might be due to the combination of their anatomy and physiology being radically different to vertebrates, the diversity of species and production sectors (wild fisheries and aquaculture), and a remarkable void in legislation, which has generated scant practical veterinary guidance.

Nevertheless, farming conditions, capture and post-capture practices impact decapod health and

R. E. Clutton, G. C. Musk (eds.), *Pain and Suffering in Farmed Animals*,
https://doi.org/10.1007/978-3-032-11407-5_22

welfare, leading to morbidity and mortality (see reviews Stoner 2012a; Albalat et al. 2022a; Browning et al. 2025). The direct effect on fishery and farm economics and indirect public perception of supply chain ethics, have moved crustacean welfare up industry and governmental agendas. In particular, the available evidence for sentience (Birch et al. 2021a; Browning et al. 2025) has encouraged commercial (SAGB 2021), scientific (Rowe 2018; Wuertz et al. 2023), regulatory (Crump 2022) and societal (British Veterinary Association 2022; Crustacean Compassion 2023; Loeb 2025) reconsideration of decapod health and welfare.

In this chapter, we summarise the evidence for and attitudes to pain in decapods. We also provide general background on their production and welfare challenges in commercial settings, recommendations for health inspections and describe the most up-to-date welfare indicators, which could prove useful for animal health practitioners inexperienced with these animals. Finally, we summarise legislation and industry standards and map knowledge gaps and challenges in the emerging topic of decapod welfare.

2 Evidence and Attitudes to Pain in Decapods

Decapod crustaceans are covered by at least some animal welfare legislation in several countries and regions, including Norway, Switzerland and New Zealand. However, despite growing scientific interest in the question, there is no consensus about whether decapods can feel pain.

Scepticism towards pain perception in animals is not specific to decapods, but generally questions the ability of humans to comprehend or demonstrate pain in any animal. Pain is an inherently private, subjective experience, whereas science studies observable, objective phenomena. Humans can describe their experiences, verbally communicating when they are in pain and its severity. However, other animals cannot use language, so some researchers argue that the scientific method cannot address whether they feel pain (at least given current knowledge; e.g. Dawkins 2022).

Yet the case of newborn human babies should make us reflect. As late as the 1980s, there were surgeons using similar sceptical arguments to justify operating on human babies without anaesthetics (De Lima et al. 1996). This is recognised to have been a serious mistake. Behavioural and physiological evidence (e.g. evidence that neonates display a behavioural stress response five times *stronger* than the adult response during surgery; Anand and Hickey 1987) provided strong reasons to err on the side of caution. Vertebrates may vocalise for many reasons, including as a consequence of pain (Olivier and Olivier 2024). Whilst decapods can produce sound, current research suggests any association is limited to courtship, defence and resource guarding (Radford and Stanley 2023).

Other objections to decapod pain are more specific to invertebrates (Diggles et al. 2024). Decapod brains are radically different from mammal brains, with decapods lacking a cortex or midbrain—structures linked to pain in humans. If these specific structures are universally essential for pain, then decapods would not have the capability to feel pain. Moreover, comparatively few studies have explicitly studied pain in decapods (Woodruff 2022). Those that have are focused on a few taxa, especially the true crabs (Brachyura), hermit crabs (Anomura) and lobsters and crayfish (Astacidea; Diggles 2019a). This lack of research limits our understanding of stress and pain, both within and across the decapods.

Other authors have argued that there is a strong case for pain in decapod crustaceans, based on neural and behavioural findings. First, the neural evidence. There is no a priori reason to think that only the mammalian cortex and midbrain might support pain. As an analogy, decapods have no visual cortex—a brain region essential for conscious vision in humans—but they can nonetheless see (Elwood et al. 2009). Decapods could also have achieved a different neural implementation of pain (Mallatt and Feinberg 2021).

Whilst there is enormous neural variation across species, many decapods have large brain regions that integrate information from different senses, facilitate learning and might conceivably support pain. Example brain regions include the well-developed hemiellipsoid bodies of true crabs (infraorder Brachyura), hermit crabs (Anomura), and caridean shrimps (Caridea; Strausfeld et al. 2020) and the large accessory lobes of lobsters and crayfish (Astacidea and Achelata; Sandeman et al. 2014).

Anaesthetics and analgesics also act on decapods behaviour in ways consistent with the hypothesis that they feel pain. For example, anaesthetising shrimp before severing their eyestalks reduced erratic swimming (Taylor et al. 2004) and rubbing (Diarte-Plata et al. 2012). However, these responses may indicate a general reduction in activity, rather than reduced pain.

Nociception could be defined as rapid detection and response to noxious stimuli, with electrophysiological observations suggesting nociceptors exist in shore crabs, *Carcinus maenas* (Kasiouras et al. 2024). Simple nociceptive reflexes, however, do not indicate pain, because humans perform them subconsciously. We, therefore, need evidence of more complex behaviours, which require integrative, centralised decision-making (Elwood 2023). Elwood and colleagues showed that crab responses to electric shocks were not simple reflexes, because they were traded off against competing motivations. When shocked in their shells, hermit crabs *Pagurus bernhardus* evacuated non-preferred *Gibbula* shells faster and more often than preferred *Littorina* shells (Appel and Elwood 2009; Elwood and Appel 2009). Shock avoidance was traded off against shell quality, so it was not a simple nociceptive reflex. Similarly, shore crabs traded the desire to enter refuges against avoiding electric shock (Barr and Elwood 2024a).

Another behavioural pain indicator is directed wound-tending. Some evidence suggests that, after acid administration, glass prawns (*Palaemon elegans*) and shore crabs (*Carcinus maenas*) will groom and rub a body part following noxious stimulation (Barr et al. 2008; Elwood et al. 2017). Further studies on these species have investigated the effect of anaesthetics and analgesics on noxious stimulation and found some modulatory behaviours (Barr and Elwood 2024b, c). However, a study on crayfish (*Procambarus clarkii*), white shrimp (*Litopenaeus setiferus*) and grass shrimp (*Palaemonetes* sp.) observed no wound-tending in response to hydrochloric acid or sodium hydroxide (Puri and Faulkes 2010). This inconsistency may reflect genuine taxonomic or irritant differences.

Further decapod studies have found evidence relevant to different feelings, such as anxiety-like states (Fossat et al. 2014, 2015). In these studies, crayfish (*Procambarus clarkii*) could explore a cross-shaped arena with two light arms and two dark arms. Stressed crayfish preferred dark arms, since wild crayfish hide from predators in dark shelters. Preference for dark arms was linked to the neurotransmitter serotonin, which underpins anxiety in humans. Injecting unstressed crayfish with serotonin caused them to spend longer in the dark arms (Fossat et al. 2014). Moreover, an anxiolytic drug, chlordiazepoxide, reduced the preference for dark arms (Fossat et al. 2015). This finding suggests a state functionally like anxiety in crayfish, modulated by some of the same neurotransmitters.

Different decapod species may also vary in their capacity for pain and other feelings. Birch et al. (2021a) developed eight criteria to evaluate evidence for pain in decapod crustaceans (based on Smith and Boyd 1991; see also Sneddon 2015; Crump et al. 2022a, b, c). Four criteria addressed the neural mechanisms that might underpin pain; the other four covered behaviours that may indicate pain. According to this framework, true crabs (Brachyura) satisfied five criteria, anomuran crabs (Anomura) and astacid lobsters/crayfish (Astacidea) satisfied three, caridean shrimps (Caridea) satisfied two and penaeid shrimps (Penaeidae) satisfied one. This is not a reason to think pain is absent in penaeid shrimps, because it largely reflects limited relevant research on the taxon, a confounding effect that is not reflected in the criteria scoring.

There are nonetheless, biological reasons to think that pain may be less plausible in some decapods. The hemiellipsoid bodies of penaeid shrimp brains have shrunk throughout evolution, so they are smaller than those of crabs, lobsters and caridean shrimps (Strausfeld et al. 2020). This has led to a suggestion that caridean shrimps may feel pain, whereas penaeids may not (e.g. Comstock 2022). However, it is not known what the lower threshold of brain size necessary for pain might be, so with current knowledge it is not possible to rule it out in penaeid shrimps.

To summarise, science cannot definitively reveal whether decapods feel pain. However, several studies have shown that decapod responses to injury are not simple reflexes, and their responses to analgesics and anxiolytics are consistent with pain and consequential behaviour. Thus, the potential for decapods to experience pain-like sensations cannot be plausibly ruled out (Farnsworth and Elwood 2023; Elwood 2025a).

Therefore, specially where threats of harm (e.g. tissue damage) are substantial, it is recommended to follow 'the precautionary principle'. We consider the precautionary principle should also guide our thinking about animal welfare risks (Birch 2017; Browning et al. 2025). Given the potential harms of ignoring signs of suffering in decapods, they should be managed as though they experience pain, despite the scientific uncertainty.

3 Commercial Use and Production of Decapods

Decapods are mostly used for food and animal feed (Jennings et al. 2016; Malaweera and Wijesundra 2013), with a small proportion collected for personal and public aquaria (Narshi et al. 2022), recreational fishing bait (Deidun et al. 2018) and biotechnical or research purposes (Hamed et al. 2016; Passantino et al. 2021). Decapods, such as lobsters, have likely been harvested from the wild by hunter-gatherers for millennia (Spanier et al. 2015) and currently support valuable commercial fisheries all over the world. However, from a total production perspective, decapod aquaculture (ca. 11 million tonnes per annum) has surpassed decapod fisheries (static at ca. 6 million tonnes per annum) (FAO 2024).

At the species level, decapod aquaculture is dominated by Pacific whiteleg shrimp (*Penaeus vannamei*), with 6.8 million tonnes produced in 2022 (FAO 2024). This species is followed by the red swamp crayfish (*Procambarus clarkii*; 2.9 million tonnes), the Chinese mitten crab (*Eriocheir sinensis*; 0.81 million tonnes), and the giant tiger prawn (*P. monodon*; 0.71 million tonnes). Production is striking if one considers production volumes in terms of individual animals. For *P. vannamei*, annual production equates to at least 300 billion individuals harvested annually (Waldhorn and Autric 2023).

The dominance of *P. vannamei* is a relatively new development; black tiger shrimp (*P. monodon*) dominated production until the early 2000s. This change was mainly driven by disease outbreaks in *P. monodon* caused by white spot syndrome virus in the 1990s (Funge-Smith and Briggs 2003) coupled with *P. vannamei*'s lower protein nutrient requirements, amenability to higher stocking densities, tolerance to wider environmental ranges, ease of domestication and subsequent development of enhanced breeding lines and the availability of specific pathogen-free (SPF) post-larvae.

Shrimp production is centred on a few countries in Asia (China, India, Vietnam and Indonesia) and Latin America, mainly driven by production in Ecuador. Farming system and management varies from super-intensive (150–300 shrimp/m^2) (Emerenciano et al. 2022) and intensive farms, which are primarily found in Asia, to extensive farms (typically <30 shrimps/m^2), which are more common in Latin America (Davis and Boyd 2021).

Shrimp farming is probably the most widespread aquaculture sector to have completed the lifecycle of a decapod group. The process consists of an intensive hatchery which produces numerous larvae from captive broodstock, and a subsequent ongrowing stage producing various size grades of shrimp for harvest and sale (FAO 2023). Iterative retention of high-quality offspring and selection of broodstock over many generations has increased production. For instance, due to the pressures of successive endemic viral diseases and an inability to immunise, other preventative measures (biosecurity and breeding programs) were investigated soon after the industry emerged (US Marine Shrimp Farming Program). This has included a selection of shrimp strains with known disease (SP, Specific Pathogen) provenance: SPR (Resistant) and SPT (Tolerant) statuses refer to stocks' defined genetic characteristics in response to pathogens and disease. SPF (Free) refers to the health status of a shrimp stock, but only indicates the sanitary status of a stock and does not guarantee stock susceptibility, resistance or tolerance to infection (Alday-Sanz et al. 2018). Nevertheless, regardless of target species, disease is a major welfare challenge and has seriously damaged the sector (see reviews Hooper et al. 2023; Kumar et al. 2021; Thitamadee et al. 2016; Stentiford et al. 2012) in some cases with devastating effects (Flegel 2012).

Decapod fisheries are more evenly distributed across different species of shrimp and crabs, which are associated with specific geographical areas (FAO 2024). Disease has been reported in wild populations of several commercially important species (Rowley and Coates 2023; Davies et al. 2019), although assessing disease as a mortality driver is far more challenging in these populations than in farmed systems (Stentiford et al. 2012; Coates and Rowley 2022). Therefore, in wild-caught decapods, welfare challenges relate to capture, impounding, transport and dispatch, as we have no or very little control over their welfare before capture.

4 Welfare Challenges in Commercial Use

Because the aquaculture and fishery sectors are fundamentally different, we must consider welfare challenges within each sector and then focus on species-specific traits. What do we mean by 'good' welfare? One of the most accepted animal welfare frameworks is the 'Five Domains Model' (Mellor and Reid 1994; Mellor et al. 2020). This conceptualises welfare as four input domains (nutrition, environment, health and behaviour) that affect one output domain (mental state). Good welfare, thus, requires meeting the animal's needs in all five domains. When translating the Five Domains framework to decapods, one obstacle is that most efforts have focused on different species' nutritional, environmental and health needs. We, therefore, focus on these three domains here. However, for a complete picture of decapod welfare, research is needed on captive decapods' behaviour needs and preferences and how to assess their mental state.

4.1 Decapod Aquaculture

Most farmed decapod research has focused on the nutritional, health/disease and environmental needs of commercially important species, because these are directly linked to survival and growth—critical from a production perspective. There is a wealth of knowledge on nutritional requirements in farm systems (Tacon et al. 2002; Glencross 2009), and some information regarding environmental tolerance for the most commercially relevant species (Chen et al. 2016; Ern et al. 2015; Gonzalez et al. 2010). Survival, disease incidence and growth depend on appropriate environmental conditions (e.g. temperature, salinity, alkalinity, pH, dissolved oxygen, NH_3-ammonia, N/P ratio and the presence of xenobiotics).

Farming approaches vary between different species and within species, from super-intensive to extensive systems. While production comparisons can be drawn, the welfare implications of different systems have not been evaluated. One possible reason is our partial understanding of the role that other parameters (such as stocking density) play in decapod welfare. Another challenge is assessing the combination of welfare hazards across an animal's overall life cycle. This has not yet been achieved for farmed decapods.

Below, however, we highlight commercial practices that have been reported to impact health and welfare. Other general husbandry practices also probably lead to suboptimal nutritional, environmental and health provision.

Eyestalk ablation In female broodstock, antagonistic hormones (gonad inhibiting hormone, GIH/MO-IH) downregulate ovarian maturation and moulting. A gland in the eyestalk (the 'X-organ') produces these hormones. Hatcheries might, therefore, sever one or both eyestalks of the female broodstock to accelerate and synchronise ovarian maturation. This eyestalk ablation is performed via pinching, slitting, cauterising or ligation. Given the practice's detrimental effects on females (summarised in Albalat et al. 2022b), many commercial operations are investigating and implementing alternative methods, such as changing nutritional profile, husbandry practices or novel chemotherapies (Uawisetwathana et al. 2011; Zacarias et al. 2019; Laphyai et al. 2021).

Spermatophore expression and artificial insemination Shrimp reproduction is external, following male courtship. The male places one of two sperm packages (spermatophores) on the female's ventral surface, which fertilises the eggs as they are extruded. To combat poor spermatophore placement and improve larval yield, spermatophores may be extruded from the males. This requires brief but stressful handling and can cause somatic tissue damage and poor sperm quality (Diamond et al. 2008; Beirão et al. 2019).

Inappropriate harvesting, stunning and killing Currently, only a few countries regulate farmed decapod stunning and slaughter (notably New Zealand, Switzerland, Norway, Germany and some Italian provinces; Conte et al. 2021a). However, very few decapods are farmed in these countries (e.g. Switzerland and Germany combined farm around 150 tonnes of shrimp per annum; Euroshrimp 2020). In countries where most decapods are farmed, there is no legislation and little information on harvesting, stunning and killing protocols or codes of practice.

Slurry ice is the primary commercial method of stunning and killing. There is limited data on this method's effectiveness, but it can be reversible and responses are species-specific (Weineck et al. 2018). When submerged in ice slurry, *P. vannamei* do not appear to lose sensory-CNS-skeletal muscle activity, because they tail-flip and muscle contractions are visible in about 1 to 2 min (personal observation and reported in Weineck et al. 2018). Cold sedation protocols in their current form, even for tropical species, are unlikely to deliver an 'immediate effective stun' (i.e. render the animal unconscious within one second) and thus other alternatives such as electrical stunning are being studied. Any new technology would require to be tested and optimised by analysing cessation of nervous activity to confirm insensibility, alongside calibration with easily observable behavioural indicators (Neil et al. 2024).

4.2 Commercial Fishing

Decapods are caught using an array of fishing gear, which can be separated into two basic types. Active or mobile gear (e.g. towed nets) collects pelagic shrimp and benthic decapods, and may damage and stress crustaceans (Ridgway et al. 2006a; Milligan et al. 2009; Larssen et al. 2013). Physical damage is caused by abrasion between animals and pressure during hauling (ICES 1994; Lancaster and Fridd 2002; Milligan et al. 2009). Physiological stress, meanwhile, is driven by increased exercise during the capture process (Newland et al. 1992; Gornik et al. 2010) and oxygen deprivation due to emersion (Ridgway et al. 2006b; Bernasconi and Uglow 2008; Gornik et al. 2010). This metabolic stress is potentially reversible (Uglow et al. 1986; Harris and Andrews 2005). Indeed, improved practices have increased the proportion of trawl-caught Norway lobsters (*Nephrops norvegicus*) that can be successfully transported alive (Albalat et al. 2010, 2017; Neil 2012). In other cases, however, the stress response is irreversible, leading to mortality or the onset of idiopathic muscle necrosis (Stentiford and Neil 2000).

Alternatively, static or immobile gear (e.g. static nets, pots, cages or SCUBA) is more size- and species-selective, as well as less damaging and energy-demanding for the animals (Albalat et al. 2009). Nonetheless, storage and live transport can lead to emersion and stress (e.g. creel-caught decapods kept out of water for long periods; Adey 2007). Appropriate environmental conditions, such as temperature (cooling before and during transport) and salinity (maintained at sufficient levels), are crucial for survival (Adey 2007).

Fishers, seafood merchants and transport operators have developed interventions to limit exposure to stressors and subsequent mortality and morbidity. These generally involve reduced handling, physical protection, separation of individuals, cooling and maintenance of humid conditions (Powell et al. 2020). In addition, the following practices may be detrimental to welfare during commercial fishing:

Banding Decapods are often aggressive and can damage each other by puncturing the exoskeleton or causing autotomy. If they cannot be kept separately, strong rubber bands are placed around the chelae to prevent flexion. This does little if any damage in the brief period before sale, although escaped or released American lobsters (*Homarus americanus*) have been recaught banded, suggesting the rubber can persist (Stebbing et al. 2012).

Nicking Although straightforward for lobsters, banding is more challenging for crabs, which have physically stout chelae. Brown crabs (*Cancer pagurus*) also have non-tapering claws, so producers report bands falling off. An alternative to banding, nicking requires severing the apodeme tendon. This causes haemolymph (blood equivalent) loss and a potential entry point for infection (Jacklin and Combes 2007). Nicking increased brown crab mortality in vivier systems after 2 weeks and was linked to significant changes in haemolymph chemistry and immune parameters compared to intact control crabs (Johnson et al. 2016).

V-notching Many fisheries prohibit landing ovigerous ('berried') lobsters (*Homarus* spp.) to promote larval recruitment. However, 'scraping' describes illegally removing the external egg mass before sale. V-notching is a semi-permanent, visual method to mark ovigerous females as illegal to sell. The process removes a small V-shaped section in the tail, which apparently does not affect growth and survival (DeAngelis et al. 2010; Mallol et al. 2014). From a welfare perspective, V-notching may be neutral.

Declawing Crab meat can be either brown or white. White meat is more valuable and located in the appendages, particularly the chelae, which are often larger in males. Declawing describes removing one or both claws for sale, before returning the crab to the sea. This practice should be strongly discouraged as it reduces survival and growth, disadvantages crabs in social and anti-predator behaviour, and increases recapture rates (e.g. Goldstein and Carloni 2021).

Holding conditions For industry, live storing decapods is critical. As an example, 45–50% of *H. americanus* landed in the United States and Canada are stored live (>136,000 tons; Thériault et al. 2013). These lobsters are usually live held in customised facilities or vivier systems, which are tightly regulated by HACCP legislation (Estrella 2002; Food and Drug Administration 2021). Live holding takes place at low temperatures (3–5 °C) without feeding (McLeese and Wilder 1964). However, EU legislation is less rigorous, with no requirement for European lobsters (*Homarus gammarus*) to be held in customised facilities, and no stipulations on the feeding regime, holding temperature or maximum confinement duration. Nevertheless, even if these systems met the animals' nutritional, environmental and health needs, it is hard to see how they would satisfy lobsters' behavioural needs, because they are designed and managed to maintain many animals in a small space.

Live transport Decapods can typically be transported either immersed in recirculating water containers (vivier trucks) or in semi-dry conditions (Jacklin and Combes 2007). Poor conditions lead to high mortality, plus low oxygen levels in both methods can trigger anaerobic metabolism. Regardless of the transport method, low temperatures are essential (Lorenzon et al. 2008; Barrento et al. 2012; Lu et al. 2016). Other key factors are animal condition before transport (namely moulting, vitality, and fasting) and transport duration (Estrella 2002; Albalat et al. 2017), which again is important for both transport methods. In addition, water quality is important in vivier tanks, and appropriate humidity (near 100%) and avoiding vibration matter for semi-dry conditions (Durand et al. 2000; Powell et al. 2017).

Stunning and slaughter in processing plants and tailing The live market's importance for wild-sourced species means that death occurs near the consumer end of the supply chain. As reviewed by Conte et al. (2021a) and Birch et al. (2021a), the least humane killing methods are probably boiling, separating the thorax from the abdomen ('tailing'), and placing live marine crustaceans in fresh water. So far, most welfare research has concentrated on optimising pre-slaughter stunning in processing plants (Roth and Grimsbø 2016). Equipment is also commercially available for electrically stunning lobsters and crabs in domestic or retail settings (Albalat et al. 2022a).

5 Disease Control, Health Inspections, and Welfare Indicators

5.1 Disease Control: Penaeid Aquaculture

Preventative actions Penaeid shrimp do not have lymphocyte-mediated immune memory, which has limited commercial interest in vaccine development. Despite some observations of specific immunity, for example, after exposure to a *Vibrio* sp. vaccine (Powell et al. 2011; Pope et al. 2011), the molecular mechanisms have not been fully determined (Rowley and Powell 2007; Rowley and Pope 2012).

Therefore, in shrimp aquaculture, most preventative disease control entails minimising risk during management and husbandry. Examples include sourcing stock of high provenance (e.g. specific pathogen free, SPR), enacting biosecurity and quarantining (reducing vermin and other disease vectors) and maintaining hygiene between growth cycles (such as 'dry out', removal of sediment, morts and application of caustic disinfectants). Additionally, reducing stressors during production lowers the risk of infection and disease spread (e.g. limiting stocking density, providing an adequate feeding regime and maintaining high water quality). Farmer support groups can also promote optimal management practices (Kumaran et al. 2015).

The importance of surveilling shrimp diseases has been slowly recognised internationally, to avoid epizootics and introduction to other decapod species (Yu et al. 2022). Rapid pond-side diagnosis is essential to control disease and limit regional spread, with portable, molecular screening tools currently under development (Flegel 2019).

Therapeutic treatment Commercial immune stimulants, pre and probiotics are commonly added to shrimp feed to prime or enhance immune response. However, the mechanism, screening and efficacy of such preparations remain largely unknown (Hauton et al. 2015; Knipe et al. 2020). Non-viral outbreaks (i.e. bacterial infections) are often treated with antibiotics—an increasingly problematic situation given the scale of shrimp production, low regulation and the threat of antimicrobial resistance (Thornber et al. 2020).

Zoonotic diseases Decapods cause few food-borne human illnesses. Most occur when humans eat undercooked decapods, leading to bacterial infections in the gastrointestinal tract. Infection typically arises from *Vibrio parahaemolyticus*, but less common bacterial and parasitic zoonotic infectious agents exist (Behringer and Duermit-Moreau 2021).

5.2 Health Inspection of Large Decapods: A Primer

Stage 1 *External examination* can be divided into: (a) exoskeleton integrity and (b) behaviour (Fig. 1).

(a) Exoskeleton integrity. Inspect the carapace (dorsum and ventrum) for signs of fractures, fouling/epibionts, discolouration (e.g. melanisation), soft areas and dysecydsis (moulting aberrations); the clawed and walking limbs for signs of damage, regrowth from autotomy or any missing (conspecifics held in captivity can remove limbs and cause physical trauma/fracturing). Unsurprisingly, diseases often negatively impact the carapace architecture or appearance. Ectoparasites, such as *Sacculina* spp. (Rowley et al. 2020), protrude from crabs' ventral reproductive opening or leave scars from previous infections. Gill parasites can accumulate on the surface and between the filaments. Shell disease syndromes—epizootic in American lobsters (*H. americanus*) and enzootic in European lobsters (*H. americanus*) and edible crabs (*Cancer pagurus*)—can present themselves as localised, superficial black spots. These may progress into extensive carapace erosion, pitted lesions, and ulceration, which provide an entry point for infection (Rowley and Coates 2023). At late stages of shell disease, necro-

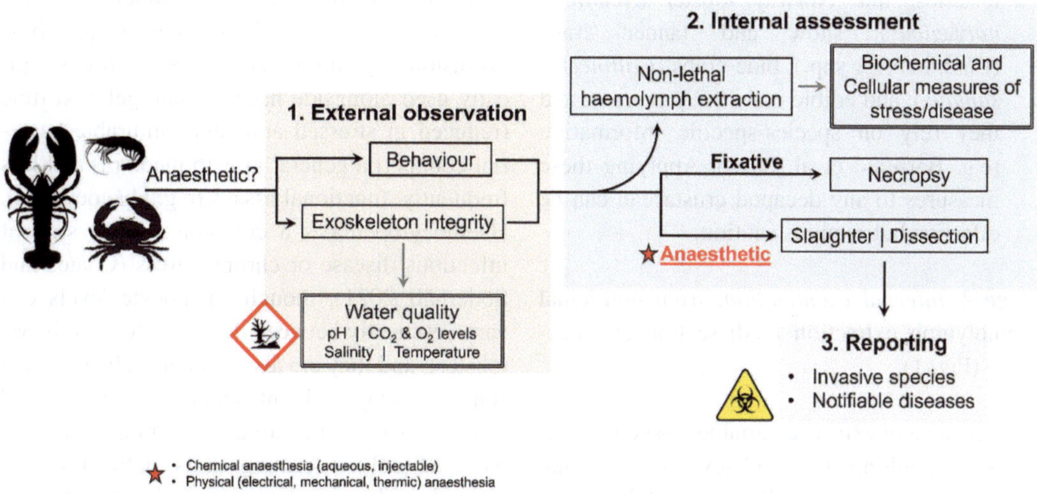

Fig. 1 General procedure for crustacean health inspection*First*, take note of any altered appearance and level of vitality. If closer inspection of the carapace is needed, then an anaesthetic should be considered to avoid handling stress, but this is not always necessary (especially if the animal is lethargic or missing claws). Check the exoskeleton carefully, note missing or damaged limbs, surface fractures, ectoparasites, fouling agents (epibionts), etc. If animals are held in tanks for experimentation, display aquaria, or under culture conditions, then water quality parameters should always be monitored; insufficient O_2 levels or over-accumulation of nitrogenous compounds can cause stress. Avoid sudden changes in water temperature. *Second*, a non-lethal haemolymph sample can be extracted to gauge potential indicators of stress/disease (e.g., raised L-lactate levels, or depleted haemocyte numbers). A physical (thermic, electrical) anaesthetic can be used here to avoid stress, which is unlikely to alter the gross haemolymph profile. Depending on the animal's size, several hundred microlitres of haemolymph can be removed for downstream processing (clean the needle entry point with 70% ethanol pre- and post-extraction). If the decapod is moribund or destined to be killed for a defined purpose, then a range of anaesthetic agents is available. Note, however, that this may render the animal unsuitable for consumption. *Third*, contact the appropriate agencies if non-native species (e.g. Chinese mitten crab, *Eriocheir sinensis*) or notifiable disease (e.g. crayfish plague, *Aphanomyces astaci*) are found

sis and melanisation can occur in the underlying muscle and viscera.

(b) Behaviour. Decapods should be lively and responsive. Lobsters, crabs and crayfish will raise their claws, move rapidly and seek shelter. Shrimp and prawns tend to flip their tails alongside escape swimming. If decapods are anorexic, lethargic and do not respond to handling (i.e. try to escape or raise claws), then they are usually morbid, stressed, carrying an infection or metabolically exhausted (e.g. limp lobster disease; Tall et al. 2003). These are signs (not symptoms) of compromised health. In fishery, culture and captive settings, such observations may be more formally assessed using damage, stamina and vigour indices or reflex action mortality predictors (RAMP; Stoner 2012a; Albalat et al. 2016). A perceived drawback of these indices is that they were developed for lucrative fisheries, like Norway lobster (*Nephrops norvegicus*), snow and tanner crabs (*Chionoecetes* spp.), blue crabs (*Callinectes sapidus*), and edible crabs (*C. pagurus*), and they rely on species-specific information (e.g. Barrento et al. 2010). Applying these measures to any decapod crustacean can be effective but requires caution.

Stage 2 *Internal examination*, from non-lethal haemolymph extraction to dissection and necropsy (Fig. 1).

Intrinsic and extrinsic variables—such as animal size, moulting stage and sex; environmental conditions (season and pollution); and the nature of an injury (physiological and mechanical)—all influence stress outcomes (poor welfare, disease susceptibility and mortality; Stoner 2012a). Of course, decapod crustaceans cannot relay symptoms of stress or disease and, without altered behaviours or integumental abnormalities (i.e. clinical signs), they are asymptomatic (Coates and Söderhäll 2021). There is no consensus among crustacean handlers or scientists for operational welfare indicators, or common biomarkers of health/disease (though recent efforts have sought to rectify this; Albalat et al. 2022a). The crustacean immune system is innate and unable to produce antibodies (clonal immunoglobulins). As such, there are fewer biochemical and cellular measures to interpret when compared to aquatic vertebrates. Useful indicators in vertebrates like cortisol levels and haematocrits (packed cell volume) are not applicable, as decapods do not have cortisol, red blood cells or intracellular haemoglobin. Additionally, decapod crustaceans show no traditional signs of inflammation, no redness as they use haemocyanin (blue pigment) for gas exchange (Coates and Costa-Paiva 2020), and little swelling due to the semi-closed circulatory system (some exceptions exist, like oedema in the gills or distended tissues due to mass haemocyte (immune cell) infiltration; reviewed by Coates 2022).

There is a series of recurrent haemolymph-based methods for monitoring decapod stress and/or infection. The primary cell-based measure is total (and differential) haemocyte counts, sometimes with a vital dye (e.g. Trypan blue exclusion; e.g. Jussila et al. 1997). This is typically used alongside haemolymph gelation time (reduced in stressed animals), culturable bacterial counts (on general growth medium) and less frequently, functional assays (e.g. phagocytosis). Haemocytopenia is a common clinical sign of infectious disease or chronic stress (Coates and Söderhäll 2021), though haemocyte levels can vary drastically between decapods (e.g. crabs vs. lobster), and they are also influenced by sex, season, moulting, ambient temperature, nutritional state and so on. Increased numbers of cultivable bacterial colony-forming units (CFUs) in the haemolymph can indicate sepsis or spoilage—an obvious complication for interpreting results is that the haemolymph in healthy animals is not axenic (symbionts present), containing 10s–100s of microorganisms per mL (Lorgen-Ritchie et al., 2023).

As for biochemical measures, crustacean hyperglycaemic hormone (CHH) is a candidate decapod stress hormone (i.e. cortisol-like; Chang 2005). According to Stoner (2012a), CHH is a possible mortality predictor, but levels can be variable, and quantification is expensive and technical. Increased lactate levels and an associated decrease in haemolymph pH are candidate universal indicators of stress (Conneely and Coates 2024). Across a diversity of stressors—emersion, handling and physical injury—L-lactate levels increase as more oxygen is consumed. This happens as the host switches from oxidative phosphorylation to glycolysis, thereby fuelling tissue repair and restoring homeostasis during and after stress exposure. Metabolic by-products like lactate can accumulate in solid tissues (e.g. muscle) and spill over into the haemolymph, leading to acidosis. Therefore, decreased haemolymph pH is another possible mortality predictor (or indirect measure of L-lactate). Total protein levels and/or defined enzymatic activity (lysozyme, phenoloxidase and superoxide dismutase) link immunity-stress-nutrition, but show no consistent patterns for healthy and compromised decapods (Johnson et al. 2016; Albalat et al. 2019, 2022a, b; Coates and Söderhäll 2021). Work is needed to establish reference intervals (see Scott et al. 2018), baseline and induced levels of metabolites and immune factors in large decapod crustaceans, as none are fixed. Should haemolymph inspection fail to diagnose a suspected underlying condition, then gill biopsies can be taken. In extremis, animals can be sacrificed/dissected for multi-tissue histological inspection (Fig. 1; Frasca et al. 2020).

Stage 3 *Reporting*

If examination reveals a notifiable disease-causing agent, then the local/national agencies should be informed (e.g. CEFAS UK). At the time of writing, the World Organisation for Animal Health (founded as OIE) lists 11 crustacean diseases that require heightened surveillance to prevent their transboundary spread, including the infamous viruses affecting penaeid shrimp (e.g. white spot syndrome virus; Flegle 2019). A similar approach should be followed if a non-native (alien), invasive species is found, such as an American lobster (*H. americanus*) in Europe. Depending on the country and region, there are defined protocols for recording, capturing and destruction (e.g. through biodiversity databases, national eradication plans and no-return policies).

5.3 Operational Welfare Indicators

Operational welfare indicators are needed for both farmed and wild-sourced crustaceans. These should be minimally or non-invasive, fit for purpose and robust. Welfare indicators may be either direct and animal-based (e.g. physical health, physiological parameters and behavioural measures; Campbell and Lee 2025) or indirect and resource-based (e.g. environmental parameters, husbandry procedures and holding conditions; Butterworth 2009). For example, both animal- and resource-based welfare indicators have been suggested for penaeid shrimps (Albalat 2022a; Pedrazzani et al. 2023, 2024). We envisage this research area expanding in the next few years, especially as welfare indicators must be species-specific and context-relevant. Moreover, as mentioned above, more work is needed on quantifying the combination of welfare hazards across an animal's life cycle.

6 Anaesthesia and Analgesia

Unless otherwise stated, we refer to anaesthesia as the administration of a chemical or concoction to induce a controlled, temporary loss of sensation, rendering the animal unconscious. Analgesia refers to a managed state of pain relief without loss of consciousness. For decapod crustaceans, anaesthesia and analgesia are relatively poorly understood. Nonetheless, Rotllant et al. (2023a) developed an online decision support tool that lists anaesthesia methods for named crustacean species, effective doses and recovery (distinguishing between perceived suitable and unsuitable approaches).

Here, we present a procedural case for using anaesthetics on large decapod crustaceans: crabs, shrimp and prawns and lobsters. Some means of inducing neural insensibility or physical inactivity are non-chemical, using low temperatures to promote a hypometabolic or torpor-like state (Weineck et al. 2018). Slow movement and docility from cold-shock may be due to muscle performance, a Q10 effect involving a decrease of 10 °C that halves enzymatic activity (akin to Arrhenius' equation of temperature dependence). However, some argue that hypothermic approaches are unsuitable for anaesthesia because there is no analgesic effect and the muscles do not relax (Wahltinez et al. 2023a).

A non-invasive approach to chemical anaesthesia in decapods is to add the agent to the water, so it can be ingested or taken in across the gills. Essential oils like clove, and its isolated constituents (e.g. eugenol), have been employed in this manner. Adding 0.06–3 mL L^{-1} (v/v) of clove oil to decapods' water can induce anaesthesia in as little as 5 minutes (e.g. 30 minutes for Dungeness crab, *Cancer magister*; Morgan et al. 2001). Cowing et al. (2015) monitored induction (sedation) and recovery (revival) in Norway lobsters (*N. norvegicus*) using passive behaviours and reaction to stimuli. The lobsters were fully anaesthetised by ≤0.9 mL L^{-1} eugenol within 5 min, and recovered in ~10 min. Moreover, depending on the decapod and dosage, immersion in isobutyl alcohol can be both an effective sedative and general anaesthetic (low to high doses [1.5 < 7 mL L^{-1}]; (Wahltinez et al. 2023a). While eugenol and isobutyl have been successful in numerous decapods (de Souza Valente 2022a), other agents may be unsuitable for general use (e.g. tricaine methanesulfonate, 2-phenoxyethanol, and magnesium salts; Robertson et al. 2018).

Alternatively, invasive intrahaemocoelic or intramuscular injection of parenteral anaesthetics can cause gross insensibility. For this process, a small gauge hypodermic needle is recommended (>27 for injection; <23 for haemolymph extraction to avoid gelation and/or haemocyte rupture). Various injectable anaesthetics have been suggested for decapods, including alfaxalone, lidocaine and ketamine in blue crabs (*C. sapidus*; Quesada et al. 2011; Minter et al. 2013). Broadly, anaesthesia induction time, duration and recovery are dose-dependent. Reported side-effects of chemical anaesthetics include an excitatory period immediately post-injection, cardiac dysrhythmias (bradycardia and extrasystole), limb autotomy (e.g. intravascular injection of 20 mg/kg propofol), and *in exitus* (e.g. prolonged exposure to clove oil can be fatal; Gardner 1997). Anaesthetics used for fish or mammals are not always transferable to decapods, and tend to lack sufficient experimental validation and pharmacokinetics. For example, intracardiac injection of 20 mg/kg ketamine led to 80% mortality in *C. sapidus* (four out of five crabs; Quesada et al. 2011).

7 Regulation

European (and subsequent national) legislation for animal welfare operates at farming (EU 1998), transport (EU 2005), and slaughter (EU 2009) stages of the supply chain. However, this tends to have broader considerations, such as food safety (Lundmark et al. 2014). Typically, decapods are not necessarily defined as 'animals' in legal terms (Carder 2017; Conte et al. 2021a). However, the United Kingdom's recent recognition of decapod sentience may influence the scope of other countries' animal welfare legislation (Birch et al. 2021a; Rowan et al. 2021; Powell et al. 2024), with some additional precedent in Switzerland, Norway, and Australasia (Conte et al. 2021a).

Specifically for the decapod (shrimp) aquaculture sector, standards are variable and focus on environmental and social issues, with a lack of specific animal welfare requirements (Aquaculture Stewardship Council 2023; Asian Seafood Improvement Collaborative 2023; Global Seafood Alliance 2023). To evaluate welfare, one option would be to introduce operational welfare indicators, currently used for several species of fish, into the Penaeid shrimp industry (Albalat et al. 2022a).

The use of animals in scientific research often demands veterinary care in the licensing and project management system (Zurlo et al. 2009), although decapods are currently excluded from legislation in the EU and most countries (European Union 2010). The topic is controversial, with proponents advocating their inclusion (Rowe 2018; Passantino et al. 2021) and recommending humane slaughter methods (Conte et al. 2021a).

For both the commercial fishing and science sectors, handbooks and best practice advice are available (e.g., Ingle 1995; Jacklin and Combes 2007; Crump et al. 2024). In the United Kingdom commercial capture fishery and supply chain, industry-led codes of practice were published in 2024, an initiative driven by public and industrial stakeholders (Seafish 2024). This is timely given the change in United Kingdom law, growing interest in decapod welfare and sustainability from retailers (Powell et al. 2024; Loeb 2025).

8 Conclusions

Decapods encompass a huge variety of species, which support very valuable aquaculture and fisheries operations. Their diversity complicates our understanding of their capacity for pain, suffering and distress, as well as the development of good welfare practices. So far, our knowledge is limited to a few species and their basic needs, mostly relating to production and product quality. Nonetheless, a science of decapod pain and welfare is emerging. Moreover, given the growing scientific and public concern, industry and other stakeholders are increasingly trying to improve decapod welfare. This is encouraging, but good research must underpin any resulting legislation and changes in industry practices. We, therefore, need more research on the welfare needs of these fascinating animals.

References

Adey JM (2007) Aspects of the sustainability of creel fishing for Norway lobster, Nephrops norvegicus (L.), on the west coast of Scotland. PhD thesis, University of Glasgow

Albalat A, Sinclair S, Laurie J, Taylor AC, Neil DM (2010) Targeting the live market: recovery of Norway lobsters *Nephrops norvegicus* (L.) from trawl-capture as assessed by stress-related parameters and nucleotide breakdown. J Exp Mar Biol Ecol 395:206–214

Albalat A, Collard A, Brucem C, Coates CJ, Fox CJ (2016) Physiological condition, short-term survival, and predator avoidance behavior of discarded Norway lobsters (Nephrops norvegicus). J Shellfish Res 35(4):1053–1065

Albalat A, Gornik SG, Atkinson RJ, Coombs GH, Neil DM (2009) Effect of capture method on the physiology and nucleotide breakdown products in the Norway lobster (Nephrops norvegicus). Mar Biol Res 5(5):441–450

Albalat A, Gornik SG, Muangnapoh C, Neil DM (2022a) Effectiveness and quality evaluation of electrical stunning versus chilling in Norway lobsters (Nephrops norvegicus). Food Control 138:108930

Albalat A, Johnson L, Coates CJ, Dykes GC, Hitte F, Morro B et al (2019) The effect of temperature on the physiological condition and immune-capacity of European lobsters (Homarus gammarus) during long-term starvation. Front Mar Sci 6:281

Albalat A, Sinclair S, Neil D (2017) Validation of a vigour index for trawl-caught Norway lobsters (Nephrops norvegicus) destined for the live market: underlying links to both physiological condition and survivability. Fish Res 191:25–29

Albalat A, Zacarias S, Coates CJ, Neil DM, Planellas SR (2022b) Welfare in farmed decapod crustaceans, with particular reference to Penaeus vannamei. Front Mar Sci 9:886024

Alday-Sanz V, Brock JA, Flegel TW, McIntosh R, Bondad-Reantaso MG, Salazar M, Subasinghe RP (2018) Facts, truths and myths about SPF shrimp in aquaculture. Rev Aquac 12:76–84

Anand KJ, Hickey PR (1987) Pain and its effects in the human neonate and fetus. N Engl J Med 317(21):1321–1329

Appel M, Elwood RW (2009) Motivational trade-offs and potential pain experience in hermit crabs. Appl Anim Behav Sci 119(1–2):120–124

Aquaculture Stewardship Council (2023). https://asc-aqua.org/producers/farm-standards/shrimp/

Asian Seafood Improvement Collaborative (2023). https://www.asicollaborative.org/asic-shrimp

Barr S, Laming PR, Dick JT, Elwood RW (2008) Nociception or pain in a decapod crustacean? Anim Behav 75(3):745–751

Barr S, Elwood RW (2024a) Trade-offs between avoidance of noxious electric shock and avoidance of bright light in shore crabs are consistent with predictions of pain. Animals 14:770

Barr S, Elwood RW (2024b) Effects of acetic acid and morphine in shore crabs, Carcinus maenas: implications for the possibility of pain in Decapods. Animals 14:1705

Barr S, Elwood RW (2024c) The effects of caustic soda and benzocaine on directed grooming to the eyestalk in the glass prawn, *Palaemon elegans*, are consistent with the idea of pain in decapods. Animals 14:364

Barrento S, Marques A, Vaz-Pires P, Nunes ML (2010) Live shipment of immersed crabs Cancer pagurus from England to Portugal and recovery in stocking

tanks: stress parameter characterization. ICES J Mar Sci 67(3):435–443
Barrento S, Marques A, Vaz-Pires P, Nunes ML (2012) Physiological changes during simulated live transport of Cancer pagurus and recovery in holding tanks. Aquac Res 43(10):1415–1426
Behringer DC, Duermit-Moreau E (2021) Crustaceans, one health and the changing ocean. J Invertebr Pathol 186:107500
Beirão J, Boulais M, Gallego V, O'Brien JK, Peixoto S, Robeck TR, Cabrita E (2019) Sperm handling in aquatic animals for artificial reproduction. Theriogenology 133:161–178
Bernasconi CJ, Uglow RF (2008) Effects of emersion-induced hypoxia on some haemolymph constituents of Nephrops norvegicus. Dis Aquat Org 82(2):135–143
Birch J (2017) Animal sentience and the precautionary principle. Animal Sentience 2(16)
Birch J, Burn C, Schnell A, Browning H, Crump A (2021a) Review of the evidence of sentience in cephalopod molluscs and decapod crustaceans. Defra
Browning H, Burn C, Schnell AK, Crump A, Birch J (2025) Animal welfare risks from commercial practices involving cephalopod molluscs and decapod crustaceans. Anim Welf 34:e24
Butterworth A (2009) Animal welfare indicators and their use in society. Welfare of production animals: assessment and management of risks. In: Food safety assurance and veterinary public health series, vol 5. Wageningen Academic Publishers, The Netherlands, pp 371–389
British Veterinary Association (2022) BVA policy position on the recognition of animals as sentient beings. Available online: https://www.bva.co.uk/media/4052/bva-policy-position-on-the-recognition-of-animals-as-sentient-beings.pdf
Campbell DL, Lee C (2025) A review of behavioral testing in decapod shrimp (Caridea) and prawns (Dendrobranchiata) with applications for welfare assessment in aquaculture. PeerJ 13:e18883
Carder G (2017) A preliminary investigation into the welfare of lobsters in the UK. Animal Sentience 16:19
Chang ES (2005) Stressed-out lobsters: crustacean hyperglycemic hormone and stress proteins. Integr Comp Biol 45(1):43–50
Chen J, Zhou F, Huang J, Ma Z, Jiang S, Qiu L, Qin JG (2016) Ammonia and salinity tolerance of Penaeus monodon across eight breeding families. Springerplus 5:1–4
Coates CJ (2022) Host defences of invertebrates to non-communicable diseases. In: Rowley AF, Coates CJ, Whitten MMA (eds) Invertebrate pathology. Oxford University Press, London, pp 41–60
Coates CJ, Costa-Paiva EM (2020) Multifunctional roles of hemocyanins. Vertebrate and invertebrate respiratory proteins, lipoproteins and other body fluid proteins. pp 233–250
Coates CJ, Söderhäll K (2021) The stress–immunity axis in shellfish. J Invertebr Pathol 186:107492
Coates CJ, Rowley AF (2022) Emerging diseases and epizootics in crabs under cultivation. Front Mar Sci 8:809759
Comstock G (2022) Pain in Pleocyemata, but not in Dendrobranchiata? Animal Sentience 7(32):13
Conte F, Voslarova E, Vecerek V, Elwood RW, Coluccio P, Pugliese M, Passantino A (2021a) Humane slaughter of edible decapod crustaceans. Animals 11(4):1089
Conneely EA, Coates CJ (2024) Meta-analytic assessment of physiological markers for decapod crustacean welfare. Fish Fish 25(1):134–150
Cowing D, Powell A, Johnson M (2015) Evaluation of different concentration doses of eugenol on the behaviour of Nephrops norvegicus. Aquaculture 442:78–85
Council Directive 98/58/EC of 20 July 1998 concerning the protection of animals kept for farming purposes. https://eur-lex.europa.eu/eli/dir/1998/58/oj/eng
Council Regulation (EC) No 1099/2009 of 24 September 2009 on the protection of animals at the time of killing. https://eur-lex.europa.eu/eli/reg/2009/1099/oj/eng
Crump A (2022) Animal sentience science and policy. Animal Sentience 6(31):15
Crump A, Browning H, Schnell AK, Burn C, Birch J (2022a) Sentience in decapod crustaceans: a general framework and review of the evidence. Animal Sentience 7(32):1
Crump A, Browning H, Schnell AK, Burn C, Birch J (2022b) Animal sentience research: synthesis and proposals. Anim Sentience 7
Crump A, Browning H, Schnell AK, Burn C, Birch J (2022c) Invertebrate sentience and sustainable seafood. Nat Food 3(11):884–886
Crump A, Fischer B, Arnott G, Birch J, Briffa M, Browning H, Coates C, Elwood R, Khan N, Thakrar U, Barrett M (2024) Guidelines for protecting and promoting decapod crustacean welfare in research. Insect Welfare Research Society. Retrieved from: insectwelfare.com/research-guidelines. https://pearl.plymouth.ac.uk/bms-research/2163/
Crustacean Compassion (2023) Sea to plate: the welfare journey of decapod crustaceans (non-peer reviewed report). Available at: https://www.crustaceancompassion.org/reports. Accessed 18 Jul 2023
Davies CE, Batista FM, Malkin SH, Thomas JE, Bryan CC, Crocombe P et al (2019) Spatial and temporal disease dynamics of the parasite Hematodinium sp. in shore crabs, Carcinus maenas. Parasit Vectors 12:1–15
Davis RP, Boyd CE (2021) A comparison of the technical efficiency of aquaculture stewardship council certified shrimp farms to non-certified farms. Curr Res Environ Sustain 3:100069
Dawkins MS (2022) Unconscious humans, autonomous machines and the difficulty of knowing which animals are sentient. Anim Sentience 32(20)
De Grave S, Pentcheff ND, Ahyong ST, Chan TY, Crandall KA, Dworschak PC et al (2009) A classification of living and fossil genera of decapod crustaceans. Raffles Bull Zool 1

De Grave S, Decock W, Dekeyzer S, Davie PJ, Fransen CH, Boyko CB, Santos S (2023) Benchmarking global biodiversity of decapod crustaceans (Crustacea: Decapoda). J Crustac Biol 43(3):ruad042

De Lima J, Lloyd-Thomas AR, Howard RF, Sumner E, Quinn TM (1996) Infant and neonatal pain: anaesthetists' perceptions and prescribing patterns. BMJ 313(7060):787

de Souza Valente C (2022a) Anaesthesia of decapod crustaceans. Vet Anim Sci 16:100252

DeAngelis BM, Cooper R, Clancy M, Cooper C, Angell T, Olszewski S, Colburn W, Catena J (2010) Impacts of V-notching the American lobster. J Shellfish Res 29(2):489–496

Deidun A, Sciberras A, Formosa J, Zava B, Insacco G, Corsini-Foka M, Crandall KA (2018) Invasion by non-indigenous freshwater decapods of Malta and Sicily, Central Mediterranean Sea. J Crustac Biol 38(6):748–753

Diamond S, Powell A, Shields RJ, Rowley AF (2008) Is spermatophore melanisation in captive shrimp (Litopenaeus vannamei) a result of an auto-immune response? Aquaculture 285(1–4):14–18

Diarte-Plata G, Sainz-Hernández JC, Aguiñaga-Cruz JA, Fierro-Coronado JA, Polanco-Torres A, Puente-Palazuelos C (2012) Eyestalk ablation procedures to minimize pain in the freshwater prawn Macrobrachium americanum. Appl Anim Behav Sci 140(3–4):172–178

Diggles BK (2019a) Review of some scientific issues related to crustacean welfare. ICES J Mar Sci 76(1):66–81

Diggles BK, Arlinghaus R, Browman HI, Cooke SJ, Cooper RL, Cowz IG et al (2024) Reasons to be skeptical about sentience and pain in fishes and aquatic invertebrates. Rev Fish Sci Aquacult 32(1):127–150

Durand F, Devillers N, Lallier FH, Regnault M (2000) Nitrogen excretion and changes in blood components during emersion of the subtidal spider crab Maia squinado (L.). Comp Biochem Physiol A Mol Integr Physiol 127(3):259–271

Elwood RW, Appel M (2009) Pain experience in hermit crabs? Anim Behav 77(5):1243–1246

Elwood RW, Barr S, Patterson L (2009) Pain and stress in crustaceans? Appl Anim Behav Sci 118(3–4):128–136

Elwood RW, Dalton N, Riddell G (2017) Aversive responses by shore crabs to acetic acid but not to capsaicin. Behav Process 140:1–5

Elwood RW (2023) Behavioural indicators of pain and suffering in arthropods and might pain bite back? Animals 13(16):2602

Elwood RW (2025a) A history of pain studies and changing attitudes to the welfare of crustaceans. Animals 15(3):445

Emerenciano MGC, Rombenso AN, Vieira FN, Martins MA, Coman GJ, Truong HH, Noble TH, Simon CJ (2022) Intensification of Penaeid Shrimp culture: an applied review of advances in production systems, nutrition and breeding. Animals 12(3):236

Ern R, Huong do TT, Phuong NT, Madsen PT, Wang T, Bayley M (2015) Some like it hot: thermal tolerance and oxygen supply capacity in two eurythermal crustaceans. Sci Rep 5:10743. https://doi.org/10.1038/srep10743. PMID: 26030412; PMCID: PMC5377234

Estrella BT (2002) Techniques for live storage & shipping of American lobster. Massachusetts Department of Fisheries, Wildlife & Environmental Law Enforcement Division of Marine Fisheries

European Union (2010) Directive 2010/63/EU of the European Parliament and of the Council of 22 September 2010 on the protection of animals used for scientific purposes. EC Council Directive 2010/63/EU. Available at: https://eur-lex.europa.eu/legal-content/EN/TXT/?uri=CELEX%3A32010L0063. Accessed 10 Jul 23

Euroshrimp (2020) Euroshrimp newsletter 14. Available at: https://www.euroshrimp.net/14-european-shrimp-production-in-2020/

Farnsworth KD, Elwood RW (2023) Why it hurts: with freedom comes the biological need for pain. Anim Cogn 26(4):1259–1275

FAO (2024) The state of world fisheries and aquaculture 2024. Towards Blue Transformation. Rome. Available at: https://doi.org/10.4060/cd0683en

FAO (2023) Penaeus vannamei. Cultured aquatic species information programme. Text by Briggs, M. Fisheries and Aquaculture Division [online]. Rome. https://www.fao.org/fishery/en/culturedspecies/litopenaeus_vannamei/en

Flegel TW (2012) Historic emergence, impact and current status of shrimp pathogens in Asia. J Invertebr Pathol 110(2):166–173

Flegel TW (2019) A future vision for disease control in shrimp aquaculture. J World Aquacult Soc 50(2):249–266

Food and Drug Administration (2021) Seafood HACCP and the FDA Food Safety Modernization Act: Guidance for Industry

Fossat P, Bacqué-Cazenave J, De Deurwaerdère P, Cattaert D, Delbecque JP (2015) Serotonin, but not dopamine, controls the stress response and anxiety-like behavior in the crayfish Procambarus clarkii. J Exp Biol 218(17):2745–2752

Fossat P, Bacqué-Cazenave J, De Deurwaerdère P, Delbecque JP, Cattaert D (2014) Anxiety-like behavior in crayfish is controlled by serotonin. Science 344(6189):1293–1297

Frasca JS, Gast RJ, Bogomolni AL, Szczepanek SM (2020) Diagnosing marine diseases. In: Lafferty KD, Silliman BR (eds) Marine disease ecology. Oxford University Press, New York

Funge-Smith S, Briggs M (2003) The introduction of Penaeus Vannamei and P Stylirostris into the Asia-Pacific Region. In: International mechanisms for the control and responsible use of Alien Species in aquatic ecosystems. Food and Agriculture Organisation of the United Nations (FAO), Rome, pp 26–29

Gardner C (1997) Options for humanely immobilizing and killing crabs. J Shellfish Res 16(1):219–224

Glencross BD (2009) Exploring the nutritional demand for essential fatty acids by aquaculture species. Rev Aquac 1:71–124

Global Seafood Alliance (2023). https://www.bapcertification.org/Standards

Goldstein JS, Carloni JT (2021) Assessing the implications of live claw removal on Jonah crab (*Cancer borealis*), an emerging fishery in the Northwest Atlantic. Fish Res 243:106046

González RA, Díaz F, Licea AF, Re AD, Sánchez LN, Garcia-Esquivel Z (2010) Thermal preference, tolerance and oxygen consumption of adult white shrimp Litopenaeus vannamei (Boone) exposed to different acclimation temperatures. J Therm Biol 35:218–224

Gornik SG, Albalat A, Atkinson RJ, Coombs GH, Neil DM (2010) The influence of defined ante-mortem stressors on the early post-mortem biochemical processes in the abdominal muscle of the Norway lobster, Nephrops norvegicus (Linnaeus, 1758). Mar Biol Res 6(3):223–238

Hamed I, Özogul F, Regenstein JM (2016) Industrial applications of crustacean by-products (chitin, chitosan, and chitooligosaccharides): a review. Trends Food Sci Technol 48:40–50

Harris RR, Andrews MB (2005) Physiological changes in the Norway lobster Nephrops norvegicus (L.) escaping and discarded from commercial trawls on the West Coast of Scotland: II. Disturbances in haemolymph respiratory gases, tissue metabolites and swimming performance after capture and during recovery. J Exp Mar Biol Ecol 320(2):195–210

Hauton C, Hudspith M, Gunton L (2015) Future prospects for prophylactic immune stimulation in crustacean aquaculture - the need for improved metadata to address immune system complexity. Dev Comp Immunol 48(2):360–368

Hooper C, Debnath PP, Stentiford GD, Bateman KS, Salin KR, Bass D (2023) Diseases of the giant river prawn Macrobrachium rosenbergii: a review for a growing industry. Rev Aquac 15(2):738–758

ICES (1994) Report from the sub-group on methodology of fish survival experiments. ICES CM 1994/B:8

Ingle RW (1995) UFAW handbook on the care and management of decapod crustaceans in captivity. Universities Federation for Animal Welfare

Jacklin M, Combes J (2007) The good practice guide to handling and storing live Crustacea. Seafish Technical Report

Jennings S, Stentiford GD, Leocadio AM, Jeffery KR, Metcalfe JD, Katsiadaki I et al (2016) Aquatic food security: insights into challenges and solutions from an analysis of interactions between fisheries, aquaculture, food safety, human health, fish and human welfare, economy and environment. Fish Fish 17(4):893–938

Johnson L, Coates CJ, Albalat A, Todd K, Neil D (2016) Temperature-dependent morbidity of 'nicked' edible crab, Cancer pagurus. Fish Res 175:127–131

Jussila J, Jago J, Tsvetnenko E, Dunstan B, Evans LH (1997) Total and differential haemocyte counts in western rock lobsters (Panulirus cygnus George) under post-harvest stress. Mar Freshw Res 48(8):863–868

Kasiouras E, Hubbard PC, Gräns A, Sneddon LU (2024) Putative nociceptive responses in a decapod crustacean: the shore crab (*Carcinus maenas*). Biology (Basel) 13(11):851

Knipe H, Temperton B, Lange A, Bass D, Tyler CR (2020) Probiotics and competitive exclusion of pathogens in shrimp aquaculture. Rev Aquacult 13:324–352

Kumar V, Roy S, Behera BK, Bossier P, Das BK (2021) Acute Hepatopancreatic Necrosis Disease (AHPND): virulence, pathogenesis and mitigation strategies in shrimp aquaculture. Toxins 13(8):524

Kumaran M, Ravichandran P, Ponniah AG, Kumar JA, Vimala DD, Chandrasekaran VS (2015) Extension methodology for assessing effectiveness of shrimp farmer groups. Indian J Fish 62:116–123

Lancaster J, Frid CL (2002) The fate of discarded juvenile brown shrimps (Crangon crangon) in the Solway Firth UK fishery. Fish Res 58(1):95–107

Laphyai P, Kruangkum T, Chotwiwatthanakun C, Semchuchot W, Thaijongrak P, Sobhon P, Tsai PS, Vanichviriyakit R (2021) Suppression of a novel Vitellogenesis-inhibiting hormone significantly increases ovarian vitellogenesis in the Black Tiger Shrimp, Penaeus monodon. Front Endocrinol 12:760538

Larssen WE, Dyb JE, Woll A, Kennedy J (2013) Factors that affect vitality of Northern Shrimp (Pandalus borealis, Krøyer 1838) during capture and storage that are destined for live trade. J Shellfish Res 32(3):807–813

Liao I-C (1985) A brief review of the larval rearing techniques of penaeid prawns. In: Taki Y, Primavera JH, Llobrera JA (eds) Proceedings of the First International Conference on the Culture of Penaeid Prawns/Shrimps, 4–7 December 1984. Aquaculture Department, Southeast Asian Fisheries Development Center, Iloilo City, pp 65–78

Loeb J (2025) Supermarket commits to prawn welfare. Vet Rec 196:166

Lorenzon S, Giulianini PG, Libralato S, Martinis M, Ferrero EA (2008) Stress effect of two different transport systems on the physiological profiles of the crab Cancer pagurus. Aquaculture 278(1–4):156–163

Lorgen-Ritchie M, Webster TU, McMurtrie J, Bass D, Tyler CR, Rowley A, Martin SA (2023) Microbiomes in the context of developing sustainable intensified aquaculture. Front Microbiol 14:1200997

Lu Y, Wang F, Li L, Dong S (2016) Responses of metabolism and haemolymph ions of swimming crab Portunus trituberculatus to thermal stresses: a comparative study between air and water. Aquac Res 47(9):2989–3000

Lundmark F, Berg C, Schmid O, Behdadi D, Röcklinsberg H (2014) Intentions and values in animal welfare legislation and standards. J Agric Environ Ethics 27(6):991–1017

Malaweera BO, Wijesundara WNM (2013) Use of seafood processing by-products in the animal feed industry. In: Seafood processing by-products: trends and applications. Springer New York, New York, pp 315–339

Mallatt J, Feinberg TE (2021) Multiple routes to animal consciousness: constrained multiple realizability rather than modest identity theory. Front Psychol 12:3895

Mallol S, Díaz D, Sobrado F, Goñi R (2014) First V-notching experience of a spiny lobster: V-notch recovery and impact on health and growth of Palinurus elephas. J Crustac Biol 34(1):25–30

McLeese D, Wilder D (1964) Lobster storage and shipment. Fish Res Bd Canada Bull 147:69

Mellor DJ, Reid CSW (1994) Concepts of animal well-being and predicting the impact of procedures on experimental animals. https://www.wellbeingintlstudiesrepository.org/exprawel/7/

Mellor DJ, Beausoleil NJ, Littlewood KE, McLean AN, McGreevy PD, Jones B, Wilkins C (2020) The 2020 five domains model: including human–animal interactions in assessments of animal welfare. Animals 10(10):1870

Milligan RJ, Albalat A, Atkinson RJA, Neil DM (2009) The effects of trawling on the physical condition of the Norway lobster Nephrops norvegicus in relation to seasonal cycles in the Clyde Sea area. ICES J Mar Sci 66(3):488–494

Minter LJ, Harms CA, Archibald KE, Broadhurst H, Bailey KM, Christiansen EF et al (2013) Efficacy of alfaxalone for intravascular anesthesia and euthanasia in blue crabs (Callinectes sapidus). J Zoo Wildl Med 44(3):694–699

Morgan J, Cargill C, Groot E (2001) The efficacy of clove oil as an anesthetic for decapod crustaceans. Bull-Aquac Assoc Canada 3:27–31

Narshi TM, Free D, Justice WSM, Smith SJ, Wolfensohn S (2022) Welfare assessment of invertebrates: Adapting the Animal Welfare Assessment Grid (AWAG) for zoo decapods and cephalopods. Animals 12(13):1675

Neil DM (2012) Ensuring crustacean product quality in the post-harvest phase. J Invertebr Pathol 110(2):267–275

Neil DM, Putyora E, Albalat A (2024) Towards the humane slaughter of decapod crustaceans: identifying the most effective indicators of insensibility following electrical stunning. Front Anim Sci 5:1378350

Newland PL, Neil DM, Chapman CJ (1992) Escape swimming in the Norway lobster. J Crustac Biol 12(3):342–353

Olivier IS, Olivier A (2024) Pain in context: indicators and expressions of animal pain. In: Glover MJ, Mitchell L (eds) Animals as experiencing entities, The Palgrave Macmillan Animal Ethics Series. Palgrave Macmillan, Cham

Passantino A, Elwood RW, Coluccio P (2021) Why protect decapod crustaceans used as models in biomedical research and in ecotoxicology? Ethical and Legislative Considerations. Animals 11(1):73

Pedrazzani AS, Cozer N, Quintiliano MH, Tavares CPDS, da Silva UDAT, Ostrensky A (2023) Non-invasive methods for assessing the welfare of farmed White-leg Shrimp (Penaeus vannamei). Animals 13(5):807

Pedrazzani AS, Cozer N, Quintiliano MH, Ostrensky A (2024) Insights into decapod sentience: applying the general welfare index (GWI) for whiteleg shrimp (*Penaeus vannamei*—Boone, 1931) reared in aquaculture grow-out ponds. Fishes 9(11):440

Pope EC, Powell A, Roberts EC, Shields RJ, Wardle R, Rowley AF (2011) Enhanced cellular immunity in shrimp (Litopenaeus vannamei) after 'vaccination'. PLoS One 6:e20960

Powell A, Barrento S, Cowing DM (2020) Management and handling of commercial crustaceans. In: Lovrich G, Thiel M (eds) The natural history of the Crustacea: fisheries and aquaculture. Oxford University Press, pp 495–523

Powell A, Pope EC, Eddy FE, Shields RJ, Francis MJ, Smith P, Topps S, Reid J, Rowley AF (2011) Enhanced immune defences in Pacific white shrimp (Litopenaeus vannamei) post-exposure to a Vibrio vaccine. J Invertebr Pathol 107:95–99

Powell A, Cowing DM, Eriksson SP, Johnson ML (2017) Stress response according to transport protocol in Norway lobster, Nephrops norvegicus. Crustacean Res 46:17–24

Powell A, Agnalt AL, Heasman K, Albalat A (2024) Considerations for implementing regulation of decapods in science. Anim Welf 33:e55

Puri S, Faulkes Z (2010) Do decapod crustaceans have nociceptors for extreme pH? PLoS One 5(4):e10244

Quesada RJ, Smith CD, Heard DJ (2011) Evaluation of parenteral drugs for anesthesia in the blue crab (Callinectes sapidus). J Zoo Wildl Med 42(2):295–299

Radford CA, Stanley JA (2023) Sound detection and production mechanisms in aquatic decapod and stomatopod crustaceans. J Exp Biol 226(10):jeb243537

Regulation (EC) No 1/2005 on the protection of animals during transport and related operations. https://eur-lex.europa.eu/EN/legal-content/summary/eu-rules-on-the-protection-of-animals-during-transport.html

Ridgway ID, Taylor AC, Atkinson RJA, Chang ES, Neil DM (2006a) Impact of capture method and trawl duration on the health status of the Norway lobster, *Nephrops norvegicus*. J Exp Mar Biol Ecol 339(2):135–147

Ridgway ID, Taylor AC, Atkinson RJA, Stentiford GD, Chang ES, Chang SA, Neil DM (2006b) Morbidity and mortality in Norway lobsters, Nephrops norvegicus: physiological, immunological and pathological effects of aerial exposure. J Exp Mar Biol Ecol 328(2):251–264

Robertson JD, Delorme NJ, Hickey A, Jeffs AG (2018) Assessment of the potential of the anesthetic AQUI-S for live transportation of the southern rock lobster, Jasus edwardsii. Bull Mar Sci 94(3):1137–1151

Roth B, Grimsbø E (2016) Electrical stunning of edible crabs (Cancer pagurus): from single experiments to commercial practice. Anim Welf 25(4):489–497

Rotllant G, Llonch P, García del Arco JA, Chic Ò, Flecknell P, Sneddon LU (2023a) Methods to induce analgesia and anesthesia in crustaceans: a supportive decision tool. Biology 12(3):387

Rowan AN, D'Silva JM, Duncan IJH, Palmer N (2021) Animal sentience: history, science, and politics. Anim Sentience 31(1)

Rowe A (2018) Should scientific research involving decapod crustaceans require ethical review? J Agric Environ Ethics 31(5):625–634

Rowley AF, Pope EC (2012) Vaccines and crustacean aquaculture—a mechanistic exploration. Aquaculture 334–337:1–11

Rowley AF, Powell A (2007) Invertebrate immune systems-specific, quasi-specific, or nonspecific? J Immunol 179:7209–7214

Rowley AF, Coates CJ (2023) Shell disease syndromes of decapod crustaceans. Environ Microbiol 25(5):931–947

Rowley AF, Davies CE, Malkin SH, Bryan CC, Thomas JE, Batista FM, Coates CJ (2020) Prevalence and histopathology of the parasitic barnacle, Sacculina carcini in shore crabs, Carcinus maenas. J Invertebr Pathol 171:107338

SAGB (2021) Advisory note on crustacea storage and transport. Available at: https://www.shellfish.org.UK/files/Literature/Misc/Crustacea-Storage-and-Transport-SAGB-Advisory-Note-2021.pdf. Accessed 18 Jul 23

Sandeman DC, Kenning M, Harzsch S (2014) Adaptive trends in malacostracan brain form and function related to behavior. In: Derby C, Thiel M (eds) Crustacean nervous system and their control of behaviour. Oxford University Press, pp 11–48

Scott K, Harsanyi P, Lyndon AR (2018) Baseline measurements of physiological and behavioural stress markers in the commercially important decapod Cancer pagurus (L.). J Exp Mar Biol Ecol 507:1–7

Seafish (2024) Codes of practice for the welfare of crabs, lobsters, crawfish and *Nephrops*. Available at https://www.seafish.org/responsible-sourcing/animal-sentience-and-welfare/

Smith J, Boyd K (eds) (1991) Lives in the balance: the ethics of using animals in biomedical research. Oxford University Press, New York

Spanier E, Lavalli KL, Goldstein JS, Groeneveld JC, Jordaan GL, Jones CM, Phillips BF, Bianchini ML, Kibler RD, Díaz D, Mallol S, Goñi R, van Der Meeren GI, Agnalt A-L, Behringer DC, Keegan WF, Jeffs A (2015) A concise review of lobster utilization by worldwide human populations from prehistory to the modern era. ICES J Mar Sci 72:7–21

Stebbing PD, Johnson P, Delahunty A, Clark PF, Mccollin T, Hale C, Clark S (2012) Reports of American lobsters, *Homarus americanus* (H. Milne Edwards, 1837), in British waters. BioInvasions Rec 1:17–23

Stentiford GD, Neil DM (2000) A rapid onset, post-capture muscle necrosis in the Norway lobster, Nephrops norvegicus (L.) from the West coast of Scotland. J Fish Dis 23(4):251–263

Stentiford GD, Neil DM, Peeler EJ, Shields JD, Small HJ, Flegel TW, Vlak JM, Jones B, Morado F, Moss S, Lotz J, Bartholomay L, Behringer DC, Hauton C, Lightner DV (2012) Disease will limit future food supply from the global crustacean fishery and aquaculture sectors. J Invertebr Pathol 110(2):141–157

Stoner AW (2012a) Assessing stress and predicting mortality in economically significant crustaceans. Rev Fish Sci 20(3):111–135

Strausfeld NJ, Wolff GH, Sayre ME (2020) Mushroom body evolution demonstrates homology and divergence across Pancrustacea. elife 9:e52411

Tacon AG, Cody J, Conquest LD, Divakaran S, Forster IP, Decamp O (2002) Effect of culture system on the nutrition and growth performance of Pacific white shrimp Litopenaeus vannamei (Boone) fed different diets. Aquacult Nutr 8:121–137

Tall BD, Fall S, Pereira MR, Ramos-Valle M, Curtis SK, Kothary MH et al (2003) Characterization of Vibrio fluvialis-like strains implicated in limp lobster disease. Appl Environ Microbiol 69(12):7435–7446

Taylor J, Vinatea L, Ozorio R, Schuweitzer R, Andreatta ER (2004) Minimizing the effects of stress during eyestalk ablation of Litopenaeus vannamei females with topical anesthetic and a coagulating agent. Aquaculture 233(1–4):173–179

Thériault G, Hanlon J, Creed L (2013) Report of the Maritime Lobster Panel. Available at https://novascotia.ca/fish/documents/Maritime-Lobster-Panel-Report-NOV1.pdf. Accessed Oct 11 2018

Thitamadee S, Prachumwat A, Srisala J et al (2016) Review of current disease threats for cultivated Penaeid shrimp in Asia. Aquaculture 452:69–87

Thornber K, Verner-Jeffreys D, Hinchliffe S, Rahman MM, Bass D, Tyler CR (2020) Evaluating antimicrobial resistance in the global shrimp industry. Rev Aquac 12(2):966–986

Uawisetwathana U, Leelatanawit R, Klanchui A, Prommoon J, Klinbunga S, Karoonuthaisiri N (2011) Insights into eyestalk ablation mechanism to induce ovarian maturation in the black tiger shrimp. PLoS One 6(9):e24427

Uglow RF, Hosie DA, Johnson IT, MacMullen PH (1986) Live handling and transport of crustacean shellfish: an investigation on mortalities. Sea Fish Industry Authority. Edinburgh. Technical Report 280

Wahltinez SJ, Harms CA, Lewbart GA (2023a) Anesthesia and analgesia in invertebrates. In: Anesthesia and analgesia in laboratory animals. Academic Press, pp 647–671

Waldhorn DR, Autric E (2023) Shrimp: the animals most commonly used and killed for food production; rethink priorities, vol 37. OSF Preprints, San Francisco

Weineck K, Ray AJ, Fleckenstein LJ, Medley M, Dzubuk N, Piana E, Cooper RL (2018) Physiological changes as a measure of crustacean welfare under different standardized stunning techniques: cooling and electroshock. Animals 8(9):158

Woodruff ML (2022) Sentience in decapods: difficulties to surmount. Anim Sentience 32(11)

Wuertz S, Bierbach D, Bögner M (2023) Welfare of decapod crustaceans with special emphasis on stress physiology. Aquac Res 2023:1307684

Yu YB, Choi JH, Kang JC, Kim HJ, Kim JH (2022) Shrimp bacterial and parasitic disease listed in the OIE: a review. Microb Pathog 166:105545

Zacarias S, Carboni S, Davie A, Little DC (2019) Reproductive performance and offspring quality of non-ablated Pacific white shrimp (Litopenaeus vannamei) under intensive commercial scale conditions. Aquaculture 503:460–466

Zurlo J, Bayne K, Clark JM (2009) Adequate veterinary care for animals in research: a comparison of guidelines from around the world. ILAR J 50(1):85–88

Recommended Reading

Evidence for Pain

Crump A, Browning H, Schnell A, Burn C, Birch J (2022) Sentience in decapod crustaceans: a general framework and review of the evidence. Animal Sentience 7(32):1

Diggles BK (2019b) Review of some scientific issues related to crustacean welfare. ICES J Marine Sci 76(1):66–81

Elwood RW (2025b) A history of pain studies and changing attitudes to the welfare of crustaceans. Animals 15(3):445

Sneddon LU (2015) Pain in aquatic animals. J Exp Biol 218(Pt 7):967–976. https://doi.org/10.1242/jeb.088823.25833131

Welfare in Commercial Settings

Albalat A, Zacarias S, Coates CJ, Neil DM, Planellas SR (2022b) Welfare in farmed decapod crustaceans, with particular reference to Penaeus vannamei. Front Marine Sci 9:886024

Birch J, Burn C, Schnell A, Browning H, Crump A (2021b) Review of the evidence of sentience in cephalopod molluscs and decapod crustaceans. Defra

Conte F, Voslarova E, Vecerek V, Elwood RW, Coluccio P, Pugliese M, Passantino A (2021b) Humane slaughter of edible decapod crustaceans. Animals 11(4):1089

Listed Approaches to Anaesthesia and Analgesia

de Souza Valente C (2022b) Anaesthesia of decapod crustaceans. Vet Anim Sci 16:100252

Rotllant G, Llonch P, García del Arco JA, Chic Ò, Flecknell P, Sneddon LU (2023b) Methods to induce analgesia and anesthesia in crustaceans: a supportive decision tool. Biology 12(3):387

Wahltinez SJ, Harms CA, Lewbart GA (2023b) Anesthesia and analgesia in invertebrates. In: Anesthesia and analgesia in laboratory animals. Academic Press, pp 647–671

Stress, Immunity, and Their Biomarkers

Coates CJ, Söderhäll K (2021) The stress–immunity axis in shellfish. J Invertebr Pathol 186:107492

Stoner AW (2012b) Assessing stress and predicting mortality in economically significant crustaceans. Rev Fish Sci 20(3):111–135

Disease Diagnosis

Bass D, Rowley AF, Coates CJ (2022) Chapter 3, Diagnostic approaches in invertebrate pathology. In: Rowley AF, Coates CJ, Whitten MMA (eds) Invertebrate pathology. Oxford University Press, London, pp 61–72

Frasca S Jr, Gast RJ, Bogomolni AL, Szczepanek SM (2020) Diagnosing marine diseases. In: Lafferty KD, Silliman BR (eds) Marine disease ecology. Oxford University Press, Oxford

Hancock-Ronemus AL, Buote M, Gregory A (2022) Chapter 16, Crustaceans. In: Lewbart GA (ed) Invertebrate medicine, 3rd edn. Wiley-Blackwell

Insects

Meghan Barrett

Abstract

Insects are an extremely diverse class of mainly terrestrial arthropods (arthropods are invertebrates in the ecdysozoa, with an exoskeleton made of chitin), with a million named species and more to be discovered. At least 1.2 trillion individual insects are farmed each year for food and feed around the globe, with trillions (perhaps quadrillions) affected by human activity in other contexts. The question of insect pain is an open one, i.e. without broad scientific or philosophical consensus. This chapter examines the history of scientific interest in the subject and presents the neurobiological and behavioural evidence related to insect pain. The potential pain and suffering experienced by farmed insects during rearing, transport and slaughter are reviewed while acknowledging the uncertainty of insect pain perception and the lack of validated tools for assessing it. The chapter also reviews analgesics and anaesthetics for use in insects and examines the attitudes of consumers, producers, academics and others to insect pain. The diversity, and contexts of use, of arthropods other than insects and crustaceans are briefly reviewed before concluding with the complexity of moral caution in the insect case.

Keywords

Pain · Insect · Analgesia · Anaesthesia · Welfare · Black Soldier Fly · Mealworms · Crickets · Ethics

1 Introduction

How does one distinguish between what is "painful" and what is "unpleasant"...? What constitutes a symptom of pain in insects? Who is qualified to judge? Can we assume insects feel no pain because they exhibit no signs of suffering?—entomologist Harry Weiss (1914)

There are one million named species of insects (Zhang 2011) and 5.5 million species estimated to exist (Stork 2018). Dedicating only a single chapter to their possible pain and suffering feels utterly inadequate; and yet, dedicating a whole chapter to their possible pain and suffering also feels entirely radical.

This chapter is one of questions. *Can insects feel pain? What could cause them to suffer—and how would that suffering be recognized? How many insects are directly affected by human activity, particularly farming? What are the public and scientific perceptions of insect pain? Are they changing?* Given the widespread uncer-

M. Barrett (✉)
Department of Biology, Indiana University
Indianapolis, Indianapolis, IN, USA
e-mail: meghbarr@iu.edu

R. E. Clutton, G. C. Musk (eds.), *Pain and Suffering in Farmed Animals*,
https://doi.org/10.1007/978-3-032-11407-5_23

tainty: *What should we do?* Many these questions are unprecedented, despite the issue of insect pain having concerned entomologists and philosophers for over 100 years.

This chapter aims to help develop an understanding of insects and their world. Seemingly alien because of their (typically) small body size, strange behaviours and inexpressive exoskeleton, many people lack familiarity with insects. However, an increasing body of information is likely to improve humanity's understanding of these animals which, after all, evolved from an ancestor common to *Homo sapiens.*

2 Insect Biodiversity

This section examines insect diversity, life history, development, physiology, cognition and behaviours. It is not exhaustive but intends to foster an appreciation of the sheer breadth of variation among and the unique capabilities of insects. The scale of insect use and management in the wild, production facilities, laboratories and classrooms, zoos and insectariums, entertainment settings, art, medicine, forensics and as pets are also described in order to demonstrate that the scale of insect use provides a strong justification for examining the question of insect pain.

2.1 Biodiversity

> *No [one] can be truly called an entomologist… the subject is too vast for any single human intelligence to grasp.*
>
> —Oliver Wendell Holmes Sr. (1872)[1, 2]

There are over thirty orders of insects (Fig. 1), with beetles (Coleoptera) and bees, ants, sawflies and wasps (Hymenoptera) being the most speciose (Forbes et al. 2018). Insects in orders that are less well-known by non-entomologists include: angel insects, lerp insects, scale insects, moss bugs, firebrats, ice crawlers, fringe wings, treehoppers, caddisflies, snakeflies, stoneflies, dobsonflies, whiteflies and scorpionflies (none of which belong to the order of the more-familiar true flies, i.e. Diptera).

Insects are found on every continent and in most environments: silver ants forage in the midday heat of the Sahara (70 °C), while a single midge species survives the Antarctic cold (−15 °C); sea skaters bob alongside cargo ships on the open seas. Most insect species are entirely terrestrial, but some live in aquatic environments, particularly in the juvenile life stage, e.g. mosquitoes or dragonflies.

Two *major* developmental strategies occur in insects: (i) 'holometabolous' or complete metamorphosis, which is employed by most species and (ii) the more ancestral 'hemimetabolous' or incomplete metamorphosis. Holometabolous insects, such as butterflies, bees, flies and beetles, begin life as a grub-like larval form that occupies distinctly different ecological niches from the adult. At the end of juvenile development, the larvae undergo pupation, a period during which the entire anatomy, including the nervous system, metamorphosizes into the adult form. In hemimetabolous insects, such as cockroaches or crickets, there is no complete pupation event. Juveniles, which usually look like miniature adults without wings or reproductive anatomy and which are often called nymphs or naiads—moult directly into adults. Other, rarer developmental strategies have been described by Vea and Minakuchi (2021).

Insect diversity is largely undescribed. By remaining undiscovered or becoming extinct, many existing species may never be described. Studies suggest both insect abundance and diversity are in decline due to climate change, habitat loss, agricultural intensification (including habitat simplification, pesticide usage) and other factors (Sánchez-Bayo and Wyckhuys 2019;

[1] The author's background is in the social and solitary Hymenoptera and other social arthropods: She has worked with bees and ants, and, to a lesser extent, wasps, termites, *Cryptocercus* wood roaches, flies, mealworms, mantids and spiders (which, notably, are not insects) throughout her career. As such, you may expect this overview to be heavily biased towards those arthropods. Yet even peering through such a small window into the world of insects as her own knowledge provides: what incredible diversity there is to see.

[2] Oliver Wendell Holmes Sr also gave the word 'anaesthesia' to the world.

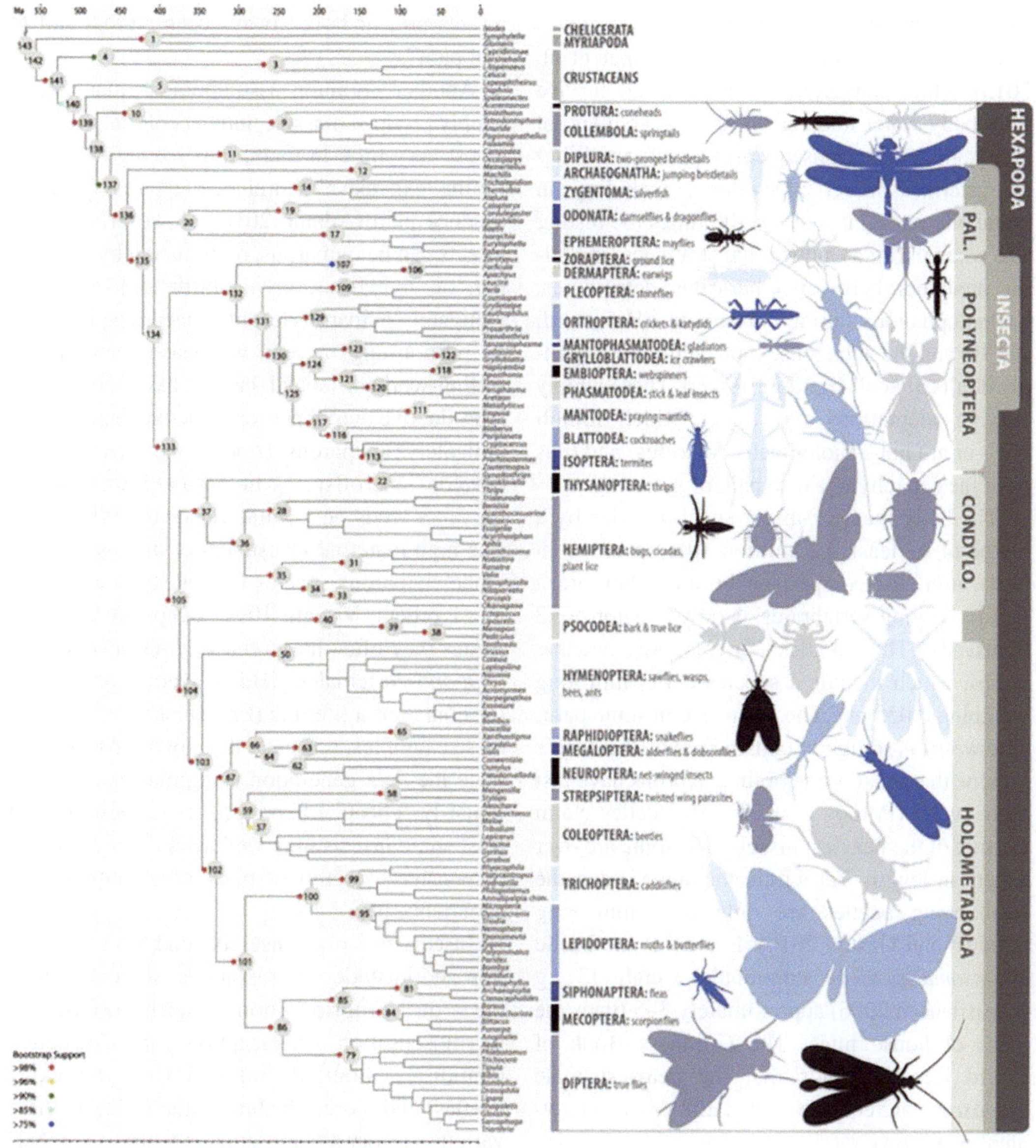

Fig. 1 Phylogeny of the insects. (From Misof et al. 2014)

Montgomery et al. 2020), a phenomenon often referred to as the 'insect apocalypse'.

2.2 Life History, Morphological and Physiological Variation

Insect life history, morphology and physiology are extremely diverse. For example, mayflies (Ephemeroptera) spend around a year as aquatic, gilled nymphs before emerging as a winged subimago and then an adult that lives for about a day. Some mayfly species are even more ephemeral, living as adults for as little as five minutes. Periodical cicadas spend 17 years underground as nymphs before emerging as adults to live for 4–6 weeks. In contrast, some queen ants have lived for at least 28 years as captive adults (Keller and Genoud 1997). Some insect species are adapted to extreme environmental conditions.

The haemolymph of cold-adapted species makes a biological form of 'antifreeze' (Kuiper et al. 2015), while anhydrobiotic midges can survive 97% body water loss for up to 17 years (Thorat and Nath 2018). Consequently, great variability characterizes insect physiology, including in metabolic rate (this may be 400 times the resting rate when in flight), nutritional strategies (including predators, herbivores, parasites, detritivores, haematophagous, urine-consumers (Hu et al. 2018b) and the aphagous: some adult moths simply don't eat at all (Kellogg 1907)). These widely varying adaptations may affect, which stimuli are—or are not—biologically 'noxious' and thus have the capacity to cause pain or suffering.

The body mass of insect species varies by a factor of at least 5.2 million (the variation in birds, used for comparison because they are a group of flying vertebrates, is only a factor of 72 thousand). The smallest insects are parasitic wasps, which include a species 0.139 mm long weighing 0.025 mg. The brain size in some parasitic wasps is so tightly constrained by the head's volume that most of the brain's neurons have lost their nuclei (Polilov et al. 2023). Beetles claim the title of the heaviest insects, and some are over 11 cm in length. Yet within the same order, the featherwing beetles are only 0.25 mm long (Chown and Gaston 2010). Larvae of the beetle *Megasoma actaeon* commonly weigh 131 g (Christiansen 2006) approximately 3–5 times the mass of house mice. The Guinness Book of World Records holds that the heaviest male *Megasoma* on record weighed 228 g—the equivalent of a brown rat.

These facts undermine a common assumption that insects are all too small to potentially experience pain. In any case, sentience evolved in the ancestors of modern animals which, in the case of many insects, were considerably larger than their extant descendants. High atmospheric oxygen levels during the Carboniferous era (approximately 300 million years ago) are believed to have supported the existence of "gigantic" insects in the late Palaeozoic, with wingspans up to 71 cm (Chown and Gaston 2010). These facts do not support the notion that insects are, or have always been, "too small to feel pain".

Another common assumption is that insects simply produce massive numbers of offspring and invest very little in them, thus having no reason for the 'expensive' neural architecture required to produce pain (Adamo 2016). However, different insects can be oviparous, reproducing by egg-laying, or viviparous, giving birth to live young. Obligately oviparous termite queens can lay up to 11 million eggs a year, with each cared for by sibling-workers (social insects have the highest reproductive output per reproductive female). The obligately viviparous *Hippobosca variegata* fly develops one offspring in utero at a time, where the larva feeds on a 'milk gland' until live birth. The high maternal investment of these and many other viviparous species (a strategy documented in 11 orders; Visagie 1992; Benoit et al. 2015) means they often have relatively fewer offspring. *Hippobosca* females yield an average of 4.5 over the course of a lifetime (Englemann 1970). From these examples, it is clear that not *all* insects populate the next generation using masses of evolutionarily 'cheap' offspring; this suggests that at least some insects may well need the presumably 'expensive' cognitive machinery required to experience pain.

Insects possess senses beyond vision, olfaction and gustation. For example, some are capable of proprioception, mechanosensation, thermosensation, nociception and detecting humidity, vibrations, infrared (in fire-seeking beetles) and more. The brains that integrate these various sensory streams contain neurons numbering from a few thousand to nearly two million in some wasps (Godfrey et al. 2021; Polilov et al. 2023). However, the number may be greater in the brains of the aforementioned large-bodied insects (which are yet to be studied). Incidentally, neuron numbers are the same between some solitary wasp species and some smaller-bodied vertebrates, such as geckos (reviewed in Barrett and Fischer 2025). These neurons are presumably important for powering the fascinating behaviours and cognitive abilities of insects covered in the next section.

2.3 Behavioural and Cognitive Variation

The extraordinary behavioural and cognitive abilities of insects represent a vast and largely unexplored field of study, making any brief description inevitably inadequate. Insects may live alone, in large or small eusocial colonies, or use a range of subsocial strategies such as communal or aggregative living with co-operative behaviours (Costa 2006). Some insects are extremely polyandrous: female *Apis dorsata* queen bees can mate with over a hundred males (Wattanachaiyingcharoen et al. 2003). Conversely, termite pairs live monogamously for their entire lifespan, i.e. up to 20 years (Shellman-Reeve 1999). Monogamous, biparental care is also found in non-eusocial insects such as wood roaches (Suzuki 2013), some of which practice extended care throughout the entire, multi-year development of their young.

Some 'single' insects deposit and abandon their eggs while others (of either sex) care for their offspring. For example, male water bugs carry fertilized eggs on their backs, which renders the parent vulnerable to predation, hinders their foraging and prevents further matings until the eggs have hatched (Crowl and Alexander Jr. 1989). Female earwigs groom their eggs regularly to prevent infections (Boos et al. 2014), while webspinners (insects, not spiders) treat, organize, add silk to and then cover their egg masses to protect them from predation (Morinaga et al. 2023). Such examples of parental investment also undermine the argument that insects necessarily produce 'cheap' offspring.

Some insects fight over established territories for mates (Switzer 2004) or engage in complex courtship rituals involving dances, 'songs', physical touch, aphrodisiac-like substances or gift-giving (Frantsevich and Gorb 2006; Salvemini et al. 2010). While these behaviours can be innate, they can also be improved through experience (Saleem et al. 2014). Female choice is an understudied but important part of the diversity of insect mating behaviour (Peretti and Aisenberg 2015). Alternatively, sexual coercion and conflict are observed in some species, which can result in lethal injuries (Stutt and Siva-Jothy 2001; Burke and Holwell 2021). Insect species may never build a nest, single nest or multiple small nests (as in the solitary desert bee *Centris pallida*; Alcock et al. 1976). Some insects will live in the same log all of their lives, while others may migrate 3000 miles in a single season (e.g. the monarch butterfly). These examples reveal that insect behaviours may be tremendously sophisticated and flexible and that consequently, insects may encounter many novel environments or stimuli where they could plausibly benefit from subjective experience, including pain.

The cognitive abilities of social insects are relatively better studied than their solitary relatives and are impressive. They include 'counting' in bees (Dacke and Srinivasan 2008), numerical cognition that enables addition and subtraction in bees (Howard et al. 2019), abstract concepts of same and different in wasps and bees (Giurfa et al. 2001; Weise et al. 2022), facial recognition in wasps (Tibbetts et al. 2021), tool use in ants (Fellers and Fellers 1976), play-like behaviours in bees and flies (Galpayage Dona et al. 2022; Triphan and Huetteroth 2023), social information transfer (von Frisch and Seeley 1967; Leadbeater and Chittka 2007; Mirwan and Kevan 2013; Bridges et al. 2023), metacognitive-like processes (Perry and Barron 2013; Giurfa 2015), selective attention (Perry et al. 2017), emotion-like states and judgment bias (Bateson et al. 2011; Perry and Baciadonna 2017; Strang and Muth 2023) and self-medication (de Roode and Hunter 2019). An increasing field of insect research involves 'personality'—defined as consistent individual behavioural differences (Jandt et al. 2014).

Some ant species rescue kin trapped under stones, on water, under attack by predators or buried in soil, a behaviour that involves the communication of distress and other rescuer activities (Miler and Turza 2021). Following termite raids, *Megaponera analis* ants will rescue conspecifics that are 'treatable' and tend their wounds in the nest, increasing their chances of survival. Lightly injured ants adjust their behaviours to appear more injured near nestmates (Frank et al. 2018). Ants are the only known animal besides humans to engage in life-saving limb amputation to combat infection (Frank et al. 2024). In contrast to their care for kin are their negative interactions with ants from other colonies, e.g. the theft of resources and offspring from colonies of their own and other species (Guénard and McGlynn 2013; Apple et al. 2014). These examples reveal that insects can engage in complex, cognitively sophisticated behaviours despite the possession of small nervous systems—suggesting that the possession of small nervous systems does not preclude complex behaviours and the possibility of feeling pain.

3 Contexts of Use and Management

Insects are used and managed in numerous ways that may expose them to noxious stimuli.

3.1 Wild Insects

The Smithsonian Institute estimates there are 10 quintillion (10,000,000,000,000,000,000) insects alive on the planet at any one time, and, while this may not be especially accurate, it makes the important point that a very large number of wild insects exist.

In addition to ubiquitous and natural sources of noxious stimulation that all animals may face (e.g. starvation, predation/parasitism and disease), there are those arising from interactions with humans. For instance, populations of disease-carrying, invasive and agricultural/residential pest species are controlled with pesticides or disease-causing agents (which may have sublethal effects on nearby non-target insect populations; Desneux et al. 2007). One report estimates 3.5 quadrillion insects live on insecticide-treated farmland in the United States, each year (Howe 2019). Another report gives low-confidence estimates that between 100 trillion and 10 quadrillion insects may be killed by agricultural pesticides, and 1–10 trillion by residential pesticides, each year globally (Rowe 2020a). While no peer-reviewed estimates currently exist, it may be assumed that these reports are thematically correct: a *lot of wild insects are affected by pesticides,* which may plausibly cause pain through their sublethal—or lethal—effects.

Insects are mass-reared and released into the wild as parasitoids or predators in insect pest control initiatives (Kruitwagen et al. 2018). For the same reason, genetically modified insects may be bred and released as part of sterile insect programmes or gene drives (Simon et al. 2018; Vargas-Terán et al. 2021). Rowe (2020b) gives low-confidence estimates of 1–100 billion insects being used for sterile insect techniques annually. (Insects used for biological control probably number in the trillions, but include other non-arthropod invertebrates.) Reports in *National Geographic* suggest that some 20 million screwworm flies are irradiated and released from just one facility in Panama each year (and over one billion released in total; Bittle 2019), though this number may grow with the recent increase in the threat posed by screwworm flies in North America. Other indirect anthropogenic harms may affect wild insects through light pollution, research and conservation practices, and traffic collisions, amongst others (Barrett et al. 2023b).

3.2 Farmed Insects: Food, Feed, Fashion and Function

Insects are mass-reared: to feed humans, livestock and pets; for use as fishing bait; to produce silk; for biocontrol; and to produce products such as shellac, wax, traditional medicine, dyes for food and cosmetics, chitosan and fertilizer. They are also mass-reared for use in laboratories as biological models or for generating recombinant proteins used in producing, amongst other things, cultured meat (Hall 2023).

There are at least 2000 insect species that humans can eat. There are presumably many more with potential as feed for other mammals, birds or fish (Liceaga 2022). A single estimate for the number of food insects reared in 2020 is 1–1.2 trillion (Rowe 2020b).[3] The

[3]The industry's rapid growth (de Jong and Nikolik 2021)

industry has since received significant investment, and major growth is projected to meet demands, mainly for replacing fishmeal in aquaculture (de Jong and Nikolik 2021). An estimated minimum of 8 trillion insects will be farmed as food and feed by 2030. Using insect protein to replace 25% of the fishmeal currently used will require the production of an estimated additional 40–80 trillion insects (Rowe 2020b). The most popular insect species farmed for food and feed are crickets, flies and beetles.

- Crickets. *Acheta domesticus* (house cricket), *Gryllus assimilis* and *G. bimaculatus* (field crickets), and *Gryllodes sigillatus* (banded cricket).
- Flies: *Hermetia illucens* (black soldier fly) and *Musca domestica* (house fly).
- Beetles: *Alphitobius diaperinus* (lesser mealworm beetle) and *Tenebrio molitor* (yellow mealworm beetle) (Rumbos and Athanassiou 2021).

Other species include:

- *Lucilia sericata* (green bottle flies).
- *Drosophila melanogaster* and *D. hydei* (fruit flies).
- *Zophobas morio* (superworm beetles).
- *Galleria mellonella* (wax moths).
- *Locusta migratoria* and *Schistocerca gregaria* (locusts).
- *Rhynchophorus* spp. (palm weevils).
- *Blaptica dubia* (Dubia roaches) (van Huis and Tomberlin 2017).

Silkworm moths (*Bombyx mori*) were amongst the first domesticated insects and are currently reared for both human consumption and the fashion industry. To these ends, estimates suggest 420 billion to one trillion silkworms were killed in 2017 (Rowe 2021).

Cochineal insects (*Dactylopius* spp.) are wild-caught and farmed to produce carmine (a red dye used in cosmetics and foods). Reports estimate 4.6–21 trillion deaths attributable to these uses (Rowe 2020c).

Shellac production involves the estimated death of 1–100 trillion lac bugs (*Kerria* spp.) per year to produce colourant, food glaze and wood finish (Rowe 2020a).[4] This is a highly imprecise estimate as little information is available on the number of individuals used for each industry or on mortality rates before slaughter.

Western honey bees (*Apis mellifera*), bumble bees (*Bombus* spp.) and a variety of stingless bees (e.g. *Melipona beecheii*) and solitary bees (e.g. *Osmia lignaria*) are managed for honey, wax and pollination services. The latter two groups are usually involved in much smaller scale rearing operations (Villanueva-Gutiérrez et al. 2013). Rowe (2020b) estimates at least one trillion insects, mostly Western honey bees, are managed industrially for pollination purposes each year. Managed honey bees face challenges including pesticide exposure, colony collapse disorder, stressful transport, low available food quality (and quantity), a range of parasites and pathogens (e.g. American foulbrood and *Varroa* mites), light pollution, heavy metal pollution and heavy machinery collisions, depending on environmental (urban or rural) factors (Garrido and Nanetti 2019; Monchanin et al. 2022). Colony-dwelling animals present unique challenges in management, as rearing and euthanasia methods must be applied in 'batches' to the whole hive (see slaughter/depopulation below).

No estimates yet exist for the number of insects farmed for cultured meat production, biocontrol or gene drives, fishing bait or chitosan/fertilizer production.

probably underestimates this (single published) value for farmed insect numbers.

[4]Rowe (2020a, c) expresses low confidence in the number of individual insects estimated to be directly affected by human management and farming practices. Furthermore, reports are not peer-reviewed as no peer-reviewed data exists, which indicates a pressing need for further research in this area. That said, there is general confidence that Rowe's work accurately points to a massive use of insects.

3.3 Insects Reared for Educational and Research Purposes

No estimates are available for the number of insects used in education. These may be wild-collected, killed, dissected or used for other purposes in undergraduate/graduate entomology courses (Trietsch and Deans 2018). Alternatively, they may be purchased 'online' to be reared in grade school classrooms. In either setting, students may not receive adequate—or any—training on insect rearing, handling, anaesthesia or euthanasia (Barrett et al. 2024a).

Insects are studied under both laboratory and natural conditions in research on pest management, biomedical neuroscience, conservation and ecology, genetics and insect behaviour. Unlike vertebrates and cephalopod molluscs, insects are not protected by any legislation or regulatory guidelines anywhere in the world. There is no legal requirement regarding the methods of euthanasia, anaesthesia and/or analgesia for insects involved in research, as there is with protected species. Accidental neglect, regular depopulation (to maintain manageable stocks), extended methods of sacrifice and traumatic manipulation without anaesthesia are not uncommon in research settings (Perl et al. 2024).

Rowe (2020a) estimates (with low confidence) that 2–15 billion insects are used in scientific research annually. Many universities worldwide maintain very large, non-active stock populations of different genetic lines (e.g. University of Sheffield 2023) of *Drosophila*. Extremely rough calculations based on the number of labs studying fruit flies (> 3,300; Bilder and Irvine 2017), the lifespan of a fly from egg hatch to death (average 4 weeks to sacrifice), and a relatively conservative stocking density for both experimental and stock populations (120 flies + larvae/vial, 4 trays of 100 vials/lab) would suggest at least 2 billion flies are used each year (probably more). Additionally, many insects are sampled from wild populations, with numbers often in the thousands for a single survey period (Spears and Ramirez 2015). In summary, billions of insects are probably used annually in research.

3.4 Insectariums and Zoos, Entertainment, Pets, Art, Medicine, Forensics, etc.

Insects are used, alive or dead, in zoos, insectariums, museums, butterfly houses and for movies or TV shows (Rivers 2017). In zoos, insects may be 'on display' or reared to feed other animals. One estimate puts the number of pupae used in butterfly houses around the world at five to ten million per annum (Boppré and Vane-Wright 2012).

Hobby entomologists rear insects as pets, to exhibit or to sell, and there are an estimated five million insect pets in South Korea alone (Kim et al. 2008). Some may constitute part of a legal (or illegal) wildlife trade business (Olsen et al. 2021). Beetle shells and other insect products are used in art and fashion. Insects are also used in forensics for estimating the time of death of corpses, and in warfare as biological vectors for human pathogens or to destroy the enemy's crops (Lockwood 2010).

Maggots (*Lucilia sericata*) are used in medicine for treating chronic wounds (Mohd Zubir et al. 2020), although the numbers involved are unknown.

3.5 Key Uncertainties and Future Directions

Ultimately, there are no peer-reviewed estimates for the number of insects used or managed globally in any context. Better methods of tracking insect use in the major contexts listed above will be essential to accurately assess the animal welfare costs of management decisions in different settings—the first step to integrating insect welfare into decision making more broadly.

4 Pain in Insects

This section reviews the empirical evidence for insect pain. While this is sparse and distributed across many taxa and life stages, the empirical evidence against pain is weak. This significant paucity of evidence strengthens the case for more research. Even if it were accepted that insects may feel pain, there are few data that allow robust conclusions to be drawn about when or how this occurs on farms and elsewhere. However, there are practices in the rearing, slaughter and transport of farmed insects that might reasonably be expected to cause pain or suffering. It is hoped that some initial ideas from this chapter may thus serve as genuine first steps for answering the most pressing questions concerning the potential for farmed insect pain and suffering.

The previous section reveals the inappropriateness of regarding insects as a single animal form. Consequently, variation in all aspects of pain biology across different orders, species or even individual insects at different stages of development should not be dismissed (Fischer and Sandall 2023; Gibbons et al. 2022a). Despite this, discussions of insect pain have thus far been conducted at the level of *Insecta,* and there are insufficient data to add additional granularity beyond the level of order. Nevertheless, it is hoped that, in time, researchers will progress beyond answering the question, 'Can [all] insects feel pain?', to something more nuanced.

4.1 Changing Attitudes to Pain in Insects

In 1872, Charles Darwin wrote that "Even insects express anger, terror, jealousy, and love by their stridulation" in *The Expression of Emotions in Man and Animals.* His statement lacked evidence, and while some entomologists were tentatively sympathetic to the inherent uncertainty of insect pain (Wheeler 1910; Weiss 1914), the idea quickly lost favour in light of subsequent observations that some insects appear completely unresponsive to traumatic injuries (Bastin 1927). By 1984, the idea that insects had a near-negligible chance of experiencing pain had become the standard view and was summarized in perhaps the most influential paper of the twentieth century on the subject (Eisemann et al. 1984).

Eisemann's case rested on six pieces of neurobiological and behavioural evidence, summarized in Barrett and Fischer (2024): "(1) insects lack nociceptors; (2) endogenous opioid peptides and their receptor sites do not necessarily indicate the capacity for pain perception; (3) because insects have simple brains with relatively few neurons compared to mammals, they cannot generate the experience of pain; (4) insects do not protect injured body parts; (5) insects often behave normally after severe injuries; (6) insect behaviour is inflexible in a way that suggests that pain would have little adaptive value for them". As we will see below, many of these pieces of evidence are no longer considered empirically or theoretically accurate, given the advances in insect neuroscience and behaviour in the last 40 years. Further, even at the time, not all authorities were in agreement (see: Lockwood 1987, 1988; Smith 1991).

In recent decades, the appearance of additional genetic, behavioural, neurobiological and physiological evidence has rekindled the debate on insect pain and weakened scientific certainty once again (Elwood 2011; Sneddon et al. 2014; Tiffin 2016; Klein and Barron 2016; Adamo 2019, 2016; Elwood 2016; Burrell 2017; Drinkwater et al. 2019; Fischer and Larson 2019; Gibbons and Sarlak 2020; Overgaard 2021; Gibbons et al. 2022a, b).

There is no current expert consensus on the probability of insect pain. However, accumulating evidence increasingly questions the assertion that the chance of pain in insects is negligible (Eisemann et al. 1984). For example, the *Oxford Handbook of Invertebrate Neurobiology* (2019) recognized the complexity of *morally-relevant* invertebrate consciousness following the conclusions of the 2012 Cambridge Declaration on Consciousness, writing:

> *If, therefore, the expression of consciousness is present in cephalopods and exists on a continuum within the invertebrates, where should we draw the line to divide those worthy of protection from those that are not? This extremely thorny question will become an increasingly urgent matter for invertebrate neuroscientists…* (Gelperin 2019; p. 18)

Similarly, the Animal Behaviour Society's Animal Care Committee guidelines recognize the ethical uncertainties of insect use, obliquely calling for further information on the ethical treatment of unregulated species, and referencing publications on insect research ethics (Fischer and Larson 2019; ASAB Ethical Committee/ ABS Animal Care Committee 2023).

The Insect Welfare Research Society created the first guidelines for protecting and promoting insect welfare in research in 2023 (Fischer et al. 2023). The Royal Entomological Society has also started a special interest group on Welfare and Ethics (Harrington 2023) and produced a statement recognizing the plausibility of insect sentience in 2025, recognizing increased need to consider their ethical treatment in research (Royal Entomological Society 2025). And in 2024, the New York Declaration on Animal Consciousness (signed by over a hundred comparative cognition experts, consciousness scientists, neuroethologists and philosophers of mind), affirmed:

> *the empirical evidence indicates at least a realistic possibility of conscious experience in all vertebrates (including reptiles, amphibians, and fishes) and many invertebrates (including, at minimum, cephalopod molluscs, decapod crustaceans, and insects)... when there is a realistic possibility of conscious experience in an animal, it is irresponsible to ignore that possibility in decisions affecting that animal. We should consider welfare risks and use the evidence to inform our responses to these risks.* (Andrews et al. 2024)

The following sections review the current status of the neurobiological and behavioural evidence for and against pain in insects. The evidence for pain is presented using the invertebrate sentience framework proposed in Crump et al. (2022), which is an adapted version of an earlier vertebrate framework (Smith and Boyd 1991). The framework is not intended to *prove that insects definitely feel pain* (see Fischer et al. 2025); rather, it summarizes the strength that an empirical case for pain might possess, based on eight categories of evidence that are consistent with the hypothesis for pain (Table 1). For the most thorough review of insect pain, citing over 350 studies across six insect orders (Blattodea, Coleoptera, Diptera, Hymenoptera, Lepidoptera, and Orthoptera)[5] at both adult and juvenile life

Table 1 Eight criteria required as evidence for animal pain perception (Crump et al. 2022)

1. The organism possesses receptors sensitive to noxious stimuli, i.e. nociceptors.
2. The organism possesses brain regions capable of integrating information from different sensory sources.
3. The organism possesses neural pathways connecting nociceptors to integrative brain regions.
4. The animal's behavioural response to noxious stimuli is modulated by chemicals affecting the nervous system in either or both of the following ways: (a) the animal possesses an endogenous neurotransmitter system that modulates (in a way consistent with the experience of pain, distress or harm) their responses to threatened or actual noxious stimuli. (b) Putative local anaesthetics, analgesics (such as opioids), anxiolytics or antidepressants modify an animal's responses to threatened or actual noxious stimuli in a way consistent with the hypothesis that these compounds attenuate the experience of pain, distress or harm.
5. The organism shows motivational trade-offs, in which the disvalue of a noxious or threatening stimulus is weighed (traded-off) against the value of an opportunity for reward, leading to flexible decision-making. Enough flexibility must be shown to indicate centralized, integrative processing of information involving an evaluative common currency.
6. The animal shows flexible self-protective behaviours, e.g. wound tending, guarding, grooming and rubbing, of a type likely to identify the location of noxious stimulation.
7. The organism shows forms of associative learning in which noxious stimuli become associated with neutral stimuli or in which novel ways of avoiding noxious stimuli are learned through reinforcement. These forms of associative learning go beyond classical conditioning in which a single-conditioned stimulus overlaps temporally with an unconditioned stimulus.
8. The organism shows that they value a putative analgesic or anaesthetic when injured, in any or all of the following ways: (a) Learning to self-administer putative analgesics or anaesthetics when injured. (b) Learning to prefer, when injured, a location at which analgesics or anaesthetics can be accessed. (c) Prioritizing the procurement of these compounds over other needs, e.g. food, when injured.

[5] Blattodea: cockroaches and termites; Coleoptera: beetles; Diptera: flies and mosquitoes; Hymenoptera: bees, ants, wasps and sawflies; Lepidoptera: butterflies and moths; Orthoptera: crickets and grasshoppers.

stages and following the same framework, see Gibbons et al. (2022a).

4.2 Neurobiological Evidence

4.2.1 Possession of Nociceptors

Eisemann et al. (1984) describe "the apparent absence from insects of any known or likely candidate nociceptors" (see Acute Pain chapter). However, the review in Gibbons et al. (2022a) confirmed the presence of nociceptors or ion channels with roles in sensing noxious stimuli in several insect orders identified using genetic, functional and imaging data (and see Borjon et al. 2025; Goldberg et al. 2024). Indeed, insect peripheral sensory cells contain ion channel types that are homologous or orthologous with vertebrates, e.g. Transient Receptor Potential [TRP] channels. The vertebrate and invertebrate genetic mechanisms for regulating acute and persistent pain perception are highly conserved (Khuong and Neely 2014).

Class IV multidendritic neurons that respond exclusively to noxious stimuli are found from the first instar, i.e. immediately after egg hatch, in larval fruit flies (Hwang et al. 2007; Lopez-Bellido et al. 2019). The branched dendritic morphology extending below the surface of the larval epidermis is reminiscent of vertebrate nociceptors (Tracey et al. 2003). Tobacco hornworm moth larvae also have anatomically similar sensory cells responding to noxious heat and mechanical stress (Grueber et al. 2001; Caron et al. 2020). Adult fruit fly sensory neurons with the ion channel *pickpocket* are responsible for mediating responses to noxious heat (Khuong et al. 2019). Altogether, Gibbons et al. (2022a) present strong evidence of nociceptive ion channels or nociceptive neurons in adult and juvenile insects of the six orders they reviewed.

Possession of Integrative Brain Regions

There is neurobiological evidence that insects are capable of integrating multiple kinds of sensory information in higher order brain regions to generate a unified stream that could be termed experience. While there is agreement that the presence of integrative brain regions raises the probability of sentience, there is disagreement about which kinds of integration matter (Mudrik et al. 2014; Barron and Klein 2016; Mashour et al. 2020). Three sensory integrative brain regions are present in all adults of the insect taxa reviewed in Gibbons et al. (2022a). These three regions are: the mushroom bodies, the central complex and the lateral horn (Thiagarajan and Sachse 2022). The mushroom bodies have roles in learning, memory and sensory integration. The central complex has roles in navigation, memory and descending control of locomotion. The lateral horn has a possible role in innate olfactory-driven behaviour but also receives other sensory inputs (Gibbons et al. 2022a).

Some hypotheses for sentience require connections *between* sensory integrative brain regions, although this is not required under the Crump et al. (2022) framework. These arguments are more advanced iterations of the 'nervous system simplicity' theme posed by Eisemann et al. (1984), who argued that insects have too simple a nervous system, with too few neurons, to form a pain experience (but see Barrett and Fischer 2025 for a rebuttal to this, and other, Eisemann claims). The 'neuron number' argument still resurfaces occasionally, but now largely focuses on comparing connectivity in insect and mammalian brains. For example, Adamo (2019) posits that insect pain is unlikely because of the "apparent lack of a higher-order network combining brain regions involved in nociception, e.g. the mushroom bodies and central complex...... there are no direct connections between them". The lack of direct connections in insects would suggest parallel, instead of integrated, processing of sensory information in these regions, where integrated information processing *across regions* is expected to be important for producing pain experiences in vertebrates; thus, parallel processing only could suggest pain experiences are implausible in insects.

However, the first connectome of the adult fruit fly mushroom bodies found direct connections between the mushroom bodies and the central complex (Li et al. 2020). Importantly, electric shocks activate neurons in both the fan-shaped body of the central complex and the mushroom bodies, with the fan-shaped body being required for escape behaviours (Hu et al. 2018a). While both sensory-integrative brain regions are activated in response to noxious stimuli *and* are directly connected, it remains to be seen if there is cross-region integration of nociceptive sensory information processing (as per the argument in Adamo 2019). However, this example reveals the hazards of accepting claims using an absence of evidence as evidence for absence.

Possession of Connections Between Nociceptors and Integrative Brain Regions

Nociception as a peripheral sensory input probably reaches higher integrative brain regions largely or exclusively via the ventral nerve cord—or the insect version of the vertebrate spinal cord (Tsubouchi et al. 2017). The connections between nociceptors and integrative brain regions have been studied in the two insect species that have become recognized models for nociception research: *Periplaneta americana* (the American cockroach) and *D. melanogaster.* In adults of both species, ascending nociceptive pathways from the periphery to the aforementioned sensory integrative brain regions have been found and largely mapped (Hu et al. 2018a; Emanuel and Libersat 2019; Khuong et al. 2019; Rana et al. 2022), which meets Crump's third criterion. And, the brain itself has been shown to be critical for modulating nocifensive responses in caterpillars (Kondakath et al. 2025).

Nocifensive Responses Amenable to Chemical Neuromodulation

Nocifensive behaviours in insects can be modulated by endogenous or exogenous chemical compounds that affect the nervous system in ways that support the possibility that insects feel pain. Endogenous neurotransmitters (such as gamma-aminobutyric acid, or GABA) as well as analgesics, anaesthetics (local), anxiolytics and antidepressants also affect the "pain behaviours" of insects (see Jang et al. 2023 for a particularly compelling example of nocifensive neuromodulation with opioids, gabapentin, and NSAIDs). Unfortunately, most data have been derived from the fruit fly, while some pharmacological data have been collected from cockroaches (Gibbons et al. 2022a). Specific analgesics and anaesthetics are reviewed later in this chapter.

4.3 Behavioural Evidence

The classical case against insect pain has often been based on the lack of vertebrate-typical behaviours in response to traumatic injury (Bastin 1927; Eisemann et al. 1984). For example, ants with 'broken' legs appear to walk normally, while male mantids continue to mate while being consumed. Cockroaches and ants have been observed eating 'normally' even when most of their body or abdomen is lost, while locusts may eat their own intestines if their abdomen is punctured. These observations are at first compelling, but ultimately depend on the reasoning behind the expectation that injured insects *should* behave like injured vertebrates. Further, some of these arguments are based on flawed methodology. As this is the strongest and *oldest* behavioural case against insect pain, it is worth considering these theoretical and methodological concerns at length (and see Barrett and Fischer 2025).[6]

1. *Body composition.* It is not unreasonable to expect that responses to traumatic injury would differ between animals with exo- or endoskeletons. While insects form scar tissue at injury sites (Burke and Holwell 2021), they do not necessarily require the continual behavioural modifications, e.g. limb conservation—manifest as "limping"—to aid in healing. Vertebrate-type responses to physical injuries should thus not always be expected in insects.
2. *Biological scaling.* Nocifensive responses probably evolved to offer a fitness-relevant protective benefit; damaged mammalian

[6] Gibbons et al. (2022a) describe further reasons, beyond those listed in this chapter, that counter the claim of behavioural insensitivity to damage as promoted by Eisemann (including behavioural manipulation by parasites; hunger-induced neuromodulation to noxious stimuli in vertebrates and invertebrates and hyperalgesia).

limbs heal more rapidly and correctly when less force is applied to them. However, many insects weigh very little. Therefore, it seems possible that the forces imposed on the damaged limb of a walking ant are insufficient to warrant protective locomotor modifications. (For similar reasons, the consequences of ants and humans falling from a great height are different: the *physics* of being small and covered in chitin greatly affects species-appropriate responses.)

3. *Pain response modulation by species-specific motivations*. Numerous examples exist of humans and mammals enduring traumatic injury yet behaving 'situationally normal' to objective observers, e.g. injured rugby players. In these species, motivational drives clearly cause nervous system modulation to the extent that other behaviours become more prominent than the expected response to injury. This occurs in other invertebrates *and* vertebrates in the context of feeding and mating (Gibbons and Sarlak 2020).
4. *Anecdotal evidence quality*. Most observations of insects not responding to injury are largely anecdotal. For example, reporting that ants 'walk normally' with leg injuries offers no data beyond visual observation. Objective measures, e.g. forces applied by the affected limb, gait analysis or motion tracking have not been used (Gibbons et al. 2022a).
5. *Prevalence of behaviours*. The prevalence of insect behaviours observed in response to injury, e.g. self-cannibalism, is generally neither reported across nor within species. In some species, these behaviours may be rare[7] but, being strange to witness, create a bias towards remembering and reporting them. These observations may not be representative of how most insects respond to injury and assumptions on the behaviour of all insects should not be based on the observed behaviour of a few. Alternatively, they may be typical responses in the observed but not other (un-observed) contexts.
6. *Insect responses to traumatic injury*. Behavioural changes *have* been observed in injured insects (Gibbons and Sarlak 2020; Gibbons et al. 2022a), and despite the widely held assumption that injured insects will continue to mate, male crickets receiving a strong pinch to the epiphallus stopped mating for an unexpectedly long time before renewing courtship attempts, compared to uninjured males (Sakai et al. 1991). In another example involving fruit flies, peripheral nerve injury caused by leg amputation resulted in neuropathic allodynia, i.e. increased sensitivity to previously innocuous temperature stimulation (see Physiology of Chronic Pain Chapter), a method now used as one of the several 'chronic pain' paradigms in insects (Khuong et al. 2019). In short, there is now evidence undermining the claim that all insects do not respond to traumatic injury in any context.
7. *Avoidance of potential injury*. Some insects appear to take measures to avoid traumatic injury. For example, male mantises of some species appear to act cautiously when approaching females, presumably to avoid the 13–28% probability of being eaten (Brown and Barry 2016). In mantis species where cannibalization is largely inevitable, males forego caution and fight back, to the extent that female abdomens may become pierced by the male's raptorial fore-tibial claws, leading to haemolymph loss and scar tissue formation while risking her (and thus his own) fitness in the process (Burke and Holwell 2021).
8. *There are many causes and kinds of pain*. Insects appear to respond to noxious injuries beyond limb trauma. Indeed, the Eisemann et al. review was preceded by evidence that insects respond nocifensively to shock and heat, if not mechanical injury (Wigglesworth

[7] For instance, in a study on humane slaughter via grinding for farmed black soldier flies by the author of this chapter (Barrett et al. 2024b), 200 individual larvae were cut in half on body segments 9–12, where they could be expected to reach their open body cavity with their mouthparts, to replicate the effects of the blade in a commercial grinder. Behaviours were observed for either 90 s (n = 160) or 5 min (n = 40) after the cut. Out of the 26,400 s of observation across 200 larvae, only 1 larva, for about 45 s, 0.17% of the total observation time, attempted to eat its own body.

1980). This includes stimuli that might be expected to be more fitness-relevant than mechanical disruption of the exoskeleton—such as noxiously high heat (see Veit 2022). Evidence of insect responses to shock, chemicals, and heat are especially well documented; in these, insects respond much more like vertebrates by recoiling immediately from, or escaping, the noxious stimuli (Im and Galko 2012; Borjon et al. 2025). Such robust, repeatable observations in many insect species and life stages form the basis of numerous experimental paradigms on which the entire field of insect nociception rests.

As demonstrated above, this initially intuitive, behavioural case (of apparent unresponsiveness to traumatic damage in insects) is no longer empirically or philosophically supportable.

4.3.1 Motivational Trade-Offs

Behavioural evidence for insect pain can be found in studies of motivational trade-offs, in which insects 'trade-off' the value of aversive and rewarding stimuli (for instance, high heat and a rewarding food source). This also represents one of the fitness benefits that pain experiences (as compared with nociceptive reflex responses) may be expected to provide, i.e. that affective states can serve as a way for animals to integrate and then flexibly modulate their responses to a complex environment containing numerous aversive and rewarding stimuli. (For more on this theory of affective states as 'common currency', see Crump et al. 2022).

There are few studies on motivational trade-offs in insects. In one, fruit flies conditioned to associate odours with a sucrose reward would "risk" being electrocuted with 100 volts, but would not be tempted by an ethanol reward when 120 V were used (Kaun et al. 2011). In another, bumble bees used memory to flexibly trade-off avoiding noxious heat with a preference for sucrose (Gibbons et al. 2022c, but see Read and Nityananda 2025). These data suggest nociceptive motivational trade-offs are possible in at least some insects and add to the case for complex affective states like pain instead of simply basic nociceptive reflexes.

Flexible, Self-Protective Behaviours Directed at the Site of Injury

Crump et al. (2022) also consider examples of flexible, self-protective behaviours related to the injury site to provide some evidence consistent with the hypothesis of pain. Some insects, e.g. the tobacco hornworm moth larvae, demonstrate wound-tending behaviours, while cockroaches groom abdominal puncture wounds (Hentschel and Penzlin 1982; Walters et al. 2001). A recent study in adult bees shows that they will groom their antennae in a site-specific manner when they receive a noxious injury from a heated probe (Gibbons et al. 2024). Whilst the fitness-relevance of these behaviours is not established, an allogrooming study in ants showed that wound-tending *can* increase survival under the non-sterile conditions in which insects evolved (i.e. not in the laboratory; Frank et al. 2018).

Nociceptive Associative Learning

With respect to nociceptive associative learning, Crump et al. (2022) accept a range of learning patterns beyond habituation or sensitization as evidence consistent with the hypothesis of pain perception. However, the kinds of associative learning that provide evidence for pain are unestablished and controversial. Some popular theories rely on instrumental conditioning, reversal learning and trace conditioning. When considering a range of classical and operant conditioning paradigms, there is much basic evidence for nociceptive associative learning in 5 of 6 orders reviewed in Gibbons et al. (2022a). For example, a study of trace conditioning in fruit flies, pairing shapes with aversive heat, found flies could learn the association with trace intervals of 20 s (though without learning the interval duration). Distracting air puffs impaired trace, but not delay, conditioning (Grover et al. 2022). Basic reversal learning with nociceptive stimuli has also been demonstrated in insects, including cockroaches and fruit flies (Gibbons et al. 2022a).

Analgesia Preference

The final piece of behavioural evidence consistent with the hypothesis for pain is when injured animals demonstrate they value, and thus seek out, putative analgesics, for example, through self-medication or conditioned place preference paradigms. Only one study, Groening et al. (2017), has attempted to test analgesia preference in injured insects. Anaesthetized bees had a leg cut off and, after some period of recovery, were simultaneously offered morphine-sucrose and sucrose-only solutions; there was no difference in consumption of either solution in injured bees compared to controls. However, there are several reasons why this does not provide strong evidence against analgesia preference in insects. First, insects do not possess opioid receptors, which means that opioids may not be analgesic in insects (Mirabeau and Joly 2013). Second, bees find the taste of morphine aversive and thus simply may have preferred the sucrose solution over a 'bitter' morphine solution. Third, there had been no prior conditioning, meaning that bees may not have learned to associate analgesic effects with drinking from the morphine feeder. Indeed, an association may have formed with analgesia and the sucrose feeder, given morphine's slow onset times. The question of analgesia preference deserves further empirical attention in insects.

Table 2 summarizes the neurobiological and behavioural evidence for adults from Gibbons et al. (2022a). Overall, the table shows that no adult insect convincingly fails to meet any criterion. Furthermore, the evidence consistent with the hypothesis of pain is strongest in adult flies and cockroaches, which meet six out of eight criteria to a high or very high degree of confidence.

In insects, as in vertebrates, it is difficult to identify at which point during development sentience, and thus the capacity to feel pain and suffer, arises. In insects, the divergent developmental strategies discussed earlier in this chapter are relevant: holometabolous insect larvae have radically different nervous systems than the adults of their species and occupy different ecological niches. In contrast, the brains and behaviours of hemimetabolous insect juveniles (nymphs) are, largely, smaller versions of the adults of their species. In general, the data for insect juveniles summarized by Gibbons et al. (2022a) was incomplete and rarely included the entire development period from first to last instar. These data are summarized in Table 3, which reveals there is less evidence consistent with the hypothesis for pain perception in juveniles compared with adults of all orders though a lack of research attention also played a large role.

4.4 Key Uncertainties and Future Research

Tables 2 and 3 reveal the extent of uncertainty that exists in response to the question 'do insects plausibly feel pain?'. All cells stating 'no research found' represent an area in which future research is needed and weaken the confidence with which it can be affirmed that insects may—or may not be—sentient and feel pain. Future research on taxa that are strongly divergent in developmental strategy, neurobiology or behaviours could benefit an understanding of how broadly distributed the capacities reviewed in Gibbons et al. (2022a) actually are amongst insects. Still, generalizing from a few well-studied cases to other animals in the same class is common practice in precautionary animal welfare policy or practice. Therefore, it is not necessary for all this data to be existence for every insect species before precautionary actions can be taken to prevent potential insect pain in some contexts of use.

Meeting the eight criteria of Crump et al. (2022) suggests that insects may perceive pain but does not definitively prove pain or sentience. Studies, like those outlined in Mason and Lavery (2022) on self-report, forms of operant conditioning that seem to require conscious processing, working memory and episodic memory/self-recognition, in the context of noxious stimuli, would also greatly increase consensus on the plausibility of insect sentience. Further, understanding the potential for other emotions in insects might also increase confidence in the affective dimension of pain by increasing the likelihood of subjective experience or conscious awareness more broadly (for a review, see: Perry and Bacciadonna 2017). Ultimately, using several readily-determined proxies to determine the existence of subjective things that cannot be

Table 2 Summary of the evidence consistent with the hypothesis of pain perception in six insect orders at the adult life stage using the Crump et al. (2022) criteria; table from Gibbons et al. 2022a

Adults	Criterion 1: Nociception	Criterion 2: Sensory integration	Criterion 3: Integrated nociception	Criterion 4: Analgesia	Criterion 5: Motivational trade-offs	Criterion 6: Flexible self-protection	Criterion 7: Associative learning	Criterion 8: Analgesia preference
Blattodea	H	VH	VH	H	VL (no research found)	H	VH	VL (no research found)
Coleoptera	VH	VH	VL (no research found)	VL (no research found)	VL (no research found)	VL (no research found)	VL (no research found)	VL (no research found)
Diptera	VH	VH	VH	VH	H	VL (no research found)	VH	VL (no research found)
Hymenoptera	VH	VH	VL (no research found)	M	H	VL (no research found)	VH	VL
Lepidoptera	VH	VH	VL (no research found)	VL (no research found)	VL (no research found)	VL (no research found)	VH	VL (no research found)
Orthoptera	H	VH	VL (no research found)	L	VL (no research found)	VL (no research found)	VH	VL (no research found)

VH = Very high; H = High; M = Medium; L = Low; VL = Very low; VL (no research found) = Very low (no research found).

observed, as in the Crump et al. 2022 framework, is not the same as certain proof of the subjective thing itself. However, these proxy-based approaches remain the only option in all non-human animals until it is possible to measure pain and sentience directly.

5 Potential Pain and Suffering in Farmed Insects

The scale of the farmed insect industry means that, despite uncertainty concerning insects' ability to feel pain, a precautionary position is justified and, indeed, even proposed by the industry itself (IPIFF 2019). Unfortunately, attempting to assess what may cause 'pain' or 'suffering' in farmed insects is complicated by the absence of validated indicators (Johnson and Barrett 2025). Tentative progress is possible by considering a priori what may cause pain or suffering. It is known that insect nociceptors respond to acute noxious heat, cold, electrical shocks and chemicals, in all species studied to date, and that they respond to these stimuli in ways that demonstrate the stimuli are aversive (Gibbons et al. 2022a). It is thus possible that these stimuli applied during rearing or slaughter may be painful.

5.1 Rearing

Currently available comprehensive reports on farmed insect welfare involve honey bees (Garrido and Nanetti 2019), black soldier flies (Barrett et al. 2023a), black soldier flies and house flies (Kortsmit et al. 2023), yellow mealworms (Barrett et al. 2023c) and crickets (Rowe et al. 2024). However, no studies explicitly examine factors causing pain or suffering in farmed insects. The following are some potential welfare concerns affecting farmed insects, with all information sourced from the aforementioned reviews.

Table 3 Summary of the evidence consistent with the hypothesis of pain perception in six insect orders at the juvenile life stage using the Crump et al. (2022) criteria; table from Gibbons et al. 2022a

Juvenile	Criterion 1: Nociception	Criterion 2: Sensory integration	Criterion 3: Integrated nociception	Criterion 4: Analgesia	Criterion 5: Motivational trade-offs	Criterion 6: Flexible self-protection	Criterion 7: Associative learning	Criterion 8: Analgesia preference
Blattodea	H	H	H	VL (no research found)	VL (no research found)	VL (no research found)	VL (no research found)	VL (no research found)
Coleoptera	H	L	VL (no research found)	VL (no research found)	VL (no research found)	VL (no research found)	VL (no research found)	VL (no research found)
Diptera	VH	VH	VL (no research found) / M	VH	VL (no research found)	VL (no research found)	VL (no research found) / VH	VL (no research found)
Hymenoptera	H	VL / M	VL (no research found)	VL (no research found)	VL (no research found)	VL (no research found)	VL (no research found)	VL (no research found)
Lepidoptera	VH	L / M	VL (no research found) / L	VL (no research found)	VL (no research found)	VL (no research found) / H	VL (no research found) / VH	VL (no research found)
Orthoptera	H	H	VL (no research found	VL (no research found)	VL (no research found)	VL (no research found)	VL (no research found)	VL (no research found)

When cells are split for a criterion, left cell indicates the first instar and right cell indicates the last instar juvenile. VH = Very high; H = High; M = Medium; L = Low; VL = Very low; VL (no research found) = Very low (no research found).

Inadequate Nutrition/Hydration Larval black soldier flies (BSF) may be fed on diverse waste substrates, e.g. manure, restaurant/food waste, etc., and larval yellow mealworms can consume many plastics/polymers (though these are rarely provided in the feed by North American or European farmers due to safety regulations). These sources may not always represent adequate nutrition and cause slow growth, behavioural changes and mortality. While adult farmed BSF have the morphological and physiological capacity to eat and absorb nutrients (and see data on their preferences in Barrett et al. 2025), they are generally not provided food and instead starve (if not culled first). Inadequate hydration appears to be less common, but can occur in mealworms when attempts are made to rear them without wet feed, which may be done to reduce the likelihood of fungal infections that occur with moist substrates.

Cannibalism/injury/Aggression Injurious aggression is encountered in farmed crickets that compete over resources like shelters. Such injuries increase the likelihood of disease, as the intact cuticle represents the first physical barrier to pathogen infection. Cannibalism can also cause or exacerbate existing injuries and is common in both crickets and mealworms. Depending on the species, cannibalism may be caused or partially driven by inadequate hydration, nutrition or excessive stocking densities. The relationship between traumatic injury and pain in these species is unclear.

Overheating Overheating occurs when climate conditions are not carefully managed. This is challenging in large facilities where microhabitats may vary and are difficult to monitor. Overheating is most likely when large numbers of larvae are reared in moist, dense substrates that partially insulate the rapidly growing larvae. Lethal overheating has been observed in both mealworms and black soldier fly larval aggregations. Given that insects are particularly responsive to noxious heat and attempt to escape areas of high temperature, this experience is likely to induce putative suffering.

Toxins Numerous fungal, bacterial or artificial toxins that enter production facilities on agricultural products intended for insect feed can cause feed avoidance, reduced growth rates, increased oxidative stress and even death. Sublethal pesticide doses affect cognitive ability, behaviours and physiology in insects. Toxins are known to limit production and so are generally avoided by producers. However, accidental contamination remains possible. Additionally, mealworms are being explored for the reclamation of mycotoxin-contaminated grains. If the levels of specific mycotoxins are sufficiently high, they may induce prolonged death for the mealworms, but survival generally is unaffected, and sublethal effects are more common.

Suffocation/Drowning Black soldier fly larvae are reared in moist substrates to prevent desiccation. However, when substrates are too wet or deep, larvae may suffocate from a lack of air flow. Larvae generally attempt to escape substrates that provide inadequate living conditions (be it low oxygen, improper pH, etc.); however, pans are designed to prevent larval escape, which can result in larvae suffocating if there are pockets of standing water. Finally, poor air flow at high densities could lead to the accumulation of ammonia, which might also lead to hypoxic conditions for larvae. Low oxygen levels induce hyperventilation and loss of spiracular control and can lead to the accumulation of metabolic by-products; death may take hours when insects are asphyxiated due to low oxygen levels.

Disease Disease is probably amongst the greatest welfare problems facing farmed insects; they can spread quickly, and new diseases are constantly being discovered (e.g., Walt et al. 2024). Some have no detectable clinical impact. Some kill farmed insects rapidly, but others cause prolonged deaths, with organ rot, sepsis, paralysis, starvation, slowed growth, morphological defects like twisted wings and other behavioural and physiological changes occurring before death. Non-lethal diseases may also exacerbate other welfare problems—for instance, reducing activity to levels that facilitate injury, predation and cannibalism.

Handling-Stress Some insects are stressed by vibration and handling, which can mimic the cues that signal attack by a predator. Honey bees (Bateson et al. 2011) and crickets are particularly sensitive to shaking, yet are regularly transported by road for pollination services or as exotic pet food. In some cases, significant mortality can result, for instance, reports of 'shipping sickness' amongst cricket producers. Losses of 10–30% are not uncommon after the transport of honey bees in the United States (Johnson 2010).

5.2 Slaughter/Depopulation

Slaughter methods used on insects depend on species, context and the desired characteristics of the animal-based products. For example, silkworm pupae are often boiled in order to preserve the silken cocoons, while black soldier fly larvae may be sold live as poultry feed or microwaved to produce a puffed appearance and crispy texture.

Slaughter methods used for farmed insects include: freezing in air, freezing in liquid nitrogen, asphyxiation, microwaving, shredding/grinding, boiling/blanching, sand roasting, drowning and baking (in the sun or in convection ovens). Freezing in air, baking and boiling/blanching may be especially common, although there is no information on the numbers dispatched by each method. Currently, these methods are typically used without pre-stunning, analgesics, or anaesthesia. The intensity of pain experienced by insects killed with these methods is speculative, although some data exist for the effect time, i.e. from onset of technique application to death. For example, freezing in liquid nitrogen, boiling/blanching, and grinding/shredding may be instantaneous when specific standard operating procedures are followed (as in Zacarias et al. 2025). However, significant unknowns exist for the welfare of insects killed by chilling or asphyxiation. Invertebrate veterinarians do not recommend cold as an insect anaesthetic or acceptable euthanasia technique (see references in Barrett et al. 2023a for all prior information in this section, and Cooper 2011).

Large groups of insects may need to be killed (depopulated) during disease outbreaks or culled (in the case of adult breeders with waning reproductive performance). Depopulation may be complicated when (i) insects of a range of body sizes cannot be accommodated by normal slaughter procedures; (ii) difficulty is encountered separating insects from substrate, or (iii) it is preferred not to contaminate the processing equipment. As culling usually involves adult insects, increased mobility (like flight) can make it difficult to achieve "uniform death". Currently, freezing in air appears to be the most common method of culling insects, though data are sparse on actual practices.

The depopulation of managed honey bees is reasonably well-described because disease outbreaks necessitate legally-mandated hive destruction. After sealing the hive, sulfur dioxide (SO_2) or the fumes from the combustion of unleaded petrol or diesel creates anoxic conditions that will destroy the entire hive in approximately 15 min. Dry ice and 70% isopropyl alcohol also cause the rapid death of bees on contact. Soapy water disrupts the wax layer of the insect cuticle and allows large water volumes to be absorbed, resulting in osmotic shock (Mutinelli 2020). An evaluation of different euthanasia methods in honey bees, considering effect time, cost, complexity, operator risk and other disadvantages, recommended SO_2 as the best method (acting within 0–10 min). A standard methodology based on SO_2 use has been devised that veterinarians can recommend to beekeepers intent on using this method (Roy and Vidal-Naquet 2022).

5.3 Key Uncertainties and Future Research

Most studies currently investigating farmed insect welfare, pain, or suffering during rearing typically rely on mortality or growth rates as the only metrics of 'welfare'. These are imprecise and so of limited use for detecting *severe* problems in populations during, for example, disease outbreaks. Importantly, they fail to address welfare at an individual level. Consequently, there is a need to refine the quantification of pain and suffering in farmed insects and validate behavioural and physiological proxies. Potentially useful indicators for assessing pain and suffering in insects include escape behaviours, preference/avoidance behaviours, injury presence, disease/parasite loads, changes in activity levels, gene expression patterns, oxidative stress, stress hormone levels, longevity and body mass or composition changes (Johnson and Barrett 2025). Others may be discovered as the industry continues to innovate.

Additional research is needed on stunning and slaughter methods. Electrical stunning is commonly employed in vertebrate livestock slaughter and is possible in insects. Whether specific methods of slaughter or depopulation cause insects to feel pain and suffering is not understood. Freezing in air and asphyxiation both cause unconsciousness before death, but it is possible that significant suffering occurs before consciousness is lost. Establishing slaughter methods that protect insect welfare, are feasible at scale, and ensure product quality should be an industry priority, especially as there is little information available on current producer practices. Greater collaboration between welfare scientists, entomologists and producers is required to address this challenge (Barrett and Fischer 2023a).

6 Anaesthetics and Analgesics

This section reviews what is known about anaesthetics and analgesics in insects. Particular emphasis is placed on pain in farmed and experimental animals.

The analgesic potential of numerous compounds has been tested in insects although most underlying mechanisms have not. Consequently, doubt remains as to whether ostensibly effective analgesics truly obtund the insect's pain experience or whether they simply suppress the motor response.

6.1 Anaesthetics

Hypothermia (cold), CO_2, halogenated hydrocarbons (chloroform, halothane) and ethers (enflurane, isoflurane, sevoflurane, desflurane), eugenol, Freon-12 (Dichlorodifluoromethane; a colourless gas), ethyl acetate, diethyl ether and specialty products like "FlyNap®" (Carolina Biological) have been used as general anaesthetics for insects (Sinclair and Osborn 1970; Allada and Nash 1993; Campbell and Nash 1994; Duncan and Newton 2000; Loru et al. 2010; Weineck et al. 2019; Wahltinez et al. 2023). Unfortunately, few anaesthetics have been tested rigorously: pharmacological information, e.g. doses and durations of effect, or even the agent used is often unreported or not standardized in insect studies (Perl et al. 2025). Adverse behavioural effects (e.g. excitability), physiological changes (e.g. tissue damage, increased heart rate, or reduced immune function) and/or mortality during or after anesthesia occur with hypothermia, FlyNap®, ethyl acetate and CO_2 (Chen and Hillyer 2013; MacMillan et al. 2017).

Importantly, it is unknown at what time point during the application of cold the hypothermic animals are truly *unconscious* rather than conscious and simply unable to move. The latter is more likely in larger-bodied insects, where the core predictably takes longer to cool than the periphery. Consequently, most invertebrate veterinarians do not recommend hypothermia for insect anaesthesia (Cooper 2011; Lewbart and Mosley 2012; Lewbart 2022; Wahltinez et al. 2023). It is unknown how stressful or pain-inducing the process of being cooled may be for insects, as at least some insects have cold-evoked nocifensive responses (Patel et al. 2025). Despite this, some entomologists have suggested that humane anaesthesia can be achieved with cold, as insects are generally small-bodied and poikilothermic, which should make cooling rapid. Further, it is known that insect central nervous system shut down can occur with the application of prolonged cold (Robertson et al. 2023). The long-lasting behavioural and physiological effects of cold-induced anaesthesia should also be considered when using this method for insects (MacMillan et al. 2017).

Some invertebrate veterinarians also do not recommend the use of CO_2 because of a lack of information on its welfare effects: "if a procedure is considered to be potentially painful, there may be merit in using isoflurane, halothane, sevoflurane, rather than CO_2 because the extent to which the latter induces analgesia in invertebrates is not known, and its use in vertebrate animals is controversial because of concerns about its effects on the animals' health and welfare" (Cooper 2011). Carbon dioxide can cause excitability on knock-down in insects (Wahltinez et al., 2023) and also has long-lasting behavioural effects in insects during recovery (MacMillan et al. 2017).

The efficacy of other anaesthetics, i.e. halothane, desflurane, enflurane, chloroform, Freon-12, ethyl acetate and diethyl ether, has largely not been tested on insects. Sevoflurane, or isoflurane, appear to cause the fewest side-effects during post-recovery though these results need to be replicated on more species than just fruit flies (MacMillan et al. 2017).

Importantly, the answer to what provides a humane induction of unconsciousness for insects may also depend on insect body size and physiology, as some insects may be more cold- or hypoxia-tolerant than others. The response of insects to anaesthetics is likely to be affected by body size and species-related physiological adaptations. For example, many bumble bees are quite cold-tolerant as they rarely die following hypothermic episodes of several hours. In contrast, similarly-sized desert-adapted *Centris caesalpiniae* bees, being highly cold-sensitive, frequently die or show significant behavioural changes after hypothermia of only a few minutes duration (Barrett, pers. comm.). The rate of temperature change, or inhalant exposure concentration, may also affect the animal's potential experience. The doses of several anaesthetics used for insects have been reported in Wahltinez et al. (2023), and the efficacy of these doses may be life-stage dependent (Weineck et al. 2019).

The local anaesthetics lidocaine and procaine have been used in insects and act dose-dependently on insect sodium channels to prevent nerve conduction, as they do in mammalian nerves (Devaud et al. 2007; Song et al. 2011).

6.2 Analgesics

Invertebrate veterinarians do not recommend the use of cold or CO_2 as analgesics (Wahltinez et al. 2023). Several compounds appear to cause antinociceptive effects in insects. In fruit flies, gabapentin (delivered in feed at concentrations of at least 2.5 mM), non-steroidal anti-inflammatory drugs (diclofenac and ibuprofen), GABA agonists and opioids (morphine) all decreased responses to noxious compounds (Jang et al. 2023; and see Manev and Dimitrijevic 2004). Morphine's potentially analgesic effects have also been seen in other insects (Núñez et al. 1983; Zabala et al. 1984; Zabala and Gómez 1991; Gritsai et al. 2004; but see Groening et al. 2017) despite there being no opioid receptors in insects (Mirabeau and Joly 2013). Other compounds producing dose-dependent antinociceptive effects in cockroaches, honey bees and other taxa include: β-casomorphines (Gritsai et al. 2000, 2004), rubiscolin-5, dinorphin A, met-enkephalin, exorphine C (Gritsai et al. 2008, 2009), ethanol (Núñez et al. 1983; Giannoni-Guzmán et al. 2014; but see Núñez et al. 1983) and Tv1, a teretoxin Tv1 from *Terebra variegata* (Eriksson et al. 2018).

Unfortunately, the mechanisms of action for most of these analgesics have not been convincingly demonstrated, so caution in interpretation is advised. For instance, the effects of ethanol on learning and shock avoidance in bees may be due to its effects on cognition, memory or locomotion (Abramson et al. 2000; Giannoni-Guzmán et al. 2014; Black et al. 2021). Whether these drugs are producing analgesic or sedative effects is unknown (Wahltinez et al. 2023).

6.3 Key Uncertainties and Future Research

Despite adverse behavioural and physiological effects, cold and CO_2 remain the most commonly used methods for putative general anaesthesia in entomological research (Perl et al. 2025). Research conclusively demonstrating that cold and CO_2 can, or cannot, be considered humane insect anaesthetics would greatly benefit both researchers and farmers aspiring to humane research and production standards, respectively.

The mechanisms underpinning the action of most tested anaesthetics and analgesics in insects are not understood. For analgesics, this represents a particular problem because if mechanisms are unknown, it's difficult to determine whether these drugs are simply sedative or possess true analgesic effects. The lack of mechanistic understanding thus complicates the interpretation of behavioural responses.Nevertheless, GABA agonists are currently showing promise as insect analgesics, and there is a better understanding of their action. The halogenated ether anaesthetics, sevoflurane and isoflurane, also show promise—they appear to have few adverse side effects, though further research is required.

There are few "standard" techniques in insect anaesthesia. This is unsurprising given the range in body size and variation in resistance to or tolerance of hypoxia and cold that characterize *Insecta*. However, the requirements for suitable techniques would not differ in principle from those expected of an anaesthetic for vertebrates, i.e. efficacy and safety.

7 Attitudes on Insect Pain: Changing Public Perceptions?

This section reviews what is known about the attitudes of producers, academics, and the public on insect pain and welfare.

7.1 Insects as Food and Feed: Consumers, Producers and Academics

The attitudes of academics towards farmed insect pain (and thus welfare) have received scant attention, although academics from several relevant disciplines, i.e. entomology, animal welfare science, animal ethics and industry-adjacent academics in particular, regard the topic as important. For example, the current editor-in-chief of *Journal of Insects as Food and Feed* has repeatedly written about pain in the context of farmed insect welfare (van Huis 2013, 2021). In addition, the first textbook on insects as food and the feed industry (van Huis and Tomberlin 2017) included a chapter dedicated to the ethical challenges faced by the industry and refers to pain and welfare (Röcklinsberg et al. 2017).

The subject of farmed insect pain and welfare continues to receive increasing attention—though predominantly from European academics (Erens et al. 2012; de Goede et al. 2013; Gjerris et al. 2016; Boppré and Vane-Wright 2019; Pali-Schöll et al. 2019; Mutinelli 2020; Shelomi 2021; van Loon and Bovenkerk 2021; Lambert et al. 2021; Gibbons et al. 2022a; Parodi et al. 2022; Barrett and Fischer 2023a, b; Klobučar and Fisher 2023; Barrett et al. 2023a, b; Kortsmit et al. 2023; Voulgari-Kokota et al. 2023). This localized focus on insect welfare likely reflects the absence of global consensus on academic attitudes towards the importance of farmed insect pain.

The attitudes of producers have not been quantified, although qualitative information suggests at least some are interested in the subject of insect pain and welfare for moral or commercial (social acceptability) reasons. Interviewing UK insect producers, Bear (2019, 2021) found that while most did not believe insects could feel pain, others expressed discomfort because of the empirical uncertainty. Some producers claim to have adopted a precautionary principle, taking measures to limit pain during slaughter should pain indeed exist. Further, some producers have employed animal welfare scientists to study pertinent areas where information is sparse or absent (Barrett, pers. comm.). Producer websites, e.g. Protix (Protix n.d.), describe general commitments to insect welfare in the company's practice, while international trade organizations, e.g. the International Platform for Insects as Food and Feed (IPIFF), have issued basic recommendations for ensuring high welfare for farmed insects, alongside calling for additional research on the topic of insect pain specifically (IPIFF 2019).

The insect as food and feed industry is very new, and there is thus little information on how the public perceives insect pain in the industry (reviewed in Delvendahl et al. 2022, and see Dullaghan et al. 2021; Fukada et al. 2023). A recent survey of American adults found 47.2% of the general population believed insects used as feed could feel pain (Fukada et al. 2023) while an earlier one revealed 52–65% did, depending on the species (Dullaghan et al. 2021). Consumers that were opposed to industry growth were more likely to believe insects could feel pain (73.6%) and were more likely to be young and knowledgeable about the industry's practices (Fukada et al. 2023).

Not unexpectedly, most consumers reported less concern for insects', than for other farmed animals', welfare (except vegans; Elorinne et al. 2019; though see Rumpold and Langen 2019). Consumers were also confused about the role of insect protein in changing the overall balance of welfare concerns faced by farmed animals (Orkusz et al. 2020), where consuming insect protein may improve health and provide behavioural enrichment for traditional livestock like chickens (Ipema et al. 2022) or where humans could replace vertebrate animal protein with insect protein in their diets (thereby reducing the number of vertebrate animals, but increasing the number of insects, being farmed). These confusions may also occur because some people are not aware that insects are animals (Santaoja and Niva 2019), thus not factoring insects into their overall animal welfare equation. Nevertheless, the popular press has begun showing interest in farmed insect pain, particularly during slaughter (Sebo and Schukraft 2021; Reynolds 2023; Carlson 2023) with animal advocacy groups (EuroGroup for Animals 2023;

Orzechowski 2023) and legislators (Evira 2018; Ministerie van Landbouw 2018) following suit. Despite growing public interest in insect pain and welfare, no legislation currently exists to protect the welfare of insects in production or research settings (and see Meijer et al. 2025).

7.2 Insects in Research and Education: Academics, Students and the Public

Little information exists on the attitudes of academics and the public to the subject of insect pain in research and education (though see Barrett et al. 2024a). Historically, scientists have dismissed the notion that insects feel pain, although even sceptics have recommended anaesthesia before traumatizing manipulations as a precautionary measure (Eisemann et al. 1984; Adamo 2016). However, scientists are increasingly expressing uncertainty about insect pain and/or morally-relevant consciousness (Barron and Klein 2016; Klein and Barron 2016; Elwood 2016; de Mori and Normando 2019; Gibbons and Sarlak 2020; Chittka 2022; Gibbons et al. 2022a; Buchmann 2023; Klobučar and Fisher 2023; Crump et al. 2023). Recently, student members of the major North American entomological societies debated whether insects in research were treated ethically, with the winning team concluding they were not (Sandhi et al. 2023). A survey of entomologists' attitudes towards insect welfare in research and education suggests that early career entomologists believe training on this topic is important, and entomologists generally were concerned about insect welfare due to the intrinsic value of the animal as well as scientific reproducibility, public perception issues and their own moral discomfort (Barrett et al. 2024a). Similarly, a survey of animal behavior researchers indicated the (slim) majority believe at least some insects had emotions or were conscious (Zipple et al. 2024), demonstrating the divide in the field.

Academic ethicists have written on the likelihood and moral importance of pain or sentience in insects (Mikhalevich and Powell 2020; Fischer 2021; Kakrada and Colombo 2022). Others have argued that *even if* insects feel pain, humans have no moral obligations to them (Carruthers 2007), or that regulation of their welfare is not necessary, even if better practices should be employed for insects in research (Freelance 2019). The growing interest in insect pain, sentience and welfare among academics from different disciplines has resulted in the foundation of several insect welfare working groups. For instance, the Veterinary Invertebrate Society has many academics on the steering committee, over 1200 followers on Facebook and its own journal—the *Veterinary Invertebrate Society Journal,* which focuses on invertebrate (arthropod-inclusive) medicine. According to the date on their Twitter account, they were likely founded around 2017. There is also the Honeybee Veterinary Consortium (founded, 2017) the Insect Welfare Research Society (IWRS; founded, 2023) and the Royal Entomological Society special interest group on Welfare and Ethics (founded, 2023).

Beyond debating the plausibility of insect sentience, several researchers have begun promoting strategies for improving the ethical treatment of insects in research and education (Drinkwater et al. 2019; Sandall and Fischer 2019; Browning and Veit 2020). These include calls for the application of the 3Rs (Crump et al. 2023), multiple calls for less destructive sampling or insect collectors' codes of best practice (Trietsch and Deans 2018; Fischer and Larson 2019; Barrett et al. 2023b; Lövei et al. 2023), a suggestion to create an Insect Welfare Charter (Boppré and Vane-Wright 2019) and the production of guidelines for considering welfare by the IWRS (Fischer et al. 2023). The American Veterinary Medical Association first mentioned captive invertebrates in their 2013 guidelines on euthanasia and elaborated on that in 2020, although recommendations for insects remain limited (Leary et al. 2020).

The public's view of insects in research is largely unknown. A recent Canadian study suggests that the public's trust in science is undermined by the lack of ethical oversight, which includes the use of invertebrates (Brunt et al. 2022). The adverse reaction of some in the UK's press to a wasp biodiversity monitoring program that involved lethal sampling of wasps (The Big Wasp Survey; Hart 2018) prompted the scientists involved to respond via a BBC podcast episode 'Into the Killing Jar' (BBC Radio 4 2018). This indicates that at least some members of the public are concerned with the suffering of insects in research.

7.3 Key Uncertainties and Future Research

There is uncertainty about how different stakeholder groups view insect pain in different contexts. Attitudes to insect pain are likely to vary for national, cultural, religious and demographic reasons. More research globally on the attitudes of producers, researchers and the public towards insect welfare would be beneficial.

8 The "Other" Arthropods: Chelicerates, Myriapods and Other Hexapods

Arthropods are an extremely diverse and speciose group, estimated to contain over 80% (up to ten million) of all known extant animal species (Ødegaard 2000). Marine and terrestrial arthropods account for nearly half of the animal biomass on the planet, while humans account for 2% (Bar-On et al. 2018). The decapod crustaceans (the preceding chapter) and insects comprise the majority of farmed arthropods.

Other arthropods are farmed or otherwise managed less intensively, although this may change: Costa-Neto and Grabowski (2021) provide a list of edible arachnids and myriapods in the *Journal of Insects as Food and Feed.* Consequently, it may be prudent to consider these other groups in deliberations of insect (and crustacean) pain experiences. A comprehensive review of behavioural pain indicators in these groups can be found in Elwood (2023) and largely reveals that less is known about the nervous systems and behaviours of arachnids and myriapods than the insects and crustaceans.

Arthropod phylogenies are frequently revised; the phylogeny in Fig. 2 (Regier et al. 2010) roughly captures the evolutionary relationship between the three mainly-terrestrial groups featured here, i.e. Hexapoda, Myriapoda and Chelicerata. Schwentner et al. 2017 have produced an improved arthropod phylogeny.

8.1 Hexapoda

The *Hexapoda* contains the insects, as well as three groups of wingless, soil-dwelling arthropods with six legs: the Collembola, Protura and Diplura. The Collembola (springtails) are the second-most abundant group of arthropods in terrestrial environments, occurring at densities up to 50–200,000 individuals m^{-2} in mature soils (Beutel et al. 2014; Roy et al. 2018). However, other studies estimate 7–30,000 individuals m^2 (compared to 1–3,000 individuals m^{-2} of all less-abundant arthropods combined; Pearse 1946; Hoffmann et al. 1949). In particular, Collembola have been shown to be among the most pesticide-sensitive of all soil fauna present in high numbers in agroecosystems (Joimel et al. 2022), and it is possible that these pesticides could cause pain or suffering to this group. Many pest control websites advertise the use of pesticides to control outdoor populations of Collembola in humid regions. All may be studied in research settings. Extraordinarily little data are available on their nervous systems or potential nocifensive behaviors.

8.2 Myriapoda

Myriapods include approximately 14,100 known species of terrestrial millipedes and centipedes, alongside around 1,000 known species of the lesser-known symphyla and pauropoda (Adis and Harvey 2000). As soil-dwelling arthropods, they may experience pain and suffering due to pesticides. People may also use pesticides to control myriapods in their homes and gardens even though they are ecologically beneficial. No species of myriapods are cur-

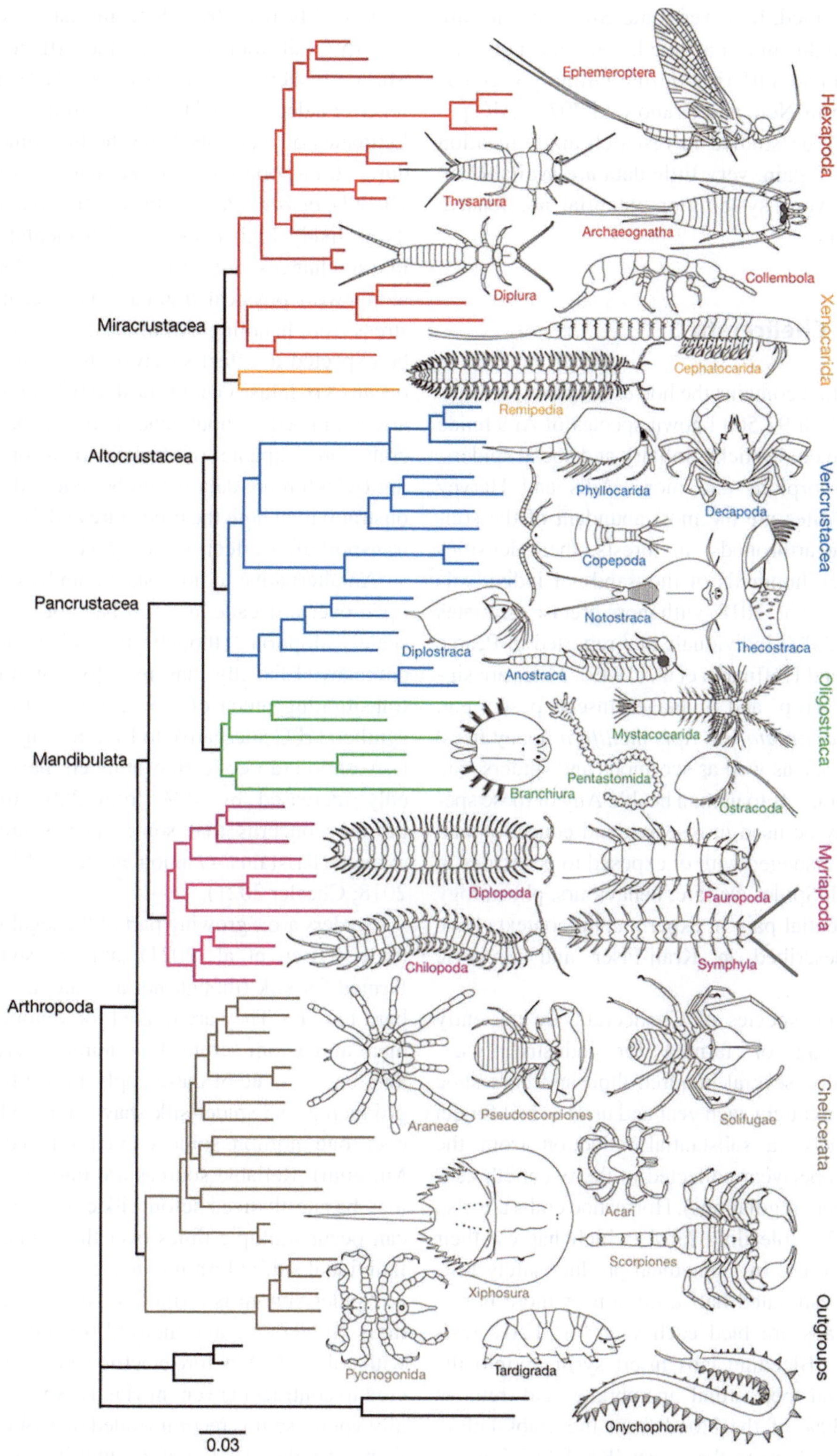

Fig. 2 Phylogeny of the Arthropoda. (From Regier et al. 2010)

rently farmed, however some edible species are wild-caught and consumed, with the potential for pain and suffering during culinary preparation (Costa-Neto and Grabowski 2021). All species may be studied in research and education settings. Again, very little data are available on their nervous systems or potential nocifensive behaviors.

8.3 Chelicerata

Chelicerata contains the horseshoe crabs, sea spiders and all 92,500 known species of Arachnida (which includes ticks, solitary and social spiders, mites, scorpions and more; Adis and Harvey 2000). Mites are the most abundant of the soil-dwelling arthropods: in forests, their densities can reach hundreds of thousands of individuals m^{-2} (Roy et al. 2018) with more precise estimates of 22–52,000 individuals m^{-2} reported in Pearse (1946) and Hoffmann et al. (1949). Mites are significant crop and managed-insect pests, e.g. *Varroa destructor* on *Apis mellifera* honey bees, while ticks, as well as scorpions and spiders, can present threats to human health. Any of these species may be used in research and education and may be managed with or exposed to pesticides in the wild. Spider welfare, behaviours, physiology and potential pain in experimental contexts have been described in Kralj-Fišer and Gregorič (2019).

Several species of Chelicerata are currently wild-caught or farmed for industrial uses. Minimally, several hundred thousand horseshoe crabs are caught each year and used as bait for eel and whelk (a substantial reduction from the millions-per-year collected in the twentieth century before regulations). Horseshoe crabs are also caught by 'bleeding labs', which harvest their blood for use in biomedical product safety testing; an estimated half a million or more horseshoe crabs are bled each year for 'LAL' tests (Fig. 3). Bleeding labs insert syringes into the heart and pericardial membrane and remove about 30% of the blood from the crabs before returning them to the ocean (Krisfalusi-Gannon et al. 2018).

Mortality rates from bleeding have exceeded the 15% allowance set by the Atlantic State Marine Fisheries Commission nearly every year for over a decade, and have increased since 2004. Estimates of mortality from the bleeding procedure, transportation and storage range from 10–30% *before return* to the ocean. Total mortality is likely higher still, as detrimental behavioural changes are often seen after bleeding, along with physical trauma from bleeding and stress from handling. All of these can reasonably be expected to affect survival *after* return to the ocean (Krisfalusi-Gannon et al. 2018). Procedures are conducted without anaesthetics. If horseshoe crabs are capable of feeling pain or suffering (which more data would be required to demonstrate) then both the procedure and the storage/transport/after-effects require revision.

An alternative to horseshoe crab blood developed over a decade ago still has not seen wide-scale adoption (though the pharmaceutical company, Eli Lilly, has recently committed to transitioning much of their practices to use the synthetic rFC alternative to LAL testing). In fact, horseshoe crab collection numbers have reportedly increased by 78% from 2004 to 2012, despite concerns that some species are going extinct (Krisfalusi-Gannon et al. 2018; Zhang 2018; Chesler 2021).

Spiders are a growing part of the legal wildlife trade (Olsen et al. 2021) and are sometimes farmed for silk (though not at scale, as they are hard to rear). They are milked for venom to produce antivenom serum. Few numbers are available on the scale of these applications; however, producing one spider-silk shawl reportedly took over one million spiders (Victoria and Albert Museum). Reliable sources are rare, but spiders may be anesthetized before silk collection, which can occur multiple times over the course of the individual's life (Terminix n.d.).

Spider venom is farmed for medical applications. Spiders are anesthetized (or immobilized) with cold or CO_2 before electric shocks are delivered to contract the venom glands. Spiders generally vomit, so it is recommended to avoid feeding them for the week before milking to reduce regurgitation volume (which can ruin the venom

Fig. 3 Horseshoe crab blood is being harvested. Still taken from a corporate video showing the bleeding of horseshoe crabs at the Charles River Laboratories facility in South Carolina. (CHARLES RIVER LABORATORIES)

if contaminated; Besson et al. 2016). When not anesthetized, spiders find electrical shocks aversive and learn to avoid places associated with shocks (Peckmezian and Taylor 2017). Spiders are generally milked many times throughout their life, which may be several years depending on the species. Following collection, venom is injected into mammals (which produce antibodies that can be collected in blood to make antivenom serum). In the future, there may be applications for spiders in the edible arthropods industry as it expands (Costa-Neto and Grabowski 2021).

8.4 Non-arthropod Invertebrates

If there may be reasons to do away with the vertebrate-invertebrate distinction when considering which animals may be capable of pain and suffering, then the invertebrate animal world offers a daunting, seemingly endless empirical and moral challenge (Andrews 2022). Non-arthropod invertebrate animals include: octopuses, squids, oysters, leeches, snails, nudibranchs, nematodes, platyhelminths, earthworms, starfish, sand dollars, sea cucumbers, corals, sponges, sea jellies, rotifera, tunicates, tardigrades, and much more (though this is not to claim that all are sentient). Altogether, invertebrates account for 95% of the animal biodiversity (Eisenhauer and Hines 2021). But, surely this speciosity and numerosity present an opportunity and not just a challenge. After all - there is nothing quite as helpful as biological diversity when it comes to making progress on the evolution, function, and mechanisms underpinning a trait. It seems plausible that many critical pieces of the puzzle that subjective experience and pain present could be hiding in plain sight -among the world's most abundant animals that are regularly forgotten.

9 Conclusion: The Complexity of Moral Caution

The challenge that nonhuman animals present, and none more so than the insects, is that *humans simply do not know.* Humans simply do not know if they are sentient or not. Humans simply do not know if they feel pain and suffer. Despite this uncertainty, humans affect quadrillions of insects every yearly in farming them for food and feed, conducting research, keeping them as companion animals, or killing them in gardens, fields, homes, and forests.

If insects feel pain—and subsequently suffer—then the scale of anthropogenically-linked suffering is immense. Irrespective of the challenges in understanding insect pain, or determining "welfare ranges" relative to other animals (Fischer 2024), the sheer numbers involved means the potential problem is huge. Risking a moral catastrophe of this magnitude, it is reasonable to employ a precautionary principle that prevents the worst cases of potential suffering while more research on their pain or sentience is completed(Birch 2017; Knutsson and Munthe 2017; Mikhalevich and Powell 2020; Andrews et al. 2024). There are several degrees of precaution to consider, ranging from strong action, e.g. regulations in research and farming, to weaker actions, such as a more attentive, personal harm-reduction approach.

Entomologists that argue the case *against* insect pain have always acknowledged the possibility that insect pain may exist and so promote the application of precautionary principles, generally recommending personal harm reduction approaches. In research, strategies like inactivating the nervous system before traumatic manipulation or reducing stress in handling (Eisemann et al. 1984; Adamo 2016) have been recommended by these sceptics. When there is *no cost* to considering insect welfare, most entomologists accept that there is no argument against this approach. However, 'no cost' scenarios may be uncommon and examples of inter-specific trade-offs and costs abound in most every context, including farming.

For example, poultry welfare is improved when the birds consume live black soldier fly larvae (Ipema et al. 2022) whose welfare may be, in turn, impoverished by inadequate nutrition, disease, behavioural restriction and—presumably—by being eaten. Alternatively, equipping insect production facilities with stunning equipment may increase the likelihood of humane death for the insects, but adds cost and time to the slaughter process, affecting sustainability and economic goals. However, the fact that there may be costs or trade-offs when insect pain is considered plausible should not prevent their consideration.

Initially, the nature of the trade-offs—real or imagined—must be identified. For instance, it is not always the case that the financial cost of a single stunning machine for each production facility is so high that profits are significantly affected. Reducing insect pain could even increase profits where it allows product differentiation, mitigates social risk or improves product quantity or quality (as has occurred previously with welfare improvements on livestock farms; Barrett and Adcock 2023, and see Niyonsaba et al. 2024). It is at least worth investigating whether the budget allocations for safeguarding insect welfare to any degree are likely to significantly alter profits or industry success. Other low-cost welfare improvements may also be feasible.

Second, even when the costs of reducing insect pain and suffering are real, profit losses are not necessarily foregone conclusions. Consumers may be prepared to pay higher prices for insects produced in welfare-prioritizing systems. Alternatively, industry financers or regulators may be persuaded to support high welfare systems with greater subsidies.

Possible solutions to these challenges are endless. But it is only when the time is taken to consider all relevant societal values, i.e. insect welfare, traditional livestock or human welfare, economic production, sustainability, etc., that it may be possible to minimize costs while maximizing benefits. Analyzing actual and not hypothetical costs under numerous conditions reveals that (1) there is a potential cost in not considering insect pain and welfare; (2) there may be creative solutions that resolve the most important trade-offs between diverse values; and (3) there can be benefits to factoring insect pain and welfare into decision-making.

Fundamentally, the complexity of morally cautious action does not preclude society from needing to consider moral caution at all. Humans must consider which costs are reasonable to pay to secure the potential welfare (and prevent the potential pain) of these animals, in research, farming, or the wild. Humanity is at the beginning of a long overdue conversation on insect welfare that involves many diverse stakeholders with different views. The difficulties in answering the question, "Do insects feel pain?" are no reason not to make the attempt nor to pay precautionary costs of creating a more ethical world. Now, that process of moral caution and fortitude begins anew, in service of a better world, for insects and humans alike.

References

Abramson CI, Stone SM, Ortez RA, Luccardi A, Vann KL, Hanig KD, Rice J (2000) The development of an ethanol model using social insects I: behavior studies of the honey bee (Apis mellifera L.). Alcohol Clin Exp Res 24:1153–1166

Adamo SA (2016) Do insects feel pain? A question at the intersection of animal behaviour, philosophy and robotics. Anim Behav 118:75–79. https://doi.org/10.1016/j.anbehav.2016.05.005

Adamo SA (2019) Is it pain if it does not hurt? On the unlikelihood of insect pain. Can Entomol 151:685–695. https://doi.org/10.4039/tce.2019.49

Adis J, Harvey M (2000) How many Arachnida and Myriapoda are there world-wide and in Amazonia? Stud Neotrop Fauna Environ 35:129–141

Alcock J, Jones C, Buchmann S (1976) The nesting behaviour of three species of Centris bees (Hymenoptera: Anthoporidae). J Kans Entomol Soc 49:469–474

Allada R, Nash HA (1993) Drosophila melanogaster as a model for study of general anesthesia: the quantitative response to clinical Anaesthetics and alkanes. Anesth Analg 77:19

Andrews K (2022) Does the sentience framework imply all animals are sentient? Anim Sentience 7. https://doi.org/10.51291/2377-7478.1737

Andrews K et al (2024) The New York Declaration on Animal Consciousness. Retrieved from: https://sites.google.com/nyu.edu/nydeclaration/declaration

Apple JL, Lewandowski SL, Levine JL (2014) Nest relocation in the slavemaking ants Formica subintegra and Formica pergandei: a response to host nest availability that increases raiding success. Insect Soc 61:347–356. https://doi.org/10.1007/s00040-014-0359-1

ASAB Ethical Committee/ABS Animal Care Committee (2023) Guidelines for the ethical treatment of nonhuman animals in behavioural research and teaching. Anim Behav 195:I–XI. https://doi.org/10.1016/j.anbehav.2022.09.006

Bar-On YM, Phillips R, Milo R (2018) The biomass distribution on earth. Proc Natl Acad Sci 115:6506–6511. https://doi.org/10.1073/pnas.1711842115

Barrett M, Adcock SJJ (2023) Animal welfare science: an integral piece of sustainable insect agriculture. J Insects Food Feed 1:1–15. https://doi.org/10.1163/23524588-20230126

Barrett M, Fischer B (2023a) Challenges in farmed insect welfare: beyond the question of sentience. Anim Welf 32:e4. https://doi.org/10.1017/awf.2022.5

Barrett M, Fischer B (2023b) Introducing ethos: a regular forum for interdisciplinary conversations about ethics in entomology. Am Entomol 69:40–44. https://doi.org/10.1093/ae/tmad016

Barrett M, Fischer B (2024) I. The Era Beyond Eisemann et al. (1984): insect pain in the 21st century. OSF Preprints. https://osf.io/preprints/osf/ng7p

Barrett and Fischer (2025) https://www.journals.uchicago.edu/doi/full/10.1086/737365

Barrett et al. (2025) https://brill.com/view/journals/jiff/aop/article-10.1163-23524588-bja10295/article-10.1163-23524588-bja10295.xml?language=en&srsltid=AfmBOoozs2G4xhzWEbjjHY6QtugNe0HdjNnNpVKJbOqAg5SzZ4ghuBhR

Barrett M, Chia S, Fischer B, Tomberlin JK (2023a) Welfare considerations for farming black soldier flies, *Hermetia illucens* (Diptera: Stratiomyidae): a model for the insects as food and feed industry. J Insects Food Feed 9:119–148. https://doi.org/10.3920/JIFF2022.0041

Barrett M, Fischer B, Buchmann S (2023b) Informing policy and practice on insect pollinator declines: tensions between conservation and animal welfare. Front Ecol Evol 10

Barrett M, Godfrey RK, Schnell A, Fischer B (2023c) Farmed yellow mealworm (Tenebrio molitor; Coleoptera: Tenebrionidae) welfare: species-specific recommendations for a global industry. J Insects Food Feed 10:903

Barrett M, Drewery M, Fischer B (2024a) Entomologists' knowledge of, and attitudes towards, insect welfare in research and education. OSF Preprints. https://resjournals.onlinelibrary.wiley.com/doi/full/10.1111/een.13415

Barrett M, Miranda C, Veloso IT, Flint C, Perl CD, Martinez A, Fischer B, Tomberlin JK (2024b) Grinding as a slaughter method for farmed black soldier fly (Hermetia illucens) larvae: empirically-supported recommendations for more instantaneous and humane practice. Anim Welf 33:e16

Brown and Barry (2016) https://royalsocietypublishing.org/doi/full/10.1098/rspb.2016.0656#:~:text=Brown%20William%20D.,org/10.1098/rspb.2016.0656

Barron AB, Klein C (2016) What insects can tell us about the origins of consciousness. PNAS 113:4900–4908

Bastin H (1927) Do insects feel pain? Sci Am 137:531–531

Bateson M, Desire S, Gartside SE, Wright GA (2011) Agitated honeybees exhibit pessimistic cognitive biases. Curr Biol 21:1070–1073. https://doi.org/10.1016/j.cub.2011.05.017

BBC Radio 4 (2018) BBC radio 4 – inside the killing jar. In: BBC. https://www.bbc.co.uk/programmes/b09qfwpt. Accessed 22 Jun 2023

Bear C (2019) Approaching insect death: understandings and practices of the UK'S edible insect farmers. Soc Anim 27:751–768. https://doi.org/10.1163/15685306-00001871

Bear C (2021) Making insects tick: responsibility, attentiveness and care in edible insect farming. Environ Plan E Nat Space 4:1010–1030. https://doi.org/10.1177/2514848620945321

Benoit JB, Attardo GM, Baumann AA, Michalkova V, Aksoy S (2015) Adenotrophic Viviparity in tsetse flies: potential for population control and as an insect model for lactation. Annu Rev Entomol 60:351–371. https://doi.org/10.1146/annurev-ento-010814-020834

Besson T, Debayle D, Diochot S, Salinas M, Lingueglia E (2016) Low cost venom extractor based on Arduino® board for electrical venom extraction from arthro-

pods and other small animals. Toxicon 118:156–161. https://doi.org/10.1016/j.toxicon.2016.05.001
Beutel R, Friedrich F, Ge S-Q, Yang X-K (2014) Insect morphology and phylogeny. De Gruyter
Bilder D, Irvine KD (2017) Taking stock of the Drosophila research ecosystem. Genetics 206:1227–1236. https://pubmed.ncbi.nlm.nih.gov/28684603/
Birch J (2017) Animal sentience and the precautionary principle. Anim Sentience 2. https://doi.org/10.51291/2377-7478.1200
Bittle J (2019) Inside the program that breeds billions of screwworm flies each year. In: Natl Geogr https://www.nationalgeographic.com/animals/article/north-american-screwworm-barrier
Black TE, Stauch KLN, Wells H, Abramson CI (2021) Effects of ethanol ingestion on aversive conditioning in honey bees (Apis mellifera L.). J Comp Psychol Wash DC 1983 135:559–567. https://doi.org/10.1037/com0000296
Boos S, Meunier J, Pichon S, Kölliker M (2014) Maternal care provides antifungal protection to eggs in the European earwig. Behav Ecol 25:754–761. https://doi.org/10.1093/beheco/aru046
Borjon (2025) https://www.cell.com/current-biology/fulltext/S0960-9822(24)01637-3
Boppré M, Vane-Wright RI (2012) The butterfly house industry: conservation risks and education opportunities. Conserv Soc 10:285–303
Boppré M, Vane-Wright RI (2019) Welfare dilemmas created by keeping insects in captivity. 18:23–67. https://doi.org/10.1007/978-3-030-13947-6_3
Bridges AD, MaBouDi H, Procenko O, Lockwood C, Mohammed Y, Kowalewska A, González JER, Woodgate JL, Chittka L (2023) Bumblebees acquire alternative puzzle-box solutions via social learning. PLoS Biol 21:e3002019. https://doi.org/10.1371/journal.pbio.3002019
Browning H, Veit W (2020) Improving invertebrate welfare. Anim Sentience 5. https://doi.org/10.51291/2377-7478.1585
Brunt MW, Kreiberg H, von Keyserlingk MAG (2022) Invertebrate research without ethical or regulatory oversight reduces public confidence and trust. Humanit Soc Sci Commun 9:1–9. https://doi.org/10.1057/s41599-022-01272-8
Buchmann S (2023) What a bee knows. Island Press
Burke NW, Holwell GI (2021) Male coercion and female injury in a sexually cannibalistic mantis. Biol Lett 17:20200811. https://doi.org/10.1098/rsbl.2020.0811
Burrell BD (2017) Comparative biology of pain: what invertebrates can tell us about how nociception works. J Neurophysiol 117:1461–1473. https://doi.org/10.1152/jn.00600.2016
Campbell DB, Nash HA (1994) Use of drosophila mutants to distinguish among volatile general anaesthetics. Proc Natl Acad Sci 91:2135–2139. https://doi.org/10.1073/pnas.91.6.2135
Carlson T (2023) Chef Joseph Yoon: edible insects are just another option for humans to consider. Fox News. https://www.foxnews.com/video/6325081705112
Caron DP, Rimniceanu M, Scibelli AE, Trimmer BA (2020) Nociceptive neurons respond to multimodal stimuli in Manduca sexta. J Exp Biol 223:jeb218859. https://doi.org/10.1242/jeb.218859
Carruthers P (2007) Invertebrate minds: a challenge for ethical theory. J Ethics 11:275–297. https://doi.org/10.1007/s10892-007-9015-6
Chen W, Hillyer JF (2013) FlyNap (Triethylamine) increases the heart rate of mosquitoes and eliminates the cardioacceleratory effect of the neuropeptide CCAP. PLoS One 8:e70414. https://doi.org/10.1371/journal.pone.0070414
Chesler C (2021) A horseshoe crab's blood is vital in testing drugs. Critics say using it endangers the ancient creature. Washington Post
Chittka L (2022) The mind of a bee. Oxford University Press
Chown SL, Gaston KJ (2010) Body size variation in insects: a macroecological perspective. Biol Rev 85:139–169. https://doi.org/10.1111/j.1469-185X.2009.00097.x
Christiansen P (2006) Somatic proportions in genus Megasoma (Scarabaeidae: Dynastinae): Megasoma actaeon. Ann Entomol Soc Am 99:342–351. https://doi.org/10.1603/0013-8746(2006)099[0342:SPIGMS]2.0.CO;2
Cooper JE (2011) Anesthesia, analgesia, and euthanasia of invertebrates. ILAR J 52:196–204. https://doi.org/10.1093/ilar.52.2.196
Costa JT (2006) The other insect societies. Belknap Press, Cambridge, MA
Costa-Neto EM, Grabowski NT (2021) Edible arachnids and myriapods worldwide – updated list, nutritional profile and food hygiene implications. J Insects Food Feed 7:261–279. https://doi.org/10.3920/JIFF2020.0046
Crowl TA, Alexander JE Jr (1989) Parental care and foraging ability in male water bugs (Belostoma flumineum). Can J Zool 67:513–515. https://doi.org/10.1139/z89-074
Crump A, Browning H, Schnell A, Burn C, Birch J (2022) Sentience in decapod crustaceans: a general framework and review of the evidence. Anim Sentience 7. https://doi.org/10.51291/2377-7478.1691
Crump A, Gibbons M, Barrett M, Birch J, Chittka L (2023) Is it time for insect researchers to consider their subjects' welfare? PLoS Biol 21:e3002138. https://doi.org/10.1371/journal.pbio.3002138
Dacke M, Srinivasan MV (2008) Evidence for counting in insects. Anim Cogn 11:683–689. https://doi.org/10.1007/s10071-008-0159-y
de Goede D, Erens J, Kapesomenou E, Peters M (2013) Large scale insect rearing and animal welfare. In: Röcklinsberg H, Sandin P (eds) The ethics of consumption. Wageningen Academic Publishers, pp 236–243
de Jong B, Nikolik G (2021) No longer crawling: insect protein to come of age in the 2020s. RaboBank
de Mori B, Normando S (2019) Is "history" repeating itself? The case of fish and arthropods' sentience and welfare. Etica Polit XXI:491–516

de Roode JC, Hunter MD (2019) Self-medication in insects: when altered behaviors of infected insects are a defense instead of a parasite manipulation. Curr Opin Insect Sci 33:1–6. https://doi.org/10.1016/j.cois.2018.12.001

Delvendahl N, Rumpold BA, Langen N (2022) Edible insects as food-insect welfare and ethical aspects from a consumer perspective. Insects 13:121. https://doi.org/10.3390/insects13020121

Desneux N, Decourtye A, Delpuech J-M (2007) The sublethal effects of pesticides on beneficial arthropods. Annu Rev Entomol 52:81–106. https://doi.org/10.1146/annurev.ento.52.110405.091440

Devaud J-M, Blunk A, Podufall J, Giurfa M, Grünewald B (2007) Using local anaesthetics to block neuronal activity and map specific learning tasks to the mushroom bodies of an insect brain. Eur J Neurosci 26:3193–3206. https://doi.org/10.1111/j.1460-9568.2007.05904.x

Drinkwater E, Robinson EJH, Hart AG (2019) Keeping invertebrate research ethical in a landscape of shifting public opinion. Methods Ecol Evol 10:1265–1273. https://doi.org/10.1111/2041-210X.13208

Dullaghan N, Wildeford P, Moss D (2021) Rethink Priorities poll: US attitudes towards insects. Rethink Priorities

Duncan FD, Newton RD (2000) The use of the anaesthetic, enflurane, for determination of metabolic rates and respiratory parameters in insects, using the ant, Camponotus maculatus (Fabricius) as the model. J Insect Physiol 46:1529–1534. https://doi.org/10.1016/S0022-1910(00)00076-7

Eisemann CH, Jorgensen WK, Merritt DJ, Rice MJ, Cribb BW, Webb PD, Zalucki MP (1984) Do insects feel pain?—a biological view. Experientia 40:164–167. https://doi.org/10.1007/BF01963580

Eisenhauer N, Hines J (2021) Invertebrate biodiversity and conservation. Curr Biol 31:R1214–R1218. https://doi.org/10.1016/j.cub.2021.06.058

Elorinne A-L, Niva M, Vartiainen O, Väisänen P (2019) Insect consumption attitudes among vegans, non-vegan vegetarians, and omnivores. Nutrients 11:292. https://doi.org/10.3390/nu11020292

Elwood R (2011) Pain and suffering in invertebrates? ILAR J 52:175–184. https://doi.org/10.1093/ilar.52.2.175

Elwood R (2016) Might insects experience pain? Anim Sentience 1. https://doi.org/10.51291/2377-7478.1156

Elwood RW (2023) Behavioural indicators of pain and suffering in arthropods and might pain bite Back? Animals 13:2602. https://doi.org/10.3390/ani13162602

Emanuel S, Libersat F (2019) Nociceptive pathway in the cockroach Periplaneta americana. Front Physiol 10:1100. https://doi.org/10.3389/fphys.2019.01100

Englemann F (1970) The physiology of insect reproduction. Pergamon Press

Erens J, Es van S, Haverkort F, Kapesomenou E, Luijben A (2012) A bug's life: large-scale insect rearing in relation to animal welfare

Eriksson A, Anand P, Gorson J, Grijuc C, Hadelia E, Stewart JC, Holford M, Claridge-Chang A (2018) Using Drosophila behavioral assays to characterize terebrid venom-peptide bioactivity. Sci Rep 8:15276. https://doi.org/10.1038/s41598-018-33215-2

EuroGroup for Animals (2023) The future of insect farming: where's the catch? EuroGroup for Animals

Evira (Finnish Food Safety Authority) (2018) Insects as food: Evira Guide 10588/2/UK. Evira, Helsinki

Fellers JH, Fellers GM (1976) Tool use in a social insect and its implications for competitive interactions. Science 192:70–72. https://doi.org/10.1126/science.192.4234.70

Fischer (2025) https://pmc.ncbi.nlm.nih.gov/articles/PMC12058431/

Fischer B (2021) Animal ethics: a contemporary introduction. Routledge

Fischer B (2024) Weighing animal welfare: comparing animal well-being across species. Oxford

Fischer B, Larson BMH (2019) Collecting insects to conserve them: a call for ethical caution. Insect Conserv Divers 12:173–182. https://doi.org/10.1111/icad.12344

Fischer B, Sandall E (2023) If adult insects matter, how much do juveniles matter? In: Rethink Priorities. https://rethinkpriorities.org/publications/how-much-do-juveniles-matter. Accessed 16 Apr 2023

Fischer B, Barrett M, Adcock S, Barron A, Browning H, Chittka L, Drinkwater E, Gibbons M, Haverkamp A, Perl C (2023) Guidelines for protecting and promoting insect welfare in research. Insect Welfare Research Society. Available at: https://www.researchgate.net/publication370441807_2023_Guidelines_for_Protecting_and_Promoting_Insect_Welfare_in_Research

Forbes AA, Bagley RK, Beer MA, Hippee AC, Widmayer HA (2018) Quantifying the unquantifiable: why Hymenoptera, not Coleoptera, is the most speciose animal order. BMC Ecol 18:21. https://doi.org/10.1186/s12898-018-0176-x

Frank ET, Wehrhahn M, Linsenmair KE (2018) Wound treatment and selective help in a termite-hunting ant. Proc R Soc B Biol Sci 285:20172457. https://doi.org/10.1098/rspb.2017.2457

Frank ET, Buffat D, Liberti J, Aibekova L, Economo EP, Keller L (2024) Wound-dependent leg amputations to combat infections in an ant society. Curr Biol 34:P3273–P3278

Frantsevich L, Gorb S (2006) Courtship dances in the flies of the genus lispe (Diptera: Muscidae): from the fly's viewpoint. Arch Insect Biochem Physiol 62:26–42. https://doi.org/10.1002/arch.20118

Freelance CB (2019) To regulate or not to regulate? The future of animal ethics in experimental research with insects. Sci Eng Ethics 25:1339–1355. https://doi.org/10.1007/s11948-018-0066-9

Fukada EP, Carrasco MN, Perez AJ, Fischer B, Drewery ML (2023) U.S. consumer perceptions of insects as livestock feed: ethical considerations for insects

Galpayage Dona HS, Solvi C, Kowalewska A, Mäkelä K, MaBouDi H, Chittka L (2022) Do bumble bees play? Anim Behav 194:239–251. https://doi.org/10.1016/j.anbehav.2022.08.013

Garrido C, Nanetti A (2019) Welfare of managed honey bees. In: Carere C, Mather J (eds) The welfare of invertebrate animals. Springer, pp 69–104

Gelperin A (2019) Recent trends in invertebrate neuroscience. In: Byrne J (ed) The Oxford handbook of invertebrate neuroscience. Oxford University Press, pp 3–30

Giannoni-Guzmán MA, Giray T, Agosto-Rivera JL, Stevison BK, Freeman B, Ricci P, Brown EA, Abramson CI (2014) Ethanol-induced effects on sting extension response and punishment learning in the western honey bee (Apis mellifera). PLoS One 9:e100894. https://doi.org/10.1371/journal.pone.0100894

Gibbons M, Sarlak S (2020) Inhibition of pain or response to injury in invertebrates and vertebrates. Anim Sentience 5. https://www.wellbeingintlstudiesrepository.org/animsent/vol5/iss29/34/

Gibbons M, Crump A, Barrett M, Sarlak S, Birch J, Chittka L (2022a) Chapter three – can insects feel pain? A review of the neural and behavioural evidence. In: Advances in insect physiology. Academic, pp 155–229

Gibbons M, Sarlak S, Chittka L (2022b) Descending control of nociception in insects? Proc R Soc B Biol Sci 289:20220599. https://doi.org/10.1098/rspb.2022.0599

Gibbons M, Versace E, Crump A, Baran B, Chittka L (2022c) Motivational trade-offs and modulation of nociception in bumblebees. Proc Natl Acad Sci 119:e2205821119. https://doi.org/10.1073/pnas.2205821119

Gibbons M, Pasquini E, Kowalewska A, Read E, Gibson S, Crump A, Solvi C, Versace E, Chittka L (2024) Noxious stimulation induces self-protective behavior in bumblebees. iScience 27:110440

Giurfa M (2015) Learning and cognition in insects. WIREs Cogn Sci 6:383–395. https://doi.org/10.1002/wcs.1348

Giurfa M, Zhang S, Jenett A, Menzel R, Srinivasan MV (2001) The concepts of 'sameness' and 'difference' in an insect. Nature 410:930–933. https://doi.org/10.1038/35073582

Gjerris M, Gamborg C, Röcklinsberg H (2016) Ethical aspects of insect production for food and feed. J Insects Food Feed 2:101–110. https://doi.org/10.3920/JIFF2015.0097

Godfrey RK, Swartzlander M, Gronenberg W (2021) Allometric analysis of brain cell number in hymenoptera suggests ant brains diverge from general trends. Proc R Soc B Biol Sci 288:20210199. https://doi.org/10.1098/rspb.2021.0199

Goldberg J, Godfrey RK, Barrett M (2024) A long-read draft assembly of the Chinese mantis (Mantodea: Mantidae: Tenodera sinensis) genome reveals patterns of ion channel gain and loss across Arthropoda. G3: Genes, Genomes, Genetics 14:jkae062

Gritsai OB, Dubynin VA, Pilipenko VE, Petrov OP, Andreeva LA (2000) The effects of beta-casomorphine-7 and naloxone of the locomotor defense response of the cockroach Periplaneta americana to electrical stimulation. Dokl Biochem Proc Acad Sci USSR Biochem Sect 375:245–247. https://doi.org/10.1023/a:1026632108709

Gritsai OB, Dubynin VA, Pilipenko VE, Petrov OP (2004) Effects of peptide and non-peptide opioids on protective reaction of the cockroach Periplaneta americana in the "Hot Camera". J Evol Biochem Physiol 40:153–160. https://doi.org/10.1023/B:JOEY.0000033806.85565.3e

Gritsai OB, Pilipenko VE, Bespalova ZD, Dubynin VA (2008) Role of food-derived opioid peptides in regulation of thermosensitivity of Periplaneta americana cockroach. Mosc Univ Biol Sci Bull 63:150–155. https://doi.org/10.3103/S0096392508040032

Gritsaĭ OB, Dubynin VA, Bespalova ZD, Pilipenko VE (2009) Effects of several exorphins and endorphins on the escape reaction of the cockroach Periplaneta americana under elevated temperature conditions. Zh Evol Biokhim Fiziol 45:391–397

Groening J, Venini D, Srinivasan MV (2017) In search of evidence for the experience of pain in honeybees: a self-administration study. Sci Rep 7:45825. https://doi.org/10.1038/srep45825

Grover D, Chen J-Y, Xie J, Li J, Changeux J-P, Greenspan RJ (2022) Differential mechanisms underlie trace and delay conditioning in drosophila. Nature 603:302–308. https://doi.org/10.1038/s41586-022-04433-6

Grueber WB, Graubard K, Truman JW (2001) Tiling of the body wall by multidendritic sensory neurons in Manduca sexta. J Comp Neurol 440:271–283. https://doi.org/10.1002/cne.1385

Guénard B, McGlynn TP (2013) Intraspecific thievery in the ant Ectatomma ruidum is mediated by food availability. Biotropica 45:497–502

Hall C (2023) Future fields is turning fruit flies into bioreactors. In: TechCrunch. https://techcrunch.com/2023/02/22/future-fields-fruit-flies-bioreactors/. Accessed 16 Apr 2023

Harrington R (2023) Insect welfare and ethics special interest group. Antenna 47(3):150–151

Hart A (2018) Inside the killing jar. In: RSB. https://www.rsb.org.uk//biologist-features/inside-the-killing-jar. Accessed 22 Jun 2023

Hentschel E, Penzlin H (1982) Beeinflussung des putzverhaltens bei Periplaneta americana (L.) durch wundsetzung, naloxon-, morphin-, und met-Enkephalingaben. Zool Jahrb Abt Allg Zool Physiol Tiere 86:361–370

Hoffmann CH, Townes HK, Swift HH, Sailer RI (1949) Field studies on the effects of airplane applications of DDT on forest invertebrates. Ecol Monogr 19:1–46. https://doi.org/10.2307/1943583

Howard SR, Avarguès-Weber A, Garcia JE, Greentree AD, Dyer AG (2019) Numerical cognition in honeybees enables addition and subtraction. Sci Adv 5:eaav0961. https://doi.org/10.1126/sciadv.aav0961

Howe H (2019) Improving pest management for wild insect welfare. Wild Animal Initiative

Hu W, Peng Y, Sun J, Zhang F, Zhang X, Wang L, Li Q, Zhong Y (2018a) Fan-shaped body neurons in the drosophila brain regulate both innate and conditioned nociceptive avoidance. Cell Rep 24:1573–1584. https://doi.org/10.1016/j.celrep.2018.07.028

Hu Y, Sanders JG, Łukasik P, D'Amelio CL, Millar JS, Vann DR, Lan Y, Newton JA, Schotanus M, Kronauer DJC, Pierce NE, Moreau CS, Wertz JT, Engel P, Russell JA (2018b) Herbivorous turtle ants obtain essential nutrients from a conserved nitrogen-recycling gut microbiome. Nat Commun 9:964. https://doi.org/10.1038/s41467-018-03357-y

Hwang RY, Zhong L, Xu Y, Johnson T, Zhang F, Deisseroth K, Tracey WD (2007) Nociceptive neurons protect Drosophila larvae from parasitoid wasps. Curr Biol CB 17:2105–2116. https://doi.org/10.1016/j.cub.2007.11.029

Im SH, Galko MJ (2012) Pokes, sunburn, and hot sauce: Drosophila as an emerging model for the biology of nociception. Dev Dyn 241:16–26. https://doi.org/10.1002/dvdy.22737

Ipema AF, Bokkers EAM, Gerrits WJJ, Kemp B, Elizabeth Bolhuis J (2022) Provision of black soldier fly larvae (Hermetia illucens) in different ways benefits broiler welfare and performance, with largest effects of scattering live larvae. Physiol Behav 257:113999. https://doi.org/10.1016/j.physbeh.2022.113999

IPIFF (2019) Ensuring high standards of animal welfare in insect production. https://ipiff.org/ensuring-high-standards-of-animal-welfare-in-insect-production/

Jackson Laboratory Body Weight Information for C57BL/6J (000664). In: Jackson Lab. https://www.jax.org/jax-mice-and-services/strain-data-sheet-pages/body-weight-chart-000664. Accessed 16 Apr 2023

Jandt JM, Bengston S, Pinter-Wollman N, Pruitt JN, Raine NE, Dornhaus A, Sih A (2014) Behavioural syndromes and social insects: personality at multiple levels. Biol Rev 89:48–67. https://doi.org/10.1111/brv.12042

Jang W, Oh M, Cho E-H, Baek M, Kim C (2023) Drosophila pain sensitization and modulation unveiled by a novel pain model and analgesic drugs. PLoS One 18:e0281874. https://doi.org/10.1371/journal.pone.0281874

Johnson R (2010) Honey bee colony collapse disorder. Congr Res Serv:7–5700

Johnson and Barrett (2025)https://www.sciencedirect.com/science/article/pii/S1751731125000849

Joimel S, Chassain J, Artru M, Faburé J (2022) Collembola are among the most pesticide-sensitive soil Fauna groups: a meta-analysis. Environ Toxicol Chem 41:2333–2341. https://doi.org/10.1002/etc.5428

Kakrada E, Colombo M (2022) Extending the null hypothesis to invertebrate pain sentience. Anim Sentience 7. https://doi.org/10.51291/2377-7478.1754

Kondakath (2025) https://royalsocietypublishing.org/doi/10.1098/rsbl.2025.0124

Kaun KR, Azanchi R, Maung Z, Hirsh J, Heberlein U (2011) A drosophila model for alcohol reward. Nat Neurosci 14:612–619. https://doi.org/10.1038/nn.2805

Keller L, Genoud M (1997) Extraordinary lifespans in ants: a test of evolutionary theories of ageing. Nature 389:958–960. https://doi.org/10.1038/40130

Kellogg VL (1907) Some silkworm moth reflexes. Biol Bull 12:152–154. https://doi.org/10.2307/1535862

Khuong TM, Neely GG (2014) Conservation of pain genes across evolution. In: Pain genetics. Wiley, pp 15–36

Khuong TM, Wang Q-P, Manion J, Oyston LJ, Lau M-T, Towler H, Lin YQ, Neely GG (2019) Nerve injury drives a heightened state of vigilance and neuropathic sensitization in drosophila. Sci Adv 5:eaaw4099. https://doi.org/10.1126/sciadv.aaw4099

Kim SA, Kim K-M, Oh B-J (2008) Current status and perspective of the insect industry in Korea. Entomol Res 38:S79–S85. https://doi.org/10.1111/j.1748-5967.2008.00178.x

Klein C, Barron A (2016) Insects have the capacity for subjective experience. Anim Sentience 1. https://doi.org/10.51291/2377-7478.1113

Klobučar T, Fisher DN (2023) When do we start caring about insect welfare? Neotrop Entomol 52:5–10. https://doi.org/10.1007/s13744-022-01023-z

Knutsson S, Munthe C (2017) A virtue of precaution regarding the moral status of animals with uncertain sentience. J Agric Environ Ethics 30:213–224. https://doi.org/10.1007/s10806-017-9662-y

Kortsmit Y, van der Bruggen M, Wertheim B, Dicke M, Beukeboom LW, van Loon JJA (2023) Behaviour of two fly species reared for livestock feed: optimising production and insect welfare. J Insects Food Feed 9:149–169. https://doi.org/10.3920/JIFF2021.0214

Kralj-Fišer S, Gregorič M (2019) Spider welfare. In: Carere C, Mather J (eds) The welfare of invertebrate animals. Springer International Publishing, Cham, pp 105–122

Krisfalusi-Gannon J, Ali W, Dellinger K, Robertson L, Brady TE, Goddard MKM, Tinker-Kulberg R, Kepley CL, Dellinger AL (2018) The role of horseshoe crabs in the biomedical industry and recent trends impacting species sustainability. Front Mar Sci 5

Kruitwagen A, Beukeboom LW, Wertheim B (2018) Optimization of native biocontrol agents, with parasitoids of the invasive pest Drosophila suzukii as an example. Evol Appl 11:1473–1497. https://doi.org/10.1111/eva.12648

Kuiper MJ, Morton CJ, Abraham SE, Gray-Weale A (2015) The biological function of an insect antifreeze protein simulated by molecular dynamics. eLife 4:e05142. https://doi.org/10.7554/eLife.05142

Lambert H, Elwin A, D'Cruze N (2021) Wouldn't hurt a fly? A review of insect cognition and sentience in relation to their use as food and feed. Appl Anim Behav Sci 243:105432. https://doi.org/10.1016/j.applanim.2021.105432

Leadbeater E, Chittka L (2007) Social learning in insects—from miniature brains to consensus building. Curr Biol 17:R703–R713. https://doi.org/10.1016/j.cub.2007.06.012

Leary S, Pharmaceuticals F, Underwood W, Anthony R, Cartner S, Johnson CL, Patterson-Kane E (2020) AVMA guidelines for the euthanasia of animals: 2020 edition

Lewbart G (ed) (2022) Invertebrate medicine, 3rd edn. Wiley-Blackwell

Lewbart GA, Mosley C (2012) Clinical anesthesia and analgesia in invertebrates. J Exot Pet Med 21:59–70. https://doi.org/10.1053/j.jepm.2011.11.007

Li F, Lindsey JW, Marin EC, Otto N, Dreher M, Dempsey G, Stark I, Bates AS, Pleijzier MW, Schlegel P, Nern A, Takemura S, Eckstein N, Yang T, Francis A, Braun A, Parekh R, Costa M, Scheffer LK, Aso Y, Jefferis GS, Abbott LF, Litwin-Kumar A, Waddell S, Rubin GM (2020) The connectome of the adult Drosophila mushroom body provides insights into function. eLife 9:e62576. https://doi.org/10.7554/eLife.62576

Liceaga AM (2022) Edible insects, a valuable protein source from ancient to modern times. Adv Food Nutr Res 101:129–152. https://doi.org/10.1016/bs.afnr.2022.04.002

Lockwood JA (1987) The moral standing of insects and the ethics of extinction. Fla Entomol 70:70–89. https://doi.org/10.2307/3495093

Lockwood JA (1988) Not to harm a Fly: our ethical obligations to insects. Species 4:12

Lockwood JA (2010) Six-legged soldiers: using insects as weapons of war. Oxford University Press, Oxford/New York

Lopez-Bellido R, Himmel NJ, Gutstein HB, Cox DN, Galko MJ (2019) An assay for chemical nociception in Drosophila larvae. Philos Trans R Soc Lond Ser B Biol Sci 374:20190282. https://doi.org/10.1098/rstb.2019.0282

Loru L, Sassu A, Fois X, Pantaleoni RA (2010) Ethyl acetate: a possible alternative for anaesthetizing insects. Ann Société Entomol Fr NS 46:422–424. https://doi.org/10.1080/00379271.2010.10697677

Lövei GL, Ferrante M, Möller D, Möller G, Vincze É (2023) The need for a (non-destructive) method revolution in entomology. Biol Conserv 282:110075. https://doi.org/10.1016/j.biocon.2023.110075

MacMillan HA, Nørgård M, MacLean HJ, Overgaard J, Williams CJA (2017) A critical test of drosophila anaesthetics: isoflurane and sevoflurane are benign alternatives to cold and CO2. J Insect Physiol 101:97–106. https://doi.org/10.1016/j.jinsphys.2017.07.005

Manev H, Dimitrijevic N (2004) Drosophila model for in vivo pharmacological analgesia research. Eur J Pharmacol 491:207–208. https://doi.org/10.1016/j.ejphar.2004.03.030

Mashour GA, Roelfsema P, Changeux J-P, Dehaene S (2020) Conscious processing and the global neuronal workspace hypothesis. Neuron 105:776–798. https://doi.org/10.1016/j.neuron.2020.01.026

Mason GJ, Lavery JM (2022) What is it like to be a bass? Red herrings, fish pain and the study of animal sentience. Front Vet Sci 9

Meijer (2025) https://www.sciencedirect.com/science/article/pii/S1751731125000515

Mikhalevich I, Powell R (2020) Minds without spines: evolutionarily inclusive animal ethics. Anim Sentience 29

Miler K, Turza F (2021) "O sister, where art thou?"—a review on rescue of imperiled individuals in ants. Biology 10:1079. https://doi.org/10.3390/biology10111079

Ministerie van Landbouw N en V (2018) Invertebrates as production animals – Publication – Council on Animal Affairs. https://english.rda.nl/publications/publications/2018/09/03/the-emerging-insect-industry. Accessed 13 Apr 2023

Mirabeau O, Joly J-S (2013) Molecular evolution of peptidergic signaling systems in bilaterians. Proc Natl Acad Sci 110. https://doi.org/10.1073/pnas.1219956110

Misof (2014) https://www.science.org/doi/10.1126/science.1257570

Mirwan HB, Kevan PG (2013) Social learning in bumblebees (*Bombus impatiens*): worker bumblebees learn to manipulate and forage at artificial flowers by observation and communication within the Colony. Psyche J Entomol 2013:e768108. https://doi.org/10.1155/2013/768108

Mohd Zubir MZ, Holloway S, Mohd Noor N (2020) Maggot therapy in wound healing: a systematic review. Int J Environ Res Public Health 17:6103. https://doi.org/10.3390/ijerph17176103

Monchanin C, Gabriela de Brito Sanchez M, Lecouvreur L, Boidard O, Méry G, Silvestre J, Le Roux G, Baqué D, Elger A, Barron AB, Lihoreau M, Devaud J-M (2022) Honey bees cannot sense harmful concentrations of metal pollutants in food. Chemosphere 297:134089. https://doi.org/10.1016/j.chemosphere.2022.134089

Montgomery GA, Dunn RR, Fox R, Jongejans E, Leather SR, Saunders ME, Shortall CR, Tingley MW, Wagner DL (2020) Is the insect apocalypse upon us? How to find out. Biol Conserv 241:108327. https://doi.org/10.1016/j.biocon.2019.108327

Morinaga G, Soghigian J, Edgerly JS (2023) Macroecology and potential drivers of diversity in Webspinner maternal care (order Embioptera). Insect Syst Divers 7:1. https://doi.org/10.1093/isd/ixac031

Mudrik L, Faivre N, Koch C (2014) Information integration without awareness. Trends Cogn Sci 18:488–496. https://doi.org/10.1016/j.tics.2014.04.009

Mutinelli F (2020) Euthanasia and welfare of managed honey bee colonies. J Apic Res 62:2–10. https://doi.org/10.1080/00218839.2021.1895569

Niyonsaba HH, Hohler J, Rumpold BA, Van der Fels-Klerx HJ, Meuwissen MPM (2024) Robustness of business models for insect production for feed and food in Europe. J Insects Food Feed 11:211

Nityananda (2025) https://www.biorxiv.org/content/10.1101/2025.04.04.647054v1
Núñez J, Maldonado H, Miralto A, Balderrama N (1983) The stinging response of the honeybee: effects of morphine, naloxone and some opioid peptides. Pharmacol Biochem Behav 19:921–924. https://doi.org/10.1016/0091-3057(83)90391-x
Ødegaard F (2000) How many species of arthropods? Erwin's estimate revised. Biol J Linn Soc 71:583–597. https://doi.org/10.1111/j.1095-8312.2000.tb01279.x
Olsen MTB, Geldmann J, Harfoot M, Tittensor DP, Price B, Sinovas P, Nowak K, Sanders NJ, Burgess ND (2021) Thirty-six years of legal and illegal wildlife trade entering the USA. Oryx 55:432–441. https://doi.org/10.1017/S0030605319000541
Orkusz A, Wolańska W, Harasym J, Piwowar A, Kapelko M (2020) Consumers' attitudes facing entomophagy: polish case perspectives. Int J Environ Res Public Health 17:2427. https://doi.org/10.3390/ijerph17072427
Orzechowski K (2023) Faunalytics Index – April 2023. In: Faunalytics. https://faunalytics.org/faunalytics--index-april-2023/. Accessed 13 Apr 2023
Overgaard M (2021) Insect consciousness. Front Behav Neurosci 15
Pali-Schöll I, Binder R, Moens Y, Polesny F, Monsó S (2019) Edible insects – defining knowledge gaps in biological and ethical considerations of entomophagy. Crit Rev Food Sci Nutr 59:2760–2771. https://doi.org/10.1080/10408398.2018.1468731
Parodi A, Ipema AF, Van Zanten HHE, Bolhuis JE, Van Loon JJA, De Boer IJM (2022) Principles for the responsible use of farmed insects as livestock feed. Nat Food 3:972–974. https://doi.org/10.1038/s43016-022-00641-5
Patel (2025) https://elifesciences.org/articles/91582
Pearse AS (1946) Observations on the microfauna of the Duke Forest. Ecol Monogr 16:127–150. https://doi.org/10.2307/1943104
Peckmezian T, Taylor PW (2017) Place avoidance learning and memory in a jumping spider. Anim Cogn 20:275–284. https://doi.org/10.1007/s10071-016-1048-4
Peretti A, Aisenberg A (eds) (2015) Cryptic female choice in arthropods. Springer
Perl CD, Kissinger C, Godfrey RK, Castillo EM, Fischer B, Barrett M (2024) Identifying trends in reporting on the ethical treatment of insects in research. OSF Preprints
Perl (2025) https://journals.plos.org/plosone/article?id=10.1371/journal.pone.0328931
Perry CJ, Baciadonna L (2017) Studying emotion in invertebrates: what has been done, what can be measured and what they can provide. J Exp Biol 220:3856–3868. https://journals.biologists.com/jeb/article/220/21/3856/33729/Studying-emotion-in-invertebrates-what-has-been
Perry CJ, Barron AB (2013) Honey bees selectively avoid difficult choices. Proc Natl Acad Sci 110:19155–19159. https://doi.org/10.1073/pnas.1314571110
Perry CJ, Barron AB, Chittka L (2017) The frontiers of insect cognition. Curr Opin Behav Sci 16:111–118. https://doi.org/10.1016/j.cobeha.2017.05.011
Polilov AA, Hakimi KD, Makarova AA (2023) Extremely small wasps independently lost the nuclei in the brain neurons of at least two lineages. Sci Rep 13:4320. https://doi.org/10.1038/s41598-023-31529-4
Protix Animal Welfare Statement. In: Protix. https://protix.eu/for-our-planet/animal-welfare-statement/. Accessed 13 Apr 2023
Rana A, Emanuel S, Adams ME, Libersat F (2022) Suppression of host nocifensive behavior by parasitoid wasp venom. Front Physiol 13
Regier JC, Shultz JW, Zwick A, Hussey A, Ball B, Wetzer R, Martin JW, Cunningham CW (2010) Arthropod relationships revealed by phylogenomic analysis of nuclear protein-coding sequences. Nature 463:1079–1083. https://doi.org/10.1038/nature08742
Reynolds M (2023) Insect farming is booming. But Is It Cruel? Wired
Rivers D (2017) Insects: evolutionary success, unrivaled diversity, World Domination. Johns Hopkins University Press
Robertson RM, MacMillan HA, Andersen MK (2023) A cold and quiet brain: mechanisms of insect CNS arrest at low temperatures. Curr Opin Insect Sci 58:101055
Röcklinsberg H, Gamborg C, Gjerris M (2017) Ethical issues in insect production
Rowe A (2020a) The scale of direct human impact on invertebrates
Rowe A (2020b) Insects raised for food and feed—global scale, practices, and policy. In: Rethink Priorities. https://rethinkpriorities.org/publications/insects-raised-for-food-and-feed. Accessed 16 Apr 2023
Rowe A (2020c) Global cochineal production: scale, welfare concerns, and potential interventions. In: EA Forum. https://forum.effectivealtruism.org/posts/tDYtn4DhFsR7pR35i/global-cochineal-production-scale-welfare-concerns-and. Accessed 16 Apr 2023
Rowe A (2021) Silk production: global scale and animal welfare issues. In: Rethink Priorities. https://rethinkpriorities.org/publications/silk-production. Accessed 16 Apr 2023
Rowe E, Robles López KY, Robinson KM, Baudier KM, Barrett M (2024) Farmed cricket (Acheta domesticus, Gryllus assimilis, and Gryllodes sigillatus; Orthoptera) welfare considerations: recommendations for improving global practice. J Insects Food Feed 10:1253
Roy C, Vidal-Naquet N (2022) Euthanasia of honey-bee colonies: proposal of a standard method. Can Vet J 63:541–544
Roy S, Roy MM, Jaiswal AK, Baitha A (2018) Soil arthropods in maintaining soil health: thrust areas for sugarcane production systems. Sugar Tech 20:376–391. https://doi.org/10.1007/s12355-018-0591-5
Royal Entomological Society (2025) https://www.royensoc.co.uk/news/res-statement-on-the-ethical-treatment-of-insects/

Rumbos CI, Athanassiou CG (2021) Insects as food and feed: if you can't beat them, eat them!'—to the magnificent seven and beyond. J Insect Sci 21:9. https://doi.org/10.1093/jisesa/ieab019

Rumpold B a, Langen N (2019) Potential of enhancing consumer acceptance of edible insects via information. J Insects Food Feed 5:45–53. https://doi.org/10.3920/JIFF2018.0041

Saleem S, Ruggles PH, Abbott WK, Carney GE (2014) Sexual experience enhances Drosophila melanogaster male mating behavior and success. PLoS One 9:e96639. https://doi.org/10.1371/journal.pone.0096639

Sakai (1991) https://www.sciencedirect.com/science/article/abs/pii/002219109190037Z

Salvemini M, Polito C, Saccone G (2010) Fruitless alternative splicing and sex behaviour in insects: an ancient and unforgettable love story? J Genet 89:287–299

Sánchez-Bayo F, Wyckhuys KAG (2019) Worldwide decline of the entomofauna: a review of its drivers. Biol Conserv 232:8–27. https://doi.org/10.1016/j.biocon.2019.01.020

Sandall E, Fischer B (2019) Be a professional: attend to the insects. Am Entomol 3:176–179

Sandhi RK, Pickens V, Bello E, Elzay S, Salgado S, Hauri KC, Ternest JJ, Constancio N, Gula S, Gearner OM, Anderson M, Edeburn M, Hall B, Maille J, Toth M, Khadka A, Doherty E, Musgrove T, Silva T, Desoto A, Rampone E, Jocson D, Luppino M, Pautzke K, Wagstaff C (2023) Entomology beyond research and education: 2022 student debates. J Insect Sci 23:11. https://doi.org/10.1093/jisesa/iead036

Santaoja M, Niva M (2019) 43. The missing animal in entomophagy? Ethical, ecological and aesthetic considerations on eating insects. In: Sustainable governance and management of food systems. Wageningen Academic Publishers, pp 310–316

Schwentner M, Combosch DJ, Nelson JP, Giribet G (2017) A phylogenomic solution to the origin of insects by resolving crustacean-hexapod relationships. Curr Biol 27:1818–1824.e5. https://doi.org/10.1016/j.cub.2017.05.040

Sebo J, Schukraft J (2021) On the torment of insect minds and our moral duty not to farm them | Aeon Essays. In: Aeon. https://aeon.co/essays/on-the-torment-of-insect-minds-and-our-moral-duty-not-to-farm-them. Accessed 13 Apr 2023

Shellman-Reeve JS (1999) Courting strategies and conflicts in a monogamous, biparental termite. Proc R Soc Lond B Biol Sci 266:137–144. https://doi.org/10.1098/rspb.1999.0613

Shelomi M (2021) Pain, suffering, and euthanasia in insects. Int J Appl Philos 35:31–43. https://doi.org/10.5840/ijap2021129156

Simon S, Otto M, Engelhard M (2018) Synthetic gene drive: between continuity and novelty. EMBO Rep 19:e45760. https://doi.org/10.15252/embr.201845760

Sinclair ER, Osborn AW (1970) Freon-12 anesthesia in two insect species. J Econ Entomol 63

Smith JA (1991) A question of pain in invertebrates. ILAR J 33:25–31. https://doi.org/10.1093/ilar.33.1-2.25

Smith JA, Boyd KM (1991) Lives in the balance: the ethics of using animals in biomedical research: the report of a working Party of the Institute of Medical Ethics. Oxford University Press

Smithsonian Institution Numbers of Insects (Species and Individuals). In: Smithson. Inst. https://www.si.edu/spotlight/buginfo/bugnos. Accessed 16 Apr 2023

Sneddon LU, Elwood RW, Adamo SA, Leach MC (2014) Defining and assessing animal pain. Anim Behav 97:201–212. https://doi.org/10.1016/j.anbehav.2014.09.007

Song W, Silver KS, Du Y, Liu Z, Dong K (2011) Analysis of the action of lidocaine on insect sodium channels. Insect Biochem Mol Biol 41:36–41. https://doi.org/10.1016/j.ibmb.2010.09.010

Spears LR, Ramirez RA (2015) Learning to love leftovers: using by-catch to expand our knowledge in entomology. Am Entomol 61:168–173

Stork NE (2018) How many species of insects and other terrestrial arthropods are there on earth? Annu Rev Entomol 63:31–45. https://doi.org/10.1146/annurev-ento-020117-043348

Strang C, Muth F (2023) Judgement bias may be explained by shifts in stimulus response curves. R Soc Open Sci 10:221322. https://doi.org/10.1098/rsos.221322

Stutt AD, Siva-Jothy MT (2001) Traumatic insemination and sexual conflict in the bed bug Cimex lectularius. Proc Natl Acad Sci USA 98:5683–5687. https://doi.org/10.1073/pnas.101440698

Suzuki S (2013) Biparental care in insects: paternal care, life history, and the function of the Nest. J Insect Sci 13:131. https://doi.org/10.1673/031.013.13101

Switzer PV (2004) Fighting behavior and prior residency advantage in the territorial dragonfly, Perithemis tenera. Ethol Ecol Evol 16:71–89. https://doi.org/10.1080/08927014.2004.9522655

Terminix Milking Spiders for Their Venom | Terminix. In: Terminix. https://www.terminix.com/blog/science-nature/milking-spiders-for-their-venom/. Accessed 15 Apr 2023

Thiagarajan D, Sachse S (2022) Multimodal information processing and associative learning in the insect brain. Insects 13:332. https://doi.org/10.3390/insects13040332

Thorat L, Nath BB (2018) Insects with survival kits for desiccation tolerance under extreme water deficits. Front Physiol 9

Tibbetts EA, Pardo-Sanchez J, Ramirez-Matias J, Avarguès-Weber A (2021) Individual recognition is associated with holistic face processing in Polistes paper wasps in a species-specific way. Proc R Soc B Biol Sci 288:20203010. https://doi.org/10.1098/rspb.2020.3010

Tiffin H (2016) Do insects feel pain? Anim Stud J 5:80–96

Tracey WD, Wilson RI, Laurent G, Benzer S (2003) Painless, a Drosophila gene essential for nociception. Cell 113:261–273. https://doi.org/10.1016/s0092-8674(03)00272-1

Trietsch C, Deans AR (2018) The insect collectors' code. Am Entomol 64:156–158. https://doi.org/10.1093/ae/tmy035
Triphan T, Huetteroth W (2023) Seeking voluntary passive movement in flies is play-like behavior. https://doi.org/10.1101/2023.08.03.551880
Tsubouchi A, Yano T, Yokoyama TK, Murtin C, Otsuna H, Ito K (2017) Topological and modality-specific representation of somatosensory information in the fly brain. Science 358:615–623. https://doi.org/10.1126/science.aan4428
University of Sheffield (2023) Drosophila stocks, food and equipment. https://www.sheffield.ac.uk/flyfacility/stock. Accessed 17 Apr 2023
van Huis A (2013) Potential of insects as food and feed in assuring food security. Annu Rev Entomol 58:563–583. https://doi.org/10.1146/annurev-ento-120811-153704
van Huis A (2021) Welfare of farmed insects. J Insects Food Feed 7:573–584. https://doi.org/10.3920/JIFF2020.0061
van Huis A, Tomberlin J (eds) (2017) Insects as food and feed: from production to consumption, 1st edn. Wageningen Academic Publishers
van Loon MS, Bovenkerk B (2021) 32. The ethics and mindedness of insects. In: Justice and food security in a changing climate. Wageningen Academic Publishers, pp 218–223
Vargas-Terán M, Spradbery JP, Hofmann HC, Tweddle NE (2021) Impact of screwworm eradication programmes using the sterile insect technique. In: Sterile Insect Technique, 2nd edn. CRC Press
Vea IM, Minakuchi C (2021) Atypical insects: molecular mechanisms of unusual life history strategies. Curr Opin Insect Sci 43:46–53. https://doi.org/10.1016/j.cois.2020.09.016
Veit W (2022) Towards a comparative study of animal consciousness. Biol Theory 17:292–303. https://doi.org/10.1007/s13752-022-00409-x
Veterinary Invertebrate Society Veterinary Invertebrate Society. In: Vet. Invertebr Soc https://veterinaryinvertebratesociety.wordpress.com/. Accessed 19 Apr 2023
Victoria and Albert Museum V&A Golden spider silk. In: Vic. Albert Mus. https://www.vam.ac.uk/articles/golden-spider-silk. Accessed 15 Apr 2023
Villanueva-Gutiérrez R, Roubik DW, Colli-Ucán W, Güemez-Ricalde FJ, Buchmann SL (2013) A critical view of Colony losses in managed Mayan honey-making bees (Apidae: Meliponini) in the heart of Zona Maya. J Kans Entomol Soc 86:352–362. https://doi.org/10.2317/JKES130131.1
Visagie EJ (1992) The louse fly Lipoptena paradoxa Newstead, 1907 (Diptera: Hippoboscidae): description of its adult and puparium and biology in South Africa
von Frisch K, Seeley TD (1967) The dance language and orientation of bees. Belknap Press, Cambridge, MA
Voulgari-Kokota A, van Loon MS, Bovenkerk B (2023) Insects as mini-livestock: considering insect welfare in feed production. NJAS Impact Agric Life Sci 95:2191797. https://doi.org/10.1080/27685241.2023.2191797
Wahltinez SJ, Harms CA, Lewbart GA (2023) Chapter 26: anesthesia and analgesia in invertebrates. In: Anesthesia and analgesia in laboratory animals, 3rd edn. Academic, pp 647–671
Walters ET, Illich PA, Weeks JC, Lewin MR (2001) Defensive responses of larval Manduca sexta and their sensitization by noxious stimuli in the laboratory and field. J Exp Biol 204:457–469. https://doi.org/10.1242/jeb.204.3.457
Walt (2024) https://pubmed.ncbi.nlm.nih.gov/39205193/
Wattanachaiyingcharoen W, Oldroyd BP, Wongsiri S, Palmer K, Paar J (2003) A scientific note on the mating frequency of Apis dorsata. Apidologie 34:85–86. https://doi.org/10.1051/apido:2002044
Weineck K, Stanback A, Cooper RL (2019) The effects of eugenol as an Anaesthetic for an insect: drosophila, adults, larval heart rate, and synaptic transmission, vol 40
Weise C, Ortiz CC, Tibbetts EA (2022) Paper wasps form abstract concept of 'same and different'. Proc R Soc B Biol Sci 289:20221156. https://doi.org/10.1098/rspb.2022.1156
Weiss HB (1914) Insects and pain. Can Entomol 46:269–271. https://doi.org/10.4039/Ent46269-8
Wheeler W (1910) Ants: their structure, development and behavior. Columbia University Press
Wigglesworth C (1980) Do insects feel pain? Antenna 4:8–9
Zabala NA, Gómez MA (1991) Morphine analgesia, tolerance and addiction in the cricket pteronemobius sp. (orthoptera, insecta). Pharmacol Biochem Behav 40:887–891. https://doi.org/10.1016/0091-3057(91)90102-8
Zacarias (2025) https://brill.com/view/journals/jiff/11/12/article-p2221_11.xml
Zabala NA, Miralto A, Maldonado H, Nuñez JA, Jaffe K, Calderon LD (1984) Opiate receptor in praying mantis: effect of morphine and naloxone. Pharmacol Biochem Behav 20:683–687. https://doi.org/10.1016/0091-3057(84)90185-0
Zhang Z-Q (2011) Animal biodiversity: an outline of higher-level classification and survey of taxonomic richness (COVER). Zootaxa 3148:3–6. https://doi.org/10.11646/zootaxa.3148.1.2
Zhang S (2018) The last days of the blue-blood harvest. In: The Atlantic. https://www.theatlantic.com/science/archive/2018/05/blood-in-the-water/559229/. Accessed 15 Apr 2023

MIX
Papier aus verantwortungsvollen Quellen
Paper from responsible sources
FSC® C105338

If you have any concerns about our products,
you can contact us on
ProductSafety@springernature.com

In case Publisher is established outside the EU,
the EU authorized representative is:
Springer Nature Customer Service Center GmbH
Europaplatz 3, 69115 Heidelberg, Germany

Printed by Libri Plureos GmbH
in Hamburg, Germany